PROJECT MANAGEMENT

PROJECT MANAGEMENT

A Systems Approach to Planning, Scheduling, and Controlling

TWELFTH EDITION

HAROLD KERZNER, Ph.D.

WILEY

This book is printed on acid-free paper. ∞

Published by John Wiley & Sons, Inc., Hoboken, New Jersey.
Published simultaneously in Canada.

For general information about our other products and services, please contact our Customer Care Department within the United States at (800) 762-2974, outside the United States at (317) 572-3993 or fax (317) 572-4002.

Wiley publishes in a variety of print and electronic formats and by print-on-demand. Some material included with standard print versions of this book may not be included in e-books or in print-on-demand. If this book refers to media such as a CD or DVD that is not included in the version you purchased, you may download this material at http://booksupport.wiley.com. For more information about Wiley products, visit www.wiley.com.

Cover images: © Aeriform/Getty Images, Inc.
Cover design: Wiley

Library of Congress Cataloging-in-Publication Data:

Names: Kerzner, Harold, author.
Title: Project management : a systems approach to planning, scheduling, and controlling / Harold Kerzner.
Description: Twelfth edition. | Hoboken, New Jersey : John Wiley & Sons, Inc., 2017. | Includes bibliographical references and index.
Identifiers: LCCN 2016045434| ISBN 9781119165354 (hardback) | ISBN 9781119165361 (epub); 9781119165378 (epdf)
Subjects: LCSH: Project management. | Project management–Case studies. | BISAC: TECHNOLOGY & ENGINEERING / Industrial Engineering.
Classification: LCC HD69.P75 K47 2017 | DDC 658.4/04–dc23 LC record available at https://lccn.loc.gov/2016045434

Printed in the United States of America

V10015355_110619

To
my wife,
Jo Ellyn,
for her more than thirty years
of unending love, devotion,
and encouragement to continue
my writing of project
management books

Contents

Preface xix

1 OVERVIEW 1

1.0 Introduction 1
1.1 Understanding Project Management 2
1.2 Defining Project Success 6
1.3 Trade-Offs and Competing Constraints 7
1.4 The Entry-Level Project Manager 9
1.5 The Talent Triangle 10
1.6 Technology-Based Projects 10
1.7 The Project Manager–Line Manager Interface 11
1.8 Defining the Project Manager's Role 13
1.9 Defining the Functional Manager's Role 15
1.10 Defining the Functional Employee's Role 17
1.11 Defining the Executive's Role 17
1.12 Working with Executives 17
1.13 Committee Sponsorship/Governance 19
1.14 The Project Manager as the Planning Agent 20
1.15 Project Champions 21
1.16 Project-Driven versus Non–Project-Driven Organizations 22
1.17 Marketing in the Project-Driven Organization 24
1.18 Classification of Projects 25
1.19 Location of the Project Manager 26
1.20 Differing Views of Project Management 27
1.21 Public-Sector Project Management 28
1.22 International Project Management 31
1.23 Concurrent Engineering: A Project Management Approach 32
1.24 Added Value 32
1.25 Studying Tips for the PMI® Project Management Certification Exam 33

Problems 36

Case Study
Williams Machine Tool Company 37

2 Project Management Growth: Concepts and Definitions 39

2.0 Introduction 39
2.1 The Evolution of Project Management: 1945–2017 39
2.2 Resistance to Change 43
2.3 Systems, Programs, and Projects: A Definition 45
2.4 Product versus Project Management: A Definition 47
2.5 Maturity and Excellence: A Definition 49
2.6 Informal Project Management: A Definition 50
2.7 The Many Faces of Success 52
2.8 The Many Faces of Failure 54
2.9 Causes of Project Failure 57
2.10 Degrees of Success and Failure 59
2.11 The Stage-Gate Process 60
2.12 Project Life Cycles 61
2.13 Gate Review Meetings (Project Closure) 65
2.14 Engagement Project Management 66
2.15 Project Management Methodologies: A Definition 67
2.16 From Enterprise Project Management Methodologies to Frameworks 69
2.17 Methodologies Can Fail 70
2.18 Organizational Change Management and Corporate Cultures 71
2.19 Benefits Harvesting and Cultural Change 76
2.20 Agile and Adaptive Project Management Cultures 77
2.21 Project Management Intellectual Property 77
2.22 Systems Thinking 79
2.23 Studying Tips for the PMI® Project Management Certification Exam 82

Problems 85

Case Study
Creating a Methodology 86

3 ORGANIZATIONAL STRUCTURES 89

3.0 Introduction 89
3.1 Organizational Work Flow 90
3.2 Traditional (Classical) Organization 91
3.3 Pure Product (Projectized) Organization 93
3.4 Matrix Organizational Form 95
3.5 Modification of Matrix Structures 99
3.6 The Strong, Weak, or Balanced Matrix 101

3.7 Project Management Offices 101
3.8 Selecting the Organizational Form 103
3.9 Strategic Business Unit (SBU) Project Management 106
3.10 Transitional Management 107
3.11 Seven Fallacies that Delay Project Management Maturity 109
3.12 Studying Tips for the PMI® Project Management Certification Exam 111

Problems 113

4 ORGANIZING AND STAFFING THE PROJECT OFFICE AND TEAM 115

4.0 Introduction 115
4.1 The Staffing Environment 116
4.2 Selecting the Project Manager: an Executive Decision 117
4.3 Skill Requirements for Project and Program Managers 121
4.4 Special Cases in Project Manager Selection 125
4.5 Today's Project Managers 126
4.6 Duties and Job Descriptions 127
4.7 The Organizational Staffing Process 128
4.8 The Project Office 131
4.9 The Functional Team 133
4.10 The Project Organizational Chart 133
4.11 Selecting the Project Management Implementation Team 136
4.12 Mistakes Made by Inexperienced Project Managers 139
4.13 Studying Tips for the PMI® Project Management Certification Exam 140

Problems 142

5 MANAGEMENT FUNCTIONS 145

5.0 Introduction 145
5.1 Controlling 146
5.2 Directing 146
5.3 Project Authority 148
5.4 Interpersonal Influences 152
5.5 Barriers to Project Team Development 154
5.6 Suggestions for Handling the Newly Formed Team 157
5.7 Team Building as an Ongoing Process 158
5.8 Leadership in a Project Environment 159
5.9 Value-Based Project Leadership 160
5.10 Transformational Project Management Leadership 163
5.11 Organizational Impact 163
5.12 Employee–Manager Problems 165
5.13 General Management Pitfalls 166
5.14 Time Management Pitfalls 167

5.15 Management Policies and Procedures 171
5.16 Human Behavior Education 171
5.17 Studying Tips for the PMI® Project Management Certification Exam 174

Problems 177

Case Studies
The Trophy Project 178
McRoy Aerospace 180
The Poor Worker 182
The Prima Donna 182
The Reluctant Workers 184
Leadership Effectiveness (A) 185
Leadership Effectiveness (B) 189
Motivational Questionnaire 195

6 COMMUNICATIONS MANAGEMENT 203

6.0 Introduction 203
6.1 Modeling the Communications Environment 203
6.2 The Project Manager as a Communicator 208
6.3 Project Review Meetings 212
6.4 Project Management Bottlenecks 212
6.5 Active Listening 213
6.6 Communication Traps 214
6.7 Project Problem Solving 215
6.8 Brainstorming 223
6.9 Predicting the Outcome of a Decision 224
6.10 Facilitation 226
6.11 Studying Tips for the PMI® Project Management Certification Exam 228

Problems 230

Case Studies
Communication Failures 231
The Team Meeting 234

7 CONFLICTS 237

7.0 Introduction 237
7.1 The Conflict Environment 238
7.2 Types of Conflicts 239
7.3 Conflict Resolution 240
7.4 The Management of Conflicts 241
7.5 Conflict Resolution Modes 242

7.6 Understanding Superior, Subordinate, and Functional Conflicts 244
7.7 Studying Tips for the PMI® Project Management Certification Exam 246

Problems 248

Case Studies
Facilities Scheduling at Mayer Manufacturing 248
Telestar International 250
Handling Conflict in Project Management 251

8 SPECIAL TOPICS 257

8.0 Introduction 257
8.1 Performance Measurement 257
8.2 Financial Compensation and Rewards 262
8.3 Effective Project Management in the Small Business Organization 270
8.4 Mega Projects 271
8.5 Morality, Ethics, and the Corporate Culture 273
8.6 Professional Responsibilities 275
8.7 Internal and External Partnerships 278
8.8 Training and Education 279
8.9 Integrated Product/Project Teams 281
8.10 Virtual Project Teams 283
8.11 Managing Innovation Projects 284
8.12 Agile Project Management 287
8.13 Studying Tips for the PMI® Project Management Certification Exam 289

Problems 295

Case Study
Is It Fraud? 295

9 THE VARIABLES FOR SUCCESS 299

9.0 Introduction 299
9.1 Predicting Project Success 299
9.2 Project Management Effectiveness 302
9.3 Expectations 303
9.4 Lessons Learned 305
9.5 Understanding Best Practices 306
9.6 Studying Tips for the PMI® Project Management Certification Exam 312

Problems 313

Case Study
Radiance International 313

10 WORKING WITH EXECUTIVES 317

10.0 Introduction 317
10.1 The Project Sponsor 317
10.2 Handling Disagreements with the Sponsor 327
10.3 The Collective Belief 327
10.4 The Exit Champion 328
10.5 The In-House Representatives 329
10.6 Stakeholder Relations Management 329
10.7 Project Portfolio Management 335
10.8 Politics 337
10.9 Studying Tips for the PMI® Project Management Certification Exam 338

Problems 339

Case Studies
The Prioritization of Projects 340
The Irresponsible Sponsors 341
Selling Executives on Project Management 342

11 PLANNING 345

11.0 Introduction 345
11.1 Business Case 346
11.2 Validating the Assumptions 348
11.3 Validating the Objectives 351
11.4 General Planning 352
11.5 Life-Cycle Phases 355
11.6 Life-Cycle Milestones 356
11.7 Kickoff Meetings 358
11.8 Understanding Participants' Roles 360
11.9 Establishing Project Objectives 360
11.10 The Statement of Work 361
11.11 Project Specifications 363
11.12 Data Item Milestone Schedules 364
11.13 Work Breakdown Structure 365
11.14 Wbs Decomposition Problems 370
11.15 Work Breakdown Structure Dictionary 372
11.16 Project Selection 373
11.17 The Role of the Executive in Planning 377
11.18 Management Cost and Control System 378
11.19 Work Planning Authorization 379
11.20 Why Do Plans Fail? 380
11.21 Stopping Projects 381
11.22 Handling Project Phaseouts and Transfers 381

11.23 Detailed Schedules and Charts 383
11.24 Master Production Scheduling 385
11.25 Project Plan 386
11.26 The Project Charter 391
11.27 Project Baselines 392
11.28 Verification and Validation 395
11.29 Management Control 396
11.30 Configuration Management 397
11.31 Enterprise Project Management Methodologies 398
11.32 Project Audits 399
11.33 Studying Tips for the PMI® Project Management Certification Exam 400

Problems 404

12 NETWORK SCHEDULING TECHNIQUES 409

12.0 Introduction 409
12.1 Network Fundamentals 411
12.2 Graphical Evaluation and Review Technique (GERT) 416
12.3 Dependencies 417
12.4 Slack Time 417
12.5 Network Replanning 423
12.6 Estimating Activity Time 428
12.7 Estimating Total Project Time 429
12.8 Total PERT/CPM Planning 430
12.9 Crash Times 431
12.10 PERT/CPM Problem Areas 436
12.11 Alternative PERT/CPM Models 436
12.12 Precedence Networks 437
12.13 Lag 440
12.14 Scheduling Problems 441
12.15 The Myths of Schedule Compression 441
12.16 Project Management Software 442
12.17 Studying Tips for the PMI® Project Management Certification Exam 445

Problems 448

Case Study
The Invisible Sponsor 451

13 PRICING AND ESTIMATING 453

13.0 Introduction 453
13.1 Global Pricing Strategies 453
13.2 Types of Estimates 455
13.3 Pricing Process 458

13.4 Organizational Input Requirements 460
13.5 Labor Distributions 462
13.6 Overhead Rates 463
13.7 Materials/Support Costs 465
13.8 Pricing Out the Work 466
13.9 Smoothing Out Department Man-Hours 469
13.10 The Pricing Review Procedure 471
13.11 Systems Pricing 472
13.12 Developing the Supporting/Backup Costs 474
13.13 The Low-Bidder Dilemma 474
13.14 Special Problems 477
13.15 Estimating Pitfalls 478
13.16 Estimating High-Risk Projects 479
13.17 Project Risks 480
13.18 The Disaster of Applying the 10 Percent Solution to Project Estimates 483
13.19 Life-Cycle Costing (LCC) 484
13.20 Logistics Support 486
13.21 Economic Project Selection Criteria: Capital Budgeting 488
13.22 Payback Period 488
13.23 The Time Value of Money and Discounted Cash Flow (DCF) 489
13.24 Net Present Value (NPV) 490
13.25 Internal Rate of Return (IRR) 490
13.26 Comparing IRR, NPV, and Payback 491
13.27 Risk Analysis 492
13.28 Capital Rationing 492
13.29 Project Financing 494
13.30 Studying Tips for the PMI® Project Management Certification Exam 496

Problems 498

Case Study
The Estimating Problem 499

14 COST CONTROL 501

14.0 Introduction 501
14.1 Understanding Control 503
14.2 The Operating Cycle 506
14.3 Cost Account Codes 506
14.4 Budgets 511
14.5 The Earned Value Measurement System (EVMS) 512
14.6 Variance and Earned Value 513
14.7 The Cost Baseline 529
14.8 Justifying the Costs 531
14.9 The Cost Overrun Dilemma 532

14.10 Recording Material Costs Using Earned Value Measurement 534
14.11 Material Variances: Price and Usage 535
14.12 Summary Variances 536
14.13 Status Reporting 537
14.14 Cost Control Problems 537
14.15 Studying Tips for the PMI® Project Management Certification Exam 539

Problems 542

Case Studies
The Bathtub Period 544
Franklin Electronics 545

15 METRICS 549

15.0 Introduction 549
15.1 Project Management Information Systems 549
15.2 Enterprise Resource Planning 550
15.3 Project Metrics 550
15.4 Key Performance Indicators (KPIS) 555
15.5 Value-Based Metrics 561
15.6 Dashboards and Scorecards 566
15.7 Business Intelligence 569
15.8 Studying Tips for the PMI® Project Management Certification Exam 570

Problems 573

16 TRADE-OFF ANALYSIS IN A PROJECT ENVIRONMENT 575

16.0 Introduction 575
16.1 Methodology for Trade-Off Analysis 578
16.2 Contracts: Their Influence on Projects 593
16.3 Industry Trade-Off Preferences 594
16.4 Project Manager's Control of Trade-Offs 597
16.5 Studying Tips for the PMI® Project Management Certification Exam 597

Problems 598

17 RISK MANAGEMENT 599

17.0 Introduction 599
17.1 Definition of Risk 601
17.2 Tolerance for Risk 603
17.3 Definition of Risk Management 604
17.4 Certainty, Risk, and Uncertainty 604
17.5 Risk Management Process 610

17.6 Plan Risk Management 611
17.7 Risk Identification 612
17.8 Risk Analysis 613
17.9 Qualitative Risk Analysis 615
17.10 Quantitative Risk Analysis 616
17.11 Plan Risk Response 619
17.12 Monitor and Control Risks 621
17.13 Some Implementation Considerations 622
17.14 The Use of Lessons Learned 623
17.15 Dependencies between Risks 624
17.16 The Impact of Risk Handling Measures 628
17.17 Risk and Concurrent Engineering 631
17.18 Studying Tips for the PMI® Project Management Certification Exam 633

Problems 637

Case Studies
Teloxy Engineering (A) 640
Teloxy Engineering (B) 640
The Risk Management Department 641

18 LEARNING CURVES 643

18.0 Introduction 643
18.1 General Theory 643
18.2 The Learning Curve Concept 644
18.3 Graphic Representation 646
18.4 Key Words Associated with Learning Curves 647
18.5 The Cumulative Average Curve 648
18.6 Sources of Experience 649
18.7 Developing Slope Measures 653
18.8 Unit Costs and Use of Midpoints 654
18.9 Selection of Learning Curves 654
18.10 Follow-On Orders 655
18.11 Manufacturing Breaks 656
18.12 Learning Curve Limitations 656
18.13 Competitive Weapon 657
18.14 Studying Tips for the PMI® Project Management Certification Exam 658

Problems 659

19 CONTRACT MANAGEMENT 661

19.0 Introduction 661
19.1 Procurement 662
19.2 Plan Procurements 664

19.3 Conducting the Procurements 667
19.4 Conduct Procurements: Request Seller Responses 668
19.5 Conduct Procurements: Select Sellers 669
19.6 Types of Contracts 673
19.7 Incentive Contracts 678
19.8 Contract Type versus Risk 680
19.9 Contract Administration 680
19.10 Contract Closure 683
19.11 Using a Checklist 684
19.12 Proposal-Contractual Interaction 684
19.13 Studying Tips for the PMI® Project Management Certification Exam 686

Problems 691

Case Studies
To Bid or Not to Bid 692
The Management Reserve 693

20 QUALITY MANAGEMENT 697

20.0 Introduction 697
20.1 Definition of Quality 698
20.2 The Quality Movement 699
20.2 Quality Management Concepts 703
20.3 The Cost of Quality 707
20.4 The Seven Quality Control Tools 709
20.5 Acceptance Sampling 721
20.6 Implementing Six Sigma 722
20.7 Quality Leadership 723
20.8 Responsibility for Quality 724
20.9 Quality Circles 725
20.10 Total Quality Management (TQM) 725
20.11 Studying Tips for the PMI® Project Management Certification Exam 728

Problems 731

21 MODERN DEVELOPMENTS IN PROJECT MANAGEMENT 733

21.0 Introduction 733
21.1 The Project Management Maturity Model (PMMM) 733
21.2 Developing Effective Procedural Documentation 737
21.3 Project Management Methodologies 741
21.4 Continuous Improvement 742
21.5 Capacity Planning 743
21.6 Competency Models 745
21.7 Managing Multiple Projects 747

21.8 The Business of Scope Changes 748
21.9 End-of-Phase Review Meetings 752

Case Study
Honicker Corporation 753
Kemko Manufacturing 755

Appendix A: Solution to Leadership Exercise 759
Appendix B: Solutions to the Project Management Conflict Exercise 765
Appendix C: Dorale Products Case Studies 771
Appendix D: Solutions to the Dorale Products Case Studies 783
Appendix E: Alignment of the *PMBOK® Guide* to the Text 789

Index 795

PMBOK is a registered mark of the Project Management Institute, Inc.

Preface

Project management has evolved from a management philosophy restricted to a few functional areas and regarded as something nice to have to an enterprise project management system affecting every functional unit of the company. Simply stated, project management has evolved into a business process rather than merely a project management process. More and more companies are now regarding project management as being mandatory for the survival of the firm. Organizations that were opponents of project management are now advocates. Management educators of the past, who preached that project management could not work and would be just another fad, are now staunch supporters. Project management is here to stay. Colleges and universities are now offering undergraduate and graduate degrees in project management.

This book is addressed not only to those undergraduate and graduate students who wish to improve upon their project management skills but also to those functional managers and upper-level executives who serve as project sponsors and must provide continuous support for projects. During the past several years, management's knowledge and understanding of project management has matured to the point where almost every company is using project management in one form or another. These companies have come to the realization that project management and productivity are related, and that we are now managing our business as though it is a series of projects. Project management coursework is now consuming more of training budgets than ever before.

General reference is provided in the text to engineers. However, the reader should not consider project management as strictly engineering-related. The engineering examples are the result of the fact that project management first appeared in the engineering disciplines, and we should be willing to learn from their mistakes. Project management now resides in every profession, including information systems, healthcare, consulting, pharmaceutical, banks, and government agencies.

The text can be used for both undergraduate and graduate courses in business, information systems, and engineering. The structure of the text is based upon my belief that project management is much more behavioral than quantitative since projects are managed by people rather than tools. The first seven chapters are part of the basic core of knowledge necessary to understand project management, specifically topics related to PMI's "Talent Triangle." Chapters 8 through 10 deal with the support functions and describe factors for predicting success and management support. It may seem strange that ten chapters on organizational behavior and structuring are needed prior to the "hard-core" chapters of planning, scheduling, and controlling. These first ten chapters are needed to understand the cultural environment for all projects and systems. These chapters are necessary for the reader to understand the difficulties in achieving cross-functional cooperation on projects where team members are working on multiple projects concurrently and why the people involved, all of whom may have different backgrounds, cannot simply be forged into a cohesive work unit without friction. Chapters 11 through 20 are more of the quantitative chapters on planning, scheduling, cost control, estimating, contracting (and procurement), and quality. Chapter 21 focuses on some of the more advanced topics.

The changes that were made in the twelfth edition include:

- Updated section on the Introduction to Project Management
- Updated section on Competing Constraints
- New section on the Talent Triangle
- New section on Entry-Level Project Management
- New section on Technology-Based Projects
- Updated section on the Many Faces of Project Success
- New section on Converting Methodologies to Frameworks
- New section on the Causes of Project Failure
- New section on Degrees of Project Success and Failure
- Updated section on Knowledge Management and Data Warehouses
- Updated section on Project Management Intellectual Property
- New section on Benefits Harvesting and Cultural Change
- New section on Transformational Project Management Leadership
- Updated section on Managing Mega Projects
- Updated section on Agile Project Management
- New section on Agile and Adaptive Project Management Cultures
- Updated section on Multinational Project Management Sponsorship
- New section on Preparing a Project Business Case
- Updated section on Validating the Project's Assumptions
- Updated section on Validating the Project's Objectives
- New section on Life-Cycle Milestones
- New section on the Project Management Office
- New section on Project Portfolio Management

- Updated section on Best Practices
- Updated section on Resource Leveling Issues

The text contains case studies, multiple choice questions, and discussion questions. There is also a separate companion book of cases *(Project Management Case Studies,* fifth edition) that provides additional real-world examples. Some of the new case studies include in the case book are:

Case Study	Description
Disney (A) Imagineering Project Management	Discusses some of the different skill sets needed to be an Imagineering PM
Disney (B) Imagineering in Action: The Haunted Mansion	Discusses the challenges with evolving scope on a project
Disney (C) Theme Parks and Enterprise Environmental Factors	Discusses how important an understanding of the enterprise environmental factors are and how they can impact project success
Disney (D) The Globalization of Disney	Discusses the challenges facing the use of project management on a global scale
Disney (E) Hong Kong Ocean Park: Competing Against Disney	Discusses how one company competed against Disney by expanding the project's scope
Olympics (A) Managing Olympic Projects	Discusses how the enterprise environmental factors impact Olympic projects
Olympics (B) Olympics, Project Management and PMI's Code of Ethics and Professional Responsibility	Discusses the complexity of abiding by PMI's Code of Conduct and Professional Responsibility on some Olympic projects
Olympics (C) Feeding the Olympic Athletes	Discusses the complexities (including quality control) for feeding 23,000 Olympians, coaches and staff members
Olympics (D) Health and Safety Risks at Olympic Events	Discusses the health and safety risks when of allowing athletes to compete in environments that have known health risks
Tradeoffs (A), (B)	Discusses how the introduction of competing constraints mandated additional tradeoffs and the challenges the company faced
The Project Management Audit	Discusses the need for occasional audits on a project and what happens executives are displeased with the results
The Executive Director	Discusses how a newly appointed executive director in a government agency played the political game to prevent being blamed for any wrong-doing

The twelfth edition text, the *PMBOK® Guide* and the book of cases are ideal as self-study tools for the Project Management Institute's PMP® Certification Exam. Because of this, there are tables of cross references at the end of each chapter in the textbook detailing the sections from the book of cases and the Guide to the Project Management Body of Knowledge (*PMBOK® Guide*) that apply to that

PMBOK is a registered mark of the Project Management Institute

chapter's content. The left-hand margin of the pages in the text has side bars that identify the cross-listing of the material on that page to the appropriate section(s) of the *PMBOK® Guide*. At the end of most of the chapters is a section on study tips for the PMP® exam.

This textbook is currently used in the college market, in the reference market, and for studying for the PMP® Certification Exam. Therefore, to satisfy the needs of all markets, a compromise had to be reached on how much of the text would be aligned to the *PMBOK® Guide* and how much new material would be included without doubling the size of the text. Some colleges and universities use the textbook to teach project management fundamentals without reference to the *PMBOK® Guide*. The text does not contain all of the material necessary to support each section or process in the *PMBOK® Guide*. Therefore, to study for the PMP® Certification Exam, the *PMBOK® Guide* must also be used together with this text. The text covers material for almost all of the *PMBOK® Guide* knowledge areas but not necessarily in the depth that appears in the *PMBOK® Guide*.

An instructor's manual is available only to college and university faculty members by contacting your local Wiley sales representative or by visiting the Wiley website at www.wiley.com/kerzner. Access to the instructor's material and supporting material can be provided only through John Wiley & Sons Publishers, not the author.

One-, two-, and three-day seminars on project management and the PMP® Certification Training using the text are offered by contacting Lori Milhaven, Executive Vice President, the International Institute for Learning, at 800-325-1533, extension 5121 (e-mail address: lori.milhaven@iil.com).

The problems and case studies at the ends of the chapters cover a variety of industries. Almost all of the case studies are real-world situations taken from my consulting practice or from research. Feedback from my colleagues who are using the text has provided me with fruitful criticism, most of which has been incorporated into the twelfth edition.

The majority of the articles on project management that have become classics have been referenced in the textbook throughout the first eleven chapters. These articles were the basis for many of the modern developments in project management and are therefore identified throughout the text.

Many colleagues provided valuable criticism. In particular, I am indebted to those industrial/government training managers whose dedication and commitment to quality project management education and training have led to valuable changes in this and previous editions. In particular, I wish to thank Frank Saladis, PMP, for his constructive comments, recommendations, and assistance with the mapping of the text to the *PMBOK® Guide* as well as recommended changes to many of the chapters. I am indebted to Dr. Edmund Conrow, PMP, for more than a decade of assistance with the preparation of the risk management chapters in all of my texts. I am also indebted to Dr. Rene Rendon for his review and recommendations for changes to the chapter on Contract Management.

To the management team and employees of the International Institute for Learning, thank you all for twenty-five years of never-ending encouragement, support, and assistance with all of my project management research and writings.

Harold Kerzner
The International Institute for Learning
2017

1 Overview

1.0 INTRODUCTION

In the United States, the roots of project management date back to the Department of Defense (DOD) and heavy construction companies during the 1960s. Early use of project management focused on the completion of unique, or sometimes repetitive, projects with a heavy focus on compliance to budgets and schedules. To maintain standardization and control in the way that projects were managed, DOD established policies and procedures for gate reviews and the way that status should be reported.

In the early years, project management was seen as a part-time job rather than as a career path position. In many companies, project management existed in only a small portion of the business, which made it difficult for some projects to get total company support.

Executives began realizing the complexities of resource control and effective project staffing. In addition, the rapid rate of change in both technology and the marketplace had created enormous strains on existing organizational forms. The traditional structure, which was highly bureaucratic, showed that it could not respond rapidly enough to a changing environment. Thus, the traditional structure was replaced by project management, or other temporary management structures, that were highly organic and could respond very rapidly as situations developed inside and outside the company. The organic nature of project management practices today allow project managers to customize the project management tools and processes to adapt to a variety of different environments.

The acceptance of project management was not easy. Many executives were not willing to accept change and were inflexible when it came to adapting to a different environment and flexible organizational structures. The project management approach required a departure from the traditional business organizational form, which was basically vertical and which emphasized a strong superior–subordinate relationship. Many executives had very strong beliefs as to how a company should be run and refused to recognize or admit that project management could benefit their company.

Unfavorable economic conditions forced executives to reconsider the value that project management could bring to a firm. Some of the unfavorable conditions included the recessions of the late 1970s and early 1990s, the housing crisis that began in 2008, the European economy downturn in 2013 and 2014, and the world economic slowdown in 2015. These unfavorable conditions emphasized the need for better control of existing resources, the creation of a portfolio of projects that would maximize the value brought to the firm, and a higher percentage of project successes. It soon became apparent that project management could satisfy all of these needs and that project management is a necessity in both bad and good economic conditions. Today, the concept behind project management is being applied in such diverse industries and organizations as defense, construction, pharmaceuticals, chemicals, banking, hospitals, accounting, advertising, law, state and local governments, and the United Nations.

Almost all of today's executives are convinced that project management can and does work well. Project management is now being applied to all facets of a business rather than just parts of the business. Projects are now being aligned with corporate or strategic objectives. Simply stated, "Why work on a project that is not aligned to strategic objectives with the goal of creating business value?" In some companies such as IBM, Microsoft, and Hewlett-Packard, project management is recognized as a strategic competency necessary for the survival of the firm. This recognition of the importance of project management today permeates almost all industries and companies of all sizes.

1.1 UNDERSTANDING PROJECT MANAGEMENT

PMBOK® Guide, 6th Edition

Chapter 1 Introduction to the *PMBOK® Guide*
1.2.1 Projects
1.2.1 The Importance of Project Management
1.2.4.5 Project Management Process Groups

In order to understand project management, one must begin with the definition of a project. A project can be considered to be any series of activities and tasks that:

- Have a specific objective, with a focus on the creation of business value, to be completed within certain specifications
- Have defined start and end dates
- Have funding limits (if applicable)
- Consume human and nonhuman resources (i.e., money, people, equipment)
- Are multifunctional (i.e., cut across several functional lines)

The result or outcome of the project can be unique or repetitive, and must be achieved within a finite period of time. Because companies have very limited resources, care must be taken that the right mix of projects is approved. Given this, another outcome of a project is that it provides business value to the company as opposed to being a "pet" project for the personal whims of one person.

Project management is the application of knowledge, skills, and tools necessary to achieve the project's requirements. The knowledge, skills, and tools are usually grouped into activities or processes. PMI's *PMBOK® Guide* identifies five process groups. Some of the activities within these groups include:

- Project initiation
  - Selection of the best project given resource limits
  - Recognizing the benefits of the project

PMBOK is a registered mark of the Project Management Institute, Inc.

  - Preparation of the documents to sanction the project
  - Assigning of the project manager
- Project planning
  - Definition of the work requirements
  - Definition of the quality and quantity of work
  - Definition of the resources needed
  - Scheduling the activities
  - Evaluation of the various risks
- Project execution
  - Negotiating for the project team members
  - Directing and managing the work
  - Working with the team members to help them improve
- Project monitoring and control
  - Tracking progress
  - Comparing actual outcome to predicted outcome
  - Analyzing variances and impacts
  - Making adjustments
- Project closure
  - Verifying that all of the work has been accomplished
  - Contractual closure of the contract
  - Financial closure of the charge numbers
  - Administrative closure of the paperwork

Successful project management can then be defined as achieving a continuous stream of project objectives within time, within cost, at the desired performance/technology level, while utilizing the assigned resources effectively and efficiently, and having the results accepted by the customer and/or stakeholders. Because each project is inherently different and each customer can have different requirements, the activities included within the process groups may change from project to project. The *PMBOK® Guide* identifies industry-accepted activity regarded as best practices for each process group and these best practices can be structured to create a project management methodology that can be applied and customized to a variety of projects.

The potential benefits from effective project management are:

- Clear identification of functional responsibilities to ensure that all activities are accounted for, regardless of personnel turnover
- Minimizing the need for continuous reporting
- Identification of time limits for scheduling
- Identification of a methodology for trade-off analysis
- Measurement of accomplishment against plans
- Early identification of problems so that corrective action may follow
- Improved estimating capability for future planning
- Knowing when objectives cannot be met or will be exceeded

Unfortunately, the benefits cannot be achieved without overcoming obstacles such as project complexity, customer's special requirements and scope changes, organizational restructuring, project risks, changes in technology, and forward planning and pricing.

Project management is designed to make better use of existing resources by getting work to flow horizontally as well as vertically within the company. This approach does not really destroy the vertical, bureaucratic flow of work but simply requires that line organizations talk to one another horizontally so that horizontal and vertical work flow will be accomplished more smoothly throughout the organization and in a concurrent manner. The vertical flow of work is still the responsibility of the line managers. The horizontal flow of work is the responsibility of the project managers, and their primary effort is to communicate and coordinate activities horizontally between the line organizations.

PMBOK® Guide, 6th Edition
3.4 Project Management Competence

Figure 1–1 shows how many companies are structured. There are always "class or prestige" gaps between various levels of management. There are also functional gaps between working units of the organization. If we superimpose the management gaps on top of the functional gaps, we find that companies are made up of small operational islands that refuse to communicate with one another for fear that giving up information may strengthen their opponents. The project manager's responsibility is to get these islands to communicate crossfunctionally toward common goals and objectives.

The project manager may require a difference set of skills when working with each of the islands. The *PMBOK® Guide* identifies a talent triangle composed of technical project management, leadership and strategic and business management skills. In today's environment, strategic and business management skills are getting more attention because project managers are seen as managing part of a business rather than merely a project and, as such, are expected to make both project and business decisions.

The following is an overview definition of project management:

> Project management is the planning, organizing, directing, and controlling of company resources for a relatively short-term objective that has been established to complete specific goals and objectives. Furthermore, project management utilizes the systems approach to management by having functional personnel (the vertical hierarchy) assigned to a specific project (the horizontal hierarchy).

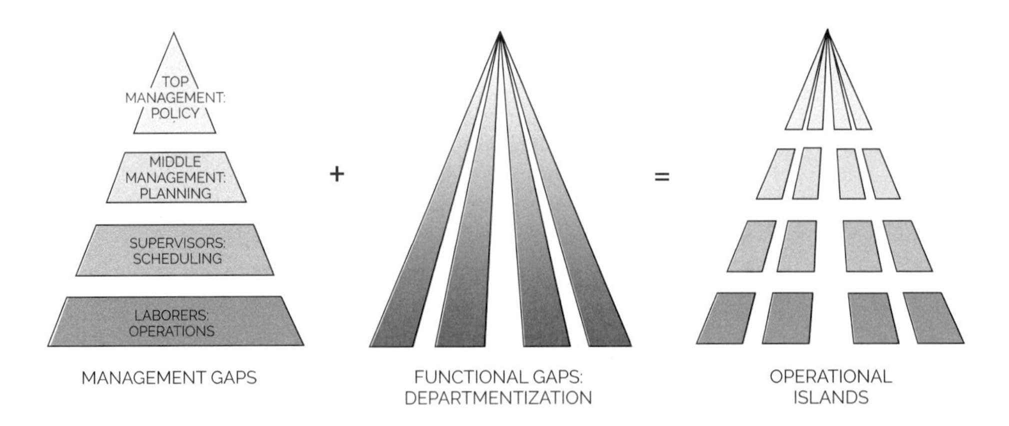

FIGURE 1–1. Organizational gaps.

PMBOK® Guide, 6th Edition
2.4 Organizational Systems

The preceding definition requires further comment. Classical management is usually considered to have five functions or principles:

- Planning
- Organizing
- Staffing
- Controlling
- Directing

You will notice that, in the definition, the staffing function has been omitted. This was intentional because the project manager does not staff the project. Staffing is a line responsibility. The project manager has the right to request specific resources, but the final decision as to what resources will be committed rests with the line managers.

We should also comment on what is meant by a "relatively" short-term project. Not all industries have the same definition for a short-term project. In engineering, the project might be for six months or two years; in construction, three to five years; in nuclear components, ten years; and in insurance, two weeks. Long-term projects, which consume resources full-time, are usually set up as a separate division (if large enough) or simply as a line organization.

Figure 1–2 is a pictorial representation of traditional project management the way it was understood in the past. The objective of the figure is to show that project management is designed to manage or control company resources on a given activity, within time, within cost, and within performance. Time, cost, and performance were considered as the

FIGURE 1–2. Overview of project management.

only constraints on the project. If the project is to be accomplished for an outside customer, then the project had a fourth constraint: good customer relations. Customers can be internal or external to the parent organization. The reader should immediately realize that it is possible to manage a project within time, cost, and performance and then also alienate the customer to such a degree that no further business will be forthcoming. Executives often select project managers based on who the customer is and what kind of customer relations will be necessary.

Projects exist to produce deliverables. The person ultimately assigned as the project manager may very well be assigned based upon the size, nature, and scope of the deliverables. Deliverables are outputs, or the end result of either the completion of the project or the end of a life-cycle phase of the project. Deliverables are measurable, tangible outputs and can take such form as:

- **Hardware Deliverables**: These are hardware items, such as a table, a prototype, or a piece of equipment.
- **Software Deliverables**: These items are similar to hardware deliverables but are usually paper products, such as reports, studies, handouts, or documentation. Some companies do not differentiate between hardware and software deliverables.
- **Interim Deliverables**: These items can be either hardware or software deliverables and progressively evolve as the project proceeds. An example is a series of interim reports leading up to the final report.

1.2 DEFINING PROJECT SUCCESS

> ***PMBOK® Guide*, 6th Edition**
> 1.2.6.4 Project Success Measures

In the previous section, we defined project success as the completion of an activity within the constraints of time, cost, and performance. This was the definition used for the past thirty to forty years or so. More recently, the definition of project success has been modified to include completion:

- Within the allocated time period
- Within the budgeted cost
- At the proper performance or specification level
- With acceptance by the customer/user
- With minimum or mutually agreed upon scope changes
- Without disturbing the main work flow of the organization
- Without changing the corporate culture

The last three elements require further explanation. Very few projects are completed within the original scope of the project. Scope changes are inevitable and have the potential to destroy not only the morale on a project, but the entire project. Scope changes *must* be held to a minimum and those that are required *must* be approved by both the project manager and the customer/user.

Project managers must be willing to manage (and make concessions/trade-offs, if necessary) such that the company's main work flow is not altered. Most project managers view themselves as self-employed entrepreneurs after project go-ahead and would like to divorce their project from the operations of the parent organization. This is not always possible. The project manager must be willing to manage within the guidelines, policies, procedures, rules, and directives of the parent organization.

All corporations have corporate cultures, and even though each project may be inherently different, the project manager should not expect his assigned personnel to deviate from cultural norms. If the company has a standard of openness and honesty when dealing with customers, then this cultural value should remain in place for all projects, regardless of who the customer/user is or how strong the project manager's desire for success is.

Excellence in project management is defined as a continuous stream of successfully managed projects. Any project can be driven to success through formal authority and strong executive meddling. But in order for a continuous stream of successful projects to occur, there must exist a strong corporate commitment to project management, and this commitment *must be visible.*

1.3 TRADE-OFFS AND COMPETING CONSTRAINTS

Although many projects are completed successfully, at least in the eyes of the stakeholders, the final criteria from which success is measured may be different from the initial criteria because of trade-offs. Trade-offs are situations where one aspect of a project may be sacrificed to gain an advantage with another aspect. As an example, additional time and money may be needed to make further improvements in the quality of the project's deliverables.

The first triangle shown in Figure 1–2 is referred to as the triple constraints on a project, namely time, cost, and performance, where performance can be scope, quality, or technology. These are considered to be the primary constraints and are often considered to be the criteria for a project against which success is measured.

Today, we realize that there can be multiple constraints on a project and, rather than use the terminology of the triple constraints, we focus our attention on competing constraints. Sometimes the constraints are referred to as primary and secondary constraints. There may be secondary factors such as risk, customer relations, image, and reputation that may cause us to deviate from our original success criteria of time, cost, and performance. These changes can occur any time during the life of a project and can then cause trade-offs in the triple constraints, thus requiring that changes be made to the success criteria. In an ideal situation, we would perform trade-offs on any or all of the competing constraints such that acceptable success criteria would still be met.

As an example, let's assume that a project was initiated using the success criteria of the triple constraints as shown in Figure 1–3. For simplicity's sake, a triangle was used for the competing constraints in Figure 1–3. However, there can be significantly more than three competing constraints in which some geometric shape other than a triangle might work best. Partway through the project, the environment changes, a new senior management team is brought in with their own agenda, or a corporate crisis occurs such

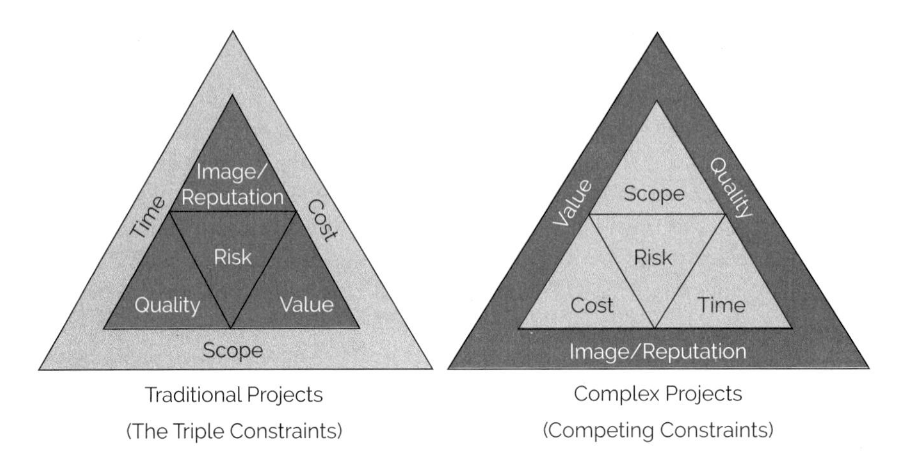

FIGURE 1–3. Competing constraints.

that the credibility of the corporation is at stake. In such a case, the competing constraints shown on the right in Figure 1–3 can be more important than the original triple constraints.

Secondary factors are also considered to be constraints and may be more important than the primary constraints. For example, years ago, in Disneyland and Disneyworld, the project managers designing and building the attractions at the theme parks had six constraints: time, cost, scope, safety, aesthetic value, and quality.

At Disney, the last three constraints of safety, aesthetic value, and quality were considered locked-in constraints that could not be altered during trade-offs. All trade-offs were made on time, cost, and scope. Some constraints simply cannot change while others may have flexibility.

Not all constraints are equal in importance. For example, in the initiation phase of a project, scope may be the critical factor and all trade-offs are made on time and cost. During the execution phase of the project, time and cost may become more important and then trade-offs will be made on scope. A more detailed discussion of trade-offs can be found in Chapter 16.

When managing a project according to the triple constraints of time, cost, and scope, we perform a juggling act and often find a way to meet all three constraints, each of which usually carries an equal degree of importance. When the number of constraints increases to five or six constraints, it may be difficult, if not impossible, to meet all of the constraints and a prioritization of constraints may be necessary.

The prioritization of constraints can change over the life of the project based upon the needs of the project manager, the client, and the stakeholders. Changing the priorities of the constraints can lead to scope changes and play havoc with the requirements and baselines. There must be a valid reason for changing the prioritization of the constraints after project go-ahead.

1.4 THE ENTRY-LEVEL PROJECT MANAGER

Too often, people desire a project management position without fully understanding what the job entails. Some people believe that they will be given a vast amount of authority, they will make any and all decisions on the project, they will have control of a small empire of workers which they personally hired, and they will interface with executives within and outside of their firm.

In reality, project management may be a lot different than some believe. Most project managers have very little real authority. The real authority may rest with the project sponsor and functional management. Some people argue that project management is actually leadership without authority.

Project managers may not have any say in staffing the project and may not even be able to fire poorly performing workers. Project staff is most commonly provided by the functional managers and only the functional managers can remove the workers. Projects managers may have no input into the wage and salary program for the employees assigned to the project. Employees assigned to the project may be working on several other projects at the same time, and the project manager may not be able to get these employees to satisfy his/her project's requirements in a timely manner. Project managers may not be allowed to communicate with personnel external to the company. This may be done by the internal project sponsor.

Today's project managers are expected to have at least a cursory understanding of the company's business model as well as the company's business processes that support project management. Project managers are now expected to make both project- and business-related decisions when necessary.

Some people believe that project managers make any and all decisions on a project. This is certainly not true. In today's high-technology environments, project managers cannot be experts in all areas. Their expertise may not be in any of the knowledge areas of the project. This is quite common when a project manager is asked to manage a technology-based project, as discussed in Section 1.5. They must therefore rely upon the governance committee and team members for support in project decision making.

The project manager may have no say or input on the imposed constraints or boundary conditions for the project. These factors may have been made by the client or the sales force during competitive bidding activities and the project manager is told that he/she must live with these conditions. It is not uncommon for the sales force to agree to unrealistic budgets and schedules just to win a contract and then tell the project manager, "This is all the time and money we could get from the client. Live with it."

Finally, the new project manager cannot take for granted that he or she fully understands the role of the participants. Because each project will be different, the roles of the players and the accompanying interface can change. This is discussed in Sections 1.6–1.10.

The characteristics of a project can change from company to company. It is important for the newly appointed or entry-level project managers to have a good understanding of what the job entails before accepting the position.

1.5 THE TALENT TRIANGLE

PMBOK® Guide, 6th Edition
3.4 Project Management Competencies

Each project is inherently different, thus possibly mandating a different set of competencies. PMI has introduced the Talent Triangle that represents the high-level skill set that global organizations consider important for project management practitioners. The Talent Triangle includes:

- Technical Project Management
- Leadership
- Strategic and Business Management

The components of the three skill areas can change between project, program, and portfolio management activities. Technical project management and leadership have been discussed briefly in this chapter and will be discussed in more depth throughout the book.

Strategic and business management is relatively new for many project managers. In some companies, the responsibility for strategic and business decisions rests solely with the project sponsor. In these situations, the project manager's primary role is to produce a deliverable and most often a technical deliverable. How the deliverable will be used and whether or not it provided value to the firm is determined by the project sponsor.

In today's world, project managers must be strategic and business oriented. Project managers today are managing more than just a project. They view themselves as managing part of a business rather than just a project and thus are expected to make both project technical and business decisions. The tools that the project manager uses, specifically the project management methodologies, are embedded with business processes rather than merely pure project management processes.

The words "business value" could become the most important words in the project manager's vocabulary. The outcome of a project is no longer just a deliverable; rather, it is now focusing more on the creation of sustainable business value. Project success is now the creation of sustainable business value rather than merely meeting certain imposed constraints. All through this chapter we focused on the importance of value, and this trend is expected to increase.

1.6 TECHNOLOGY-BASED PROJECTS

PMBOK® Guide, 6th Edition
3.4.2 Technical Project Management Skills

Technology-based projects are often considered the most difficult projects to manage, especially for entry-level project managers. There is a high degree of complexity, innovation is required, the risks are most often greater than with traditional projects, and the solution requires experimentation, iterative approaches, and creativity. According to Hans Thamhain,[1]

---

1. Hans J. Thamhain, *Managing Technology-Based Projects* (Hoboken, NJ: John Wiley & Sons, 2014), p.5.

In our highly connected world, most project managers must deal with technology. They must function in a business environment that uses technology for competitive advantage, and their projects are heavily steeped in technology. Virtually every segment of industry and government tries to leverage technology to improve effectiveness, value, and speed. Traditional linear work processes and top-down controls are no longer sufficient but are gradually being replaced by alternative organizational designs and new, more agile management techniques and business processes, such as concurrent engineering, design-build, stage-gate, and user-centered design. These techniques offer more sophisticated capabilities for cross-functional integration, resource mobility, effectiveness, and market responsiveness, but they also require more sophisticated skills to effectively deal with a broad spectrum of contemporary challenges, both technically and socially, including higher levels of conflict, change, risks, uncertainty, and a shifting attention from functional efficiency to process integration effectiveness, emphasizing organizational interfaces, human factors, and the overall business process. Taken together, technology-intensive projects can be characterized as follows:

- Value creation by applying technology
- Strong need for innovation and creativity
- High task complexities, risk, and uncertainties
- Resource constraints and tight end-date-driven schedules despite tough performance requirements
- Highly educated and skilled personnel, broad skill spectrum
- Specific technical job knowledge and competency
- Need for sophisticated people skills, ability to work across different organizational cultures and values, and to deal with organizational conflict, power, and politics
- Complex project organizations and cross-functional linkages
- Complex business processes and stakeholder communities
- Technology used as a tool for managing projects
- Replacement of labor with technology
- Advanced infrastructure
- High front-end expenditures early in the project life cycle
- Low short-term profitability in spite of large capital investment
- Fast-changing markets, technology, and regulations
- Intense global competition, open markets, and low barriers to entry
- Short product life cycles that affect time to market
- Need for quick market response
- Complex decision-making processes
- Many alliances, joint ventures, and partnerships

1.7 THE PROJECT MANAGER–LINE MANAGER INTERFACE

***PMBOK® Guide*, 6th Edition**
3.4 Project Management Competencies

We have stated that the project manager must control company resources within time, cost, and performance. Most companies have six resources:

- Money
- Employees

- Equipment
- Facilities
- Materials
- Information/technology

Actually, the project manager does *not* control any of these resources directly, except perhaps money (i.e., the project budget).[2] Resources are controlled by the line managers, functional managers, or, as they are often called, resources managers. Project managers must, therefore, negotiate with line managers for all project resources. When we say that project managers control project resources, we really mean that they control those resources (which are temporarily loaned to them) *through line managers.*

Today, we have a new breed of project manager. Years ago, virtually all project managers were engineers with advanced degrees. These people had a command of technology rather than merely an understanding of technology. If the line manager believed that the project manager did in fact possess a command of technology, then the line manager would allow the assigned functional employees to take direction from the project manager. The result was that project managers were expected to manage people.

Most project managers today have an understanding of technology rather than a command of technology. As a result, the accountability for the success of the project is now viewed as shared accountability between the project manager and all affected line managers. With shared accountability, the line managers must now have a good understanding of project management, which is why more line managers are becoming PMP® credential holders. Project managers are now expected to focus more so on managing the project's deliverables rather than providing technical direction to the project team. Management of the assigned resources is more often than not a line function.

Another important fact is that project managers are treated as though they are managing part of a business rather than simply a project, and thus are expected to make sound business decisions as well as project decisions. Project managers must understand business principles. In the future, project managers may be expected to become externally certified by PMI and internally certified by their company on the organization's business processes.

In recent years, the rapid acceleration of technology has forced the project manager to become more business oriented.

It should become obvious at this point that successful project management is strongly dependent on:

- A good daily working relationship between the project manager and those line managers who directly assign resources to projects
- The ability of functional employees to report vertically to line managers at the same time that they report horizontally to one or more project managers

These two items become critical. In the first item, functional employees who are assigned to a project manager still take technical direction from their line managers. Second,

---

2. Here we are assuming that the line manager and project manager are not the same individual. However, the terms *line manager* and *functional manager* are used interchangeably throughout the text.

employees who report to multiple managers will always favor the manager who controls their purse strings. Thus, most project managers appear always to be at the mercy of the line managers.

If we take a close look at project management, the project manager actually works for the line managers, not vice versa. Many executives do not realize this. They have a tendency to put a halo around the head of the project manager and give him a bonus at project completion when, in fact, the credit should be shared with the line managers, who are continually pressured to make better use of their resources while meeting the project's constraints. The project manager is simply the agent through whom this is accomplished. So why do some companies glorify the project management position?

When the project management–line management relationship begins to deteriorate, the project almost always suffers. Executives must promote a good working relationship between line and project management. One of the most common ways of destroying this relationship is by asking, "Who contributes to profits—the line or project manager?" Project managers feel that they control all project profits because they control the budget. The line managers, on the other hand, argue that they must staff with appropriately budgeted-for personnel, supply the resources at the desired time, and supervise performance. Actually, both the vertical and horizontal lines contribute to profits. These types of conflicts can destroy the entire project management system.

Effective project management requires an understanding of quantitative tools and techniques, organizational structures, and organizational behavior.

Most people understand the quantitative tools for planning, scheduling, and controlling work. It is imperative that project managers understand totally the operations of each line organization. In addition, project managers must understand their own job description, especially where their authority begins and ends.

Organizational behavior is important because the functional employees at the interface position find themselves reporting to more than one boss—a line manager and one project manager for each project they are assigned to. Executives must provide proper training so functional employees can report effectively to multiple managers.

1.8 DEFINING THE PROJECT MANAGER'S ROLE

***PMBOK® Guide*, 6th Edition**
2.4.3 Organizational Governance Frameworks
Chapter 3 Role of the Project Management
Chapter 4 Project Integration Management

The project manager is responsible for coordinating and integrating activities across multiple functional lines. The integration activities performed by the project manager include:

- Integrating the activities necessary to develop a project plan
- Integrating the activities necessary to execute the plan
- Integrating the activities necessary to make changes to the plan

These integrative responsibilities are shown in Figure 1–4, where the project manager must convert the inputs (i.e., resources) into outputs of products, services, and ultimately profits. In order to do this, the project manager needs strong communicative and

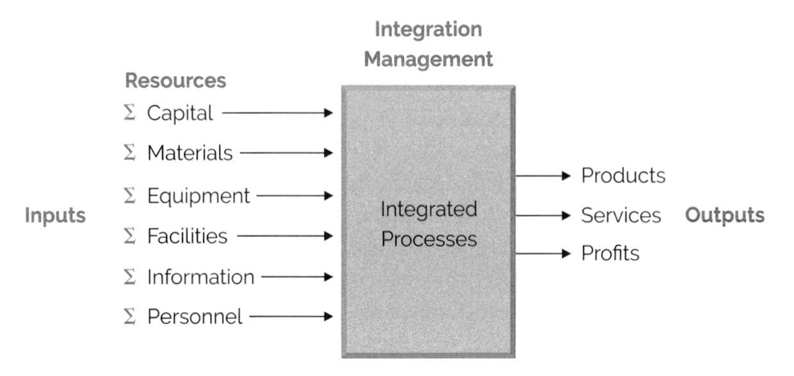

FIGURE 1–4. Integration management.

interpersonal skills, must become familiar with the operations of each line organization, and must have knowledge of the technology being used.

***PMBOK® Guide*, 6th Edition**
Chapter 4 Integration Management

An executive with a computer manufacturer stated that his company was looking externally for project managers. When asked if he expected candidates to have a command of computer technology, the executive remarked, "You give me an individual who has good communicative skills and interpersonal skills, and I'll give that individual a job. I can teach people the technology and give them technical experts to assist them in decision making. But I cannot teach somebody how to work with people."

The project manager's job is not an easy one. Project managers may have increasing responsibility, but very little authority. This lack of authority can force them to "negotiate" with upper-level management as well as functional management for control of company resources. They may often be treated as outsiders by the formal organization.

In the project environment, everything seems to revolve about the project manager.

Although the project organization is a specialized, task-oriented entity, it cannot exist apart from the traditional structure of the organization. The project manager, therefore, must walk the fence between the two organizations. The term *interface management* is often used for this role, which can be described as managing relationships:

- Within the project team
- Between the project team and the functional organizations
- Between the project team and senior management
- Between the project team and the customer's organization, whether an internal or external organization

The project manager is actually a general manager and gets to know the total operation of the company. In fact, project managers get to know more about the total operation of a company than most executives. That is why project management is often used as a training ground to prepare future general managers who will be capable of filling top management positions.

1.9 DEFINING THE FUNCTIONAL MANAGER'S ROLE

PMBOK® Guide, 5th Edition
Chapter 9 Project Resources Management
9.3 Acquire Resources

Assuming that the project and functional managers are not the same person, we can identify a specific role for the functional manager. There are three elements to this role:

- The functional manager has the responsibility to define *how* the task will be done and *where* the task will be done (i.e., the technical criteria).
- The functional manager has the responsibility to provide sufficient resources to accomplish the objective within the project's constraints (i.e., *who* will get the job done).
- The functional manager has the responsibility for the deliverable.

In other words, once the project manager identifies the requirements for the project (i.e., what work has to be done and the constraints), it becomes the line manager's responsibility to identify the technical criteria. Except perhaps in R&D efforts, the line manager should be the recognized technical expert. If the line manager believes that certain technical portions of the project manager's requirements are unsound, then the line manager has the right, by virtue of his expertise, to take exception and plead his case to a higher authority.

In Section 1.1 we stated that all resources (including personnel) are controlled by the line manager. The project manager has the right to request specific staff, but the final appointments rest with line managers. It helps if project managers understand the line manager's problems:

- Unlimited work requests (especially during competitive bidding)
- Predetermined deadlines
- All requests having a high priority
- Limited number of resources
- Limited availability of resources
- Unscheduled changes in the project plan
- Unpredicted lack of progress
- Unplanned absence of resources
- Unplanned breakdown of resources
- Unplanned loss of resources
- Unplanned turnover of personnel

Only in a very few industries will the line manager be able to identify to the project manager in advance exactly what resources will be available when the project is scheduled to begin. It is not important for the project manager to have the best available resources. Functional managers should not commit to certain people's availability. Rather, the functional manager should commit to achieving his portion of the deliverables within time, cost, and performance even if he has to use average or below-average personnel. If the project manager is unhappy with the assigned functional resources, then the project manager should closely track that portion of the project. Only if and when the project manager is convinced by the evidence that the assigned resources are unacceptable should he confront the line manager and demand better resources.

The fact that a project manager is assigned does not relieve the line manager of his functional responsibility to perform. If a functional manager assigns resources such that the constraints are not met, then *both* the project and functional managers will be blamed. Some companies are even considering evaluating line managers for merit increases and promotion based on how often they have lived up to their commitments to the project managers.

PMBOK® Guide, 6th Edition
2.4.4 Organizational Structure Types

TABLE 1–1. DUAL RESPONSIBILITY

	Responsibility	
Topic	**Project Manager**	**Line Manager**
Rewards	Give recommendation: Informal	Provide rewards: Formal
Direction	Milestone (summary)	Detailed
Evaluation	Summary	Detailed
Measurement	Summary	Detailed
Control	Summary	Detailed

Therefore, it is extremely valuable to everyone concerned to have all project commitments *made visible to all.*

Project management is designed to have shared authority and responsibility between the project and line managers. Project managers plan, monitor, and control the project, whereas functional managers perform the work. Table 1–1 shows this shared responsibility. The one exception to Table 1–1 occurs when the project and line managers are the same person. This situation, which happens more often than not, creates a conflict of interest. If a line manager has to assign resources to six projects, one of which is under his direct control, he might save the best resources for his project. In this case, his project will be a success at the expense of all of the other projects.

The exact relationship between project and line managers is of paramount importance in project management where multiple-boss reporting prevails. Table 1–2 shows that the relationship between project and line managers is not always in balance and thus, of course, has a bearing on who exerts more influence over the assigned functional employees.

PMBOK® Guide, 6th Edition
2.4.4 Organizational Structure Types

TABLE 1–2. REPORTING RELATIONSHIPS

		Project Manager (PM)/Line Manager (LM)/Employee Relationship			
Type of Project Manager	**Type of Matrix Structure***	**PM Negotiates For**	**Employees Take Technical Direction From**	**PM Receives Functional Progress From**	**Employee Performance Evaluations Made By**
Lightweight	Weak	Deliverables	LMs	Primarily LMs	LMs only with no input from PM
Heavyweight	Strong	People who report informally to PM but formally to LMs	PM and LMs	Assigned employees who report to LMs	LMs with input from PM
Tiger teams	Very strong	People who report entirely to PM full-time for duration of project	PM only	Assigned employees who now report directly to PM	PM only

*The types of organizational structures are discussed in Chapter 3.

1.10 DEFINING THE FUNCTIONAL EMPLOYEE'S ROLE

Once the line managers commit to the deliverables, it is the responsibility of the assigned functional employees to achieve the functional deliverables.

In most organizations, the assigned employees report on a "solid" line to their functional manager, even though they may be working on several projects simultaneously. The employees are usually a "dotted" line to the project but solid to their function. This places the employees in the often awkward position of reporting to multiple individuals. This situation is further complicated when the project manager has more technical knowledge than the line manager. This occurs during R&D projects.

The functional employee is expected to accomplish the following activities when assigned to projects:

- Accept responsibility for accomplishing the assigned deliverables within the project's constraints
- Complete the work at the earliest possible time
- Periodically inform both the project and line manager of the project's status
- Bring problems to the surface quickly for resolution
- Share information with the rest of the project team

1.11 DEFINING THE EXECUTIVE'S ROLE

PMBOK® Guide, 6th Edition
2.4.2 Organizational Governance Frameworks

In a project environment there are new expectations of and for the executives, as well as a new interfacing role.[3] Executives are expected to interface a project as follows:

- In project planning and objective setting
- In conflict resolution
- In priority setting
- As project sponsor[4]

Executives are expected to interface with projects very closely at project initiation and planning, but to remain at a distance during execution unless needed for priority setting and conflict resolution. One reason why executives "meddle" during project execution is that they are not getting accurate information from the project manager about project status. If project managers provide executives with meaningful status reports, then the so-called meddling may be reduced or even eliminated.

1.12 WORKING WITH EXECUTIVES

Success in project management is like a three-legged stool. The first leg is the project manager, the second leg is the line manager, and the third leg is senior management. If any of the three legs fail, the stool will topple.

3. The expectations are discussed in Section 9.3.
4. The role of the project sponsor is discussed in Section 10.1.

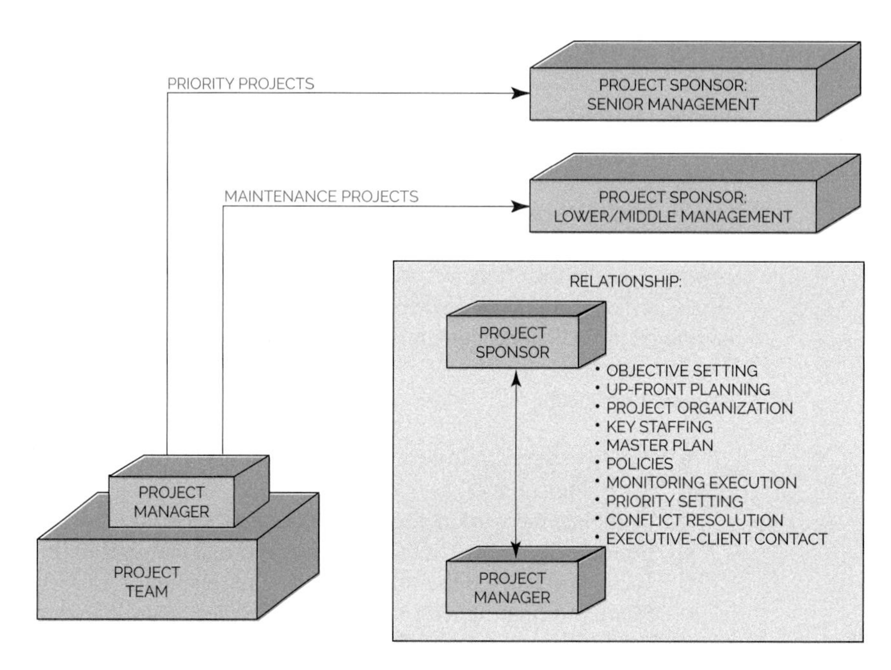

FIGURE 1–5. The project sponsor interface.

The critical node in project management is the project manager–line manager interface. At this interface, the project and line managers must view each other as equals and be willing to share authority, responsibility, and accountability. In excellently managed companies, project managers do not negotiate for resources but simply ask for the line manager's commitment to executing his portion of the work within time, cost, and performance. Therefore, it should not matter who the line manager assigns as long as the line manager lives up to his commitments.

Since the project and line managers are "equals," senior management involvement is necessary to provide advice and guidance to the project manager, as well as to provide encouragement to the line managers to keep their promises. When executives act in this capacity, they assume the role of project sponsors, as shown in Figure 1–5,[5] which also shows that sponsorship need not always be at the executive levels. The exact person appointed as the project sponsor is based on the dollar value of the project, the priority of the project, and who the customer is.

The ultimate objective of the project sponsor is to provide behind-the-scenes assistance to project personnel for projects both "internal" to the company, as well as "external," as shown in Figure 1–5. Projects can still be successful without this commitment and support, as long as all work flows smoothly. But in time of crisis, having a "big brother" available as a possible sounding board will surely help.

PMBOK® Guide, 6th Edition
2.4.2 Organizational Governance Frameworks

When an executive is required to act as a project sponsor, then the executive has the responsibility to make effective and timely project decisions. To accomplish this, the executive needs timely,

5. Section 10.1 describes the role of the project sponsor in more depth.

accurate, and complete data for such decisions. Keeping management informed serves this purpose, while the all-too-common practice of "stonewalling" prevents an executive from making effective project decisions.

It is not necessary for project sponsorship to remain exclusively at the executive levels. As companies mature in their understanding and implementation of project management, project sponsorship may be pushed down to middle-level management. Committee sponsorship is also possible.

1.13 COMMITTEE SPONSORSHIP/GOVERNANCE

All projects have the potential of getting into trouble but, in general, project management can work well as long as the project's requirements do not impose severe pressure upon the project manager and a project sponsor exists as an ally to assist the project manager when trouble does appear.

Project problems requiring executive-level support may not be able to be resolved, at least easily and in a timely manner, by a single project sponsor. These problems can be resolved using effective project governance. Project governance is actually a framework by which decisions are made. Governance relates to decisions that define expectations, accountability, responsibility, the granting of power, or verifying performance. Governance relates to consistent management, cohesive policies, and processes and decision-making rights for a given area of responsibility. Governance enables efficient and effective decision making to take place.

Every project can have different governance even if each project uses the same enterprise project management methodology. The governance function can operate as a separate process or as part of project management leadership. Governance is designed not to replace project decision making but to prevent undesirable decisions from being made.

Historically, governance was provided by a single project sponsor. Today, governance is a committee and can include representatives from each stakeholder's organization. Table 1–3

TABLE 1–3. TYPES OF PROJECT GOVERNANCE

Structure	Description	Governance
Dispersed locally	Team members can be full- or part-time. They are still attached administratively to their functional area.	Usually a single person is acting as the sponsor but may be an internal committee based upon the project's complexity.
Dispersed geographically	This is a virtual team. The project manager may never see some of the team members. Team members can be full- or part-time.	Usually governance by committee and can include stakeholder membership.
Colocated	All of the team members are physically located in close proximity to the project manager. The project manager does not have any responsibility for wage and salary administration.	Usually a single person acting as the sponsor.
Projectized	This is similar to a colocated team but the project manager generally functions as a line manager and may have wage and salary responsibilities.	May be governance by committee based upon the size of the project and the number of strategic partners.

shows various governance approaches based upon the type of project team. The membership of the committee can change from project to project and industry to industry. The membership may also vary based upon the number of stakeholders and whether the project is for an internal or external client. On long-term projects, membership can change throughout the project.

Governance on projects and programs sometimes fails because people confuse project governance with corporate governance. The result is that members of the committee are not sure what their role should be. Some of the major differences include:

- **Alignment**: Corporate governance focuses on how well the portfolio of projects is aligned to and satisfies overall business objectives. Project governance focuses on ways to keep a project on track.
- **Direction**: Corporate governance provides strategic direction with a focus on how project success will satisfy corporate objectives. Project governance is more operation direction with decisions based upon the predefined parameters on project scope, time, cost, and functionality.
- **Dashboards**: Corporate governance dashboards are based upon financial, marketing, and sales metrics. Project governance dashboards have operations metrics on time, cost, scope, quality, action items, risks, and deliverables.
- **Membership**: Corporate governance committees are composed of the seniormost levels of management. Project government membership may include some membership from middle management.

Another reason why failure may occur is when members of the project or program governance group do not understand project or program management. This can lead to micromanagement by the governance committee. There is always the question of what decisions must be made by the governance committee and what decisions the project manager can make. In general, the project manager should have the authority for decisions related to actions necessary to maintain the baselines. Governance committees must have the authority to approve scope changes above a certain dollar value and to make decisions necessary to align the project to corporate objectives and strategy.

1.14 THE PROJECT MANAGER AS THE PLANNING AGENT

PMBOK® Guide, 6th Edition
Chapter 9 Project Resource Management

The major responsibility of the project manager is planning. If project planning is performed correctly, then it is conceivable that the project manager will work himself out of a job because the project can run itself. This rarely happens, however. Few projects are ever completed without some conflict or trade-offs for the project manager to resolve.

In most cases, the project manager provides overall or summary definitions of the work to be accomplished, but the line managers (the true experts) do the detailed planning. Although project managers cannot control or assign line resources, they must make sure that the resources are adequate and scheduled to satisfy the needs of the project, not vice versa. As the architect of the project plan, the project manager must provide:

- Complete task definitions
- Resource requirement definitions (possibly skill levels)
- Major timetable milestones
- Definition of end-item quality and reliability requirements
- The basis for performance measurement
- Definition of project success

These factors, if properly established, result in:

- Assurance that functional units will understand their total responsibilities toward achieving project needs
- Assurance that problems resulting from scheduling and allocation of critical resources are known beforehand
- Early identification of problems that may jeopardize successful project completion so that effective corrective action and replanning can be taken to prevent or resolve the problems

Project managers are responsible for project administration and, therefore, must have the right to establish their own policies, procedures, rules, guidelines, and directives—provided these policies, guidelines, and so on conform to overall company policy. Companies with mature project management structures usually have rather loose company guidelines, so project managers have some degree of flexibility in how to control their projects.

Establishing project administrative requirements is part of project planning. Executives must either work with the project managers at project initiation or act as resources later. Improper project administrative planning can create a situation that requires:

- A continuous revision and/or establishment of company and/or project policies, procedures, and directives
- A continuous shifting in organizational responsibility and possible unnecessary restructuring
- A need for staff to acquire new knowledge and skills

If these situations occur simultaneously on several projects, there can be confusion throughout the organization.

1.15 PROJECT CHAMPIONS

Corporations encourage employees to think up new ideas that, if approved by the corporation, will generate monetary and nonmonetary rewards for the idea generator. One such reward is naming the individual the "project champion." Unfortunately, the project champion often becomes the project manager, and, although the idea was technically sound, the project fails.

TABLE 1–4. PROJECT MANAGER VERSUS PROJECT CHAMPIONS

Project Managers	Project Champions
• Prefer to work in groups • Committed to their managerial and technical responsibilities • Committed to the corporation • Seek to achieve the objective • Are willing to take risks • Seek what is possible • Think in terms of short time spans • Manage people • Are committed to and pursue material values	• Prefer working individually • Committed to technology • Committed to the profession • Seek to exceed the objective • Are unwilling to take risks; try to test everything • Seek perfection • Think in terms of long time spans • Manage things • Are committed to and pursue intellectual values

Table 1–4 provides a comparison between project managers and project champions. It shows that the project champions may become so attached to the technical side of the project that they become derelict in their administrative responsibilities. Perhaps the project champion might function best as a project engineer rather than the project manager.

This comparison does not mean that technically oriented project managers-champions will fail. Rather, it implies that the selection of the "proper" project manager should be based on *all* facets of the project.

1.16 PROJECT-DRIVEN VERSUS NON–PROJECT-DRIVEN ORGANIZATIONS

PMBOK® Guide, 6th Edition
2.4.1 Organizational Systems Overview

On the micro level, virtually all organizations are either marketing-, engineering-, or manufacturing-driven. But on the macro level, organizations are either project- or non–project-driven. The *PMBOK® Guide* uses the terms *project-based* and *non–project-based,* whereas in this text the terms *project-driven* and *non–project-driven* or *operational-driven* are used. In a project-driven organization, such as construction or aerospace, all work is characterized through projects, with each project as a separate cost center having its own profit-and-loss statement. The total profit to the corporation is simply the summation of the profits on all projects. In a project-driven organization, everything centers on the projects.

In the non–project-driven organization, such as low-technology manufacturing, profit and loss are measured on vertical or functional lines. In this type of organization, projects exist merely to support the product lines or functional lines. Priority resources are assigned to the revenue-producing functional line activities rather than the projects. Project management in a non–project-driven organization is generally more difficult for these reasons:

- Projects may be few and far between.
- Not all projects have the same project management requirements, and therefore they cannot be managed identically. This difficulty results from poor understanding of project management and a reluctance of companies to invest in proper training.
- Executives do not have sufficient time to manage projects themselves, yet refuse to delegate authority.

- Projects tend to be delayed because approvals most often follow the vertical chain of command. As a result, project work stays too long in functional departments.
- Because project staffing is on a "local" basis, only a portion of the organization understands project management and sees the system in action.
- There is heavy dependence on subcontractors and outside agencies for project management expertise.

Non–project-driven organizations may also have a steady stream of projects, all of which are usually designed to enhance manufacturing operations. Some projects may be customer-requested, such as:

- The introduction of statistical dimensioning concepts to improve process control
- The introduction of process changes to enhance the final product
- The introduction of process change concepts to enhance product reliability

If these changes are not identified as specific projects, the result can be:

- Poorly defined responsibility areas within the organization
- Poor communications, both internal and external to the organization
- Slow implementation
- A lack of a cost-tracking system for implementation
- Poorly defined performance criteria

Figure 1–6 shows the tip-of-the-iceberg syndrome, which can occur in all types of organizations but is most common in non–project-driven organizations. On the surface, all

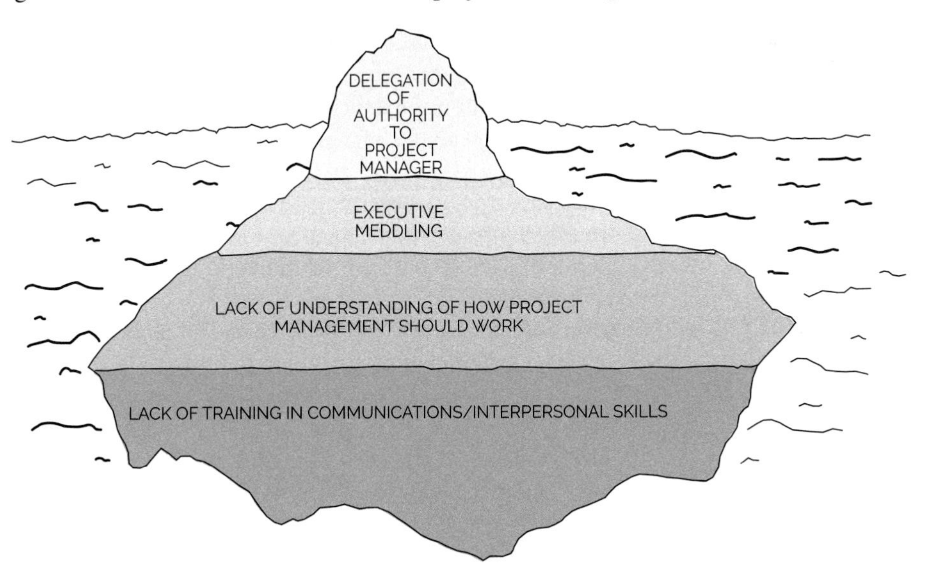

FIGURE 1–6. The tip-of-the-iceberg syndrome for matrix implementation.

we see is a lack of authority for the project manager. But beneath the surface we see the causes: there is excessive meddling due to lack of understanding of project management, which, in turn, resulted from an inability to recognize the need for proper training.

1.17 MARKETING IN THE PROJECT-DRIVEN ORGANIZATION

PMBOK® Guide, 6th Edition
1.2.3.6 Organizational Project Management (OPM) and Strategies

Getting new projects is the lifeblood of any project-oriented business. The practices of the project-oriented company are, however, substantially different from those of traditional product businesses and require highly specialized and disciplined team efforts among marketing, technical, and operating personnel, plus significant customer involvement. Projects are different from products in many respects, especially marketing. Marketing projects requires the ability to identify, pursue, and capture one-of-a-kind business opportunities, and is characterized by:

- *A systematic effort*: A systematic approach is usually required to develop a new program lead into an actual contract. The project acquisition effort is often highly integrated with ongoing programs and involves key personnel from both the potential customer and the performing organization.
- *Custom design*: While traditional businesses provide standard products and services for a variety of applications and customers, projects are custom-designed items designed to fit specific requirements of a single-customer community.
- *Project life cycle*: Project-oriented businesses have a well-defined beginning and end and are not self-perpetuating. Business must be generated on a project-by-project basis rather than by creating demand for a standard product or service.
- *Marketing phase*: Long lead times often exist between the product definition, start-up, and completion phases of a project.
- *Risks*: There are risks, especially in the research, design, and production of programs. The program manager not only has to integrate the multidisciplinary tasks and project elements within budget and schedule constraints, but also has to manage inventions and technology while working with a variety of technically oriented prima donnas.
- *The technical capability to perform*: Technical ability is critical to the successful pursuit and acquisition of a new project.

In spite of the risks and problems, profits on projects are usually very low in comparison with commercial business practices. One may wonder why companies pursue project businesses. Clearly, there are many reasons why projects are good business:

- Although immediate profits (as a percentage of sales) are usually small, the return on capital investment is often very attractive. Progress payment practices keep inventories and receivables to a minimum and enable companies to undertake projects many times larger in value than the assets of the total company.

- Once a contract has been secured and is being managed properly, the project may be of relatively low financial risk to the company. The company has little additional selling expenditure and has a predictable market over the life cycle of the project.
- Project business must be viewed from a broader perspective than motivation for immediate profits. Projects provide an opportunity to develop the company's technical capabilities and build an experience base for future business growth.
- Winning one large project often provides attractive growth potential, such as (1) growth with the project via additions and changes; (2) follow-on work; (3) spare parts, maintenance, and training; and (4) being able to compete effectively in the next project phase, such as nurturing a study program into a development contract and finally a production contract.

Customers come in various forms and sizes. For small and medium businesses particularly, it is a challenge to compete for contracts from large industrial or governmental organizations. Although the contract to a firm may be relatively small, it is often subcontracted via a larger organization. Selling to such a diversified heterogeneous customer is a marketing challenge that requires a highly sophisticated and disciplined approach.

The first step in a new business development effort is to define the market to be pursued. The market segment for a new program opportunity is normally in an area of relevant past experience, technical capability, and customer involvement. Good marketers in the program business have to think as product line managers. They have to understand all dimensions of the business and be able to define and pursue market objectives that are consistent with the capabilities of their organizations.

Program businesses operate in an opportunity-driven market. It is a common mistake, however, to believe that these markets are unpredictable and unmanageable. Market planning and strategizing is important. New project opportunities develop over periods of time, sometimes years for larger projects. These developments must be properly tracked and cultivated to form the bases for management actions such as (1) bid decisions, (2) resource commitment, (3) technical readiness, and (4) effective customer liaison.

1.18 CLASSIFICATION OF PROJECTS

The principles of project management can be applied to any type of project and to any industry. However, the relative degree of importance of these principles can vary from project to project and industry to industry. Table 1–5 shows a brief comparison of certain industries/projects.

For those industries that are project-driven, such as aerospace and large construction, the high dollar value of the projects mandates a much more rigorous project management approach. For non–project-driven industries, projects may be managed more informally than formally, especially if no immediate profit is involved. Informal project management is similar to formal project management but paperwork requirements are kept at a minimum.

TABLE 1–5. CLASSIFICATION OF PROJECTS/CHARACTERISTICS

	Type of Project/Industry					
	In-house R&D	**Small Construction**	**Large Construction**	**Aerospace/ Defense**	**MIS**	**Engineering**
Need for interpersonal skills	Low	Low	High	High	High	Low
Importance of organizational structure	Low	Low	Low	Low	High	Low
Time management difficulties	Low	Low	High	High	High	Low
Number of meetings	Excessive	Low	Excessive	Excessive	High	Medium
Project manager's supervisor	Middle management	Top management	Top management	Top management	Middle management	Middle management
Project sponsor present	Yes	No	Yes	Yes	No	No
Conflict intensity	Low	Low	High	High	High	Low
Cost control level	Low	Low	High	High	Low	Low
Level of planning/ scheduling	Milestones only	Milestones only	Detailed plan	Detailed plan	Milestones only	Milestones only

1.19 LOCATION OF THE PROJECT MANAGER

The success of project management could easily depend on the location of the project manager within the organization. Two questions must be answered:

- What salary should the project manager earn?
- To whom should the project manager report?

Figure 1–7 shows a typical organizational hierarchy. Ideally, the project manager should be at the same pay grade as the individuals with whom he must negotiate on a daily basis. A project manager earning substantially more or less money than the line manager will usually create conflict. The ultimate reporting location of the project manager (and perhaps his salary) is heavily dependent on whether the organization is project- or non–project-driven, and whether the project manager is responsible for profit or loss.

Project managers can end up reporting both high and low in an organization during the life cycle of the project. During the planning phase of the project, the project manager may report high, whereas during implementation, he may report low. Likewise, the positioning of the project manager may be dependent on the risk of the project, the size of the project, or the customer.

Finally, it should be noted that even if the project manager reports low, he should still have the right to interface with top executives during project planning, although there may be two or more reporting levels between the project manager and executives. At the

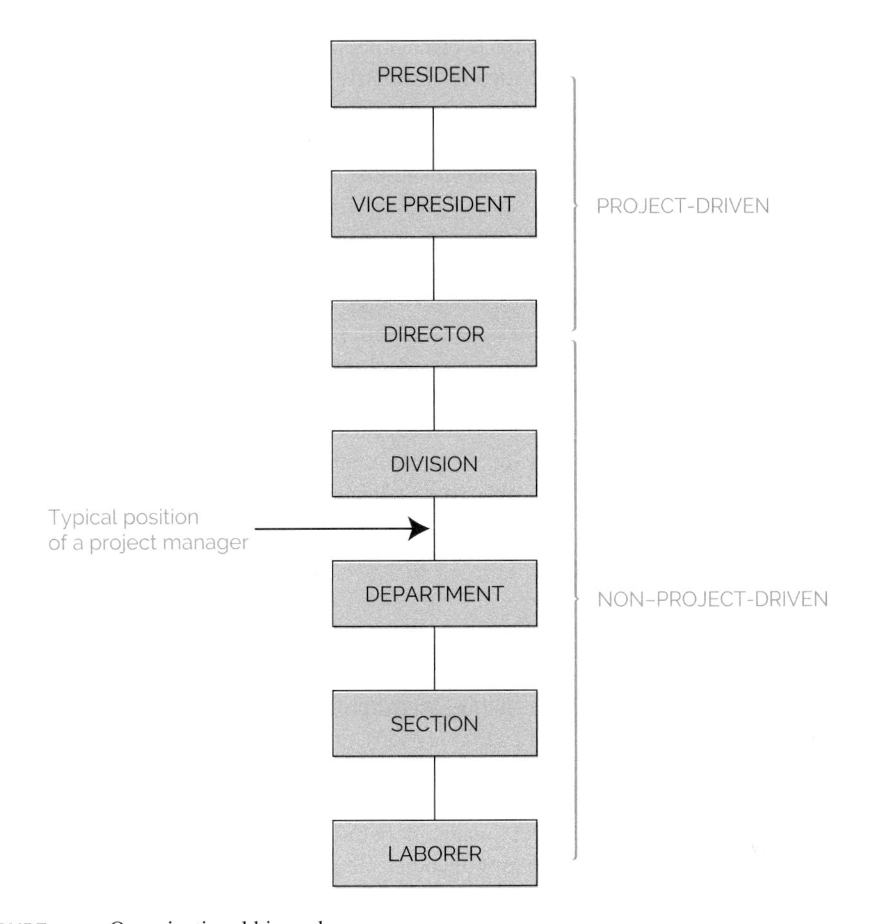

FIGURE 1–7. Organizational hierarchy.

opposite end of the spectrum, the project manager should have the right to go directly into the depths of the organization instead of having to follow the chain of command downward, especially during planning.

1.20 DIFFERING VIEWS OF PROJECT MANAGEMENT

Many companies, especially those with project-driven organizations, have differing views of project management. Some people view project management as an excellent means to achieving objectives, while others view it as a threat. In project-driven organizations, there are three career paths that lead to executive management:

- Through project management
- Through project engineering
- Through line management

A project engineer is often a person who performs project management coordination and integration activities for primarily engineering activities and may be restricted to just engineering work. There may also be manufacturing engineers that handle only that portion of the project that is in manufacturing.

In project-driven organizations, the fast-track position is in project management, whereas in a non–project-driven organization, it would be line management. Even though line managers support the project management approach, they resent the project manager because of his promotions and top-level visibility. In one construction company, a department manager was told that he had no chance for promotion above his present department manager position unless he went into project management or project engineering where he could get to know the operation of the whole company. A second construction company requires that individuals aspiring to become a department manager first spend a "tour of duty" as an assistant project manager or project engineer.

Executives may dislike project managers because more authority and control must be delegated. However, once executives realize that it is a sound business practice, it becomes important.

1.21 PUBLIC-SECTOR PROJECT MANAGEMENT

For several decades, public-sector projects were managed by contractors whose primary objective was a profit motive. Many times, contractors would make trade-offs and accompanying decisions just to support the profit motive. At the end of the project, the contractor would provide the public-sector agency with a deliverable, but the contractor would walk away with the project management best practices and lessons learned.

Today, public-sector agencies are requesting the contractor to share with them all project management intellectual property accumulated during the course of the project. Also, more agencies are becoming experienced in project management to the point where the projects are managed with internal personnel rather than contractors.

As more and more government agencies adopt the project management approach, we discover that public-sector projects can be more complex than private-sector projects and more difficult to manage.

THE CHALLENGES OF PUBLIC-SECTOR PROJECT MANAGEMENT

Private-sector project managers like to assume that their work is more demanding than projects in the public sector. They assume that their projects are more complex, subject to tougher management oversight, and mandated to move at faster speeds. Although private-sector projects can be tough, in many cases, it is easier to accomplish results in the private sector than in the public sector.

Public-sector projects can be more difficult than many private-sector projects because they:

- Operate in an environment of often-conflicting goals and outcome
- Involve many layers of stakeholders with varied interests
- Must placate political interests and operate under media scrutiny
- Are allowed little tolerance for failure
- Operate in organizations that often have a difficult time identifying outcome measures and missions
- Are required to be performed under constraints imposed by administrative rules and often-cumbersome policies and processes that can delay projects and consume project resources
- Require the cooperation and performance of agencies outside of the project team for purchasing, hiring, and other functions
- Must make do with existing staff resources more often than private-sector projects because of civil-service protections and hiring systems
- Are performed in organizations that may not be comfortable or used to directed action and project success
- Are performed in an environment that may include political adversaries

If these challenges were not tough enough, because of their ability to push the burden of paying for projects to future generations, public-sector projects have a reach deep into the future.[1] That introduces the challenges of serving the needs of stakeholders who are not yet "at the table" and whose interests might be difficult to identify. Some also cite the relative lack of project management maturity in public organizations as a challenge of public-sector projects.

In addition to these complications, public projects are often more complex than those in the private sector. For some projects, the outcome can be defined at the beginning of the project. Construction projects are one example. For other projects, the desired outcome can only be defined as the project progresses. Examples of those are organizational change projects and complex information technology projects. Although the first type of project can be difficult and require detailed planning and implementation, the second type, those whose outcomes are determined over the course of the project, are regarded as more challenging. They require more interaction with stakeholders and more openness to factors outside of the control of the project team.

Because of the multiple stakeholders involved in public-sector projects, the types of projects the public sector engages in, and the difficulty of identifying measurable outcomes in the public sector, more public-sector projects are likely to be of the latter variety and more difficult. As a result of the distinguishing characteristics of public-sector organizations, public-sector projects require the management not only of the project team but of an entire community. Little is accomplished in the public sector by lone individuals or even by teams working in isolation. Instead, public-sector projects engage

1. Project Management Institute, *Government Extension to the PMBOK® Guide Third Edition*, 2006, p. 15.

broad groups of stakeholders who not only have a stake in the project but also have a voice and an opportunity to influence outcomes. In public-sector projects, even though the project manager may be ultimately accountable, governance of the project and credit for successes must be shared.

The good news for public-sector project managers is that the community of stakeholders, which may seem to be a burden, can also be an opportunity and a source of resources and support. Many of those stakeholders stand ready to provide help to the project manager as he or she attempts to navigate the constraints affecting the project. Others can be enlisted to support the project, and their authority can make the difference between project success and failure.

THE COMING STORM

In addition to the existing challenges of public-sector projects listed previously, some factors will place soon more stress on public-sector organizations and demand even more emphasis on solid project management. Some of the emerging challenges for public-sector organizations will include:

- Modest or stagnant economic growth
- Globalization and the loss of the industrial revenue base and, increasingly, the service-sector revenue base
- A decline in real wages and pressures for tax reform
- Private-sector practices that pass the corporate safety net back to individuals, who may then look to government for such essential security mechanisms as health coverage
- Difficulty in passing on the need for government revenue to taxpayers and a general loss of confidence in government
- Structural limitations on revenue generation, such as Proposition 13 and property tax indexing
- The redirection of scarce public revenues to homeland security and defense without the imposition of war taxes
- The erosion of public-sector income as entitlement programs drain revenues in response to an aging population
- An age imbalance, with fewer workers in the workforce to support an expanding number of retirees and children
- Longer life expectancy, which further burdens entitlement and health programs
- Increasing costs of health care well beyond the level of inflation
- Long-delayed investments in our national infrastructure, including roads, bridges and water systems

In combination, these factors constitute a looming storm that will require us to question our assumptions about government operations and services. Doing far more with much less will require new thinking about how government performs its work. It will require more innovation than the development of new services. It will take radical rethinking of what government does and how it goes about getting it done.

WHY DO PUBLIC-SECTOR PROJECTS FAIL?

Public-sector projects can fail for a set of reasons related to the unique character of public-sector projects. In that regard, they:

- Run afoul of political processes
- Lack the necessary resources because of requirements to use existing staff rather than to contract for the right expertise
- Are constrained by civil-service rules that limit assignment of activities to project staff
- Lose budget authorization
- Lose support at the change of administration due to electoral cycles
- Are overwhelmed by administrative rules and required processes for purchasing and hiring
- Fail to satisfy oversight agencies
- Adopt overly conservative approaches due to the contentious nature of the project environment
- Are victimized by suboptimal vendors who have been selected by purchasing processes that are overly focused on costs or that can be influenced by factors that are not relevant to performance
- Are compromised by the bias of public-sector managers and staff toward compliance over performance
- Fail to identify project goals given the wide array of project stakeholders in the public sector and the challenges of identifying public-sector goals and metrics for success

Source: D. W. Wirick, *Public-Sector Project Management* (Hoboken, NJ: John Wiley & Sons, 2009), pp. 8–10, 18–19.

1.22 INTERNATIONAL PROJECT MANAGEMENT

As the world marketplace begins to accept project management and recognizes the need for experienced project managers, more opportunities have become available for people aspiring to become project managers. The need is there and growing. According to Thomas Grisham:[6]

6. T. W. Grisham, *International Project Management* (Hoboken, NJ: John Wiley and Sons, 2010), p. 3.

International business and project management practice have converged in the last 10 years. Organizations are tending toward hiring multitalented people who are self-motivated, intelligent, and willing to take responsibility. Some of the reasons are:

- The need for leaner and flatter organizations to reduce cost
- The need for leadership skills throughout the organizational food chain from top to bottom—lead one day, follow the next, and be comfortable personally in either role
- The need for knowledge workers throughout the organization
- Globalization and the need to improve quality while reducing cost
- Kaizen to keep quality high while reducing cost
- Diversity

Years ago, companies had three pay grades for project managers; junior project managers, project managers, and senior project managers. Today, we are adding in a fourth pay grade, namely global project managers. Unfortunately, there may be additional skills needed to be a global project manager. Some of the additional skills include managing virtual teams, understanding global cultural differences, working in an environment where politics can dictate many of the decisions, and working under committee governance rather than a single sponsor.

1.23 CONCURRENT ENGINEERING: A PROJECT MANAGEMENT APPROACH

In the past decade, organizations have become more aware of the fact that America's most formidable weapon is its manufacturing ability, and yet more and more work seems to be departing for Southeast Asia and the Far East. If America and other countries are to remain competitive, then survival may depend on the manufacturing of a quality product and a rapid introduction into the marketplace. Today, companies are under tremendous pressure to rapidly introduce new products because product life cycles are becoming shorter. As a result, organizations no longer have the luxury of performing work in series.

Concurrent or simultaneous engineering is an attempt to accomplish work in parallel rather than in series. This requires that marketing, R&D, engineering, and production are all actively involved in the early project phases and making plans even before the product design has been finalized. This concept of current engineering will accelerate product development, but it does come with serious and potentially costly risks, the largest one being the cost of rework.

Almost everyone agrees that the best way to reduce or minimize risks is for the organization to plan better. Since project management is one of the best methodologies to foster better planning, it is little wonder that more organizations are accepting project management as a way of life.

1.24 ADDED VALUE

People often wonder what project managers do with their time once the project plan is created. While it is true that they monitor and control the work being performed, they also look for ways to add value to the project. Added value can be defined as incremental improvements to the deliverable of a project such that performance is improved or a significant business advantage is obtained, and the client is willing to pay for this difference. Looking for added-value opportunities that benefit the client is a good approach whereas looking for "fictitious" added-value opportunities just to increase the cost of the project is bad.

In certain projects, such as in new product development in the pharmaceutical industry, project managers must be aware of opportunities. According to Trevor Brown and Stephen Allport:[7]

> The critical issues facing companies which understand the importance of building customer value into new products is how to incorporate this into the development process and invest appropriately to fully understand the opportunity. In practice, project teams have more opportunity than is generally realized to add, enhance, or diminish value in each of the four perspectives . . . corporate, prescriber, payer and patient. The tools at the project teams' disposal to enhance customer value include challenging and improving established processes, adopting a value-directed approach to the management of development projects, and taking advantage of tried and tested methodologies for understanding product value.

Project managers generally do not take enough time in evaluating opportunities. In such a case, either the scope change is disapproved or the scope change is allowed and suddenly the project is at risk when additional information is discovered. Opportunities must be fully understood.

Related Case Studies (from Kerzner/*Project Management Case Studies,* 5th ed.)	**Related Workbook Exercises (from Kerzner/ *Project Management Workbook and PMP®/ CAPM® Exam Study Guide,* 12th ed.)**	***PMBOK® Guide*, 6th Edition, Reference Section for the PMP® Certification Exam**
• Kombs Engineering • Williams Machine Tool Company* • Macon, Inc. • Jackson Industries • Olympics (A)	• Multiple choice exam	• Integration Management • Scope Management • Project Resource Management

*Case study also appears at end of chapter.

1.25 STUDYING TIPS FOR THE PMI® PROJECT MANAGEMENT CERTIFICATION EXAM

This section is applicable as a review of the principles or to support an understanding of the knowledge areas and domain groups in the *PMBOK® Guide*. This chapter addresses some material from the *PMBOK® Guide* knowledge areas:

- Integration Management
- Scope Management
- Project Resource Management

7. T. J. Brown and S. Allport, "Developing Products with Added Value," in P. Harpum (ed.), *Portfolio, Program, and Project Management in the Pharmaceutical and Biotechnology Industries* (Hoboken, NJ: John Wiley & Sons, 2010), p. 218.

Understanding the following principles is beneficial if the reader is using this textbook together with the *PMBOK® Guide* to study for the PMP® Certification Exam:

- Definition of a project
- Definition of the competing constraints
- Definition of successful execution of a project
- Benefits of using project management
- Responsibility of the project manager in dealing with stakeholders and how stakeholders can affect the outcome of the project
- Responsibility of the project manager in meeting deliverables
- The fact that the project manager is ultimately accountable for the success of the project
- Responsibilities of the line manager during project management staffing and execution
- Role of the executive sponsor and champion
- Difference between a project-driven and non–project-driven organization

Be sure to review the appropriate sections of the *PMBOK® Guide* and the glossary of terms at the end of the *PMBOK® Guide*.

Some multiple-choice questions are provided in this section as a review of the material. There are other sources for practice review questions that are specific for the PMP® Exam, namely:

- *Project Management IQ®* from the International Institute for Learning (iil.com)
- *PMP® Exam Practice Test and Study Guide,* by J. LeRoy Ward, PMP, editor
- *PMP® Exam Prep,* by Rita Mulcahy
- *Q & As for the PMBOK® Guide,* Project Management Institute

The more practice questions reviewed, the better prepared the reader will be for the PMP® Certification Exam.

In Appendix C, there are a series of mini case studies called Dorale Products that reviews some of the concepts. The mini cases can be used as either an introduction to the chapter or as a review of the chapter material. These mini case studies were placed in Appendix C because they can be used for several chapters in the text. For this chapter, the following are applicable:

- Dorale Products (A) [Integration and Scope Management]
- Dorale Products (B) [Integration and Scope Management]

Answers to the Dorale Products mini-cases appear in Appendix D. The following multiple-choice questions will be helpful in reviewing the above principles:

1. The traditional competing constraints on a project are:
   A. Time, cost, and profitability
   B. Resources required, sponsorship involvement, and funding
   C. Time, cost, and quality and/or scope
   D. Calendar dates, facilities available, and funding

2. Which of the following is not part of the definition of a project?
   A. Repetitive activities
   B. Constraints
   C. Consumption of resources
   D. A well-defined objective
3. Which of the following is usually not part of the criteria for project success?
   A. Customer satisfaction
   B. Customer acceptance
   C. Meeting at least 75 percent of specification requirements.
   D. Meeting the triple-constraint requirements
4. Which of the following is generally not a benefit achieved from using project management?
   A. Flexibility in the project's end date
   B. Improved risk management
   C. Improved estimating
   D. Tracking of projects
5. The person responsible for assigning the resources to a project is most often:
   A. The project manager
   B. The Human Resources Department
   C. The line manager
   D. The executive sponsor
6. Conflicts between the project and line managers are most often resolved by:
   A. The assistant project manager for conflicts
   B. The project sponsor
   C. The executive steering committee
   D. The Human Resources Department
7. Your company does only projects. If the projects performed by your company are for customers external to your company and a profit criterion exists on the project, then your organization is most likely:
   A. Project-driven
   B. Non–project-driven
   C. A hybrid
   D. All of the above are possible based upon the size of the profit margin.

ANSWERS

1. C
2. A
3. C
4. A
5. C
6. B
7. A

PROBLEMS

1–1 Because of the individuality of people, there always exist differing views of what management is all about. Below are lists of possible perspectives and a selected group of organizational members. For each individual select the possible ways that this individual might view project management:

Individuals

Upper-level manager
Project manager
Functional manager
Project team member
Scientist and consultant

Perspectives

a. A threat to established authority
b. A source for future general managers
c. A cause of unwanted change in ongoing procedures
d. A means to an end
e. A significant market for their services
f. A place to build an empire
g. A necessary evil to traditional management
h. An opportunity for growth and advancement
i. A better way to motivate people toward an objective
j. A source of frustration in authority
k. A way of introducing controlled changes
l. An area of research
m. A vehicle for introducing creativity
n. A means of coordinating functional units
o. A means of deep satisfaction
p. A way of life

1–2 Will project management work in all companies? If not, identify those companies in which project management may not be applicable and defend your answers.

1–3 What attributes should a project manager have? Can an individual be trained to become a project manager? If a company were changing over to a project management structure, would it be better to promote and train from within or hire from the outside?

1–4 What types of projects might be more appropriate for functional management rather than project management, and vice versa?

1–5 Do you think that there would be a shift in the relative degree of importance of the following terms in a project management environment as opposed to a traditional management environment?

a. Time management
b. Communications
c. Motivation

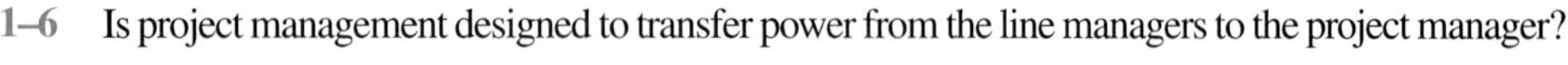
1–6 Is project management designed to transfer power from the line managers to the project manager?

1–7 Explain how career paths and career growth can differ between project-driven and non–project-driven organizations. In each organization, is the career path fastest in project management, project engineering, or line management?

CASE STUDY

WILLIAMS MACHINE TOOL COMPANY

For 85 years, the Williams Machine Tool Company had provided high-quality products to its clients, becoming the third largest U.S.-based machine tool company by 1990. The company was highly profitable and had an extremely low employee turnover rate. Pay and benefits were excellent.

Between 1980 and 1990, the company's profits soared to record levels. The company's success was due to one product line of standard manufacturing machine tools. Williams spent most of its time and effort looking for ways to improve its bread-and-butter product line rather than to develop new products. The product line was so successful that companies were willing to modify their production lines around these machine tools rather than asking Williams for major modifications to the machine tools.

By 1990, Williams Company was extremely complacent, expecting this phenomenal success with one product line to continue for 20 to 25 more years. The recession of the early 1990s forced management to realign their thinking. Cutbacks in production had decreased the demand for the standard machine tools. More and more customers were asking for either major modifications to the standard machine tools or a completely new product design.

The marketplace was changing and senior management recognized that a new strategic focus was necessary. However, lower-level management and the work force, especially engineering, were strongly resisting a change. The employees, many of them with over 20 years of employment at Williams Company, refused to recognize the need for this change in the belief that the glory days of yore would return at the end of the recession.

By 1995, the recession had been over for at least two years yet Williams Company had no new product lines. Revenue was down, sales for the standard product (with and without modifications) were decreasing, and the employees were still resisting change. Layoffs were imminent.

In 1996, the company was sold to Crock Engineering. Crock had an experienced machine tool division of its own and understood the machine tool business. Williams Company was allowed to operate as a separate entity from 1995 to 1996. By 1996, red ink had appeared on the Williams Company balance sheet. Crock replaced all of the Williams senior managers with its own personnel. Crock then announced to all employees that Williams would become a specialty machine tool manufacturer and that the "good old days" would never return. Customer demand for specialty products had increased threefold in just the last twelve months alone. Crock made it clear that employees who would not support this new direction would be replaced.

The new senior management at Williams Company recognized that 85 years of traditional management had come to an end for a company now committed to specialty products. The company culture was about to change, spearheaded by project management, concurrent engineering, and total quality management.

Senior management's commitment to product management was apparent by the time and money spent in educating the employees. Unfortunately, the seasoned 20-year-plus veterans still would not support the new culture. Recognizing the problems, management provided continuous and visible support for project management in addition to hiring a project management consultant to work with the people. The consultant worked with Williams from 1996 to 2001.

From 1996 to 2001, the Williams Division of Crock Engineering experienced losses in 24 consecutive quarters. The quarter ending March 31, 2002, was the first profitable quarter in over six years. Much of the credit was given to the performance and maturity of the project management system. In May 2002, the Williams Division was sold. More than 80% of the employees lost their jobs when the company was relocated over 1,500 miles away.

QUESTIONS

1. Why was it so difficult to change the culture of the company?
2. What could have been done differently to accelerate the change?

Project Management Growth: Concepts and Definitions

2.0 INTRODUCTION

PMBOK® Guide, 6th Edition
Chapter 2 The Project Management Environment
Chapter 4 Project Integration Management

The growth and acceptance of project management has changed significantly over the past forty years, and these changes are expected to continue well into the twenty-first century, especially in the area of global project management.

The growth of project management can be traced through topics such as roles and responsibilities, organizational structures, delegation of authority and decision making, and especially corporate profitability. Twenty years ago, companies had the choice of whether or not to accept the project management approach. Today, some companies foolishly think that they still have the choice. Nothing could be further from the truth. The survival of the firm may very well rest upon how well project management is implemented, and how quickly.

2.1 THE EVOLUTION OF PROJECT MANAGEMENT: 1945–2017

During the 1940s, line managers used the concept of over-the-fence management to manage projects. Each line manager, wearing the hat of a project manager, would perform the work necessitated by their line organization, and when completed, would throw the "ball" over the fence in hopes that someone would catch it. Once the ball was thrown over the fence, the line managers would wash their hands of any responsibility for the project because the ball was no longer in their yard. If a project failed, blame was placed on whichever line manager had the ball at that time.

PMBOK is a registered mark of the Project Management Institute, Inc.

The problem with over-the-fence management was that the customer had no single contact point for questions. The filtering of information wasted precious time for both the customer and the contractor. Customers who wanted firsthand information had to seek out the manager in possession of the ball. For small projects, this was easy. But as projects grew in size and complexity, this became more difficult.

The Cold War arms race made it clear that the traditional use of over-the-fence management would not be acceptable to the Department of Defense (DoD). The government wanted a single point of contact, namely, a project manager who had total accountability through all project phases. The use of project management was then mandated for some of the smaller weapon systems such as jet fighters and tanks. NASA mandated the use of project management for all activities related to the space program.

By the late 1950s and early 1960s, the aerospace and defense industries were using project management on virtually all projects, and they were pressuring their suppliers to use it as well. Project management was growing, but at a relatively slow rate except for aerospace and defense.

Because of the vast number of contractors and subcontractors, the government needed standardization, especially in the planning process and the reporting of information. The government established a life-cycle planning and control model and a cost monitoring system, and created a group of project management auditors to make sure that the government's money was being spent as planned. These practices were to be used on all government programs above a certain dollar value. Private industry viewed these practices as an over-management cost and saw no practical value in project management.

The growth of project management has come about more through necessity than through desire. Its slow growth can be attributed mainly to lack of acceptance of the new management techniques necessary for its successful implementation. An inherent fear of the unknown acted as a deterrent for managers.

Other than aerospace, defense, and construction, the majority of the companies in the 1960s maintained an informal method for managing projects. In informal project management, just as the words imply, the projects were handled on an informal basis whereby the authority of the project manager was minimized. Most projects were handled by functional managers and stayed in one or two functional lines, and formal communications were either unnecessary or handled informally because of the good working relationships between line managers. Many organizations today, such as low-technology manufacturing, have line managers who have been working side by side for ten or more years. In such situations, informal project management may be effective on capital equipment or facility development projects.

By 1970 and again during the early 1980s, more companies departed from informal project management and restructured to formalize the project management process, mainly because the size and complexity of their activities had grown to a point where they were unmanageable within the current structure.

By 1970, the environment began to change rapidly. Companies in aerospace, defense, and construction pioneered the implementation of project management, and other industries soon followed, some with great reluctance. NASA and the DOD "forced" subcontractors into accepting project management.

TABLE 2–1. LIFE-CYCLE PHASES FOR PROJECT MANAGEMENT MATURITY

Embryonic Phase	Executive Management Acceptance Phase	Line Management Acceptance Phase	Growth Phase	Maturity Phase
• Recognize need • Recognize benefits • Recognize applications • Recognize what must be done	• Visible executive support • Executive understanding of project management • Project sponsorship • Willingness to change way of doing business	• Line management support • Line management commitment • Line management education • Willingness to release employees for project management training	• Use of life-cycle phases • Development of a project management methodology • Commitment to planning • Minimization of "creeping scope" • Selection of a project tracking system	• Development of a management cost/schedule control system • Integrating cost and schedule control • Developing an educational program to enhance project management skills

By the 1990s, companies had begun to realize that implementing project management was a necessity, not a choice. The issue was not how to implement project management, but how fast it could be done.

Table 2–1 shows the typical life-cycle phases that an organization goes through to implement project management. There are seven driving forces that lead executives to recognize the need for project management:

- Capital projects
- Customer expectations
- Competitiveness
- Executive understanding
- New project development
- Efficiency and effectiveness
- The need for business growth

Manufacturing companies are driven to project management because of large capital projects or a multitude of simultaneous projects. Executives soon realize the impact on cash flow and that slippages in the schedule could end up idling workers.

Companies that sell products or services, including installation, to their clients must have good project management practices. These companies are usually non–project-driven but function as though they were project-driven. These companies now sell solutions to their customers rather than products. It is almost impossible to sell complete solutions to customers without having superior project management practices because what you are actually selling is your project management expertise.

The speed by which companies reach some degree of maturity in project management is most often based upon how important they perceive the driving forces to be. Non–project-driven and hybrid organizations move quickly to maturity if increased internal efficiencie and effectiveness are needed. Competitiveness is the slowest path because these typ organizations do not recognize that project management affects their competitive

***PMBOK® Guide*, 6th Edition**
1.2.3.4 Operations and Project Management
1.2.3.5 Operations Management

TABLE 2–2. BENEFITS OF PROJECT MANAGEMENT

Past View	Present View
• Project management will require more people and add to the overhead costs. • Profitability may decrease. • Project management will increase the amount of scope changes. • Project management creates organizational instability and increases conflicts. • Project management is really "eye wash" for the customer's benefit. • Project management will create problems. • Only large projects need project management. • Project management will increase quality problems. • Project management will create power and authority problems. • Project management focuses on suboptimization by looking at only the project. • Project management delivers products to a customer. • The cost of project management may make us noncompetitive.	• Project management allows us to accomplish more work in less time, with fewer people. • Profitability will increase. • Project management will provide better control of scope changes. • Project management makes the organization more efficient and effective through better organizational behavior principles. • Project management will allow us to work more closely with our customers. • Project management provides a means for solving problems. • All projects will benefit from project management. • Project management increases quality. • Project management will reduce power struggles. • Project management allows people to make good company decisions. • Project management delivers solutions. • Project management will increase our business.

directly. For project-driven organizations, the path is reversed. Competitiveness is the name of the game and the vehicle used is project management.

By the 1990s, companies finally began to recognize the benefits of project management. Table 2–2 shows the some of the benefits of project management and how our view of project management has changed.

In 2008, we saw the beginning of the housing market crunch in the United States and other nations. From 2008 to 2016, many companies saw a slowdown in company's growth. Some companies recognized that the benefits of using project management could help them develop strategic partnerships, alliances, and joint ventures, thus providing growth opportunities. But now, in addition to the traditional project management competencies, project managers would need to understand how culture, politics, and religion affect decision making in their partner's environment. In the future, a knowledge of culture, politics, and religion may be necessary to become a global project manager.

Recognizing that the organization can benefit from the implementation of project management is just the starting point. The question now becomes, "How long will it take us to achieve these benefits?" This can be partially answered from Figure 2–1. In the beginning of the implementation process, there will be added expenses to develop the project management methodology and establish the support systems for planning, scheduling, and control. Eventually, the cost will level off and become pegged. The question mark in

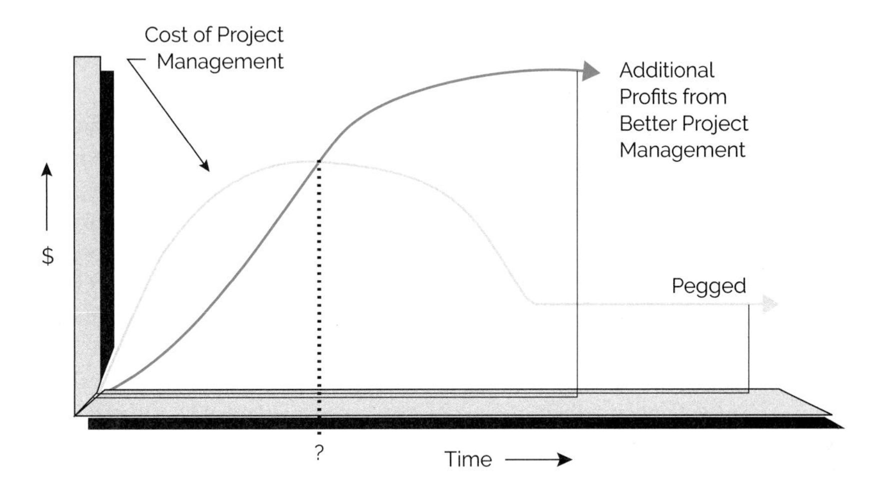

FIGURE 2–1. Project management costs versus benefits.

Figure 2–1 is the point at which the benefits equal the cost of implementation. This point can be pushed to the left through training and education.

2.2 RESISTANCE TO CHANGE

Why was project management so difficult for companies to accept and implement? The answer is shown in Figure 2–2. Historically, project management resided only in the project-driven sectors of the marketplace. In these sectors, the project managers were given the

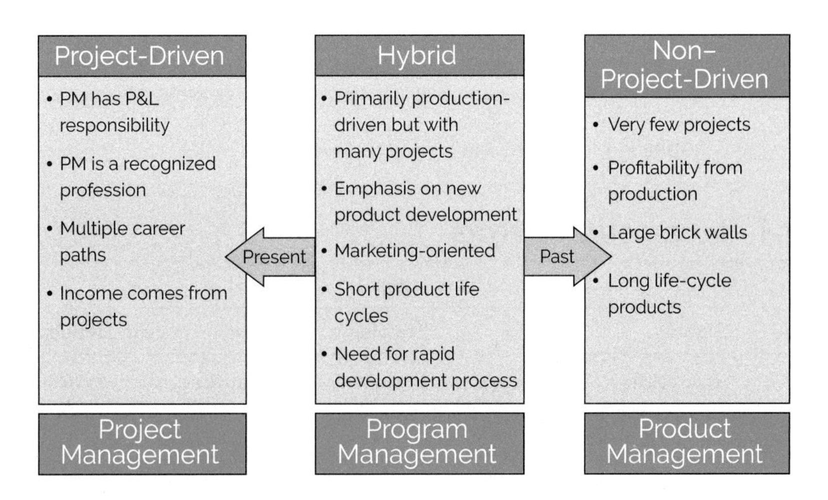

FIGURE 2–2. Industry classification (by project management utilization).

responsibility for profit and loss, which virtually forced companies to treat project management as a profession. Project managers were viewed as managing part of a business rather than just performing as a project manager.

In the non–project-driven sectors of the marketplace, corporate survival was based upon products and services, rather than upon a continuous stream of projects. Profitability was identified through marketing and sales, with very few projects having an identifiable P&L. As a result, project management in these firms was never viewed as a profession.

In reality, most firms that believed that they were non–project-driven were actually hybrids. Hybrid organizations are typically non–project-driven firms with one or two divisions that are project-driven. Historically, hybrids have functioned as though they were non–project-driven, as shown in Figure 2–2, but today they are functioning like project-driven firms. Why the change? Management has come to the realization that they can most effectively run their organization on a "management by project" basis, and thereby achieve the benefits of both a project management organization and a traditional organization. The rapid growth and acceptance of project management during the last ten years has taken place in the non–project-driven/hybrid sectors. Now, project management is being promoted by marketing, engineering, and production, rather than only by the project-driven departments (see Figure 2–3).

A second factor contributing to the acceptance of project management was the economy, specifically the recessions of 1979–1983 and 1989–1993. This can be seen from Table 2–3. By the end of the recession of 1979–1983, companies recognized the benefits of

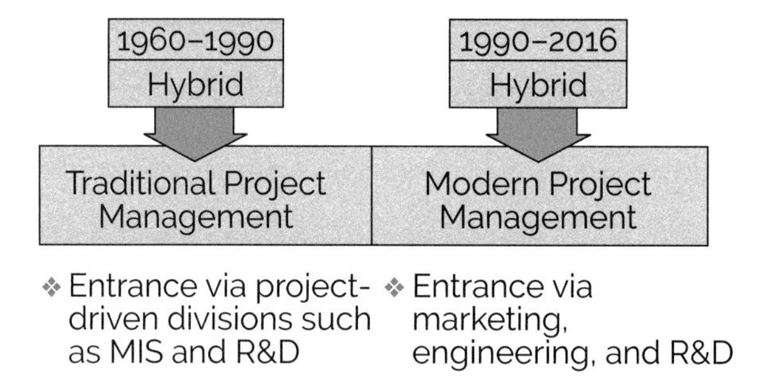

FIGURE 2–3. From hybrid to project-driven.

TABLE 2–3. RECESSIONARY EFFECTS

	Characteristics				
Recession	**Layoffs**	**R&D**	**Training**	**Solutions Sought**	**Results of the Recessions**
1979–1983	Blue collar	Eliminated	Eliminated	Short-term	• Return to status quo • No project management support • No allies for project management
1989–1993	White collar	Focused	Focused	Long-term	• Change way of doing business • Risk management • Examine lessons learned

using project management but were reluctant to see it implemented. Companies returned to the "status quo" of traditional management. There were no allies or alternative management techniques that were promoting the use of project management.

The recession of 1989–1993 finally saw the growth of project management in the non–project-driven sector. This recession was characterized by layoffs in the white collar/management ranks. Allies for project management were appearing and emphasis was being placed upon long-term solutions to problems. Project management was here to stay.

As project management continues to grow and mature, it will have more allies. In the twenty-first century, emerging world nations will come to recognize the benefits and importance of project management. Worldwide standards for project management will be established that may very well include requirements related to culture, politics, and religious aspects.

Even though project management has been in existence for more than fifty years, there are still different views and misconceptions about what it really is. Textbooks on operations research or management science still have chapters entitled "Project Management" that discuss only PERT scheduling techniques. A textbook on organizational design recognized project management as simply another organizational form.

All companies sooner or later understand the basics of project management. But companies that have achieved excellence in project management have done so through successful implementation and execution of processes and methodologies.

2.3 SYSTEMS, PROGRAMS, AND PROJECTS: A DEFINITION

In the preceding sections the word "systems" has been used rather loosely. The exact definition of a system depends on the users, environment, and ultimate goal. Business practitioners define a system as:

> A group of elements, either human or nonhuman, that is organized and arranged in such a way that the elements can act as a whole toward achieving some common goal or objective.

Systems are collections of interacting subsystems that, if properly organized, can provide a synergistic output. Systems are characterized by their boundaries or interface conditions. For example, if the business firm system were completely isolated from the environmental system, then a *closed system* would exist, in which case management would have complete control over all system components. If the business system reacts with the environment, then the system is referred to as *open.* All social systems, for example, are categorized as open systems. Open systems must have permeable boundaries.

If a system is significantly dependent on other systems for its survival, then it is an *extended system.* Not all open systems are extended systems. Extended systems are ever-changing and can impose great hardships on individuals who desire to work in a regimented atmosphere.

Military and government organizations were the first to attempt to define the boundaries of systems, programs, and projects.

PMBOK® Guide, 6th Edition
1.2.3.2 Program Management

Programs can be construed as the necessary first-level elements of a system. Additionally, they can be regarded as subsystems. However, programs are generally defined as time-phased efforts, whereas systems exist on a continuous basis.

Projects are also time-phased efforts (much shorter than programs) and are the first level of breakdown of a program. As shown in Table 2–4, the government sector tends to run efforts as programs, headed up by a program manager who hopes that their program will receive government funding year after year. Today, the majority of the industrial sector uses both project and program managers. Throughout this text, I have used the terms project and program management as being the same because they are generally regulated by the same policies, procedures, and guidelines. In general, as will be discussed in Chapter 11, projects are often considered to be the first level of subdivision of a program, and programs are often longer in duration than projects. However, there are many other significant differences, such as:

- Projects may have a single objective, whereas programs may have multiple objects with a heavy orientation toward business rather than technical objectives.
- The length of programs often makes them more susceptible to changing environmental conditions, politics, the economy, business strategy, and interest rates.
- The possibility for changing economic conditions may play havoc with pricing out long-term programs based upon estimates on forward pricing rates.
- Functional managers are often reluctant to give up their best workers that are in high demand by committing them to a single program that will run for years.
- Program governance is conducted by a committee rather than by a single individual, and the membership may change over the life of the program.
- Program funding may be on a yearly basis and changes in planned funding are based upon existing need, which may change from year to year, and economic conditions.
- Scope changes may occur more frequently and have a greater impact on the project.
- Rebaselining and replanning will occur more frequently.
- Based upon the program's length, succession planning may be necessary for workers with critical skills.
- The loss of some workers over the length of the program may be expected because of changing positions, better opportunities in another company, and retirements.
- Workers may not believe that a long-term assignment on just one program is an opportunity for career advancement.

TABLE 2–4. DEFINITION SUMMARY

Level	Sector	Title
System*	—	—
Program	Government	Program managers
Project	Industry	Project managers

*Definitions, as used here, do not include in-house industrial systems such as management information systems or shop floor control systems.

PMI has certification programs for both project and program managers and does differentiate between the two. There are textbooks written that are dedicated entirely to program management.

Once a group of tasks is selected and considered to be a project, the next step is to define the kinds of project units. There are four categories of projects:

- *Individual projects:* These are short-duration projects normally assigned to a single individual who may be acting as both a project manager and a functional manager.
- *Staff projects:* These are projects that can be accomplished by one organizational unit, say a department. A staff or task force is developed from each section involved. This works best if only one functional unit is involved.
- *Special projects:* Often special projects occur that require certain primary functions and/or authority to be assigned temporarily to other individuals or units. This works best for short-duration projects. Long-term projects can lead to severe conflicts under this arrangement.
- *Matrix or aggregate projects:* These require input from a large number of functional units and usually control vast resources.

Project management may now be defined as the process of achieving project objectives through the traditional organizational structure and over the specialties of the individuals concerned. Project management is applicable for any ad hoc (unique, one-time, one-of-a-kind) undertaking concerned with a specific end objective. In order to complete a task, a project manager must:

PMBOK® *Guide*, 6th Edition
1.2.2 The Importance of Project Management

- Set objectives
- Establish plans
- Organize resources
- Provide staffing
- Set up controls
- Issue directives
- Motivate personnel
- Apply innovation for alternative actions
- Remain flexible

The type of project will often dictate which of these functions a project manager will be required to perform.

2.4 PRODUCT VERSUS PROJECT MANAGEMENT: A DEFINITION

PMBOK® *Guide*, 6th Edition
4.1.1 Inputs to Project Charter
5.0 Project Scope Management

Some people mistakenly argue that there is no major difference between a project and a program other than the time duration. Project managers focus on the end date of their project from the day they are assigned as project manager. Program managers usually have a much longer time frame than project managers and never want to see their program come to an end. In the early years of project management with the DOD serving as the primary customer, aerospace and defense project managers were called program managers because the intent was to get follow-on government contracts each year.

But what about the definition of product management or product line management? Product managers function very much like program managers. The product manager wants his or her product to be as long-lived as possible and as profitable as possible. Even when the demand for the product diminishes, the product manager will always look for spin-offs to keep a product alive.

There is also a difference between project and product scope:

- *Project scope* defines the work that must be accomplished to produce a deliverable with specified features or functions. The deliverable can be a product, service, or other result.
- *Product scope* defines the features or functions that characterize the deliverable.

There is a relationship between project and product management. When the project is in the R&D phase, a project manager is involved. Once the product is developed and introduced into the marketplace, the product manager takes control. In some situations, the project manager can become the product manager. Product and project management can, and do, exist concurrently within companies.

Product management can operate horizontally as well as vertically. When a product is shown horizontally on the organizational chart, the implication is that the product line is not big enough to control its own resources full-time and therefore shares key functional resources. If the product line were large enough to control its own resources full-time, it would be shown as a separate division or a vertical line on the organization chart.

Based upon the nature of the project, the project manager (or project engineer) can report to a marketing-type person. The reason is that technically oriented project leaders get too involved with the technical details of the project and lose sight of when and how to "kill" a project. Remember, most technical leaders have been trained in an academic rather than a business environment. Their commitment to success often does not take into account such important parameters as return on investment, profitability, competition, and marketability. This is one of the reasons why some project plans and project business cases identify an "exit criteria" that states under what circumstances the project should be cancelled or possibly redirected toward a different business outcome.

To alleviate these problems, project managers and project engineers, especially on R&D-type projects, are now reporting to marketing so that marketing input will be included in all R&D decisions because of the high costs incurred during R&D. Executives must exercise caution with regard to this structure in which both product and project managers report to the marketing function. The marketing executive could become the focal point of the entire organization, with the capability of building a very large empire.

2.5 MATURITY AND EXCELLENCE: A DEFINITION

Some people contend that maturity and excellence in project management are the same. Unfortunately, this is not the case. Consider the following definition:

> Maturity in project management is the implementation of a standard methodology and accompanying processes such that there exists a high likelihood of repeated successes.

This definition is supported by the life-cycle phases shown in Table 2–1. Maturity implies that the proper foundation of tools, techniques, processes, and even culture, exists. When projects come to an end, there is usually a debriefing with senior management to discuss how well the methodology was used and to recommend changes. This debriefing looks at "key performance indicators," which are shared learning topics, and allows the organization to maximize what it does right and to correct what it did wrong.

The definition of excellence can be stated as:

> Organizations excellent in project management are those that create the environment in which there exists a *continuous* stream of successfully managed projects and where success is measured by what is in the best interest of *both* the company and the project (i.e., customer).

Excellence goes well beyond maturity. You must have maturity to achieve excellence. Figure 2–4 shows that once the organization completes the first four life-cycle phases in Table 2–1, it may take two years or more to reach some initial levels of maturity. Excellence, if achievable at all, may take an additional five years or more.

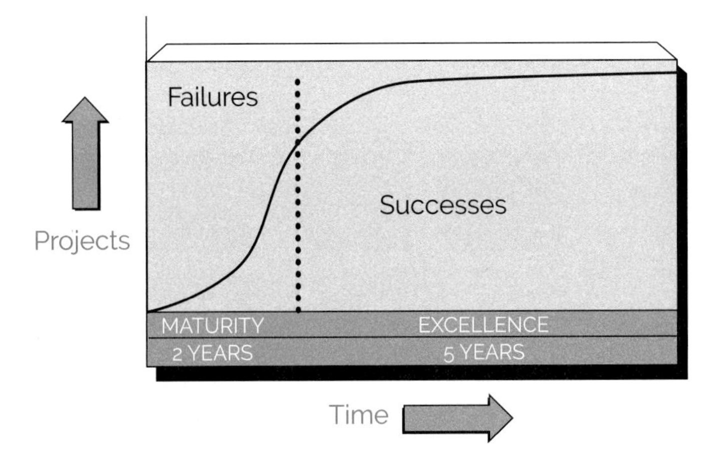

FIGURE 2–4. The growth of excellence.

Figure 2–4 also brings out another important fact. During maturity, more successes than failures occur. During excellence, we obtain a continuous stream of successful projects. Yet, even after having achieved excellence, there will still be some failures.

> Executives who always make the right decision are not making enough decisions. Likewise, organizations in which all projects are completed successfully are not taking enough risks and are not working on enough projects.

It is unrealistic to believe that all projects will be completed successfully. Some people contend that the only true project failures are the ones from which nothing is learned. Failure can be viewed as success if the failure is identified early enough so that the resources can be reassigned to other, more opportune activities.

2.6 INFORMAL PROJECT MANAGEMENT: A DEFINITION

Companies today are managing projects more informally than before. Informal project management does have some degree of formality but emphasizes managing the project with a minimum amount of paperwork. Furthermore, informal project management is based upon guidelines rather than the policies and procedures that are the basis for formal project management. This was shown previously to be a characteristic of a good project management methodology. Informal project management mandates effective communications, effective cooperation, effective teamwork, and trust.

These four elements are absolutely essential for effective informal project management. Figure 2–5 shows the evolution of project documentation over the years.

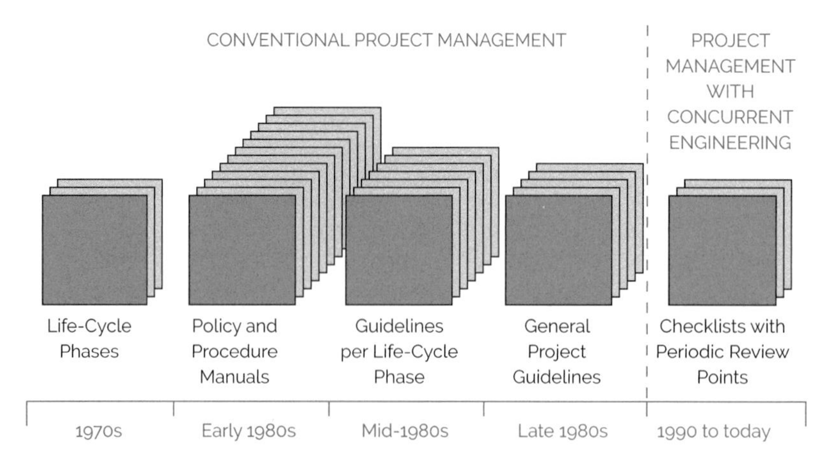

FIGURE 2–5. Evolution of policies, procedures, and guidelines.

Source: Reprinted from H. Kerzner, *In Search of Excellence in Project Management.* New York: John Wiley & Sons, 1998, p. 196.

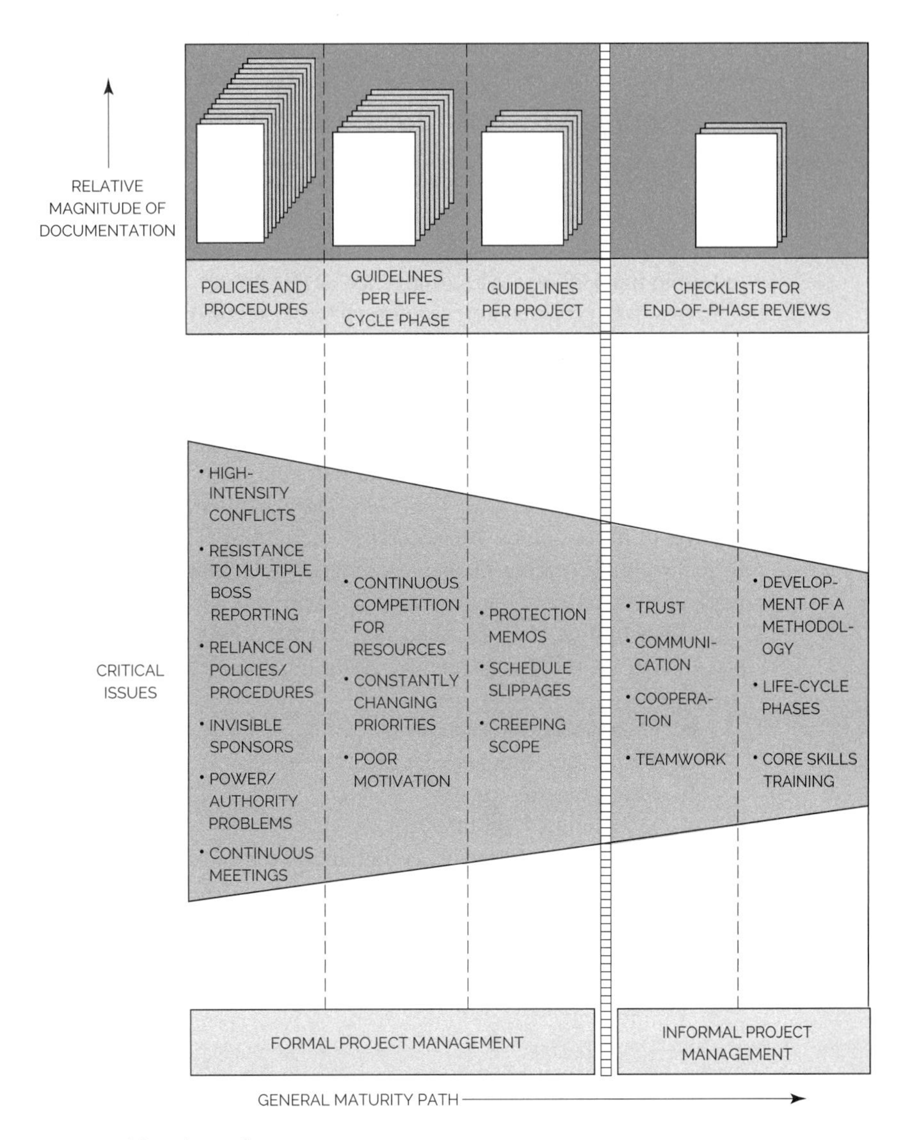

FIGURE 2–6. Maturity path.

As companies become mature in project management, emphasis is on guidelines and checklists. Figure 2–6 shows the critical issues as project management matures toward more informality.

As a final note, not all companies have the luxury of using informal project management. Customers often have a strong voice in whether formal or informal project management will be used. Dashboard reporting of project status accompanied by some degree of project management maturity has more companies leaning toward informal project management.

2.7 THE MANY FACES OF SUCCESS

PMBOK® Guide, 6th Edition
1.2.6.4 Project Success Measures

Historically, the definition of success has been meeting the customer's expectations regardless of whether the customer is internal or external. Success also includes getting the job done within the constraints of time, cost, and quality. Using this standard definition, success is defined as a point on the time, cost, quality/performance grid. But how many projects, especially those requiring innovation, are accomplished at this point?

Very few projects are ever completed without trade-offs or scope changes on time, cost, and quality. Therefore, success could still occur without exactly hitting this singular point. In this regard, success could be defined as a cube, such as seen in Figure 2–7. The singular point of time, cost, and quality would be a point within the cube, constituting the convergence of the critical success factors (CSFs) for the project.

Another factor to consider is that there may exist both primary and secondary definitions of success, as shown in Table 2–5. The primary definitions of success are seen through the eyes of the customer. The secondary definitions of success are usually internal benefits. If achieving 86 percent of the specification is acceptable to the customer and follow-on work is received, then the original project might very well be considered a success.

The definition of success can also vary according to who the stakeholder is. For example, each of the following can have his or her own definition of success on a project:

- Consumers: safety in its use
- Employees: guaranteed employment
- Management: bonuses
- Stockholders: profitability
- Government agencies: compliance with federal regulations

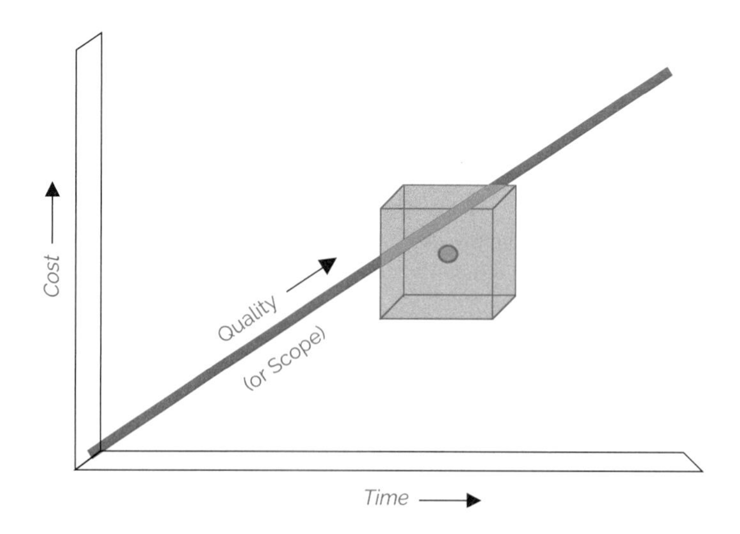

FIGURE 2–7. Success: point or cube?

TABLE 2–5. SUCCESS FACTORS

Primary	Secondary
• Within time • Within cost • Within quality limits • Accepted by the customer	• Follow-on work from this customer • Using the customer's name as a reference on your literature • Commercialization of a product • With minimum or mutually agreed upon scope changes • Without disturbing the main flow of work • Without changing the corporate culture • Without violating safety requirements • Providing efficiency and effectiveness of operations • Satisfying OSHA/EPA requirements • Maintaining ethical conduct • Providing a strategic alignment • Maintaining a corporate reputation • Maintaining regulatory agency relations

It is possible for a project management methodology to identify primary and secondary success factors (Table 2.5). This could provide guidance to a project manager for the development of a risk management plan and for deciding which risks are worth taking and which are not.

As stated in Section 1.0, projects are now being aligned to business goals and objectives. This applies to programs as well as projects. In addition to the alignment, the output of projects and programs are now expected to create sustainable business value. All of this has forced us to rethink our definitions of projects, programs, and how we will measure success. Table 2–6 shows how some of the definitions will most likely change in the future. The traditional definitions are still applicable today, but there is a need, at least in the author's opinion, to include business and value components. Consuming resources on projects and programs that are not intended to create sustainable business value may not be a good business decision.

Customers and contractors must come to an agreement on the definition of success. A project manager was managing a large project for a government agency. The project manager asked one of the vice presidents in his company, "What's our company's definition of success on the project for this government agency?" The vice president responded, "Meeting the profit margins we stated in our proposal." The project manager then responded, "Do you think the government agency has the same definition of project success as we do?" The conversation then ended.

When the customer and the contractor are working toward different definitions of success, decision making becomes suboptimal and each party makes decisions in their own best interest. In an ideal situation, the customer and the contractor will establish a mutually agreed upon definition of success that both parties can live with.

While it is possible that no such agreement can be reached, a good starting point is to view the project through the eyes of the other party. As stated by Rachel Alt-Simmons:[1]

> All too often, we take an inside-out perspective. What this means is that we see a customer's journey from how we engage with them as a company, not how they engage with

1. Rachel Alt-Simmons, *Agile by Design* (Hoboken, NJ: John Wiley, 2016), p. 33.

TABLE 2–6. TRADITIONAL AND FUTURE DEFINITIONS

Factor	Traditional Definition	Future Definition
Project	A temporary endeavor undertaken to create a unique product, service or result*	A collection of sustainable business value scheduled for realization
Program	Achieving a set of business goals through the coordinated management of interdependent projects over a finite period of time*	A collection of projects designed to achieve a business purpose and create sustainable business value within the established competing constraints
Success	Completion of the projects or programs within the triple constraints of time, cost and scope	Achieving the desired business value within the competing constraints

*These definitions are taken from the Glossary of Project Management Institute, *A Guide to the Project Management Body of Knowledge (PMBOK® Guide)*—Fifth Edition, Project Management Institute, Inc., 2013.

us as a consumer. A helpful tool in identifying how customers engage with us is by creating a customer journey map. The journey map helps identify all paths customers take in achieving their goal from start to finish. By looking at your organization through your customers' eyes, you can begin to better understand the challenges that a customer faces in doing business with your organization. The team sees the customer outside or product or functional silos and helps link pieces of a customer process across the organization. Often, teams find out that potential solutions for problems that they're identifying extend outside of their functional realm—and that's okay!

2.8 THE MANY FACES OF FAILURE

Previously we stated that success might be a cube rather than a point. If we stay within the cube but miss the point, is that a failure? Probably not! The true definition of failure is when the final results are not what were expected, even though the original expectations may or may not have been reasonable. Sometimes customers and even internal executives set performance targets that are totally unrealistic in hopes of achieving 80–90 percent. For simplicity's sake, let us define failure as unmet expectations.

With unmeetable expectations, failure is virtually assured since we have defined failure as unmet expectations. This is called a *planning failure* and is the difference between what was planned and what was, in fact, achieved. The second component of failure is poor performance or *actual failure*. This is the difference between what was achievable and what was actually accomplished.

Perceived failure is the net sum of *actual failure* and *planning failure*. Figures 2–8 and 2–9 illustrate the components of perceived failure. In Figure 2–8, *project management* has planned a level of accomplishment (C) lower than what is achievable given project circumstances and resources (D). This is a classic underplanning situation. Actual accomplishment (B), however, was less than planned.

A slightly different case is illustrated in Figure 2–9. Here, we have planned to accomplish more than is achievable. Planning failure is again assured even if no actual failure

Section 2.8 is adapted from Robert D. Gilbreath, *Winning at Project Management*. (New York: John Wiley & Sons, 1986), pp. 2–6.

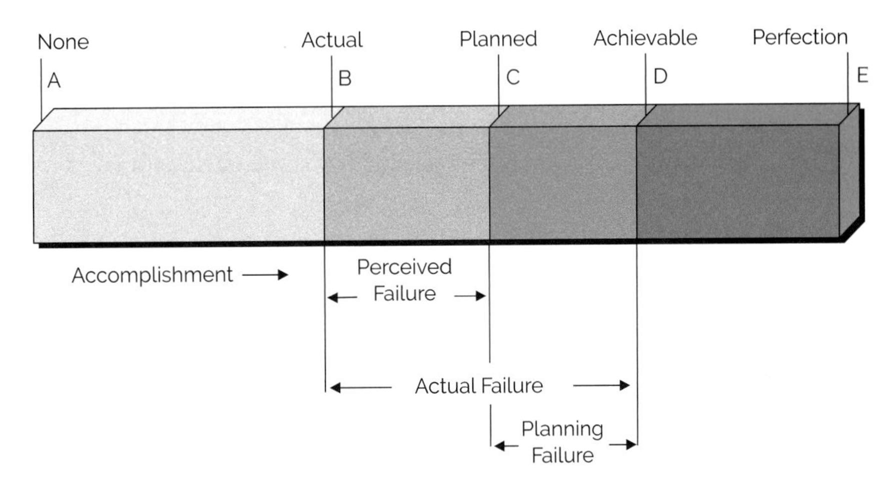

FIGURE 2–8. Components of failure (pessimistic planning).

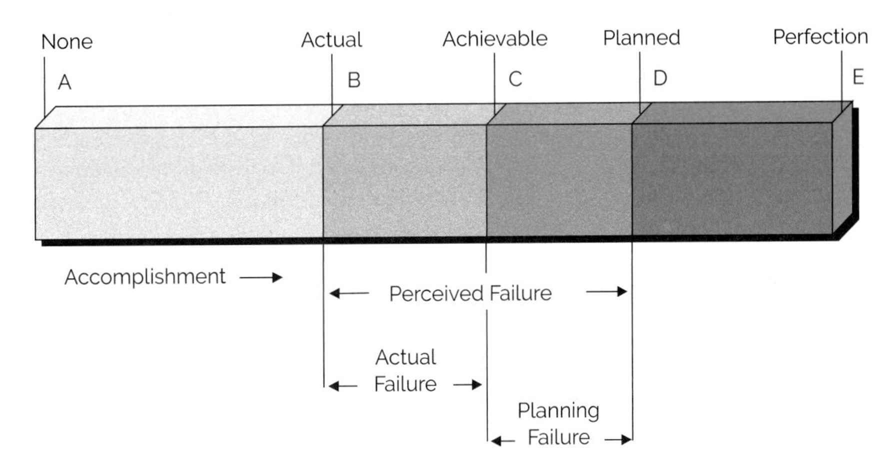

FIGURE 2–9. Components of failure (optimistic planning).

occurs. In both of these situations (overplanning and underplanning), the actual failure is the same, but the perceived failure can vary considerably.

Today, most project management practitioners focus on the *planning failure* term. If this term can be compressed or even eliminated, then the magnitude of the actual failure, should it occur, would be diminished. A good project management methodology helps to reduce this term. We now believe that the existence of this term is largely due to the project manager's inability to perform effective risk management. In the 1980s, the failure of a project was believed to be largely a quantitative failure due to:

- Ineffective planning
- Ineffective scheduling
- Ineffective estimating

- Ineffective cost control
- Project objectives being "moving targets"

During the 1990s, the view of failure changed from being quantitatively oriented to qualitatively oriented. A failure in the 1990s was largely attributed to:

- Poor morale
- Poor motivation
- Poor human relations
- Poor productivity
- No employee commitment
- No functional commitment
- Delays in problem solving
- Too many unresolved policy issues
- Conflicting priorities between executives, line managers, and project managers

Although these quantitative and qualitative approaches still hold true to some degree, today we believe that the major component of planning failure is inappropriate or inadequate risk management, or having a project management methodology that does not provide any guidance for risk management.

Sometimes, the risk management component of failure is not readily identified. For example, look at Figure 2–10. The actual performance delivered by the contractor was significantly less than the customer's expectations. Is the difference due to poor technical ability or a combination of technical inability and poor risk management? Today we believe that it is a combination.

When a project is completed, companies perform a lessons-learned review. Sometimes lessons learned are inappropriately labeled and the true reason for the risk event is not known. Figure 2–11 illustrates the relationship between the marketing personnel and technical personnel when undertaking a project to develop a new product. If the project is

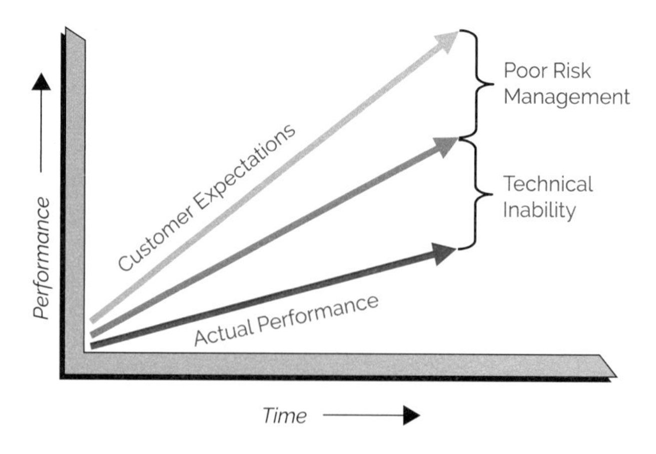

FIGURE 2–10. Risk planning.

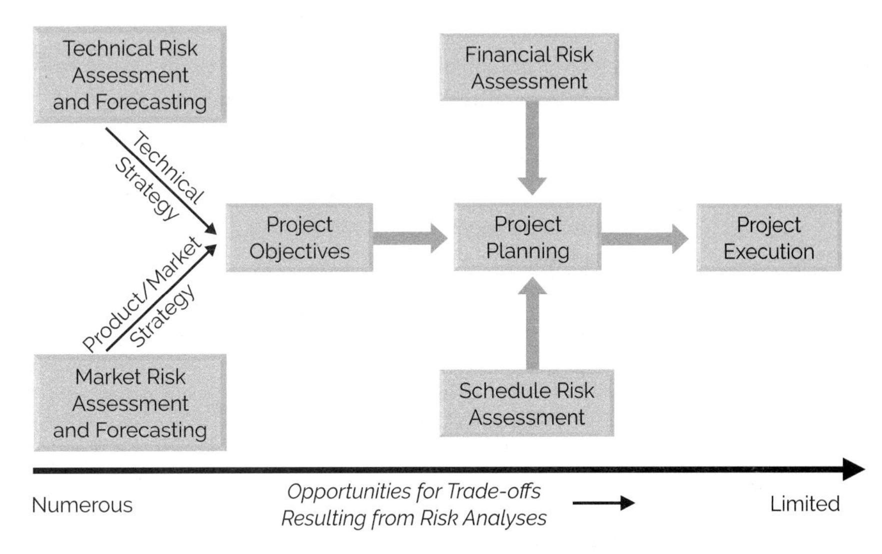

FIGURE 2–11. The relationship between marketing and technical personnel.

completed with actual performance being less than customer expectations, is it because of poor risk management by the technical assessment and forecasting personnel or poor marketing risk assessment? The relationship between marketing and technical risk management is not always clear.

Figure 2–11 also shows that opportunities for trade-offs diminish as we get further downstream on the project. There are numerous opportunities for trade-offs prior to establishing the final objectives for the project. In other words, if the project fails, it may be because of the timing when the risks were analyzed.

2.9 CAUSES OF PROJECT FAILURE

There are numerous causes of project failure, whether a partial or complete failure, and most failures are a result of more than one cause. Some cause may directly or indirectly lead to other causes. For example, business case failure can lead to planning and execution failure. For simplicity sake, project failures can be broken down into the following categories:

Planning/execution failures:

- Business case deterioration
- Business case requirements changed significantly over the life of the project
- Technical obsolescence has occurred
- Technologically unrealistic requirements
- Lack of a clear vision
- Plan asks for too much in too little time

- Poor estimates, especially financial
- Unclear or unrealistic expectations
- Assumptions, if they exist at all, are unrealistic
- Plans are based upon insufficient data
- No systemization of the planning process
- Planning is performed by a planning group
- Inadequate or incomplete requirements
- Lack of resources
- Assigned resources lack experience or the necessary skills
- Resources lack focus or motivation
- Staffing requirements are not fully known
- Constantly changing resources
- Poor overall project planning
- Established milestones are not measurable
- Established milestones are too far apart
- The environmental factors have changes causing outdated scope
- Missed deadlines and no recovery plan
- Budgets are exceeded and out of control
- Lack of replanning on a regular basis
- Lack of attention provided to the human and organizational aspects of the project
- Project estimates are best guesses and not based upon history or standards
- Not enough time provided for estimating
- No one knows the exact major milestone dates or due dates for reporting
- Team members working with conflicting requirements
- People are shuffled in and out of the project with little regard for the schedule
- Poor or fragmented cost control
- Weak project and stakeholder communications
- Poor assessment of risks if done at all
- Wrong type of contract
- Poor project management; team members possess a poor understanding of project management, especially virtual team members
- Technical objectives are more important than business objectives
- Assigning critically skilled workers, including the project manager, on a part-time basis
- Poor performance tracking metrics
- Poor risk management practices
- Insufficient organizational process assets

Governance/stakeholder failures:

- End-use stakeholders not involved throughout the project
- Minimal or no stakeholder backing; lack of ownership
- New executive team in place with different visions and goals
- Constantly changing stakeholders
- Corporate goals and/or vision not understood at the lower organizational levels

- Unclear stakeholder requirements
- Passive user stakeholder involvement after handoff
- Each stakeholder uses different organizational process assets, which may be incompatible with each other
- Weak project and stakeholder communications
- Inability of stakeholders to come to an agreement

Political failures:

- New elections resulting in a change of power
- Changes in the host country's fiscal policy, procurement policy and labor policy
- Nationalization or unlawful seizure of project assets and/or intellectual property
- Civil unrest resulting from a coup, acts of terrorism, kidnapping, ransom, assassinations, civil war and insurrection
- Significant inflation rate changes resulting in unfavorable monetary conversion policies
- Contractual failure such as license cancellation and payment failure

Failures can also be industry-specific such as IT failure or construction failure. Some failures can be corrected while other failures can lead to bankruptcy.

2.10 DEGREES OF SUCCESS AND FAILURE

Projects get terminated for one of two basic reasons; project success or project failure. Project success is considered as a natural cause for termination and is achieved when we meet the success criteria established at the onset of the project. Project failure is often the result of unnatural causes such as a sudden change in the business base, loss of critical resources, or inability to meet certain critical constraints. Previously we listed the numerous reasons why a project can get terminated. Canceling a project is a critical business decision and can have a serious impact on people, processes, materials, and money within the company. Depending on when it's canceled, it can also impact customer and partner relationships.

In an ideal situation, the business case for a project would contain a section identifying the criteria for success and also for termination. Identifying a cancellation criteria is important because too many times a project that should be cancelled on just linger on and wastes precious resources that could be assigned to other more value-driven projects.

There are degrees of project success and failure. For example, a project can come in two weeks late and still be considered as a success. A project over budget by $100,000 can also be considered as a success if the end results provide value to the client and the client accepts the deliverables. Projects can also be partial successes and partial failures. One possible way of classifying project results can be:

- *Complete success*: The project met the success criteria, value was created and all constraints were adhered to.

- *Partial success*: The project met the success criteria, the client accepted the deliverables and value was created, although one or more of the success constraints were not met.
- *Partial failure*: The project was not completed as expected and may have been canceled early on in the life cycle. However, knowledge and/or intellectual property was created that may be used on future projects.
- *Complete failure*: The project was abandoned and nothing was learned from the project.

In the future, we can expect to have more than three constraints on our projects. It is important to understand that it may not be possible to meet all of the competing constraints and therefore partial success may become the norm.

2.11 THE STAGE-GATE PROCESS

PMBOK® Guide, 6th Edition
1.2.4.1 Project and Development Life Cycles

When companies recognize the need to begin developing processes for project management, the starting point is normally the stage-gate process. The stage-gate process was created because the traditional organizational structure was designed primarily for top-down, centralized management, control, and communications, all of which were no longer practical for organizations that use project management and horizontal work flow. The stage-gate process eventually evolved into life-cycle phases.

Just as the words imply, the process is composed of stages and gates. Stages are groups of activities that can be performed either in series or parallel based upon the magnitude of the risks the project team can endure. The stages are managed by crossfunctional teams. The gates are structured decision points at the end of each stage. Good project management processes usually have no more than six gates. With more than six gates, the project team focuses too much attention on preparing for the gate reviews rather than on the actual management of the project.

Project management is used to manage the stages between the gates, and it can shorten the time between the gates. This is a critical success factor if the stage-gate process is to be used for the development and launch of new products. A good corporate methodology for project management will provide checklists, forms, and guidelines to make sure that critical steps are not omitted.

Checklists for gate reviews are critical. Without these checklists, project managers can waste hours preparing gate review reports. Good checklists focus on answering these questions:

- Where are we today (i.e., time and cost)?
- Where will we end up (i.e., time and cost)?
- What are the present and future risks?
- What assistance is needed from management?

Project managers are never allowed to function as their own gatekeepers. The gatekeepers are either individuals (i.e., sponsors) or groups of individuals designated by senior management and empowered to enforce the structured decision-making process. The gatekeepers are authorized to evaluate the performance to date against predetermined criteria and to provide the project team with additional business and technical information.

Gatekeepers must be willing to make decisions. The four most common decisions are:

- Proceed to the next gate based upon the original objectives
- Proceed to the next gate based upon revised objectives
- Delay making a gate decision until further information is obtained
- Cancel the project

Sponsors must also have the courage to terminate a project. The purpose of the gates is not only to obtain authorization to proceed, but to identify failure early enough so that resources will not be wasted but will be assigned to more promising activities.

The three major benefits of the stage-gate process are:

- Providing structure to project management
- Providing possible standardization in planning, scheduling, and control (i.e., forms, checklists, and guidelines)
- Allowing for a structured decision-making process

Companies embark upon the stage-gate process with good intentions, but there are pitfalls that may disrupt the process. These include:

- Assigning gatekeepers and not empowering them to make decisions
- Assigning gatekeepers who are afraid to terminate a project
- Denying the project team access to critical information
- Allowing the project team to focus more on the gates than on the stages

It should be recognized that the stage-gate process is neither an end result nor a self-sufficient methodology. Instead, it is just one of several processes that provide structure to the overall project management methodology.

Today, the stage-gate process appears to have been replaced by life-cycle phases. Although there is some truth in this, the stage-gate process is making a comeback. Since the stage-gate process focuses on decision making more than life-cycle phases, the stage-gate process is being used as an internal, decision-making tool within each of the life-cycle phases. The advantage is that, while life-cycle phases are the same for every project, the stage-gate process can be custom-designed for each project to facilitate decision making and risk management. The stage-gate process is now an integral part of project management, whereas previously it was used primarily for new product development efforts.

2.12 PROJECT LIFE CYCLES

PMBOK® *Guide*, 6th Edition
1.2.4.2 Project Phase

Every program, project, or product has certain phases of development known as life-cycle phases. A clear understanding of these phases permits managers and executives to better control resources to achieve goals. During the past few years, there has been at least partial agreement about the life-cycle phases of a product. They include:

- Research and development
- Market introduction
- Growth
- Maturity
- Deterioration
- Death

Today, there is no agreement among industries, or even companies within the same industry, about the life-cycle phases of a project. This is understandable because of the complex nature and diversity of projects.

The theoretical definitions of the life-cycle phases of a system can be applied to a project. These phases include:

- Conceptual
- Planning
- Testing
- Implementation
- Closure

The first phase, the conceptual phase, includes the preliminary evaluation of an idea. Most important in this phase is a preliminary analysis of risk and the resulting impact on the time, cost, and performance requirements, together with the potential impact on company resources. The conceptual phase also includes a "first cut" at the feasibility of the effort.

The second phase is the planning phase. It is mainly a refinement of the elements in the conceptual phase and requires a firm identification of the resources required and the establishment of realistic time, cost, and performance parameters. This phase also includes the initial preparation of documentation necessary to support the system. For a project based on competitive bidding, the conceptual phase would include the decision of whether to bid, and the planning phase would include the development of the total bid package (i.e., time, schedule, cost, and performance).

Because of the amount of estimating involved, analyzing system costs during the conceptual and planning phases is not an easy task. As shown in Figure 2–12, most project or system costs can be broken down into operating (recurring) and implementation (nonrecurring) categories. Implementation costs include one-time expenses such as construction of a new facility, purchasing computer hardware, or detailed planning. Operating costs include recurring expenses such as human resources. The operating costs may be reduced as shown in Figure 2–12 if personnel perform at a higher position on the learning curve. The identification of a learning curve position is vitally important during the planning phase when firm cost positions must be established. Of course, it is not always possible to know what individuals will be available or how soon they will perform at a higher learning curve position.

Once the approximate total cost of the project is determined, a cost-benefit analysis should be conducted (see Figure 2–13) to determine if the estimated value of the information obtained from the system exceeds the cost of obtaining the information. This analysis

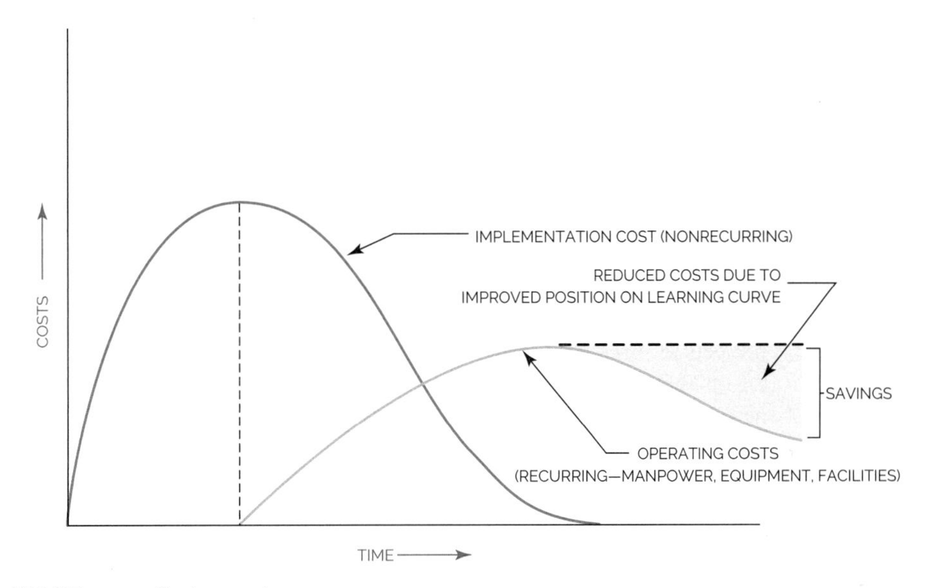

FIGURE 2–12. System costs.

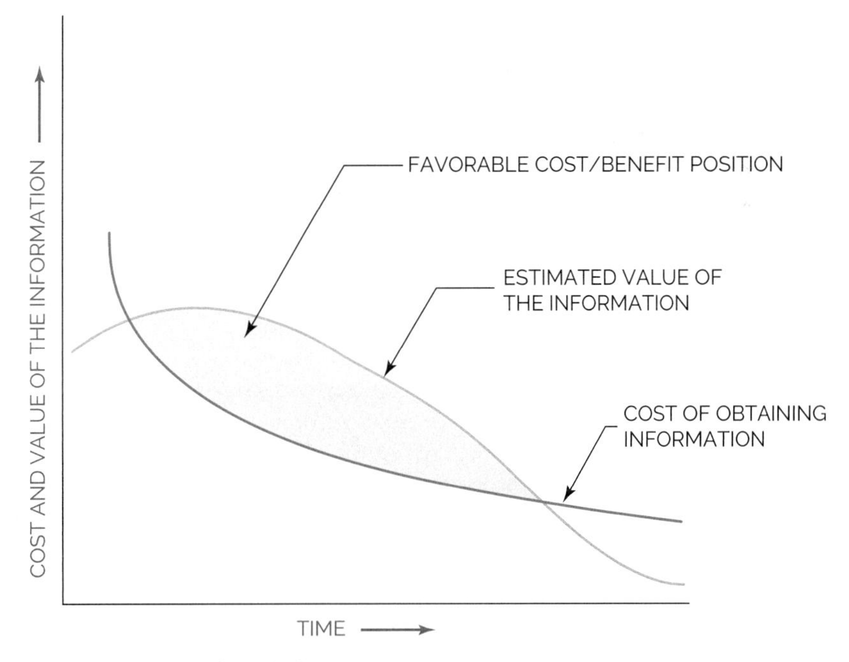

FIGURE 2–13. Cost–benefit analysis.

is often included as part of a feasibility study. There are several situations, such as in competitive bidding, where the feasibility study is actually the conceptual and definition phases. Because of the costs that can be incurred during these two phases, top-management approval is almost always necessary before the initiation of such a feasibility study.

The third phase—testing—is predominantly a testing and final standardization effort so that operations can begin. Almost all documentation must be completed in this phase.

The fourth phase is the implementation phase, which integrates the project's product or services into the existing organization. If the project was developed for establishment of a marketable product, then this phase could include the product life-cycle phases of market introduction, growth, maturity, and a portion of deterioration.

The final phase is closure and includes the reallocation of resources. Consider a company that sells products to consumers. As one product begins the deterioration and death phases of its life cycle (i.e., the divestment phase of a system), new products or projects must be established. Such a company would, therefore, require a continuous stream of projects to survive, as shown in Figure 2–14. As projects A and B begin their decline, new efforts (project C) must be developed for resource reallocation. In the ideal situation, these new projects will be established at such a rate that total revenue will increase and company growth will be clearly visible.

The closure phase evaluates the efforts of the total system and serves as input to the conceptual phases for new projects and systems. This final phase also has an impact on other ongoing projects with regard to identifying priorities.

Table 2–7 identifies the various life-cycle phases that are commonly used in different industries. However, even in mature project management industries such as construction, one could survey ten different construction companies and find ten different definitions for the life-cycle phases.

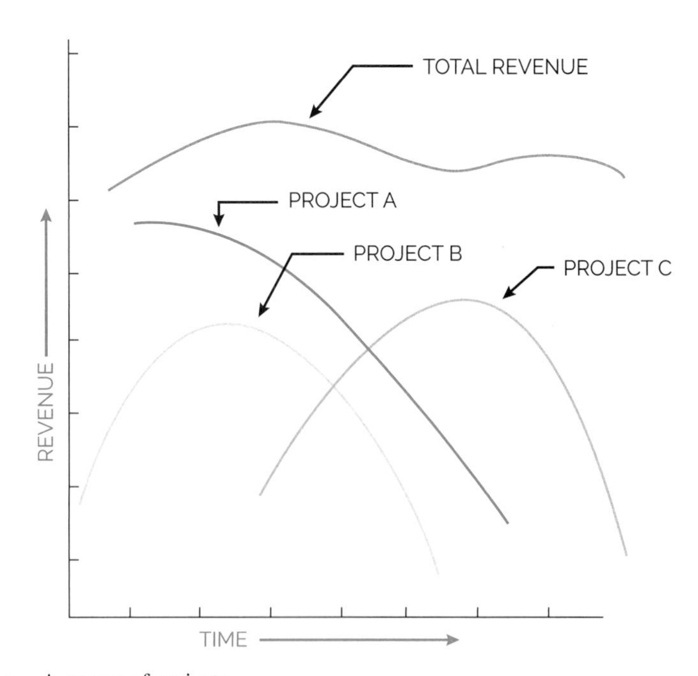

FIGURE 2–14. A stream of projects.

TABLE 2–7. LIFE-CYCLE PHASE DEFINITIONS

Engineering	Manufacturing	Computer Programming	Construction
• Start-up • Definition • Main • Termination	• Formation • Buildup • Production • Phase-out • Final audit	• Conceptual • Planning • Definition and design • Implementation • Conversion	• Planning, data gathering, and procedures • Studies and basic engineering • Major review • Detail engineering • Detail engineering/construction overlap • Construction • Testing and commissioning

Not all projects can be simply transposed into life-cycle phases (e.g., R&D). It might be possible (even in the same company) for different definitions of life-cycle phases to exist because of schedule length, complexity, or just the difficulty of managing the phases.

Top management is responsible for the periodic review of major projects. This should be accomplished, at a minimum, at the completion of each life-cycle phase.

2.13 GATE REVIEW MEETINGS (PROJECT CLOSURE)

PMBOK® Guide, 6th Edition
1.2.4.3 Phase Review

Gate review meetings are a form of project closure. Gate review meetings could result in the closure of a life-cycle phase or the closure of the entire project. Gate review meetings must be planned for, and this includes the gathering, analysis, and dissemination of pertinent information. This can be done effectively with the use of forms, templates, and checklists.

There are two forms of closure pertinent to gate review meetings: contractual closure and administrative closure. Contractual closure precedes administrative closure.

Contractual closure is the verification and signoff that all deliverables required for this phase have been completed and all action items have been fulfilled. Contractual closure is the responsibility of both the project manager and the contract administrator.

Administrative closure is the updating of all pertinent records required for both the customer and the contractor. Customers are particularly interested in documentation on any as-built or as-installed changes or deviations from the specifications. Also required is an archived trail of all scope changes agreed to during the life of the project. Contractors are interested in archived data that include project records, minutes, memos, newsletters, change management documentation, project acceptance documentation, and the history of audits for lessons learned and continuous improvement.

A subset of administrative closure is financial closure, which is the closing out of all charge numbers for the work completed. Even though contractual closure may have taken place, there may still exist open charge numbers for the repair of defects or to complete archived paperwork. Closure must be planned for, and this includes setting up a timetable and budget. Table 2–8 shows the activities for each type of closure.

TABLE 2–8. FORMS OF PROJECT CLOSURE

	Engineering	Administrative	Financial
Purpose	Customer signoff	Documentation and traceability completed	Shut down the completed work packages
When	End of the project	After contractual closure is completed	Throughout the project when work packages are completed
Activities	Verification and validation	Completion of minutes, memos, handouts, reports, and all other forms of documentation	Closing out work orders for completed work
	Conformance to acceptance criteria, including quality assurance requirements	Archiving of documentation administrative closure	Documenting results for
	Walkthroughs, testing, reviews, and audits	Capturing the lessons learned and best practices	Transferring unused funds to the management reserve or profits
	Compliance testing	Releasing resources	
	User testing		
	Review of scope changes		
	Documenting as-built changes		

2.14 ENGAGEMENT PROJECT MANAGEMENT

Companies have traditionally viewed each customer as a one-time opportunity, and after this customer's needs were met, emphasis was placed upon finding other customers. This is acceptable as long as there exists a potentially large customer base. Today, project-driven organizations, namely those that survive on the income from a continuous stream of customer-funded projects, are implementing the "engagement project management" (EPM) approach. With engagement project management, each potential new customer is approached where the contractor is soliciting a long-term relationship with the customer rather than a one-time opportunity. With this approach, contractors are selling not only deliverables and complete solutions to the client's business needs but also a willingness to make changes to the way that they manage their projects in order to receive future contracts from this client.

To maintain this level of customer satisfaction and hopefully a long-term relationship, customers are requested to provide input on how the contractor's project management methodology can be better utilized in the future. Some companies have added into their methodology a life-cycle phase entitled "Customer Satisfaction Management." This life-cycle phase takes place after administrative closure is completed. The phase involves a meeting between the client and the contractor, and in attendance are the project managers from each organization, the sponsors, selected team members and functional managers, and the sales force. The question that needs to be addressed by the contractor is, "What can we do better on the next project we perform for you?"

How much freedom should a client be given in making recommendations for changes to a contractor's EPM system? How much say should a customer have in how a contractor manages projects? What happens if this allows customers to begin telling contractors how to do their job? Obviously there are risks to be considered for this level of customer satisfaction.

If the project manager is expected to manage several projects for this client, then the project manager must understand the nature of the client's business and the environment in which the client does business. This is essential in order to identify and mitigate the risks associated with these projects. Some companies maintain an engagement manager and a project manager for each client. The engagement manager functions like an account executive for that client and may provide the project manager with the needed business information.

2.15 PROJECT MANAGEMENT METHODOLOGIES: A DEFINITION

Achieving project management excellence, or maturity, is more likely with a repetitive process that can be used on each and every project. This repetitive process is referred to as the project management methodology.

If possible, companies should maintain and support a single methodology for project management. Good methodologies integrate other processes into the project management methodology, as shown in Figure 2–15. Companies have all five of these processes integrated into their project management methodology.

In the coming years, companies can be expected to integrate more of their business processes in the project management methodology. This is shown in Figure 2–16. Managing off of a single methodology lowers cost, reduces resource requirements for support, minimizes paperwork, and eliminates duplicated efforts.

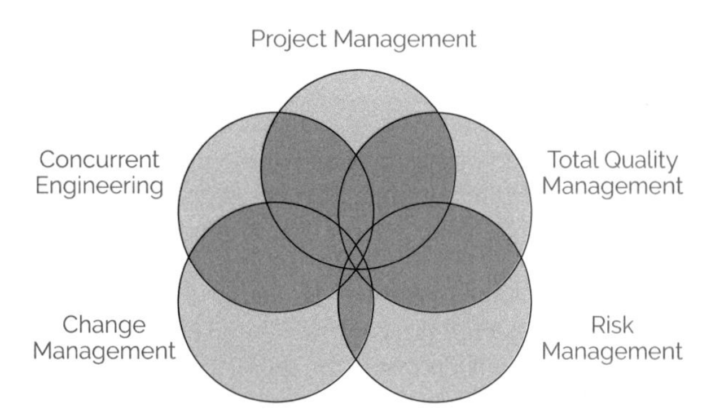

FIGURE 2–15. Integrated processes for the twenty-first century.

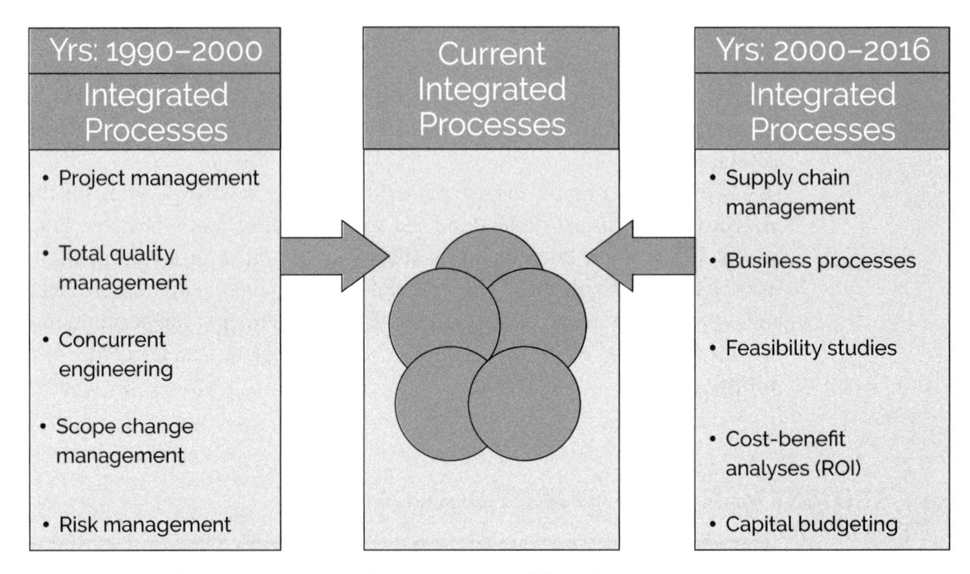

FIGURE 2–16. Integrated processes (past, present, and future).

The characteristics of a good methodology based upon integrated processes include:

- A recommended level of detail
- Use of templates
- Standardized planning, scheduling, and cost control techniques
- Standardized reporting format for both in-house and customer use
- Flexibility for application to all projects
- Flexibility for rapid improvements
- Easy for the customer to understand and follow
- Readily accepted and used throughout the entire company
- Use of standardized life-cycle phases (which can overlap) and end of phase reviews (Section 2.15)
- Based upon guidelines rather than policies and procedures (Section 2.6)
- Based upon a good work ethic

Methodologies do not manage projects; people do. It is the corporate culture that executes the methodology. Senior management must create a corporate culture that supports project management and demonstrates faith in the methodology. If this is done successfully, then the following benefits can be expected:

- Faster "time to market" through better control of the project's scope
- Lower overall project risk
- Better decision-making process
- Greater customer satisfaction, which leads to increased business
- More time available for value-added efforts, rather than internal politics and internal competition

2.16 FROM ENTERPRISE PROJECT MANAGEMENT METHODOLOGIES TO FRAMEWORKS

When the products, services, or customers have similar requirements that are reasonably well defined and do not require significant customization or numerous scope changes, companies develop often inflexible methodologies to provide some degree of consistency in the way that projects are managed. These types of methodologies are often based on rigid policies and procedures with limited flexibility but can be successful especially on large, complex, long-term projects. These "rigid" approaches are commonly called waterfall approaches, where work is done sequentially and can be easily represented by Gantt charts. The waterfall approach begins with well-defined requirements from which we must determine the budget and schedule to produce the deliverables. This approach thrives on often massive and costly documentation requirements. The approval of scope changes can be slow because rapid customer involvement may not be possible.

For some types of projects, for example in software development, the waterfall approach may not work well because the requirements may not be fully understood at the beginning of a project. We may not have a clear picture of the approach/solution we must take to create the deliverables. We may need some degree of experimentation which could lead to a significant number of scope changes. Customer involvement must occur throughout the project in order to address changes quickly, which mandates collaborative involvement by all participants, including stakeholders. In this case, we may start out with a fixed budget and schedule, and then have to decide how much work can be done within the time and cost constraints. The requirements can evolve over the life of the project. For these types of projects, a more flexible or agile approach is needed. The agile project management methodology will be discussed in more depth in Section 8.11.

As companies become reasonably mature in project management and recognize the need for a more agile approach on some projects, the policies and procedures are replaced by forms, guidelines, templates, and checklists. This provides more flexibility for the project manager in how to apply the methodology to satisfy a specific customer's requirements. This leads to a more informal or agile approach to project management.

Today, most projects are managed with an approach that is neither extremely agile nor extremely rigid; it is an in between approach with some degree of flexibility and more informal than formal. Because the amount of flexibility can change for each project, this approach is sometimes called a framework. A framework is a basic conceptual structure that is used to address an issue, such as a project. It includes a set of assumptions, concepts, templates, values, and processes that provide the project manager with a means for viewing what is needed to satisfy a customer's requirements. A framework is a skeleton support structure for building the project's deliverables. Frameworks work well as long as the project's requirements do not impose severe pressure upon the project manager. Unfortunately, in today's chaotic environment, this pressure exists and appears to be increasing. Project managers need framework methodologies to have the freedom to meet the customer's needs.

2.17 METHODOLOGIES CAN FAIL

Most companies today seem to recognize the need for one or more project management methodologies but either create the wrong methodologies or misuse the methodologies that have been created. It may not be possible to create a single enterprise-wide methodology that can be applied to each and every project. Some companies have been successful doing this, but there are still many companies that successfully maintain more than one methodology. Unless the project manager is capable of tailoring the enterprise project management methodology to his or her needs, perhaps by using a framework approach, more than one methodology may be necessary.

There are several reasons why good intentions often go astray. At the executive levels, methodologies can fail if the executives have a poor understanding of what a methodology is and believe that a methodology is:

- A quick fix
- A silver bullet
- A temporary solution
- A cookbook approach for project success[2]

At the working levels, methodologies can also fail if they:

- Are abstract and high level
- Contain insufficient narratives to support these methodologies
- Are not functional or do not address crucial areas
- Ignore the industry standards and best practices
- Look impressive but lack real integration into the business
- Use nonstandard project conventions and terminology
- Compete for similar resources without addressing this problem
- Don't have any performance metrics
- Take too long to complete because of bureaucracy and administration[3]

Other reasons why methodologies can lead to project failure include:

- The methodology must be followed exactly even if the assumptions and environmental input factors have changed.
- The methodology focuses on linear thinking.
- The methodology does not allow for out-of-the-box thinking.
- The methodology does not allow for value-added changes that are not part of the original requirements.
- The methodology does not fit the type of project.
- The methodology uses nonstandard terminology.
- The methodology is too abstract (rushing to design it).

2. J. Charvat, *Project Management Methodologies* (Hoboken, NJ: John Wiley & Sons,), 2003, p. 4.
3. Charvat, *Project Management*, p. 5.

- The methodology development team neglects to consider bottlenecks and concerns of the user community.
- The methodology is too detailed.
- The methodology takes too long to use.
- The methodology is too complex for the market, clients, and stakeholders to understand.
- The methodology does not have sufficient or correct metrics.

2.18 ORGANIZATIONAL CHANGE MANAGEMENT AND CORPORATE CULTURES

PMBOK® Guide, 6th Edition
Chapter 4 Integration Management
4.6 Perform Integrated Change Control
1.2.1 Projects

It has often been said that the most difficult projects to manage are those that involve the management of change. Figure 2–17 shows the four basic inputs needed to develop a project management methodology. Each has a "human" side that may require that people change.

Successful development and implementation of a project management methodology requires:

- Identification of the most common reasons for change in project management
- Identification of the ways to overcome the resistance to change
- Application of the principles of organizational change management to ensure that the desired project management environment will be created and sustained

For simplicity's sake, resistance can be classified as professional resistance and personal resistance to change. Professional resistance occurs when each functional unit as a whole feels threatened by project management. This is shown in Figure 2–18. Examples include:

- *Sales:* The sales staff's resistance to change arises from fear that project management will take credit for corporate profits, thus reducing the year-end bonuses for

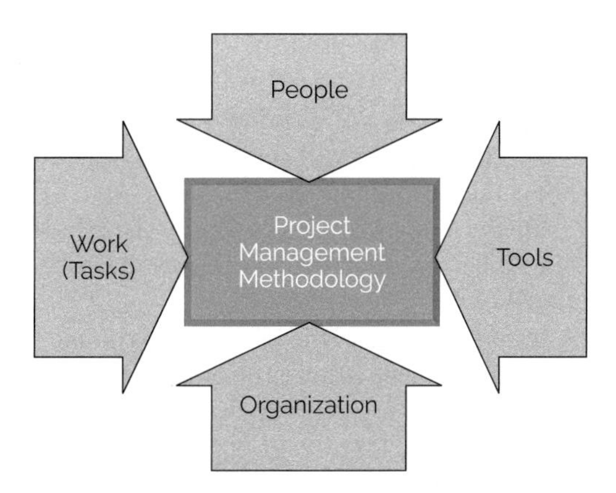

FIGURE 2–17. Methodology inputs.

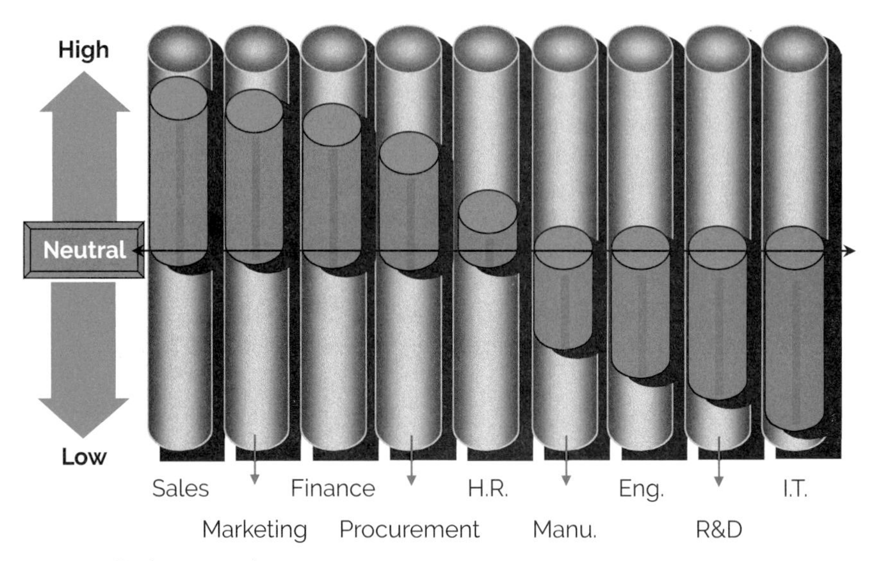

FIGURE 2–18. Resistance to change.

the sales force. Sales personnel fear that project managers may become involved in the sales effort, thus diminishing the power of the sales force.

- *Marketing:* Marketing people fear that project managers will end up working so closely with customers that project managers may eventually be given some of the marketing and sales functions. This fear is not without merit because customers often want to communicate with the personnel managing the project rather than those who may disappear after the sale is closed.
- *Finance (and Accounting):* These departments fear that project management will require the development of a project accounting system (such as earned value measurement) that will increase the workload in accounting and finance, and that they will have to perform accounting both horizontally (i.e., in projects) and vertically (i.e., in line groups).
- *Procurement:* The fear in this group is that a project procurement system will be implemented in parallel with the corporate procurement system, and that the project managers will perform their own procurement, thus bypassing the procurement department.
- *Human Resources Management:* The HR department may fear that a project management career path ladder will be created, requiring new training programs. This will increase their workloads.
- *Manufacturing:* Little resistance is found here because, although the manufacturing segment is not project-driven, there are numerous capital installation and maintenance projects which will have required the use of project management.
- *Engineering, R&D, and Information Technology:* These departments are almost entirely project-driven with very little resistance to project management.

Getting the support of and partnership with functional management can usually overcome the functional resistance. However, the individual resistance is usually more complex and more difficult to overcome. Individual resistance can stem from:

- Potential changes in work habits
- Potential changes in the social groups
- Embedded fears
- Potential changes in the wage and salary administration program

Tables 2–2 through 2–12 show the causes of resistance and possible solutions. Workers tend to seek constancy and often fear that new initiatives will push them outside their comfort zones. Most workers are already pressed for time in their current jobs and fear that new programs will require more time and energy.

TABLE 2–9. RESISTANCE: WORK HABITS

Cause of Resistance	Ways to Overcome
• New guidelines/processes • Need to share "power" information • Creation of a fragmented work environment • Need to give up established work patterns (learn new skills) • Change in comfort zones	• Dictate mandatory conformance from above • Create new comfort zones at an acceptable pace • Identify tangible/intangible individual benefits

TABLE 2–10. RESISTANCE: SOCIAL GROUPS

Cause of Resistance	Ways to Overcome
• Unknown new relationships • Multiple bosses • Multiple, temporary assignments • Severing of established ties	• Maintain existing relationships • Avoid cultural shock • Find an acceptable pace for rate of change

TABLE 2–11. RESISTANCE: EMBEDDED FEARS

Cause of Resistance	Ways to Overcome
• Fear of failure • Fear of termination • Fear of added workload • Fear or dislike of uncertainty/unknowns • Fear of embarrassment • Fear of a "we/they" organization	• Educate workforce on benefits of changes to the individual/corporation • Show willingness to admit/accept mistakes • Show willingness to pitch in • Transform unknowns into opportunities • Share information

TABLE 2–12. RESISTANCE: WAGE AND SALARY ADMINISTRATION

Causes of Resistance	Ways to Overcome
• Shifts in authority and power • Lack of recognition after the changes • Unknown rewards and punishment • Improper evaluation of personal performance • Multiple bosses	• Link incentives to change • Identify future advancement opportunities/career path

Some companies feel compelled to continually undertake new initiatives, and people may become skeptical of these programs, especially if previous initiatives have not been successful. The worst case scenario is when employees are asked to undertake new initiatives, procedures, and processes that they do not understand.

It is imperative that we understand resistance to change. If individuals are happy with their current environment, there will be resistance to change. But what if people are unhappy? There will still be resistance to change unless (1) people believe that the change is possible, and (2) people believe that they will somehow benefit from the change.

Management is the architect of the change process and must develop the appropriate strategies so the organization can change. This is done best by developing a shared understanding with employees by doing the following:

- Explaining the reasons for the change and soliciting feedback
- Explaining the desired outcomes and rationale
- Championing the change process
- Empowering the appropriate individuals to institutionalize the changes
- Investing in training necessary to support the changes

For most companies, the change management process will follow the pattern shown in Figure 2–19. Employees initially refuse to admit the need for change. As management begins pursuing the change, the support for the change diminishes and pockets of resistance crop up. Continuous support for the change by management encourages employees

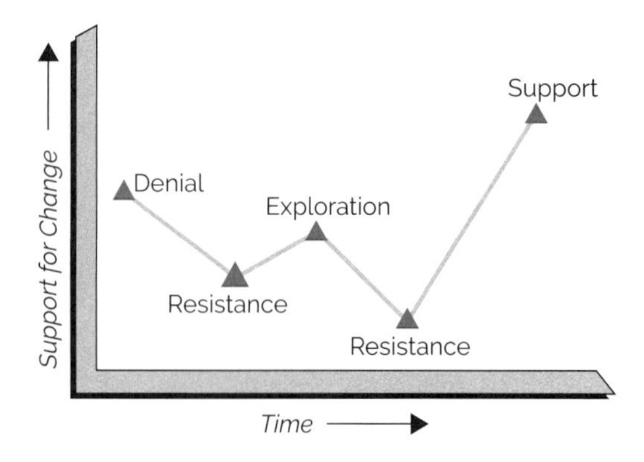

FIGURE 2–19. Change process.

to explore the potential opportunities that will result from the change about to take place. Unfortunately, this exploration often causes additional negative information to surface, thus reinforcing the resistance to change. As pressure by management increases, and employees begin to recognize the benefits of the proposed change, support begins to grow.

The ideal purpose of change management is to create a superior culture. There are different types of project management cultures based upon the nature of the business, the amount of trust and cooperation, and the competitive environment. Typical types of cultures include:

- *Cooperative cultures:* These are based upon trust and effective communications, internally and externally.
- *Noncooperative cultures:* In these cultures, mistrust prevails. Employees worry more about themselves and their personal interests than what's best for the team, company, or customer.
- *Competitive cultures:* These cultures force project teams to compete with one another for valuable corporate resources. In these cultures, project managers often demand that the employees demonstrate more loyalty to the project than to their line managers. This can be disastrous when employees are working on many projects at the same time.
- *Isolated cultures:* These occur when a large organization allows functional units to develop their own project management cultures and can result in a culture-within-a-culture environment.
- *Fragmented cultures:* These occur when part of the team is geographically separated from the rest of the team. Fragmented cultures also occur on multinational projects, where the home office or corporate team may have a strong culture for project management but the foreign team has no sustainable project management culture.

Some of the facets for an effective project management culture are shown in Figure 2–20.

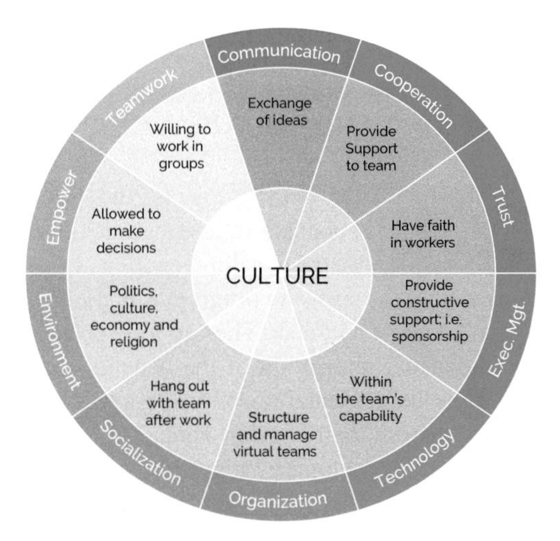

FIGURE 2–20. Facets of a project management culture.

TABLE 2–13. TRUST IN CUSTOMER-CONTRACTOR RELATIONSHIPS

Without Trust	With Trust
Continuous competitive bidding	Long-term contracts, repeat business, single- and sole-source contract awards
Massive project documentation	Minimal documentation
Excessive number of customer–contractor team meetings	Minimal number of team meetings
Team meeting with excessive documentation	Team meeting without documentation or minimal documentation
Sponsorship at the executive levels	Sponsorship at lower and middle levels of management

The critical facets of a good culture are teamwork, trust communications, and cooperation. Some project management practitioners argue that communications and cooperation are the essential ingredients for teamwork and trust. In companies with excellent cultures, teamwork is exhibited by:

- Employees and managers sharing ideas with each other and establishing high levels of innovation and creativity in work groups
- Employees and managers trusting each other and demonstrating loyalty to each other and the company
- Employees and managers being committed to the work they do and the promises they make
- Employees and managers sharing information freely
- Employees and managers consistently being open and honest with each other

When teamwork exists, trust usually follows, and this includes trust among the workers within the company and trust in dealing with clients. When trust occurs between the buyer and the seller, both parties eventually benefit, as shown in Table 2–13.

2.19 BENEFITS HARVESTING AND CULTURAL CHANGE

PMBOK® Guide, 6th Edition
Chapter 4 Integration Management

On some projects, the true benefits and resulting value are not obtained until sometime after the project is over. An example might be the development of a new software program where the benefits are not achieved until the software program is implemented and being used. This is often called the "go live" stage of a project. The "go live" stage is often called benefits harvesting stages which is the actual realization of the benefits and accompanying value. Harvesting may necessitate the implementation of an organizational change management plan that may remove people from their comfort zone. Full benefit realization may face resistance from managers, workers, customers, suppliers, and partners. There may be an inherent fear that change will be accompanied by loss of promotion prospects, less authority and responsibility, and possible loss of respect from peers.

Benefits harvesting may also increase the benefits realization costs because of:

- Hiring and training new recruits
- Changing the roles of existing personnel and providing training
- Relocating existing personnel
- Providing additional or new management support
- Updating computer systems
- Purchasing new software
- Creating new policies and procedures
- Renegotiating union contracts
- Developing new relationships with suppliers, distributors, partners, and joint ventures

2.20 AGILE AND ADAPTIVE PROJECT MANAGEMENT CULTURES

PMBOK® Guide, 6th Edition
1.2.5.1 Project Management Tailoring
1.2.5.2 Project Management Methodology Tailoring

One of the reasons why agile project management has been successful is because executive are now placing more trust in the hands of the project managers to make the correct project and business decisions. Years ago, project management methodologies were created based upon rigid policies and procedures with the mistaken belief that only through project management standardization on every project can we get repeatable project success. Tailoring the project management methodology to a particular project or client was rarely allowed.

Agile project management practices have demonstrated that project management tailoring can work. Most methodologies today are made of forms, guidelines, templates, and checklists. The project manager then selects what is appropriate for a particular client and creates a flexible methodology or framework that can be unique for each client. We live in a world of adaptive environments. This is particularly important for external clients that would prefer that the framework be adapted to their business model and way of doing business rather than how your parent company does business. Framework success can lead to repeat business.

2.21 PROJECT MANAGEMENT INTELLECTUAL PROPERTY

PMBOK® Guide, 6th Edition
1.2.4.7 Project Management Data and Information
1.2.6 Project Management Business Documents
2.3.2 Corporate Knowledge Repositories

We believe today that we are managing our business by projects. As such, project managers are expected to make business decisions as well as project decisions. Throughout the life of a project, there is a significant amount of data that must be collected including information related to the project business case, project benefits realization plan, project charter, and project plan.

When a project comes to an end, the focus is now on capturing lessons learned and best practices. We must capture not only project-related best practices, but business best practices as well. But as we capture business best practices, we begin replacing the project management best practices library with a knowledge repository that includes both project management and business-related best practices. This is shown in Figure 2–21.

Another reason for the growth in intellectual property is because of the benchmarking activities that companies are performing, most likely using the project management office. Figure 2–22 shows typical benchmarking activities and the types of information being sought.

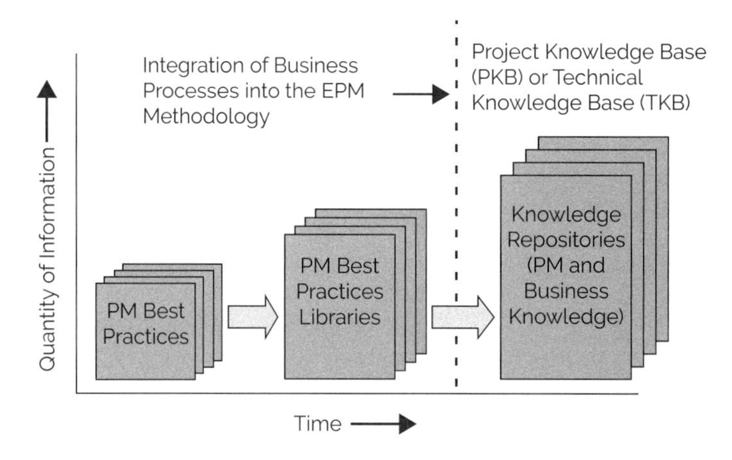

FIGURE 2–21. Growth of knowledge management.

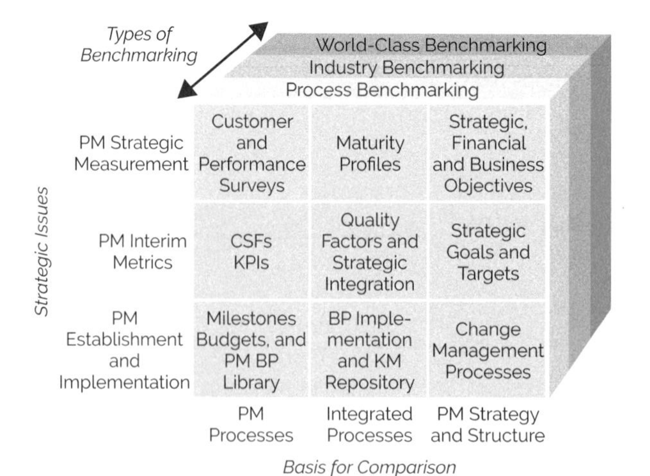

FIGURE 2–22. PM benchmarking and knowledge management (KM).

The growth in knowledge repositories and cloud computing has provided companies with the opportunities for data warehouses. According to Melik:

> Many organizations use diverse applications and information systems, each having its own database. The data from disparate systems can be merged into one single database (centralized data) in a process known as *data warehousing*. For example, a company could use a customer relations management (CRM) solution from Vendor A, a project management system from Vendor B, and an enterprise resource planning (ERP) or accounting system from Vendor C; data warehousing would be used to aggregate the data from these three sources. Business intelligence and reporting tools are then used to perform detailed analysis on all of the data. Data warehouse reports are usually not real time, since the data aggregation takes time to complete and is typically scheduled for once per week, month, or even quarter.[4]

2.22 SYSTEMS THINKING

Ultimately, all decisions and policies are made on the basis of judgments; there is no other way, and there never will be. In the end, analysis is but an aid to the judgment and intuition of the decision maker. These principles hold true for project management as well as for systems management.

The systems approach may be defined as a logical and disciplined process of problem solving. The word *process* indicates an active ongoing system that is fed by input from its parts. The systems approach:

- Forces review of the relationship of the various subsystems
- Is a dynamic process that integrates all activities into a meaningful total system
- Systematically assembles and matches the parts of the system into a unified whole
- Seeks an optimal solution or strategy in solving a problem

The systems approach to problem-solving has phases of development similar to traditional life-cycle phases. These phases are defined as follows:

- *Translation:* Terminology, problem objective, and criteria and constraints are defined and accepted by all participants.
- *Analysis:* All possible approaches to or alternatives to the solution of the problem are stated.
- *Trade-off:* Selection criteria and constraints are applied to the alternatives to meet the objective.
- *Synthesis:* The best solution in reaching the objective of the system is the result of the combination of analysis and trade-off phases.

4. Rudolf Melik, *The Rise of the Project Workforce* (Hoboken, NJ: John Wiley & Sons, 2007), p. 238.

Other terms essential to the systems approach are:

- *Objective:* The function of the system or the strategy that must be achieved
- *Requirement:* A partial need to satisfy the objective
- *Alternative:* One of the selected ways to implement and satisfy a requirement
- *Selection criteria:* Performance factors used in evaluating the alternatives to select a preferable alternative
- *Constraint:* An absolute factor that describes conditions that the alternatives *must* meet

A common error by potential decision makers (those dissatisfied individuals with authority to act) who base their thinking solely on subjective experience, judgment, and intuition is that they fail to recognize the existence of alternatives. Subjective thinking is inhibited or affected by personal bias.

Objective thinking, on the other hand, is a fundamental characteristic of the systems approach and is exhibited or characterized by emphasis on the tendency to view events, phenomena, and ideas as external and apart from self-consciousness. Objective thinking is unprejudiced.

The systems analysis process, as shown in Figure 2–23, begins with systematic examination and comparison of those alternative actions that are related to the accomplishment of the desired objective. The alternatives are then compared on the basis of the resource costs and the associated benefits. The loop is then completed using feedback to determine how compatible each alternative is with the objectives of the organization.

The above analysis can be arranged in steps:

- Input data to mental process
- Analyze data
- Predict outcomes
- Evaluate outcomes and compare alternatives
- Choose the best alternative
- Take action
- Measure results and compare them with predictions

The systems approach is most effective if individuals can be trained to be ready with alternative actions that directly tie in with the prediction of outcomes. The basic tool is the outcome array, which represents the matrix of all possible circumstances. This outcome array can be developed only if the decision maker thinks in terms of the wide scope of possible outcomes. Outcome descriptions force the decision maker to spell out clearly just what he is trying to achieve (i.e., his objectives).

Systems thinking is vital for the success of a project. Project management systems urgently need new ways of strategically viewing, questioning, and analyzing project needs

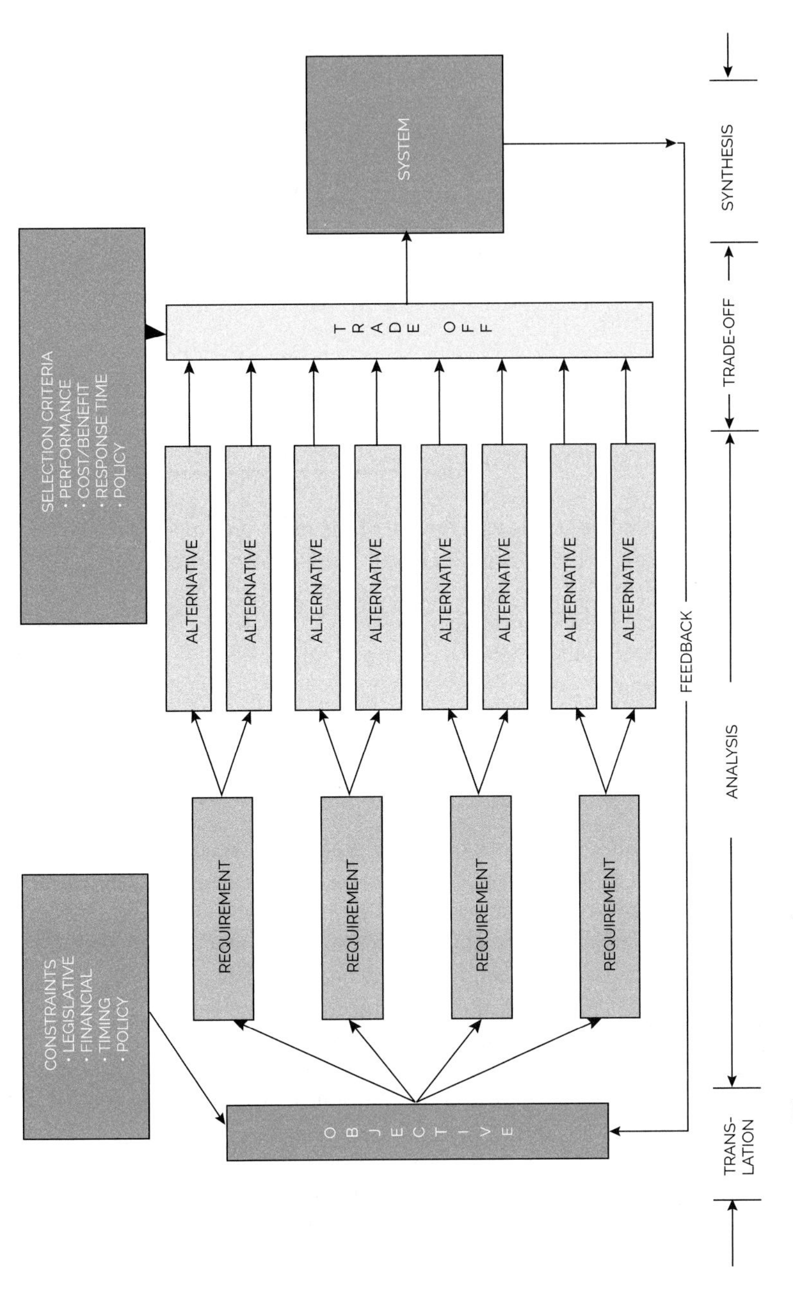

FIGURE 2–23. The systems approach.

for alternative nontechnical and technical solutions. The ability to analyze the total project, rather than the individual parts, is essential for successful project management.

Related Case Studies (from Kerzner/*Project Management Case Studies*, 5th ed.)	**Related Workbook Exercises (from Kerzner/*Project Management Workbook and PMP®/CAPM® Exam Study Guide*, 12th ed.)**	***PMBOK® Guide*, 6th Edition, Reference Section for the PMP® Certification Exam**
• Cordova Research Group • Cortez Plastics • Apache Metals, Inc. • Haller Specialty Manufacturing • Creating a Methodology* • Disney (A), (B)	• Multiple Choice Exam	• Integration Management • Scope Management

*Case study appears at end of the chapter.

2.23 STUDYING TIPS FOR THE PMI® PROJECT MANAGEMENT CERTIFICATION EXAM

This section is applicable as a review of the principles to support the knowledge areas and domain groups in the *PMBOK® Guide*. This chapter addresses:

- Integration Management
- Scope Management
- Closure

Understanding the following principles is beneficial if the reader is using this text to study for the PMP® Certification Exam:

- Brief historical background of project management
- That, early on, project managers were assigned from engineering
- Benefits of project management
- Barriers to project management implementation and how to overcome them
- Differences between a program and a project
- What is meant by informal project management
- How to identify success and failure in project management
- Project life-cycle phases
- What is meant by closure to a life-cycle phase or to the entire project
- What is meant by a project management methodology
- What is meant by critical success factors (CSFs) and key performance indicators (KPIs)

PMP is a registered mark of the Project Management Institute, Inc.

In Appendix C, the following Dorale Products mini–case studies are applicable:

- Dorale Products (A) [Integration and Scope Management]
- Dorale Products (B) [Integration and Scope Management]
- Dorale Products (C) [Integration and Scope Management]
- Dorale Products (D) [Integration and Scope Management]
- Dorale Products (E) [Integration and Scope Management]
- Dorale Products (F) [Integration and Scope Management]

The following multiple-choice questions will be helpful in reviewing the principles of this chapter:

1. A structured process for managing a multitude of projects is most commonly referred to as:
   A. Project management policies
   B. Project management guidelines
   C. Industrywide templates
   D. A project management methodology

2. The most common terminology for a reusable project management methodology is:
   A. Template
   B. Concurrent scheduling technique
   C. Concurrent planning technique
   D. Skeleton framework document

3. The major behavioral issue in getting an organization to accept and use a project management methodology effectively is:
   A. Lack of executive sponsorship
   B. Multiple boss reporting
   C. Inadequate policies and procedures
   D. Limited project management applications

4. The major difference between a project and a program is usually:
   A. The role of the sponsor
   B. The role of the line manager
   C. The time frame
   D. The specifications

5. Projects that remain almost entirely within one functional area are best managed by the:
   A. Project manager
   B. Project sponsor
   C. Functional manager
   D. Assigned functional employees

6. Large projects are managed by:
   A. The executive sponsor
   B. The project or program office for that project
   C. The manager of project managers
   D. The director of marketing
7. The most common threshold limits on when to use the project management methodology are:
   A. The importance of the customer and potential profitability
   B. The size of the project (i.e., $) and duration
   C. The reporting requirements and position of the sponsor
   D. The desires of management and functional boundaries crossed
8. A grouping of projects is called a:
   A. Program
   B. Project template
   C. Business template
   D. Business plan
9. Project management methodologies often work best if they are structured around:
   A. Rigid policies
   B. Rigid procedures
   C. Minimal forms and checklists
   D. Life-cycle phases
10. One way to validate the successful implementation of project management is by looking at the number and magnitude of the conflicts requiring:
   A. Executive involvement
   B. Customer involvement
   C. Line management involvement
   D. Project manager involvement
11. Standardization and control are benefits usually attributed to:
   A. Laissez-faire management
   B. Project management on R&D efforts
   C. Use of life cycle-phases
   D. An organization with weak executive sponsorship
12. The most difficult decision for an executive sponsor to make at the end-of-phase review meeting is to:
   A. Allow the project to proceed to the next phase based upon the original objective.
   B. Allow the project to proceed to the next phase based upon a revised objective.
   C. Postpone making a decision until more information is processed.
   D. Cancel the project.

13. Having too many life-cycle phases may be detrimental because:
    A. Executive sponsors will micromanage.
    B. Executive sponsors will become "invisible."
    C. The project manager will spend too much time planning for gate review meetings rather than managing the phases.
    D. The project manager will need to develop many different plans for each phase.
14. A project is terminated early because the technology cannot be developed, and the resources are applied to another project that ends up being successful. Which of the following is true concerning the first project?
    A. The first project is regarded as a failure.
    B. The first project is a success if the termination is done early enough before additional resources are squandered.
    C. The first project is a success if the project manager gets promoted.
    D. The first project is a failure if the project manager gets reassigned to a less important project.
15. Which of the following would *not* be regarded as a secondary definition of project success?
    A. The customer is unhappy with the deliverable, but follow-on business is awarded based on effective customer relations.
    B. The deliverables are met but OSHA and EPA laws are violated.
    C. The customer is displeased with the performance, but you have developed a new technology that could generate many new products.
    D. The project's costs were overrun by 40 percent, but the customer funds an enhancement project.

ANSWERS

1. D
2. A
3. B
4. C
5. C
6. B
7. B
8. A
9. D
10. A
11. C
12. D
13. C
14. B
15. B

PROBLEMS

2–1 Do you think that someone could be a good systems manager but a poor project manager? What about the reverse situation? State any assumptions that you may have to make.

2–2 For each of the following projects, state whether we are discussing an open, closed, or extended system:

a. A high-technology project
b. New product R&D

c. An online computer system for a bank
d. Construction of a chemical plant
e. Developing an in-house cost accounting reporting system

2–3 What impact could the product life cycle have on the selection of the project organizational structure?

2–4 In the development of a system, what criteria should be used to determine where one phase begins and another ends and where overlap can occur?

2–5 Can a company be successful at project management without having or using a project management methodology?

2–6 Who determines how many life-cycle phases should be part of a project management methodology?

2–7 Under what conditions can a project be considered as both a success and a failure at the same time?

2–8 Is it possible to attain an informal project management approach without first going through formalized project management?

CASE STUDY

CREATING A METHODOLOGY

Background

John Compton, the president of the company, expressed his feelings quite bluntly at the executive staff meeting:

We are no longer competitive in the marketplace. Almost all of the Requests for Proposal (RFP) that we want to bid on have a requirement that we must identify in the proposal the project management methodology we will use on the contract should we be awarded the contract. We have no project management methodology. We have just a few templates we use based upon the *PMBOK® Guide*. All of our competitors have methodologies, but not us.

I have been asking for a methodology to be developed for more than a year now, and all I get are excuses. Some of you are obviously afraid that you might lose power and authority once the methodology is up and running. That may be true, but losing some power and authority is obviously better than losing your job. In six months I want to see a methodology in use on all projects or I will handle the situation myself. I simply cannot believe that my executive staff is afraid to develop a project management methodology.

Critical Issues

The executive staff knew this day was inevitable; they had to take the initiative in the implementation of a project management methodology. Last year, a consultant was brought in to conduct a morning three-hour session on the benefits of project management and the value of an enterprise project management methodology (EPM). As part of the session, the consultant explained that the time needed to develop and implement an EPM system can be shortened if the company has a project management office (PMO) in place to take the lead role. The consultant also explained that whichever executive gets control of the PMO may become more powerful than other executives because he or she now controls all of the project management intellectual property. The executive staff fully understood the implication of this and therefore became reluctant to visibly support project management until they could see how their organization would be affected. In the meantime, project management suffered.

Reluctantly, a PMO was formed reporting to the chief information officer. The PMO comprised a handful of experienced project managers who could hopefully take the lead in the development of a methodology. The PMO concluded that there were five steps that had to be done initially. After the five steps were done, the executive committee would receive a final briefing on what had been accomplished. The final briefing would be in addition to the monthly updates and progress reports. The PMO believed that getting executive support and sign-offs in a timely manner would be difficult.

The first step that needed to be done was the establishment of the number of life-cycle phases. Some people interviewed wanted ten to twelve life-cycle phases. That meant that there would be ten to twelve gate review meetings and the project managers would spend a great deal of time preparing paperwork for the gate review meetings rather than managing the project. The decision was then made to have no more than six life-cycle phases.

The second step was to decide whether the methodology should be designed around rigid policies and procedures or go the more informal route of using forms, guidelines, checklists, and templates. The PMO felt that project managers needed some degree of freedom in dealing with clients and therefore the more informal approach would work best. Also, clients were asking to have the methodology designed around the client's business needs and the more informal approach would provide the flexibility to do this.

The third step was to see what could be salvaged from the existing templates and checklists. The company had a few templates and checklists but not all of the project managers used them. The decision was made to develop a standardized set of documents in accordance with the information in the *PMBOK® Guide*. The project managers could then select whatever forms, guidelines, templates, and checklists were appropriate for a particular project and client.

The fourth step would be to develop a means for capturing best practices using the EPM system. Clients were now requiring in their RFP that best practices on a project must be captured and shared with the client prior to the closeout of the project. Most of the people in the PMO believed that this could be done using forms or checklists at the final project debriefing meeting.

The fifth step involved education and training. The project managers and functional organizations that would staff the projects would need to be trained in the use of the new methodology. The PMO believed that a one-day training program would suffice and the functional organizations could easily release their people for a one-day training session.

QUESTIONS

1. What can you determine about the corporate culture from the fact that they waited this long to consider the development of an EPM system?
2. Can a PMO accelerate the implementation process?
3. Is it acceptable for the PMO to report to the chief information officer or to someone else?
4. Why is it best to have six or fewer life-cycle phases in an EPM system?
5. Is it best to design an EPM system around flexible or inflexible elements? Generally, when first developing an EPM system, do companies prefer to use formality or informality in the design?
6. Should an EPM system have the capability of capturing best practices?

3 Organizational Structures

3.0 INTRODUCTION

PMBOK® Guide, 6th Edition
2.4 Organizational Systems
2.4.2 Organizational Structure Types
Chapter 9 Project Resource Management

During the past fifty years there has been a so-called hidden revolution in the introduction and development of new organizational structures. Management has come to realize that organizations must be dynamic in nature; that is, they must be capable of rapid restructuring should environmental conditions so dictate. These environmental factors evolved from the increasing competitiveness of the market, changes in technology, and a requirement for better control of resources for multiproduct firms.

Much has been written about how to identify and interpret those signs that indicate that a new organizational form may be necessary. Some signs include underutilization of talent, frequent inability to meet the constraints and the lack of a cooperative culture.

Unfortunately, many companies do not realize the necessity for organizational change until it is too late. Management looks externally (i.e., to the environment) rather than internally for solutions to problems. A typical example would be that new product costs are rising while the product life cycle may be decreasing. Should emphasis be placed on lowering costs or developing new products?

If we assume that an organizational system is composed of both human and nonhuman resources, then we must analyze the sociotechnical subsystem whenever organizational changes are being considered. The social system is represented by the organization's personnel and their group behavior. The technical system includes the technology, materials, and machines necessary to perform the required tasks.

Behaviorists contend that there is no one best structure to meet the challenges of tomorrow's organizations. The structure used, however, must be one that optimizes company performance by achieving a balance between the social and the technical requirements.

Even the simplest type of organizational change can induce major conflicts. The creation of a new position, the need for better planning, the lengthening or shortening of the span of control, the need for additional technology (knowledge), and centralization or decentralization can result in major changes in the sociotechnical subsystem.

Organizational restructuring is a compromise between the traditional (classical) and the behavioral schools of thought; management must consider the needs of individuals as well as the needs of the company. Is the organization structured to manage people or to manage work?

There is a wide variety of organizational forms for restructuring management. The exact method depends on the people in the organization, the company's product lines, and management's philosophy. A poorly restructured organization can sever communication channels that may have taken months or years to cultivate; cause a restructuring of the informal organization, thus creating new power, status, and political positions; and eliminate job satisfaction and motivational factors to such a degree that complete discontent results.

In the sections that follow, a variety of organizational forms will be presented. Obviously, it is an impossible task to describe all possible organizational structures. Each form describes how the project management organization evolved from the classical theories of management. Advantages and disadvantages are listed for technology and social systems.

The answers to these questions are not easy. For the most part, they are a matter of the judgment exercised by organizational and behavioral managers.

3.1 ORGANIZATIONAL WORK FLOW

Organizations are continually restructured to meet the demands imposed by the environment. Restructuring can change the role of individuals in the formal and the informal organization. Many researchers believe that the greatest usefulness of behaviorists lies in their ability to help the informal organization adapt to changes and resolve the resulting conflicts. Unfortunately, behaviorists cannot be totally effective unless they have input into the formal organization as well. Whatever organizational form is finally selected, formal channels must be developed so that each individual has a clear description of the authority, responsibility, and accountability necessary for the work to proceed.

In the discussion of organizational structures, the following definitions will be used:

- *Authority* is the power granted to individuals (possibly by their position) so that they can make final decisions.
- *Responsibility* is the obligation incurred by individuals in their roles in the formal organization to effectively perform assignments.
- *Accountability* is being answerable for the satisfactory completion of a specific assignment. (Accountability = authority + responsibility)

Authority and responsibility can be delegated to lower levels in the organization, whereas accountability usually rests with the individual. Yet many executives refuse to delegate and argue that an individual can have total accountability just through responsibility.

Even with these clearly definable divisions of authority, responsibility, and accountability, establishing good relationships between project and functional managers can take a great deal of time, especially during the conversion from a traditional to a project organizational form. Trust is the key to success here.

3.2 TRADITIONAL (CLASSICAL) ORGANIZATION

The traditional management structure has survived for more than two centuries. However, recent business developments, such as the rapid rate of change in technology and increased stockholder demands, have created strains on existing organizational forms. Fifty years ago companies could survive with only one or two product lines. The classical management organization, as shown in Figure 3–1, was satisfactory for control, and conflicts were minimal.[1]

However, with the passing of time, companies found that survival depended on multiple product lines (i.e., diversification) and vigorous integration of technology into the existing organization. As organizations grew and matured, managers found that company activities were not being integrated effectively, and that new conflicts were arising in the well-established formal and informal channels. Managers began searching for more innovative organizational forms that would alleviate these problems.

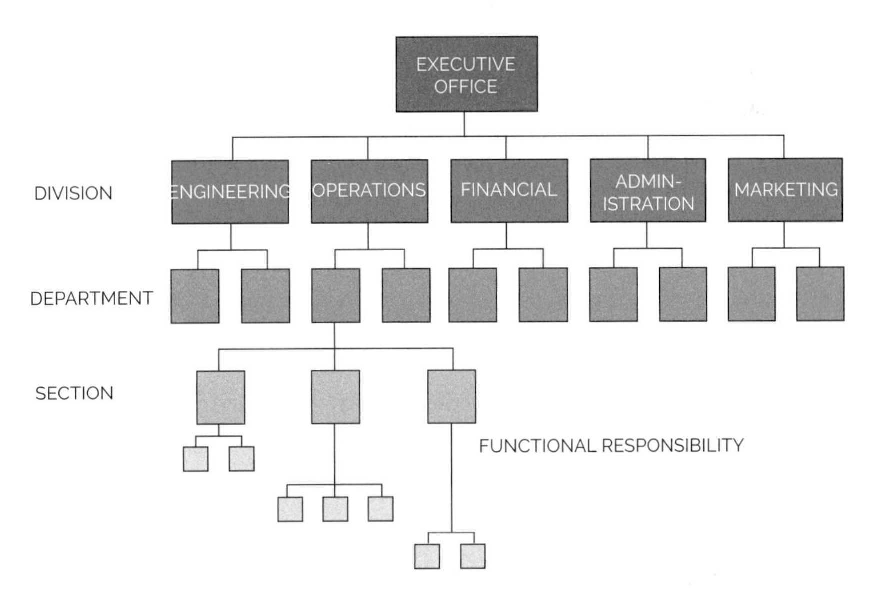

FIGURE 3–1. The traditional management structure.

1. Many authors refer to classical organizations as pure functional organizations. This can be seen from Figure 3–1. Also note that the department level is below the division level. In some organizations these titles are reversed.

TABLE 3–1. ADVANTAGES OF THE TRADITIONAL (CLASSICAL) ORGANIZATION

- Easier budgeting and cost control are possible.
- Better technical control is possible.
  - Specialists can be grouped to share knowledge and responsibility.
  - Personnel can be used on many different projects.
  - All projects will benefit from the most advanced technology (better utilization of scarce personnel).
- Flexibility in the use of manpower.
- A broad manpower base to work with.
- Continuity in the functional disciplines; policies, procedures, and lines of responsibility are easily defined and understandable
- Admits mass production activities within established specifications.
- Good control over personnel, since each employee has one and only one person to report to.
- Communication channels are vertical and well established.
- Quick reaction capability exists, but may be dependent upon the priorities of the functional managers.

Before a valid comparison can be made with the newer forms, the advantages and disadvantages of the traditional structure must be shown. Table 3–1 lists the advantages of the traditional organization. As seen in Figure 3–1, the general manager has all of the functional entities necessary to perform R&D or develop and manufacture a product. All activities are performed within the functional groups and are headed by a department (or, in some cases, a division) head. Each department maintains a strong concentration of technical expertise. Since all projects must flow through the functional departments, each project can benefit from the most advanced technology, thus making this organizational form well suited to mass production. Functional managers can hire a wide variety of specialists and provide them with easily definable paths for career progression.

The functional managers maintain absolute control over the budget. They establish their own budgets, on approval from above, and specify requirements for additional personnel. Because the functional manager has manpower flexibility and a broad base from which to work, most projects are normally completed within cost.

Both the formal and informal organizations are well established, and levels of authority and responsibility are clearly defined. Because each person reports to only one individual, communication channels are well structured.

Yet, for each advantage, there is almost always a corresponding disadvantage (see Table 3–2). The majority of these disadvantages are related to the absence of a strong

TABLE 3–2. DISADVANTAGES OF THE TRADITIONAL (CLASSICAL ORGANIZATION)

- No one individual is directly responsible for the total project (i.e., no formal authority; committee solutions).
- Does not provide the project-oriented emphasis necessary to accomplish the project tasks.
- Coordination becomes complex, and additional lead time is required for approval of decisions.
- Decisions normally favor the strongest functional groups.
- No customer focal point.
- Response to customer needs is slow.
- Difficulty in pinpointing responsibility; this is the result of little or no direct project reporting, very little project-oriented planning, and no project authority.
- Motivation and innovation are decreased.
- Ideas tend to be functionally oriented with little regard for ongoing projects.

central authority or individual responsible for the total project. As a result, integration of activities that cross functional lines becomes difficult, and top-level executives must get involved with the daily routine. Conflicts occur as each functional group struggles for power. Ideas may remain functionally oriented with very little regard for ongoing projects, and the decision-making process will be slow and tedious.

Because there is no customer focal point, all communications must be channeled through upper-level management. Upper-level managers then act in a customer-relations capacity and refer all complex problems down through the vertical chain of command to the functional managers. The response to the customer's needs therefore becomes a slow and aggravating process.

Projects have a tendency to fall behind schedule in the classical organizational structure. Incredibly large lead times are required. Functional managers attend to those tasks that provide better benefits to themselves and their subordinates first.

With the growth of project management in the late 1960s, executives began to realize that many of the problems were the result of weaknesses in the traditional structure.

3.3 PURE PRODUCT (PROJECTIZED) ORGANIZATION

As the traditional organizational structure began to evolve and develop with varying degrees of success, the need for project management structures were conceived. The pure product organization, as shown in Figure 3–2, develops as a division within a division.

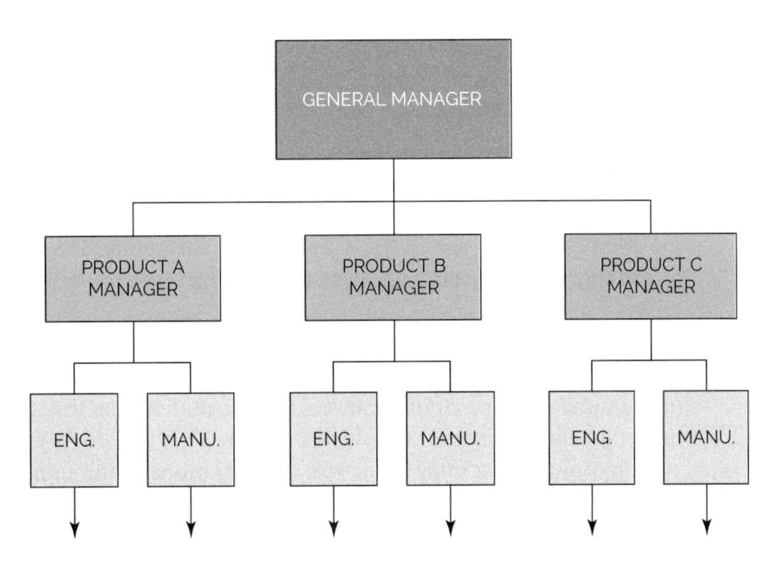

FIGURE 3–2. Pure product or projectized structure.

As long as there exists a continuous flow of projects, work is stable and conflicts are at a minimum. The major advantage of this organizational flow is that one individual, the program manager, maintains complete line authority over the entire project. Not only does he assign work, but he also conducts merit reviews. Because each individual reports to only one person, strong communication channels develop that result in a very rapid reaction time.

In pure product organizations, long lead times became a thing of the past. Trade-off studies could be conducted as fast as time would permit without the need to look at the impact on other projects (unless, of course, identical facilities or equipment were required). Functional managers were able to maintain qualified staffs for new product development without sharing personnel with other programs and projects.

The responsibilities attributed to the project manager were entirely new. First, his authority was now granted by the vice president and general manager. The program manager handled all conflicts, both those within his organization and those involving other projects. Interface management was conducted at the program manager level. Upper-level management was now able to spend more time on executive decision making than on conflict arbitration.

The major disadvantage with the pure project form is the cost of maintaining the organization. There is no chance for sharing an individual with another project in order to reduce costs. Personnel are usually attached to these projects long after they are needed because once an employee is given up, the project manager might not be able to get him back. Motivating personnel becomes a problem. At project completion, functional personnel do not "have a home" to return to. Many organizations place these individuals into an overhead labor pool from which selection can be made during new project development. People remaining in the labor pool may be laid off. As each project comes to a close, people become uneasy and often strive to prove their worth to the company by overachieving, a condition that is only temporary. It is very difficult for management to convince key functional personnel that they do, in fact, have career opportunities in this type of organization.

In pure functional (traditional) structures, technologies are well developed, but project schedules often fall behind. In the pure project structure, the fast reaction time keeps activities on schedule, but technology suffers because without strong functional groups, which maintain interactive technical communication, the company's outlook for meeting the competition may be severely hampered. The engineering department for one project might not communicate with its counterpart on other projects, resulting in duplication of efforts.

The last major disadvantage of this organizational form lies in the control of facilities and equipment. The most frequent conflict occurs when two projects require use of the same piece of equipment or facilities at the same time. Upper-level management must then assign priorities to these projects. This is normally accomplished by defining certain projects as strategic, tactical, or operational—the same definitions usually given to plans.

Table 3–3 summarizes the advantages and disadvantages of this organizational form.

TABLE 3–3. ADVANTAGES AND DISADVANTAGES OF THE PRODUCT ORGANIZATIONAL FORM

Advantages

- Provides complete line authority over the project (i.e., strong control through a single project authority).
- Participants work directly for the project manager. Unprofitable product lines are easily identified and can be eliminated.
- Strong communications channels.
- Staffs can maintain expertise on a given project without sharing key personnel.
- Very rapid reaction time is provided.
- Personnel demonstrate loyalty to the project; better morale with product identification.
- A focal point develops for out-of-company customer relations.
- Flexibility in determining time (schedule), cost, and performance trade-offs.
- Interface management becomes easier as unit size is decreased.
- Upper-level management maintains more free time for executive decision making.

Disadvantages

- Cost of maintaining this form in a multiproduct company would be prohibitive due to duplication of effort, facilities, and personnel; inefficient usage.
- A tendency to retain personnel on a project long after they are needed. Upper-level management must balance workloads as projects start up and are phased out.
- Technology suffers because, without strong functional groups, outlook of the future to improve company's capabilities for new programs would be hampered (i.e., no perpetuation of technology).
- Control of functional (i.e., organizational) specialists requires top-level coordination.
- Lack of opportunities for technical interchange between projects.
- Lack of career continuity and opportunities for project personnel.

3.4 MATRIX ORGANIZATIONAL FORM

PMBOK® Guide, 6th Edition
2.4.4 Organizational Structure Types

The matrix organizational form is an attempt to combine the advantages of the pure functional structure and the product organizational structure. This form is ideally suited for "project-driven" companies. Figure 3–3 shows a typical matrix structure. Each project manager reports directly to the vice president and general manager. Since each project represents a potential profit center, the power and authority used by the project manager come directly from the general manager. The project manager has total responsibility and accountability for project success. The functional departments, on the other hand, have functional responsibility to maintain technical excellence on the project. Each functional unit is headed by a department manager whose prime responsibility is to ensure that a unified technical base is maintained and that all available information can be exchanged for each project. Department managers must also keep their people aware of the latest technical accomplishments in the industry.

Project management is a "coordinative" function, whereas matrix management is a collaborative function division of project management. In the coordinative or project organization, work is generally assigned to specific people or units who "do their own thing." In the collaborative or matrix organization, information sharing may be mandatory, and several people may be required for the same piece of work. In a project organization,

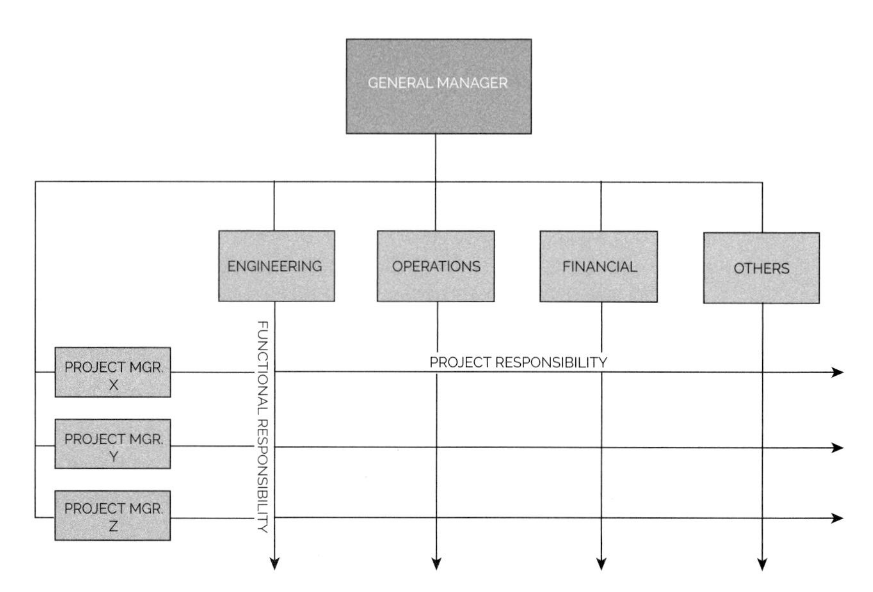

FIGURE 3–3. Typical matrix structure.

authority for decision making and direction rests with the project leader, whereas in a matrix it rests with the team.

Certain ground rules exist for matrix development:

- Participants are full time on the project; this ensures a degree of loyalty.
- Horizontal as well as vertical channels must exist for making commitments.
- There must be quick and effective methods for conflict resolution.
- There must be good communication channels and free access between managers.
- All managers must have input into the planning process.
- Both horizontally and vertically oriented managers must be willing to negotiate for resources.
- The horizontal line must be permitted to operate as a separate entity except for administrative purposes.

The basis for the matrix approach is an attempt to create synergism through shared responsibility between project and functional management. Yet this is easier said than done. *No two working environments are the same, and, therefore, no two companies will have the same matrix design.* The following questions must be answered before a matrix structure can be successful:

- If each functional unit is responsible for one aspect of a project, and other parts are conducted elsewhere (possibly subcontracted to other companies), how can a synergistic environment be created?
- Who decides which element of a project is most important?
- How can a functional unit (operating in a vertical structure) answer questions and achieve project goals and objectives that are compatible with other projects?

The answers to these questions depend on mutual understanding between the project and functional managers. Since both individuals maintain some degree of authority, responsibility, and accountability on each project, they must continuously negotiate. Unfortunately, the program manager might only consider what is best for his project (disregarding all others), whereas the functional manager might consider his organization more important than each project.

In order to get the job done, project managers need organizational status and authority. A corporate executive contends that the organization chart shown in Figure 3–6 can be modified to show that the project managers have adequate organizational authority by placing the department manager boxes at the tip of the functional responsibility arrowheads. With this approach, the project managers appear to be higher in the organization than their departmental counterparts but are actually equal in status. Executives who prefer this method must exercise caution because the line and project managers may not feel that there is still a balance of power.

Problem solving in this environment is fragmented and diffused. The project manager acts as a unifying agent for project control of resources and technology. He must maintain open channels of communication to prevent suboptimization of individual projects.

In many situations, functional managers have the power to make a project manager look good, if they can be motivated to think about what is best for the project. Unfortunately, this is not always accomplished. As stated by Mantell:[2]

> There exists an inevitable tendency for hierarchically arrayed units to seek solutions and to identify problems in terms of scope of duties of particular units rather than looking beyond them. This phenomenon exists without regard for the competence of the executive concerned. It comes about because of authority delegation and functionalism.

The project environment and functional environment cannot be separated; they must interact. The location of the project and functional unit interface is the focal point for all activities.

The functional manager controls departmental resources (i.e., people). This poses a problem because, although the project manager maintains the maximum control (through the line managers) over all resources including cost and personnel, the functional manager must provide staff for the project's requirements. It is therefore inevitable that conflicts occur between functional and project managers.

The matrix structure provides us with the best of two worlds: the traditional structure and the matrix structure. The advantages of the matrix structure eliminate almost all of the disadvantages of the traditional structure. The word "matrix" often brings fear to the hearts of executives because it implies radical change, or at least they think that it does. If we take a close look at Figure 3–3, we can see that the traditional structure is still there. The matrix is simply horizontal lines superimposed over the traditional structure. The horizontal lines will come and go as projects start up and terminate, but the traditional structure will remain.

Table 3–4 summarizes the advantages and disadvantages of this.

---

2. Leroy H. Mantell, "The Systems Approach and Good Management." Reprinted with permission from *Business Horizons,* October 1972 (p. 50). Copyright © 1972 by the Board of Trustees at Indiana University.

TABLE 3–4. ADVANTAGES AND DISADVANTAGES OF A PURE MATRIX ORGANIZATIONAL FORM

Advantages

- The project manager maintains maximum project control (through the line managers) over all resources, including cost and personnel.
- Policies and procedures can be set up independently for each project, provided that they do not contradict company policies and procedures.
- The project manager has the authority to commit company resources, provided that scheduling does not cause conflicts with other projects.
- Rapid responses are possible to changes, conflict resolution, and project needs (as technology or schedule).
- The functional organizations exist primarily as support for the project.
- Each person has a "home" after project completion. People are susceptible to motivation and end-item identification. Each person can be shown a career path.
- Because key people can be shared, the program cost is minimized. People can work on a variety of problems; that is, better people control is possible.
- A strong technical base can be developed, and much more time can be devoted to complex problem solving. Knowledge is available for all projects on an equal basis.
- Conflicts are minimal, and those requiring hierarchical referrals are more easily resolved.
- There is a better balance among time, cost, and performance.
- Rapid development of specialists and generalists occurs.
- Authority and responsibility are shared.
- Stress is distributed among the team (and the functional managers).

Disadvantages

- Multidimensional information flow.
- Multidimensional work flow.
- Dual reporting.
- Continuously changing priorities.
- Management goals different from project goals.
- Potential for continuous conflict and conflict resolution.
- Difficulty in monitoring and control.
- Company-wide, the organizational structure is not cost-effective because more people than necessary are required, primarily administrative.
- Each project organization operates independently. Care must be taken that duplication of efforts does not occur.
- More effort and time are needed initially to define policies and procedures, compared to traditional form.
- Functional managers may be biased according to their own set of priorities.
- Balance of power between functional and project organizations must be watched.
- Balance of time, cost, and performance must be monitored.
- Although rapid response time is possible for individual problem resolution, the reaction time can become quite slow.
- Employees and managers are more susceptible to role ambiguity than in traditional form.
- Conflicts and their resolution may be a continuous process (possibly requiring support of an organizational development specialist).
- People do not feel that they have any control over their own destiny when continuously reporting to multiple managers.

We should note that with proper executive-level planning and control, all of the disadvantages can be eliminated. This is the only organizational form where such control is possible. But companies must resist creating more positions in executive management than are actually necessary as this will drive up overhead rates. However, there is a point where the matrix will become mature and fewer people will be required at the top levels of management.

Matrix implementation requires:

- Training in matrix operations
- Training in how to maintain open communications

- Training in problem solving
- Compatible reward systems
- Role definitions

3.5 MODIFICATION OF MATRIX STRUCTURES

The matrix can take many forms, but there are basically three common varieties. Each type represents a different degree of authority attributed to the program manager and indirectly identifies the relative size of the company. As an example, in the matrix of Figure 3–3, all program managers report directly to the general manager. This type of arrangement works best for small companies that have few projects and assumes that the general manager has sufficient time to coordinate activities between his project managers. In this type of arrangement, all conflicts between projects are referred to the general manager for resolution.

As companies grow in size and the number of projects, the general manager will find it increasingly difficult to act as the focal point for all projects. A new position must be created, that of director of programs, or manager of programs or projects, who is responsible for all program management. See Figure 3–4.

Finally, we must discuss the characteristics of a project engineer. In Figure 3–5, most people would place the project manager to the right of center with stronger human

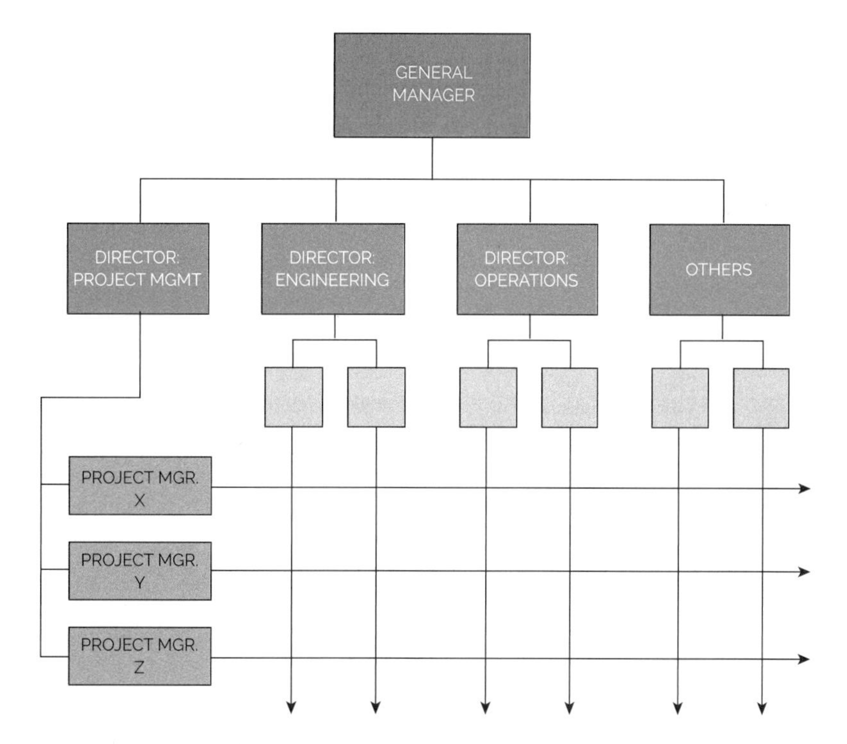

FIGURE 3–4. Development of a director of project management.

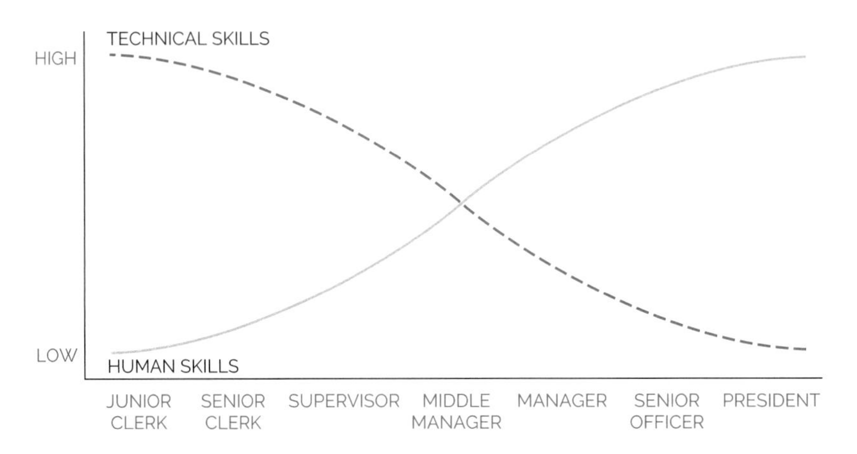

FIGURE 3–5. Philosophy of management.

skills than technical skills, and the project engineer to the left of center with stronger technical skills than human skills. How far from the center point will the project manager and project engineer be? Today, many companies are merging project management and project engineering into one position. This can be seen in Table 3–5. The project manager and project engineer have similar functions above the line but different ones below the line.[3]

The main reason for separating project management from project engineering is so that the project engineer will remain "solid" to the director of engineering in order to have the full authority to give technical direction to engineering.

TABLE 3–5. PROJECT MANAGEMENT COMPARED TO PROJECT ENGINEERING

Project Management	Project Engineering
• Total project planning • Cost control • Schedule control • System specifications • Logistics support • Contract control • Report preparation and distribution • Procurement • Identification of reliability and maintainability requirements • Staffing • Priority scheduling • Management information systems	• Total project planning • Cost control • Schedule control • System specifications • Logistics support • Configuration control • Fabrication, testing, and production technical leadership support

3. Procurement, reliability, and maintainability may fall under the responsibility of the project engineer in some companies.

3.6 THE STRONG, WEAK, OR BALANCED MATRIX

PMBOK® Guide, 6th Edition
2.4.4.1 Organizational Structure Types
2.4.4.2 Factors in Organizational Structure Selection

Matrix structures can be strong, weak, or balanced. The strength of the matrix is based upon who has more influence over the daily performance of the workers: project manager or line managers. If the project manager has more influence over the worker, then the matrix structure functions as a strong matrix as seen through the eyes of the project manager. If the line manager has more influence than does the project manager, then the organization functions as a weak matrix as seen by the project manager.

The most common differentiator between a strong and weak matrix is where the command of technology resides: project manager or line managers. If the project manager has a command of technology and is recognized by the line managers and the workers as being a technical expert, then the line managers will allow the workers to take technical direction from the project manager. This will result in a strong matrix structure. Workers will seek solutions to their problems from the project manager first and the line managers second. The reverse is true for a weak matrix. Project managers in a strong matrix generally possess more authority than in a weak matrix.

When a company desires a strong matrix, the project manager is generally promoted from within the organization and may have had assignments in several line functions throughout the organization. In a weak matrix, the company may hire from outside the organization but should at least require that the person selected understand the technology and the industry.

3.7 PROJECT MANAGEMENT OFFICES

PMBOK® Guide, 6th Edition
2.3.2 Corporate Knowledge Repositories
2.4.4.3 Project Management Office

In project-driven companies, the creation of a project management division is readily accepted as a necessity to conduct business. Organizational restructuring can quite often occur based on environmental changes and customer needs. In non–project-driven organizations, employees are less tolerant of organizational change. Power, authority, and turf become important. The implementation of a separate division for project management is extremely difficult. Resistance can become so strong that the entire project management process can suffer.

Over the past two decades, non–project-driven as well as project-driven companies have created project management offices (PMOs) which previously were called centers for project management expertise. These PMOs are not necessarily formal line organizations, but may function more as informal committees whose membership may come from each functional unit of the company. This is dependent upon the size of the company. The assignment to the PMO can be part-time or full-time; it may be for only six months to a year; and it may or may not require the individual to manage projects.

As companies begin to recognize the favorable effect that project management has on profitability, emphasis is placed upon achieving professionalism in project management using the project office concept. The concept of a PMO could very well be the most important project management activity in this decade. With this recognition of importance comes strategic planning for both project management and the project office. Maturity and excellence in project management do *not* occur simply by using project management over a prolonged period of time. Rather, it comes through strategic planning for project management.

Usually, the PMO has as its charter:

- To develop and update a methodology for project management. The methodology usually advocates informal or flexible project management and treated as a framework rather than as a methodology.
- To act as a facilitator or trainer in conducting project management training programs.
- To provide project management assistance to any employee who is currently managing projects and requires support in planning, scheduling, and controlling projects.
- To develop or maintain files on "lessons learned" and to see that this information is made available to all project managers.
    - To provide standardization in estimating
    - To provide standardization in planning
    - To provide standardization in scheduling
    - To provide standardization in control
    - To provide standardization in reporting
    - To provide clarification of project management roles and responsibilities
    - Provide Human Resources with advice for the preparation of job descriptions for project management
    - Preparation of archive data on lessons learned
    - Benchmarking continuously
    - Developing project management templates
    - Developing a project management methodology
    - Recommending and implementing changes and improvements to the existing methodology
    - Identifying project standards
    - Identifying best practices
    - Performing strategic planning for project management
    - Establishing a project management problem-solving hotline
    - Coordinating and/or conducting project management training programs
    - Transferring knowledge through coaching and mentorship
    - Developing a corporate resource capacity/utilization plan
    - Supporting portfolio management activities
    - Assessing risks
    - Planning for disaster recovery
    - Auditing the use of the project management methodology
    - Auditing the use of best practices

PMOs have become commonplace in the corporate hierarchy. Although the majority of activities assigned to the PMO had not changed, there was now a new mission for the PMO:

- The PO now has the responsibility for maintaining all intellectual property related to project management and to actively support corporate strategic planning.

The PO was now servicing the corporation, especially the strategic planning activities for project management, rather than focusing on a specific customer. The PO was transformed into a corporate center for control of project management intellectual property. This was a necessity as the magnitude of project management information grew almost exponentially throughout the organization.

All of the benefits of using a PMO are either directly or indirectly related to project management intellectual property. To maintain the project management intellectual property, the PMO must maintain the vehicles for capturing the data and then disseminating the data to the various stakeholders. These vehicles include the company project management intranet, project websites, project databases, and project management information systems. Since much of this information is necessary for both project management and corporate strategic planning, then there must exist strategic planning for the PMO.

3.8 SELECTING THE ORGANIZATIONAL FORM

PMBOK® Guide, 6th Edition
2.4.4.2 Factors in Organizational Structure Selection
Chapter 4 Integration Management

Project management has matured as an outgrowth of the need to develop and produce complex and/or large projects in the shortest possible time, within anticipated cost, with required reliability and performance, and (when applicable) to realize a profit. Granted that organizations have become so complex that traditional organizational structures and relationships no longer allow for effective management, how can executives determine which organizational form is best, especially since some projects last for only a few weeks or months while others may take years?

To answer this question, we must first determine whether the necessary characteristics exist to warrant a project management organizational form. Generally speaking, the project management approach can be effectively applied to a onetime undertaking that is[4]:

- Definable in terms of a specific goal
- Infrequent, unique, or unfamiliar to the present organization
- Complex with respect to interdependence of detailed tasks
- Critical to the company

Once a group of tasks is selected and considered to be a project, the next step is to define the kinds of projects, described in Section 2.3. These include individual, staff, special, and matrix or aggregate projects.

Unfortunately, many companies do not have a clear definition of what a project is. As a result, large project teams are often constructed for small projects when they could be handled more quickly and effectively by some other structural form. All structural forms have their advantages and disadvantages, but the project management approach appears to be the best possible alternative.

4. John M. Stewart, "Making Project Management Work." Reprinted with permission from Business Horizons, Fall 1965 (p. 54). Copyright © 1964 by the Board of Trustees at Indiana University.

The basic factors that influence the selection of a project organizational form are:

- Typical project size
- Typical project length
- Span of control
- Typical project cost
- Experience with project management organization
- Philosophy and visibility of upper-level management
- Project location
- Available resources
- Unique aspects of the project

This last item requires further comment. Project management (especially with a matrix) usually works best for the control of human resources and thus may be more applicable to labor-intensive projects rather than capital-intensive projects. Labor-intensive organizations have formal project management, whereas capital-intensive organizations may use informal project management.

Four fundamental parameters must be analyzed when considering implementation of a project organizational form:

- Integrating devices
- Authority structure
- Influence distribution
- Information system

Project management is a means of integrating all company efforts, especially research and development, by selecting an appropriate organizational form. This can be accomplished formally or informally.

Informal integration works best if, and only if, effective collaboration can be achieved between conflicting units. Without any clearly defined authority, the role of the integrator is simply to act as an exchange medium across the interface of two functional units. As the size of the organization increases, formal integration positions must exist, especially in situations where intense conflict can occur (e.g., research and development).

Not all organizations need a pure matrix structure to achieve this integration. Many problems can be solved simply through the chain of command, depending on the size of the organization and the nature of the project. The organization needed to achieve project control can vary in size from one person to several thousand people. The organizational structure needed for effective project control is governed by the desires of top management and project circumstances.

Top management must decide on the authority structure that will control the integration mechanism. The authority structure can range from pure functional authority (traditional management), to product authority (product management), and finally to dual authority (matrix management). From a management point of view, organizational forms are often selected based on how much authority top management wishes to delegate or surrender.

Integration of activities across functional boundaries can also be accomplished by influence. Influence includes such factors as participation in budget planning and approval, design changes, location and size of offices, salaries, and so on. Influence can also cut administrative red tape and develop a much more unified informal organization.

Information systems also play an important role. They are designed to get the right information to the right person at the right time in a cost-effective manner. Organizational functions must facilitate the flow of information through the management network.

Galbraith has described additional factors that can influence organizational selection. These factors are[5]:

- Diversity of product lines
- Rate of change of the product lines
- Interdependencies among subunits
- Level of technology
- Presence of economies of scale
- Organizational size

A diversity of project lines requires both top-level and functional managers to maintain knowledge in all areas. A diversity is needed because the rate of change in customer demands will necessitate changes in the product lines. Diversity makes it more difficult for managers to make realistic estimates concerning resource allocations and the control of time, cost, schedules, and technology. The systems approach to management requires sufficient information and alternatives to be available so that effective trade-offs can be established. For diversity in a high-technology environment, the organizational choice might, in fact, be a trade-off between the flow of work and the flow of information. Diversity tends toward strong product authority and control.

Many functional organizations consider themselves companies within a company and pride themselves on their independence. This attitude poses a severe problem in trying to develop a synergistic atmosphere. Successful project management requires that functional units recognize the interdependence that must exist in order for technology to be shared and schedule dates to be met. Interdependency is also required in order to develop strong communication channels and coordination.

The use of new technologies poses a serious problem in that technical expertise must be established in all specialties, including engineering, production, material control, and safety. Maintaining technical expertise works best in strong functional disciplines, provided the information is not purchased outside the organization. The main problem, however, is how to communicate this expertise across functional lines. Independent R&D units can be established, as opposed to integrating R&D into each functional department's routine efforts. Organizational control requirements are much more difficult in high-technology industries with ongoing research and development than with pure production groups.

Economies of scale and size can also affect organizational selection. The economies of scale are most often controlled by the amount of physical resources that a company has available. The larger the economies of scale, the more the organization tends to favor pure functional management.

---

5. Jay R. Galbraith, "Matrix Organization Designs." Reprinted with permission from *Business Horizons,* February 1971, pp. 29–40. Copyright © 1971 by the Board of Trustees at Indiana University.

The size of the organization is important in that it can limit the amount of technical expertise in the economies of scale. While size may have little effect on the organizational structure, it does have a severe impact on the economies of scale. Small companies, for example, cannot maintain large specialist staffs and, therefore, incur a larger cost for lost specialization and lost economies of scale.

The way in which companies operate their project organization is bound to affect the organization, both during the operation of the project and after the project has been completed and personnel have been disbanded. The overall effects on the company must be looked at from a personnel and cost control standpoint. This will be accomplished, in depth, in later chapters. Although project management is growing, the creation of a project organization does not necessarily ensure that an assigned objective will be accomplished successfully. Furthermore, weaknesses can develop in the areas of maintaining capability and structural changes.

Although the project organization is a specialized, task-oriented entity, it seldom, if ever, exists apart from the traditional structure of the organization. All project management structures overlap the traditional structure. Furthermore, companies can have more than one project organizational form in existence at one time. A major steel product, for example, has a matrix structure for R&D and a product structure elsewhere.

Accepting a project management structure is a giant step from which there may be no return. The company may have to create more management positions without changing the total employment levels. In addition, incorporation of a project organization is almost always accompanied by the upgrading of jobs. In any event, management must realize that whichever project management structure is selected, a dynamic state of equilibrium will be necessary.

3.9 STRATEGIC BUSINESS UNIT (SBU) PROJECT MANAGEMENT

During the past several years, large companies have restructured into strategic business units (SBUs). An SBU is a grouping of functional units that have the responsibility for profit (or loss) of part of the organization's core businesses. Figure 3–6 shows how one of the automotive suppliers restructured into three SBUs; one each for Ford, Chrysler, and General Motors. Each strategic business unit is large enough to maintain its own project and program managers. The executive in charge of the strategic business unit may act as the sponsor for all the program and project managers with the SBU. The major benefit of these types of project management SBUs is that it allows the SBU to work more closely with the customer. It is a customer-focused organizational structure.

It is possible for some resources to be shared across several SBUs. Manufacturing plants can end up supporting more than one SBU. Also, corporate may provide the resources for cost accounting, human resource management, and training.

A more recent organizational structure, and a more complex one, is shown in Figure 3–7. In this structure, each SBU may end up using the same platform (i.e., powertrain, chassis, and other underneath components). The platform managers are responsible for the design and enhancements of each platform, whereas the SBU program managers

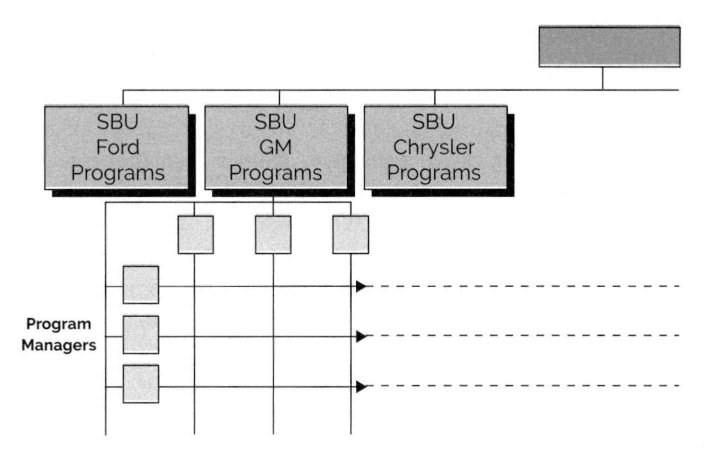

FIGURE 3–6. Strategic business unit project management.

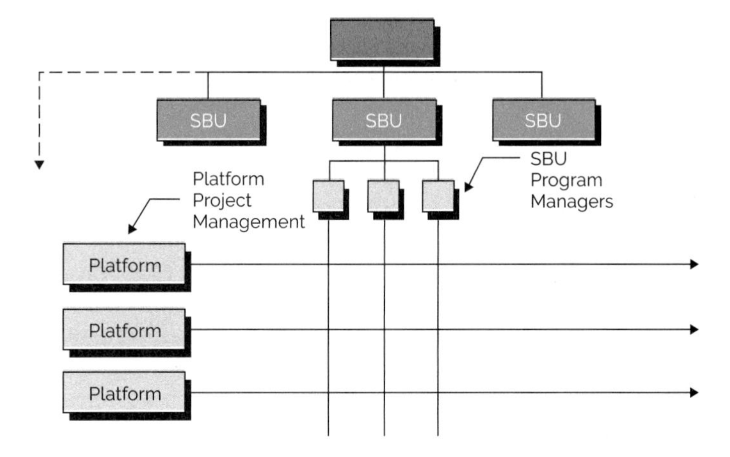

FIGURE 3–7. SBU project management using platform management.

must adapt this platform to a new model car. This type of matrix is multidimensional inasmuch as each SBU could already have an internal matrix. Also, each manufacturing plant could be located outside of the continental United States, making this structure a multinational, multidimensional matrix.

3.10 TRANSITIONAL MANAGEMENT

Organizational redesign is occurring at a rapid rate because of shorter product life cycles, rapidly changing environments, accelerated development of sophisticated information systems, and increased marketplace competitiveness. Because of these factors, more companies are considering project management organizations as a solution.

Why have some companies been able to implement this change in a short period of time while other companies require years? The answer is that successful implementation requires good transitional management.

Transitional management is the art and science of managing the conversion period from one organizational design to another. Transitional management necessitates an understanding of the new goals, objectives, roles, expectations, and employees' fears.

A survey was conducted of executives, managers, and employees in thirty-eight companies that had implemented matrix management. Almost all executives felt that the greatest success could be achieved through proper training and education, both during and after transition. In addition to training, executives stated that the following fifteen challenges must be accounted for during transition:

- *Transfer of power.* Some line managers will find it extremely difficult to accept someone else managing their projects, whereas some project managers will find it difficult to give orders to workers who belong to someone else.
- *Trust.* The secret to a successful transition without formal executive authority will be trust between line managers, between project managers, and between project and line managers. It takes time for trust to develop. Senior management should encourage it throughout the transition life cycle.
- *Policies and procedures.* The establishment of well-accepted policies and procedures is a slow and tedious process. Trying to establish rigid policies and procedures at project initiation will lead to difficulties.
- *Hierarchical consideration.* During transition, every attempt should be made to minimize hierarchical considerations that could affect successful organizational maturity.
- *Priority scheduling.* Priorities should be established only when needed, not on a continual basis. If priority shifting is continual, confusion and disenchantment will occur.
- *Personnel problems.* During transition there will be personnel problems brought on by moving to new locations, status changes, and new informal organizations. These problems should be addressed on a continual basis.
- *Communications.* During transition, new channels of communications should be built but not at the expense of old ones. Transition phases should show employees that communication can be multidirectional, for example, a project manager talking directly to functional employees.
- *Project manager acceptance.* Resistance to the project manager position can be controlled through proper training. People tend to resist what they do not understand.
- *Competition.* Although some competition is healthy within an organization, it can be detrimental during transition. Competition should not be encouraged at the expense of the total organization.
- *Tools.* It is common practice for each line organization to establish its own tools and techniques. During transition, no attempt should be made to force the line organizations to depart from their current practices. Rather, it is better for the project managers to develop tools and techniques that can be integrated with those in the functional groups.
- *Contradicting demands.* During transition and after maturity, contradicting demands will be a way of life. When they first occur during transition, they should be handled in a "working atmosphere" rather than a crisis mode.

- *Reporting.* If any type of standardization is to be developed, it should be for project status reporting, regardless of the size of the project.
- *Teamwork.* Systematic planning with strong functional input will produce teamwork. Using planning groups during transition will not obtain the necessary functional and project commitments.
- *Theory X–Theory Y.* During transition, functional employees may soon find themselves managed under either Theory X or Theory Y approaches. People must realize (through training) that this is a way of life in project management, especially during crises.
- *Overmanagement costs.* A mistake often made by executives is thinking that projects can be managed with fewer resources. This usually leads to disaster because undermanagement costs may be an order of magnitude greater than overmanagement costs when problems arise.

Transition to a project-driven matrix organization is not easy. Managers and professionals contemplating such a move should know:

- Proper planning and organization of the transition on a life-cycle basis will facilitate a successful change.
- Training of the executives, line managers, and employees in project management knowledge, skills, and attitudes is critical to a successful transition and probably will shorten the transition time.
- Employee involvement and acceptance may be the single most important function during transition.
- The strongest driving force of success during transition is a demonstration of commitment to and involvement in project management by senior executives.
- Organizational behavior becomes important during transition.
- Commitments made by senior executives prior to transition must be preserved during and following transition.
- Major concessions by senior management will come slowly.
- Schedule or performance compromises are not acceptable during transition; cost overruns may be acceptable.
- Conflict among participants increases during transition.
- If project managers are willing to manage with only implied authority during transition, then the total transition time may be drastically reduced.
- It is not clear how long transition will take.

Making the transition from a classical or product organization to a project-driven organization is not easy. With proper understanding, training, demonstrated commitment, and patience, the transition will have a good chance for success.

3.11 SEVEN FALLACIES THAT DELAY PROJECT MANAGEMENT MATURITY

All too often, companies embark upon a journey to implement project management only to discover that the path they thought was clear and straightforward is actually filled with obstacles and fallacies. Without sufficient understanding of the looming roadblocks and how to overcome them, an organization may never reach a high level of project management maturity. Their competitors, on the other hand, may require only a few years to implement an organization-wide strategy that predictably and consistently delivers successful projects.

One key obstacle to project management maturity is that implementation activities are often spearheaded by people in positions of authority within an organization. These people often have a poor understanding of project management yet are unwilling to attend training programs, even short ones, to capture a basic understanding of what is required to successfully bring project management implementation to maturity. A second key obstacle is that these same people often make implementation decisions based upon personal interests or hidden agendas. Both obstacles cause project management implementation to suffer.

The fallacies affecting the maturity of a project management implementation do not necessarily prevent project management from occurring. Instead, these mistaken beliefs elongate the implementation time frame and create significant frustration in the project management ranks. The seven most common fallacies are explained below.

Fallacy 1: Our ultimate goal is to implement project management. Wrong goal! The ultimate goal must be the progressive development of project management systems and processes that consistently and predictably result in a continuous stream of successful projects. A successful implementation occurs in the shortest amount of time and causes no disruption to the existing work flow. Anyone can purchase a software package and implement project management piecemeal. But effective project management systems and processes do not necessarily result. And successfully completing one or two projects does not mean that only successfully managed projects will continue.

Fallacy 2: We need to establish a mandatory number of forms, templates, guidelines, and checklists by a certain point in time. Wrong criteria! Project management maturity can be evaluated only by establishing time-based levels of maturity and by using assessment instruments for measurement. While it is true that forms, guidelines, templates, and checklists are necessities, maximizing their number or putting them in place does not equal project management maturity. Many project management practitioners believe that project management maturity can be accelerated if the focus is on the development of an organization-wide project management methodology that everyone buys into and supports.

Methodologies should be designed to streamline the way the organization handles projects. For example, when a project is completed, the team should be debriefed to capture lessons learned and best practices. The debriefing session often uncovers ways to minimize or combine processes and improve efficiency and effectiveness without increasing costs.

Fallacy 3: We need to purchase project management software to accelerate the maturity process. Wrong approach! Purchasing software just for the sake of having it is a bad idea. Too often, decision makers purchase project management software based upon the "bells and whistles" that are packaged with it, believing that a larger project management software package can accelerate maturity.

The goal of software selection must be the benefits to the project and the organization, such as cost reductions through efficiency, effectiveness, standardization, and consistency. A $500 software package can, more often than not, reduce project costs just as effectively as a $200,000 package. What is unfortunate is that the people who order the software focus more on the number of packaged features than on how much money will be saved by using the software.

Fallacy 4: We need to implement project management in small steps with a small breakthrough project that everyone can track. Wrong method! This works if time is not a constraint. The best bet is to use a large project as the breakthrough project. A successfully managed large project implies that the same processes can work on small projects, whereas the reverse is not necessarily true.

On small breakthrough projects, some people will always argue against the implementation of project management and find numerous examples why it will not work. Using a large project generally comes with less resistance, especially if project execution proceeds smoothly.

Fallacy 5: We need to track and broadcast the results of the breakthrough project. Wrong course of action! Expounding a project's success benefits only that project rather than the entire company. Illuminating how project management caused a project to succeed benefits the entire organization. People then understand that project management can be used on a multitude of projects.

Fallacy 6: We need executive support. Almost true! We need *visible* executive support. People can easily differentiate between genuine support and lip service. Executives must *walk the talk*. They must hold meetings to demonstrate their support of project management and attend various project team meetings. They must maintain an open-door policy for problems that occur during project management implementation.

Fallacy 7: We need a project management course so our workers can become Project Management Professionals (PMPs). Getting closer! What is really needed is lifelong education in project management. Becoming a PMP® is just the starting point. There is life beyond the *PMBOK® Guide*. Continuous organization-wide project management education is the fastest way to accelerate maturity in project management.

Needless to say, significantly more fallacies than discussed here are out there waiting to block your project management implementation and delay its maturity. What is critical is that your organization implements project management through a well-thought-out plan that receives organization-wide buy-in and support. Fallacies create unnecessary delays. Identifying and overcoming faulty thinking can help fast-track your organization's project management maturity.

Related Case Studies (from Kerzner/*Project Management Case Studies*, 5th ed.)	**Related Workbook Exercises (from Kerzner/*Project Management Workbook and PMP®/CAPM® Exam Study Guide*, 12th ed.)**	***PMBOK® Guide*, 6th Edition, Reference Section for the PMP® Certification Exam**
• Quasar Communications, Inc. • Fargo Foods	• The Struggle with Implementation • Multiple Choice Exam	• Human Resource Management

3.12 STUDYING TIPS FOR THE PMI® PROJECT MANAGEMENT CERTIFICATION EXAM

This section is applicable as a review of the principles to support the knowledge areas and domain groups in the *PMBOK® Guide*. This chapter addresses:

- Project Resource Management
- Planning

PMP and CAPM are registered marks of the Project Management Institute, Inc.

Understanding the following principles is beneficial if the reader is using this text to study for the PMP® Certification Exam:

- Different types of organizational structures
- Advantages and disadvantages of each structure
- In which structure the project manager possesses the greatest amount of authority
- In which structure the project manager possesses the least amount of authority
- Three types of matrix structures

In Appendix C, the following Dorale Products mini–case studies are applicable:

- Dorale Products (H) [Human Resources Management]
- Dorale Products (J) [Human Resources Management]
- Dorale Products (K) [Human Resources Management]

The following multiple-choice questions will be helpful in reviewing the principles of this chapter:

1. In which organizational form is it most difficult to integrate project activities?
   A. Classical/traditional
   B. Projectized
   C. Strong matrix
   D. Weak matrix
2. In which organization form would the project manager possess the greatest amount of authority?
   A. Classical/traditional
   B. Projectized
   C. Strong matrix
   D. Weak matrix
3. In which organizational form does the project manager often have the least amount of authority?
   A. Classical/traditional
   B. Projectized
   C. Strong matrix
   D. Weak matrix
4. In which organizational form is the project manager least likely to share resources with other projects?
   A. Classical/traditional
   B. Projectized
   C. Strong matrix
   D. Weak matrix

5. In which organizational form do project managers have the greatest likelihood of possessing reward power and have a wage-and-salary administration function? (The project and line manager are the same person.)
   A. Classical/traditional
   B. Projectized
   C. Strong matrix
   D. Weak matrix
6. In which organizational form is the worker in the greatest jeopardy of losing his or her job if the project gets canceled?
   A. Classical/traditional
   B. Projectized
   C. Strong matrix
   D. Weak matrix
7. In which type of matrix structure would a project manager most likely have a command of technology?
   A. Strong matrix
   B. Balanced matrix
   C. Weak matrix
   D. Cross-cultural matrix

ANSWERS

1. A
2. B
3. D
4. B
5. A
6. B
7. A

PROBLEMS

3–1 One of the most difficult problems facing management is that of how to minimize the transition time between changeover from a purely traditional organizational form to a project organizational form. Managing the changeover is difficult in that management must consistently "provide individual training on teamwork and group problem solving; also, provide the project and functional groups with assignments to help build teamwork."

3–2 Which organizational form would be best for the following corporate strategies?

A. Developing, manufacturing, and marketing many diverse but interrelated technological products and materials
B. Having market interests that span virtually every major industry
C. Becoming multinational with a rapidly expanding global business
D. Working in a business environment of rapid and drastic change, together with strong competition

3–3 In deciding to go to a new organizational form, what impact should the capabilities of the following groups have on your decision?

A. Top management
B. Middle management
C. Lower-level management

3–4 Below are three statements that are often used to describe the environment of a matrix. Do you agree or disagree? Defend your answer.

A. Project management in a matrix allows for fuller utilization of personnel.
B. The project manager and functional manager must agree on priorities.
C. Decision making in a matrix requires continual trade-offs on time, cost, technical risk, and uncertainty.

3–5 Some organizational structures are considered to be "project-driven." Define what is meant by "project-driven." Which organizational forms described in this chapter would fall under your definition?

3–6 The internal functioning of an organization must consider the demands imposed on the organization by task complexity, available technology, the external environment, and the needs of the organizational membership.

Considering these facts, should an organization search for the one best way to organize under all conditions? Should managers examine the functioning of an organization relative to its needs, or vice versa?

3–7 Defend or attack the following two statements concerning the operation of a matrix:

- There should be no disruption due to dual accountability.
- A difference in judgment should not delay work in progress.

3–8 A company has fifteen projects going on at once. Three projects are over $5 million, seven projects are between $1 million and $3 million, and five projects are between $500,000 and $700,000. Each project has a full-time project manager. Just based upon this information, which organizational form would be best? Can all the project managers report to the same person?

3–9 A major insurance company is considering the implementation of project management. The majority of the projects in the company are two weeks in duration, with very few existing beyond one month. Can project management work here?

3–10 A company has decided to go to full project management utilizing a matrix structure. Can the implementation be done in stages? Can the matrix be partially implemented, say, in one portion of the organization, and then gradually expanded across the rest of the company?

Organizing and Staffing the Project Office and Team

4.0 INTRODUCTION

PMBOK® *Guide*, 6th Edition
Chapter 9 Project Resource Management

Successful project management, regardless of the organizational structure, is only as good as the individuals and leaders who are managing the key functions. Project management is not a one-person operation; it requires a group of individuals dedicated to the achievement of a specific goal. Project management includes:

- A project manager
- Assistant project managers if necessary
- A project (home) office
- A project team

Large projects may require a project office (PO) for the management of a single project. The PO should not be confused with the PMO discussed in Chapter 3. Generally, project office personnel are assigned full-time to the project and work out of the project office, whereas the project team members work out of the functional units and may spend only a small percentage of their time on the project. Normally, project office personnel report directly to the project manager, but they may still be solid to their line function just for administrative control. A project office usually is not required on small projects, and sometimes the project can be accomplished by just one person who may fill all of the project office positions.

Before the staffing function begins, five basic questions are usually considered:

1. What are the requirements for an individual to become a successful project manager?
2. Who should be a member of the project team?

PMBOK is a registered mark of the Project Management Institute, Inc.

3. Who should be a member of the project office?
4. What problems can occur during recruiting activities?
5. What can happen downstream to cause the loss of key team members?

On the surface, these questions may not seem especially complex. But when we apply them to a project environment (which is by definition a "temporary" situation), where a constant stream of projects is necessary for corporate growth, the staffing problems become complex, especially if the organization is understaffed or lacks workers with the necessary skills.

4.1 THE STAFFING ENVIRONMENT

PMBOK® Guide, 6th Edition
9.1 Plan Resource Management

To understand the problems that occur during staffing, we must first investigate the characteristics of project management, including the project environment, the project management process, and the project manager.

Two major kinds of problems are related to the project environment: personnel performance problems and personnel policy problems. Performance is difficult for many individuals in the project environment because it represents a change in the way of doing business. Individuals, regardless of how competent they are, find it difficult to adapt continually to a changing situation in which they report to multiple managers.

On the other hand, many individuals thrive on temporary assignments because it gives them a "chance for glory." Unfortunately, some employees might consider the chance for glory more important than the project. For example, an employee may pay no attention to the instructions of the project manager and instead perform the task his own way. In this situation, the employee wants only to be recognized as an achiever and really does not care if the project is a success or failure, as long as he still has a functional home to return to where he will be identified as an achiever with good ideas.

The second major performance problem lies in the project–functional interface, where an individual suddenly finds himself reporting to two bosses, the functional manager and the project manager. If the functional manager and the project manager are in agreement about the work to be accomplished, then performance may not be hampered. But if conflicting directions are received, then the individual may let his performance suffer because of his compromising position. In this case, the employee will "bend" in the direction of the manager who controls his purse strings.

Personnel policy problems can create havoc in an organization, especially if the "grass is greener" in a project environment than in the functional environment. Functional organizations normally specify grades and salaries for employees. Project offices, on the other hand, have no such requirements and can promote and pay according to achievement. Bonuses are also easier to obtain in the project office but may create conflict and jealousy between the horizontal and vertical elements.

Because each project is different, the project management process allows each project to have its own policies, procedures, rules, and standards, provided they fall within broad company guidelines. Each project must be recognized as a project by top management so that the project manager has the delegated authority necessary to enforce the policies, procedures, rules, and standards.

Project management is successful only if the project manager and his team are totally dedicated to the successful completion of the project. This requires each team member of the project team and office to have a good understanding of the project requirements.

Ultimately, the person with the greatest influence during the staffing phase is the project manager. The personal attributes and abilities of project managers will either attract or deter highly desirable individuals. Project managers must exhibit honesty and integrity to foster an atmosphere of trust. They should not make impossible promises, such as immediate promotions for everyone if a follow-on contract is received. Also, on temporarily assigned activities, such as a project, managers cannot wait for personnel to iron out their own problems because time, cost, and performance requirements will not be satisfied.

Project managers should have both business management and technical expertise. They must understand the fundamental principles of management, especially those involving the rapid development of temporary communication channels. Project managers must understand the technical implications of a problem, since they are ultimately responsible for all decision making. However, many good technically oriented managers have failed because they have become too involved with the technical side of the project rather than the management side. There are strong arguments for having a project manager who has more than just an understanding of the necessary technology.

Because a project has a relatively short time duration, decision making must be rapid and effective. Managers must be alert and quick in their ability to perceive "red flags" that can eventually lead to serious problems. They must demonstrate their versatility and toughness in order to keep subordinates dedicated to goal accomplishment. Executives must realize that the project manager's objectives during staffing are to:

- Acquire the best available assets and try to improve them
- Provide a good working environment for all personnel
- Make sure that all resources are applied effectively and efficiently so that all constraints are met, if possible

4.2 SELECTING THE PROJECT MANAGER: AN EXECUTIVE DECISION

PMBOK® Guide, 6th Edition
9.3 Acquire Resources
9.4.2.3 Interpersonal and Team Skills

Probably the most difficult decision facing upper-level management is the selection of project managers. Some managers work best on long-duration projects where decision making can be slow; others may thrive on short duration projects that can result in a constant-pressure environment.

The selection process for project managers is not easy. Five basic questions must be considered:

1. What are the internal and external sources?
2. How do we select?
3. How do we provide career development in project management?
4. How can we develop project management skills in a reasonable time frame?
5. How do we evaluate project management performance?

Project management cannot succeed unless a good project manager is at the controls. It is far more likely project managers will succeed if it is obvious to the subordinates the general manager has appointed them. Usually, a brief memo to the line managers will suffice. The major responsibilities of the project manager include:

- To produce the end-item with the available resources and within the constraints of time, cost, and performance/technology
- To meet contractual profit objectives
- To make all required decisions whether they be for alternatives or termination
- To act as the customer (external) and upper-level and functional management (internal) communications focal point
- To "negotiate" with all functional disciplines for accomplishment of the necessary work packages within the constraints of time, cost, and performance/technology
- To resolve all conflicts

In order for project managers to fulfill their responsibilities successfully, they are constantly required to demonstrate their skills in interface, resource, and planning and control management. These implicit responsibilities are:

- Interface Management
  - Product interfaces
    - Performance of parts or subsections
    - Physical connection of parts or subsections
  - Project interfaces
  - Customer
  - Management (functional and upper-level)
  - Change of responsibilities
  - Information flow
  - Material interfaces (inventory control)
- Resource Management
  - Time (schedule)
  - Manpower
  - Money
  - Facilities
  - Equipment

- Material
- Information/technology
- Planning and Control Management
  - Increased equipment utilization
  - Increased performance efficiency
  - Reduced risks
  - Identification of alternatives to problems
  - Identification of alternative resolutions to conflicts

Finding the person with the right qualifications is not an easy task because the selection of project managers is based more on personal characteristics than on the job description. Russell Archibald defines a broader range of desired personal characteristics[1]:

PMBOK® *Guide*, 6th Edition
9.4 Develop Team

- Flexibility and adaptability
- Preference for significant initiative and leadership
- Aggressiveness, confidence, persuasiveness, verbal fluency
- Ambition, activity, forcefulness
- Effectiveness as a communicator and integrator
- Broad scope of personal interests
- Poise, enthusiasm, imagination, spontaneity
- Able to balance technical solutions with time, cost, and human factors
- Well organized and disciplined
- A generalist rather than a specialist
- Able and willing to devote most of his time to planning and controlling
- Able to identify problems
- Willing to make decisions
- Able to maintain proper balance in the use of time

The best project managers are willing and able to identify their own shortcomings and know when to ask for help.

So far we have discussed the personal characteristics of the project manager. There are also job-related questions to consider, such as:

- Are feasibility and economic analyses necessary?
- Is complex technical expertise required? If so, is it within the individual's capabilities?
- If the individual is lacking expertise, will there be sufficient backup strength in the line organizations?
- Is this the company's or the individual's first exposure to this type of project and/or client? If so, what are the risks to be considered?
- What is the priority for this project, and what are the risks?
- With whom must the project manager interface, both inside and outside the organization?

---

1. Russell D. Archibald, *Managing High-Technology Programs and Projects* (New York: John Wiley & Sons, 1976), p. 55.

While there may sometimes be some degree of commonality about the leadership qualities that every project manager should possess, industry and company requirements play a dominant role. According to Anthony Walker[2]:

> These qualities can be split into characteristics and skills. Project manager's characteristics will in many cases determine how they will deploy their skills. Examples of the characteristics which help to form good leaders in construction project management are:
>
> - Integrity
> - Preferred leadership style (tending towards democratic)
> - Self-confidence
> - Ability to delegate and trust others
> - Ability to cope with stress
> - Decisiveness
> - Judgment
> - Consistency and stability
> - Personal motivation and dedications
> - Determination
> - Positive thinking
> - Excellent health
> - Openness and the ability to hear what others say
> - Ease in social interactions with many types of people
>
> In terms of skills, the following are important:
>
> - Persuasive ability
> - Negotiation skills
> - Commercial expertise
> - 'Political' awareness
> - Breadth of vision
> - Integrative skills
> - Ability to set clear objectives
> - Communication skills
> - Management of meetings
> - Early warning antennae
> - Skills of diplomacy
> - The skill of discriminating important information

One of the most important but often least understood characteristics of good project managers is the ability to know their own strengths and weaknesses and those of their employees. Managers must understand that in order for employees to perform efficiently:

---

2. Anthony Walker, *Project Management in Construction* (Hoboken, NJ: John Wiley & Sons, 2015), pp. 245-246.

- They must know what they are supposed to do.
- They must have a clear understanding of authority and its limits.
- They must know what their relationship with other people is.
- They should know what constitutes a job well done in terms of specific results.
- They should know where and when they are falling short.
- They must be made aware of what can and should be done to correct unsatisfactory results.
- They must feel that their superior has an interest in them as individuals.
- They must feel that their superior believes in them and wants them to succeed.

4.3 SKILL REQUIREMENTS FOR PROJECT AND PROGRAM MANAGERS

PMBOK® Guide, 6th Edition
Chapter 9 Project Resources Management
9.4.2.3 Interpersonal and Team Skills
1.2.3.2 Program Management

To get results, the project manager must relate to (1) the people to be managed, (2) the task to be done, (3) the tools available, (4) the organizational structure, and (5) the organizational environment, including the customer community.

With an understanding of the interaction of corporate organization and behavior elements, the manager can build an environment conducive to the working team's needs. In addition, the project manager must understand the culture and value system of the organization he is working with. Effective project management is directly related to proficiency in the following ten skills:

1. Team building
2. Leadership
3. Conflict resolution
4. Technical expertise
5. Planning
6. Organization
7. Entrepreneurship
8. Administration
9. Management support
10. Resource allocation

The days of the manager who gets by with technical expertise alone or pure administrative skills are gone.

Team-Building Skills

Building the project team is one of the prime responsibilities of the project manager. Team building involves a whole spectrum of management skills required to identify, commit, and integrate the various task groups from the traditional functional organization into a single project management system.

To be effective, the project manager must provide an atmosphere conducive to teamwork. A climate with the following characteristics must be nurtured:

- Team members committed to the project
- Good interpersonal relations and team spirit
- The necessary expertise and resources
- Clearly defined goals and project objectives
- Involved and supportive top management
- Good project leadership
- Open communication among team members and support organizations
- A low degree of detrimental interpersonal and intergroup conflict

Three major considerations are involved in all of the above factors: (1) effective communications, (2) sincere interest in the professional growth of team members, and (3) commitment to the project.

Leadership Skills

A prerequisite for project success is the project manager's ability to lead the team within a relatively unstructured environment. It involves dealing effectively with managers and supporting personnel across functional lines and the ability to collect and filter relevant data for decision making in a dynamic environment. It involves the ability to integrate individual demands, requirements, and limitations into decisions and to resolve intergroup conflicts.

As with a general manager, quality leadership depends heavily on the project manager's personal experience and credibility within the organization. An effective management style might be characterized this way:

- Clear project leadership and direction
- Assistance in problem solving
- Facilitating the integration of new members into the team
- Ability to handle interpersonal conflict
- Facilitating group decisions
- Capability to plan and elicit commitments
- Ability to communicate clearly
- Presentation of the team to higher management
- Ability to balance technical solutions against economic and human factors

The personal traits desirable and supportive of the above skills are:

- Project management experience
- Flexibility and change orientation
- Innovative thinking
- Initiative and enthusiasm
- Charisma and persuasiveness
- Organization and discipline

Conflict Resolution Skills

Conflict is fundamental to complex task management. Understanding the determinants of conflicts is important to the project manager's ability to deal with conflicts effectively. When conflict becomes dysfunctional, it often results in poor project decision making, lengthy delays over issues, and a disruption of the team's efforts, all negative influences to project performance. However, conflict can be beneficial when it produces involvement and new information and enhances the competitive spirit.

To successfully resolve conflict and improve overall project performance, project managers must:

- Understand interaction of the organizational and behavioral elements in order to build an environment conducive to their team's motivational needs. This will enhance active participation and minimize unproductive conflict.
- Communicate effectively with all organizational levels regarding both project objectives and decisions. Regularly scheduled status review meetings can be an important communication vehicle.
- Recognize the determinants of conflict and their timing in the project life cycle. Effective project planning, contingency planning, securing of commitments, and involving top management can help to avoid or minimize many conflicts before they impede project performance.

Technical Skills

The project manager rarely has all the technical, administrative, and marketing expertise needed to direct the project single-handedly. It is essential, however, for the project manager to understand the technology, the markets, and the environment of the business. Without this understanding, the consequences of local decisions on the total project, the potential growth ramifications, and relationships to other business opportunities cannot be foreseen by the manager. Further technical expertise is necessary to evaluate technical concepts and solutions, to communicate effectively in technical terms with the project team, and to assess risks and make trade-offs between cost, schedule, and technical issues. Frequently, the project begins with an exploratory phase leading into a proposal. This is normally an excellent testing ground for the future project manager. It also allows top management to judge the new candidate's capacity for managing the technological innovations and integration of solutions.

Planning Skills

Planning skills are helpful for any undertaking; they are absolutely essential for the successful management of large complex projects. The project plan is the road map that defines how to get from the start to the final results. Project planning is an ongoing activity at all organizational levels. However, the preparation of a project summary plan, prior to project start, is the responsibility of the project manager. Effective project planning requires particular skills far beyond writing a document with schedules and budgets. It requires communication and information processing skills to define the actual resource requirements and administrative support necessary. It requires the ability to negotiate the necessary resources and commitments from key personnel in various support organizations with little or no formal authority.

In addition, the project manager must assure that the plan remains a viable document. Changes in project scope and depth are inevitable. The plan should reflect necessary changes through formal revisions and should be the guiding document throughout the life cycle of the project. An obsolete or irrelevant plan is useless.

Finally, project managers need to be aware that planning can be overdone. If not controlled, planning can become an end in itself and a poor substitute for innovative work. It is the responsibility of the project manager to build flexibility into the plan and police it against misuse.

Organizational Skills

The project manager must be a social architect; that is, he must understand how the organization works and how to work with the organization. Organizational skills are particularly important during project formation and start-up when the project manager is integrating people from many different disciplines into an effective work team. It requires defining the reporting relationships, responsibilities, lines of control, and information needs. A good project plan and a task matrix are useful organizational tools. In addition, the organizational effort is facilitated by clearly defined project objectives, open communication channels, good project leadership, and senior management support.

Entrepreneurial Skills

The project manager also needs a general management perspective. For example, economic considerations affect the organization's financial performance, but objectives often are much broader than profits. Customer satisfaction, future growth, cultivation of related market activities, and minimum organizational disruptions of other projects might be equally important goals. The effective project manager is concerned with all these issues. Entrepreneurial skills are developed through actual experience. However, formal MBA-type training, special seminars, and cross-functional training projects can help to develop the entrepreneurial skills needed by project managers.

Administrative Skills

Administrative skills are essential. The project manager must be experienced in planning, staffing, budgeting, scheduling, and other control techniques. In dealing with technical personnel, the problem is seldom to make people understand administrative techniques such as budgeting and scheduling, but to impress on them that costs and schedules are just as important as elegant technical solutions.

Some helpful tools for the manager in the administration of his project include: (1) the meeting, (2) the report, (3) the review, and (4) the budget and schedule controls. Project managers must be thoroughly familiar with these available tools and know how to use them effectively.

Management Support

The project manager is surrounded by a myriad of organizations that Building Skills either support him or control his activities. An understanding of these interfaces is important to project managers as it enhances their ability to build favorable relationships with senior management. Project organizations are shared-power systems with personnel of many diverse interests and "ways of doing things." Only a strong leader backed by senior management can prevent the development of unfavorable biases.

Four key variables influence the project manager's ability to create favorable relationships with senior management: (1) his ongoing credibility, (2) the visibility of his project, (3) the priority of his project relative to other organizational undertakings, and (4) his own accessibility.

Resource Allocation Skills A project organization has many bosses. Functional lines often shield support organizations from direct financial control by the project office. Once a task has been authorized, it is often impossible to control the personnel assignments, priorities, and indirect manpower costs. In addition, profit accountability is difficult owing to the interdependencies of various support departments and the often changing work scope and contents.

Effective and detailed project planning may facilitate commitment and reinforce control. Part of the plan is the "Statement of Work," which establishes a basis for resource allocation. It is also important to work out specific agreements with all key contributors and their superiors on the tasks to be performed and the associated budgets and schedules. Measurable milestones are not only important for hardware components, but also for the "invisible" project components such as systems and software tasks.

4.4 SPECIAL CASES IN PROJECT MANAGER SELECTION

Thus far we have assumed that the project is large enough for a full-time project manager to be appointed. This is not always the case. There are four major problem areas in staffing projects:

- Part-time versus full-time assignments
- Several projects assigned to one project manager
- Projects assigned to functional managers
- The project manager role retained by the general manager

The first problem is generally related to the size of the project. If the project is small (in time duration or cost), a part-time project manager may be selected. Many executives have fallen into the trap of letting line personnel act as part-time project managers while still performing line functions. If the employee has a conflict between what is best for the project and what is best for his line organization, the project will suffer. It is only natural that the employee will favor the place the salary increases come from.

It is a common practice for one project manager to control several projects, especially if they are either related, similar, small in size and may not justify a full time project manager. Problems come about when the projects have drastically different priorities. The low-priority efforts will be neglected.

If the project is a high-technology effort that requires specialization and can be performed by one department, then it is not unusual for the line manager to take on a dual role and act as project manager as well. This can be difficult to do, especially if the project manager is required to establish the priorities for the work under his supervision. The line manager may keep the best resources for the project, regardless of the priority. Then that project will be a success at the expense of every other project he must supply resources to.

Probably the worst situation is that in which an executive fills the role of project manager for a particular effort. The executive may not have the time necessary for total dedication to the achievement of the project. He cannot make effective decisions as a project manager while still discharging normal duties. Additionally, the executive may hoard the best resources for his project.

4.5 TODAY'S PROJECT MANAGERS

The skills needed to be an effective, twenty-first-century project manager have changed from those needed during the 1980s. As project management began to grow and mature, the project manager was converted from a technical manager to a business manager. The primary skills needed to be an effective project manager in the twenty-first century are:

- Knowledge of the business
- Risk management
- Integration skills

The critical skill is risk management. However, to perform risk management effectively, a sound knowledge of the business is required. Figure 4–1 shows the changes in project management skills needed between 1985 and 2016.

As projects become larger, the complexities of integration management become more pronounced. Figure 4–2 illustrates the importance of integration management. In 1985, project managers spent most of their time planning and replanning with their team. This was necessary because the project manager was the technical expert. Today, the project manager's efforts are heavily oriented toward integration of the function plans into a total project

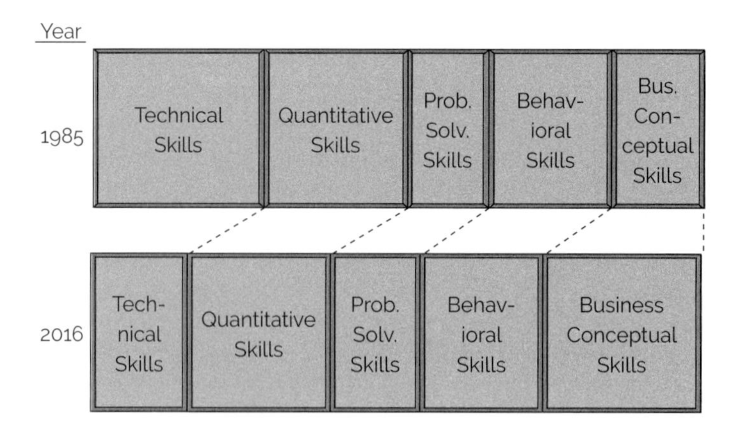

FIGURE 4–1. Project management skills.

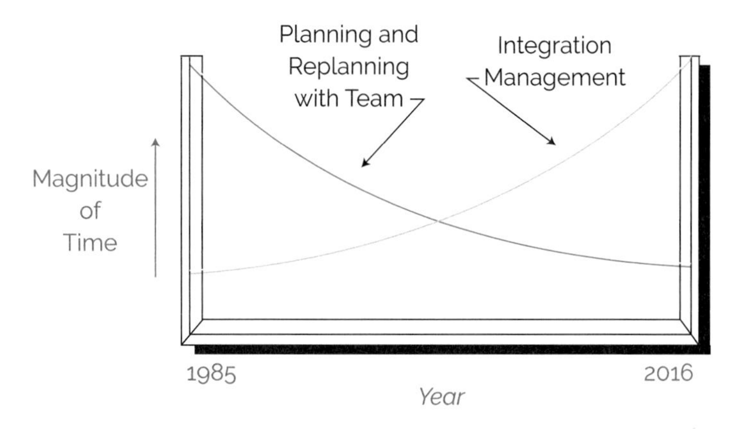

FIGURE 4–2. How do project managers spend their time?

plan. Some people contend that, with the increased risks and complexities of integration management, the project manager of the future will become an expert in damage control.

4.6 DUTIES AND JOB DESCRIPTIONS

PMBOK® Guide, 6th Edition
Chapter 9 Project Resource Management

Since projects, environments, and organizations differ from company to company as well as project to project, it is not unusual for companies to struggle to provide reasonable job descriptions of the project manager and associated personnel.

Because of the potential overlapping nature of job descriptions in a project management environment, some companies try to define responsibilities for each project management position, as shown in Table 4–1.

TABLE 4–1. PROJECT MANAGEMENT POSITIONS AND RESPONSIBILITIES

Project Management Position	Typical Responsibility	Skill Requirements
• Project Administrator • Project Coordinator • Technical Assistant	Coordinating and integrating of subsystem tasks. Assisting in determining technical and manpower requirements, schedules, and budgets. Measuring and analyzing project performance regarding technical progress, schedules, and budgets.	• Planning • Coordinating • Analyzing • Understanding the organization
• Task Manager • Project Engineer • Assistant Project Manager	Same as above, but stronger role in establishing and maintaining project requirements. Conducting trade-offs. Directing the technical implementation according to established schedules and budgets.	• Technical expertise • Assessing trade-offs • Managing task implementation • Leading task specialists

(continues)

TABLE 4–1. PROJECT MANAGEMENT POSITIONS AND RESPONSIBILITIES *(Continued)*

Project Management Position	Typical Responsibility	Skill Requirements
• Project Manager • Program Manager	Same as above, but stronger role in project planning and controlling. Coordinating and negotiating requirements between sponsor and performing organizations. Bid proposal development and pricing. Establishing project organization and staffing. Overall leadership toward implementing project plan. Project profit. New business development.	• Overall program leadership • Team building • Resolving conflict • Managing multidisciplinary tasks • Planning and allocating resources • Interfacing with customers/ • sponsors
• Executive Program Manager	Title reserved for very large programs relative to host organization. Responsibilities same as above. Focus is on directing overall program toward desired business results. Customer liaison. Profit performance. New business development. Organizational development.	• Business leadership • Managing overall program businesses • Building program organizations • Developing personnel • Developing new business
• Director of Programs • V.P. Program Development	Responsible for managing multiprogram businesses via various project organizations, each led by a project manager. Focus is on business planning and development, profit performance, technology development, establishing policies and procedures, program management guidelines, personnel development, organizational development.	• Leadership • Strategic planning • Directing and managing program businesses • Building organizations • Selecting and developing key personnel • Identifying and developing new business

4.7 THE ORGANIZATIONAL STAFFING PROCESS

PMBOK® Guide, **6th Edition**
Chapter 9 Human Resource Management
9.1.3 Resource Management Plan
9.3 Acquire Team

Staffing the project organization can become a long and tedious effort, especially on large and complex engineering projects. Three major questions must be answered:

- What people resources are required?
- Where will the people come from?
- What type of project organizational structure will be best?

To determine the people resources required, the types of individuals (possibly job descriptions) must be decided on, as well as how many individuals from each job category are necessary and when these individuals will be needed. Other factors to be considered include the cost of the resources, their availability over the duration of the project, their skill level, training needs and your previous experience working with them. The organizational staffing process time can be reduced if a resource management plan is created to address these three questions.

Consider the following situation: As a project manager, you have an activity that requires three separate tasks, all performed within the same line organization. The line manager promises you the best available resources right now for the first task but cannot make any commitments beyond that. The line manager may have only below-average workers available for the second and third tasks. However, the line manager is willing to make a deal with you. He can give you an employee who can do the work but will only give an average performance. If you accept the average employee, the line manager will guarantee that the employee will be available to you for all three tasks. How important is continuity to you? There is no clearly definable answer to this question. Some people will always want the best resources and are willing to fight for them, whereas others prefer continuity and dislike seeing new people coming and going.

Mutual trust between project and line managers is crucial, especially during staffing sessions. Once a project manager has developed a good working relationship with employees, the project manager would like to keep those individuals assigned to his activities. There is nothing wrong with a project manager requesting the same administrative and/or technical staff as before. Line managers realize this and usually agree to it.

There must also be mutual trust between the project managers themselves. Project managers must work as a team, recognize each other's needs, and be willing to make decisions that are in the best interest of the company.

Once the resources are defined as in the resource management plan, the next question must be whether staffing will be from within the existing organization or from outside sources, such as new hires or consultants. Outside consultants are advisable if, and only if, internal manpower resources are being fully utilized on other projects, or if the company does not possess the required project skills. The answer to this question will indicate which organizational form is best for achievement of the objectives. The form might be a virtual team, matrix, product, or staff project management structure.

Selecting the project manager is the beginning of the organizational staffing process. The next step, selecting the project office personnel and team members, can be a time-consuming chore. The project office consists of personnel who are usually assigned as full-time members of the project. The evaluation process should include active project team members, functional team members available for promotion or transfer, and outside applicants.

Upon completion of the evaluation process, the project manager meets with upper-level management. This coordination is required to assure that:

- All assignments fall within current policies on rank, salary, and promotion.
- The individuals selected can work well with both the project manager (formal reporting) and upper-level management (informal reporting).
- The individuals selected have good working relationships with the functional personnel.

Good project office personnel usually have experience with several types of projects and are self-disciplined.

If the resources needed are currently assigned on other projects, then a meeting is held between the project manager, upper-level management, and the project manager on whose project the requested individuals are currently assigned. Project managers are very reluctant to give up qualified personnel to other projects, but unfortunately, this procedure is a way of life in a project environment. Upper-level management attends these meetings to show all negotiating parties that top management is concerned with maintaining the best possible mix of individuals from available resources and to help resolve staffing conflicts. Staffing from within is a negotiation process in which upper-level management establishes the ground rules and priorities.

Figure 4–3 shows the typical staffing pattern as a function of time. Staff is provided from functional areas or from workers being released from other projects. People should be brought on board as needed, but most project managers would like to get their project staffed quickly for fear of losing the workers to another project. In an ideal situation the workers would be released from the project as early as possible for work on other projects. But as seen by the low slope in Figure 4–3, project managers tend to release workers at a much slower rate than bringing them on board because they want to be absolutely sure that the workers are no longer needed.

Thus far we have discussed staffing the project on the assumption that the workers are performing as expected. Unfortunately, there are situations in which employees must be terminated from the project because of:

- Nonacceptance of rules, policies, and procedures
- Nonacceptance of established formal authority
- Professionalism being more important to them than company loyalty
- Focusing on technical aspects at the expense of the budget and schedule
- Incompetence

There are three possible solutions for working with incompetent personnel. First, the project manager can provide an on-the-spot appraisal of the employee. This includes identification of weaknesses, corrective action to be taken, and threat of punishment if

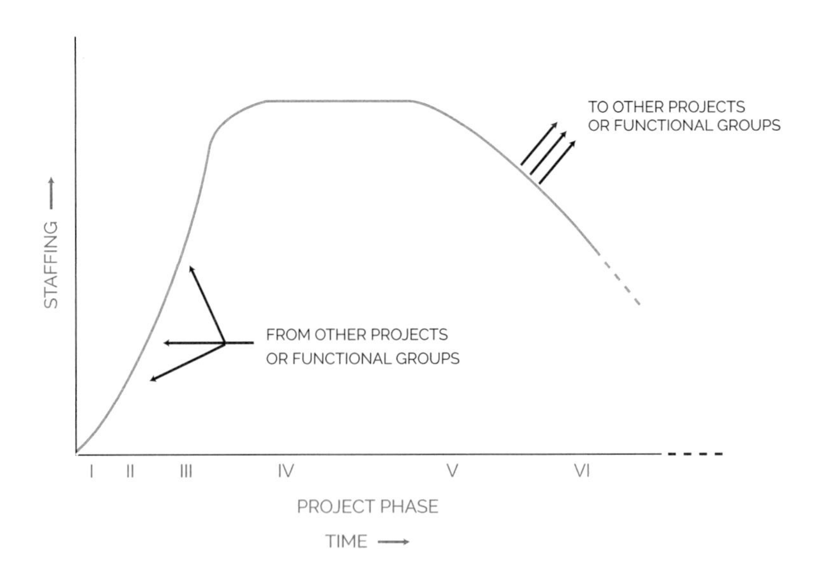

FIGURE 4–3. Staffing pattern versus time.

the situation continues. A second solution is reassignment of the employee to less critical activities. This solution is usually not preferred by project managers. The third and most frequent solution is the removal of the employee.

Although project managers can get project office people (who report to the project manager) removed directly, the removal of a line employee is an indirect process and must be accomplished through the line manager. The removal of the line employee should be made to look like a transfer; otherwise, the project manager will be branded as an individual who fires people.

4.8 THE PROJECT OFFICE

PMBOK® Guide, 6th Edition
9.3 Acquire Resources

The project team is a combination of the project office and functional employees, as shown in Figure 4–4. Although the figure identifies the project office personnel as assistant project managers, some employees may not have any such title. The advantage of such a title is that it entitles the employee to speak directly to the customer. The title is important because when the assistant project manager speaks to the customer, he represents the company, whereas the functional employee represents himself.

The project office is an organization developed to support the project manager in carrying out his duties. Project office personnel must have the same dedication toward the project as the project manager and must have good working relationships with both the project and functional managers. The responsibilities of the project office include:

- Acting as the focal point of information for both in-house control and customer reporting
- Controlling time, cost, and performance to adhere to contractual requirements

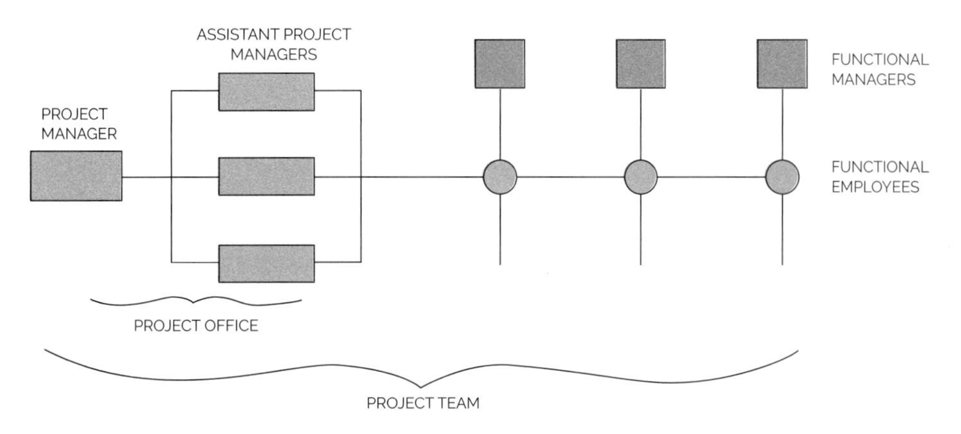

FIGURE 4–4. Project organization.

- Ensuring that all work required is documented and distributed to all key personnel
- Ensuring that all work performed is both authorized and funded by contractual documentation

The major responsibility of the project manager and the project office personnel is the integration of work across the functional lines of the organization. Functional units, such as engineering, R&D, and manufacturing, together with extra-company subcontractors, must work toward the same specifications, designs, and even objectives. The lack of proper integration of these functional units is the most common cause of project failure. The team members must be dedicated to all activities required for project success, not just their own functional responsibilities. The problems resulting from lack of integration can best be solved by full-time membership and participation of project office personnel. Not all team members are part of the project office. Functional representatives, performing at the interface position, also act as integrators but at a closer position to where the work is finally accomplished (i.e., the line organization).

One of the biggest challenges facing project managers is determining the size of the project office. The optimal size is determined by a trade-off between the maximum number of members necessary to assure compliance with requirements and the maximum number for keeping the total administrative costs under control. Membership is determined by factors such as project size, internal support requirements, type of project (e.g., R&D, qualification, production), level of technical competency required, and customer support requirements. Membership size is also influenced by how strategic management views the project to be. There is a tendency to enlarge project offices if the project is considered strategic, especially if follow-on work is possible.

On large projects, and even on some smaller efforts, it is often impossible to achieve project success without permanently assigned personnel. The four major activities of the project office, shown below, indicate the need for using full-time people:

- Integration of activities
- In-house and out-of-house communication
- Scheduling with risk and uncertainty
- Effective control

These four activities require continuous monitoring by trained project personnel. The training of good project office members may take weeks or even months, and can extend beyond the time allocated for a project.

Many executives have a misconception concerning the makeup and usefulness of the project office. People who work in the project office should be individuals whose first concern is project management, not the enhancement of their technical expertise. It is almost impossible for individuals to perform for any extended period of time in the project office without becoming cross-trained in a second or third project office function. For example, the project manager for cost could acquire enough expertise eventually to act as the assistant to the assistant project manager for procurement. This technique of project office cross-training is an excellent mechanism for creating good project managers.

4.9 THE FUNCTIONAL TEAM

PMBOK® Guide, 6th Edition
Chapter 9 Project Resource Management
9.2 Estimating Activity Resources

The project team consists of the project manager, the project office (whose members may or may not report directly to the project manager), and the functional or interface members (who must report horizontally as well as vertically for information flow). Functional team members are often shown on organizational charts as project office team members. This is normally done to satisfy customer requirements.

Upper-level management can have an input into the selection process for functional team members but should not take an active role unless the project and functional managers cannot agree. Functional management must be represented at all staffing meetings because functional staffing is directly dependent on project requirements and because:

- Functional managers generally have more expertise and can identify high-risk areas.
- Functional managers must develop a positive attitude toward project success. This is best achieved by inviting their participation in the early activities of the planning phase.
- Functional team members are not always full-time. They can be full-time or part-time for either the duration of the project or only specific phases.

The selection process for both the functional team member and the project office must include evaluation of any special requirements. The most common special requirements develop from:

- Changes in technical specifications
- Special customer requests
- Organizational restructuring because of deviations from existing policies
- Compatibility with the customer's project office

A typical project office may include between ten and thirty members, whereas the total project team may be in excess of a hundred people, causing information to be shared slowly. For large projects, it is desirable to have a full-time functional representative from each major division or department assigned permanently to the project, and perhaps even to the project office. Both the project manager and team members must understand fully the responsibilities and functions of each team member so that total integration can be achieved rapidly and effectively.

When employees are attached to a project, the project manager must identify the "star" employees. These are the employees who are vital for the success of the project and who can either make or break the project manager. Most of the time, star employees are found in the line organization, not the project office.

4.10 THE PROJECT ORGANIZATIONAL CHART

One of the first requirements of the project start-up phase is to develop the organizational chart for the project and determine its relationship to the parent organizational structure. Figure 4–5 shows, in abbreviated form, the six major programs at Dalton Corporation. It is

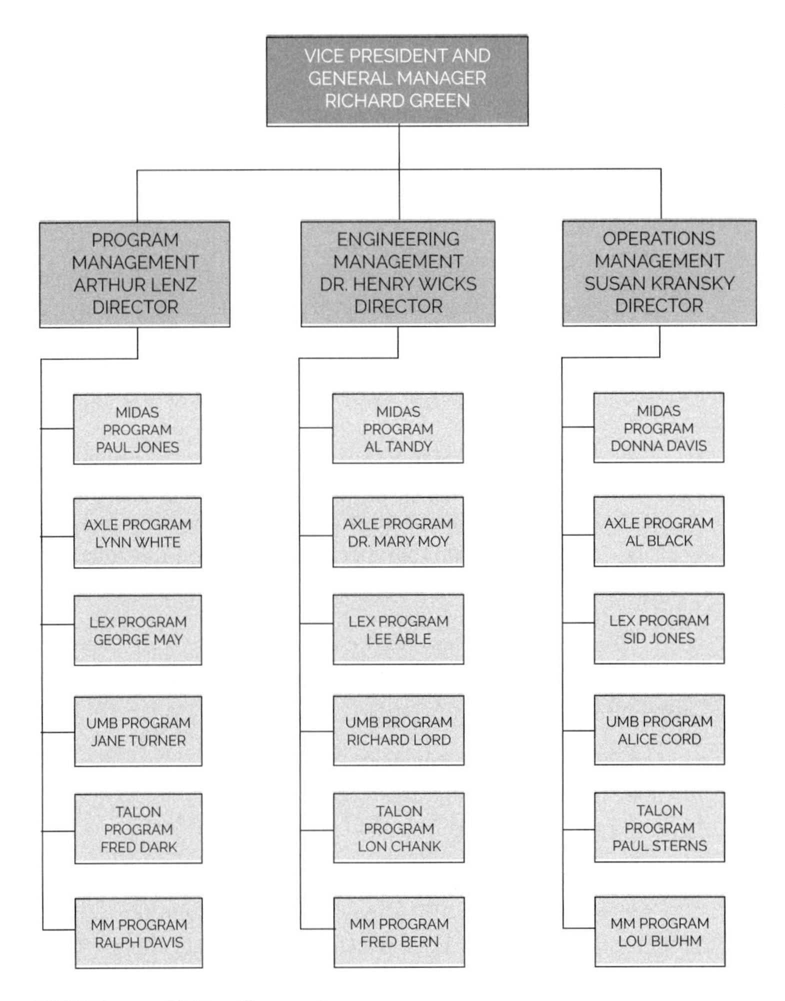

FIGURE 4–5. Dalton Corporation.

more common to see organization charts like Figure 4–5 represent programs, but they can be used for projects as well. Our concern is with the Midas Program. Although the Midas Program may have the lowest priority of the six programs, it is placed at the top, and in boldface, to give the impression that it is the top priority. This type of representation usually makes the client or customer feel that his program is important to the contractor.

The employees shown in Figure 4–5 may be part-time or full-time, depending upon the project's requirements. Perturbations on Figure 4–5 might include one employee's name identified on two or more vertical positions (e.g., the project engineer on two projects) or the same name in two horizontal boxes (e.g., for a small project, the same person could be the project manager and project engineer). Remember, this type of chart is for the customer's benefit and may not show the true "dotted/solid" reporting relationships in the company.

The next step is to show the program office structure, as illustrated in Figure 4–6. Note that the chief of operations and the chief engineer have dual reporting responsibility; they

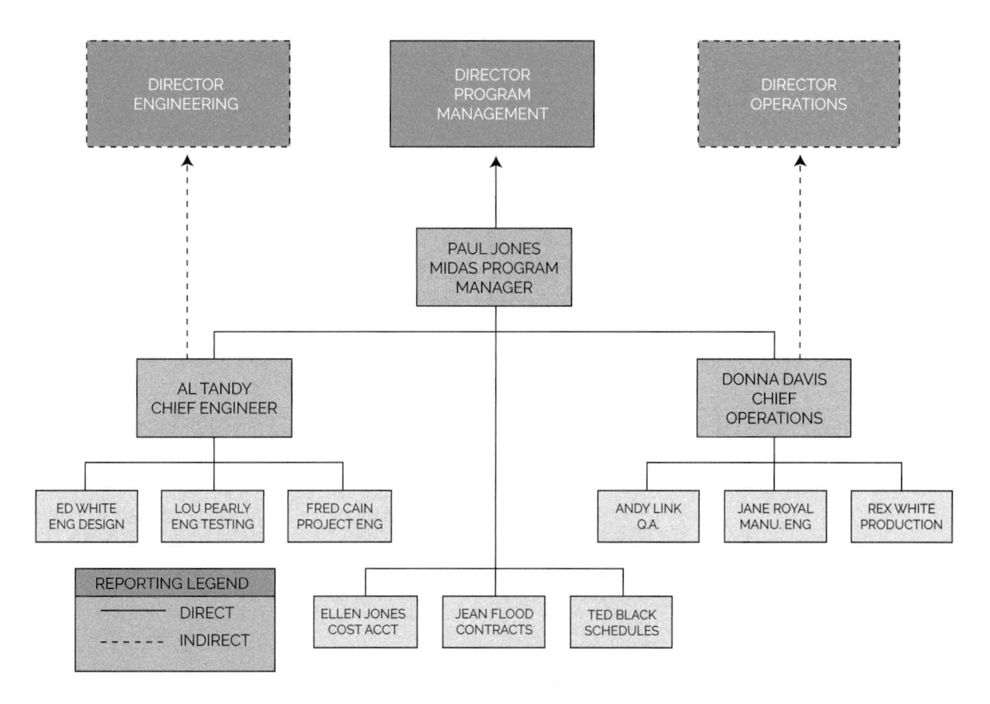

FIGURE 4–6. Midas Program Office.

report directly to the program manager and indirectly to the directors. Again, this may be just for the customer's benefit, with the real reporting structure being reversed. Beneath the chief engineer, there are three positions. Although these positions appear as solid lines, they might actually be dotted lines. For example, Ed White might be working only part-time on the Midas Program but is still shown on the chart as a permanent program office member. Jean Flood, under contracts, might be spending only ten hours per week on the Midas Program.

If the function of two positions on the organizational chart takes place at different times, then both positions may be shown as manned by the same person. For example, Ed White may have his name under both engineering design and engineering testing if the two activities are far enough apart that he can perform them independently.

The people shown in the project office organizational chart, whether full-time or part-time, may not be physically sitting in the project office. For full-time, long-term assignments, as in construction projects, the employees may be physically sitting side by side, whereas for part-time assignments, it may be imperative for them to sit in their functional group. Remember, these types of charts may simply be eyewash for the customer.

Most customers realize that the top-quality personnel may be shared with other programs and projects. Project manning charts, such as the one shown in Figure 4–7, can be used for this purpose. These manning charts are also helpful in preparing the management volume of proposals to show the customer that key personnel will be readily available on his project.

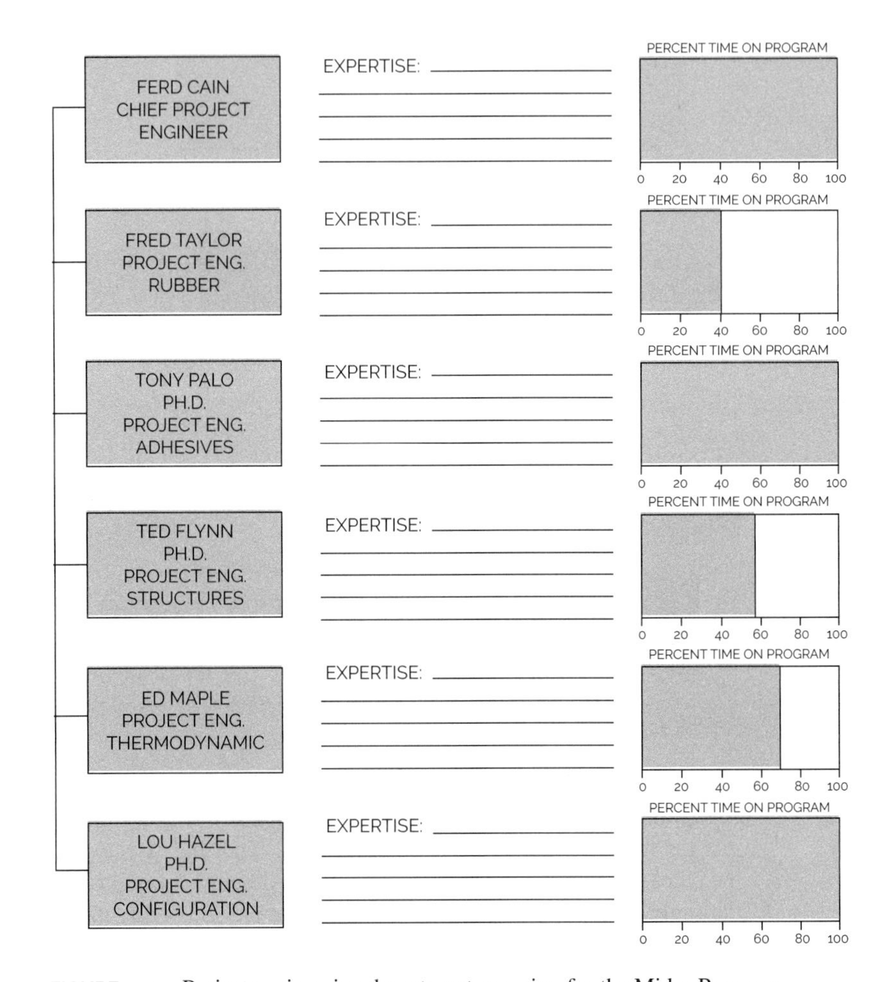

FIGURE 4–7. Project engineering department manning for the Midas Program.

4.11 SELECTING THE PROJECT MANAGEMENT IMPLEMENTATION TEAM

> ***PMBOK® Guide*, 6th Edition**
> Chapter 9 Project Resource Management
> 9.3 Acquire Resources

The implementation of project management within an organization requires strong executive support and an implementation team that is dedicated to making project management work. Selecting the wrong team players can either lengthen the implementation process or reduce employee morale. Some employees may play destructive roles on a project team. These roles, which undermine project management implementation, are shown in Figure 4–8 and described below:

- The aggressor
  - Criticizes everybody and everything on project management
  - Deflates the status and ego of other team members
  - Always acts aggressively

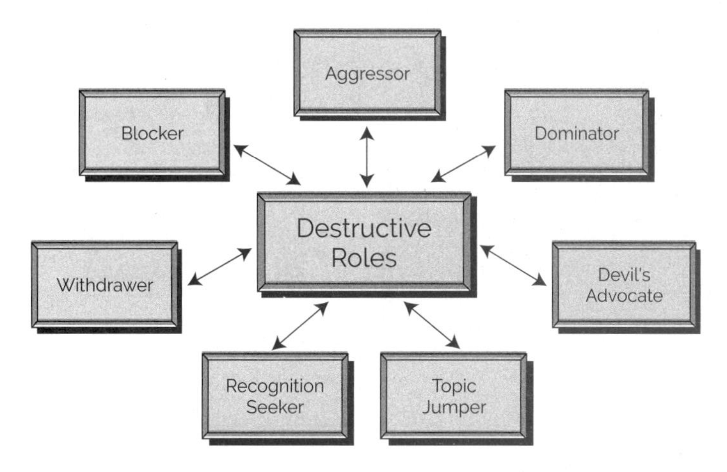

FIGURE 4–8. Roles people play that undermine project management implementation.

- The dominator
  - Always tries to take over
  - Professes to know everything about project management
  - Tries to manipulate people
  - Will challenge those in charge for leadership role
- The devil's advocate
  - Finds fault in all areas of project management
  - Refuses to support project management unless threatened
  - Acts more of a devil than an advocate
- The topic jumper
  - Must be the first one with a new idea/approach to project management
  - Constantly changes topics
  - Cannot focus on ideas for a long time unless it is his/her idea
  - Tries to keep project management implementation as an action item forever
- The recognition seeker
  - Always argues in favor of his/her own ideas
  - Always demonstrates status consciousness
  - Volunteers to become the project manager if status is recognized
  - Likes to hear himself/herself talk
  - Likes to boast rather than provide meaningful information
- The withdrawer
  - Is afraid to be criticized
  - Will not participate openly unless threatened
  - May withhold information
  - May be shy
- The blocker
  - Likes to criticize
  - Rejects the views of others
  - Cites unrelated examples and personal experiences
  - Has multiple reasons why project management will not work

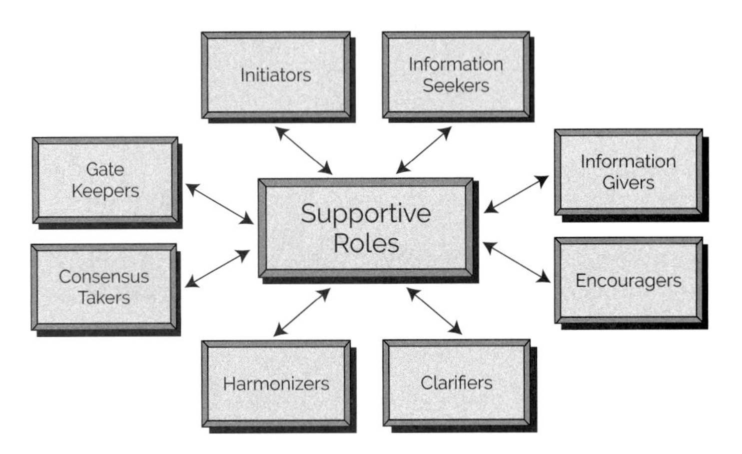

FIGURE 4–9. Roles people play that support project management implementation.

These types of people should not be assigned to project management implementation teams. The types of people who should be assigned to implementation teams are shown in Figure 4–9 and described below. Their roles are indicated by their words:

- The initiators
  - "Is there a chance that this might work?"
  - "Let's try this."
- The information seekers
  - "Have we tried anything like this before?"
  - "Do we know other companies where this has worked?"
  - "Can we get this information?"
- The information givers
  - "Other companies found that . . ."
  - "The literature says that . . ."
  - "Benchmarking studies indicate that . . ."
- The encouragers
  - "Your idea has a lot of merit."
  - "The idea is workable, but we may have to make small changes."
  - "What you said will really help us."
- The clarifiers
  - "Are we saying that . . . ?"
  - "Let me state in my own words what I'm hearing from the team."
  - "Let's see if we can put this into perspective."
- The harmonizers
  - "We sort of agree, don't we?"
  - "Your ideas and mine are close together."
  - "Aren't we saying the same thing?"

- The consensus takers
  - "Let's see if the team is in agreement."
  - "Let's take a vote on this."
  - "Let's see how the rest of the group feels about this."
- The gatekeepers
  - "Who has not given us their opinions on this yet?"
  - "Should we keep our options open?"
  - "Are we prepared to make a decision or recommendation, or is there additional information to be reviewed?"

4.12 MISTAKES MADE BY INEXPERIENCED PROJECT MANAGERS

We are all prone to making mistakes as a project manager or team member. Project managers are not infallible. The list below shows twenty of the most common mistakes that young or inexperienced project managers make. Obviously, there are more than twenty mistakes, and many of these may be unique to specific industries. However, the list is a good starting point for understanding why many project managers get into trouble because of their own doing.[3]

- Believing that excessive detail is needed to be an effective leader
- Pretending to know more than you actually do, thus alienating the true subject matter experts
- Trying to impress people by preparing an ambitious schedule that line managers may find difficulty in supporting
- Having an overreliance on repeatable processes that lack flexibility
- Ignoring problems in the belief that they will go away
- Failing to share accountability for success and failure with functional managers
- Gold-plating the deliverables by adding in unnecessary functionality
- Failing to understand what stakeholders and sponsors want to hear
- Not fully understanding requirements
- Refusing to ask for help
- Ignoring problems that are the responsibility of the project manager to resolve
- Believing in saviors and miracles rather than effective leadership
- Trying to motivate by making promises that cannot be kept
- Failing to see dependencies between your project and other company projects
- Refusing to tell the client that they are wrong
- Continuously reminding everyone who's the boss
- Failing to understand the effects on the project resulting from internal and external politics

---

3. For additional information, see H. Kerzner, "Twenty Common Mistakes Made by Inexperienced Project Managers," learningcenter.iil.com/Saba/Web/Main/goto/Catalog, ©2012 by the International Institute for Learning, New York City. Reproduced by permission.

- Unwilling to say "no"
- Unable to determine which battles are worth fighting and when

Related Case Studies (from Kerzner/*Project Management Case Studies*, 5th ed.)	**Related Workbook Exercises (from Kerzner/*Project Management Workbook and PMP®/CAPM® Exam Study Guide*, 12th ed.)**	***PMBOK® Guide*, 6th Edition, Reference Section for the PMP® Certification Exam**
• Government Project Management • Falls Engineering • White Manufacturing • Martig Construction Company • Ducor Chemical • The Carlson Project	• The Bad Apple • Multiple Choice Exam	• Project Resource Management

4.13 STUDYING TIPS FOR THE PMI® PROJECT MANAGEMENT CERTIFICATION EXAM

This section is applicable as a review of the principles to support the knowledge areas and domain groups in the *PMBOK® Guide*. This chapter addresses:

- Human Resources Management
- Planning
- Project Staffing

Understanding the following principles is beneficial if the reader is using this text to study for the PMP® Certification Exam:

- What is meant by a project team
- Staffing process and environment
- Role of the line manager in staffing
- Role of the executive in staffing
- Skills needed to be a project manager
- That the project manager is responsible for helping the team members grow and learn while working on the project

In Appendix C, the following Dorale Products mini case studies are applicable:

- Dorale Products (H) [Human Resources Management]
- Dorale Products (I) [Human Resources Management]
- Dorale Products (J) [Human Resources Management]
- Dorale Products (K) [Human Resources Management]

PMP and CAPM are registered marks of the Project Management Institute, Inc.

The following multiple-choice questions will be helpful in reviewing the principles of this chapter:

1. During project staffing, the *primary* role of senior management is in the selection of the:
   A. Project manager
   B. Assistant project managers
   C. Functional team
   D. Executives do not get involved in staffing.
2. During project staffing, the *primary* role of line management is:
   A. Approving the selection of the project manager
   B. Approving the selection of assistant project managers
   C. Assigning functional resources based upon who is available
   D. Assigning functional resources based upon availability and the skill set needed
3. A project manager is far more likely to succeed if it is obvious to everyone that:
   A. The project manager has a command of technology.
   B. The project manager is a higher pay grade than everyone else on the team.
   C. The project manager is over 45 years of age.
   D. Executive management has officially appointed the project manager.
4. Most people believe that the best way to train someone in project management is through:
   A. On-the-job training
   B. University seminars
   C. Graduate degrees in project management
   D. Professional seminars and meeting
5. In staffing negotiations with the line manager, you identify a work package that requires a skill set of a grade 7 worker. The line manager informs you that he will assign a grade 6 and a grade 8 worker. You should:
   A. Refuse to accept the grade 6 because you are not responsible for training
   B. Ask for two different people
   C. Ask the sponsor to interfere
   D. Be happy! You have two workers.
6. You priced out a project at 1,000 hours assuming a grade 7 employee would be assigned. The line manager assigns a grade 9 employee. This will result in a significant cost overrun. The project manager should:
   A. Reschedule the start date of the project based upon the availability of a grade 7
   B. Ask the sponsor for a higher priority for your project
   C. Reduce the scope of the project
   D. See if the grade 9 can do the job in less time

7. As a project begins to wind down, the project manager should:
   A. Release all nonessential personnel so that they can be assigned to other projects
   B. Wait until the project is officially completed before releasing anyone
   C. Wait until the line manager officially requests that the people be released
   D. Talk to other project managers to see who wants your people

ANSWERS

1. A
2. D
3. D
4. A
5. D
6. D
7. A

PROBLEMS

4–1 David Cleland made the following remarks:

> His [project manager's] staff should be qualified to provide personal administrative and technical support. He should have sufficient authority to increase or decrease his staff as necessary throughout the life of the project. This authorization should include selective augmentation for varying periods of time from the supporting functional areas.[4]

Do you agree or disagree with these statements? Should the type of project or type of organization play a dominant role in your answer?

4–2 Some people believe that a project manager functions, in some respects, like a physician. Is there any validity in this?

4–3 Paul is a project manager for an effort that requires twelve months. During the seventh, eighth, and ninth months he needs two individuals with special qualifications. The functional manager has promised that these individuals will be available two months before they are needed. If Paul does not assign them to his project at that time, they will be assigned elsewhere and he will have to do with whomever will be available later. What should Paul do? Do you have to make any assumptions in order to defend your answer?

4–4 Frank Boone is the most knowledgeable piping engineer in the company. For five years, the company has turned down his application for transfer to project engineering and project management, stating that he is too valuable to the company in his current position. If you were a project manager, would you want this individual as part of your functional team? How should an organization cope with this situation?

---

4. David Cleland, "Why Project Management?" *Business Horizons*, Winter 1964, p. 85.

4–5 For each of the organizational forms shown below, who determines what resources are needed, when they are needed, and how they will be employed? Who has the authority and responsibility to mobilize these resources?

A. Traditional organization
B. Matrix organization
C. Product line organization
D. Line/staff project organization

4–6 Do you agree or disagree that project organizational forms encourage peer-to-peer communications and dynamic problem solving?

4–7 You are the project engineer on a program similar to one that you directed previously. Should you attempt to obtain the same administrative and/or technical staff that you had before?

4–8 A person assigned to your project is performing unsatisfactorily. What should you do? Will it make a difference if he is in the project office or a functional employee?

4–9 Can a project manager create dedication and a true winning spirit and still be hated by all?

4–10 Can anyone be trained to be a project manager?

4–11 Sometimes, project office personnel report dotted (i.e., indirectly) to the project manager and remain a solid line reportee to their functional manager. Can this work effectively if it were reversed and the personnel are solid to the project manager and dotted to their functional manager?

4–12 Most organizations have "star" people who are usually identified as those individuals who are the key to success. How does a project manager identify these people? Can they be in the project office, or must they be functional employees or managers?

4–13 A major utility company is worried about the project manager's upgrading functional employees. On an eight-month project that employs four hundred full-time project employees, the department managers have set up "check" people whose responsibility is to see that functional employees do not have unauthorized (i.e., not approved by the functional manager) work assignments above their current grade level. Can this system work? What if the work is at a position below their grade level?

Management Functions

5.0 INTRODUCTION

PMBOK® Guide, 6th Edition
2.4.4.3 Project Management Office
3.4 Project Manager Competencies
9.4.2.3 Interpersonal and Team Skills

The project manager measures his success by how well he can negotiate with both upper-level and functional management for the resources necessary to achieve the project objective. Moreover, the project manager may have a great deal of delegated authority but very little power. Hence, the managerial skills he requires for successful performance may be drastically different from those of his functional management counterparts.

The difficult aspect of the project management environment is that individuals at the project–functional interface must report to two bosses. Functional managers and project managers, by virtue of their different authority levels and responsibilities, treat their people in different fashions depending on their "management school" philosophies. This imposes hardships on both the project managers and functional representatives. The project manager must motivate functional representatives toward project dedication on the horizontal line, often with little regard for the employee. After all, the employee might be assigned for a very short-term effort, whereas the end-item is the most important objective. The functional manager, however, expresses more concern for the individual needs of the employee.

Modern practitioners still tend to identify management responsibilities and skills in terms of the principles and functions developed in the early management schools, namely:

- Planning
- Organizing
- Staffing
- Controlling
- Directing

PMBOK is a registered mark of the Project Management Institute, Inc.

Although these management functions have generally been applied to traditional management structures, they have recently been redefined for temporary management positions. Their fundamental meanings remain the same, but the applications are different.

5.1 CONTROLLING

Controlling is a three-step process of measuring progress toward an objective, evaluating what remains to be done, and taking the necessary corrective action to achieve or exceed the objectives. These three steps—measuring, evaluating, and correcting—are defined as follows:

1. *Measuring:* determining through formal and informal reports the degree to which progress toward objectives is being made
2. *Evaluating:* determining cause of and possible ways to act on significant deviations from planned performance
3. *Correcting:* taking control action to correct an unfavorable trend or to take advantage of an unusually favorable trend

The project manager is responsible for ensuring the accomplishment of group and organizational goals and objectives. To effect this, he or she must have a thorough knowledge of standards and cost-control policies and procedures so that a comparison is possible between operating results and preestablished standards. The project manager must then take the necessary corrective actions. Later chapters provide a more in-depth analysis of control, especially the cost control function.

In Chapter 1, we stated that project managers must understand organizational behavior in order to be effective and must have strong interpersonal skills. This is especially important during the controlling function. Line managers may have the luxury of time to build up relationships with each of their workers. But for a project manager time is a constraint, and it is not always easy to predict how well or how poorly an individual will interact with a group, especially if the project manager has never worked with this employee previously. Understanding the physiological and social behavior of how people perform in a group cannot happen overnight.

5.2 DIRECTING

Directing is the implementing and carrying out (through others) of those approved plans that are necessary to achieve or exceed objectives. Directing involves such steps as:

- *Staffing:* seeing that a qualified person is selected for each position
- *Training:* teaching individuals and groups how to fulfill their duties and responsibilities

- *Supervising:* giving others day-to-day instruction, guidance, and discipline as required so that they can fulfill their duties and responsibilities
- *Delegating:* assigning work, responsibility, and authority so others can make maximum utilization of their abilities
- *Motivating:* encouraging others to perform by fulfilling or appealing to their needs
- *Counseling:* holding private discussions with another about how he might do better work, solve a personal problem, or realize his ambitions
- *Coordinating:* seeing that activities are carried out in relation to their importance and with a minimum of conflict

Directing subordinates is not an easy task because of both the short time duration of the project and the fact that employees might still be assigned to a functional manager while temporarily assigned to your effort. The luxury of getting to "know" one's subordinates may not be possible in a project environment.

Project managers must be decisive and move forward rapidly whenever directives are necessary. It is better to decide an issue and be 10 percent wrong than it is to wait for the last 10 percent of a problem's input and cause a schedule delay and improper use of resources. Directives are most effective when the KISS (keep it simple, stupid) rule is applied. Directives should be written with one simple and clear objective so that subordinates can work effectively and get things done right the first time. Orders must be issued in a manner that expects immediate compliance. Whether people will obey an order depends mainly on the amount of respect they have for you. Therefore, never issue an order that you cannot enforce. Oral orders and directives should be disguised as suggestions or requests. The requestor should ask the receiver to repeat the oral orders so that there is no misunderstanding.

Motivating employees so that they feel secure on the job is not easy, especially since a project has a finite lifetime. Specific methods for producing security in a project environment include:

- Letting people know why they are where they are
- Making individuals feel that they belong where they are
- Placing individuals in positions for which they are properly trained
- Letting employees know how their efforts fit into the big picture

Since project managers cannot motivate by promising material gains, they must appeal to each person's pride. The guidelines for proper motivation are:

- Adopt a positive attitude
- Do not criticize management
- Do not make promises that cannot be kept
- Circulate customer reports
- Give each person the attention he or she requires

There are several ways of motivating project personnel. Some effective ways include:

- Giving assignments that provide challenges
- Clearly defining performance expectations

- Giving proper criticism as well as credit
- Giving honest appraisals
- Providing a good working atmosphere
- Developing a team attitude
- Providing a proper direction

5.3 PROJECT AUTHORITY

***PMBOK® Guide*, 6th Edition**
9.4 Develop Teams
9.4.2.3 Interpersonal and Team Skills

Project management structures create a web of relationships that can cause chaos in the delegation of authority and the internal authority structure. Four questions must be considered in describing project authority:

- What is project authority?
- What is power, and how is it achieved?
- How much project authority should be granted to the project manager?
- Who settles project authority interface problems?

One form of the project manager's authority can be defined as the legal or rightful power to command, act, or direct the activities of others. Authority can be delegated from one's superiors. Power, on the other hand, is granted to an individual by his subordinates and is a measure of their respect for him. A manager's authority is a combination of his power and influence such that subordinates, peers, and associates willingly accept his judgment.

In the traditional structure, the power spectrum is realized through the hierarchy, whereas in the project structure, power comes from credibility, expertise, or being a sound decision maker.

Authority is the key to the project management process. The project manager must manage across functional and organizational lines by bringing together activities required to accomplish the objectives of a specific project. Project authority provides the way of thinking required to unify all organizational activities toward accomplishment of the project regardless of where they are located. The project manager who fails to build and maintain his alliances will soon find opposition or indifference to his project requirements.

The amount of authority granted to the project manager varies according to project size, management philosophy, and management interpretation of potential conflicts with functional managers.

Generally speaking, a project manager should have more authority than his responsibility calls for, the exact amount of authority usually depending on the amount of risk that the project manager must take. The greater the risk, the greater the amount of authority. A good project manager knows where his authority ends and does not hold an employee responsible for duties that he (the project manager) does not have the authority to enforce. Some projects are directed by project managers who have only monitoring authority. These project managers are referred to as influence project managers.

Failure to establish authority relationships can result in:

- Poor communication channels
- Misleading information

- Antagonism, especially from the informal organization
- Poor working relationships with superiors, subordinates, peers, and associates
- Surprises for the customer

The following are the most common sources of power and authority problems in a project environment:

- Poorly documented or no formal authority
- Power and authority perceived incorrectly
- Dual accountability of personnel
- Two bosses (who often disagree)
- The project organization encouraging individualism
- Subordinate relations stronger than peer or superior relationships
- Shifting of personnel loyalties from vertical to horizontal lines
- Group decision making based on the strongest group
- Ability to influence or administer rewards and punishment
- Sharing resources among several projects

The project manager does not have unilateral authority in the project effort. He frequently negotiates with the functional manager. The project manager has the authority to determine the "when" and "what" of the project activities, whereas the functional manager has the authority to determine "how the support will be given." The project manager accomplishes his objectives by working with personnel who are largely professional. For professional personnel, project leadership must include explaining the rationale of the effort as well as the more obvious functions of planning, organizing, directing, and controlling.

Certain ground rules exist for authority control through negotiations:

- Negotiations should take place at the lowest level of interaction.
- Definition of the problem must be the first priority:
  - The issue
  - The impact
  - The alternative
  - The recommendations
- Higher-level authority should be used if, and only if, agreement cannot be reached.

The critical stage of any project is planning. This includes more than just planning the activities to be accomplished; it also includes the planning and establishment of the authority relationships that must exist for the duration of the project. Because the project management environment is an ever-changing one, each project establishes its own policies and procedures, a situation that can ultimately result in a variety of authority relationships. It is therefore possible for functional personnel to have different responsibilities on different projects, even if the tasks are the same.

During the planning phase the project team develops a responsibility assignment matrix (RAM) that contains such elements as:

- General management responsibility
- Operations management responsibility

- Specialized responsibility
- Who must be consulted
- Who may be consulted
- Who must be notified
- Who must approve

The responsibility matrix is often referred to as a linear responsibility chart (LRC) or responsibility assignment matrix (RAM). Linear responsibility charts identify the participants, and to what degree an activity will be performed or a decision will be made. The LRC attempts to clarify the authority relationships that can exist when functional units share common work.

Figure 5–1 shows a typical linear responsibility chart. The rows, which indicate the activities, responsibilities, or functions required, can be all of the tasks in the work breakdown structure. The columns identify either positions, titles, or the people themselves. If the chart will be given to an outside customer, then only the titles should appear, or the customer will call the employees directly without going through the project manager. The symbols indicate the degrees of authority or responsibility existing between the rows and columns.

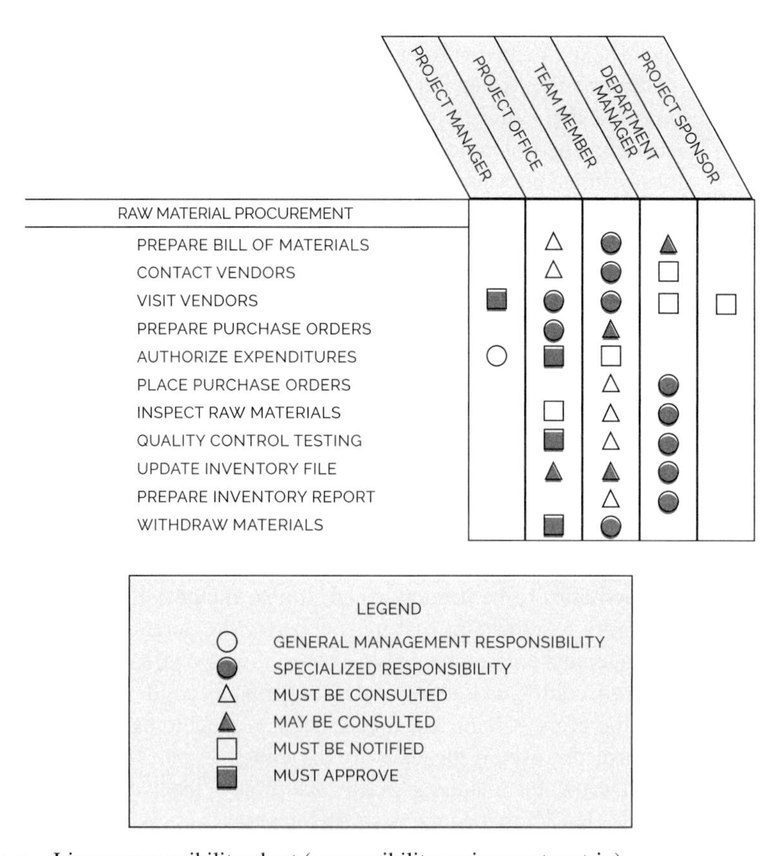

FIGURE 5–1. Linear responsibility chart (responsibility assignment matrix).

Several key factors affect the delegation of authority and responsibility, both from upper-level management to project management and from project management to functional management. These key factors include:

- The maturity of the project management function
- The size, nature, and business base of the company
- The size and nature of the project
- The life cycle of the project
- The capabilities of management at all levels

Once agreement has been reached as to the project manager's authority and responsibility, the results must be documented to clearly delineate his role in regard to:

- His focal position
- Conflict between the project manager and functional managers
- Influence to cut across functional and organizational lines
- Participation in major management and technical decisions
- Collaboration in staffing the project
- Control over allocation and expenditure of funds
- Selection of subcontractors
- Rights in resolving conflicts
- Voice in maintaining integrity of the project team
- Establishment of project plans
- Providing a cost-effective information system for control
- Providing leadership in preparing operational requirements
- Maintaining prime customer liaison and contact
- Promoting technological and managerial improvements
- Establishment of project organization for the duration
- Cutting red tape

Perhaps the best way to document the project manager's authority is through the project charter, which is one of the three methods, shown in Figure 5–2, by which project managers attain authority. Documenting the project manager's authority is necessary because:

- All interfacing must be kept as simple as possible.
- The project manager must have the authority to "force" functional managers to depart from existing standards and possibly incur risk.
- The project manager must gain authority over those elements of a program that are not under his control. This is normally achieved by earning the respect of the individuals concerned.
- The project manager should not attempt to fully describe the exact authority and responsibilities of his project office personnel or team members. Instead, he should encourage problem solving rather than role definition.

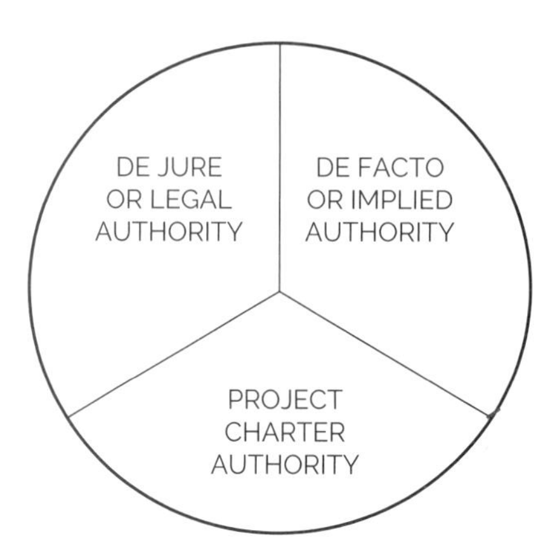

FIGURE 5–2. Types of project authority.

5.4 INTERPERSONAL INFLUENCES

PMBOK® Guide, 6th Edition
3.4.4.3 Politics, Power and Getting Things Done
9.4.2.3 Interpersonal and Team Skills

There exist a variety of relationships (although they are not always clearly definable) between power and authority. These relationships are usually measured by "relative" decision power as a function of the authority structure and are strongly dependent on the project organizational form.

Project managers are generally known for having a lot of delegated authority but very little formal power. They must, therefore, get jobs done through the use of interpersonal influences. There are five such interpersonal influences:

- *Legitimate power:* the ability to gain support because project personnel perceive the project manager as being officially empowered to issue orders
- *Reward power:* the ability to gain support because project personnel perceive the project manager as capable of directly or indirectly dispensing valued organizational rewards (i.e., salary, promotion, bonus, future work assignments)
- *Penalty power:* the ability to gain support because the project personnel perceive the project manager as capable of directly or indirectly dispensing penalties that they wish to avoid; usually derives from the same source as reward power, with one being a necessary condition for the other
- *Expert power:* the ability to gain support because personnel perceive the project manager as possessing special knowledge or expertise (that functional personnel consider as important)
- *Referent power:* the ability to gain support because project personnel feel personally attracted to the project manager or his project

Expert and referent power are examples of personal power that comes from the personal qualities or characteristics to which team members are attracted. Legitimate, reward, and penalty power are often referred to as examples of position power, which is directly related to one's position within the organization. Line managers generally possess a great amount of position power. But in a project environment, position power may be difficult to achieve. According to Magenau and Pinto[1]:

> Within the arena of project management, the whole issue of position power becomes more problematic. Project managers in many organizations operate outside the standard functional hierarchy. While that position allows them a certain freedom of action without direct oversight, it has some important concomitant disadvantages, particularly as they pertain to positional power. First, because cross-functional relationships between the project man-ager and other functional departments can be ill-defined, project managers discover rather quickly that they have little or no legitimate power to simply force their decisions through the organizational system. Functional departments usually do not have to recognize the rights of the project managers to interfere with functional responsibilities; consequently, novice project managers hoping to rely on positional power to implement their projects are quickly disabused.
>
> As a second problem with the use of positional power, in many organizations, project managers have minimal authority to reward team members who, because they are temporary subordinates, maintain direct ties and loyalties to their functional departments. In fact, project managers may not even have the opportunity to complete a performance evaluation on these temporary team members. Likewise, for similar reasons, project managers may have minimal authority to punish inappropriate behavior. Therefore, they may discover that they have the ability to neither offer the carrot nor threaten the stick. As a result, in addition to positional power, it is often necessary that effective project managers seek to develop their personal power bases.

Like relative power, interpersonal influences can be identified with various project organizational forms as to their relative value. This is shown in Figure 5–3.

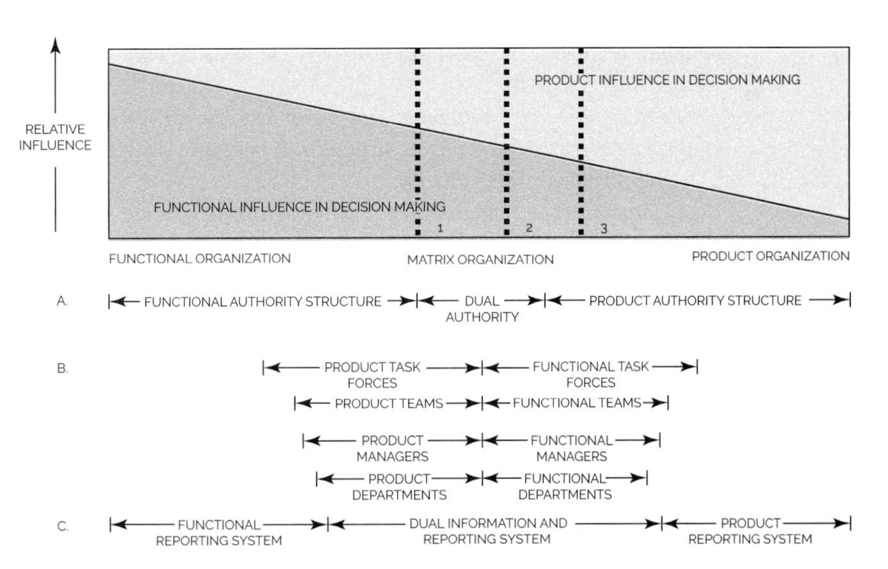

FIGURE 5–3. The range of alternatives. Source: Jay R. Galbraith, "Matrix Organization Designs." Reprinted with permission from *Business Horizons,* February 1971, p. 37. Copyright © 1971 by the Board of Trustees at Indiana University.

1. John M. Magenau, and Jeffrey K. Pinto, "Power, Influence, and Negotiation in Project Management," Peter W. G. Morris and Jeffrey Pinto, eds., *Project Organization and Project Management Competencies* (Hoboken, NJ: John Wiley & Sons, 2007), p. 91.

For any temporary management structure to be effective, there must exist a rational balance of power between functional and project management. Unfortunately, a balance of equal power is often impossible to obtain because each project is inherently different from others, and the project managers possess different leadership abilities.

Regardless of how much authority and power a project manager develops over the course of the project, the ultimate factor in his ability to get the job done is usually his leadership style. Developing bonds of trust, friendship, and respect with the functional workers can promote success.

5.5 BARRIERS TO PROJECT TEAM DEVELOPMENT

PMBOK® Guide, 6th Edition
9.4 Develop Project Team

Most people within project-driven and non–project-driven organizations have differing views of project management. These differing views can create severe barriers to successful project management operations.

The understanding of barriers to project team building can help in developing an environment conducive to effective teamwork. The following barriers are typical for many project environments.

Differing outlooks, priorities, and interests. A major barrier exists when team members have professional objectives and interests that are different from the project objectives. These problems are compounded when the team relies on support organizations that have different interests and priorities.

Role conflicts. Team development efforts are thwarted when role conflicts exist among the team members, such as ambiguity over who does what within the project team and in external support groups.

Project objectives/outcomes not clear. Unclear project objectives frequently lead to conflict, ambiguities, and power struggles. It becomes difficult, if not impossible, to define roles and responsibilities clearly.

Dynamic project environments. Many projects operate in a continual state of change. For example, senior management may keep changing the project scope, objectives, and resource base. In other situations, regulatory changes or client demands can drastically affect the internal operations of a project team.

Competition over team leadership. This barrier most likely occurs in the early phases of a project or if the project runs into severe problems. Obviously, such cases of leadership challenge can result in barriers to team building. Frequently, these challenges are covert challenges to the project leader's ability.

Lack of team definition and structure. Many senior managers complain that teamwork is severely impaired because it lacks clearly defined task responsibilities and reporting structures. A common pattern is that a support department is charged with a task but no one leader is clearly delegated the responsibility. As a consequence, some personnel are working on the project but are not entirely clear on the extent of their responsibilities. In other cases, problems result when a project is supported by several departments without interdisciplinary coordination.

Team personnel selection. This barrier develops when personnel feel unfairly treated or threatened during the staffing of a project. In some cases, project personnel are assigned to a team by functional managers, and the project manager has little or no input into the selection process. This can impede team development efforts, especially when the project leader is given available personnel versus the best, hand-picked team members. The assignment of "available personnel" can result in several problems (e.g., low motivation levels, discontent, and uncommitted team members). The more power the project leader has over the selection of his team members, and the more negotiated agreement there is over the assigned task, the more likely it is that team-building efforts will be fruitful.

Credibility of project leader. Team-building efforts are hampered when the project leader suffers from poor credibility within the team or from other managers. In such cases, team members are often reluctant to make a commitment to the project or the leader. Credibility problems may come from poor managerial skills, poor technical judgments, or lack of experience relevant to the project.

Lack of team member commitment. Lack of commitment can have several sources. Team members having professional interests elsewhere, the feeling of insecurity that is associated with projects, the unclear nature of the rewards that may be forthcoming upon successful completion, and intense interpersonal conflicts within the team can all lead to lack of commitment.

Lack of team member commitment may result from suspicious attitudes existing between the project leader and a functional support manager, or between two team members from two warring functional departments. Finally, low commitment levels are likely to occur when a "star" on a team "demands" too much effort from other team members or too much attention from the team leader. It's a motivation killer.

Communication problems. Not surprisingly, poor communication is a major enemy to effective team development. Poor communication exists on four major levels: problems of communication among team members, between the project leader and the team members, between the project team and top management, and between the project leaders and the client. Often the problem is caused by team members simply not keeping others informed on key project developments. Yet the "whys" of poor communication patterns are far more difficult to determine. The problem can result from low motivation levels, poor morale, or carelessness. It was also discovered that poor communication patterns between the team and support groups result in severe team-building problems, as does poor communication with the client. Poor communication practices often lead to unclear objectives and poor project control, coordination, and work flow.

Lack of senior management support. If senior management support and commitment is unclear and subject to waxing and waning over the project life cycle, it can result in an uneasy feeling among team members and lead to low levels of enthusiasm and project commitment. Two other common problems are that senior management often does not help set the right environment for the project team at the outset, nor do they give the team timely feedback on their performance and activities during the life of the project.

Project managers who are successfully performing their role not only recognize these barriers but also know when in the project life cycle they are most likely to occur. Moreover, these managers take preventive actions and usually foster a work environment that is conducive to effective teamwork. The effective team builder is usually a social architect who understands the interaction of organizational and behavior variables and can foster a climate of active participation and minimal conflict. This requires carefully developed skills in leadership, administration, organization, and technical expertise on the project. However, besides the delicately balanced management skills, the project manager's sensitivity to the basic issues underlying each barrier can help to increase success in developing an effective project team. Specific suggestions for managing team building barriers are advanced in Table 5–1.

TABLE 5–1. BARRIERS TO EFFECTIVE TEAM BUILDING AND SUGGESTED HANDLING APPROACHES

Barrier	Suggestions for Effectively Managing Barriers (How to Minimize or Eliminate Barriers)
Differing outlooks, priorities, interests, and judgments of team members	Make effort early in the project life cycle to discover these conflicting differences. Fully explain the scope of the project and the rewards that may be forthcoming on successful project completion. Sell "team" concept and explain responsibilities. Try to blend individual interests with the overall project objectives.
Role conflicts	As early in a project as feasible, ask team members where they see themselves fitting into the project. Determine how the overall project can best be divided into subsystems and subtasks (e.g., the work breakdown structure). Assign/negotiate roles. Conduct regular status review meetings to keep team informed on progress and watch for unanticipated role conflicts over the project's life.
Project objectives/ outcomes not clear	Assure that all parties understand the overall and interdisciplinary project objectives. Clear and frequent communication with senior management and the client becomes critically important. Status review meetings can be used for feedback. Finally, a proper team name can help to reinforce the project objectives.
Dynamic project environments	The major challenge is to stabilize external influences. First, key project personnel must work out an agreement on the principal project direction and "sell" this direction to the total team. Also educate senior management and the customer on the detrimental consequences of unwarranted change. It is critically important to forecast the "environment" within which the project will be developed. Develop contingency plans.
Competition over team leadership	Senior management must help establish the project manager's leadership role. On the other hand, the project manager needs to fulfill the leadership expectations of team members. Clear role and responsibility definition often minimizes competition over leadership.
Lack of team definition and structure	Project leaders need to sell the team concept to senior management as well as to their team members. Regular meetings with the team will reinforce the team notion as will clearly defined tasks, roles, and responsibilities. Also, visibility in memos and other forms of written media as well as senior management and client participation can unify the team.
Project personnel selection	Attempt to negotiate the project assignments with potential team members. Clearly discuss with potential team members the importance of the project, their role in it, what rewards might result on completion, and the general "rules of the road" of project management. Finally, if team members remain uninterested in the project, then replacement should be considered.
Credibility of project leader	Credibility of the project leader among team members is crucial. It grows with the image of a sound decision maker in both general management and relevant technical expertise. Credibility can be enhanced by the project leader's relationship to other key managers who support the team's efforts.
Lack of team member commitment	Try to determine lack of team member commitment early in the life of the project and attempt to change possible negative views toward the project. Often, insecurity is a major reason for the lack of commitment; try to determine why insecurity exists, then work on reducing the team members' fears. Conflicts with other team members may be another reason for lack of commitment. It is important for the project leader to intervene and mediate the conflict quickly. Finally, if a team member's professional interests lie elsewhere, the project leader should examine ways to satisfy part of the team member's interests or consider replacement.
Communication problems	The project leader should devote considerable time communicating with individual team members about their needs and concerns. In addition, the leader should provide a vehicle for timely sessions to encourage communications among the individual team contributors. Tools for enhancing communications are status meetings, reviews, schedules, reporting system, and colocation. Similarly, the project leader should establish regular and thorough communications with the client and senior management. Emphasis is placed on written and oral communications with key issues and agreements in writing.

Lack of senior management support	Senior management support is an absolute necessity for dealing effectively with interface groups and proper resource commitment. Therefore, a major goal for project leaders is to maintain the continued interest and commitment of senior management in their projects. We suggest that senior management become an integral part of project reviews. Equally important, it is critical for senior management to provide the proper environment for the project to function effectively. Here the project leader needs to tell management at the onset of the program what resources are needed. The project manager's relationship with senior management and ability to develop senior management support is critically affected by his own credibility and the visibility and priority of his project.

5.6 SUGGESTIONS FOR HANDLING THE NEWLY FORMED TEAM

A major problem faced by many project leaders is managing the anxiety that usually develops when a new team is formed. The anxiety experienced by team members is normal and predictable, but is a barrier to getting the team quickly focused on the task.

This anxiety may come from several sources. For example, if the team members have never worked with the project leader, they may be concerned about his leadership style. Some team members may be concerned about the nature of the project and whether it will match their professional interests and capabilities, or help or hinder their career aspirations. Further, team members can be highly anxious about life-style/work-style disruptions. As one project manager remarked, "Moving a team member's desk from one side of the room to the other can sometimes be just about as traumatic as moving someone from Chicago to Manila."

Another common concern among newly formed teams is whether there will be an equitable distribution of the workload among team members and whether each member is capable of pulling his own weight. In some newly formed teams, members not only must do their own work, but also must train other team members. Within reason this is bearable, but when it becomes excessive, anxiety increases.

Certain steps taken early in the life of a team can minimize these problems. First, the project leader must talk with each team member one-to-one about the following:

1. What the objectives are for the project.
2. Who will be involved and why.
3. The importance of the project to the overall organization or work unit.
4. Why the team member was selected and assigned to the project and what role that person will perform.
5. What rewards might be forthcoming if the project is successfully completed.
6. What problems and constraints are likely to be encountered.
7. The rules of the road that will be followed in managing the project (e.g., regular status review meetings).
8. What suggestions the team member has for achieving success.
9. What the professional interests of the team member are.
10. What challenge the project will present to individual members and the entire team.
11. Why the team concept is so important to project management success and how it should work.

Dealing with these anxieties and helping team members feel that they are an integral part of the team can yield rich dividends. First, team members are more likely to openly share their ideas and approaches. Second, it is more likely that the team will be able to develop effective decision-making processes. Third, the team is likely to develop more effective project control procedures, including those traditionally used to monitor project performance (PERT/CPM, networking, work breakdown structures, etc.) and those in which team members give feedback to each other regarding performance.

5.7 TEAM BUILDING AS AN ONGOING PROCESS

While proper attention to team building is critical during early phases of a project, it is a never-ending process. The project manager is continually monitoring team functioning and performance to see what corrective action may be needed to prevent or correct various team problems. Several barometers (summarized in Table 5–2) provide good clues of potential team dysfunction. First, noticeable changes in performance levels for the team and/or for individual team members should always be investigated. Such changes can be symptomatic of more serious problems (e.g., conflict, lack of work integration, communication problems, and unclear objectives). Second, the project leader and team members must be aware of the changing energy levels of team members. These changes, too, may signal more serious problems or that the team is tired and stressed. Sometimes changing the work pace or taking time off can reenergize team members. Third, verbal and nonverbal clues from team members may be a source of information on team functioning. It is important to hear the needs and concerns of team members (verbal clues) and to observe how they act in carrying out their responsibilities (nonverbal clues). Finally, detrimental behavior of one team member toward another can be a signal that a problem within the team warrants attention.

TABLE 5–2. EFFECTIVENESS-INEFFECTIVENESS INDICATORS

The Effective Team's Likely Characteristics	The Ineffective Team's Likely Characteristics
• High performance and task efficiency • Innovative/creative behavior • Commitment • Professional objectives of team members coincident with project requirements • Team members highly interdependent, interface effectively • Capacity for conflict resolution, but conflict encouraged when it can lead to beneficial results • Effective communication • High trust levels • Results orientation • Interest in membership • High energy levels and enthusiasm • High morale • Change orientation	• Low performance • Low commitment to project objectives • Unclear project objectives and fluid commitment levels from key participants • Unproductive gamesmanship, manipulation of others, hidden feelings, conflict avoidance at all costs • Confusion, conflict, inefficiency • Subtle sabotage, fear, disinterest, or foot-dragging • Cliques, collusion, isolation of members • Lethargy/unresponsiveness

Project leaders should hold regular meetings to evaluate overall team performance and deal with team functioning problems. The focus of these meetings can be directed toward "what we are doing well as a team" and "what areas need our team's attention." This approach often brings positive surprises in that the total team is informed of progress in diverse project areas (e.g., a breakthrough in technology development, a subsystem schedule met ahead of the original target, or a positive change in the client's behavior toward the project). After the positive issues have been discussed the review session should focus on actual or potential problem areas. The meeting leader should ask each team member for his observations and then open the discussion to ascertain how significant the problems really are. Assumptions should, of course, be separated from the facts of each situation. Next, assignments should be agreed on for best handling these problems. Finally, a plan for problem follow-up should be developed. The process should result in better overall performance and promote a feeling of team participation and high morale.

5.8 LEADERSHIP IN A PROJECT ENVIRONMENT

PMBOK® Guide, 6th Edition
3.4.4 Leadership Skills

Leadership can be defined as a style of behavior designed to integrate both the organizational requirements and one's personal interests into the pursuit of some objective. All managers have some sort of leadership responsibility. If time permits, successful leadership techniques and practices can be developed.

Leadership is composed of several complex elements, the three most common being:

- The person leading
- The people being led
- The situation (i.e., the project environment)

Project managers are often selected or not selected because of their leadership styles. The most common reason for not selecting an individual is his inability to balance the technical and managerial project functions.

There have been several surveys to determine what leadership techniques are best. The following are the results of a survey by Richard Hodgetts.[2] The results of the survey are still applicable for many project management environments.

- Human relations–oriented leadership techniques
  - "The project manager must make all the team members feel that their efforts are important and have a direct effect on the outcome of the program."
  - "The project manager must educate the team concerning what is to be done and how important its role is."
  - "Provide credit to project participants."

2. Richard M. Hodgetts, "Leadership Techniques in Project Organizations," *Academy of Management Journal*, Vol. 11, pp. 211–219, 1968.

- "Project members must be given recognition and prestige of appointment."
    - "Make the team members feel and believe that they play a vital part in the success (or failure) of the team."
    - "By working extremely closely with my team I believe that one can win a project loyalty while to a large extent minimizing the frequency of authority-gap problems."
    - "I believe that a great motivation can be created just by knowing the people in a personal sense. I know many of the line people better than their own supervisor does. In addition, I try to make them understand that they are an indispensable part of the team."
    - "I would consider the most important technique in overcoming the authority gap to be understanding as much as possible the needs of the individuals with whom you are dealing and over whom you have no direct authority."
- Formal authority–oriented leadership techniques
    - "Point out how great the loss will be if cooperation is not forthcoming."
    - "Put all authority in functional statements."
    - "Apply pressure beginning with a tactful approach and minimum application warranted by the situation and then increasing it."
    - "Threaten to precipitate high-level intervention and do it if necessary."
    - "Convince the members that what is good for the company is good for them."
    - "Place authority on full-time assigned people in the operating division to get the necessity work done."
    - "Maintain control over expenditures."
    - "Utilize implicit threat of going to general management for resolution."
    - "It is most important that the team members recognize that the project manager has the charter to direct the project."

5.9 VALUE-BASED PROJECT LEADERSHIP

PMBOK® Guide, 6th Edition
3.4.4 Leadership Skills

The importance of value has had a significant impact on the leadership style of today's project managers. Historically, project management leadership was perceived as the inevitable conflict between individual values and organizational values. Today, companies are looking for ways to get employees to align their personal values with the organization's values. In doing so, companies have created cultures that support project management and many of the cultures are driven by a change in perceived values.

Table 5–3, adapted from Hultman and Gellerman, shows how our concept of value has changed over the years.[3] If you look closely at the items in Table 5–3, you can see that the changing values affect more than just individual versus organization values. Instead, it is more likely to be a conflict of four groups, namely the project manager,

3. K. Hultman and B. Gellerman, *Balancing Individual and Organizational Values* (Jossey-Bass/Pfeiffer: San Francisco, 2002). pp. 105–106.

TABLE 5–3. CHANGING VALUES

Moving away from: Ineffective Values	Moving toward: Effective Values
Mistrust	Trust
Job descriptions	Competency models
Power and authority	Teamwork
Internal focus	Stakeholder focus
Security	Taking risks
Conformity	Innovation
Predictability	Flexibility
Internal competition	Internal collaboration
Reactive management	Proactive management
Bureaucracy	Boundaryless
Traditional education	Lifelong education
Hierarchical leadership	Multidirectional leadership
Tactical thinking	Strategic thinking
Compliance	Commitment
Meeting standards	Continuous improvements

the project team, the parent organization, and the stakeholders. The needs of each group might be:

- Project manager:
  - Accomplishment of objectives
  - Demonstration of creativity
  - Demonstration of innovation
- Team members:
  - Achievement
  - Advancement
  - Ambition
  - Credentials
  - Recognition
- Organization
  - Continuous improvement
  - Learning
  - Quality
  - Strategic focus
  - Morality and ethics
  - Profitability
  - Recognition and image
- Stakeholders
  - Organizational stakeholders: job security
  - Product/market stakeholders: high-quality performance and product usefulness
  - Capital markets: financial growth

There are several reasons why the role of the project manager and the accompanying leadership style has changed. Some reasons are:

- We are now managing our business as though it is a series of projects.
- Project management is now viewed as a full-time profession.
- Project managers are now viewed as both business managers and project managers and are expected to make decisions in both areas.
- The value of a project is measured more so in business terms rather than solely technical terms.
- Project management is now being applied to parts of the business that traditionally haven't used project management.

The nontraditional types of projects have made it clear why traditional project management must change. Here are areas that necessitate changes:

- New projects have become:
  - Highly complex and with greater acceptance of risks that may not be fully understood during project approval
  - More uncertain in the outcomes of the projects and with no guarantee of value at the end
  - Pressed for speed-to-market irrespective of the risks
- The statement of work (SOW) is:
  - Not always well defined, especially on long-term projects
  - Based upon possibly flawed, irrational, or unrealistic assumptions
  - Inconsiderate of unknown and rapidly changing economic and environmental conditions
  - Based upon a stationary rather than moving target for final value
- The management cost and control systems [enterprise project management methodologies (EPM)] focus on:
  - An ideal situation (as in the *PMBOK® Guide*)
  - Theories rather than the understanding of the workflow
  - Inflexible processes
  - Periodically reporting time at completion and cost at completion but not value (or benefits) at completion
  - Project continuation rather than canceling projects with limited or no value

Over the years, small steps have been taken to plan for the use of project management on nontraditional projects. This included:

- Project managers are provided with more business knowledge and are allowed to provide an input during the project selection process.
- Because of the above item, project managers are brought on board the project at the beginning of the initiation phase rather than the end of the initiation phase.
- Projects managers now seem to have more of an understanding of technology rather than a command of technology.

5.10 TRANSFORMATIONAL PROJECT MANAGEMENT LEADERSHIP

PMBOK® Guide, 6th Edition
3.4.4.1 Dealing with People

There have been numerous books written on effective project management leadership. Most books seem to favor situational leadership where the leadership style that the project manager selects is based upon the size and nature of the project, the importance of the deliverables, the skill level of the project team members, the project manager's previous experience working with these team members, and the risks associated with the project. Historically, project managers perceived themselves as being paid to produce deliverables rather than managing people. Team leadership was important to some degree as long as what was expected in the way of employee performance and behavior was consistent with the desires of the employee's functional manager that conducted the employee's performance review. In the past, project managers were expected to provide leadership in a manner that improves the employee's performance and skills, and allows the employee to grow while working on project teams. Today, project managers are being asked to function as managers of organizational change on selected projects. Organizational change requires that people change. This mandates that project managers possess a set of skills that may be different than what was appropriate for managing projects. This approach is now being called transformational project management leadership.

There are specific situations where transformational leadership must be used and employees must be removed from their previous comfort zones. As an example, not all projects come to an end once the deliverables are created. Consider a multinational company that establishes an IT project to create a new, high-security company-wide email system. Once the software is developed, the project is ready to "go live." Historically, the person acting as the project manager to develop the software moves on to another project at "go live," and the responsibility for implementation goes to the functional managers or someone else. Today, companies are asking the project manager to remain on board the project and act as the change agent for full, corporate-wide implementation of the change-over to the new system. In these situations, the project manager must adopt a transformational leadership style.

Transformational project management is heavily focused upon the people side of the change and is a method for managing the resistance to the change, whether the change is in processes, technology, acquisitions, targets, or organizational restructuring. People need to understand the change and buy into it. Imposing change upon people is an invitation for prolonged resistance especially if people see their job threatened. Transformational projects can remove people from their comfort zones.

5.11 ORGANIZATIONAL IMPACT

In most companies, whether or not project-oriented, the impact of management emphasis on the organization is well known. In the project environment there also exists a definite impact due to leadership emphasis. The leadership emphasis is best seen by employee contributions, organizational order, employee performance, and the project manager's performance:

- Contributions from People
  - A good project manager encourages active cooperation and responsible participation. The result is that both good and bad information is contributed freely.
  - A poor project manager maintains an atmosphere of passive resistance with only responsive participation. This results in information being withheld.
- Organizational Order
  - A good project manager develops policy and encourages acceptance. A low price is paid for contributions.
  - A poor project manager goes beyond policies and attempts to develop procedures and measurements. A high price is normally paid for contributions.
- Employee Performance
  - A good project manager keeps people informed and satisfied (if possible) by aligning motives with objectives. Positive thinking and cooperation are encouraged. A good project manager is willing to give more responsibility to those willing to accept it.
  - A poor project manager keeps people uninformed, frustrated, defensive, and negative. Motives are aligned with incentives rather than objectives. The poor project manager develops a "stay out of trouble" atmosphere.
- Performance of the Project Manager
  - A good project manager assumes that employee misunderstandings can and will occur, and therefore blames himself. A good project manager constantly attempts to improve and be more communicative. He relies heavily on moral persuasion.
  - A poor project manager assumes that employees are unwilling to cooperate and therefore blames subordinates. The poor project manager demands more through authoritarian attitudes and relies heavily on material incentives.

Management emphasis also impacts the organization. The following four categories show this management emphasis resulting for both good and poor project management:

- Management Problem Solving
  - A good project manager performs his own problem solving at the level for which he is responsible through delegation of problem-solving responsibilities.
  - A poor project manager will do subordinate problem solving in known areas. For areas that he does not know, he requires that his approval be given prior to idea implementation.
- Organizational Order
  - A good project manager develops, maintains, and uses a single integrated management system in which authority and responsibility are delegated to the subordinates. In addition, he knows that occasional slippages and overruns will occur, and simply tries to minimize their effect.

- A poor project manager delegates as little authority and responsibility as possible, and runs the risk of continual slippages and overruns. A poor project manager maintains two management information systems: one informal system for himself and one formal (eyewash) system simply to impress his superiors.

- Performance of People
  - A good project manager finds that subordinates willingly accept responsibility, are decisive in their attitude toward the project, and are satisfied.
  - A poor project manager finds that his or her subordinates are reluctant to accept responsibility, are indecisive in their actions, and seem frustrated.
- Performance of the Project Manager
  - A good project manager assumes that his key people can "run the show." He exhibits confidence in those individuals working in areas in which he has no expertise, and exhibits patience with people working in areas where he has a familiarity. A good project manager is never too busy to help his people solve personal or professional problems.
  - A poor project manager considers himself indispensable, is overcautious with work performed in unfamiliar areas, and becomes overly interested in work he knows. A poor project manager is always tied up in meetings.

5.12 EMPLOYEE–MANAGER PROBLEMS

PMBOK® Guide, 6th Edition
3.4.4.1 Dealing with People

The two major problem areas in the project environment are the "who has what authority and responsibility" question, and the resulting conflicts associated with the individual at the project–functional interface. Almost all project problems in some way or another involve these two major areas. Other problem areas found in the project environment include:

- The pyramidal structure
- Superior–subordinate relationships
- Departmentalization
- Scalar chain of command
- Organizational chain of command
- Power and authority
- Planning goals and objectives
- Decision making
- Reward and punishment
- Span of control

The two most common employee problems involve the assignment and resulting evaluation processes.

On the manager level, the two most common problems involve personal values and conflicts. Personal values are often attributed to the "changing of the guard." New managers have a different sense of values from that of the older, more experienced managers.

Previously, we defined one of the attributes of a project manager as liking risks. Unfortunately, the amount of risk that today's managers are willing to accept varies not only with their personal values but also with the impact of current economic conditions and top management philosophies. If top management views a specific project as vital for the growth of the company, then the project manager may be directed to assume virtually no risks during the execution of the project. In this case the project manager may attempt to pass all responsibility to higher or lower management claiming that "his hands are tied."

The amount of risk that managers will accept also varies with age and experience. Older, more experienced managers tend to take few risks, whereas the younger, more aggressive managers may adopt a risk-lover policy in hopes of achieving a name for themselves.

Conflicts exist at the project–functional interface regardless of how hard we attempt to structure the work.

Major conflicts can also arise during problem resolution sessions because the time constraints imposed on the project often prevent both parties from taking a logical approach. One of the major causes of prolonged problem solving is a lack of pertinent information. The following information should be reported by the project manager[4]:

- The problem
- The cause
- The expected impact on schedule, budget, profit, or other pertinent area
- The action taken or recommended and the results expected of that action
- What top management can do to help

5.13 GENERAL MANAGEMENT PITFALLS

The project environment offers numerous opportunities for project managers and team members to get into trouble. Common types of management pitfalls are:

- Lack of self-control (knowing oneself): Knowing oneself, especially one's capabilities, strengths, and weaknesses, is the first step toward successful project management. Too often, managers will assume that they are jacks-of-all-trades, will "bite off more than they can chew," and then find that insufficient time exists for training additional personnel.
- Activity traps: Activity traps result when the means become the end, rather than the means to achieve the end. The most common activity traps are team meetings, customer–technical interchange meetings, and the development of special schedules and charts that cannot be used for customer reporting but are used to inform upper-level management of project status. Sign-off documents are another activity trap, and managers must evaluate whether all this paperwork is worth the effort.
- Managing versus doing: There often exists a very fine line between managing and doing. As an example, consider a project manager who was asked by one of

4. Russell D. Archibald, *Managing High-Technology Programs and Projects* (New York: John Wiley & Sons, 1976), p. 230.

his technical people to make a telephone call to assist him in solving a problem. Simply making the phone call is doing work that should be done by the project team members or even the functional manager. However, if the person being called requires that someone in absolute authority be included in the conversation, then this can be considered managing instead of doing. There are several other cases where one must become a doer in order to be an effective manager and command the loyalty and respect of subordinates.

- People versus task skills: Another major pitfall is the decision to utilize either people skills or task skills. Is it better to utilize subordinates with whom you can obtain a good working relationship or to employ highly skilled people simply to get the job done? Obviously, the project manager would like nothing better than to have the best of both worlds, but this is not always possible.
- Ineffective communications.
- Time management: It is often said that a good project manager must be willing to work sixty to eighty hours a week to get the job done. This might be true if he is continually fighting fires or if budgeting constraints prevent employing additional staff. The major reason, however, is the result of ineffective time management. Prime examples might include the continuous flow of paperwork, unnecessary meetings, unnecessary phone calls, and acting as a tour guide for visitors.
- Management bottlenecks.

5.14 TIME MANAGEMENT PITFALLS

Managing projects within time, cost, and performance is easier said than done. The project management environment is extremely turbulent and is composed of numerous meetings, report writing, conflict resolution, continuous planning and replanning, communications with the customer, and crisis management. Ideally, the effective project manager is a manager, not a doer, but in the "real world," project managers often compromise their time by doing both, a fact that makes effective project management leadership difficult.

Disciplined time management is one of the keys to effective project management. It is often said that if the project manager cannot control his own time, then he will control nothing else on the project. For most people, time is a resource that, when lost or misplaced, is gone forever. For a project manager, however, time is more of a constraint, and effective time management principles must be employed to make it a resource. Experienced personnel soon learn to delegate tasks and to employ effective time management principles. The following questions should help managers identify problem areas:

- Do you have trouble completing work within the allocated deadlines?
- How many interruptions are there each day?

Section 5.13 is adapted from David Cleland and Harold Kerzner, *Engineering Team Management* (Melbourne, Florida: Krieger, 1986), Chapter 8.

- Do you have a procedure for handling interruptions?
- If you need a large block of uninterrupted time, is it available? With or without overtime?
- How do you handle drop-in visitors and phone calls?
- How is incoming mail handled?
- Do you have established procedures for routine work?
- Are you accomplishing more or less than you were three months ago? Six months ago?
- How difficult is it for you to say no?
- How do you approach detail work?
- Do you perform work that should be handled by your subordinates?
- Do you have sufficient time each day for personal interests?
- Do you still think about your job when away from the office?
- Do you make a list of things to do? If yes, is the list prioritized?
- Does your schedule have some degree of flexibility?

The project manager who can deal with these questions has a greater opportunity to convert time from a constraint to a resource.

TIME ROBBERS

PMBOK® Guide, 6th Edition
Chapter 6 Project Schedule Management
Chapter 11 Project Risk Management

The most challenging problem facing the project manager is his inability to say no. Consider the situation in which an employee comes into your office with a problem. The employee may be sincere when he says that he simply wants your advice but, more often than not, the employee wants to take the monkey off of his back and put it onto yours. The employee's problem is now *your* problem.

To handle such situations, first screen out the problems with which you do not wish to get involved. Second, if the situation does necessitate your involvement, then you must make sure that when the employee leaves your office, he realizes that the problem is still his, not yours. Third, if you find that the problem will require your continued attention, remind the employee that all future decisions will be joint decisions and that the problem will still be on the employee's shoulders. Once employees realize that they cannot put their problems on your shoulders, they learn how to make their own decisions.

There are numerous time robbers in the project management environment. These include:

- Incomplete work
- Lack of a job description
- A job poorly done that must be done over
- Too many people involved in minor decision making
- Telephone calls, mail, and email
- Lack of technical knowledge
- Lack of adequate responsibility and commensurate authority
- Lack of authorization to make decisions
- Too many changes to the plan
- Poor functional status reporting
- Work overload
- Waiting for people

- Unreasonable time constraints
- Failure to delegate, or unwise delegation
- Too much travel
- Lack of adequate project management
- Poor retrieval systems tools
- Lack of information in a ready-to-use format
- Departmental "buck passing"
- Company politics
- Day-to-day administration
- Going from crisis to crisis
- Union grievances
- Conflicting directives
- Having to explain "thinking" to everyone
- Bureaucratic roadblocks ("ego")
- Empire-building line managers
- Too many levels of review
- No communication between sales and engineering
- Casual office conversations
- Misplaced information
- Excessive paperwork
- Shifting priorities
- Lack of clerical/administrative support
- Indecision at any level
- Procrastination
- Dealing with unreliable subcontractors
- Setting up appointments
- Too many meetings
- Personnel not willing to take risks
- Monitoring delegated work
- Demand for short-term results
- Unclear roles/job descriptions
- Lack of long-range planning
- Executive meddling
- Learning new company systems
- Budget adherence requirements
- Poor lead time on projects
- Poorly educated customers
- Documentation (reports/red tape)
- Not enough proven managers
- Large number of projects
- Vague goals and objectives
- Desire for perfection
- Lack of project organization
- Shifting of functional personnel
- Constant pressure
- Lack of employee discipline
- Constant interruptions
- Lack of qualified manpower

EFFECTIVE TIME MANAGEMENT

There are several techniques that project managers can practice in order to make better use of their time:

- Delegate.
- Follow the schedule.
- Decide fast.
- Decide who should attend.
- Learn to say no.
- Start now.
- Do the tough part first.
- Travel light.
- Work at travel stops.
- Avoid useless memos.
- Refuse to do the unimportant.
- Look ahead.
- Ask: Is this trip necessary?
- Know your energy cycle.

- Control telephone and email time.
- Send out the meeting agenda.
- Overcome procrastination.
- Manage by exception.

To be effective, the project manager must establish time management rules and then ask himself four questions:

- What am I doing that I don't have to do at all?
- What am I doing that can be done better by someone else?
- What am I doing that could be done as well by someone else?
- Am I establishing the right priorities for my activities?

STRESS AND BURNOUT

The factors that serve to make any occupation especially stressful are responsibility without the authority or ability to exert control, a necessity for perfection, the pressure of deadlines, role ambiguity, role conflict, role overload, the crossing of organizational boundaries, responsibility for the actions of subordinates, and the necessity to keep up with the information explosions or technological breakthroughs. Project managers have all of these factors in their jobs.

A project manager has his resources controlled by line management, yet the responsibilities of bringing a project to completion by a prescribed deadline are his. A project manager may be told to increase the work output, while the work force is simultaneously being cut. Project managers are expected to get work out on schedule, but are often not permitted to pay overtime.

Project managers are subject to stress due to several different facets of their jobs. This can manifest itself in a variety of ways, such as:

- Being tired
- Feeling depressed
- Being physically and emotionally exhausted
- Burned out
- Being unhappy
- Feeling trapped
- Feeling worthless
- Feeling resentful and disillusioned about people
- Feeling hopeless
- Feeling rejected
- Feeling anxious

Stress is not always negative, however. Without certain amounts of stress, reports would never get written or distributed, deadlines would never be met, and no one would even get to work on time. In a project environment, with continually changing requirements, impossible deadlines, and each project being considered as a unique entity in itself, we must ask, "How much prolonged stress can a project manager handle comfortably?"

The stresses of project management may seem excessive for whatever rewards the position may offer. However, the project manager who is aware of the stresses inherent in the job and knows stress management techniques can face this challenge objectively and make it a rewarding experience.

5.15 MANAGEMENT POLICIES AND PROCEDURES

Although project managers have the authority and responsibility to establish project policies and procedures, they must fall within the general guidelines established by top management. Table 5–4 identifies sample top-management guidelines. Guidelines can also be established for planning, scheduling, controlling, and communications.

5.16 HUMAN BEHAVIOR EDUCATION

If there is a weakness in some of the project management education programs, it lies in the area of human behavior education. The potential problem is that there is an abundance of courses on planning, scheduling, and cost control but not very many courses on behavioral sciences that are directly applicable to a project management environment. All too often, lectures on human behavior focus upon application of the theories and principles based upon a superior (project manager) to subordinate (team member) relationship. This approach fails because:

- Team members can be at a higher pay grade than the project manager.
- The project manager most often has little overall authority.
- The project manager most often has little formal reward power.
- Team members may be working on multiple projects at the same time.
- Team members may receive conflicting instructions from the project managers and their line manager.
- Because of the project's duration, the project manager may not have the time necessary to adequately know the people on the team on a personal basis.
- The project manager may not have any authority to have people assigned to the project team or removed.

Topics that managers and executives believe should be covered in more depth in the behavioral courses include:

- Conflict management with all levels of personnel
- Facilitation management
- Counseling skills
- Mentorship skills
- Negotiation skills
- Communication skills with all stakeholders
- Presentation skills

TABLE 5–4. PROJECT GUIDELINES

Program Manager	Functional Manager	Relationship
The program manager is responsible for overall program direction, control, and coordination; and is the principal contact with the program management of the customer.	The functional organization managers are responsible for supporting the program manager in the performance of the contract(s) and in accordance with the terms of the contract(s) and are accountable to their cognizant managers for the total performance.	The program manager determines what will be done: he obtains, through the assigned program team members, the assistance and concurrence of the functional support organizations in determining the definitive requirements and objectives of the program.
To achieve the program objectives, the program manager utilizes the services of the functional organizations in accordance with the prescribed division policies and procedures affecting the functional organizations.		The functional organizations determine *how the work will be done.*
The program manager establishes program and technical policy as defined by management policy.	The functional support organizations perform all work within their functional areas for all programs within the cost, schedule, quality, and specifications established by contract for the program so as to assist the program manager in achieving the program objectives.	The program manager operates within prescribed division policies and procedures except where requirements of a particular program necessitate deviations or modifications as approved by the general manager. The functional support organizations provide strong, aggressive support to the program managers.
The program manager is responsible for the progress being made as well as the effectiveness of the total program. Integrates research, development, production, procurement, quality assurance, product support, test, and financial and contractual aspects.		
	The functional support organization management seeks out or initiates innovations, methods, improvements, or other means that will enable that function to better schedule commitments, reduce cost, improve quality, or otherwise render exemplary performance as approved by the program manager.	
Approves detailed performance specifications, pertinent physical characteristics, and functional design criteria to meet the program's development or operational requirements.		The program manager relies on the functional support program team members for carrying out specific program assignments.
		Program managers and the functional support program team members are jointly responsible for ensuring that unresolved conflicts between requirements levied on functional organizations by different program managers are brought to the attention of management.

Ensures preparation of, and approves, overall plan, budgets, and work statements essential to the integration of system elements. Directs the preparation and maintenance of a time, cost, and performance schedule to ensure the orderly progress of the program.	
Coordinates and approves subcontract work statement, schedules, contract type, and price for major "buy" items.	
	Program managers do not make decisions that are the responsibility of the functional support organizations as defined in division policies and procedures and/or as assigned by the general manager.
Coordinates and approves vendor evaluation and source selections in conjunction with procurement representative to the program team.	
	Functional organization managers do not request decisions of a program manager that are not within the program manager's delineated authority and responsibility and that do not affect the requirements of the program.
Program decision authority rests with the program manager for all matters relating to his assigned program, consistent with division policy and the responsibilities assigned by the general manager.	
	Functional organizations do not make program decisions that are the responsibility of the program manager.
	Joint participation in problem solution is essential to providing satisfactory decisions that fulfill overall program and company objectives, and is accomplished by the program manager and the assigned program team members. In arriving at program decisions, the program manager obtains the assistance and concurrence of cognizant functional support managers, through the cognizant program team member, since they are held accountable for their support of each program and for overall division functional performance.

The problem may emanate from the limited number of textbooks on human behavior applications directly applicable to the project management environment. One of the best books in the marketplace was written by Steven Flannes and Ginger Levin.[5] The book stresses application of project management education by providing numerous examples from the authors' project management experience.

Related Case Studies (from Kerzner/*Project Management Case Studies*, 5th ed.)	**Related Workbook Exercises (from Kerzner/*Project Management Workbook and PMP®/CAPM® Exam Study Guide*, 12th ed.)**	***PMBOK® Guide*, Sixth Edition, Reference Section for the PMP® Certification Exam**
• The Trophy Project* • McRoy Aerospace* • The Poor Worker* • The Prima Donna* • The Reluctant Workers*	• Multiple Choice Exam • Crossword Puzzle on Project Resource Management	

*Case Study also appears at end of chapter.

5.17 STUDYING TIPS FOR THE PMI® PROJECT MANAGEMENT CERTIFICATION EXAM

This section is applicable as a review of the principles to support the knowledge areas and domain groups in the *PMBOK® Guide*. This chapter addresses:

- Human Resources Management
- Communications Management
- Closure

Understanding the following principles is beneficial if the reader is using this text to study for the PMP® Certification Exam:

- Various leadership styles
- Different types of power
- Different types of authority
- Need to document authority
- Importance of human resources management in project management
- Need to clearly identify each team member's role and responsibility
- Various ways to motivate team members
- That both the project manager and the team are expected to solve their own problems
- Barriers to encoding and decoding
- Need for communication feedback

5. Steven W. Flannes and Ginger Levin, *People Skills for Project Managers* (Vienna, VA: Management Concepts, 2001).
PMP is a registered mark of the Project Management Institute, Inc.

- Various communication styles
- Types of meetings

In Appendix C, the following Dorale Products mini case study is applicable:

- Dorale Products (I) [Human Resources and Communications Management]

The following multiple-choice questions will be helpful in reviewing the principles of this chapter:

1. Which of the following is not one of the sources of authority for a project manager?
   A. Project charter
   B. Job description for a project manager
   C. Delegation from senior management
   D. Delegation from subordinates
2. Which form of power do project managers that have a command of technology and are leading R&D projects most frequently use?
   A. Reward power
   B. Legitimate power
   C. Expert power
   D. Referent power
3. If a project manager possesses penalty (or coercive) power, he or she most likely also possesses:
   A. Reward power
   B. Legitimate power
   C. Expert power
   D. Referent power
4. A project manager with a history of success in meeting deliverables and in working with team members would most likely possess a great deal of:
   A. Reward power
   B. Legitimate power
   C. Expert power
   D. Referent power
5. Most project managers are motivated by which level of Maslow's hierarchy of human needs?
   A. Safety
   B. Socialization
   C. Self-esteem
   D. Self-actualization
6. You have been placed in charge of a project team. The majority of the team members have less than two years of experience working on project teams and most of the

people have never worked with you previously. The leadership style you would most likely select would be:

A. Telling
B. Selling
C. Participating
D. Delegating

7. You have been placed in charge of a new project team and are fortunate to have been assigned the same people that worked for you on your last two projects. Both previous projects were very successful and the team performed as a high-performance team. The leadership style you would most likely use on the new project would be:

   A. Telling
   B. Selling
   C. Participating
   D. Delegating

8. A project manager provides a verbal set of instructions to two team members on how to perform a specific test. Without agreeing or disagreeing with the project manager, the two employees leave the project manager's office. Later, the project manager discovers that the tests were not conducted according to his instructions. The most probable cause of failure would be:

   A. Improper encoding
   B. Improper decoding
   C. Improper format for the message
   D. Lack of feedback on instructions

9. A project manager who allows workers to be actively involved with the project manager in making decisions would be using which leadership style?

   A. Passive
   B. Participative/democratic
   C. Autocratic
   D. Laissez-faire

10. A project manager who dictates all decisions and does not allow for any participation by the workers would be using which leadership style?

    A. Passive
    B. Participative/democratic
    C. Autocratic
    D. Laissez-faire

11. A project manager that allows the team to make virtually all of the decisions without any involvement by the project manager would be using which leadership style?

    A. Passive
    B. Participative/democratic
    C. Autocratic
    D. Laissez-faire

ANSWERS

1. D
2. C
3. A
4. D
5. D
6. A
7. D
8. D
9. B
10. C
11. D

PROBLEMS

5–1 A project manager finds that he does not have direct reward power over salaries, bonuses, work assignments, or project funding for members of the project team with whom he interfaces. Does this mean that he is totally deficient in reward power? Explain your answer.

5–2 For each of the remarks made below, what types of interpersonal influences could exist?

a. "I've had good working relations with department X. They like me and I like them. I can usually push through anything ahead of schedule."
b. A research scientist was temporarily promoted to project management for an advanced state-of-the-art effort. He was overheard making the following remark to a team member: "I know it's contrary to department policy, but the test must be conducted according to these criteria or else the results will be meaningless."

5–3 Do you agree or disagree that scientists and engineers are likely to be more creative if they feel that they have sufficient freedom in their work? Can this condition backfire?

5–4 Should the amount of risk and uncertainty in the project have a direct bearing on how much authority is granted to a project manager?

5–5 Some projects are directed by project managers who have only monitoring authority. These individuals are referred to as influence project managers. What kind of projects would be under their control? What organizational structure might be best for this?

5–6 What kind of working relationships would result if the project manager had more reward power than the functional managers?

5–7 What is the correct way for a project manager to invite line managers to attend team meetings?

5–8 How do you handle a project manager or project engineer who continually tries to "bite off more than he can chew"? If he were effective at doing this, at least temporarily, would your answer change?

5–9 Manpower requirements indicate that a specific functional pool will increase sharply from eight to seventeen people over the next two weeks and then drop back to eight people. Should you question this?

5–10 Below are several sources from which legal authority can be derived. State whether each source provides the project manager with sufficient authority from which he can effectively manage the project.

a. The project or organizational charter
b. The project manager's position in the organization
c. The job description and specifications for project managers

d. Policy documents
e. The project manager's "executive" rank
f. Dollar value of the contract
g. Control of funds

5–11 Below are three broad statements describing the functions of management. For each statement, are we referring to upper-level management, project management, or functional management?

a. Acquire the best available assets and try to improve them.
b. Provide a good working environment for all personnel.
c. Make sure that all resources are applied effectively and efficiently such that all constraints are met, if possible.

5–12 Is it possible for a product manager to have the same degree of tunnel vision that a project manager has? If so, under what circumstances?

5–13 Are there situations in which a project manager can wait for long-term changes instead of an immediate response to actions?

5–14 Should a project manager encourage the flow of problems to him? If yes, should he be selective in which ones to resolve?

5–15 If all projects are different, should there exist a uniform company policies and procedures manual?

5–16 Should time robbers be added to direct labor standards for pricing out work?

5–17 Is it possible for a project manager to improve his time management skills by knowing the "energy cycle" of his people? Can this energy cycle be a function of the hour of the day, day of the week, or whether overtime is required?

CASE STUDIES

THE TROPHY PROJECT

The ill-fated Trophy Project was in trouble right from the start. Reichart, who had been an assistant project manager, was involved with the project from its conception. When the Trophy Project was accepted by the company, Reichart was assigned as the project manager. The program schedules started to slip from day one, and expenditures were excessive. Reichart found that the functional managers were charging direct labor time to his project but working on their own "pet" projects. When Reichart complained of this, he was told not to meddle in the functional manager's allocation of resources and budgeted expenditures. After approximately six months, Reichart was requested to make a progress report directly to corporate and division staffs.

Reichart took this opportunity to bare his soul. The report substantiated that the project was forecasted to be one complete year behind schedule. Reichart's staff, as supplied by the line managers, was inadequate to stay at the required pace, let alone make up any time that had already been lost. The estimated cost at completion at this interval showed a cost overrun of at least 20 percent. This was Reichart's first opportunity to tell his story to people who were in a position to correct the situation. The result of Reichart's frank, candid evaluation of the Trophy Project was very predictable. Nonbelievers finally saw the light, and the line managers realized that they had a role to play in the completion of the project. Most of the problems were now out in the open and could be corrected by providing adequate staffing and resources. Corporate staff ordered immediate remedial action and staff support to provide Reichart a chance to bail out his program.

The results were not at all what Reichart had expected. He no longer reported to the project office; he now reported directly to the operations manager. Corporate staff's interest in the project became very intense, requiring a 7:00 A.M. meeting every Monday morning for complete review of the project status and plans for recovery. Reichart found himself spending more time preparing paperwork, reports, and projections for his Monday morning meetings than he did administering the Trophy Project. The main concern of corporate was to get the project back on schedule. Reichart spent many hours preparing the recovery plan and establishing manpower requirements to bring the program back onto the original schedule.

Group staff, in order to closely track the progress of the Trophy Project, assigned an assistant program manager. The assistant program manager determined that a sure cure for the Trophy Project would be to computerize the various problems and track the progress through a very complex computer program. Corporate provided Reichart with twelve additional staff members to work on the computer program. In the meantime, nothing changed. The functional managers still did not provide adequate staff for recovery, assuming that the additional manpower Reichart had received from corporate would accomplish that task.

After approximately $50,000 was spent on the computer program to track the problems, it was found that the program objectives could not be handled by the computer. Reichart discussed this problem with a computer supplier and found that $15,000 more was required for programming and additional storage capacity. It would take two months for installation of the additional storage capacity and the completion of the programming. At this point, the decision was made to abandon the computer program.

Reichart was now a year and a half into the program with no prototype units completed. The program was still nine months behind schedule with the overrun projected at 40 percent of budget. The customer had been receiving his reports on a timely basis and was well aware of the fact that the Trophy Project was behind schedule. Reichart had spent a great deal of time with the customer explaining the problems and the plan for recovery. Another problem that Reichart had to contend with was that the vendors who were supplying components for the project were also running behind schedule.

One Sunday morning, while Reichart was in his office putting together a report for the client, a corporate vice president came into his office. "Reichart," he said, "in any project I look at the top sheet of paper and the man whose name appears at the top of the sheet is the one I hold responsible. For this project your name appears at the top of the sheet. If you cannot bail this thing out, you are in serious trouble in this corporation." Reichart did not know which way to turn or what to say. He had no control over the functional managers who were creating the problems, but he was the person who was being held responsible.

After another three months the customer, becoming impatient, realized that the Trophy Project was in serious trouble and requested that the division general manager and his entire staff visit the customer's plant to give a progress and "get well" report within a week. The division general manager called Reichart into his office and said, "Reichart, go visit our customer. Take three or four functional line people with you and try to placate him with whatever you feel is necessary." Reichart and four functional line people visited the customer and gave a four-and-a-half-hour presentation defining the problems and the progress to that point. The customer was very polite and even commented that it was an excellent presentation, but the content was totally unacceptable. The program was still six to eight months late, and the customer demanded progress reports on a weekly basis. The customer made arrangements to assign a representative in Reichart's department to be "on-site" at the project on a daily basis and to interface with Reichart and his staff as required. After this turn of events, the program became very hectic.

The customer representative demanded constant updates and problem identification and then became involved in attempting to solve these problems. This involvement created many changes in the program and the product in order to eliminate some of the problems. Reichart had trouble with the customer and did not agree with the changes in the program. He expressed his disagreement vocally when, in many cases, the customer felt the changes were at no cost. This caused a deterioration of the relationship between client and producer.

One morning Reichart was called into the division general manager's office and introduced to Mr. "Red" Baron. Reichart was told to turn over the reins of the Trophy Project to Red immediately. "Reichart, you will be temporarily reassigned to some other division within the corporation. I suggest you start looking outside the company for another job." Reichart looked at Red and asked, "Who did this? Who shot me down?"

Red was program manager on the Trophy Project for approximately six months, after which, by mutual agreement, he was replaced by a third project manager. The customer reassigned his local program manager to another project. With the new team the Trophy Project was finally completed one year behind schedule and at a 40 percent cost overrun.

QUESTIONS

1. Why did the project get into trouble?
2. What mistakes did Reichart make?
3. What could Reichart have done differently?
4. Did Reichart have support from a project sponsor?

MCROY AEROSPACE

McRoy Aerospace was a highly profitable company building cargo planes and refueling tankers for the armed forces. It had been doing this for more than fifty years and was highly successful. But because of a downturn in the government's spending on these types of planes, McRoy decided to enter the commercial aviation aircraft business, specifically wide-body planes that would seat up to 400 passengers, and compete head on with Boeing and Airbus Industries.

During the design phase, McRoy found that the majority of the commercial airlines would consider purchasing its plane provided that the costs were lower than the other aircraft manufacturers. While the actual purchase price of the plane was a consideration for the buyers, the greater interest was in the life-cycle cost of maintaining the operational readiness of the aircraft, specifically the maintenance costs.

Operations and support costs were a considerable expense and maintenance requirements were regulated by the government for safety reasons. The airlines make money when the planes are in the air rather than sitting in a maintenance hangar. Each maintenance depot maintained an inventory of spare parts so that, if a part did not function properly, the part could be removed and replaced with a new part. The damaged part would be sent to the manufacturer for repairs or replacement. Inventory costs could be significant but were considered a necessary expense to keep the planes flying.

One of the issues facing McRoy was the mechanisms for the eight doors on the aircraft. Each pair of doors had their own mechanisms, which appeared to be restricted by their location in the plane. If McRoy could come up with a single design mechanism for all four pairs of doors, it would significantly lower the inventory costs for the airlines as well as the necessity to train mechanics on one set of mechanisms rather than four. On the cargo planes and refueling tankers, each pair of doors had a unique mechanism. For commercial aircrafts, finding one design for all doors would be challenging.

Mark Wilson, one of the department managers at McRoy's design center, assigned Jack, the best person he could think of to work on this extremely challenging project. If anyone could accomplish it, it was Jack. If Jack could not do it, Mark sincerely believed it could not be done.

The successful completion of this project would be seen as a value-added opportunity for McRoy's customers and could make a tremendous difference from a cost and efficiency standpoint. McRoy would be seen as an industry leader in life-cycle costing, and this could make the difference in getting buyers to purchase commercial planes from McRoy Aerospace.

The project was to design an opening/closing mechanism that was the same for all of the doors. Until now, each door could have a different set of open/close mechanisms, which made the design, manufacturing, maintenance, and installation processes more complex, cumbersome, and costly.

Without a doubt, Jack was the best—and probably the only—person to make this happen even though the equipment engineers and designers all agreed that it could not be done. Mark put all of his cards on the table when he presented the challenge to Jack. He told him wholeheartedly that his only hope was for Jack to take on this project and explore it from every possible, out-of-the-box angle he could think of. But Jack said right off the bat that this might not be possible. Mark was not happy hearing Jack say this right away, but he knew Jack would do his best.

Jack spent two months looking at the problem and simply could not come up with the solution needed. Jack decided to inform Mark that a solution was not possible. Both Jack and Mark were disappointed that a solution could not be found.

"I know you're the best, Jack," stated Mark. "I can't imagine anyone else even coming close to solving this critical problem. I know you put forth your best effort and the problem was just too much of a challenge. Thanks for trying. But if I had to choose one of your coworkers to take another look at this project, who might have even half a chance of making it happen? Who would you suggest? I just want to make sure that we have left no stone unturned," he said rather glumly.

Mark's words caught Jack by surprise. Jack thought for a moment and you could practically see the wheels turning in his mind. Was Jack thinking about who could take this project on and waste more time trying to find a solution? No, Jack's wheels were turning on the subject of the challenging problem itself. A glimmer of an idea whisked through his brain and he said, "Can you give me a few days to think about some things, Mark?" he asked pensively.

Mark had to keep the little glimmer of a smile from erupting full force on his face. "Sure, Jack," he said. "Like I said before, if anyone can do it, it's you. Take all the time you need." A few weeks later, the problem was solved and Jack's reputation rose to even higher heights than before.

QUESTIONS

1. Was Mark correct in what he said to get Jack to continue investigating the problem?
2. Should Mark just have given up on the idea rather than what he said to Jack?
3. Should Mark have assigned this to someone else rather than giving Jack a second chance, and if so, how might Jack have reacted?

4. What should Mark have done if Jack still was not able to resolve the problem?
5. Would it have made sense for Mark to assign this problem to someone else, if Jack could not solve the problem the second time around?
6. What other options, if any, were now available to Mark?

THE POOR WORKER

Paula, the project manager, was reasonably happy the way that work was progressing on the project. The only issue was the work being done by Frank. Paula knew from the start of the project that Frank was a mediocre employee and often regarded as a troublemaker. The tasks that Frank was expected to perform were not overly complex and the line manager assured Paula during the staffing function that Frank could do the job. The line manager also informed Paula that Frank demonstrated behavioral issues on other projects and sometimes had to be removed from the project. Frank was a chronic complainer and found fault with everything and everybody. But the line manager also assured Paula that Frank's attitude was changing and that the line manager would get actively involved if any of these issues began to surface on Paula's project. Reluctantly, Paula agreed to allow Frank to be assigned to her project.

Unfortunately, Frank's work on the project was not being performed according to Paula's standards. Paula had told Frank on more than one occasion what she expected from him, but Frank persisted in doing his own thing. Paula was now convinced that the situation was getting worse. Frank's work packages were coming in late and sometimes over budget. Frank continuously criticized Paula's performance as a project manager and Frank's attitude was beginning to affect the performance of some of the other team members. Frank was lowering the morale of the team. It was obvious that Paula had to take some action.

QUESTIONS

1. What options are available to Paula?
2. If Paula decides to try to handle the situation first by herself rather than approach the line manager, what should Paula do and in what order?
3. If all of Paula's attempts fail to change the worker's attitude and the line refuses to remove the worker, what options are available to Paula?
4. What rights, if any, does Paula have with regard to wage and salary administration regarding this employee?

THE PRIMA DONNA

Ben was placed in charge of a one-year project. Several of the work packages had to be accomplished by the Mechanical Engineering Department and required three people to be assigned full time for the duration of the project. When the project was originally proposed, the Mechanical Engineering Department manager estimated that he would assign three of his grade

7 employees to do the job. Unfortunately, the start date of the project was delayed by three months and the department manager was forced to assign the resources he planned to use on another project. The resources that would be available for Ben's project at the new starting date were two grade 6s and a grade 9.

The department manager assured Ben that these three employees could adequately perform the required work and that Ben would have these three employees full-time for the duration of the project. Furthermore, if any problems occurred, the department manager made it clear to Ben that he personally would get involved to make sure that the work packages and deliverables were completed correctly.

Ben did not know any of the three employees personally. But since a grade 9 was considered a senior subject matter expert pay grade, Ben made the grade 9 the lead engineer representing his department on Ben's project. It was common practice for the senior-most person assigned from each department to act as the lead and even as an assistant project manager. The lead was often allowed to interface with the customers at information exchange meetings.

By the end of the first month of the project, work was progressing as planned. Although most of the team seemed happy to be assigned to the project and team morale was high, the two grade 6 team members in the Mechanical Engineering Department were disenchanted with the project. Ben interviewed the two grade 6 employees to see why they were somewhat unhappy. One of the two employees stated:

> The grade 9 wants to do everything himself. He simply does not trust us. Every time we use certain equations to come up with a solution, he must review everything we did in microscopic detail. He has to approve everything. The only time he does not micromanage us is when we have to make copies of reports. We do not feel that we are part of the team.

Ben was unsure how to handle the situation. Resources are assigned by the department managers and usually cannot be removed from a project without the permission of the department managers. Ben met with the mechanical engineering department manager, who stated:

> The grade 9 that I assigned is probably the best worker in my department. Unfortunately, he's a prima donna. He trusts nobody else's numbers or equations other than his own. Whenever co-workers perform work, he feels obligated to review everything that they have done. Whenever possible, I try to assign him to one-person activities so that he will not have to interface with anyone. But I have no other one-person assignments right now, which is why I assigned him to your project. I was hoping he would change his ways and work as a real team member with the two grade 6 workers, but I guess not. Don't worry about it. The work will get done, and get done right. We'll just have to allow the two grade 6 employees to be unhappy for a little while.

Ben understood what the department manager said but was not happy about the situation. Forcing the grade 9 to be removed could result in the assignment of someone with lesser capabilities, and this could impact the quality of the deliverables from the Mechanical Engineering Department. Leaving the grade 9 in place for the duration of the project would alienate the two grade 6 employees and their frustration and morale issues could infect other team members.

QUESTIONS

1. What options are available to Ben?
2. Is there a risk in leaving the situation as is?
3. Is there a risk in removing the grade 9?

THE RELUCTANT WORKERS

Tim Aston had changed employers three months ago. His new position was project manager. At first, he had had stars in his eyes about becoming the best project manager that his company had ever seen. Now, he wasn't sure if project management was worth the effort. He made an appointment to see Phil Davies, director of project management.

Tim Aston: "Phil, I'm a little unhappy about the way things are going. I just can't seem to motivate my people. Every day, at 4:30 P.M., all of my people clean off their desks and go home. I've had people walk out of late afternoon team meetings because they were afraid that they'd miss their car pool. I have to schedule morning team meetings."

Phil Davies: "Look, Tim. You're going to have to realize that in a project environment, people think that they come first and that the project is second. This is a way of life in our organizational form."

Tim Aston: "I've continually asked my people to come to me if they have problems. I find that the people do not think that they need help and, therefore, do not want it. I just can't get my people to communicate more."

Phil Davies: "The average age of our employees is about forty-six. Most of our people have been here for twenty years. They're set in their ways. You're the first person that we've hired in the past three years. Some of our people may just resent seeing a thirty-year-old project manager."

Tim Aston: "I found one guy in the accounting department who has an excellent head on his shoulders. He's very interested in project management. I asked his boss if he'd release him for a position in project management, and his boss just laughed at me, saying something to the effect that as long as that guy is doing a good job for him, he'll never be released for an assignment elsewhere in the company. His boss seems more worried about his personal empire than he does in what's best for the company.

"We had a test scheduled for last week. The customer's top management was planning on flying in for firsthand observations. Two of my people said that they had programmed vacation days coming, and that they would not change, under any conditions. One guy was going fishing and the other guy was planning to spend a few days working with fatherless children in our community. Surely, these guys could change their plans for the test."

Phil Davies: "Many of our people have social responsibilities and outside interests. We encourage social responsibilities and only hope that the outside interests do not interfere with their jobs.

"There's one thing you should understand about our people. With an average age of forty-six, many of our people are at the top of their pay grades and have no place to go. They must look elsewhere for interests. These are the people you have to work with and motivate. Perhaps you should do some reading on human behavior."

LEADERSHIP EFFECTIVENESS (A)

Instructions

This tabulation form on page 188 is concerned with a comparison of personal supervisory styles. Indicate your preference to the two alternatives after each item by writing appropriate figures in the blanks. Some of the alternatives may seem equally attractive or unattractive to you. Nevertheless, please attempt to choose the alternative that is relatively more characteristic of you. For each question given, you have three (3) points that you may distribute in any of the following combinations:

A. If you agree with alternative (a) and disagree with (b), write 3 in the top blank and 0 in bottom blank.
  a. 3
  b. 0

B. If you agree with (b) and disagree with (a), write:
  a. 0
  b. 3

C. If you have a slight preference for (a) over (b), write:
  a. 2
  b. 1

D. If you have a slight preference for (b) over (a), write:
  a. 1
  b. 2

Important: Use only the combinations shown above. Try to relate each item to your own personal experience. Please make a choice from every pair of alternatives.

1. On the job, a project manager should make a decision and
   a. ______ tell his team to carry it out.
   b. ______ "tell" his team about the decision and then try to "sell" it.

2. After a project manager has arrived at a decision
   a. ______ he should try to reduce the team's resistance to his decision by indicating what they have to gain.
   b. ______ he should provide an opportunity for his team to get a fuller explanation of his ideas.

3. When a project manager presents a problem to his subordinates
   a. ______ he should get suggestions from them and then make a decision.
   b. ______ he should define it and request that the group make a decision.

4. A project manager
   a. ______ is paid to make all the decisions affecting the work of his team.
   b. ______ should commit himself in advance to assist in implementing whatever decision his team selects when they are asked to solve a problem.

5. A project manager should
   a. ______ permit his team an opportunity to exert some influence on decisions but reserve final decisions for himself.
   b. ______ participate with his team in group decision making but attempt to do so with a minimum of authority.

6. In making a decision concerning the work situation, a project manager should
   a. ______ present his decision and ideas and engage in a "give-and-take" session with his team to allow them to fully explore the implications of the decision.
   b. ______ present the problem to his team, get suggestions, and then make a decision.

7. A good work situation is one in which the project manager
   a. ______ "tells" his team about a decision and then tries to "sell" it to them.
   b. ______ calls his team together, presents a problem, defines the problem, and requests they solve the problem with the understanding that he will support their decision(s).

8. A well-run project will include
   a. ______ efforts by the project manager to reduce the team's resistance to his decisions by indicating what they have to gain from them.
   b. ______ "give-and-take" sessions to enable the project manager and team to explore more fully the implications of the project manager's decisions.

9. A good way to deal with people in a work situation is
   a. ______ to present problems to your team as they arise, get suggestions, and then make a decision.
   b. ______ to permit the team to make decisions, with the understanding that the project manager will assist in implementing whatever decision they make.

10. A good project manager is one who takes
    a. ______ the responsibility for locating problems and arriving at solutions, then tries to persuade his team to accept them.
    b. ______ the opportunity to collect ideas from his team about problems, then he makes his decision.

11. A project manager
    a. ______ should make the decisions in his organization and tell his team to carry them out.
    b. ______ should work closely with his team in solving problems, and attempt to do so with a minimum of authority.

12. To do a good job, a project manager should
    a. ______ present solutions for his team's reaction.
    b. ______ present the problem and collect from the team suggested solutions, then make a decision based on the best solution offered.

13. A good method for a project manager is
    a. ______ to "tell" and then try to "sell" his decision.
    b. ______ to define the problem for his team, then pass them the right to make decisions.

14. On the job, a project manager
    a. ______ need not give consideration to what his team will think or feel about his decisions.
    b. ______ should present his decisions and engage in a "give-and-take" session to enable everyone concerned to explore, more fully, the implications of the decisions.

15. A project manager
    a. ______ should make all decisions himself.
    b. ______ should present the problem to his team, get suggestions, and then make a decision.

16. It is good
    a. ______ to permit the team an opportunity to exert some influence on decisions, but the project manager should reserve final decisions for himself.
    b. ______ for the project manager to participate with his team in group decision making with as little authority as possible.

17. The project manager who gets the most from his team is the one who
    a. ______ exercises direct authority.
    b. ______ seeks possible solutions from them and then makes a decision.

18. An effective project manager should
    a. ______ make the decisions on his project and tell his team to carry them out.
    b. ______ make the decisions and then try to persuade his team to accept them.

19. A good way for a project manager to handle work problems is to
    a. ______ implement decisions without giving any consideration to what his team will think or feel.
    b. ______ permit the team an opportunity to exert some influence on decisions but reserve the final decision for himself.

20. Project managers
    a. ______ should seek to reduce the team's resistance to their decisions by indicating what they have to gain from them.
    b. ______ should seek possible solutions from their team when problems arise and then make a decision from the list of alternatives.

LEADERSHIP QUESTIONNAIRE
Tabulation Form

	1	2	3	4	5
1	_____ a	_____ b			
2		_____ a	_____ b		
3				_____ a	_____ b
4	_____ a				_____ b
5			_____ a		_____ b
6			_____ a	_____ b	
7		_____ a			_____ b
8		_____ a	_____ b		
9				_____ a	_____ b
10	_____ a		_____ b		
11	_____ a				_____ b
12			_____ a	_____ b	
13		_____ a			_____ b
14	_____ a		_____ b		
15	_____ a			_____ b	
16			_____ a		_____ b
17	_____ a			_____ b	
18	_____ a	_____ b			
19	_____ a		_____ b		
20		_____ a		_____ b	
TOTAL	_____	_____	_____	_____	_____

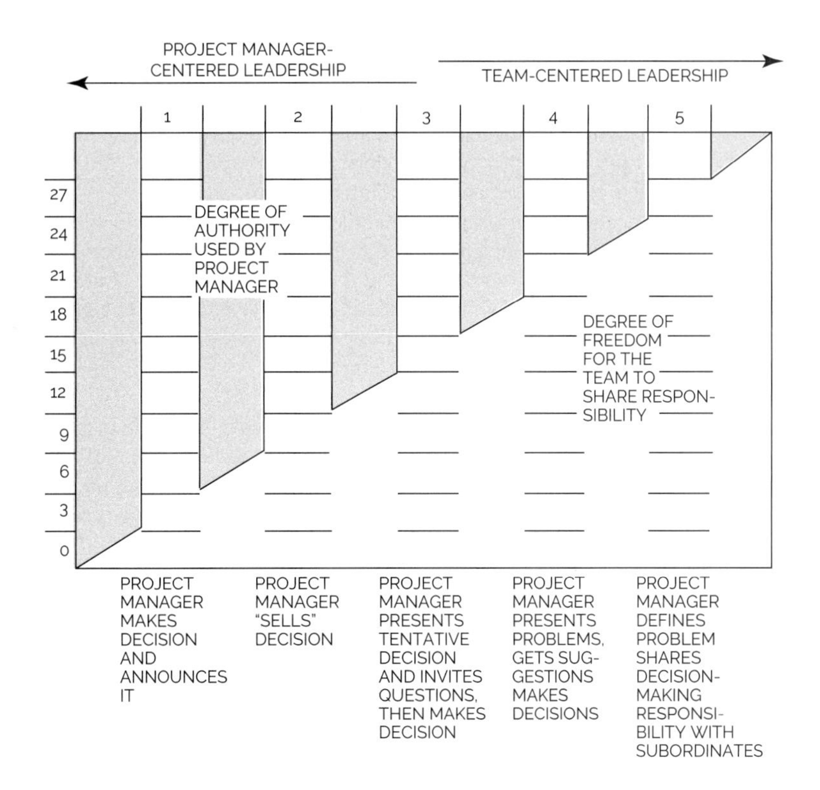

LEADERSHIP EFFECTIVENESS (B)

The Project

> **PMBOK® Guide, 5th Edition**
> Chapter 9 Human Resources Management
> Chapter 10 Communications Management

Your company has just won a contract for an outside customer. The contract is for one year, broken down as follows: R&D, six months; prototype testing, one month; manufacturing, five months. In addition to the risks involved in the R&D stage, both your management and the customer have stated that there will be absolutely no trade-offs on time, cost, or performance.

When you prepared the proposal six months ago, you planned and budgeted for a full-time staff of five people, in addition to the functional support personnel. Unfortunately, due to limited resources, your staff (i.e., the project office) will be as follows:

Tom: An excellent engineer, somewhat of a prima donna, but has worked very well with you on previous projects. You specifically requested Tom and were fortunate to have him assigned, although your project is not regarded as a high priority. Tom is recognized as both a technical leader and expert, and is considered perhaps the best engineer in the company. Tom will be full-time for the duration of the project.

Bob: Started with the company a little over a year ago, and may be a little "wet behind the ears." His line manager has great expectations for him in the future but, for the time being, wants you to give him on-the-job-training as a project office team member. Bob will be full-time on your project.

Carol: She has been with the company for twenty years and does an acceptable job. She has never worked on your projects before. She is full-time on the project.

George: He has been with the company for six years, but has never worked on any of your projects. His superior tells you that he will be only half-time on your project until he finishes a crash job on another project. He should be available for full-time work in a month or two. George is regarded as an outstanding employee.

Management informs you that there is nobody else available to fill the fifth position. You'll have to spread the increased workload over the other members. Obviously, the customer may not be too happy about this.

In each situation that follows, circle the best answer. The grading system will be provided later. Remember: These staff individuals are "dotted" to you and "solid" to their line manager, although they are in your project office.

Situation 1: The project office team members have been told to report to you this morning. They have all received your memo concerning the time and place of the kickoff meeting. However, they have not been provided any specific details concerning the project except that the project will be at least one year in duration. For your company, this is regarded as a long-term project. A good strategy for the meeting would be:

A. The team must already be self-motivated or else they would not have been assigned. Simply welcome them and assign homework.
B. Motivate the employees by showing them how they will benefit: esteem, pride, and self-actualization. Minimize discussion on specifics.
C. Explain the project and ask them for their input. Try to get them to identify alternatives and encourage group decision making.
D. Identify the technical details of the project: the requirements, performance standards, and expectations.

Situation 2: You give the team members a copy of the winning proposal and a "confidential" memo describing the assumptions and constraints you considered in developing the proposal. You tell your team to review the material and be prepared to perform detailed planning at the meeting you have scheduled for the following Monday. During Monday's planning meeting, you find that Tom (who has worked with you before) has established a take-charge role and has done some of the planning that should have been the responsibility of other team members. You should:

A. Do nothing. This may be a beneficial situation. However, you may wish to ask if the other project office members wish to review Tom's planning.
B. Ask each team member individually how he or she feels about Tom's role. If they complain, have a talk with Tom.
C. Ask each team member to develop his or her own schedules and then compare results.
D. Talk to Tom privately about the long-term effects of his behavior.

Situation 3: Your team appears to be having trouble laying out realistic schedules that will satisfy the customer's milestones. They keep asking you pertinent questions and seem to be making the right decisions, but with difficulty.

A. Do nothing. If the team is good, they will eventually work out the problem.
B. Encourage the team to continue but give some ideas as to possible alternatives. Let them solve the problem.

C. Become actively involved and help the team solve the problem. Supervise the planning until completion.
D. Take charge yourself and solve the problem for the team. You may have to provide continuous direction.

Situation 4: Your team has taken an optimistic approach to the schedule. The functional managers have reviewed the schedules and have sent your team strong memos stating that there is no way that they can support your schedules. Your team's morale appears to be very low. Your team expected the schedules to be returned for additional iterations and trade-offs, but not with such harsh words from the line managers. You should:

A. Take no action. This is common to these types of projects and the team must learn to cope.
B. Call a special team meeting to discuss the morale problem and ask the team for recommendations. Try to work out the problem.
C. Meet with each team member individually to reinforce his or her behavior and performance. Let members know how many other times this has occurred and been resolved through trade-offs and additional iterations. State your availability to provide advice and support.
D. Take charge and look for ways to improve morale by changing the schedules.

Situation 5: The functional departments have begun working, but are still criticizing the schedules. Your team is extremely unhappy with some of the employees assigned out of one functional department. Your team feels that these employees are not qualified to perform the required work. You should:

A. Do nothing until you are absolutely sure (with evidence) that the assigned personnel cannot perform as needed.
B. Sympathize with your team and encourage them to live with this situation until an alternative is found.
C. Assess the potential risks with the team and ask for their input and suggestions. Try to develop contingency plans if the problem is as serious as the team indicates.
D. Approach the functional manager and express your concern. Ask to have different employees assigned.

Situation 6: Bob's performance as a project office team member has begun to deteriorate. You are not sure whether he simply lacks the skills, cannot endure the pressure, or cannot assume part of the additional work that resulted from the fifth position in the project being vacant. You should:

A. Do nothing. The problem may be temporary and you cannot be sure that there is a measurable impact on the project.
B. Have a personal discussion with Bob, seek out the cause, and ask him for a solution.
C. Call a team meeting and discuss how productivity and performance are decreasing. Ask the team for recommendations and hope Bob gets the message.
D. Interview the other team members and see if they can explain Bob's actions lately. Ask the other members to assist you by talking to Bob.

Situation 7: George, who is half-time on your project, has just submitted for your approval his quarterly progress report for your project. After your signature has been attained, the report is sent to senior management and the customer. The report is marginally acceptable and not at all what you would have expected from George. George apologizes to you for the report and blames it on his other project, which is in its last two weeks. You should:

A. Sympathize with George and ask him to rewrite the report.
B. Tell George that the report is totally unacceptable and will reflect on his ability as a project office team member.
C. Ask the team to assist George in redoing the report since a bad report reflects on everyone.
D. Ask one of the other team members to rewrite the report for George.

Situation 8: You have completed the R&D stage of your project and are entering phase II: prototype testing. You are entering month seven of the twelve-month project. Unfortunately, the results of phase I R&D indicate that you were too optimistic in your estimating for phase II and a schedule slippage of at least two weeks is highly probable. The customer may not be happy. You should:

A. Do nothing. These problems occur and have a way of working themselves out. The end date of the project can still be met.
B. Call a team meeting to discuss the morale problem resulting from the slippage. If morale is improved, the slippage may be overcome.
C. Call a team meeting and seek ways of improving productivity for phase II. Hopefully, the team will come up with alternatives.
D. This is a crisis and you must exert strong leadership. You should take control and assist your team in identifying alternatives.

Situation 9: Your rescheduling efforts have been successful. The functional managers have given you adequate support and you are back on schedule. You should:

A. Do nothing. Your team has matured and is doing what they are paid to do.
B. Try to provide some sort of monetary or nonmonetary reward for your team (e.g., management-granted time off or a dinner team meeting).
C. Provide positive feedback/reinforcement for the team and search for ideas for shortening phase III.
D. Obviously, your strong leadership has been effective. Continue this role for the phase III schedule.

Situation 10: You are now at the end of the seventh month and everything is proceeding as planned. Motivation appears high. You should:

A. Leave well enough alone.
B. Look for better ways to improve the functioning of the team. Talk to them and make them feel important.
C. Call a team meeting and review the remaining schedules for the project. Look for contingency plans.
D. Make sure the team is still focusing on the goals and objectives of the project.

Situation 11: The customer unofficially informs you that his company has a problem and may have to change the design specifications before production actually begins. This would be a catastrophe for your project. The customer wants a meeting at your plant within the next seven days. This will be the customer's first visit to your plant. All previous meetings were informal and at the customer's facilities, with just you and the customer. This meeting will be formal. To prepare for the meeting, you should:

A. Make sure the schedules are updated and assume a passive role since the customer has not officially informed you of his problem.
B. Ask the team to improve productivity before the customer's meeting. This should please the customer.
C. Call an immediate team meeting and ask the team to prepare an agenda and identify the items to be discussed.
D. Assign specific responsibilities to each team member for preparation of handout material for the meeting.

Situation 12: Your team is obviously not happy with the results of the customer interface meeting because the customer has asked for a change in design specifications. The manufacturing plans and manufacturing schedules must be developed anew. You should:

A. Do nothing. The team is already highly motivated and will take charge as before.
B. Reemphasize the team spirit and encourage your people to proceed. Tell them that nothing is impossible for a good team.
C. Roll up your shirt sleeves and help the team identify alternatives. Some degree of guidance is necessary.
D. Provide strong leadership and close supervision. Your team will have to rely on you for assistance.

Situation 13: You are now in the ninth month. While your replanning is going on (as a result of changes in the specifications), the customer calls and asks for an assessment of the risks in canceling this project right away and starting another one. You should:

A. Wait for a formal request. Perhaps you can delay long enough for the project to finish.
B. Tell the team that their excellent performance may result in a follow-on contract.
C. Call a team meeting to assess the risks and look for alternatives.
D. Accept strong leadership for this and with *minimum,* if any, team involvement.

Situation 14: One of the functional managers has asked for your evaluation of all of his functional employees currently working on your project (excluding project office personnel). Your project office personnel appear to be working more closely with the functional employees than you are. You should:

A. Return the request to the functional manager since this is not part of your job description.
B. Talk to each team member individually, telling them how important their input is, and ask for their evaluations.
C. As a team, evaluate each of the functional team members, and try to come to some sort of agreement.
D. Do not burden your team with this request. You can do it yourself.

Situation 15: You are in the tenth month of the project. Carol informs you that she has the opportunity to be the project leader for an effort starting in two weeks. She has been with the company for twenty years and this is her first opportunity as a project leader. She wants to know if she can be released from your project. You should:

A. Let Carol go. You do not want to stand in the way of her career advancement.
B. Ask the team to meet in private and conduct a vote. Tell Carol you will abide by the team vote.
C. Discuss the problem with the team since they must assume the extra workload, if necessary. Ask for their input into meeting the constraints.
D. Counsel her and explain how important it is for her to remain. You are already shorthanded.

Situation 16: Your team informs you that one of the functional manufacturing managers has built up a brick wall around his department and all information requests must flow through him. The brick wall has been in existence for two years. Your team members are having trouble with status reporting, but always get the information after catering to the functional manager. You should:

A. Do nothing. This is obviously the way the line manager wants to run his department. Your team is getting the information they need.
B. Ask the team members to use their behavioral skills in obtaining the information.
C. Call a team meeting to discuss alternative ways of obtaining the information.
D. Assume strong leadership and exert your authority by calling the line manager and asking for the information.

Situation 17: The executives have given you a new person to replace Carol for the last two months of the project. Neither you nor your team have worked with this man before. You should:

A. Do nothing. Carol obviously filled him in on what he should be doing and what is involved in the project.
B. Counsel the new man individually, bring him up to speed, and assign him Carol's work.
C. Call a meeting and ask each member to explain his or her role on the project to the new man.
D. Ask each team member to talk to this man as soon as possible and help him come on board. Request that individual conversations be used.

Situation 18: One of your team members wants to take a late-afternoon course at the local college. Unfortunately, this course may conflict with his workload. You should:

A. Postpone your decision. Ask the employee to wait until the course is offered again.
B. Review the request with the team member and discuss the impact on his performance.
C. Discuss the request with the team and ask for the team's approval. The team may have to cover for this employee's workload.
D. Discuss this individually with each team member to make sure that the task requirements will still be adhered to.

Situation 19: Your functional employees have used the wrong materials in making a production run test. The cost to your project was significant, but absorbed in a small "cushion" that you saved for emergencies such as this. Your team members tell you that the test will be rerun without any slippage of the schedule. You should:

A. Do nothing. Your team seems to have the situation well under control.
B. Interview the employees that created this problem and stress the importance of productivity and following instructions.
C. Ask your team to develop contingency plans for this situation should it happen again.
D. Assume a strong leadership role for the rerun test to let people know your concern.

Situation 20: All good projects must come to an end, usually with a final report. Your project has a requirement for a final report. This final report may very well become the basis for follow-on work. You should:

A. Do nothing. Your team has things under control and knows that a final report is needed.
B. Tell your team that they have done a wonderful job and there is only one more task to do.
C. Ask your team to meet and provide an outline for the final report.
D. You must provide some degree of leadership for the final report, at least the structure. The final report could easily reflect on your ability as a manager.

Fill in the table below. The answers appear in Appendix B.

Situation	Answer	Points		Situation	Answer	Points
1				11		
2				12		
3				13		
4				14		
5				15		
6				16		
7				17		
8				18		
9				19		
10				20		
					Total	

MOTIVATIONAL QUESTIONNAIRE

On the next several pages, you will find forty statements concerning what motivates you and how you try to motivate others. Beside each statement, circle the number that corresponds to your opinion. In the example below, the choice is "Slightly Agree."

–3	Strongly Disagree
–2	Disagree
–1	Slightly Disagree
0	No Opinion
(+1)	Slightly Agree
+2	Agree
+3	Strongly Agree

Part 1

The following twenty statements involve *what motivates you.* Please rate each of the statements as honestly as possible. Circle the rating that you think is correct, *not* the one you think the instructor is looking for:

1. My company pays me a reasonable salary for the work that I do.	– 3	– 2	– 1	0	+ 1	+ 2	+ 3
2. My company believes that every job that I do can be considered as a challenge.	– 3	– 2	– 1	0	+ 1	+ 2	+ 3
3. The company provides me with the latest equipment (hardware, software, etc.) so I can do my job effectively.	– 3	– 2	– 1	0	+ 1	+ 2	+ 3
4. My company provides me with recognition for work well done.	– 3	– 2	– 1	0	+ 1	+ 2	+ 3
5. Seniority on the job, job security, and vested rights are provided by the company.	– 3	– 2	– 1	0	+ 1	+ 2	+ 3
6. Executives provide managers with feedback of strategic or long-range information that may affect the manager's job.	– 3	– 2	– 1	0	+ 1	+ 2	+ 3
7. My company provides off-hour clubs and organizations so that employees can socialize, as well as sponsoring social events.	– 3	– 2	– 1	0	+ 1	+ 2	+ 3
8. Employees are allowed either to set their own work/performance standards or at least to approve/review standards set for them by management.	– 3	– 2	– 1	0	+ 1	+ 2	+ 3

9. Employees are encouraged to maintain membership in professional societies and/or attend seminars and symposiums on work related subjects.	–3	–2	–1	0	+1	+2	+3
10. The company often reminds me that the only way to have job security is to compete effectively in the marketplace.							
11. Employees who develop a reputation for "excellence" are allowed to further enhance their reputation, if job related.	–3	–2	–1	0	+1	+2	+3
12. Supervisors encourage a friendly, cooperative working environment for employees.	–3	–2	–1	0	+1	+2	+3
13. My company provides me with a detailed job description, identifying my role and responsibilities.	–3	–2	–1	0	+1	+2	+3
14. My company gives automatic wage and salary increases for the employees.	–3	–2	–1	0	+1	+2	+3
15. My company gives me the opportunity to do what I do best.	–3	–2	–1	0	+1	+2	+3
16. My job gives me the opportunity to be truly creative, to the point where I can solve complex problems.	–3	–2	–1	0	+1	+2	+3
17. My efficiency and effectiveness is improving because the company provided me with better physical working conditions (e.g., lighting, low noise, temperature, restrooms)	–3	–2	–1	0	+1	+2	+3
18. My job gives me constant self-development.	–3	–2	–1	0	+1	+2	+3
19. Our supervisors have feelings for employees rather than simply treating them as "inanimate tools."	–3	–2	–1	0	+1	+2	+3
20. Participation in the company's stock option/retirement plan is available to employees.	–3	–2	–1	0	+1	+2	+3

Part 2

Statements 21–40 involve how project managers motivate team members. Again, it is important that your ratings honestly reflect the way you think that *you,* as project manager, try to motivate employees. Do *not* indicate the way others or the instructor might recommend motivating the employees. Your thoughts are what are important in this exercise.

21. Project managers should encourage employees to take advantage of company benefits such as stock option plans and retirement plans.	–3	–2	–1	0	+1	+2	+3
22. Project managers should make sure that team members have a good work environment (e.g., heat, lighting, low noise, restrooms, cafeteria).	–3	–2	–1	0	+1	+2	+3
23. Project managers should assign team members work that can enhance each team member's reputation.	–3	–2	–1	0	+1	+2	+3
24. Project managers should create a relaxed, cooperative environment for the team members.	–3	–2	–1	0	+1	+2	+3
25. Project managers should continually remind the team that job security is a function of competitiveness, staying within constraints, and good customer relations.	–3	–2	–1	0	+1	+2	+3
26. Project managers should try to convince team members that each new assignment is a challenge.	–3	–2	–1	0	+1	+2	+3
27. Project managers should be willing to reschedule activities, if possible, around the team's company and out-of-company social functions.	–3	–2	–1	0	+1	+2	+3
28. Project managers should continually remind employees of how they will benefit, monetarily, by successful performance on your project.	–3	–2	–1	0	+1	+2	+3

29. Project managers should be willing to "pat people on the back" and provide recognition where applicable.	–3	–2	–1	0	+1	+2	+3
30. Project managers should encourage the team to maintain constant self-development with each assignment.	–3	–2	–1	0	+1	+2	+3
31. Project managers should allow team members to set their own standards, where applicable.	–3	–2	–1	0	+1	+2	+3
32. Project managers should assign work to functional employees according to seniority on the job.	–3	–2	–1	0	+1	+2	+3
33. Project managers should allow team members to use the informal, as well as formal, organization to get work accomplished.	–3	–2	–1	0	+1	+2	+3
34. As a project manager, I would like to control the salaries of the full-time employees on my project.	–3	–2	–1	0	+1	+2	+3
35. Project managers should share information with the team. This includes project information that may not be directly applicable to the team member's assignment.	–3	–2	–1	0	+1	+2	+3
36. Project managers should encourage team members to be creative and to solve their own problems.	–3	–2	–1	0	+1	+2	+3
37. Project managers should provide detailed job descriptions for team members, outlining the team member's role and responsibility.	–3	–2	–1	0	+1	+2	+3
38. Project managers should give each team member the opportunity to do what the team member can do best.	–3	–2	–1	0	+1	+2	+3
39. Project managers should be willing to interact informally with the team members and get to know them, as long as there exists sufficient time on the project.	–3	–2	–1	0	+1	+2	+3
40. Most of the employees on my project earn a salary commensurate with their abilities.	–3	–2	–1	0	+1	+2	+3

Part 1 Scoring Sheet (What Motivates You?)

Place your answers (the numerical values you circled) to questions 1–20 in the corresponding spaces in the chart below.

Basic Needs	*Safety Needs*	*Belonging Needs*
#1 _____	#5 _____	#7 _____
#3 _____	#10 _____	#9 _____
#14 _____	#13 _____	#12 _____
#17 _____	#20 _____	#19 _____
Total _____	Total _____	Total _____

Esteem/Ego Needs	*Self-Actualization Needs*
#4 _____	#2 _____
#6 _____	#15 _____
#8 _____	#16 _____
#11 _____	#18 _____
Total _____	Total _____

Transfer your total score in each category to the table on page 201 by placing an "X" in the appropriate area for motivational needs.

Part 2 Scoring Sheet (How Do You Motivate?)

Place your answers (the numerical values you circled) to questions 21–40 in the corresponding spaces in the chart below.

Basic Needs	*Safety Needs*	*Belonging Needs*
#22 _____	#21 _____	#24 _____
#28 _____	#25 _____	#27 _____
#34 _____	#32 _____	#33 _____
#40 _____	#37 _____	#39 _____
Total _____	Total _____	Total _____

Esteem/Ego Needs	*Self-Actualization Needs*
#23 _____	#26 _____
#29 _____	#30 _____
#31 _____	#36 _____
#35 _____	#38 _____
Total _____	Total _____

Transfer your total score in each category to the table on page 201 by placing an "X" in the appropriate area for motivational needs.

QUESTIONS 1–20

Needs	–12	–11	–10	–9	–8	–7	–6	–5	–4	–3	–2	–1	0	+1	+2	+3	+4	+5	+6	+7	+8	+9	+10	+11	+12
Self-Actualization																									
Esteem/Ego																									
Belonging																									
Safety																									
Basic																									

QUESTIONS 21–40

Needs	–12	–11	–10	–9	–8	–7	–6	–5	–4	–3	–2	–1	0	+1	+2	+3	+4	+5	+6	+7	+8	+9	+10	+11
Self-Actualization																								
Esteem/Ego																								
Belonging																								
Safety																								
Basic																								

Communications Management

6.0 INTRODUCTION

PMBOK® Guide,
6th Edition
Chapter 10 Project Communications Management

Effective project communications ensure that we get the right information to the right person at the right time and in a cost-effective manner. Proper communication is vital to the success of a project. Typical definitions of effective communication include:

- An exchange of information
- An act or instance of transmitting information
- A verbal or written message
- A technique for expressing ideas effectively
- A process by which meanings are exchanged between individuals through a common system of symbols

In the previous chapter, we could argue that most of the leadership and management pitfalls are directly or indirectly attributed to a communications failure. Some people argue that the most important skill a project manager can possess is the ability to communicate effectively to everyone.

6.1 MODELING THE COMMUNICATIONS ENVIRONMENT

The communications environment can be regarded as a network of channels. Most channels are two-way channels. The number of two-way channels, N, can be calculated from the formula

$$N = \frac{X(X-1)}{2}$$

PMBOK is a registered mark of the Project Management Institute, Inc.

In this formula, X represents the number of people communicating with each other. For example, if four people are communicating (i.e., $X = 4$), then there are six two-way channels.

Sometimes it does not matter whether we have a few or many channels of communication. Breakdowns can occur. When a breakdown in communications occurs, disaster may follow, as Figure 6–1 demonstrates.

Figures 6–2 and 6–3 show typical communications patterns. Some people consider Figure 6–2 "politically incorrect" because project managers should not be identified as talking "down" to people. Most project managers communicate laterally, whereas line managers communicate vertically downward to subordinates. Sometimes it is politically incorrect to use the word "subordinates" because these people may be a higher pay grade than the project manager.

Figure 6–3 shows typical communications patterns between customers and contractors. The informal channels of communication, especially the employee to employee relation, can be troublesome if these people agree to scope modifications that are not called out contractually.

Figure 6–4 shows the complete communication model. The screens or barriers are from one's perception, personality, attitudes, emotions, and prejudices.

FIGURE 6–1. A breakdown in communications. (Source unknown)

PMBOK® *Guide*, 6th Edition
10.1 Plan Communication Management

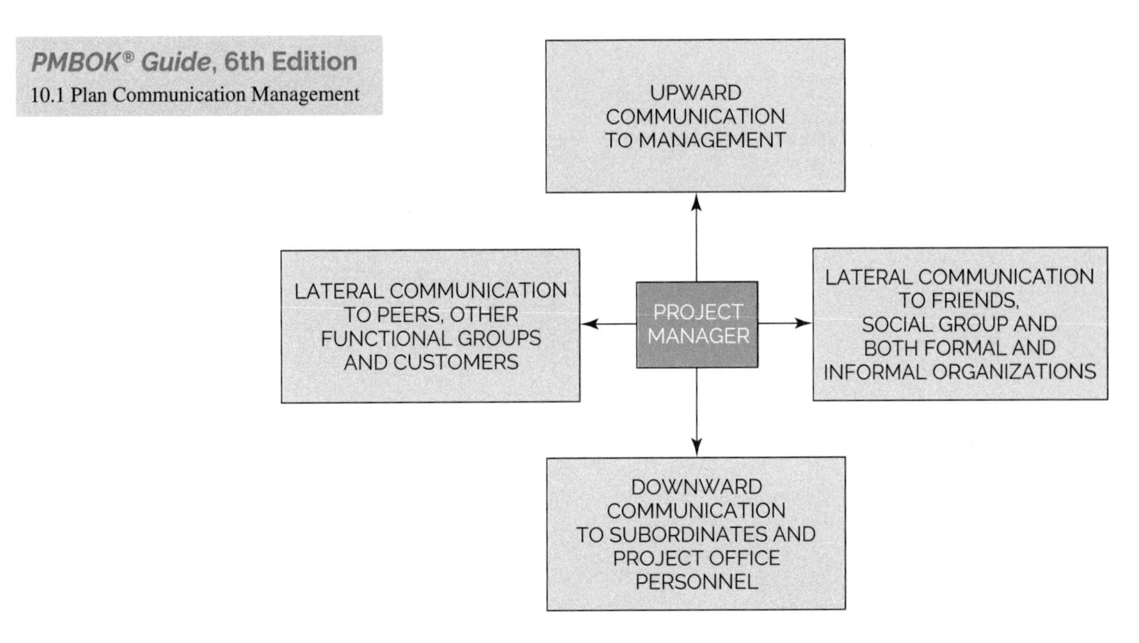

FIGURE 6–2. Communication channels. Source: D. I. Cleland and H. Kerzner, *Engineering Team Management* (Melbourne, Florida: Krieger, 1986), p. 39.

PMBOK® *Guide*, 6th Edition
10.1 Plan Communication Management

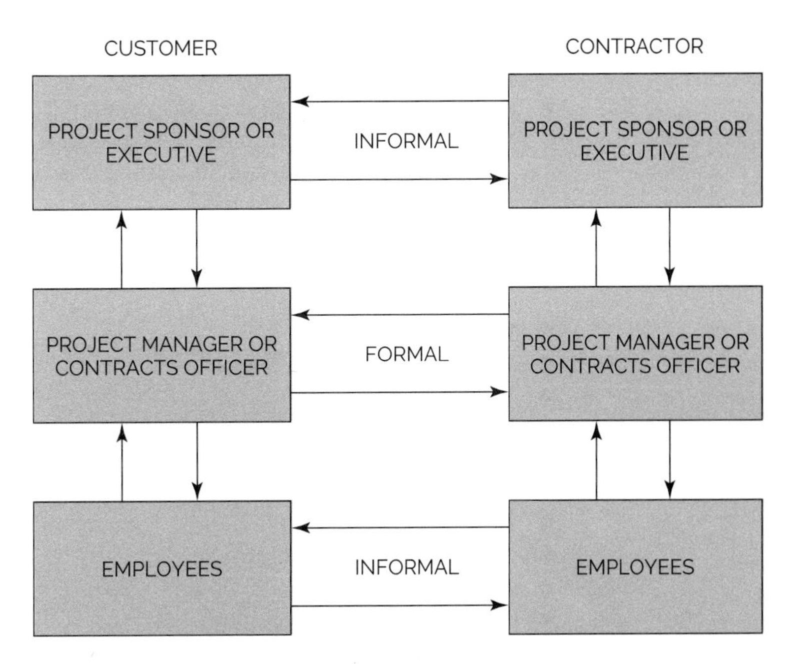

FIGURE 6–3. Customer communications. Source: D. I. Cleland and H. Kerzner, *Engineering Team Management* (Melbourne, Florida: Krieger, 1986), p. 64.

***PMBOK® Guide*, 6th Edition**
10.2.2 Manage Communications: Tools & Techniques

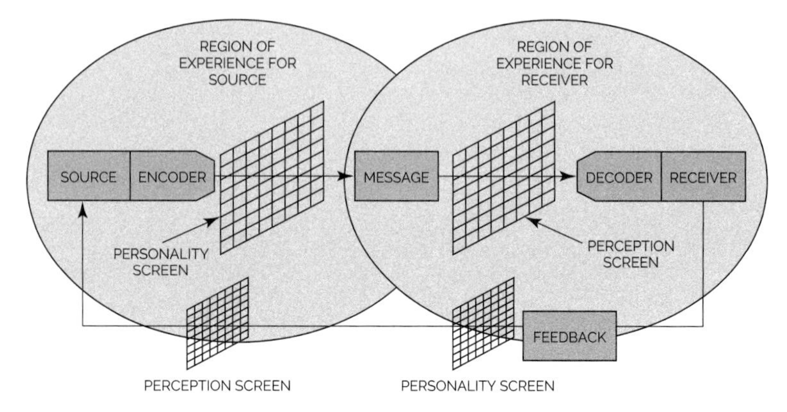

FIGURE 6–4. Total communication process. Source: D. I. Cleland and H. Kerzner, *Engineering Team Management* (Melbourne, Florida: Krieger, 1986), p. 46.

***PMBOK® Guide*, 6th Edition**
Figure 10.2.2 Manage Communications: Tools and Techniques

- *Perception barriers* occur because individuals can view the same message in different ways. Factors influencing perception include the individual's level of education and region of experience.
- Perception problems can be minimized by using words that have precise meaning.
- *Personality and interests,* such as the likes and dislikes of individuals, affect communications. People tend to listen carefully to topics of interest but turn a deaf ear to unfamiliar or boring topics.
- *Attitudes, emotions, and prejudices* warp our sense of interpretation. Individuals who are fearful or have strong love or hate emotions will tend to protect themselves by distorting the communication process. Strong emotions rob individuals of their ability to comprehend.

Typical barriers that affect the encoding process include:

- Communication goals
- Communication skills
- Frame of reference
- Sender credibility
- Needs
- Personality and interests
- Interpersonal sensitivity
- Attitude, emotion, and self-interest
- Position and status
- Assumptions (about receivers)
- Existing relationships with receivers

Typical barriers that affect the decoding process include:

- Evaluative tendency
- Preconceived ideas
- Communication skills
- Frame of reference
- Needs
- Personality and interest
- Attitudes, emotion, and self-interest
- Position and status
- Assumptions about sender
- Existing relationship with sender
- Lack of responsive feedback
- Selective listening

The receiving of information can be affected by the way the information is received. The most common ways include:

- Hearing activity
- Reading skills
- Visual activity
- Tactile sensitivity
- Olfactory sensitivity
- Extrasensory perception

The communications environment is controlled by both the internal and external forces, which can act either individually or collectively. These forces can either assist or restrict the attainment of project objectives.

Typical internal factors include:

- Power games
- Withholding information
- Management by memo
- Reactive emotional behavior
- Mixed messages
- Indirect communications
- Stereotyping
- Transmitting partial information
- Blocking or selective perception

Typical external factors include:

- The business environment
- The political environment
- The economic climate
- Regulatory agencies
- The technical state-of-the-art

The communications environment is also affected by:

- Logistics/geographic separation
- Personal contact requirements
- Group meetings
- Telephone
- Correspondence (frequency and quantity)

Noise tends to distort or destroy the information within the message. Noise results from our own personality screens, which dictate the way we present the message, and perception screens, which may cause us to "perceive" what we thought was said. Noise therefore can cause ambiguity:

- Ambiguity causes us to hear what we want to hear.
- Ambiguity causes us to hear what the group wants.
- Ambiguity causes us to relate to past experiences without being discriminatory.

6.2 THE PROJECT MANAGER AS A COMMUNICATOR

In a project environment, a project manager may very well spend 90 percent or more of his or her time communicating. Typical functional applications include:

- Providing project direction, including decision making, authorizing work, directing activities, negotiating, and reporting (including briefings)
- Attending meetings
- Overall project management
- Marketing and selling
- Public relations
- Records management, including minutes, memos/letters/newsletters, reports, specifications, and contract documents

Because of the time spent in a communications mode, the project manager may very well have as his or her responsibility the process of *communications management.* Communications management is the formal or informal process of conducting or supervising the exchange of information either upward, downward, laterally, or diagonally. There appears to be a direct correlation between the project manager's ability to manage the communications process and project performance.

The communications process is more than simply conveying a message; it is also a source for control. Proper communications let the employees in on the act because employees need to know and understand. Communication must convey both information and motivation. The problem, therefore, is how to communicate. Below are six simple steps:

- Think through what you wish to accomplish.
- Determine the way you will communicate.

- Appeal to the interest of those affected.
- Give feedback on ways others communicate to you.
- Get feedback on what you communicate.
- Test the effectiveness of your communication by relying on others to carry out your instructions.

Knowing how to communicate does not guarantee that a clear message will be generated. There are techniques that can be used to improve communications. These techniques include:

- Obtaining feedback, possibly in more than one form
- Establishing multiple communications channels
- Using face-to-face communications if possible
- Determining how sensitive the receiver is to your communications
- Being aware of symbolic meaning such as expressions on people's faces
- Communicating at the proper time
- Reinforcing words with actions
- Using a simple language
- Using redundancy (i.e., saying it two different ways) whenever possible

With every effort to communicate there are always barriers. The barriers include:

- Receiver hearing what he wants to hear. This results from people doing the same job so long that they no longer listen.
- Sender and receiver having different perceptions. This is vitally important in interpreting contractual requirements, statements of work, and proposal information requests.
- Receiver evaluating the source before accepting the communications.
- Receiver ignoring conflicting information and doing as he pleases.
- Words meaning different things to different people.
- Communicators ignoring nonverbal cues.
- Receiver being emotionally upset.

The scalar chain of command can also become a barrier with regard to in-house communications. The project manager must have the authority to go to the general manager or counterpart to communicate effectively. Otherwise, filters can develop and distort the final message.

PMBOK® Guide, 6th Edition
Chapter 10 Project Communications Management

Three important conclusions can be drawn about communications techniques and barriers:

- Don't assume that the message you sent will be received in the form you sent it.
- The swiftest and most effective communications take place among people with common points of view. The manager who fosters good relationships with his associates will have little difficulty in communicating with them.
- Communications must be established early in the project.

In a project environment, communications are often filtered. There are several reasons for the filtering of upward communications:

- Unpleasantness for the sender
- Receiver cannot obtain information from any other source
- To embarrass a superior
- Lack of mobility or status for the sender
- Insecurity
- Mistrust

Communication is also listening. Good project managers must be willing to listen to their employees, both professionally and personally. The advantages of listening properly are that:

- Subordinates know you are sincerely interested.
- You obtain feedback.
- Employee acceptance is fostered.

The successful manager must be willing to listen to an individual's story from beginning to end, without interruptions, and to see the problem through the eyes of the subordinate. Finally, before making a decision, the manager should ask the subordinate for his or her solutions to the problem.

PMBOK® Guide, 6th Edition
Chapter 10 Project Communications Management

The project manager's communication skills and personality screen often dictates the communication style. Typical communication styles include:

- Authoritarian: gives expectations and specific guidance
- Promotional: cultivates team spirit
- Facilitating: gives guidance as required, noninterfering
- Conciliatory: friendly and agreeable, builds compatible team
- Judicial: uses sound judgment
- Ethical: honest, fair, by the book
- Secretive: not open or outgoing (to project detriment)
- Disruptive: breaks apart unity of group, agitator
- Intimidating: "tough guy," can lower morale
- Combative: eager to fight or be disagreeable

PMBOK® Guide, 6th Edition
Chapter 10 Project Communications Management
10.2.2.7 Meetings

Team meetings are often used to exchange valuable and necessary information. The following are general guides for conducting more effective meetings:

- Start on time. If you wait for people, you reward tardy behavior.
- Develop agenda "objectives." Generate a list and proceed; avoid getting hung up on the order of topics.
- Conduct one piece of business at a time.
- Allow each member to contribute in his own way. Support, challenge, and counter; view differences as helpful; dig for reasons or views.

- Silence does not always mean agreement. Seek opinions.
- Be ready to confront the verbal member: "Okay, we've heard from Mike on this matter; now how about some other views?"
- Test for readiness to make a decision.
- Make the decision.
- Test for commitment to the decision.
- Assign roles and responsibilities (only after decision making).
- Agree on follow-up or accountability dates.
- Indicate the next step for this group.
- Set the time and place for the next meeting.
- End on time.
- Ask yourself if the meeting was necessary.

Many times, company policies and procedures can be established for the development of communications channels. Table 6–1 illustrates such communications guidelines.

TABLE 6–1. COMMUNICATIONS POLICY

Program Manager	Functional Manager	Relationship
The program manager utilizes existing authorized communications media to the maximum extent rather than create new ones.		Communications up, down, and laterally are essential elements to the success of programs in a multiprogram organization, and to the morale and motivation of supporting functional organizations. In principle, communication from the program manager should be channeled through the program team member to functional managers.
Approves program plans, subdivided work description, and/or work authorizations, and schedules defining specific program requirements.	Assures his organization's compliance with all such program direction received.	Program definition must be within the scope of the contract as expressed in the program plan and work breakdown structure.
Signs correspondence that provides program direction to functional organizations. Signs correspondence addressed to the customer that pertains to the program except that which has been expressly assigned by the general manager, the function organizations, or higher management in accordance with division policy.	Assures his organization's compliance with all such program direction received. Functional manager provides the program manager with copies of all "Program" correspondence released by his organization that may affect program performance. Ensures that the program manager is aware of correspondence with unusual content, on an exception basis, through the cognizant program team member or directly if such action is warranted by the gravity of the situation.	In the program manager's absence, the signature authority is transferred upward to his reporting superior unless an acting program manager has been designated. Signature authority for correspondence will be consistent with established division policy.
Reports program results and accomplishments to the customer and to the general manager, keeping them informed of significant problems and events.	Participates in program reviews, being aware of and prepared in matters related to his functional specialty. Keeps his line or staff management and cognizant program team member informed of significant problems and events relating to any program in which his personnel are involved.	Status reporting is the responsibility of functional specialists. The program manager utilizes the specialist organizations. The specialists retain their own channels to the general manager but must keep the program manager informed.

6.3 PROJECT REVIEW MEETINGS

PMBOK® Guide, 6th Edition
Chapter 10 Project Communications Management
10.2.2.7 Meetings

Project review meetings are necessary to show that progress is being made on a project. There are three types of review meetings:

- Project team review meetings: Most projects have weekly, bimonthly, or monthly meetings in order to keep the project manager and his team informed about the project's status. These meetings are flexible and should be called only if they will benefit the team.
- Executive management review meetings: Executive management has the right to require monthly status review meetings. However, if the project manager believes that other meeting dates are better (because they occur at a point where progress can be identified), then he or she should request them
- Customer project review meetings: Customer review meetings are often the most critical and most inflexibly scheduled. Project managers must allow time to prepare handouts and literature well in advance of the meeting.

6.4 PROJECT MANAGEMENT BOTTLENECKS

Poor communications can easily produce communications bottlenecks. The most common bottleneck occurs when all communications between the customer and the parent organization must flow through the project office. Requiring that all information pass through the project office may be necessary but slows reaction times. Regardless of the qualifications of the project office members, the client always fears that the information he receives will be "filtered" prior to disclosure.

Customers not only like firsthand information, but also prefer that their technical specialists be able to communicate directly with the parent organization's technical specialists. Many project managers dislike this arrangement, because they fear that the technical specialists may say or do something contrary to project strategy or thinking. These fears can be allayed by telling the customer that this situation will be permitted if, and only if, the customer realizes that the remarks made by the technical specialists do not, in any way, shape, or form, reflect the position of the project office or company.

For long-duration projects the customer may require that the contractor have an established customer representative office in the contractor's facilities. The idea behind this is sound in that all information to the customer must flow through the customer's project office at the contractor's facility. This creates a problem in that it attempts to sever direct communications channels between the customer and contractor project managers. The result is the establishment of a local project office to satisfy contractual requirements, while actual communications go from customer to contractor as though the local project office did not exist. This creates an antagonistic local customer project office.

Another bottleneck occurs when the customer's project manager considers himself to be in a higher position than the contractor's project manager and, therefore, seeks some higher authority with which to communicate. Project managers who seek status can often

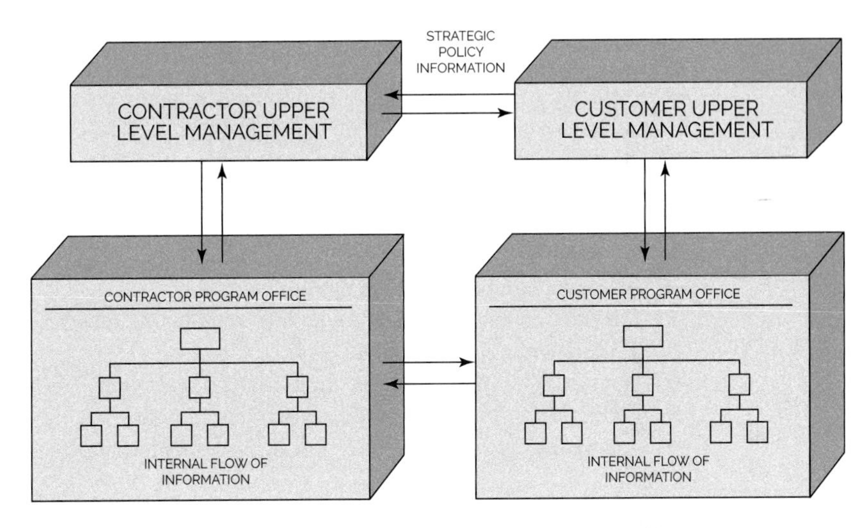

FIGURE 6–5. Information flow pattern from contractor program office.

jeopardize the success of the project by creating rigid communications channels. Almost always, there will exist a minimum of two paths for communications flow to and from the customer, which can cause confusion. Figure 6–5 identifies why communications bottlenecks such as these occur.

6.5 ACTIVE LISTENING

Part of good communication skills is effective or active listening. Improper listening can result in miscommunication, numerous and costly mistakes, having to repeat work, schedule delays, and the creation of a poor working environment. The result of poor listening is often more team meetings than originally thought and an abundance of action items. This can happen in any of the domain areas of the *PMBOK® Guide*.

Active listening involves more than just listening to the words that the speaker says. It also involves reading body language. Sometimes, the person's body language provides the listener with a much more accurate understanding of the intent of the message.

All elements of communication, including active listening, may be affected by barriers that can impede the flow of conversation. Such barriers include distractions, trigger words, poor choice of vocabulary, and limited attention span. Listening barriers may be psychological (e.g., emotions) or physical (e.g., noise and visual distraction). Cultural differences, including speakers' accents, vocabulary, and misunderstandings due to cultural assumptions, often obstruct the listening process. Frequently, the listener's personal interpretations, attitudes, biases, and prejudices lead to ineffective communication.

Sometimes the barriers to active listening are created by the speaker. This can occur when the speaker continuously changes subjects, uses words and expressions that confuse the listener, distracts the listener with improper or unnecessary body language, and neglects to solicit feedback as to whether the listener truly understood the message.

Typical active listening barriers created by the speaker include:

- Creating a communications environment where excessive note-taking is required such that the listener never gets to digest the material or see the body language
- Allowing constant interruptions to take place, which can lead to conflicts and arguments or allowing the interruptions to get the actual subject of the communications way off track
- Allowing people to cut you off, change subjects, and/or defend their positions
- Allowing competitive interruptions
- Speaking in an environment where there may be excessive noise or distractions
- Talking without pauses or talking too fast
- Neglecting to paraphrase or summarize critical points
- Failing to solicit feedback by asking the right questions
- Answering questions with responses that are slightly off

Typical active listening barriers created by the listener include:

- Looking at distractions rather than focusing on the speaker
- Letting your mind wander and looking off in the distance rather than staying focused
- Failing to ask for clarification of information that you do not understand
- Multitasking; doing some task, such as reading, while the speaker is presenting his or her message
- Not trying to see the information through the eyes of the speaker
- Allowing your emotions to cloud your thinking and listening
- Being anxious for your turn to speak
- Being anxious for the meeting to be over

Some techniques for active listening effectiveness might include:

- Always face the speaker.
- Maintain eye contact.
- Look at the speaker's body language.
- Minimize distractions, whether internal or external.
- Focus on what the speaker is saying without evaluating the message or defending your position.
- Keep an open mind on what is being discussed and try to empathize with the speaker even if you disagree.

6.6 COMMUNICATION TRAPS

> ***PMBOK® Guide*, 6th Edition**
> Chapter 10 Communications Management
> 10.1.3.1 Communications Management Plan

Projects are run by communications. The work is defined by the communications tool known as the work breakdown structure. Actually, this is the easy part of communications, where everything is well defined. Unfortunately, project managers cannot document everything they wish to say or relate to other people, regardless of the level in the company. The worst possible situation occurs when an outside customer loses faith in the contractor. When a situation of mistrust prevails, the logical sequence of events would be:

- More documentation
- More interchange meetings
- Customer representation on your site

In each of these situations, the project manager becomes severely overloaded with work. This situation can also occur in-house when a line manager begins to mistrust a project manager, or vice versa. There may suddenly appear an exponential increase in the flow of paperwork, and everyone is writing "protection" memos.

Communication traps occur most frequently with customer–contractor relationships but also occur between the project office and line managers.

- Do not interrupt the speaker even though you have a different position.

6.7 PROJECT PROBLEM SOLVING

When we have problems and make decisions in our personal lives, we usually adopt a "let's live with it" attitude. If the decision is wrong, we may try to change the decision. But in a project environment, changing a decision is more complicated and there may be a significant cost associated with the change. Some project decisions are irreversible. But there's one thing we know for sure in a business environment: anybody who always makes the right decision probably isn't making enough decisions. Expecting always to make the right decision is wishful thinking.

Problem solving and decision making go hand in hand. Problem solving involves understanding the problem, gathering the facts, and developing alternatives. Decisions are made when we select the appropriate alternative. There is a strong argument that decision making is also needed and used as part of identifying the problem and developing alternatives.

Today, there seems to be an abundance of information available to everyone. We all seem to suffer from information overload thanks to advances in information system technologies. Our main problem is being able to discern what information is critical to understand the problem. For simplicity's sake, information can be broken down into primary and secondary information. Primary information is information that is readily available to us. This is information that we can directly access from our desktop or laptop. Information that is company sensitive or considered as proprietary information may be password protected but still accessible. Secondary information is information that must be collected from someone else.

Even with information overload, project managers generally do not have all of the information they need to solve a problem and make a timely decision. This is largely due to the complexity of our projects as well as the complexity of the problems that need to be resolved. We generally rely upon a problem-solving team to provide us with the secondary information. The secondary information is often more critical for decision making than the primary information. Many times the secondary information is controlled by the subject matter experts, and they must tell us what information is directly pertinent to this problem.

Constraints play havoc with both the problem-solving and the decision-making processes. The time constraint probably has the greatest impact on decision making. Time is not a luxury. The decision may have to be made even though the project manager has only partial information and may not fully comprehend the problem. Making decisions with complete information is usually not a luxury that the project team will possess. And to make matters worse, we often have little knowledge on what the impact of the decisions will be.

Identifying the Problem

To understand problem solving, we must first understand what is meant by a problem. A problem is a deviation between an actual and desired situation. It is an obstacle, impediment, difficulty, or challenge, or any situation that invites resolution, the resolution of which is recognized as a solution or contribution toward a known purpose or goal. The problem could be to add something that is currently absent but desired, to remove something that is potentially bad, or to correct something that is not performing as expected. Therefore, problems can be formulated in a positive or negative manner. Problems are formulated in a positive manner if the problem is to determine how to take advantage of an opportunity.

We tend to identify alternatives as being good or bad choices. If the decision maker has all of the alternatives labeled as good or bad, then the job of the decision maker or project manager would be easy. Unfortunately, a problem implies that there exists doubt or uncertainty or else a problem would not exist. This uncertainty can happen on all projects and therefore makes it difficult to classify all alternatives as only good or bad.

Problems imply that some alternatives exist. Problems that have no alternatives are called open problems. Not all problems can be solved or should be solved. There are projects that may need to be created to solve problems that require compliance to government regulations. For these projects, which are almost always very costly, all of the alternatives are often considered poor choices. When forced to comply, we select the best of the worst. But more often than not, we leave them as open problems until the very last minute, hoping the problem will be forgotten or will disappear.

Companies encourage all project team members to bring forth all problems quickly. The quicker the problem is exposed, the more time is available for finding a solution, the more alternatives are usually available, and the greater the number of resources that can assist in the solution. Unfortunately, some people simply may not want to identify the problem with the hope that they can resolve it by themselves before anyone finds out about it. This is true for people who may have been involved in creating the problem.

In such cases, people try to solve the problem by themselves, in secret, before anyone finds out about the problem. In reality, the problem is often hard to hide. Sometimes the entire problem-solving team is in collusion in hiding the problem. Unfortunately, problem-solving sessions clearly identify that a problem exists and this alone could make it difficult to hide a problem. It is easier for one person to try to solve a problem secretly than for an entire team to do so. There are several reasons for the team wanting the problem to be resolved quietly:

- The client and/or the stakeholders may overreact to the problem and dictate the solution.
- The client and/or stakeholders may overreact to the problem and remove financial support.
- The client may cancel the project.
- Problem resolution requires the discussion of proprietary or classified information.
- Open identification of the problem may cause people to be fired.
- Open identification of the problem may cause damage to your company's image and reputation.
- Open identification of the problem can result in potential lawsuits.
- The cause of the problem is unknown.
- The problem can be resolved quickly without any impact on the competing constraints and the deliverables.

Problem Data

Numerous techniques are available for data gathering about a problem. The selection of the technique is based upon the information being sought out, the timing of the information, who will provide the information, the criticality of the information, and the type of decisions that the information must support. Each technique comes with strengths and weaknesses. Some data-gathering techniques can be done quickly. Table 6–2 illustrates some of the most commonly used techniques.[1]

Using just one technique may not suffice. For the most part, data-gathering techniques are time consuming. It may be necessary to use several techniques in order to capture all of the required data. Effective data gathering requires an understanding of what questions to ask. While it is true that the questions will be predicated upon the type of problem, typical questions might include:

- Are there any other resources or subject matter experts that can help us with this problem?
- How many problems do we have?
- Are there hidden problems that are below the surface?
- What is the extent of the problem?
- Is the problem getting worse, getting better, or remaining stable?
- Did this problem exist previously?
- Can the problem be quantified?
- Can we determine the severity of the problem?
- What physical evidence exists to identify the problem?
- Who identified the problem?
- To whom was it first reported?
- Is there an action plan to collect additional information?
- Do we have the right team members addressing this problem?

---

1. Adapted from R. K. Wysocki, *Effective Project Management: Traditional, Agile, Extreme*, 5th ed. (Hoboken, NJ: John Wiley & Sons, 2009), pp. 62–63.

TABLE 6–2. STRENGTHS AND RISKS OF VARIOUS TECHNIQUES

Method	Strength	Risks
Root-cause analysis	Looks for root cause rather than just symptoms Designed to prevent recurrence of the problem Provides in-depth analysis of the problem	Highly systematic Iterative process Very time consuming
Facilitated group sessions	Excellent for cross-functional processes Detailed requirements can be documented and verified immediately	Use of untrained facilitators can lead to a negative response from users The time and cost of the planning and/or executing sessions can be high
Panels of experts	Resolves issues with an impartial facilitator Selection of the best of the best in people Good when there is no one correct solution	Personalities can influence decisions Too much defense of one's own position Relationship between interviewer and interviewee may hide the truth People may be afraid that the truth will be held against them
Interviews	Best if one-on-one sessions Allows for follow-up interviews if necessary Interviewees usually speak freely	
Surveys	Can be formal or informal Responses are usually honest People will defend their positions	Often difficult to get enough people to respond Expensive to design a questionnaire Statistical reliability may be necessary
Observations	Specific and complete descriptions of actions are provided Effective when routine activities are difficult to describe	Documenting and videotaping may be time consuming and expensive Confusing or conflicting information must be clarified Can lead to misinterpretation of what is observed
Requirements reuse	Requirements are quickly generated and refined Redundant efforts are reduced Client satisfaction is enhanced by previous proof Quality is increased Reinventing the wheel is minimized	Requires a significant investment for developing archives, maintenance, and library functions May violate intellectual property rights Similarity of an archived requirement to a new requirement may be misunderstood
Business process diagramming	Excellent for cross-functional processes Visual communication through process diagrams Verification of "what is" and "what is not"	Implementation of improvement is dependent on an organization being open to changes Good facilitation, data gathering, and interpretation are required Time consuming
Prototypes	Innovated ideas can be generated Users clarify what they want Users identify requirements that may be missed Client focused Stimulates thought processes	Client may want to implement the prototype Difficult to know when to stop Specialized skills are required Absence of documentation
Use case scenarios	The state of the system is described before the client first interacts with that system Complete scenarios are used to describe state of systems The normal flow of events and/or exceptions is revealed Improved client satisfaction and design	Newness has resulted in some inconsistencies Information may still be missing from scenario description Long interaction is required Training is expensive
Review of performance data	Available from archives Data are usually reliable for the situation at hand	May not describe how the data were collected Data may be outdated

We generally believe that most problems are real and need to be resolved. But that is not always the case. Some problems are created because of the personalities of the individuals involved. Some people create problems unnecessarily as long as they can somehow benefit, perhaps by being the only person capable of solving the problem.

Meetings

Problems are not resolved in a vacuum. Meetings are needed and the hard part is to determine who should attend. If people are not involved in the problem or the problem is unrelated to the work they do, then having them attend these meetings may be a waste of their time. This holds true for some of the team members as well. As an example, if the problem is with procurement, then it may not be necessary for the drafting personnel to be in attendance.

For simplicity's sake, we shall consider just two types of meetings: problem solving and decision making. The purpose of the problem-solving meeting is to obtain a clear understanding of the problem, collect the necessary data, and develop a list of workable alternatives accompanies by recommendations. More than one meeting will probably be required.

Sending out an agenda is important. The agenda should include a problem statement that clearly explains why the meeting is being called. If people know about the problem in advance, they will have a chance to think about the problem and bring the necessary information, thus reducing some of the time needed for data gathering. It is also possible that the information gathered will identify that the real problem is quite different from what was considered to be the problem at first.

It is essential that subject matter experts familiar with the problem be in attendance. These subject matter experts may not be part of the original project team but may be brought in just to resolve this problem. The subject matter experts may also be contractors hired to assist with the problem. The people brought in for the identification of the problem and data gathering usually remain for the development of the alternatives. But there are situations where additional people may participate just for the evaluation of alternatives.

The decision-making meeting can be different from the problem-solving meeting. In general, all of the participants who were involved in the problem-solving meeting will most likely be in attendance in the decision-making meeting, but there may be a significant number of other participants. Project team members should have the ability to resolve problems, but not all of the team members have the authority to make decisions for their functional units. It is normally a good idea at the initiation of the project for the project manager to determine which team members possess this authority and which do not. Team members who do not possess decision-making authority will still be allowed to attend the decision-making sessions but may need to be accompanied by their respective functional managers when decisions are required and voting takes place.

Stakeholder attendance is virtually mandatory at the decision-making meetings. The people making the decisions must have the authority to commit resources to the solution of the problem. The commitment could involve additional funding or the assignment of subject matter experts and higher-pay-grade employees. Project managers are responsible for the implementation of the solution. Therefore, the project manager must have the authority to obtain the resources needed for a timely solution to the problem.

Team meetings that involve problem solving and decision making often get people to act in an irrational manner, especially if the outcome of the meeting can have a negative impact on them personally. This is particularly true for people who are closely identified with the cause of the problem. You may also be inviting people you have never worked with previously, and you have no idea how they will react to the problem or the solution.

Alternatives

A major part of problem solving and ultimately the decision making involves the identification and analysis of a finite set of alternatives described in terms of some evaluative criteria. These criteria may be benefit or cost in nature or the criteria could simply be the adherence to the cost, schedule, and scope baselines of the project. Then the problem might be to rank these alternatives in terms of how attractive they are to the decision maker(s) when all the criteria are considered simultaneously. Another goal might be to just find the best alternative or to determine the relative total priority of each alternative.

The number of alternatives is often limited by the constraints imposed upon the project. For example, if the schedule is exceeding the baseline schedule, then the project manager may have five alternatives: overtime, performing some work in parallel rather than in series, adding more resources to the project, outsourcing some of the work to a lower cost supplier, or reducing the scope of the project. Each alternative will be accompanied by advantages and disadvantages. If the goal is to lower the costs, then there may be only one viable alternative, namely reducing the scope.

There are several variables that must be considered when identifying and selecting alternatives. The variables are usually project-specific and based upon the size, nature, and complexity of the problem. However, we can identify a list of core variables that usually apply to the identification and evaluation of most alternatives:

- **Cost:** There is a cost associated with each alternative. This includes not only the cost of implementing the alternative but also the financial impact on the remaining work on the project.
- **Schedule:** Implementing an alternative takes time. If the implementation time is too long or cannot be done in parallel with the other project work, then there may be a significant impact on the end date of the project.
- **Quality:** Care must be taken that the speed to resolve a problem does not result in a degradation of quality in the project's deliverables.
- **Resources:** Implementing a solution requires resources. The problem is that the people needed with the necessary skills may not be available.
- **Feasibility:** Some alternatives may seem plausible on paper but may be unfeasible when needed to be implemented. Feasibility or complexity of the alternative must be considered. Otherwise, you could make the problem worse.
- **Risks:** Some alternatives expose the company to increased risks. These may be future risks (or even opportunities) that will appear well after the project is completed.

We must also look at the features that make up the alternatives. Many times there are several features that can be included in each of the alternatives and we may have a choice on whether to include these features. Part of understanding the boundary conditions is to know the importance of each feature. The features can be classified as:

- **Must have:** Any alternative that does not include this feature should be discarded.
- **Should have:** These are features that in most situations should be included in the alternatives that are being considered. Failure to consider these could result in a degradation of performance. Some of these features may be omitted if including them results in unfavorable consequences when trying to satisfy the competing constraints.
- **Might have:** These are usually add-ons to enhance performance but not necessarily part of the project's requirements. These are nice-to-have items but not a necessity when deciding upon a final solution. Might-have features are often characterized as bells and whistles that are part of gold-plating efforts.

After looking at the variables and evaluating all of the alternatives, the conclusion may be that none of the alternatives are acceptable. In this case, the project manager may be forced to select the "best of the worst."

Creativity

It is possible that after evaluating the alternatives the best approach might just be a combination of alternatives. This is referred to as a hybrid alternative. Alternative A might be a high risk but a low cost of implementation. Alternative B might be a low risk but a high cost of implementation. By combining alternatives A and B, we may be able to come up with a hybrid alternative with an acceptable cost and risk factor.

Sometimes, creativity is needed to develop alternatives. Not all people are creative even if they are at the top of their pay grade. People can do the same repetitive task for so long that they are considered subject matter experts. They can rise to the top of their pay grade based upon experience and years of service. But that alone does not mean that they have creativity skills. Most people think that they are creative when, in fact, they are not. Companies also do not often provide their workers training in creative thinking.

In a project environment, creativity is the ability to use one's imagination to come up with new and original ideas or things to meet requirements and/or solve problems. People are assigned to project teams based upon experience. It is impossible for the project manager, and sometimes even the functional managers, to know whether these people have the creativity skills needed to solve problems that can arise during a project. Unless you have worked with these people previously, it is difficult to know if they have imagination, inspiration, ingenuity, inventiveness, vision, and resourcefulness, all common characteristics of creativity.

Innovation

Creativity is the ability to think up ideas to produce something new through imaginative skills, whether a new solution to a problem or a new method or device. Innovation is the ability to solve the problem by converting the idea into reality, whether it is a product, service, or any form of deliverable for the client. Innovation goes beyond creative thinking. Creativity and innovation do not necessarily go hand in hand. Any problem-solving team can come up with creative solutions that cannot be implemented. Any engineering team can design a product (or a modification to a product) that manufacturing cannot build.

Innovation is more than simply turning an idea into reality. It is a process that creates value. Clients are paying for something of value. Whatever solution is arrived at must be recognized by the client as possessing value. The best of all possibilities is when the real value can be somehow shared between the client's needs and your company's strategy. The final alternative selected might increase or decrease the value of the end deliverables, as seen by the client, but there must always be some recognizable value in the solution selected.

Because of constraints and limitations, some solutions to a problem may necessitate a reduction in value compared to the original requirements of the project. This is referred to as negative innovation. In such cases, innovation for a solution that reduces value can have a negative or destructive effect upon the team. People could see negative innovation as damage to their reputation and career.

If the innovation risks are too great, the project team may recommend some form of open innovation. Open innovation is a partnership with those outside your company by sharing the risks and rewards of the outcome. Many companies have creative ideas for solving problems but lack the innovative talent to implement a solution. Partnerships and joint ventures may be the final solution.

There are four types of innovation. Each type comes with advantages:

- **Add-ons, enhancements, product/quality improvements, and cost reduction efforts:** This type of innovation may be able to be accomplished quickly and with the existing resources in the company. The intent is to solve a problem and add incremental value to the end result.
- **Radical breakthrough in technology:** This type of innovation has risks. You may not be able to determine when the breakthrough will be made and the accompanying cost. Even if the breakthrough can be made, there is no guarantee that the client will receive added value from this solution. If the breakthrough cannot be made, the client may still be happy with the partial solution. This type of innovation may require the skills of only one or two people.
- **New family members:** This is the creation of new products and may require a technological breakthrough.
- **Totally complex system or platform (next-generation projects):** This is the solution with the greatest risk. If the complex system cannot be developed, then the project will probably be considered a total loss. A large number of highly talented resources are needed for this form of innovation.

Downsizing

Sometimes, we start out projects with the best of intentions and later discover that some problem has occurred that could result in the cancellation of the project. Rather than cancel the project outright, the solution might be to downsize the project and readjust our innovation attempts. Factors that can lead to a readjustment in innovation include:

- The market for the deliverable has shrunk.
- The deliverable will be overpriced and demand will not be there.
- The technical breakthrough cannot be achieved in a timely manner.
- There is a loss of faith and enthusiasm by the team and they no longer believe this solution is workable.

- There is possible loss of interest by top management and the client.
- There are insurmountable technical obstacles.
- There is a significant decrease in the likelihood of success.

If these factors exist, then it is entirely possible that another alternative must be selected in order to salvage the project. As long as the client is willing to accept a possible reduction in final value, the project may be allowed to continue.

6.8 BRAINSTORMING

Throughout the life of any project, the team will be tested on their ability to find the best possible solution to a problem within the imposed limitations and boundaries. This could occur in the planning phase of the project where we must come up with the best possible plan or it could happen in any later phases where problems arise and the best solution must be found. These are situations where brainstorming techniques may not be appropriate. Most people seem to have heard about brainstorming but very few have been part of brainstorming teams. Although we normally discuss brainstorming as a means for identifying alternative solutions to a problem, brainstorming can also be used for root-cause identification of the problem.

There are four basic rules in brainstorming. These rules are intended to stimulate idea generation and increase overall creativity of the group while minimizing the inhibitions people may have about working in groups.

- **Focus on quantity:** This rule focuses on the maximization of possible ideas, both good and bad. The assumption made is that the greater the number of ideas, the greater the chance of finding the optimal solution to a problem.
- **Withhold criticism:** In brainstorming, criticism of ideas creates conflict and wastes valuable time needed to generate the maximum number of ideas. When people see ideas being criticized, they tend to withhold their own ideas to avoid being criticized. Criticism should take place but after the brainstorming session is completed. Typical brainstorming sessions last about an hour or less.
- **Welcome unusual ideas:** All ideas should be encouraged, whether good or bad. People must be encouraged to think "outside the box," and this may generate new perspectives and a new way of thinking. Sometimes, what appears as a radical solution initially may be the best possible solution in the end.
- **Combine and improve ideas:** The best possible solution may be a combination of ideas. New ideas should be encouraged from the combination of ideas already presented.

The process of conducting a brainstorming session includes the following:

The process

- Participants who have ideas but were unable to present them are encouraged to write down the ideas and present them later.

- The idea collector should number the ideas, so that the chairperson can use the number to encourage an idea generation goal, for example: *We have 14 ideas now, let's get it to 20!*
- The idea collector should repeat the idea in the words he or she has written verbatim to confirm that it expresses the meaning intended by the originator.
- When many participants are having ideas, the one with the most associated idea should have priority. This is to encourage elaboration on previous ideas.
- During a brainstorming session, managers and other superiors may be discouraged from attending, since it may inhibit and reduce the effect of the four basic rules, especially the generation of unusual ideas.

Evaluation

Brainstorming is not just about generating ideas for others to evaluate and select. Usually the group itself will, in its final stage, evaluate the ideas and select one as the solution to the problem proposed to the group.

- The solution should not require resources or skills the members of the group do not have or cannot acquire.
- If acquiring additional resources or skills is necessary, that needs to be the first part of the solution.
- There must be a way to measure progress and success.
- The steps to carry out the solution must be clear to all and amenable to being assigned to the members so that each will have an important role.
- There must be a common decision-making process to enable a coordinated effort to proceed and to reassign tasks as the project unfolds.
- There should be evaluations at milestones to decide whether the group is on track toward a final solution.
- There should be incentives to participation so that participants maintain their efforts.

6.9 PREDICTING THE OUTCOME OF A DECISION

Problem solving and decision making require the project manager to predict how those impacted by the decision will react to the alternative selected. Soliciting feedback prior to the implementation of the solution seems nice to do, but the real impact of the decision may not be known until after full implementation of the solution. As an example, as part of developing a new product, marketing informs the project manager that the competition has just come out with a similar product, and marketing believes that we must add in some additional features into the product we are developing. The project team adds a significant number of "bells and whistles," to the point where the product's selling price is higher than that of the competition and the payback period is now elongated. When the product is eventually launched, the consumers do not believe that the added features are worth the additional cost.

TABLE 6–3. CONSEQUENCE TABLE

Competing Constraints					
Alternative	**Time**	**Cost**	**Quality**	**Safety**	**Overall Impact**
1	A	C	B	B	B
2	A	C	A	C	B
3	A	C	C	C	C
4	B	A	C	A	B
5	A	B	A	A	A

Notes: A = high impact, B = moderate impact, C = low impact.

It is not always possible to evaluate or predict the impact of a decision when making a choice among alternatives. But soliciting feedback prior to full implementation is helpful.

A useful tool for assisting in the selection of alternatives is a consequence table, as shown in Table 6–3. For each alternative, the consequences are measured against a variety of factors such as each of the competing constraints. For example, an alternative could have a favorable consequence on quality but an unfavorable consequence on time and cost. Most consequence tables have the impacts identified quantitatively rather than qualitatively. Risk is also a factor that is considered, but the impact on risk is usually defined qualitatively rather than quantitatively.

If there are three alternatives and five constraints, then there may be fifteen rows in the consequence table. Once all fifteen consequences are identified, they are ranked. They may be ranked according to either favorable or unfavorable consequences. If none of the consequences is acceptable, then it may be necessary to perform trade-offs on the alternatives.

This could become an iterative process until an agreed-upon alternative is found. The table could be prepared quantitatively or qualitatively. With a quantitative table, weighting factors can be used for the relative importance of each of the competing constraints.

The people preparing the table are the people who make up the project team rather than possible outsiders that were brought in as subject matter experts for a particular problem. Project team members know the estimating techniques as well as the tools that are part of the organization process assets that can be used for determining impacts.

It is nice to have several possible alternatives for the solution to a problem.

Unfortunately, the alternative that is finally selected must be implemented, and that can create additional problems. One of the ways to analyze the impact is to create an impact implementation matrix, as shown in Figure 6–6. Each alternative considered could have a high or low impact on the project. Likewise, the implementation of each alternative could be easy or hard.

Each alternative is identified in its appropriate quadrant. The most obvious choice would be the alternatives that have a low impact and are easy to implement. But in reality, we often do not find very many alternatives in this quadrant.

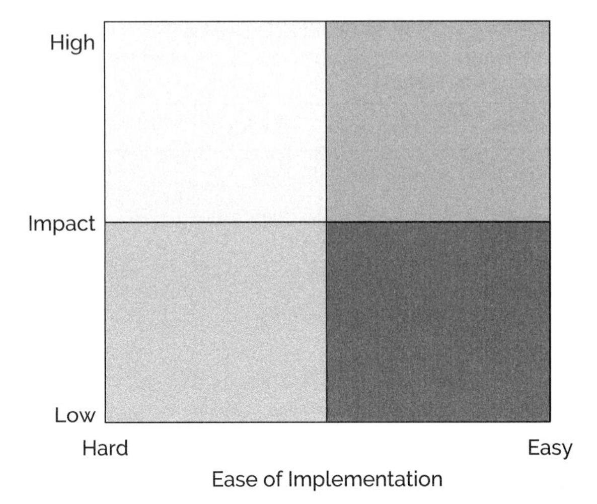

FIGURE 6–6. Impact analysis matrix.

6.10 FACILITATION

A good facilitator helps people understand their common objectives and assists them in planning to achieve these objectives but without taking a particular position in the discussion. This may be difficult for a project manager to do. Some people identify a facilitator as:

- An enabler
- Someone who helps people communicate and work together
- A person who adds structure and process to decision making and problem solving
- A person who creates synergy in the decision-making process
- A person who can get everyone to do their best thinking
- Someone who can tap into each person's creative potential

The facilitator will not lead the group toward the answer that he or she thinks is best even if they possess an opinion on the subject matter. The facilitator does not evaluate ideas. The facilitator's role is to make it easier for the group to arrive at its own answer, decision, solution, or deliverable. The facilitator sometimes acts as a resource for the group in the area of data analysis tools and problem-solving techniques. The facilitator must be comfortable with team-building techniques and group processes in order to assist the group in performing tasks and maintaining roles essential to team building. The facilitator intervenes to help the group stay focused and build cohesiveness, getting the job done with excellence, while developing the final product.

To keep the meeting on track, the facilitator must remain aware of the agenda, the time, and the flow of work. Facilitation skills are used to ensure total participation. Facilitators observe group development, noting both task and maintenance roles, and encourage group members to perform them. Facilitators handle inappropriate participant behaviors with skill and sensitivity.

The basic skills of a facilitator are about following good meeting practices: timekeeping, following an agreed-upon agenda, and keeping a clear record. Facilitators also need a variety of crossover skills that include active listening skills, the ability to paraphrase, draw people into the discussion, balance participation, and make space for more reticent group members. It is critical to the facilitator's role to have the knowledge and skill to be able to intervene in a way that adds to the group's creativity rather than taking away from it.

Some of the readily apparent skills of good facilitators include:

- Knowing how to deal with difficult people
- Knowing how to minimize or prevent gamesmanship during meetings
- Knowing when and how to use intervention effectively
- Knowing the importance of a good environment for the meeting
- Being able to identify when participants are becoming, lazy, bored, or frustrated
- Being able to protect team members from attack

Good facilitators are able to see not only the obvious but also what else is happening that may not be quite apparent to the rest of the people in the meeting. In other words, facilitators must be experts in identifying negative dynamics or actions by people who can disrupt the intent of the meeting.

The facilitator is the protector of the processes used in group dynamics, more specifically as they relate to the project's crossover skills requirements. The facilitator's toolkit is a set of techniques, knowledge, and experience that they apply to protect the process the group is working through. The function of facilitation is to keep a meeting focused and moving and to ensure even participation. The facilitator makes sure these things occur, either by doing it or by monitoring the group and intervening as needed. The facilitator is the keeper of the task and doesn't influence the content or product of the group. The facilitator pays attention to the way the group works—the process. The facilitator helps to create the process, adjust it, keep it heading in the right direction, and most importantly keep the people attached to it.

More projects today are being managed with virtual project teams. The people occupy positions within and outside of organizations. They reside throughout the hierarchy and they come from different functional areas. When they are assigned as part of the team, they bring with them, in addition to their knowledge, their backgrounds, beliefs, organizational culture, technical jargon, and personal behaviors. The facilitator may never see them all face to face, yet they may be part of the meeting. Understanding cultural diversity is essential. This may include:

- Understanding that each person learns differently
- Having tolerance for ambiguity and recognizing the need to explain things perhaps more than once
- Demonstrating a sense of humility when needed
- Understanding cultural differences
- Demonstrating patience
- Demonstrating interpersonal sensitivity
- Possessing a sense of humor

There are risks when the project manager assumes the role of the facilitator. The project manager may try to lead the discussion toward the answer that the project manager wants. This is dangerous if the project manager acts as the facilitator and has preconceived expectations. Good facilitators do not have preconceived expectations or an axe to grind. Another problem might occur if people are afraid to contribute ideas because the project manager is leading the discussion and acting as the facilitator. The situation becomes more complicated if the project manager has wage and salary responsibilities for people in the meeting when acting as the facilitator.

Unfortunately, project budgets do not always allow for the cost of a facilitator whenever a meeting is needed. Project managers must learn facilitation skills to be effective. The project management office may have people assigned with facilitation skills and these people may be able to provide some support to various project teams.

Related Case Studies (from Kerzner/*Project Management Case Studies*, 5th ed.)	**Related Workbook Exercises (from Kerzner/*Project Management Workbook and PMP®/CAPM® Exam Study Guide*, 12th ed.)**	***PMBOK® Guide*, Sixth Edition, Reference Section for the PMP® Certification Exam**
• Time Management Exercise • Communication Failures* • The Team Meeting*	• Multiple Choice Exam • The Communication Problem • Meetings, Meetings, and Meetings • Crossword Puzzle on Communications Management	• Human Resource Management • Risk Management

*Case study also appears at end of chapter.

6.11 STUDYING TIPS FOR THE PMI® PROJECT MANAGEMENT CERTIFICATION EXAM

This section is applicable as a review of the principles to support the knowledge areas and domain groups in the *PMBOK® Guide*. This chapter addresses:

- Project Resource Management
- Project Risk Management
- Execution

Understanding the following principles is beneficial if the reader is using this text to study for the PMP® Certification Exam:

- How stress can affect the way that the project manager works with the team
- How stress affects the performance of team members

The following multiple-choice questions will be helpful in reviewing the principles of this chapter:

1. Which of the following leadership styles most frequently creates "additional" time robbers for a project manager?
   A. Telling
   B. Selling
   C. Participating
   D. Delegating
2. Which of the following leadership styles most frequently creates "additional" time robbers for the project team?
   A. Telling
   B. Selling
   C. Participating
   D. Delegating
3. Which of the following time robbers would a project manager most likely want to handle by himself or herself rather than through delegation to equally qualified team members?
   A. Approval of procurement expenditures
   B. Status reporting to a customer
   C. Conflicting directives from the executive sponsor
   D. Earned-value status reporting
4. Five people are in attendance in a meeting and are communicating with one another. How many two-way channels of communication are present?
   A. 4
   B. 5
   C. 10
   D. 20

ANSWERS

1. A
2. D
3. C
4. C

PROBLEMS

6–1 Is it possible for functional employees to have performed a job so long or so often that they no longer listen to the instructions given by the project or functional managers?

6–2 Below are eight common methods that project and functional employees can use to provide communications:

a. Counseling sessions
b. Telephone conversation
c. Individual conversation
d. Formal letter
e. Project office memo
f. Project office directive
g. Project team meeting
h. Formal report

For each of the following actions, select one and only one means of communication from the above list that you would utilize in accomplishing the action:

1. Defining the project organizational structure to functional managers
2. Defining the project organizational structure to team members
3. Defining the project organizational structure to executives
4. Explaining to a functional manager the reasons for conflict between his employee and your assistant project managers
5. Requesting overtime because of schedule slippages
6. Reporting an employee's violation of company policy
7. Reporting an employee's violation of project policy
8. Trying to solve a functional employee's grievance
9. Trying to solve a project office team member's grievance
10. Directing employees to increase production
11. Directing employees to perform work in a manner that violates company policy
12. Explaining the new indirect project evaluation system to project team members
13. Asking for downstream functional commitment of resources
14. Reporting daily status to executives or the customer
15. Reporting weekly status to executives or the customer
16. Reporting monthly or quarterly status to executives or the customer
17. Explaining the reason for the cost overrun
18. Establishing project planning guidelines
19. Requesting a vice president to attend your team meeting
20. Informing functional managers of project status
21. Informing functional team members of project status

22. Asking a functional manager to perform work not originally budgeted for
23. Explaining customer grievances to your people
24. Informing employees of the results of customer interchange meetings
25. Requesting that a functional employee be removed from your project because of incompetence

6–3 How does a project manager find out if the project team members from the functional departments have the authority to make decisions?

6–4 Below are several problems that commonly occur in project organizations. State, if possible, the effect that each problem could have on communications and time management:

a. People tend to resist exploration of new ideas.
b. People tend to mistrust each other in temporary management situations.
c. People tend to protect themselves.
d. Functional people tend to look at day-to-day activities rather than long-range efforts.
e. Both functional and project personnel often look for individual rather than group recognition.
f. People tend to create win-or-lose positions.

6–5 What is meant by polarization of communications? What are the most common causes?

6–6 The customer has asked to have a customer representative office set up in the same building as the project office. As project manager, you put the customer's office at the opposite end of the building from where you are, and on a different floor. The customer states that he wants his office next to yours. Should this be permitted, and, if so, under what conditions?

6–7 Is it possible for a project manager to hold too few project review meetings?

CASE STUDIES

COMMUNICATION FAILURES

Background

Herb had been with the company for more than eight years and had worked on various R&D and product enhancement projects for external clients. He had a Ph.D. in engineering and had developed a reputation as a subject matter expert. Because of his specialized skills, he worked by himself most of the time and interfaced with the various project teams only during project team meetings. All of that was about to change. Herb's company had just won a two-year contract from one of its best customers. The first year of the contract would be R&D and the second year would be manufacturing. The company made the decision that the best person qualified to be the project manager was Herb because of his knowledge of R&D and manufacturing. Unfortunately, Herb had never taken any courses in project management, and because of his limited involvement with previous project teams, there were risks in assigning him as the project manager. But management believed he could do the job.

The team is formed

Herb's team consisted of fourteen people, most of whom would be full-time for at least the first year of the project. The four people who Herb would be interfacing with on a daily basis were Alice, Bob, Betty, and Frank.

- Alice was a seasoned veteran who worked with Herb in R&D. Alice had been with the company longer than Herb and would coordinate the efforts of the R&D personnel.
- Bob also had been with the company longer than Herb and had spent his career in engineering. Bob would coordinate the engineering efforts and drafting.
- Betty was relatively new to the company. She would be responsible for all reports, records management, and procurements.
- Frank, a five-year employee with the company, was a manufacturing engineer. Unlike Alice, Bob, and Betty, Frank would be part-time on the project until it was time to prepare the manufacturing plans.

For the first two months of the program, work seemed to be progressing as planned. Everyone understood their role on the project and there were no critical issues.

Friday the 13th

Herb held weekly teams meetings every Friday from 2:00 to 3:00 P.M. Unfortunately, the next team meeting would fall on Friday the 13th, and that bothered Herb because he was somewhat superstitious. He was considering canceling the team meeting just for that week but decided against it.

At 9:00 A.M. on Friday the 13th, Herb met with his project sponsor as he always did in the past. Two days before, Herb casually talked to his sponsor in the hallway and the sponsor told Herb that on Friday the sponsor would like to discuss the cash flow projections for the next six months and have a discussion on ways to reduce some of the expenditures. The sponsor had seen some expenditures that bothered him. As soon as Herb entered the sponsor's office, the sponsor said:

> It looks like you have no report with you. I specifically recall asking you for a report on the cash flow projections.

Herb was somewhat displeased over this. He specifically recalled that this was to be a discussion only and no report was requested. But Herb knew that "rank has its privileges" and questioning the sponsor's communication skills would be wrong. Obviously, this was not a good start to Friday the 13th.

At 10:00 A.M., Alice came into Herb's office and he could see from the expression on her face that she was somewhat distraught. Alice then spoke:

> Herb, last Monday I told you that the company was considering me for promotion and the announcements would be made this morning. Well, I did not get promoted. How come you never wrote a letter of recommendation for me?

Herb remembered the conversation vividly. Alice did say that she was being considered for promotion but never asked him to write a letter of recommendation. Did Alice expect Herb to read between the lines and try to figure out what she really meant?

Herb expressed his sincere apologies for what happened. Unfortunately, this did not make Alice feel any better as she stormed out of Herb's office. Obviously, Herb's day was getting worse and it was Friday the 13th.

No sooner had Alice exited the doorway to Herb's office when Bob entered. Herb could tell that Bob had a problem. Bob then stated:

> In one of our team meetings last month, you stated that you had personally contacted some of my engineering technicians and told them to perform this week's tests at 70°F, 90°F, and 110°F. You and I know that the specifications called for testing at 60°F, 80°F, and 100°F. That's the way it was always done and you were asking them to perform the tests at different intervals than the specifications called for.
>
> Well, it seems that the engineering technicians forgot the conversation you had with them and did the tests according to the specification criteria. I assumed that you had followed up your conversation with them with a memo, but that was not the case. It seems that they forgot.
>
> When dealing with my engineering technicians, the standard rule is, "If it's not in writing, then it hasn't been said." From now on, I would recommend that you let me provide the direction to my engineering technicians. My responsibility is engineering and all requests of my engineering personnel should go through me.

Yes, Friday the 13th had become a very bad day for Herb. What else could go wrong, Herb thought. It was now 11:30 A.M. and almost time for lunch. Herb was considering locking his office door so that nobody could find him and then disconnecting his phone. But in walked Betty and Frank, and once again he could tell by the expressions on their faces that they had a problem. Frank spoke first:

> I just received confirmation from procurement that they purchased certain materials that we will need when we begin manufacturing. We are a year away from beginning manufacturing and, if the final design changes in the slightest, we will be stuck with costly raw materials that cannot be used. Also, my manufacturing budget did not have the cash flow for early procurement. I should be involved in all procurement decisions involving manufacturing. I might have been able to get it cheaper that Betty did. So, how was this decision made without me?

Before Herb could say anything, Betty spoke up:

> Last month, Herb, you asked me to look into the cost of procuring these materials. I found a great price at one of the vendors and made the decision to purchase them. I thought that this was what you wanted me to do. This is how we did it in the last company I worked for.

Herb then remarked:

> I just wanted you to determine what the cost would be, not to make the final procurement decision, which is not your responsibility.

Friday the 13th was becoming possibly the worst day in Herb's life. Herb decided not to take any further chances. As soon as Betty and Frank left, Herb immediately sent out emails to all of the team members canceling the team meeting scheduled for 2:00 to 3:00 P.M. that afternoon.

QUESTIONS

1. How important are communication skills in project management?
2. Was Herb the right person to be assigned as the project manager?
3. There were communications issues with Alice, Bob, Betty, and Frank. For each communication issue, where was the breakdown in communications: encoding, decoding, feedback, and so on?

Source:

THE TEAM MEETING

Background

Every project team has team meetings. The hard part is deciding when during the day to have the team meeting.

Know Your Energy Cycle

Vince had been a "morning person" ever since graduating from college. He enjoyed getting up early. He knew his own energy cycle and the fact that he was obviously more productive in the morning than in the afternoon.

Vince would come into work at 6:00 A.M., 2 hours before the normal work force would show up. Between 6:00 A.M. and noon, Vince would keep his office door closed and often would not answer the phone. This prevented people from robbing Vince of his most productive time. Vince considered time robbers such as unnecessary phone calls lethal to the success of the project. This gave Vince 6 hours of productive time each day to do the necessary project work. After lunch, Vince would open his office door and anyone could then talk with him.

A Tough Decision

Vince's energy cycle worked well, at least for Vince. But Vince had just become the project manager on a large project. Vince knew that he might have to sacrifice some of his precious morning time for team meetings. It was customary for each project team to have a weekly team meeting, and most project team meetings seemed to be held in the morning. Initially, Vince decided to go against tradition and hold team meetings between 2:00 and 3:00 P.M. This would allow Vince to keep his precious morning time for his own productive work. Vince was somewhat disturbed when there was very little discussion on some of the critical issues and it appeared that people were looking at their watches. Finally, Vince understood the problem. A large portion of Vince's team members were manufacturing personnel that started work as early as 5:00 A.M. The manufacturing personnel were ready to go home at 2:00 P.M. and were tired.

The following week Vince changed the team meeting time to 11:00 A.M. to 12:00 P.M. It was evident to Vince that he had to sacrifice some of his morning time. But once again, during the team meetings there really wasn't very much discussion about some of the critical issues on the project and the manufacturing personnel were looking at their watches. Vince was disappointed and, as he exited the conference room, one of the manufacturing personnel commented to Vince, "Don't you know that the manufacturing people usually go to lunch around 11:00 A.M.?"

Vince came up with a plan for the next team meeting. He sent out emails to all of the team members stating that the team meeting would be at 11:00 A.M. to noon as before but the project would pick up the cost for providing lunch in the form of pizzas and salads. Much to Vince's surprise, this worked well. The atmosphere in the team meeting improved significantly. There were meaningful discussions and decisions were being made instead of creating action items for future team meetings. It suddenly became an informal rather than a formal team meeting.

While Vince's project could certainly incur the cost of pizzas, salads, and soft drinks for team meetings, this might set a bad precedent if this were to happen at each team meeting. At the next team meeting, the team decided that it would be nice if this could happen once or twice a month. For the other team meetings, it was decided to leave the time for the team meetings the same at 11:00 A.M. to noon, but they would be "brown bag" team meetings where the team members would bring their lunches and the project would provide only the soft drinks and perhaps some cookies or brownies.

QUESTIONS

1. How should a project manager determine when (i.e., time of day) to hold a team meeting? What factors should be considered?
2. What mistakes did Vince make initially?
3. If you were an executive in this company, would you allow Vince to continue doing this?

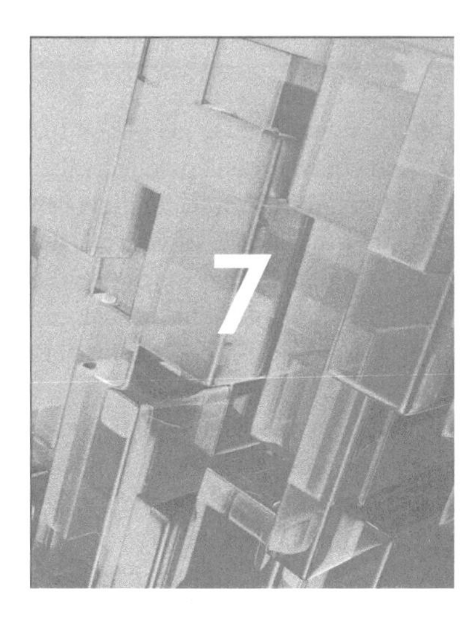

7 Conflicts

7.0 INTRODUCTION

PMBOK® Guide, 6th Edition
9.5 Manage Project Team
9.5.2.1 Interpersonal and Team Skills

In discussing the project environment, we have purposely avoided discussion of what may be its single most important characteristic: conflicts. Opponents of project management assert that the major reason why many companies avoid changeover to a project management organizational structure is either fear or an inability to handle the resulting conflicts. Conflicts are a way of life in a project structure and can generally occur at any level in the organization, usually as a result of conflicting objectives.

The project manager has often been described as a conflict manager. In many organizations the project manager continually fights fires and crises evolving from conflicts, and delegates the day-to-day responsibility of running the project to the project team members. Although this is not the best situation, it cannot always be prevented, especially after organizational restructuring or the initiation of projects requiring new resources. The ability to handle conflicts requires an understanding of why they occur. Asking and answering these four questions may help handle and prevent conflicts.

What are the project objectives and are they in conflict with other projects?

- Why do conflicts occur?
- How do we resolve conflicts?
- Is there any type of analysis that could identify possible conflicts before they occur?

7.1 THE CONFLICT ENVIRONMENT

In the project environment, conflicts are inevitable. Conflicts occur because people on the project team may have different values, interests, feelings, and goals. Project managers that cannot resolve these conflicts in a timely manner are doomed to failure. Some conflicts can be resolved quickly while other conflicts may take much longer to resolve. In general, the fewer the number of people involved in the conflict, the less time is needed to resolve the issues. Determining the amount of time needed to resolve an issue is difficult. Resolving conflicts with direct reportees is easier than resolving conflicts with those team members that are still attached administratively to other functional managers.

There are several causes of conflicts. First, project managers have historically been brought on board the project after the business case has been prepared. As a result, the business case, schedule, cost, assumptions, and other constraints are imposed upon the project team. All of this happens well before a detailed project plan is prepared. Once the project plan is finally prepared, it is often the case that the deliverables cannot be achieved in a timely manner within the imposed requirements and constraints.

Second, companies often approve projects without any consideration being given to capacity planning and whether or not qualified resources will be available once the project begins. This is particularly true for companies that survive on competitive bidding. These companies may have no idea how many contracts they will win, if any. The result is usually a shortage of qualified resources.

Third, projects are often approved and added to the queue without knowing when the project will begin. High-level schedules are established from a go-ahead date rather than a calendar date and, once again, with little regard for available or qualified resources. Once the project officially begins, the qualifications or work habits of the assigned project team members may not fit the needs of the project. And, as expected, you are then told that these are the only resources that are available.

Fourth, your project must be accomplished without disrupting the ongoing business of your company and other projects being performed. If your project has a low priority, then you must expect that your most critical resources may be temporarily removed to put out fires elsewhere in the company. These conflicts are highly probable in non–project-driven companies.

Fifth, the type of organizational structure can create conflicts. As an example, line managers that perform in a matrix structure are under tremendous pressure to staff a multitude of projects possibly at the same time. A delay on one project could result is a late release of personnel needed to staff new projects about to begin.

Here, we described five common causes of conflicts that can occur as the project begins. There are also numerous other conflicts that can occur during project execution. Ginger Levin provides a good discussion of the types of conflicts that can exist in each life-cycle phase as well as ways to handle them.[1]

Good project managers understand that conflicts will happen and try to plan for their resolution.

---

1. G. Levin, *Interpersonal Skills for Portfolio, Program, and Project Managers* (Leesburg Pike, VA: Management Concepts, 2010), Chapter 8.

7.2 TYPES OF CONFLICTS

It is impossible to develop a list of all of the different types of conflicts that can exist on each and every project. All projects differ in size, scope, and complexity. The most common types of conflicts involve:

- Staffing resources
- Equipment and facilities
- Capital expenditures
- Costs
- Technical opinions and trade-offs
- Priorities
- Administrative procedures
- Scheduling
- Responsibilities
- Personality clashes

Each of these conflicts can vary in relative intensity over the life cycle of a project. However, project managers believe that the most frequently occurring conflicts are over schedules but the potentially damaging conflicts can occur over personality clashes. The relative intensity can vary as a function of:

- Getting closer to project constraints
- Having only two constraints instead of three (i.e., time and performance, but not cost)
- The project life cycle itself
- The person with whom the conflict occurs

Sometimes conflict is "meaningful" and produces beneficial results. These meaningful conflicts should be permitted to continue as long as project constraints are not violated and beneficial results are being received. An example of this is two technical specialists arguing that each has a better way of solving a problem, and each trying to find additional supporting data for his hypothesis.

Conflicts can occur with anyone and over anything. Some people contend that personality conflicts are the most difficult to resolve.

Ideally, the project manager should report to someone high enough up to get timely assistance in resolving conflicts. Unfortunately, this is easier said than done. Therefore, project managers must plan for conflict resolution. As examples of this:

- The project manager might wish to concede on a low-intensity conflict if he knows that a high-intensity conflict is expected to occur at a later point in the project.
- Jones Construction Company has recently won a $120 million effort for a local company. The effort includes three separate construction projects, all beginning at the same time. Two of the projects are twenty-four months in duration, and the third is thirty-six months. Each project has its own project manager. When resource conflicts occur between the projects, the customer is usually called in.

- Richard is a department manager who must supply resources to four different projects. Although each project has an established priority, the project managers continually argue that departmental resources are not being allocated effectively. Richard now holds a monthly meeting with all four of the project managers and lets them determine how the resources should be allocated.

Many executives feel that the best way of resolving conflicts is by establishing priorities. This may be true as long as priorities are not continually shifted around.

The most common factors influencing the establishment of project priorities include:

- The technical risks in development
- The risks that the company will incur, financially or competitively
- The nearness of the delivery date and the urgency
- The penalties that can accompany late delivery dates
- The expected savings, profit increase, and return on investment
- The amount of influence that the customer possesses, possibly due to the size of the project
- The impact on other projects or product lines
- The impact on affiliated organizations

The ultimate responsibility for establishing priorities rests with top-level management. Yet even with priority establishment, conflicts still develop.

7.3 CONFLICT RESOLUTION

PMBOK® *Guide*, 6th Edition
9.5.2.1 Interpersonal and Team Skills

Although each project within the company may be inherently different, the company may wish to have the resulting conflicts resolved in the same manner. The four most common methods are:

1. *The development of company-wide conflict resolution policies and procedures.* Many companies have attempted to develop company-wide policies and procedures for conflict resolution, but this method is often doomed to failure because each project and conflict is different. Furthermore, project managers, by virtue of their individuality, and sometimes differing amounts of authority and responsibility, prefer to resolve conflicts in their own fashion.
2. *The establishment of project conflict resolution procedures during the early planning activities.* One method that is often very effective is to "plan" for conflicts during the planning activities. This can be accomplished through the use of linear responsibility charts. Planning for conflict resolution is similar to the first method except that each project manager can develop his or her own policies, rules, and procedures.
3. *The use of hierarchical referral.* In theory, this appears as the best method because neither the project manager nor the functional manager will dominate. Under this arrangement, the project and functional managers agree that for a proper balance to exist

their common superior must resolve the conflict to protect the company's best interest. Unfortunately, this is not realistic because the common superior cannot be expected to continually resolve lower-level conflicts, and it gives the impression that the functional and project managers cannot resolve their own problems.

4. *The requirement of direct contact.* This is direct contact in which conflicting parties meet face to face and resolve their disagreement. Unfortunately, this method does not always work and, if continually stressed, can result in conditions where individuals will either suppress the identification of problems or develop new ones during confrontation.

Many conflicts can be either reduced or eliminated by constant communication of the project objectives to the team members. This continual repetition may prevent individuals from going too far in the wrong direction.

7.4 THE MANAGEMENT OF CONFLICTS

PMBOK® *Guide*, 6th Edition
9.5.2.1 Interpersonal and Team Skills

Good project managers realize that conflicts are inevitable, but that good procedures or techniques can help resolve them. Once a conflict occurs, the project manager must:

- Study the problem and collect all available information
- Develop a situational approach or methodology
- Set the appropriate atmosphere or climate

If a confrontation meeting is necessary between conflicting parties, then the project manager should be aware of the logical steps and sequence of events that should be taken. These include:

- Setting the climate: establishing a willingness to participate
- Analyzing the images: how do you see yourself and others, and how do they see you?
- Collecting the information: getting feelings out in the open
- Defining the problem: defining and clarifying all positions
- Sharing the information: making the information available to all
- Setting the appropriate priorities: developing working sessions for setting priorities and timetables
- Organizing the group: forming cross-functional problem-solving groups
- Problem-solving: obtaining cross-functional involvement, securing commitments, and setting the priorities and timetable
- Developing the action plan: getting commitment
- Implementing the work: taking action on the plan
- Following up: obtaining feedback on the implementation for the action plan

---

The majority of Section 7.4, including the figures, was adapted from Seminar in Project Management Workbook, ©1977 by Hans J. Thamhain. Reproduced by permission of Dr. Hans J. Thamhain.

The project manager or team leader should also understand conflict minimization procedures. These include:

- Pausing and thinking before reacting
- Building trust
- Trying to understand the conflict motives
- Keeping the meeting under control
- Listening to all involved parties
- Maintaining a give-and-take attitude
- Educating others tactfully on your views
- Being willing to say when you were wrong
- Not acting as a superman and leveling the discussion only once in a while

Thus, the effective manager, in conflict problem-solving situations:

- Knows the organization
- Listens with understanding rather than evaluation
- Clarifies the nature of the conflict
- Understands the feelings of others
- Suggests the procedures for resolving differences
- Maintains relationships with disputing parties
- Facilitates the communications process
- Seeks resolutions

7.5 CONFLICT RESOLUTION MODES

PMBOK® *Guide*, 6th Edition
9.5.2.1 Interpersonal and Team Skills

The management of conflicts places the project manager in the precarious situation of having to select a conflict resolution mode (previously defined in Section 7.3). Based upon the situation, the type of conflict, and whom the conflict is with, any of these modes could be justified.

Confronting (or Collaborating) With this approach, the conflicting parties meet face to face and try to work through their disagreements. This approach should focus more on solving the problem and less on being combative. This approach is collaboration and integration where both parties need to win. This method should be used:

- When you and the conflicting party can both get at least what you wanted and maybe more
- To reduce cost
- To create a common power base
- To attack a common foe
- When skills are complementary

- When there is enough time
- When there is trust
- When you have confidence in the other person's ability
- When the ultimate objective is to learn

Compromising

To compromise is to bargain or to search for solutions so both parties leave with some degree of satisfaction. Compromising is often the result of confrontation. Some people argue that compromise is a "give-and-take" approach, which leads to a "win-win" position. Others argue that compromise is a "lose-lose" position, since neither party gets everything he/she wants or needs. Compromise should be used:

- When both parties need to be winners
- When you can't win
- When others are as strong as you are
- When you haven't time to win
- To maintain your relationship with your opponent
- When you are not sure you are right
- When you get nothing if you don't
- When stakes are moderate
- To avoid giving the impression of "fighting"

Smoothing (or Accommodating)

Smoothing is an attempt to reduce the emotions that exist in a conflict. This is accomplished by emphasizing areas of agreement and deemphasizing areas of disagreement. An example of smoothing would be to tell someone, "We have agreed on three of the five points and there is no reason why we cannot agree on the last two points." Smoothing does not necessarily resolve a conflict, but tries to convince both parties to remain at the bargaining table because a solution is possible. In smoothing, one may sacrifice one's own goals in order to satisfy the needs of the other party. Smoothing should be used:

- To reach an overarching goal
- To create obligation for a trade-off at a later date
- When the stakes are low
- When liability is limited
- To maintain harmony
- When any solution will be adequate
- To create goodwill (be magnanimous)
- When you'll lose anyway
- To gain time

Forcing (or Competing, Being Assertive)

This is what happens when one party tries to impose the solution on Being Uncooperative, the other party. Conflict resolution works best when resolution is achieved at the lowest possible levels. The higher up the conflict goes, the greater the tendency for the conflict to be forced, with the result being a "win-lose" situation in which one party wins at the expense of the other. Forcing should be used:

- When you are right
- When a do-or-die situation exists
- When stakes are high
- When important principles are at stake
- When you are stronger (never start a battle you can't win)
- To gain status or to gain power
- In short-term, one-shot deals
- When the relationship is unimportant
- When it's understood that a game is being played
- When a quick decision must be made

Avoiding (or Withdrawing) Avoidance is often regarded as a temporary solution to a problem. The problem and the resulting conflict can come up again and again. Some people view avoiding as cowardice and an unwillingness to be responsive to a situation. Avoiding should be used:

- When you can't win
- When the stakes are low
- When the stakes are high, but you are not ready yet
- To gain time
- To unnerve your opponent
- To preserve neutrality or reputation
- When you think the problem will go away
- When you win by delay

7.6 UNDERSTANDING SUPERIOR, SUBORDINATE, AND FUNCTIONAL CONFLICTS

PMBOK® Guide, 5th Edition
9.5.2.1 Interpersonal and Team Skills

In order for the project manager to be effective, he must understand how to work with the various employees who interface with the project. These employees include upper-level management, subordinate project team members, and functional personnel. Quite often, the project manager must demonstrate an ability for continuous adaptability by creating a different working environment with each group of employees. The need for this was shown in the previous section by the fact that the relative intensity of conflicts can vary in the life cycle of a project.

The type and intensity of conflicts can also vary with the type of employee, as shown in Figure 7–1. Both conflict causes and sources are rated according to relative conflict intensity.

The majority of Section 7.6, including the figures, was adapted from *Seminar in Project Management Workbook,* © 1977 by Hans J. Thamhain. Reproduced by permission of Dr. Hans J. Thamhain.

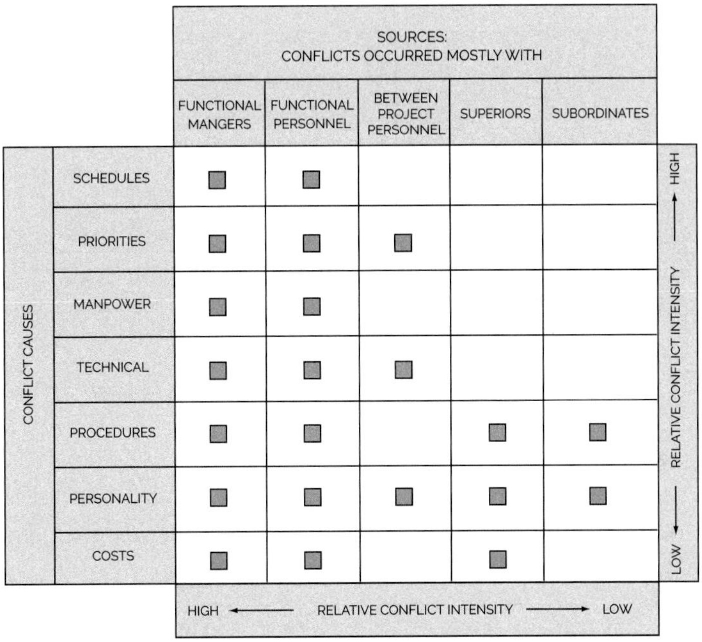

CONFLICT CAUSES	SOURCES: CONFLICTS OCCURRED MOSTLY WITH				
	FUNCTIONAL MANGERS	FUNCTIONAL PERSONNEL	BETWEEN PROJECT PERSONNEL	SUPERIORS	SUBORDINATES
SCHEDULES	■	■			
PRIORITIES	■	■	■		
MANPOWER	■	■			
TECHNICAL	■	■	■		
PROCEDURES	■	■		■	■
PERSONALITY	■	■	■	■	■
COSTS	■	■		■	

HIGH ← RELATIVE CONFLICT INTENSITY → LOW (across the columns)

LOW ← RELATIVE CONFLICT INTENSITY → HIGH (down the rows, from HIGH at top to LOW at bottom)

FIGURE 7–1. Relationship between conflict causes and sources.

The specific resolution mode a project manager will use might easily depend on whom the conflict is with, as shown in Figure 7–2. The data in Figure 7–2 do not necessarily show the modes that project managers would prefer, but rather identify the modes that will increase or decrease the potential conflict intensity. For example, although project managers consider, in general, that withdrawal is their least favorite mode, it can be used quite effectively with functional managers. In dealing with superiors, project managers would rather be ready for an immediate compromise than for face-to-face confrontation that could favor upper-level management.

Figure 7–3 identifies the various influence styles that project managers find effective in helping to reduce potential conflicts. Penalty power, authority, and expertise are considered as strongly unfavorable associations with respect to low conflicts. As expected, work challenge and promotions (if the project manager has the authority) are strongly favorable.

Related Case Studies (from Kerzner/*Project Management Case Studies*, 5th ed.)	**Related Workbook Exercises (from Kerzner/*Project Management Workbook and PMP®/CAPM® Exam Study Guide*, 12th ed.)**	***PMBOK® Guide*, Sixth Edition, Reference Section for the PMP® Certification Exam**
• Facilities Scheduling at Mayer Manufacturing* • Scheduling the Safety Lab • Telestar International* • The Problem with Priorities	• Multiple Choice Exam	• Human Resource Management

*Case study also appears at end of chapter.

PMP and CAPM are registered marks of the Project Management Institute, Inc.

(The figure shows only those associations which are statistically significant at the 95 percent level)

INTENSITY OF CONFLICT PERCEIVED BY PROJECT MANAGERS (P.M.)	ACTUAL CONFLICT RESOLUTION STYLE				
	FORCING	CONFRONTATION	COMPROMISE	SMOOTHING	WITHDRAWAL
BETWEEN P.M. AND HIS PERSONNEL	■	△	△	△	■
BETWEEN P.M. AND HIS SUPERIOR		■	△		
BETWEEN P.M. AND FUNCTIONAL SUPPORT DEPARTMENTS	■	■			△

△ STRONGLY FAVORABLE ASSOCIATION WITH REGARD TO LOW CONFLICT (− τ)

■ STRONGLY UNFAVORABLE ASSOCIATION WITH REGARD TO LOW CONFLICT(+ τ)

• KENDALL − τ CORRELATION

FIGURE 7–2. Association between perceived intensity of conflict and mode of conflict resolution.

(The figure shows only those associated which are statistically significant at the 95 percent level)

INTENSITY OF CONFLICT PERCEIVED BY PROJECT MANAGER (P.M.)	INFLUENCE METHODS AS PERCEIVED BY PROJECT MANAGERS						
	EXPERTISE	AUTHORITY	WORK CHALLENGE	FRIENDSHIP	PROMOTION	SALARY	PENALTY
BETWEEN P.M. AND HIS PERSONNEL	■	■	△		△		■
BETWEEN P.M. AND HIS SUPERIOR			△				■
BETWEEN P.M. AND FUNCTIONAL SUPPORT DEPARTMENTS		■					■

△ STRONGLY FAVORABLE ASSOCIATION WITH REGARD TO LOW CONFLICT (− τ)

■ STRONGLY UNFAVORABLE ASSOCIATION WITH REGARD TO LOW CONFLICT(+ τ)

• KENDALL τ CORRELATION

FIGURE 7–3. Association between influence methods of project managers and their perceived conflict intensity.

7.7 STUDYING TIPS FOR THE PMI® PROJECT MANAGEMENT CERTIFICATION EXAM

This section is applicable as a review of the principles to support the knowledge areas and domain groups in the *PMBOK® Guide*. This chapter addresses:

- Project Resource Management
- Execution

Understanding the following principles is beneficial if the reader is using this text to study for the PMP® Certification Exam:

- Different types of conflicts that can occur in a project environment
- Different conflict resolution modes and when each one should be used

The following multiple-choice questions will be helpful in reviewing the principles of this chapter:

1. Project managers believe that the most commonly occurring conflict is:
   A. Priorities
   B. Schedules
   C. Personalities
   D. Resources
2. The conflict that generally is the most damaging to the project when it occurs is:
   A. Priorities
   B. Schedules
   C. Personalities
   D. Resources
3. The most commonly preferred conflict resolution mode for project managers is:
   A. Compromise
   B. Confrontation
   C. Smoothing
   D. Withdrawal
4. Which conflict resolution mode is equivalent to problem solving?
   A. Compromise
   B. Confrontation
   C. Smoothing
   D. Withdrawal
5. Which conflict resolution mode avoids a conflict temporarily rather than solving it?
   A. Compromise
   B. Confrontation
   C. Smoothing
   D. Withdrawal

ANSWERS

1. B
2. C
3. B
4. B
5. D

PROBLEMS

7–1 Is it possible to establish formal organizational procedures (either at the project level or company-wide) for the resolution of conflicts? If a procedure is established, what can go wrong?

7–2 If a situation occurs that can develop into meaningful conflict, should the project manager let the conflict continue as long as it produces beneficial contributions, or should he try to resolve it as soon as possible?

7–3 For each part below there are two statements; one represents the traditional view and the other the project organizational view. Identify each one.

A. Conflict should be avoided; conflict is part of change and is therefore inevitable.

B. Conflict is the result of troublemakers and egoists; conflict is determined by the structure of the system and the relationship among components.

C. Conflict may be beneficial; conflict is bad.

7–4 Would you agree or disagree with the statement that "Conflict resolution through collaboration needs trust; people must rely on one another"?

7–5 Determine the best conflict resolution mode for each of the following situations:

a. Two of your functional team members appear to have personality clashes and almost always assume opposite points of view during decision making.

b. R&D quality control and manufacturing operations quality control continually argue as to who should perform testing on an R&D project. R&D postulates that it's their project, and manufacturing argues that it will eventually go into production and that they wish to be involved as early as possible.

c. Two functional department managers continually argue as to who should perform a certain test. You know that this situation exists, and that the department managers are trying to work it out themselves, often with great pain. However, you are not sure that they will be able to resolve the problem themselves.

7–6 One of the most common conflicts in an organization occurs with raw materials and finished goods. Why would finance/accounting, marketing/sales, and manufacturing have disagreements?

7–7 Explain how the relative intensity of a conflict can vary as a function of:

A. Getting closer to the actual constraints

B. Having only two constraints instead of three (i.e., time and performance, but not cost)

C. The project life cycle

D. The person with whom the conflict occurs

CASE STUDIES

FACILITIES SCHEDULING AT MAYER MANUFACTURING

Eddie Turner was elated with the good news that he was being promoted to section supervisor in charge of scheduling all activities in the new engineering research laboratory. The new laboratory was a necessity for Mayer Manufacturing. The engineering, manufacturing, and quality

Mayer Manufacturing organizational structure

V.P.

OTHERS
ENGINEERING
MANUFACTURING
QUALITY CONTROL
DIVISION MANAGERS
DEPARTMENT MANAGERS
GARY WHITEHEAD
SECTION SUPERVISORS
EDDIE TURNER

control directorates were all in desperate need of a new testing facility. Upper-level management felt that this new facility would alleviate many of the problems that previously existed.

The new organizational structure (as shown in the illustration) required a change in policy over use of the laboratory. The new section supervisor, on approval from his department manager, would have full authority for establishing priorities for the use of the new facility. The new policy change was a necessity because upper-level management felt that there would be inevitable conflict between manufacturing, engineering, and quality control.

After one month of operations, Eddie Turner was finding his job impossible, so Eddie has a meeting with Gary Whitehead, his department manager.

Eddie: "I'm having a hell of a time trying to satisfy all of the department managers. If I give engineering prime-time use of the facility, then quality control and manufacturing say that I'm playing favorites. Imagine that! Even my own people say that I'm playing favorites with other directorates. I just can't satisfy everyone."

Gary: "Well, Eddie, you know that this problem comes with the job. You'll get the job done."

Eddie: "The problem is that I'm a section supervisor and have to work with department managers. These department managers look down on me like I'm their servant. If I were a department manager, then they'd show me some respect. What I'm really trying to say is that I would like you to send out the weekly memos to these department managers telling them of the new priorities. They wouldn't argue with you like they do with me. I can supply you with all the necessary information. All you'll have to do is to sign your name."

Gary: "Determining the priorities and scheduling the facilities is your job, not mine. This is a new position and I want you to handle it. I know you can because I selected you. I do not intend to interfere."

During the next two weeks, the conflicts got progressively worse. Eddie felt that he was unable to cope with the situation by himself. The department managers did not respect the authority delegated to him by his superiors. For the next two weeks, Eddie sent memos to Gary in the early part of the week asking whether Gary agreed with the priority list. There was no response to the two memos. Eddie then met with Gary to discuss the deteriorating situation.

Eddie: "Gary, I've sent you two memos to see if I'm doing anything wrong in establishing the weekly priorities and schedules. Did you get my memos?"

Gary: "Yes, I received your memos. But as I told you before, I have enough problems to worry about without doing your job for you. If you can't handle the work let me know and I'll find someone who can."

Eddie returned to his desk and contemplated his situation. Finally, he made a decision.

Next week he was going to put a signature block under his for Gary to sign, with carbon copies for all division managers. "Now, let's see what happens," remarked Eddie.

QUESTIONS

1. What was Eddie's problem?
2. Did Eddie create the problem himself or did others create it for him?
3. Was there a failure of project sponsorship or is this the way it typically works in companies?
4. Were there better solutions to the conflict, and if so, what are they?

*TELESTAR INTERNATIONAL**

On November 15, 2008, the Department of Energy Resources awarded Telestar a $475,000 contract for the developing and testing of two waste treatment plants. Telestar had spent the better part of the last two years developing waste treatment technology under its own R&D activities. This new contract would give Telestar the opportunity to "break into a new field"—that of waste treatment.

The contract was negotiated at a firm-fixed price. Any cost overruns would have to be incurred by Telestar. The original bid was priced out at $847,000. Telestar's management, however, wanted to win this one. The decision was made that Telestar would "buy in" at $475,000 so that they could at least get their foot into the new marketplace.

The original estimate of $847,000 was very "rough" because Telestar did not have any good man-hour standards, in the area of waste treatment, on which to base their man-hour projections. Corporate management was willing to spend up to $400,000 of their own funds in order to compensate the bid of $475,000.

By February 15, 2009, costs were increasing to such a point where overrun would be occurring well ahead of schedule. Anticipated costs to completion were now $943,000. The project manager decided to stop all activities in certain functional departments, one of which was structural analysis. The manager of the structural analysis department strongly opposed the closing out of the work order prior to the testing of the first plant's high-pressure pneumatic and electrical systems.

Structures Manager: "You're running a risk if you close out this work order. How will you know if the hardware can withstand the stresses that will be imposed during the test? After all, the test is scheduled for next month and I can probably finish the analysis by then."

Project Manager: "I understand your concern, but I cannot risk a cost overrun. My boss expects me to do the work within cost. The plant design is similar to one that we have tested before, without any structural problems being detected. On this basis I consider your analysis unnecessary."

Structures Manager: "Just because two plants are similar does not mean that they will be identical in performance. There can be major structural deficiencies."

Project Manager: "I guess the risk is mine."

Structures Manager: "Yes, but I get concerned when a failure can reflect on the integrity of my department. You know, we're performing on schedule and within the time and money budgeted. You're setting a bad example by cutting off our budget without any real justification."

Project Manager: "I understand your concern, but we must pull out all the stops when overrun costs are inevitable."

* Revised, 2015.

Structures Manager: "There's no question in my mind that this analysis should be completed. However, I'm not going to complete it on my overhead budget. I'll reassign my people tomorrow. Incidentally, you had better be careful; my people are not very happy to work for a project that can be canceled immediately. I may have trouble getting volunteers next time."

Project Manager: "Well, I'm sure you'll be able to adequately handle any future work. I'll report to my boss that I have issued a work stoppage order to your department."

During the next month's test, the plant exploded. Post analysis indicated that the failure was due to a structural deficiency.

QUESTIONS

1. Who is at fault?
2. Should the structures manager have been dedicated enough to continue the work on his own?
3. Can a functional manager, who considers his organization as strictly support, still be dedicated to total project success?

HANDLING CONFLICT IN PROJECT MANAGEMENT

The next several pages contain a six-part case study in conflict management. Read the instructions in Appendix A carefully on how to keep score and use the boxes in the table on page 252 as the worksheet for recording your choice and the group's choice; after the case study has been completed, your instructor will provide you with the proper grading system for recording your scores.

Part 1: Facing the Conflict

As part of his first official duties, the new department manager informs you by memo that he has changed his input and output requirements for the MIS project (on which you are the project manager) because of several complaints by his departmental employees. This is contradictory to the project plan that you developed with the previous manager and are currently working toward. The department manager states that he has already discussed this with the vice president and general manager, a man to whom both of you report, and feels that the former department manager made a poor decision and did not get sufficient input from the employees who would be using the system as to the best system specifications. You telephone him and try to convince him to hold off on his request for change until a later time, but he refuses.

Changing the input–output requirements at this point in time will require a major revision and will set back total system implementation by three weeks. This will also affect other department managers who expect to see this system operational according to the original schedule. You can explain this to your superiors, but the increased project costs will be hard to absorb. The potential cost overrun might be difficult to explain at a later date.

At this point you are somewhat unhappy with yourself at having been on the search committee that found this department manager and especially at having recommended him for this position. You know that something must be done, and the following are your alternatives:

A. You can remind the department manager that you were on the search committee that recommended him and then ask him to return the favor, since he "owes you one."

B. You can tell the department manager that you will form a new search committee to replace him if he doesn't change his position.

C. You can take a tranquilizer and then ask your people to try to perform the additional work within the original time and cost constraints.

D. You can go to the vice president and general manager and request that the former requirements be adhered to, at least temporarily.

E. You can send a memo to the department manager explaining your problem and asking him to help you find a solution.

F. You can tell the department manager that your people cannot handle the request and his people will have to find alternate ways of solving their problems.

G. You can send a memo to the department manager requesting an appointment, at his earliest convenience, to help you resolve your problem.

H. You can go to the department manager's office later that afternoon and continue the discussion further.

I. You can send the department manager a memo telling him that you have decided to use the old requirements but will honor his request at a later time.

Line	Part	Personal		Group	
		Choice	Score	Choice	Score
1	1. Facing the Conflict				
2	2. Understanding Emotions				
3	3. Establishing Communications				
4	4. Conflict Resolution				
5	5. Understanding Your Choices				
6	6. Interpersonal Influences				
	TOTAL				

Although other alternatives exist, assume that these are the only ones open to you at the moment. Without discussing the answer with your group, record the letter representing your choice in the appropriate space on line 1 of the worksheet under "Personal."

As soon as all of your group have finished, discuss the problem as a group and determine that alternative that the group considers to be best. Record this answer on line 1 of the worksheet under "Group." Allow ten minutes for this part.

Part 2: Understanding Emotions

Never having worked with this department manager before, you try to predict what his reactions will be when confronted with the problem. Obviously, he can react in a variety of ways:

A. He can *accept* your solution in its entirety without asking any questions.

B. He can discuss some sort of justification in order to *defend* his position.

C. He can become extremely annoyed with having to discuss the problem again and demonstrate *hostility*.

D. He can demonstrate a willingness to *cooperate* with you in resolving the problem.

E. He can avoid making any decision at this time by *withdrawing* from the discussion.

	Your Choice					Group Choice				
	Acc.	**Def.**	**Host.**	**Coop.**	**With.**	**Acc.**	**Def.**	**Host.**	**Coop.**	**With.**
A. I've given my answer. See the general manager if you're not happy.										
B. I understand your problem. Let's do it your way.										
C. I understand your problem, but I'm doing what is best for my department.										
D. Let's discuss the problem. Perhaps there are alternatives.										
E. Let me explain to you why we need the new requirements.										
F. See my section supervisors. It was their recommendation.										
G. New managers are supposed to come up with new and better ways, aren't they?										

In the table above are several possible statements that could be made by the department manager when confronted with the problem. Without discussion with your group, place a check mark beside the appropriate emotion that could describe this statement. When each member of the group has completed his or her choice, determine the group choice. Numerical values will be assigned to your choices in the discussion that follows. Do not mark the worksheet at this time. Allow ten minutes for this part.

Part 3: Establishing Communications

Unhappy over the department manager's memo and the resulting follow up phone conversation, you decide to walk in on the department manager. You tell him that you will have a problem trying to honor his request. He tells you that he is too busy with his own problems of restructuring his department and that your schedule and cost problems are of no concern to him at this time. You storm out of his office, leaving him with the impression that his actions and remarks are not in the best interest of either the project or the company. The department manager's actions do not, of course, appear to be those of a dedicated manager. He should be more concerned about what's in the best interest of the company. As you contemplate the situation, you wonder if you could have received a better response from him had you approached him differently. In other words, what is your best approach to opening up communications between you and the department manager? From the list of alternatives shown below, and working alone, select the alternative that best represents how you would handle this situation. When all members of the group have selected their personal choices, repeat the process and make a group choice. Record your personal and group choices on line 3 of the worksheet. Allow ten minutes for this part.

A. Comply with the request and document all results so that you will be able to defend yourself at a later date in order to show that the department manager should be held accountable.
B. Immediately send him a memo reiterating your position and tell him that at a later time you will reconsider his new requirements. Tell him that time is of utmost importance, and you need an immediate response if he is displeased.
C. Send him a memo stating that you are holding him accountable for all cost overruns and schedule delays.
D. Send him a memo stating you are considering his request and that you plan to see him again at a later date to discuss changing the requirements.
E. See him as soon as possible. Tell him that he need not apologize for his remarks and actions, and that you have reconsidered your position and wish to discuss it with him.
F. Delay talking to him for a few days in hopes that he will cool off sufficiently and then see him in hopes that you can reopen the discussions.
G. Wait a day or so for everyone to cool off and then try to see him through an appointment; apologize for losing your temper, and ask him if he would like to help you resolve the problem.

Part 4: Conflict Resolution Modes

Having never worked with this manager before, you are unsure about which conflict resolution mode would work best. You decide to wait a few days and then set up an appointment with the department manager without stating what subject matter will be discussed. You then try to determine what conflict resolution mode appears to be dominant based on the opening remarks of the department manager. Neglecting the fact that your conversation with the department manager might already be considered as confrontation, for each statement shown below, select the conflict resolution mode that the *department manager* appears to prefer. After each member of the group has recorded his personal choices in the table below, determine the group choices. Numerical values will be attached to your answers at a later time. Allow ten minutes for this part.

A. *Withdrawal* is retreating from a potential conflict.
B. *Smoothing* is emphasizing areas of agreement and deemphasizing areas of disagreement.
C. *Compromising* is the willingness to give and take.
D. *Forcing* is directing the resolution in one direction or another, a win-or-lose position.
E. *Confrontation* is a face-to-face meeting to resolve the conflict.

	Personal Choice					Group Choice				
	With.	**Smooth.**	**Comp.**	**Forc.**	**Conf.**	**With.**	**Smooth.**	**Comp.**	**Forc.**	**Conf.**
A. The requirements are my decision, and we're doing it my way.										
B. I've thought about it and you're right. We'll do it your way.										
C. Let's discuss the problem. Perhaps there are alternatives.										
D. Let me again explain why we need the new requirements.										
E. See my section supervisors; they're handling it now.										
F. I've looked over the problem and I might be able to ease up on some of the requirements.										

Part 5: Understanding Your Choices

Assume that the department manager has refused to see you again to discuss the new requirements. Time is running out, and you would like to make a decision before the costs and schedules get out of hand. From the list below, select your personal choice and then, after each group member is finished, find a group choice.

A. Disregard the new requirements, since they weren't part of the original project plan.
B. Adhere to the new requirements, and absorb the increased costs and delays.
C. Ask the vice president and general manager to step in and make the final decision.
D. Ask the other department managers who may realize a schedule delay to try to convince this department manager to ease his request or even delay it.

Record your answer on line 5 of the worksheet. Allow five minutes for this part.

Part 6: Interpersonal Influences

Assume that upper-level management resolves the conflict in your favor. In order to complete the original work requirements you will need support from this department manager's organization. Unfortunately, you are not sure as to which type of interpersonal influence to use. Although you are considered an expert in your field, you fear that this manager's functional employees may have a strong allegiance to the department manager and may not want to adhere to your requests. Which of the following interpersonal influence styles would be best under the given set of conditions?

A. You threaten the employees with penalty power by telling them that you will turn in a bad performance report to their department manager.
B. You can use reward power and promise the employees a good evaluation, possible promotion, and increased responsibilities on your next project.
C. You can continue your technique of trying to convince the functional personnel to do your bidding because you are the expert in the field.
D. You can try to motivate the employees to do a good job by convincing them that the work is challenging.
E. You can make sure that they understand that your authority has been delegated to you by the vice president and general manager and that they must do what you say.
F. You can try to build up friendships and off-work relationships with these people and rely on referent power.

Record your personal and group choices on line 6 of the worksheet. Allow ten minutes for completion of this part.

The solution to this exercise appears in Appendix A.

Special Topics

8.0 INTRODUCTION

There are several situations or special topics that deserve attention. These include:

- Performance measurement
- Compensation and rewards
- Managing small projects
- Managing mega projects
- Morality, ethics, and the corporate culture
- Professional responsibility
- Internal and external partnerships
- Training and education
- Integrated project teams
- Virtual teams
- Innovation projects
- Agile project management

8.1 PERFORMANCE MEASUREMENT

> ***PMBOK® Guide*, 5th Edition**
> 9.4.3.1 Team Performance Assessments
> 9.5 Manage Team

A good project manager will make it immediately clear to all new functional employees that, if they perform well in the project, then he or she (the project manager) will inform the functional manager of their progress and achievements.

This assumes that the functional manager is not providing close supervision over the functional employees and is, instead, passing on some of the responsibility to the project manager—a common situation in project management organization structures.

Many good projects, as well as project management structures, have failed because of the inability of the system to evaluate properly the functional employee's performance. In a project management structure, there are basically six ways that a functional employee can be evaluated on a project:

- *The project manager prepares a written, confidential evaluation and gives it to the functional manager.* The functional manager will evaluate the validity of the project manager's comments and prepare his or her own evaluation. Only the line manager's evaluation is shown to the employee. The use of confidential forms is not preferred because it may be contrary to government regulations and it does not provide the feedback necessary for an employee to improve.
- *The project manager prepares a nonconfidential evaluation and gives it to the functional manager.* The functional manager prepares his or her own evaluation form and shows both evaluations to the functional employee. This is the technique preferred by most project and functional managers. However, there are several major difficulties with this technique. If the functional employee is an average or below-average worker, and if this employee is still to be assigned to that project after his or her evaluation, then the project manager might rate the employee as above average simply to prevent any sabotage or bad feelings downstream. In this situation, the functional manager might want a confidential evaluation instead, knowing that the functional employee will see both evaluation forms. Functional employees tend to blame the project manager if they receive a below-average merit pay increase, but give credit to the functional manager if the increase is above average. The best bet here is for the project manager periodically to tell the functional employees how well they are doing, and to give them an honest appraisal. Several companies that use this technique allow the project manager to show the form to the line manager first (to avoid conflict later) and then show it to the employee.
- *The project manager provides the functional manager with an oral evaluation of the employee's performance.* Although this technique is commonly used, most functional managers prefer documentation on employee progress. Again, lack of feedback may prevent the employee from improving.
- *The functional manager makes the entire evaluation without any input from the project manager.* In order for this technique to be effective, the functional manager must have sufficient time to supervise each subordinate's performance on a continual basis. Unfortunately, most functional managers do not have this luxury because of their broad span of control and must therefore rely heavily on the project manager's input.
- *The project manager makes the entire evaluation for the functional manager.* This technique can work if the functional employee spends 100 percent of his time on one project, or if he is physically located at a remote site where he cannot be observed by his functional manager.

PERFORMANCE FACTORS	EXCELLENT (1 OUT OF 15)	VERY GOOD (3 OUT OF 15)	GOOD (8 OUT OF 15)	FAIR (2 OUT OF 15)	UNSATISFACTORY (1 OUT OF 15)
	FAR EXCEEDS JOB REQUIREMENTS	EXCEEDS JOB REQUIREMENTS	MEETS JOB REQUIREMENTS	NEEDS SOME IMPROVEMENT	DOES NOT MEET MINIMUM STANDARDS
QUALITY	LEAPS TALL BUILDINGS WITH A SINGLE BOUND	MUST TAKE RUNNING START TO LEAP OVER TALL BUILDING	CAN ONLY LEAP OVER A SHORT BUILDING OR MEDIUM ONE WITHOUT SPIRES	CRASHES INTO BUILDING	CANNOT RECOGNIZE BUILDINGS
TIMELINESS	IS FASTER THAN A SPEEDING BULLET	IS AS FAST AS A SPEEDING BULLET	NOT QUITE AS FAST AS A SPEEDING BULLET	WOULD YOU BELIEVE A SLOW BULLET?	WOUNDS HIMSELF WITH THE BULLET
INITIATIVE	IS STRONGER THAN A LOCOMOTIVE	IS STRONGER THAN A BULL ELEPHANT	IS STRONGER THAN A BULL	SHOOTS THE BULL	SMELLS LIKE A BULL
ADAPTABILITY	WALKS ON WATER CONSISTENTLY	WALKS ON WATER IN EMERGENCIES	WASHES WITH WATER	DRINKS WATER	PASSES WATER IN EMERGENCIES
COMMUNICATIONS	TALKS WITH GOD	TALKS WITH ANGELS	TALKS TO HIMSELF	ARGUES WITH HIMSELF	LOSES THE ARGUMENT WITH HIMSELF

FIGURE 8–1. Guide to performance appraisal.

- *All project and functional managers jointly evaluate all project functional employees at the same time.* This technique should be limited to small companies with fewer than fifty or so employees; otherwise the evaluation process might be time-consuming for key personnel. A bad evaluation will be known by everyone.

Figure 8–1 represents, in a humorous way, how project personnel perceive the evaluation form. Unfortunately, the evaluation process is very serious and can easily have a severe impact on an individual's career path with the company even though the final evaluation rests with the functional manager.

Figure 8–2 shows a simple type of evaluation form on which the project manager identifies the best description of the employee's performance. This type of form is generally used whenever the employee is up for evaluation.

Figure 8–3 shows another typical form that can be used to evaluate an employee. In each category, the employee is rated on a subjective scale. In order to minimize time and paperwork, it is also possible to have a single evaluation form at project termination for evaluation of all employees. This is shown in Figure 8–4. All employees are rated in each category on a scale of 1 to 5. Totals are obtained to provide a relative comparison of employees.

Obviously, evaluation forms such as that shown in Figure 8–4 have severe limitations, as a one-to-one comparison of all project functional personnel is of little value if the

EMPLOYEE'S NAME	DATE
PROJECT TITLE	JOB NUMBER
EMPLOYEE ASSIGNMENT	
EMPLOYEE'S TOTAL TIME TO DATE ON PROJECT	EMPLOYEE'S REMAINING TIME ON PROJECT

TECHNICAL JUDGMENT:				
☐ Quickly reaches sound conclusions	☐ Usually makes sound conclusions	☐ Marginal decision-making ability	☐ Needs technical assistance	☐ Makes faulty conclusions
WORK PLANNING:				
☐ Good planner	☐ Plans well with help	☐ Occasionally plans well	☐ Needs detailed instructions	☐ Cannot plan at all
COMMUNICATIONS:				
☐ Always understands instructions	☐ Sometimes needs clarification	☐ Always needs clarifications	☐ Needs follow-up	☐ Needs constant instruction
ATTITUDE:				
☐ Always job interested	☐ Shows interest most of the time	☐ Shows no job interest	☐ More interested in other activities	☐ Does not care about job
COOPERATION:				
☐ Always enthusiastic	☐ Works well until job is completed	☐ Usually works well with others	☐ Works poorly with others	☐ Wants it done his/her way
WORK HABITS:				
☐ Always project oriented	☐ Most often project oriented	☐ Usually consistent with requests	☐ Works poorly with others	☐ Always works alone

ADDITIONAL COMMENTS: ______________________________

____

____

FIGURE 8–2. Description-based evaluation form.

employees are from different departments. How can a project engineer be compared to a cost accountant?

Even though the project manager fills out an evaluation form, there is no guarantee that the functional manager will believe the project manager's evaluation. There are always situations in which the project and functional managers disagree as to either quality or direction of work.

Another problem may exist in the situation where the project manager is a "generalist," say at a grade-7 level, and requests that the functional manager assign his best employee to the project. The functional manager agrees to the request and assigns his best employee, a grade-10 specialist. One solution to this problem is to have the project manager evaluate the expert only in certain categories such as communications, work habits, and problem solving, but not in the area of his technical expertise.

As a final note, it is sometimes argued that functional employees should have some sort of indirect input into a project manager's evaluation. This raises rather interesting questions as to how far we can go with the indirect evaluation procedure.

EMPLOYEE'S NAME	DATE
PROJECT TITLE	JOB NUMBER
EMPLOYEE ASSIGNMENT	
EMPLOYEE'S TOTAL TIME TO DATE ON PROJECT	EMPLOYEE'S REMAINING TIME ON PROJECT

	EXCELLENT	ABOVE AVERAGE	AVERAGE	BELOW AVERAGE	INADEQUATE
TECHNICAL JUDGMENT					
WORK PLANNING					
COMMUNICATIONS					
ATTITUDE					
COOPERATION					
WORK HABITS					
PROFIT CONTRIBUTION					

ADDITIONAL COMMENTS: ____________________

FIGURE 8–3. Rate-based evaluation form.

PROJECT TITLE	JOB NUMBER
EMPLOYEE ASSIGNMENT	DATE

CODE:

EXCELLENT = 5
ABOVE AVERAGE = 4
AVERAGE = 3
BELOW AVERAGE = 2
INADEQUATE = 1

NAMES	TECHNICAL JUDGMENT	WORK PLANNING	COMMUNICATIONS	ATTITUDE	COOPERATION	WORK HABITS	PROFIT CONTRIBUTION	SELF MOTIVATION	TOTAL POINTS

FIGURE 8–4. Project termination evaluation form.

From a top-management perspective, the indirect evaluation process brings with it several headaches. Wage and salary administrators readily accept the necessity for using different evaluation forms for white-collar and blue-collar workers. But now, we have a situation in which there can be more than one type of evaluation system for white-collar workers alone. Those employees who work in project-driven functional departments will be evaluated directly and indirectly, but based on formal procedures. Employees who charge their time to overhead accounts and non–project-driven departments might simply be evaluated by a single, direct evaluation procedure.

Many wage and salary administrators contend that they cannot live with a white-collar evaluation system and therefore have tried to combine the direct and indirect evaluation forms into one, as shown in Figure 8–5. Some administrators have even gone so far as to adopt a single form company-wide, regardless of whether an individual is a white- or blue-collar worker.

8.2 FINANCIAL COMPENSATION AND REWARDS

PMBOK® Guide, 6th Edition
9.4.2.4 Recognition and Rewards

Proper financial compensation and rewards are important to the morale and motivation of people in any organization. However, there are several issues that often make it necessary to treat compensation practices of project personnel separately from the rest of the organization:

- *Job classification and job descriptions* for project personnel are usually not compatible with those existing for other professional jobs. It is often difficult to pick an existing classification and adapt it to project personnel. Without proper adjustment, the small amount of formal authority of the project and the small number of direct reports may distort the position level of project personnel in spite of their broad range of business responsibilities.
- *Dual accountability* and dual reporting relationships of project personnel raise the question of who should assess performance and control the rewards.
- *Bases for financial rewards* are often difficult to establish, quantify, and administer. The criteria for "doing a good job" are difficult to quantify.
- *Special compensations* for overtime, extensive travel, or living away from home should be considered in addition to bonus pay for preestablished results. Bonus pay is a particularly difficult and delicate issue because often many people contribute to the results of such incentives. Discretionary bonus practices can be demoralizing to the project team.

Some specific guidelines are provided here to help managers establish compensation systems for their project organizations. The foundations of these compensation practices are based on four systems: (1) job classification, (2) base pay, (3) performance appraisals, and (4) merit increases.

I. EMPLOYEE INFORMATION:

1. NAME ______ 2. DATE OF EVALUATION ______
3. JOB ASSIGNMENT ______ 4. DATE OF LAST EVALUATION ______
5. PAY GRADE ______
6. EMPLOYEE'S IMMEDIATE SUPERVISOR ______
7. SUPERVISOR'S LEVEL: ☐ SECTION ☐ DEPT. ☐ DIVISION ☐ EXECUTIVE

II. EVALUATOR'S INFORMATION:

1. EVALUATOR'S NAME ______
2. EVALUATOR'S LEVEL: ☐ SECTION ☐ DEPT. ☐ DIVISION ☐ EXECUTIVE
3. RATE THE EMPLOYEE ON THE FOLLOWING:

	EXCELLENT	VERY GOOD	GOOD	FAIR	POOR
ABILITY TO ASSUME RESPONSIBILITY					
WORKS WELL WITH OTHERS					
LOYAL ATTITUDE TOWARD COMPANY					
DOCUMENTS WORK WELL AND IS BOTH COST AND PROFIT CONSCIOUS					
RELIABILITY TO SEE JOB THROUGH					
ABILITY TO ACCEPT CRITCISM					
WILLINGNESS TO WORK OVERTIME					
PLANS JOB EXECUTION CAREFULLY					
TECHNICAL KNOWLEDGE					
COMMUNICATIVE SKILLS					
OVERALL RATING					

4. RATE THE EMPLOYEE IN COMPARISON TO HIS CONTEMPORARIES:

LOWER 10%	LOWER 25%	LOWER 40%	MIDWAY	UPPER 40%	UPPER 25%	UPPER 10%

5. RATE THE EMPLOYEE IN COMPARISON TO HIS CONTEMPORARIES:

SHOULD BE PROMOTED AT ONCE	PROMOTABLE NEXT YEAR	PROMOTABLE ALONG WITH CONTEMPORARIES	NEEDS TO MATURE IN GRADE	DEFINITELY NOT PROMOTABLE

6. EVALUATOR'S COMMENTS: ______

SIGNATURE ______

III. CONCURRENCE SECTION:

1. NAME ______
2. POSITION: ☐ DEPARTMENT ☐ DIVISION ☐ EXECUTIVE
3. CONCURRENCE ☐ AGREE ☐ DISAGREE
4. COMMENTS: ______

SIGNATURE ______

IV. PERSONNEL SECTION: (to be completed by the Personnel Department only)

6/15
6/14
6/13
6/12
6/11
6/10
6/09
6/08
6/07
6/06

LOWER 10% LOWER 25% LOWER 40% MIDWAY UPPER 40% UPPER 25% UPPER 10%

V. EMPLOYEE'S SIGNATURE: ______ DATE: ______

FIGURE 8–5. Job evaluation.

Job Classifications and Job Descriptions

Every effort should be made to fit the new classifications for project personnel into the existing standard classification that has already been established for the organization.

The first step is to define job titles for various project personnel and their corresponding responsibilities. Titles are noteworthy because they imply certain responsibilities, position power, organizational status, and pay level. Furthermore, titles may indicate certain functional responsibilities, as does, for example, the title of task manager.[1] Therefore, titles should be carefully selected and each of them should be supported by a formal job description.

The job description provides the basic charter for the job and the individual in charge of it. A good job description is brief and concise, not exceeding one page. Typically, it is broken down into three sections: (1) overall responsibilities, (2) specific duties, and (3) qualifications. A sample job description is given in Table 8–1.

TABLE 8–1. SAMPLE JOB DESCRIPTION

Job Description: Lead Project Engineer of Processor Development

Overall Responsibility

Responsible for directing the technical development of the new Central Processor including managing the technical personnel assigned to this development. The Lead Project Engineer has dual responsibility: (1) to his/her functional superior for the technical implementation and engineering quality and (2) to the project manager for managing the development within the established budget and schedule.

Specific Duties and Responsibilities

1. Provide necessary program direction for planning, organizing, developing, and integrating the engineering effort, including establishing the specific objectives, schedules, and budgets for the processor subsystem.
2. Provide technical leadership for analyzing and establishing requirements, preliminary designing, designing, prototyping, and testing of the processor subsystem.
3. Divide the work into discrete and clearly definable tasks. Assign tasks to technical personnel within the Lead Engineer's area of responsibility and other organizational units.
4. Define, negotiate, and allocate budgets and schedules according to the specific tasks and overall program requirements.
5. Measure and control cost, schedule, and technical performance against program plan.
6. Report deviations from program plan to program office.
7. Replan trade-off and redirect the development effort in case of contingencies such as to best utilize the available resources toward the overall program objectives.
8. Plan, maintain, and utilize engineering facilities to meet the long-range program requirements.

Qualifications

1. Strong technical background in state-of-the-art central processor development.
2. Prior task management experience with proven record for effective cost and schedule control of multidisciplinary technology-based task in excess of SIM.
3. Personal skills to lead, direct, and motivate senior engineering personnel.
4. Excellent communication skills, both orally and in writing.

1. In most organizations the title of task manager indicates responsibility for managing the technical content of a project subsystem within a functional unit, having dual accountabilities to the functional superior and the project office.

Base-Pay Classifications and Incentives

After the job descriptions have been developed, one can delineate pay classes consistent with the responsibilities and accountabilities for business results. If left to the personnel specialist, these pay scales may slip toward the lower end of an equitable compensation. This is understandable because, on the surface, project positions look less senior than their functional counterparts, as formal authority over resources and direct reports is often less necessary for project positions than for traditional functional positions. The impact of such a skewed compensation system is that the project organization will attract less qualified personnel and may be seen as an inferior career path.

Many companies that have struggled with this problem have solved it by (1) working out compensation schemes as a team of senior managers and personnel specialists, and (2) applying criteria of responsibility and business/profit accountability to setting pay scales for project personnel in accord with other jobs in their organization. Managers who are hiring can choose a salary from the established range based on their judgment of actual position responsibilities, the candidate's qualifications, the available budget, and other considerations.

Performance Appraisals

Traditionally, the purpose of the performance appraisal is to:

- Assess the employee's work performance, preferably against preestablished objectives
- Provide a justification for salary actions
- Establish new goals and objectives for the next review period
- Identify and deal with work-related problems
- Serve as a basis for career discussions

In reality, however, the first two objectives are in conflict. As a result, traditional performance appraisals essentially become a salary discussion with the objective to justify subsequent managerial actions. In addition, discussions dominated by salary actions are usually not conducive to future goal setting, problem solving, or career planning.

In order to get around this dilemma, many companies have separated the salary discussion from the other parts of the performance appraisal. Moreover, successful managers have carefully considered the complex issues involved and have built a performance appraisal system solidly based on content, measurability, and source of information.

The first challenge is in content, that is, to decide "what to review" and "how to measure performance." Modern management practices try to individualize accountability as much as possible. Furthermore, subsequent incentive or merit increases are tied to profit performance. Although most companies apply these principles to their project organizations, they do it with a great deal of skepticism. Practices are often modified to assure balance and equity for jointly performed responsibilities. A similar dilemma exists in the area of profit accountability. Acknowledging the realities, organizations are measuring performance of their *project managers,* in at least two areas:

- *Business results* as measured by profits, contribution margin, return on investment, new business, and income; also, on-time delivery, meeting contractual requirements, and within-budget performance.

- *Managerial performance* as measured by overall project management effectiveness, organization, direction and leadership, and team performance

The first area applies only if the project manager is indeed responsible for business results such as contractual performance or new business acquisitions. Many project managers work with company-internal sponsors, for example, on a company-internal new product development or a feasibility study. In these cases, producing the results within agreed-on schedule and budget constraints becomes the primary measure of performance. The second area is clearly more difficult to assess. Moreover, if handled improperly, it will lead to manipulation and game playing. Table 8–2 provides some specific measures of project management performance. Whether the sponsor is company-internal or external, project managers are usually being assessed on how long it took to organize the team, whether the project is moving along according to agreed-on schedules and budgets, and how closely they meet the global goals and objectives set by their superiors.

On the other side of the project organization, resource managers or project personnel are being assessed primarily on their ability to direct the implementation of a specific project subsystem:

- *Technical implementation* as measured against requirements, quality, schedules, and cost targets
- *Team performance* as measured by ability to staff, build an effective task group, interface with other groups, and integrate among various functions

PMBOK® Guide, 6th Edition
9.4 Manage Team
9.5.1.4 Team Performance Assessments

TABLE 8–2. PERFORMANCE MEASURES FOR PROJECT MANAGERS

Who Performs Appraisal
Functional superior of project manager

Source of Performance Data
Functional superior, resource managers, general managers

Primary Measures
1. Project manager's success in leading the project toward preestablished global objectives
   - Target costs
   - Key milestones
   - Profit, net income, return on investment, contribution margin
   - Quality
   - Technical accomplishments
   - Market measures, new business, follow-on contract
2. Project manager's effectiveness in overall project direction and leadership during all phases, including establishing:
   - Objectives and customer requirements
   - Budgets and schedules
   - Policies
   - Performance measures and controls
   - Reporting and review system

(continues)

TABLE 8–2. PERFORMANCE MEASURES FOR PROJECT MANAGERS (*Continued*)

Secondary Measures

1. Ability to utilize organizational resources
   - Overhead cost reduction
   - Working with existing personnel
   - Cost-effective make-buy decisions
2. Ability to build effective project team
   - Project staffing
   - Interfunctional communications
   - Low team conflict complaints and hassles
   - Professionally satisfied team members
   - Work with support groups
3. Effective project planning and plan implementation
   - Plan detail and measurability
   - Commitment by key personnel and management
   - Management involvement
   - Contingency provisions
   - Reports and reviews
4. Customer/client satisfaction
   - Perception of overall project performance by sponsor
   - Communications, liaison
   - Responsiveness to changes
5. Participation in business management
   - Keeping management informed of new project/product/business opportunities
   - Bid proposal work
   - Business planning, policy development

Additional Considerations

1. Difficulty of tasks involved
   - Technical tasks
   - Administrative and organizational complexity
   - Multidisciplinary nature
   - Staffing and start-up
2. Scope of the project
   - Total project budget
   - Number of personnel involved
   - Number of organizations and subcontractors involved
3. Changing work environment
   - Nature and degree of customer changes and redirections
   - Contingencies

Specific performance measures are shown in Table 8–3. In addition, the actual project performance of both project managers and their resource personnel should be assessed on the conditions under which it was achieved: the degree of task difficulty, complexity, size, changes, and general business conditions.

Finally, one needs to decide who is to perform the performance appraisal and to make the salary adjustment. Where dual accountabilities are involved, good practices call for inputs from both bosses. Such a situation could exist for project managers who report functionally to one superior but are also accountable for specific business results to another person. While dual accountability of project managers is an exception for most organizations, it is common for project resource personnel who are responsible to their functional superior for the quality of the work and to their project manager for meeting the

TABLE 8–3. PERFORMANCE MEASURES FOR PROJECT PERSONNEL

Who Performs Appraisal
Functional superior of project person

Source of Performance Data
Project manager and resource managers

Primary Measures

1. Success in directing the agreed-on task toward completion
   - Technical implementation according to requirements
   - Quality
   - Key milestones/schedules
   - Target costs, design-to-cost
   - Innovation
   - Trade-offs
2. Effectiveness as a team member or team leader
   - Building effective task team
   - Working together with others, participation, involvement
   - Interfacing with support organizations and subcontractors
   - Interfunctional coordination
   - Getting along with others
   - Change orientation
   - Making commitments

Secondary Measures

1. Success and effectiveness in performing functional tasks in addition to project work in accordance with functional charter
   - Special assignments
   - Advancing technology
   - Developing organization
   - Resource planning
   - Functional direction and leadership
2. Administrative support services
   - Reports and reviews
   - Special task forces and committees
   - Project planning
   - Procedure development
3. New business development
   - Bid proposal support
   - Customer presentations
4. Professional development
   - Keeping abreast in professional field
   - Publications
   - Liaison with society, vendors, customers, and educational institutions

Additional Considerations

1. Difficulty of tasks involved
   - Technical challenges
   - State-of-the-art considerations
   - Changes and contingencies
2. Managerial responsibilities
   - Task leader for number of project personnel
   - Multifunctional integration
   - Budget responsibility
   - Staffing responsibility
   - Specific accountabilities
3. Multiproject involvement
   - Number of different projects
   - Number and magnitude of functional task and duties
   - Overall workload

requirements within budget and schedule. Moreover, resource personnel may be shared among many projects. Only the functional or resource manager can judge overall performance of resource personnel.

Merit Increases, Bonuses, and Incentive Plans

PMBOK® Guide, 6th Edition
9.5.2.4 Recognition and Rewards

Under inflationary conditions, pay adjustments seldom keep up with cost-of-living increases. To deal with this salary compression and to give incentive for management performance, many companies have introduced bonus structures. The problem is that standard plans for merit increases and bonuses are based on individual accountability while project personnel work in teams with shared accountabilities, responsibilities, and controls. It is usually very difficult to credit project success or failure to a single individual or a small group.

Most managers with these dilemmas have turned to the traditional remedy of the performance appraisal. If done well, the appraisal should provide particular measures of job performance that assess the level and magnitude at which the individual has contributed to the success of the project, including the managerial performance and team performance components. Therefore, a properly designed and executed performance appraisal that includes input from all accountable management elements, and the basic agreement of the employee with the conclusions, is a sound basis for future salary reviews.

Project team incentives are important because team members expect appropriate rewards and recognition for work well done. These incentive plans usually have some combination of these basic measures:

- Project Milestones: Hit a milestone, on budget and on time, and all team members earn a defined amount. Although sound in theory, there are inherent problems in tying financial incentives to hitting milestones. Milestones often change for good reason (technological advances, market shifts, other developments) and you don't want the team and management to get into a negotiation on slipping dates to trigger the incentive. Unless milestones are set in stone and reaching them is simply a function of the team doing its normal, everyday job, it's generally best to use recognition-after-the-fact celebration of reaching milestones—rather than tying financial incentives to it.
- Rewards need not always be time-based, such that when the team hits a milestone by a certain date it earns a reward. If, for example, a product development team debugs a new piece of software on time, that's not necessarily a reason to reward it. But if it discovers and solves an unsuspected problem or writes better code before a delivery date, rewards are due.
- Project Completion: All team members earn a defined amount when they complete the project on budget and on time (or to the team champion's quality standards).
- Value Added: This award is a function of the value added by a project, and depends largely on the ability of the organization to create and track objective measures. Examples include reduced turnaround time on customer requests, improved cycle times for product development, cost savings due to new process efficiencies, or incremental profit or market share created by the product or service developed or implemented by the project team.

8.3 EFFECTIVE PROJECT MANAGEMENT IN THE SMALL BUSINESS ORGANIZATION

Here, we are discussing project management in both small companies and small organizations within a larger corporation. In small organizations, major differences from large companies must be accounted for:

- *In small companies, the project manager has to wear multiple hats and may have to act as a project manager and line manager at the same time.* Large companies may have the luxury of a single full-time project manager for the duration of a project. Smaller companies may not be able to afford a full-time project manager and therefore may require that functional managers wear two hats. This poses a problem in that the functional managers may be more dedicated to their own functional unit than to the project, and the project may suffer. There is also the risk that when the line manager also acts as project manager, the line manager may keep the best resources for his own project. The line manager's project may be a success at the expense of all the other projects that he must supply resources for.

In the ideal situation, the project manager works horizontally and has project dedication, whereas the line manager works vertically and has functional (or company) dedication. If the working relationship between the project and functional managers is a good one, then decisions will be made in a manner that is in the best interest of both the project and the company. Unfortunately, this may be difficult to accomplish in small companies when an individual wears multiple hats.

- *In a small company, the project manager handles multiple projects, perhaps each with a different priority.* In large companies, project managers normally handle only one project at a time. Handling multiple projects becomes a serious problem if the priorities are not close together. For this reason, many small companies avoid the establishment of priorities for fear that the lower-priority activities will never be accomplished.
- *In a small company, the project manager has limited resources.* In a large company, if the project manager is unhappy with resources that are provided, he may have the luxury of returning to the functional manager to either demand or negotiate for other resources. In a small organization, the resources assigned may be simply the only resources available.
- *In a small company, project managers must generally have a better understanding of interpersonal skills than in a larger company.* This is a necessity because a project manager in the small company has limited resources and must provide the best motivation that he can.
- *In the smaller company, the project manager generally has shorter lines of communications.* In small organizations project managers almost always report to a top-level executive, whereas in larger organizations the project managers can report to any level of management. Small companies tend to have fewer levels of management.
- *Small companies do not have a project office.* Large companies, especially in aerospace or construction, can easily support a project office of twenty to thirty people, whereas in the smaller company the project manager may have to be the entire project

office. This implies that the project manager in a small company may be required to have more general and specific information about all company activities, policies, and procedures than his counterparts in the larger companies.

- *In a small company, there may be a much greater risk to the total company with the failure of as little as one project.* Large companies may be able to afford the loss of a multimillion-dollar program, whereas the smaller company may be in serious financial trouble. Thus many smaller companies avoid bidding on projects that would necessitate hiring additional resources or giving up some of its smaller accounts.
- *In a small company, there might be tighter monetary controls but with less sophisticated control techniques.* Because the smaller company incurs greater risk with the failure (or cost overrun) of as little as one project, costs are generally controlled much more tightly and more frequently than in larger companies. However, smaller companies generally rely on manual or partially computerized systems, whereas larger organizations rely heavily on sophisticated software packages.
- *In a small company, there is usually more upper-level management interference.* This is expected because in the small company there is a much greater risk with the failure of a single project. In addition, executives in smaller companies "meddle" more than executives in larger companies, and quite often delegate as little as possible to project managers.
- *Evaluation procedures for individuals are usually easier in a smaller company.* This holds true because the project manager gets to know the people better, and, as stated above, there exists a greater need for interpersonal skills on the horizontal line in a smaller company.
- *In a smaller company, project estimating is usually more precise and based on either history or standards.* This type of planning process is usually manual as opposed to computerized. In addition, functional managers in a small company usually feel obligated to live up to their commitments, whereas in larger companies, much more lip service is given.

8.4 MEGA PROJECTS

Mega projects may have a different set of rules and guidelines from those of smaller projects. For example, in large projects:

- Vast numbers of people may be required, often for short or intense periods of time.
- Continuous organizational restructuring may be necessary as each project goes through a different life-cycle phase.
- The matrix and project organizational form may be used interchangeably.

The following elements are critical for success.

- Training in project management
- Rules and procedures clearly defined

- Communications at all levels
- Quality front-end planning

Virginia Greiman has developed a list of 25 attributes common to most mega projects:[2]

1. Long duration
2. Scale and dimension
3. Type of industry and purpose
4. Design and construction complexity
5. Sponsorship and financing
6. Life cycle
7. Long, complex, and critical front end
8. High public profile
9. Public scrutiny
10. Pursuit of large-scale policy making
11. Project delivery and procurement
12. Continuity of management
13. Technology and procedural complexity
14. Organizational structure
15. High degree of regulation
16. Multiple stakeholders
17. Dynamic governance structures
18. Ethical dilemmas and challenges
19. Consistent cost underestimation and poor performance
20. Risk management in complex projects
21. Socioeconomic impacts
22. Cultural dimension
23. Systems methodology complexity
24. Environmental impact
25. Collaborative contracting, integration and partnering

Many companies dream of winning mega project contracts only to find disaster rather than a pot of gold. Perhaps the most common difficulty in managing mega projects stems mainly from resource restraints:

- Lack of available on-site workers (or local labor forces)
- Lack of skilled workers
- Lack of properly trained on-site supervision
- Lack of raw materials

As a result of such problems, the company immediately assigns its best employees to the mega project, thus creating severe risks for the smaller projects, many of which could lead to substantial follow-on business. Overtime is usually required, on a prolonged basis, and this results in lower efficiency and unhappy employees.

As the project schedule slips, management hires additional home-office personnel to support the project. By the time that the project is finished, the total organization is overstaffed, many smaller customers have taken their business elsewhere, and the company

2. Virginia A. Greiman, *Mega Project Management* (Hoboken, NJ: John Wiley & Sons, 2013), pp. 12–24.

finds itself in the position of needing another mega project in order to survive and support the existing staff.

Mega projects are not always as glorious as people think they are. Organizational stability, accompanied by a moderate growth rate, may be more important than quantum steps to mega projects. The lesson here is that mega projects should be left to those companies that have the facilities, expertise, resources, and management know-how to handle the situation.

8.5 MORALITY, ETHICS, AND THE CORPORATE CULTURE

***PMBOK® Guide*, 6th Edition**

1.1.3 Code of Ethics and Professional Conduct

Professional Responsibility and the PMP® Code of Conduct

Companies that promote morality and ethics in business usually have an easier time developing a cooperative culture than those that encourage unethical or immoral behavior. The adversity generated by unethical acts can be either internally or externally driven. Internally driven adversity occurs when employees or managers in your own company ask you to take action that may be in the best interest of your company but violates your own moral and ethical beliefs. Typical examples might include:

You are asked to lie to the customer in a proposal in order to win the contract.

You are asked to withhold bad news from your own management.

You are asked to withhold bad news from the customer.

You are instructed to ship a potentially defective unit to the customer in order to maintain production quotas.

You are ordered to violate ethical accounting practices to make your numbers "look good" for senior management.

You are asked to cover up acts of embezzlement or use the wrong charge numbers.

You are asked to violate the confidence of a private personal decision by a team member.

External adversity occurs when your customers ask you to take action that may be in the customer's best interest (and possibly your company's best interest), but once again violates your personal moral and ethical beliefs. Typical examples might include:

You are asked to hide or destroy information that could be damaging to the customer during legal action against your customer.

You are asked to lie to consumers to help maintain your customer's public image.

You are asked to release unreliable information that would be damaging to one of your customer's competitors.

The customer's project manager asks you to lie in your proposal so that he/she will have an easier time in approving contract award.

PMP is a registered mark of the Project Management Institute, Inc.

Project managers are often placed in positions where an action must be taken for the best interest of the company and its customers, and yet the same action could be upsetting to the workers. Consider the following example as a positive way to handle this:

A project had a delivery date where a specific number of completed units had to be on the firm's biggest customer's receiving dock by January 5. This customer represented 30% of the firm's sales and 33% of its profits. Because of product development problems and slippages, the project could not be completed early. The employees, many of whom were exempt, were informed that they would be expected to work 12-hour days, including Christmas and New Year's, to maintain the schedule. The project manager worked the same hours as his manufacturing team and was visible to all. The company allowed family members to visit the workers during the lunch and dinner hours during this period. After delivery was accomplished, the project manager arranged for all of the team members to receive two weeks of paid time off. At completion of the project, the team members were volunteering to work again for this project manager.

The project manager realized that asking his team to work these days might be viewed as immoral. Yet, because he also worked, his behavior reinforced the importance of meeting the schedule. The project manager's actions actually strengthened the cooperative nature of the culture within the firm.

Not all changes are in the best interest of both the company and the workers. Sometimes change is needed simply to survive, and this could force employees to depart from their comfort zones. The employees might even view the change as immoral. Consider the following example:

Because of a recession, a machine tool company switched from a non–project-driven to a project-driven company. Management recognized the change and tried to convince employees that customers now wanted specialty products rather than standard products, and that the survival of the firm might be at stake. The company hired a project management consulting company to help bring in project management since the business was now project-driven. The employees vigorously resisted both the change and the training with the mistaken belief that, once the recession ended, the customers would once again want the standard, off-the-shelf products and that project management was a waste of time. The company is no longer in business and, as the employees walked out of the plant for the last time, they blamed project management for the loss of employment.

Some companies develop "standard practice manuals," which describe in detail what is meant by ethical conduct in dealing with customers and suppliers. Yet, even with the existence of these manuals, well-meaning individuals may create unintended consequences that wreak havoc. Consider the following example:

The executive project sponsor on a government-funded R&D project decided to "massage" the raw data to make the numbers look better before presenting the data to a customer. When the customer realized what had happened, their relationship, which had been based upon trust and open communications, became based upon mistrust and formal documentation. The entire project team suffered because of the self-serving conduct of one executive.

Sometimes, project managers find themselves in situations where the outcome most likely will be a win-lose position rather than a win-win situation. Consider the following three situations:

An assistant project manager, Mary, had the opportunity to be promoted and manage a new large project that was about to begin. She needed her manager's permission to accept the new assignment, but if she left, her manager would have to perform her work in addition to his own for at least three months. The project manager refused to release her, and the project manager developed a reputation of preventing people from being promoted while working on his projects.

In the first month of a twelve-month project, the project manager realized that the end date was optimistic, but he purposely withheld information from the customer in hopes that a miracle would occur. Ten months later, the project manager was still withholding information, waiting for the miracle. In the eleventh month, the customer was told the truth. People then labeled the project manager as an individual who would rather lie than tell the truth because it was easier.

To maintain the customer's schedule, the project manager demanded that employees work excessive overtime, knowing that this often led to more mistakes. The company fired a tired worker who inadvertently withdrew the wrong raw materials from inventory, resulting in a $55,000 manufacturing mistake.

In all three situations, the project manager believed that his decision was in the best interest of the company at that time. Yet the final result in each case was that the project manager was labeled unethical or immoral.

It is often said that "money is the root of all evil." Sometimes companies believe that recognizing the achievements of an individual through a financial reward system is appropriate without considering the impact on the culture. Consider the following example:

At the end of a highly successful project, the project manager was promoted, and given a $5,000 bonus and a paid vacation. The team members who were key to the project's success, and who earned minimum wage, went to a fast food restaurant to celebrate their contribution to the firm and their support of each other. The project manager celebrated alone.

The company failed to recognize that project management is a team effort. The workers viewed management's reward policy as immoral and unethical because the project manager was successful due to the efforts of the entire team.

Moral and ethical conduct by project managers, project sponsors, and line managers can improve the corporate culture. Likewise, poor decisions can destroy a culture, often in much less time than it took for the culture to be developed.

8.6 PROFESSIONAL RESPONSIBILITIES

***PMBOK® Guide*, 6th Edition**
Professional Responsibilities; The PMP® Code of Conduct
1.1.3 Code of Ethics and Professional Conduct

Professional responsibilities for project managers have become increasingly important because of the unfavorable publicity on some of the dealings of corporate America. These have been with us for some time, especially in dealing with government agencies. Professional responsibilities for a project manager are both broad-based and encompassing.

Professional responsibilities cover two major areas: our responsibilities to the profession of project management and our responsibilities to the customers and the public.

In 2010, PMI released the finding of the latest version of the Role Delineation Study (RDS). In addition, PMI updated its Code of Professional Conduct and renamed it the Code of Ethics and Professional Conduct. PMI has also eliminated the Professional Responsibility Domain Area from the PMP® exam and included the professional responsibility questions in the other domain areas where appropriate.

The PMI® Code of Ethics and Professional Conduct applies to everyone working in a project environment, not merely the project manager. Therefore, the code emphasizes that PMP credential holders must function as "role models" and exhibit characteristics such as honesty, integrity, and ethical behavior. The code guides members of the project management profession on how to handle ethical issues as well as providing customers and stakeholders with some degree of assuredness that project personnel will make the right decisions.

The Code of Ethics and Professional Conduct is divided into four sections:

- **Responsibility:** This includes taking ownership for the decisions we make, or failure to make decisions, as well as the consequences, whether they are good or bad.
- **Respect:** This includes the way we regard ourselves as well as the resources provided to us.
- **Fairness:** This focuses on the way we make decisions. Decisions should be made in an impartial and objective manner and be free from conflicts of interest, favoritism, and prejudices.
- **Honesty:** This states that we should act in a truthful manner.

Each of the four sections includes both aspirational standards and mandatory standards. An aspirational standard is an agreed-upon, repeatable way of doing something. It can be a published document designed to be used consistently as a general rule, guideline, or definition. Aspirational standards are often a summary of good and best practices rather than general practices. Although we tend to say that abiding by aspirational standards is not optional, the aspirational standards are actually designed for voluntary use in the way that they are acted upon and may not impose any regulations. However, laws and regulations may refer to certain aspirational standards and make compliance with them compulsory.

There are numerous situations that can create problems for project managers in dealing with professional responsibilities expectations. These situations include:

- Maintaining professional integrity
- Adhering to ethical standards
- Recognizing diversity
- Avoiding/reporting conflicts of interest
- Not making project decisions for personal gains
- Receiving gifts from customers and vendors
- Providing gifts to customers and vendors
- Truthfully reporting information
- Willing to identify violations
- Balancing stakeholder needs
- Succumbing to stakeholder pressure
- Managing your firm's intellectual property
- Managing your customer's intellectual property

- Adhering to security and confidentiality requirements
- Abiding by the Code of Ethics and Professional Conduct
- Report omissions and errors
- Living up to commitments
- Not engaging in deceptive practices
- Truthful and accurate communications
- Appropriate use of power and authority
- Demonstrating honest behavior

Several of these topics are explained below.

Conflict of Interest

A conflict of interest is a situation where the individual is placed in a compromising position where the individual can gain personally based upon the decisions made. This is also referred to as personal enrichment. There are numerous situations where a project manager is placed in such a position. Examples include:

- Insider knowledge that the stock will be going up or down
- Being asked to improperly allow employees to use charge numbers on your project even though they are not working on your project
- Receiving or giving inappropriate (by dollar value) gifts
- Receiving unjustified compensation or kickbacks
- Providing the customer with false information just to keep the project alive

Project managers are expected to abide by the PMI® Code of Professional Conduct, which makes it clear that project managers should conduct themselves in an ethical manner. Unjust compensation or gains not only are frowned upon but are unacceptable. Unless these conflict-of-interest situations are understood, the legitimate interests of both the customer and the company may not be forthcoming.

Inappropriate Connections

Not all stakeholders are equal in their ability to influence the decisions made by the project manager. Some stakeholders can provide inappropriate influence/compensation, such as:

- A loan with a very low interest rate
- Ability to purchase a product/service at a price that may appear equivalent to a gift
- Ability to receive free gifts such as airline tickets, tickets to athletic events, free meals and entertainment, or even cash

Another form of inappropriate connections can occur with family or friends. These individuals may provide you with information or influence by which you could gain personally in a business situation. Examples of affiliation connections might be:

- Receiving insider information
- Receiving privileged information
- Opening doors that you could not open by yourself, at least without some difficulty

Acceptance of Gifts

Today, all companies have rules concerning the acceptance of gifts and their disclosure. While it may be customary in some countries to give or accept gifts, the standard rule is usually to avoid all gifts. Some companies may stipulate limits on when gifts are permitted and the appropriateness of the gift. The gifts might be cash, free meals, or other such items.

Responsibility to Your Company (and Stakeholders)

Companies today, more than ever before, are under pressure to maintain ethical practices with customers and suppliers. This could be interpreted as a company code of ethics that stipulates the professional behavior expected from the project manager and the team members. This applies specifically to the actions of both the project manager as well as the team members. Some companies even go so far as to develop "standard practice manuals" on how to act in a professional manner. Typical sections of such manuals might be:

- Truthful representation of all information
- Full disclosure of all information
- Protection of company-proprietary information
- Responsibility to report violations
- Full compliance with groups auditing violations
- Full disclosure, and in a timely manner, of all conflicts of interest
- Ensure that all of the team members abide by the above items

8.7 INTERNAL AND EXTERNAL PARTNERSHIPS

A partnership is a group of two or more individuals working together to achieve a common objective. In project management, maintaining excellent, working relations with internal partners is essential. Internally, the critical relationship is between the project and line manager.

In the early days of project management, the selection of the individual to serve as the project manager was most often dependent upon who possessed the greatest command of technology. The result was a very poor working relationship between the project and line manager.

As the magnitude and technical complexity of the projects grew, it became obvious that the project managers could not maintain a command of technology in all aspects of the project.

As the partnership between the project and line managers developed, management recognized that partnerships worked best on a peer-to-peer basis. Project and line managers began to view each other as equals and share in the authority, responsibility, and accountability needed to ensure project success. Good project management methodologies emphasize the cooperative working relationship that must exist between the project and line managers.

> ***PMBOK® Guide*, 6th Edition**
> 2.4.2 Organizational Governance Framework
> Chapter 13 Project Stakeholder Management

Project management methodologies also emphasize the working relationships with external organizations such as suppliers. Outsourcing has become a major trend because it allows companies to bring their products and services to the market faster and often at a more competitive price. Therefore, external partnerships can be beneficial for both the suppliers and the customers. If properly managed, they can provide significant long-term benefits to both the customer and supplier.

8.8 TRAINING AND EDUCATION

Given that most companies use the same basic tools as part of their methodology, what makes one company better than another? The answer lies in the execution of the methodology. Training and education can accelerate not only the project management maturity process but also the ability to execute the methodology.

Actual learning takes place in three areas, as shown in Figure 8–6: on-the-job experience, education, and knowledge transfer. Ideal project management knowledge would be obtained by allowing each employee to be educated on the results of the company's lessons-learned studies, including risk management, benchmarking, and continuous improvement efforts. Unfortunately, this is rarely done and ideal learning is hardly ever reached. To make matters worse, actual learning is less than most people believe because of lost knowledge. This lost knowledge is shown in Figure 8–7 and will occur even in companies that maintain low employee turnover ratios. These two figures also illustrate the importance of maintaining the same personnel on the project for the duration of the effort.

Companies often find themselves in a position of having to provide a key initiative for a multitude of people, or simply specialized training to a program team about to embark

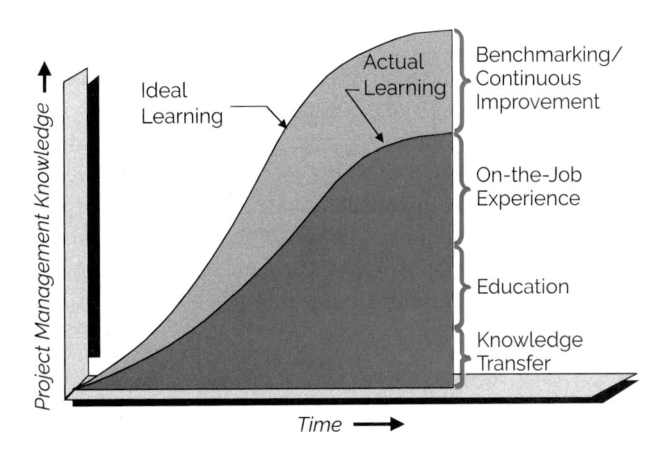

FIGURE 8–6. Project management learning curve.

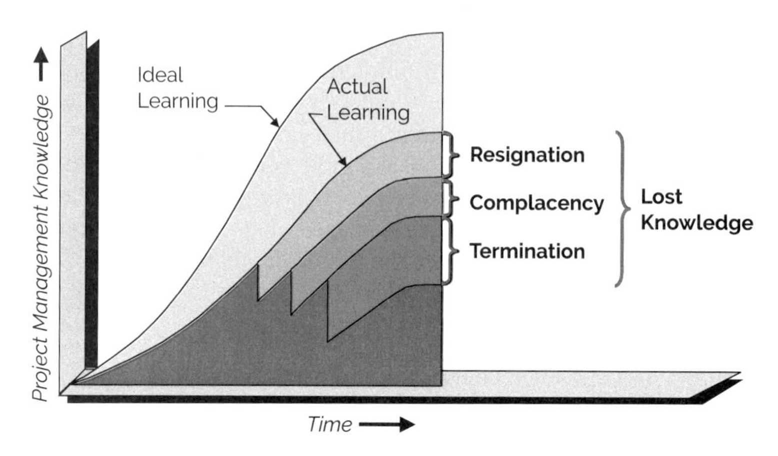

FIGURE 8–7. Lost knowledge within the learning curve.

upon a new long-term effort. In such cases, specialized training is required, with targeted goals and results that are specifically planned for. The elements common to training on a key initiative or practice include[3]:

- A front-end analysis of the program team's needs and training requirements
- Involvement of the program teams in key decisions
- Customized training to meet program team's specific needs
- Targeted training for the implementation of specific practices
- Improved training outcomes, including better course depth, timeliness, and reach

The front-end analysis is used to determine the needs and requirements of the program office implementing the practice. The analysis is also used to identify and address barriers each program office faces when implementing new practices. According to the director of the benchmarking forum for the American Society of Training and Development, this type of analysis is crucial for an organization to be able to institute performance-improving measures. Using information from the front-end analysis, the training organizations customize the training to ensure that it directly assists program teams in implementing new practices. To ensure that the training will address the needs of the program teams, the training organizations involve the staff in making important training decisions. Program staff help decide the amount of training to be provided for certain job descriptions, course objectives, and depth of course coverage. Companies doing this believe their training approach, which includes program staff, has resulted in the right amount of course depth, timeliness, and coverage of personnel.

3. Adapted from *DoD Training Can Do More to Help Weapon System Programs Implement Best Practices,* Best Practices Series, GAO/NSIAD-99-206, Government Accounting Office, August 1999, pp. 40–41, 51.

8.9 INTEGRATED PRODUCT/PROJECT TEAMS

PMBOK® Guide, 6th Edition
Chapter 9 Project Resource Management
Chapter 4 Project Integration Management

In recent years, there has been an effort to substantially improve the formation and makeup of teams required to develop a new product or implement a new practice. These teams have membership from across the management and through the entire organization and are called integrated product/project teams (IPTs).

The IPT consists of a sponsor, program manager, and the core team. For the most part, members of the core team are assigned full-time to the team but may not be on the team for the duration of the entire project.

The skills needed to be a member of the core team include:

- Self-starter ability
- Work without supervision
- Good communication skills
- Cooperative
- Technical understanding
- Willing to learn backup skills
- Able to perform feasibility studies and cost/benefit analyses
- Able to perform or assist in market research studies
- Able to evaluate asset utilization
- Decision maker
- Knowledgeable in risk management
- Understand the need for continuous validation

Each IPT is given a project charter that identifies the project's mission and identifies the assigned project manager. However, unlike traditional charters, the IPT charter can also identify the key members of the IPT by name or job responsibility.

Unlike traditional project teams, the IPT thrives on sharing information across the team and collective decision making. IPTs eventually develop their own culture and, as such, can function in either a formal or informal capacity. Since the concept of an IPT is well suited to large, long-term projects, it is no wonder that the Department of Defense has been researching best practices for an IPT.[4] The government looked at four projects, in both the public and private sectors, which were highly successful using the IPT approach and four government projects that had less than acceptable results. The successful IPT projects are shown in Table 8–4. The unsuccessful IPT projects are shown in Table 8–5. In analyzing the data, the government came up with the results shown in Figure 8–8. Each vertical line in Figure 8–8 is a situation where the IPT must go outside of its own domain to seek information and approvals. Each time this happens, it is referred to as a "hit." The

4. *DoD Teaming Practices Not Achieving Potential Results,* Best Practices Series, GOA-01-501, Government Accounting Office, April 2001.

TABLE 8–4. EFFECTIVE IPTS

Program	Cost Status	Schedule Status	Performance Status
Daimler-Chrysler	Product cost was lowered	Decreased development cycle months by 50 percent	Improved vehicle designs
Hewlett-Packard	Lowered cost by over 60 percent	Shortened development schedule by over 60 percent	Improved system integration and product design
3M	Outperformed cost goals	Product deliveries shortened by 12 to 18 months	Improved performance by 80 percent
Advanced Amphibious Assault Vehicle	Product unit cost lower than original estimate	Ahead of original development schedule	Demonstrated fivefold increase in speed

TABLE 8–5. INEFFECTIVE IPTS

Program	Cost Status	Schedule Status	Performance Status
CH-60S Helicopter	Increased cost but due to additional purchases	Schedule delayed	Software and structural difficulties
Extended Range Guided Munitions	Increases in development costs	Schedule slipped three years	Redesigning due to technical difficulties
Global Broadcast Service	Experiencing cost growth	Schedule slipped 1.5 years	Software and hardware design shortfalls
Land Warrior	Cost increase of about 50 percent	Schedule delayed four years	Overweight equipment, inadequate battery power and design

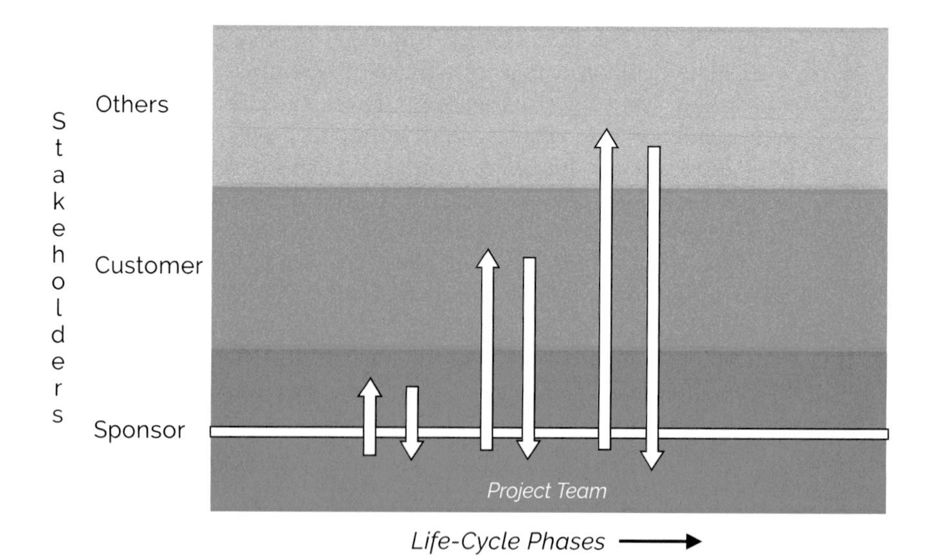

FIGURE 8–8. Knowledge and authority.

government research indicated that the greater the number of hits, the more likely it is that the time, cost, and performance constraints will not be achieved. The research confirmed that if the IPT has the knowledge necessary to make decisions, and also has the authority to make the decisions, then the desired performance would be achieved. Hits will delay decisions and cause schedule slippages.

8.10 VIRTUAL PROJECT TEAMS

PMBOK® Guide, 6th Edition
9.3.2.4 Virtual Teams

Historically, project management was a face-to-face environment where team meetings involved all players convening together in one room. The team itself may even be co-located. Today, because of the size and complexity of projects, it is rare to find all team members located under one roof. Other possible characteristics of a virtual team are shown in Table 8–6.

Duarte and Snyder define seven types of virtual teams.[5] These are shown in Table 8–7. Culture and technology can have a major impact on the performance of virtual teams. Duarte and Snyder have identified some of these relationships in Table 8–8.

TABLE 8–6. CHARACTERISTICS OF TRADITIONAL AND VIRTUAL TEAMS

Characteristic	Traditional Teams	Virtual Teams
Membership	Team members are all from the same company.	Team members may be multinational and all from different companies and countries.
Proximity	Team members work in close proximity with each other.	Team members may never meet face-to-face.
Methodology usage	One approach exists, perhaps an enterprise project management methodology.	Each unit can have their own methodology
Methodology structure	One approach, which is based upon either policies and procedures, or forms guidelines, templates, and checklists.	Each unit's methodology can have its own structure.
Trust	Very little trust may exist.	Trust is essential.
Authority	Leadership may focus on authority.	Leadership may focus on influence power.

TABLE 8–7. TYPES OF VIRTUAL TEAMS

Type of Team	Description
Network	Team membership is diffuse and fluid; members come and go as needed. Team lacks clear boundaries within the organization.
Parallel	Team has clear boundaries and distinct membership. Team works in the short term to develop recommendation for an improvement in a process or system.
Project or product development	Team has fluid membership, clear boundaries, and a defined customer base, technical requirement, and output. Longer-term team task is nonroutine, and the team has decision-making authority.
Work or production	Team as distinct membership and clear boundaries. Members perform regular and outgoing work, usually in one functional area.
Service Management	Team has distinct membership and supports ongoing customer network activity. Team has distinct membership and works on a regular basis to lead corporate activities.
Action	Team deals with immediate action, usually in an emergency situation. Membership may be fluid or distinct.

5. D. L. Duarte and N. T. Snyder, *Mastering Virtual Teams* (San Francisco: Jossey-Bass, an imprint of John Wiley & Sons, 2001), p. 10.

TABLE 8–8. TECHNOLOGY AND CULTURE

Cultural Factor	Technological Considerations
Power distance	Members from high-power-distance cultures may participate more freely with technologies that are asynchronous and allow anonymous input. These cultures sometimes use technology to indicate status differences between team members.
Uncertainty avoidance	People from cultures with high uncertainty avoidance may be slower adopters of technology. They may also prefer technology that is able to produce more permanent records of discussions and decisions.
Individualism–collectivism	Members from highly collectivistic cultures may prefer face-to-face interactions.
Masculinity–femininity	People from cultures with more "feminine" orientations are more prone to use technology in a nurturing way, especially during team startups.
Context	People from high-content cultures may prefer more information-rich technologies, as well as those that offer opportunities for the feeling of social presence. They may resist using technologies with low social presence to communicate with people they have never met. People from low-context cultures may prefer more asynchronous communications.

Source: D.L. Duarte and N. Tennant Snyder, *Mastering Virtual Teams* (San Francisco: Jossey-Bass, 2001), p. 60.

The importance of culture cannot be understated. Duarte and Snyder identify four important points to remember concerning the impact of culture on virtual teams:[6]

1. There are national cultures, organizational cultures, functional cultures, and team cultures. They can be sources of competitive advantages for virtual teams that know how to use cultural differences to create synergy. Team leaders and members who understand and are sensitive to cultural differences can create more robust outcomes than can members of homogeneous teams with members who think and act alike. Cultural differences can create distinctive advantages for teams if they are understood and used in positive ways.
2. The most important aspect of understanding and working with cultural differences is to create a team culture in which problems can be surfaced and differences can he discussed in a productive, respectful manner.
3. It is essential to distinguish between problems that result from cultural differences and problems that are performance based.
4. Business practices and business ethics vary in different parts of the world. Virtual teams need to clearly articulate approaches to these that every member understands and abides by.

8.11 MANAGING INNOVATION PROJECTS

Understanding Innovation

Innovation is generally regarded as a new way of doing something. The new way of doing something should be substantially different from the way it was done before rather than a small incremental change such as with continuous improvement activities. The ultimate goal of innovation is to create hopefully long-lasting additional value for the company, the users, and the deliverable itself. Innovation can be viewed as the conversion of an idea into cash or a cash equivalent. While the goal of

6. See Duarte and Snyder, p. 70, note 11.

successful innovation is to add value, the effect can be negative or even destructive if it results in poor team morale, an unfavorable cultural change, or a radical departure from existing ways of doing work. The failure of an innovation project can lead to demoralizing the organization and causing talented people to be risk avoiders in the future rather than risk takers.

Not all project managers are capable of managing projects involving innovation. Characteristics of innovation projects include an understanding that:

- Specific innovation tools and decision-making techniques may be necessary.
- It may be impossible to prepare a detailed schedule showing when innovation will actually occur.
- It may be impossible to determine a realistic budget for innovation.
- Innovation may not be possible and there comes a time when one must simply "give up."
- The deliverable from the innovation project may not need extra "bells and whistles" that would make it too costly to users.

Failure is an inevitable part of many innovation projects. The greater the degree of innovation desired, the greater the need for effective risk management practices to be in place. Without effective risk management, it may be impossible within a reasonable time period to "pull the plug" on a project that is a cash drain and with no likelihood of achieving success.

Standard project management methodologies do not necessarily lend themselves to projects requiring innovation. Frameworks may be more appropriate. Methodologies work well when there exists a well-defined statement of work and reasonable estimates. Schedules and WBS development are normally based upon rolling wave or progressive planning since it is unlikely that we can develop a detailed plan and schedule for the entire project.

Project Selection

Your company has a list of twenty projects that they would like to accomplish, and each project may or may not require some different degree of innovation. If current funding will support only a few projects, then how does a company decide which of the twenty projects to work on first? This is the project selection and prioritization process. Project selection is a necessity for all projects but usually more difficult to perform if some degree of innovation is required. Predicting the degree of complexity is difficult. Based upon the company's financial health, the company may select projects not requiring innovation but have a high expectation of generating short-term cash flow. Project selection decisions are not made in a vacuum. The decision is usually related to other factors such as funding limitations, timing of the funding, criticality of the project and its alignment with the strategic plan, timing of cash flow expected from the completed project, fit with other projects in the portfolio, availability or the required resources, and, perhaps most important, the availability of qualified project managers and team members. The selection of an innovation project could be based upon the completion of other projects that would release resources needed for the new project. Also, the project selected may be constrained by the completion date such that resources must be released to other activities. In any event, some form of selection process is needed, and this may very well involve the PMO. What is unfortunate is that project managers are generally brought on board after the selection has been made, therefore having very little to say in the selection process.

Project Selection Obstacles

Project selection decision makers frequently have much less information to evaluate possible innovation projects than they would wish, especially if they do not consult the project manager. Uncertainties often surround the success likelihood of a project and market response, the ultimate market value of the project, its total cost to completion, and the probability of success and/or a technical breakthrough. This lack of an adequate information base often leads to another difficulty: the lack of a systematic approach to project selection and evaluation. Consensus criteria and methods for assessing each candidate project against these criteria are essential for rational decision making. Though most organizations have established organizational goals and objectives, these are usually not detailed enough to be used as criteria for project selection decision making. However, they are an essential starting point.

Project selection and evaluation decisions are often confounded by several behavioral and organizational factors. Departmental loyalties, conflicts in desires, differences in perspectives, and an unwillingness to openly share information can stymie the project selection and evaluation process. Adding to these, the uncertainties of innovation and possibly a lack of understanding of the complexities of the innovation project can make decision making riskier than for projects where innovation may be unnecessary. Much project evaluation data and information are necessarily subjective in nature. Thus, the willingness of the parties to openly share and put trust in each other's opinions becomes an important factor.

The risk-taking climate or culture of an organization can also have a decisive bearing on the project selection process as well as creating additional problems for the project manager during project execution. If the climate is risk averse, then high-risk projects requiring innovation may never surface. Attitudes within the organization toward ideas and the volume of ideas being generated will influence the quality of the projects selected. In general, the greater the number of creative ideas generated, the greater the chances of selecting projects that will yield the greatest value.

Identification of Projects

Because the number of potential ideas can be large, some sort of classification system is needed. There are three common methods of classification. The first method is to place the projects into two major categories such as survival and growth. The sources and types of funds for these two categories can and will be different. The second method comes from typical R&D strategic planning models. Using this approach, projects to develop new products or services are classified as either offensive or defensive projects. Offensive projects are designed to capture new markets or expand market share within existing markets. Offensive projects mandate the continuous development of new products and services predicated upon innovation. Defensive projects are designed to extend the life of existing products or services. This could include add-ons or enhancements geared toward keeping present customers or finding new customers for one's existing products or services. Defensive projects are usually easier to manage than offensive projects and have a higher probability of success. Some other methods for classifying projects are:

- Radical technical breakthrough projects
- Next-generation projects

- New family members
- Add-ons and enhancement projects

Radical technological breakthrough projects are the most difficult to manage because of the need for possibly extensive innovation. These innovation projects, if successful, can lead to profits that are many times larger than the original development costs. Unsuccessful innovation projects can lead to equally dramatic losses, which is one of the reasons why senior management must exercise due caution in approving innovation projects. Care must be taken to identify and screen out inferior candidate projects before committing significant resources to them.

There is no question that innovation projects are the most costly and most difficult to manage. Some companies mistakenly believe that the solution is to minimize or limit the total number of ideas for new projects or to limit the number of ideas in each category. This could be a costly mistake for long-term growth.

8.12 AGILE PROJECT MANAGEMENT

When companies first get involved in project management, there is a tendency to focus on a formalized project management approach structured around a project management methodology that contains rigid policies and procedures as to how project work must take place. Deviations from the plan were discouraged because there was an inherent fear at the executive levels that there might be a loss of control without formalized processes.

As companies become reasonably mature in project management, there is a tendency to go from formal to informal project management, minimize the need for excessive documentation (possibly even hoping for paperless project management), and trust that the project team will make the right decisions. In order to support the existence of the more informal approach to project management, techniques such as agile project management have surfaced. There are several forms of agile project management. Most of these forms are designed to alleviate some of the problems with traditional project management, as seen in Table 8–9.

TABLE 8–9. COMPARISON OF TRADITIONAL VERSUS AGILE PROJECT MANAGEMENT

Factor	Traditional Project Management	Agile Project Management
Structured focus	Tools and processes	People
Completion focus	Paperwork and contractual documentation	Results and deliverables
Leadership style	Authoritarian	Participative
Amount of documentation	Extremely heavy	Minimal
Trust	Mistrust may prevail	Trust
Customer interfacing	Negotiation	Collaboration
Customer feedback	Minimal, perhaps only at project termination	Throughout the project
Project direction	Follow the plan exactly	Respond to changes
Project solution	Follow the contractual requirements exactly	Constantly evolving solution
Delivery	Often a late delivery	Shorter delivery time
Unused features	Too much "gold-plating"	Minimal
Number of features	Too many	What the client needs
Acceptance	Often a high rejection of deliverables	Minimal number of rejected deliverables
Best practices and lessons learned	Discovered from successes	Discovered from successes and failures

While agile project management practices may eventually replace traditional project management practices, there is the fear that implementation failure may occur. Therefore, it may be difficult to go directly to agile project management without first having some form of traditional project management in place.

While the benefits of agile are quite apparent, there is a risk that the benefits may not be achieved if the company rushing into an adoption of an agile approach without having at least a cursory understanding of project management. Companies that are experienced in project management practices seem to have an easier time implementing agile and achieving the benefits. Precautions must be taken because, while traditional project management knowledge is necessary to develop project management skills in general, the same knowledge may not be directly applicable in an agile environment. Charles Cobb states:[7]

> Agile is causing us to broaden our vision of what project management is, and this will have a dramatic impact on the potential roles that a project manager can play in an agile project. The image of a project manager is typically very heavily focused on a plan-driven approach.
>
> All of these things are good things for a traditional project manager to know, but many of them become irrelevant in an agile environment, which calls for a very different approach. In an agile environment, you may not have enough information upfront to develop detailed project plans and work breakdown structures, and it may be very difficult, if not impossible, to predict how all of the project activities will be organized in advance in terms of PERT [networking] charts and Gantt charts. Agile calls for a much more fluid and dynamic approach that is optimized around dealing with a much more unpredictable and uncertain environment.
>
> That doesn't mean that the typical plan-driven approach to project management is obsolete and no longer useful, but it certainly should not be the only way to run a project. The challenge for project managers is going beyond a traditional, plan-driven approach and learning a much broader range of project management practices in a wider range of environments with potentially much higher levels of uncertainty. It is much more of a multidimensional and adaptive approach to project management.

Related Case Studies (from Kerzner/*Project Management Case Studies*, 5th ed.)	**Related Workbook Exercises (from Kerzner/*Project Management Workbook and PMP®/CAPM® Exam Study Guide*, 12th ed.)**	***PMBOK® Guide*, Sixth Edition, Reference Section for the PMP® Certification Exam**
• American Electronics International • The Tylenol Tragedies • Photolite Corporation (A) • Photolite Corporation (B) • Photolite Corporation (C) • Photolite Corporation (D) • Jackson Industries • Is It Fraud?*	• The Potential Problem Audit • The Situational Audit • Multiple Choice Exam	• Integration Management • Human Resource Management • Project Management • Roles and Responsibilities

*Case study appears at end of chapter.

7. Charles G. Cobb, *The Project Manager's Guide to Mastering Agile* (Hoboken, NJ: John Wiley & Sons, 2015), p. 115.

8.13 STUDYING TIPS FOR THE PMI® PROJECT MANAGEMENT CERTIFICATION EXAM

This section is applicable as a review of the principles to support the knowledge areas and domain groups in the *PMBOK® Guide*. This chapter addresses:

- Project Resource Management
- Professional Responsibility
- Planning
- Execution

Understanding the following principles is beneficial if the reader is using this text to study for the PMP® Certification Exam:

- Principles and tasks included under professional responsibility
- Factors that affect professional responsibility such as conflicts of interest and gifts
- PMI® Code of Professional Conduct (this can be downloaded from the PMI website pmi.org)
- That personnel performance reviews, whether formal or informal, are part of a project manager's responsibility
- Differences between project management in a large company and project management in a small company

The following multiple-choice questions will be helpful in reviewing the principles of this chapter:

1. You have been sent on a business trip to visit one of the companies bidding on a contract to be awarded by your company. You are there to determine the validity of the information in its proposal. They take you to dinner one evening at a very expensive restaurant. When the bill comes, you should:
   A. Thank them for their generosity and let them pay the bill.
   B. Thank them for their generosity and tell them that you prefer to pay for your own meal.
   C. Offer to pay for the meal for everyone and put it on your company's credit card.
   D. Offer to pay the bill, put it on your company's credit card, and make the appropriate adjustment in the company's bid price to cover the cost of the meals.
2. You are preparing a proposal in response to a request for proposal (RFP) from a potentially important client. The salesperson in your company working on the proposal tells you to "lie" in the proposal to improve the company's chance of winning the contract. You should:
   A. Do as you are told.
   B. Refuse to work on the proposal.
   C. Report the matter to your superior, the project sponsor, or the corporate legal group.
   D. Resign from the company.

3. You are preparing for a customer interface meeting and your project sponsor asks you to lie to the customer about certain test results. You should:
   A. Do as you are told.
   B. Refuse to work on the project from this point forth.
   C. Report the matter to either your superior or the corporate legal group for advice.
   D. Resign from the company.
4. One of the project managers in your company approaches you with a request to use some of the charge numbers from your project (which is currently running under budget) for work on their project (which is currently running over budget). Your contract is a cost-reimbursable contract for a client external to your company. You should:
   A. Do as you are requested.
   B. Refuse to do this unless the project manager allows you to use his charge numbers later on.
   C. Report the matter to your superior, the project sponsor, or the corporate legal group.
   D. Ask the project manager to resign from the company.
5. You have submitted a proposal to a client as part of a competitive bidding effort. One of the people evaluating your bid informs you that it is customary to send them some gifts in order to have a better chance of winning the contract. You should:
   A. Send them some gifts.
   B. Do not send any gifts and see what happens.
   C. Report the matter to your superior, the project sponsor, or the corporate legal group for advice.
   D. Withdraw the proposal.
6. You just discovered that the company where your brother-in-law is employed has submitted a proposal to your company. Your brother-in-law has asked you to do everything possible to make sure that his company will win the contract because his job may be in jeopardy. You should:
   A. Do what your brother-in-law requests.
   B. Refuse to look into the matter and pretend it never happened.
   C. Report the conflict of interest to your superior, the project sponsor, or the corporate legal group.
   D. Hire an attorney for advice.
7. As part of a proposal evaluation team, you have discovered that the contract will be awarded to Alpha Company and that a formal announcement will be made in two days. The price of Alpha Company's stock may just skyrocket because of this contract award. You should:
   A. Purchase as much Alpha Company stock as you can within the next two days.
   B. Tell family members to purchase the stock.
   C. Tell employees in the company to purchase the stock.
   D. Do nothing about stock purchases until after the formal announcement has been made.
8. Your company has decided to cancel a contract with Beta Company. Only a handful of employees know about this upcoming cancellation. The announcement of the cancellation will

be made in about two days. You own several shares of Beta Company stock and know full well that the stock will plunge on the bad news. You should:

A. Sell your stock as quickly as possible.

B. Sell your stock and tell others whom you know own the stock to do the same thing.

C. Tell the executives to sell their shares if they are stockowners.

D. Do nothing until after the formal announcement is made.

9. You are performing a two-day quality audit of one of your suppliers. The supplier asks you to remain a few more days so that they can take you out deep-sea fishing and gambling at the local casino. You should:

A. Accept as long as you complete the audit within two days.

B. Accept but take vacation time for fishing and gambling.

C. Accept their invitation but at a later time so that it does not interfere with the audit.

D. Gracefully decline their invitation.

10. You have been assigned as the project manager for a large project in the Pacific Rim. This is a very important project for both your company and the client. In your first meeting with the client, you are presented with a very expensive gift for yourself and another expensive gift for your spouse. You were told by your company that this is considered an acceptable custom when doing work in this country. You should:

A. Gracefully accept both gifts.

B. Gracefully accept both gifts but report only your gift to your company.

C. Gracefully accept both gifts and report both gifts to your company.

D. Gracefully refuse both gifts.

11. Your company is looking at the purchase of some property for a new plant. You are part of the committee making the final decision. You discover that the owner of a local auto dealership from whom you purchase family cars owns one of the properties. The owner of the dealership tells you in confidence that he will give you a new model car to use for free for up to three years if your company purchases his property for the new plant. You should:

A. Say thank you and accept the offer.

B. Remove yourself from the committee for conflict of interest.

C. Report the matter to your superior, the project sponsor, or the corporate legal group for advice.

D. Accept the offer as long as the car is in your spouse's name.

12. Your company has embarked upon a large project (with you as project manager) and as an output from the project there will be some toxic waste as residue from the manufacturing operations. A subsidiary plan has been developed for the containment and removal of the toxic waste and no environmental danger exists. This information on toxic waste has not been made available to the general public as yet, and the general public does not appear to know about this waste problem. During an interview with local newspaper personnel you are discussing the new project and the question of environmental concerns comes up. You should:

A. Say there are no problems with environmental concerns.

B. Say that you have not looked at the environmental issues problems as yet.

C. Say nothing and ask for the next question.

D. Be truthful and reply as delicately as possible.

13. As a project manager, you establish a project policy that you, in advance of the meeting, review all handouts presented to your external customer during project status review meetings. While reviewing the handouts, you notice that one slide contains company confidential information. Presenting this information to the customer would certainly enhance good will. You should:
    A. Present the information to the customer.
    B. Remove the confidential information immediately.
    C. Discuss the possible violation with senior management and the legal department before taking any action.
    D. First discuss the situation with the team member that created the slide and then discuss the possible violation with senior management and the legal department before taking any action.
14. You are managing a project for an external client. Your company developed a new testing procedure to validate certain properties of a product and the new testing procedure was developed entirely with internal funds. Your company owns all of the intellectual property rights associated with the new test. The workers who developed the new test used one of the components developed for your current customer as part of the experimental process. The results using the new test showed that the component would actually exceed the customer's expectations. You should:
    A. Show the results to the customer but do not discuss the fact that it came from the new test procedure.
    B. Do not show the results of the new test procedure since the customer's specifications call for use of the old test procedures.
    C. First change the customer's specifications and then show the customer the results.
    D. Discuss the release of this information with your legal department and senior management before taking any action.
15. Using the same scenario as in the previous question, assume that the new test procedure that is expected to be more accurate than the old test procedure indicates that performance will not meet customer specifications, whereas the old test indicates that customer specifications will be barely met. You should:
    A. Present the old test results to the customer showing that specification requirements will be met.
    B. Show both sets of test results and explain that the new procedure is unproven technology.
    C. First change the customer's specifications and then show the customer the results.
    D. Discuss the release of this information with your legal department and senior management before taking any action.
16. Your customer has demanded to see the "raw data" test results from last week's testing. Usually the test results are not released to customers until after the company reaches a conclusion on the meaning of the test results. Your customer has heard through the grapevine that the testing showed poor results. Management has left the entire decision up to you. You should:
    A. Show the results and explain that it is simply raw data and that your company's interpretation of the results will be forthcoming.
    B. Withhold the information until after the results are verified.
    C. Stall for time even if it means lying to the customer.
    D. Explain to the customer your company's policy of not releasing raw data.

17. One of your team members plays golf with your external customer's project manager. You discover that the employee has been feeding the customer company-sensitive information. You should:
    A. Inform the customer that project information from anyone other than the project manager is not official until released by the project manager.
    B. Change the contractual terms and conditions and release the information.
    C. Remove the employee from your project team.
    D. Explain to the employee the ramifications of his actions and that he still represents the company when not at work; then report this as a violation.
18. Your company has a policy that all company-sensitive material must be stored in locked filing cabinets at the end of each day. One of your employees has received several notices from the security office for violating this policy. You should:
    A. Reprimand the employee.
    B. Remove the employee from your project.
    C. Ask the Human Resources Group to have the employee terminated.
    D. Counsel the employee as well as other team members on the importance of confidentiality and the possible consequences for violations.
19. You have just received last month's earned-value information that must be shown to the customer in the monthly status review meeting. Last month's data showed unfavorable variances that exceeded the permissible threshold limits on time and cost variances. This was the result of a prolonged power outage in the manufacturing area. Your manufacturing engineer tells you that this is not a problem and next month you will be right on target on time and cost as you have been in the last five months. You should:
    A. Provide the data to the customer and be truthful in the explanation of the variances.
    B. Adjust the variances so that they fall within the threshold limits since this problem will correct itself next month.
    C. Do not report any variances this month.
    D. Expand the threshold limits on the acceptable variances but do not tell the customer.
20. You are working in a foreign country where it is customary for a customer to present gifts to the contractor's project manager throughout the project as a way of showing appreciation. Declining the gifts would be perceived by the customer as an insult. Your company has a policy on how to report gifts received. The *best* way to handle this situation would be to:
    A. Refuse all gifts.
    B. Send the customer a copy of our company's policy on accepting gifts.
    C. Accept the gifts and report the gifts according to policy.
    D. Report all gifts even though the policy says that some gifts need not be reported.
21. You are interviewing a candidate to fill a project management position in your company. On her resume, she states that she is a PMP® credential holder. One of your workers who knows the candidate informs you that she is not a PMP® credential holder yet but is planning to take the test next month and certainly expects to pass. You should:
    A. Wait until she passes the exam before interviewing her.
    B. Interview her and ask her why she lied.

C. Inform PMI of the violation.

D. Forget about it and hire her if she looks like the right person for the job.

22. You are managing a multinational project from your office in Chicago. Half of your project team are from a foreign country but are living in Chicago while working on your project. These people inform you that two days during next week are national religious holidays in their country and they will be observing the holiday by not coming into work. You should:

A. Respect their beliefs and say nothing.

B. Force them to work because they are in the United States where their holiday is not celebrated.

C. Tell them that they must work noncompensated overtime when they return to work in order to make up the lost time.

D. Remove them from the project team if possible.

23. PMI® informs you that one of your team members who took the PMP® exam last week and passed may have had the answers to the questions in advance provided to him by some of your other team members who are also PMP® credential holders and were tutoring him. PMI is asking for your support in the investigation. You should:

A. Assist PMI in the investigation of the violation.

B. Call in the employee for interrogation and counseling.

C. Call in the other team members for interrogation and counseling.

D. Tell PMI® that it is their problem, not your problem.

24. One of your team members has been with you for the past year since her graduation from college. The team member informs you that she is now a PMP® credential holder and shows you her certificate from PMI acknowledging this. You wonder how she was qualified to take the exam since she had no prior work experience prior to joining your company one year ago. You should:

A. Report this to PMI as a possible violation.

B. Call in the employee for counseling.

C. Ask the employee to surrender her PMP credentials.

D. Do nothing.

25. Four companies have responded to your RFP. Each proposal has a different technical solution to your problem and each proposal states that the information in the proposal is company-proprietary knowledge and not to be shared with anyone. After evaluation of the proposals, you discover that the best technical approach is from the highest bidder. You are unhappy about this. You decide to show the proposal from the highest bidder to the lowest bidder to see if the lowest bidder can provide the same technical solution but at a lower cost. This situation is:

A. Acceptable since once the proposals are submitted to your company, you have unlimited access to the intellectual property in the proposals

B. Acceptable since all companies do this

C. Acceptable as long as you inform the high bidder that you are showing their proposal to the lowest bidder

D. Unacceptable and is a violation of the Code of Professional Conduct

ANSWERS

1. B	10. C	19. A
2. C	11. C	20. D
3. C	12. D	21. C
4. C	13. D	22. A
5. C	14. D	23. A
6. C	15. D	24. A
7. D	16. A	25. D
8. D	17. D	
9. D	18. D	

PROBLEMS

8–1 Beta Company has decided to modify its wage and salary administration program whereby line managers are evaluated for promotion and merit increases based on how well they have lived up to the commitments that they made to the project managers. What are the advantages and disadvantages of this approach?

8–2 Does a functional employee have the right to challenge any items in the project manager's nonconfidential evaluation form?

8–3 Some people contend that functional employees should be able to evaluate the effectiveness of the project manager after project termination. Design an evaluation form for this purpose.

8–4 As a functional employee, the project manager tells you, "Sign these prints or I'll fire you from this project." How should this situation be handled?

8–5 How efficient can project management be in a unionized, immobile manpower environment?

8–6 Corporate salary structures and limited annual raise allocations often prevent proper project management performance rewards. Explain how each of the following could serve as a motivational factor:

A. Job satisfaction
B. Personal recognition
C. Intellectual growth

CASE STUDY

IS IT FRAUD?

Background

Paul was a project management consultant and often helped the Judge Advocate General's Office (JAG) by acting as an expert witness in lawsuits filed by the U.S. government against defense contractors. While most lawsuits were based upon unacceptable performance by the contractors, this lawsuit was different; it was based upon supposedly superior performance.

Meeting with Colonel Jensen

Paul sat in the office of Army colonel Jensen listening to the colonel's description of the history behind this contract. Colonel Jensen stated:

"We have been working with the Welton Company for almost ten years. This contract was one of several contracts we have had with them over the years. It was a one-year contract to produce 1,500 units for the Department of the Navy. Welton told us during contract negotiations that they needed two quarters to develop their manufacturing plans and conduct procurement. They would then ship the Navy 750 units at the end of the third quarter and the remaining 750 units at the end of the fourth quarter. On some other contracts, manufacturing planning and procurement was done in less than one quarter.

"On other contracts similar to this one, the Navy would negotiate a firm-fixed-price contract because the risk to both the buyer and seller was quite low. The government's proposal statement of work also stated that this would be a firm-fixed-price contract. But during final contract negotiations, Welton became adamant in wanting this contract to be an incentive-type contract with a bonus for coming in under budget and/or ahead of schedule.

"We were somewhat perplexed about why they wanted an incentive contract. Current economic conditions in the United States were poor during the time we did the bidding and companies like Welton were struggling to get government contracts and keep their people employed. Under these conditions, we believed that they would want to take as long as possible to finish the contract just to keep their people working.

"Their request for an incentive contract made no sense to us, but we reluctantly agreed to it. We often change the type of contract based upon special circumstances. We issued a fixed-price-incentive-fee contract with a special incentive clause for a large bonus should they finish the work early and ship all 1,500 units to the Navy. The target cost for the contract, including $10 million in procurement, was $35 million with a sharing ratio of 90%–10% and a profit target of $4 million. The point of total assumption was at a contract price of $43.5 million.

"Welton claimed that they finished their procurement and manufacturing plans in the first quarter of the year. They shipped the Navy 750 units at the end of the second quarter and the remaining 750 units at the end of the third quarter. According to their invoices, which we audited, they spent $30 million in labor in the first nine months of the contract and $10 million in procurement. The government issued them checks totaling $49.5 million. That included $43.5 million plus the incentive bonus of $6 million for early delivery of the units.

"The JAG office believes that Welton took advantage of the Department of the Navy when they demanded and received a fixed-price-incentive-fee contract. We want you to look over their proposal and what they did on the contract and see if anything looks suspicious."

Consultant's Audit

The first thing that Paul did was to review the final costs on the contract.

Labor:	$30,000.000
Material:	$10,000,000
	$40,000,000
Cost overrun:	$5,000,000
Welton's cost:	$500,000
Final profit:	$3,500,000

Welton completed the contract exactly at the contract price ceiling, also the point of total assumption, of $43.5 million.

The cost overrun of $5 million was entirely in labor. Welton originally expected to do the job in twelve months for $25 million in labor. That amounted to an average monthly labor expenditure of $2,083,333. But Welton actually spent $30 million in labor over nine months, which amounted to an average monthly labor cost of $3,333,333. Welton was spending about $1.25 million more per month than planned for during the first nine months. Welton explained that part of the labor overrun was due to overtime and using more people than anticipated.

It was pretty clear in Paul's mind what Welton had done. Welton overspent the labor by $5 million and only $500,000 of the overrun was paid by Welton because of the sharing ratio. In addition, Welton received a $6 million bonus for early delivery. Simply stated, Welton received $6 million for a $500,000 investment.

Paul knew that believing this to be true was one thing, but being able to prove this in court would require more supporting information. Paul's next step was to read the proposal that Welton submitted. On the bottom of the first page of the proposal was a paragraph entitled "Truth of Negotiations" which stated that everything in the proposal was the truth. The letter was signed by a senior officer at Welton. Paul then began reading the management section of the proposal. In the management section, Welton bragged about previous contracts almost identical to this one with the Department of the Navy and other government organizations. Welton also stated that most of the people used on this contract had worked on the previous contracts. Paul found other statements in the proposal that implied that the manufacturing plans for this contract were similar to those of other contracts and Paul now wondered why two quarters were needed to develop the manufacturing plans for this project. Paul was now convinced that something was wrong.

QUESTIONS

1. What information does Paul have to support his belief that something is wrong?
2. Knowing that you are not an attorney, does it appear from a project management perspective that sufficient information exists for a possible lawsuit to recover all or part of the incentive bonus for early delivery?
3. How do you think this case study ended? (It is a factual case and the author was the consultant.)

The Variables for Success

9.0 INTRODUCTION

Project management cannot succeed unless the project manager is willing to employ the systems approach to project management by analyzing those variables that lead to success and failure. This chapter briefly discusses the dos and don'ts of project management and provides a "skeleton" checklist of the key success variables. The following five topics are included:

- Predicting project success
- Project management effectiveness
- Expectations
- Lessons learned
- Best practices

9.1 PREDICTING PROJECT SUCCESS

PMBOK® Guide, 6th Edition
1.2.6.4 Measure Success

One of the most difficult tasks is predicting whether the project will be successful. Most goal-oriented managers look only at the time, cost, and performance parameters. If an out-of-tolerance condition exists, then additional analysis is required to identify the cause of the problem. Looking only at time, cost, and performance might identify immediate contributions to profits, but will not identify whether the project itself was managed correctly. This takes on paramount importance if the survival of the organization is based on a steady stream of successfully managed projects. Once or twice a project manager might be able to force a project to success. After a while, however, either the effect of the big bat will become tolerable, or people will avoid working on his projects.

PMBOK is a registered mark of the Project Management Institute, Inc.

Project success is often measured by the "actions" of three groups: the project manager and team, the parent organization, and the customer's organization. There are certain actions that the project manager and team can take in order to stimulate project success. These actions include:

- Insist on the right to select key project team members.
- Select key team members with proven track records in their fields.
- Develop commitment and a sense of mission from the outset.
- Seek sufficient authority and organizational form that supports project management.
- Coordinate and maintain a good relationship with the client, parent, and team.
- Seek to enhance the public's image of the project.
- Have key team members assist in decision making and problem solving.
- Develop realistic cost, schedule, and performance estimates and goals.
- Have backup strategies in anticipation of potential problems.
- Provide a team structure that is appropriate, yet flexible and flat.
- Go beyond formal authority to maximize influence over people and key decisions.
- Employ a workable set of project planning and control tools.
- Avoid overreliance on one type of control tool.
- Stress the importance of meeting cost, schedule, and performance goals.
- Give priority to achieving the mission or function of the end item.
- Keep changes under control.
- Seek to find ways of assuring job security for effective project team members.

In Chapter 4 we stated that a project cannot be successful unless it is recognized as a project and has the support of top-level management. Top-level management must be willing to commit company resources and provide the necessary administrative support so that the project easily adapts to the company's day-to-day routine of doing business. Furthermore, the parent organization must develop an atmosphere conducive to good working relationships between the project manager, parent organization, and client organization.

With regard to the parent and client organizations, there exist a number of variables that can be used to evaluate parent organization support. These variables are included in Table 9–1.

TABLE 9–1. VARIABLES TO PARENT AND CLIENT ORGANIZATIONS SUPPORT

Parent Organization Variables	Client Organization Variables
• A willingness to coordinate efforts • A willingness to maintain structural flexibility • A willingness to adapt to change • Effective strategic planning • Rapport maintenance • Proper emphasis on past experience • External buffering • Prompt and accurate communications • Enthusiastic support • Identification to all concerned parties that the project does, in fact, contribute to parent capabilities	• A willingness to coordinate efforts • Rapport maintenance • Establishment of reasonable and specific goals and criteria • Well-established procedures for changes • Prompt and accurate communications • Commitment of client resources • Minimization of red tape • Providing sufficient authority to the client contact (especially for decision making)

The mere identification and existence of these variables do not guarantee project success. Instead, they imply that there exists a good foundation with which to work so that if the project manager and team, and the parent and client organizations, take the appropriate actions, project success is likely.

Within the parent organization, the following actions must be taken:

- Select at an early point, a project manager with a proven track record of technical skills, human skills, and administrative skills (not necessarily in that order) to lead the project team.
- Develop clear and workable guidelines for the project manager.
- Delegate sufficient authority to the project manager, and let him make important decisions in conjunction with key team members.
- Demonstrate enthusiasm for and commitment to the project and team.
- Develop and maintain short and informal lines of communication.
- Avoid excessive pressure on the project manager to win contracts.
- Avoid arbitrarily slashing or ballooning the project team's cost estimate.
- Avoid "buy-ins."
- Develop close, not meddling, working relationships with the principal client contact and project manager.

Both the parent organization and the project team must employ proper managerial techniques to ensure that judicious and adequate, but not excessive, use of planning, controlling, and communications systems can be made. These proper management techniques must also include preconditioning, such as clearly established specifications and designs, realistic schedules, realistic cost estimates, avoidance of "buy-ins," and avoidance of overoptimism.

The client organization can have a great deal of influence on project success by minimizing team meetings, making rapid responses to requests for information, and simply letting the contractor "do his thing" without any interference.

With the client organization variables stated in Table 9–1 as the basic foundation, it should be possible for the project management team to:

- Encourage openness and honesty from the start from all participants.
- Create an atmosphere that encourages healthy competition, but not cutthroat situations or "liars'" contests.
- Plan for adequate funding to complete the entire project.
- Develop clear understandings of the relative importance of cost, schedule, and technical performance goals.
- Develop short and informal lines of communication and a flat organizational structure.
- Delegate sufficient authority to the principal client contact, and allow prompt approval or rejection of important project decisions.
- Reject "buy-ins."
- Make prompt decisions regarding contract award or go-ahead.
- Develop close, not meddling, working relationships with project participants.
- Avoid arm's-length relationships.
- Avoid excessive reporting schemes.
- Make prompt decisions regarding changes.

By combining the relevant actions of the project team, parent organization, and client organization, we can identify the fundamental lessons for management. These include:

- When starting off in project management, plan to go all the way.
  - Recognize authority conflicts—resolve.
  - Recognize change impact—be a change agent.
- Match the right people with the right jobs.
  - No system is better than the people who implement it.
- Allow adequate time and effort for laying out the project groundwork and defining work:
  - Work breakdown structure
  - Network planning
- Ensure that work packages are the proper size:
  - Manageable, with organizational accountability
  - Realistic in terms of effort and time
- Establish and use planning and control systems as the focal point of project implementation:
  - Know where you're going.
  - Know when you've gotten there.
- Be sure information flow is realistic:
  - Information is the basis for problem solving and decision making.
  - Communication "pitfalls" are the greatest contributor to project difficulties.
- Be willing to replan—do so:
  - The best-laid plans can often go astray.
  - Change is inevitable.
- Tie together responsibility, performance, and rewards:
  - Management by objectives
  - Key to motivation and productivity
- Long before the project ends, plan for its end:
  - Disposition of personnel
  - Disposal of material and other resources
  - Transfer of knowledge
  - Closing out work orders
  - Customer/contractor financial payments and reporting

9.2 PROJECT MANAGEMENT EFFECTIVENESS

PMBOK® Guide, 6th Edition
Chapter 4 Project Integration Management
Chapter 9 Project Resource Management

Project managers interact continually with upper-level management, perhaps more so than with functional managers. Not only the success of the project, but even the career path of the project manager can depend on the working relationships and expectations established with upper-level management. There are four key variables in

Section 9.2 and Section 9.3 are adapted from *Seminar in Project Management Workbook,* copyright 1977 by Hans J. Thamhain. Reproduced by permission of Dr. Hans J. Thamhain.

measuring the effectiveness of dealing with upper-level management. These variables are credibility, priority, accessibility, and visibility:

- Credibility
  - Credibility comes from being a sound decision maker.
  - It is normally based on experience in a variety of assignments.
  - It is refueled by the manager and the status of his project.
  - Making success visible to others increases credibility.
  - To be believable, emphasize facts rather than opinions.
  - Give credit to others; they may return this favor.
- Priority
  - Sell the specific importance of the project to the objectives of the total organization.
  - Stress the competitive aspect, if relevant.
  - Stress changes for success.
  - Secure testimonial support from others—functional departments, other managers, customers, independent sources.
  - Emphasize "spin-offs" that may result from projects.
  - Anticipate "priority problems."
  - Sell priority on a one-to-one basis.
- Accessibility
  - Accessibility involves the ability to communicate directly with top management.
  - Show that your proposals are good for the total organization, not just the project.
  - Weigh the facts carefully; explain the pros and cons.
  - Be logical and polished in your presentations.
  - Become personally known by members of top management.
  - Create a desire in the "customer" for your abilities and your project.
  - Make curiosity work for you.
- Visibility
  - Be aware of the amount of visibility you really need.
  - Make a good impact when presenting the project to top management.
  - Adopt a contrasting style of management when feasible and possible.
  - Use team members to help regulate the visibility you need.
  - Conduct timely "informational" meetings with those who count.
  - Use available publicity media.

9.3 EXPECTATIONS

PMBOK® Guide, 6th Edition
Chapter 9 Project Human Resource Management
Chapter 10 Project Communications

In the project management environment, the project managers, team members, and upper-level managers each have expectations of what their relationships should be with the other parties. To illustrate this, top management expects project managers to:

- Assume total accountability for the success or failure to provide results
- Provide effective reports and information

- Provide minimum organizational disruption during the execution of a project
- Present recommendations, not just alternatives
- Have the capacity to handle most interpersonal problems
- Demonstrate a self-starting capacity
- Demonstrate growth with each assignment

At first glance, it may appear that these qualities are expected of all managers, not necessarily project managers. But this is not true. The first four items are different. The line managers are not accountable for total project success, just for that portion performed by their line organization. Line managers can be promoted on their technical ability, not necessarily on their ability to write effective reports. Line managers cannot disrupt an entire organization, but the project manager can. Line managers do not necessarily have to make decisions, just provide alternatives and recommendations.

Just as top management has expectations of project managers, project managers have certain expectations of top management. Project management expects top management to:

- Provide clearly defined decision channels
- Take actions on requests
- Facilitate interfacing with support departments
- Assist in conflict resolution
- Provide sufficient resources/charter
- Provide sufficient strategic/long-range information
- Provide feedback
- Give advice and stage-setting support
- Define expectations clearly
- Provide protection from political infighting
- Provide the opportunity for personal and professional growth

The project team also has expectations of their leader, the project manager. The project team expects the project manager to:

- Assist in the problem-solving process by coming up with ideas
- Provide proper direction and leadership
- Provide a relaxed environment
- Interact informally with team members
- Stimulate the group process
- Facilitate adoption of new members
- Reduce conflicts
- Defend the team against outside pressure
- Resist changes
- Act as the group spokesperson
- Provide representation with higher management

In order to provide high task efficiency and productivity, a project team should have certain traits and characteristics. A project manager expects the project team to:

- Demonstrate membership self-development
- Demonstrate the potential for innovative and creative behavior

- Communicate effectively
- Be committed to the project
- Demonstrate the capacity for conflict resolution
- Be results oriented
- Be change oriented
- Interface effectively and with high morale

Team members want, in general, to fill certain primary needs. The project manager should understand these needs before demanding that the team live up to his expectations. Members of the project team need:

- A sense of belonging
- Interest in the work itself
- Respect for the work being done
- Protection from political infighting
- Job security and job continuity
- Potential for career growth

Project managers must remember that team members may not always be able to verbalize these needs, but they exist nevertheless.

9.4 LESSONS LEARNED

Lessons can be learned from each and every project, even if the project is a failure. But many companies do not document lessons learned because employees are reluctant to sign their names to documents that indicate they made mistakes. Thus employees end up repeating the mistakes that others have made.

Today, there is increasing emphasis on documenting lessons learned. Boeing maintains diaries of lessons learned on each airplane project. Another company conducts a post-implementation meeting where the team is required to prepare a three- to five-page case study documenting the successes and failures on the project. The case studies are then used by the training department in preparing individuals to become future project managers. Some companies even mandate that project managers keep project notebooks documenting all decisions as well as a project file with all project correspondence. On large projects, this may be impractical.

Most companies seem to prefer post-implementation meetings and case study documentation. The problem is when to hold the post-implementation meeting. One company uses project management for new product development and production. When the first production run is complete, the company holds a post-implementation meeting to discuss what was learned. Approximately six months later, the company conducts a second post-implementation meeting to discuss customer reaction to the product. There have been situations where the reaction of the customer indicated that what the company thought they did right turned out to be a wrong decision. A follow-up case study is now prepared during the second meeting.

9.5 UNDERSTANDING BEST PRACTICES

PMBOK® Guide, 6th Edition
4.4.3.1 Lessons Learned register
Chapter 9 Project Resource Management
9.5 Manage Project Team

One of the benefits of understanding the variables of success is that it provides you with a means for capturing and retaining best practices. Unfortunately this is easier said than done. There are multiple definitions of a best practice, such as:

- Something that works
- Something that works well
- Something that works well on a repetitive basis
- Something that leads to a competitive advantage
- Something that can be identified in a proposal to generate business

Ultimately, *best practices are those actions or activities undertaken by the company or individuals that lead to a sustained competitive advantage in project management.*

It has only been in recent years that the importance of best practices has been recognized. In the early years of project management, there were misconceptions concerning project management.

As project management evolved, best practices became important. Best practices can be learned from both successes and failures. In the early years of project management, private industry focused on learning best practices from successes. The government, however, focused on learning about best practices from failures. When the government finally focused on learning from successes, the knowledge on best practices came from their relationships with both their prime contractors and the subcontractors. Some of the best practices that emerged included:

- Use of life-cycle phases
- Standardization and consistency
- Use of templates for planning, scheduling, control, and risk
- Providing military personnel in project management positions with extended tours of duty at the same location
- Use of integrated project teams (IPTs)
- Control of contractor-generated scope changes
- Use of earned-value measurement (discussed in Chapter 14)

Critical Questions

There are several questions that must be addressed before an activity is recognized as a best practice. Three frequently asked questions are:

- Who determines that an activity is a best practice?
- How do you properly evaluate what you think is best practice to validate that in fact it is a true best practice?
- How do you get executives to recognize that best practices are true value-added activities and should be championed by executive management?

Some organizations have committees that report to senior management that have as their primary function the evaluation of potential best practices. Other organizations use the PMO to perform this work.

There is a difference between lessons learned and best practices. Lessons learned can be favorable or unfavorable, whereas best practices are usually favorable outcomes.

Evaluating whether or not something is a best practice is not time-consuming, but it is complex. Simply believing that an action is a best practice does not mean that it is a best practice, but rather a proven practice. PMOs are currently developing templates and criteria for determining whether an activity may qualify as a best practice. Some items that might be included in the template are:

- Is it a measurable metric?
- Does it identify measurable efficiency?
- Does it identify measurable effectiveness?
- Does it add value to the company?
- Does it add value to the customers?
- Is it transferable to other projects?
- Does it have the potential for longevity?
- Is it applicable to multiple users?
- Does it differentiate us from our competitors?
- Will the best practice require governance?
- Will the best practice require employee training?
- Is the best practice company proprietary knowledge?

One company had two unique characteristics in its best practices template:

- Helps to avoid failure
- In a crisis, helps to resolve a critical situation

Executives must realize that these best practices are, in fact, intellectual property to benefit the entire organization. If the best practice can be quantified, then it is usually easier to convince senior management.

Best Practices versus Proven Practices

For more than a decade, companies have become fascinated with the expression "best practices." But now, after a decade or more of use, we are beginning to scrutinize the term and perhaps better expressions exist.

A best practice begins with an idea that there is a technique, process, method, or activity that can be more effective at delivering an outcome than any other approach and provides us with the desired outcome with fewer problems and unforeseen complications. As a result, we end up with the most efficient and effective way of accomplishing a task based upon a repeatable process that has been proven over time for a large number of people and/or projects.

But once this idea has been proven to be effective, we normally integrate the best practice into our processes so that it becomes a standard way of doing business. Therefore, after acceptance and proven use of the idea, the better expression possibly should be a "proven practice" rather than a best practice.

This is just one argument why best practice may be just a buzzword and should be replaced by proven practice.

Another argument is that the identification of a best practice may lead some to believe that we were performing some activities incorrectly in the past, and that may not have been the case. This may simply be a more efficient and effective way of achieving a deliverable. Another issue is that some people believe that best practices imply that there is one and only one way of accomplishing a task. This also may be a faulty interpretation.

Perhaps in the future the expression best practices will be replaced by proven practices. However, for the remainder of this text, we will refer to the expression as best practices, but the reader must understand that other terms may be more appropriate.

Levels of Best Practices

Best practices can be discovered anywhere within or outside an organization. Figure 9–1 shows various levels of best practices. The bottom level is the professional standards level, which would include professional standards as defined by PMI. The professional standards level contains the greatest number of best practices, but they are general rather than specific and have a low level of complexity.

The industry standards level identifies best practices related to performance within the industry. For example, the automotive industry has established standards and best practices specific to the auto industry.

As we progress to the individual best practices in Figure 9–1, the complexity of the best practices goes from general to very specific applications and, as expected, the quantity of best practices is less. An example of a best practice at each level might be (from general to specific):

- Professional standards: Preparation and use of a risk management plan, including templates, guidelines, forms, and checklists for risk management.
- Industry-specific: The risk management plan includes industry best practices such as the best way to transition from engineering to manufacturing.

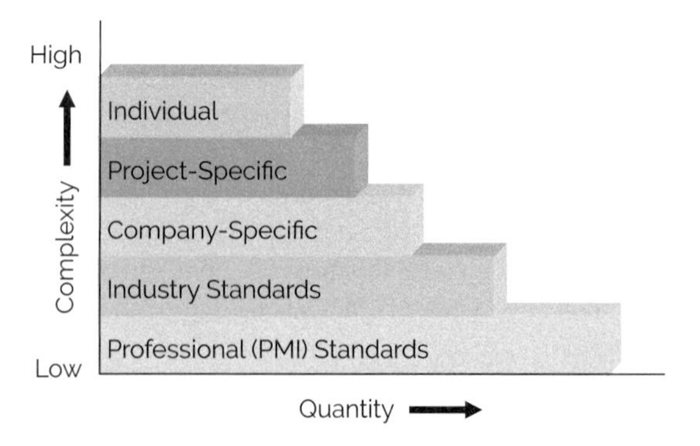

FIGURE 9–1. Levels of best practices.

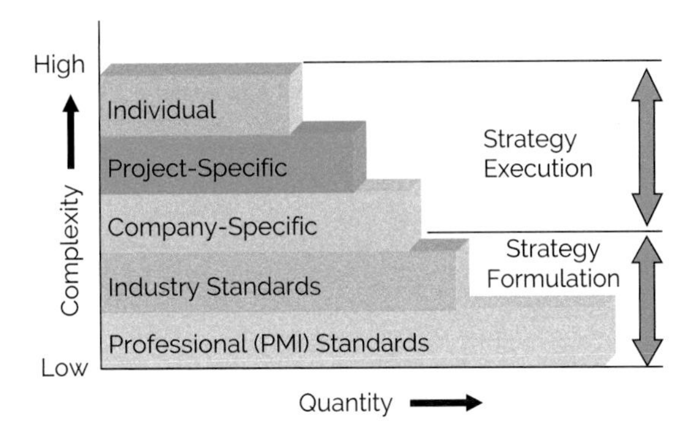

FIGURE 9–2. Usefulness of best practices.

- Company-specific: The risk management plan identifies the roles and interactions of engineering, manufacturing, and quality assurance groups during transition.
- Project-specific: The risk management plan identifies the roles and interactions of affected groups as they relate to a specific product/service for a customer.
- Individual: The risk management plan identifies the roles and interactions of affected groups based upon their personal tolerance for risk, possibly through the use of a responsibility assignment matrix prepared by the project manager.

Best practices can be extremely useful during strategic planning activities. As shown in Figure 9–2, the bottom two levels may be more useful for project strategy formulation whereas the top three levels are more appropriate for the execution of a strategy.

Common Beliefs

There are several common beliefs concerning best practices. A partial list includes:

- Because best practices can be interrelated, the identification of one best practice can lead to the discovery of another best practice, especially in the same category or level of best practices.
- Because of the dependencies that can exist between best practices, it is often easier to identify categories of best practices rather than individual best practices.
- Best practices may not be transferable. What works well for one company may not work for another company.
- Even though some best practices seem simplistic and common sense in most companies, the constant reminder and use of these best practices lead to excellence and customer satisfaction.
- Best practices are not limited exclusively to companies in good financial health.

Companies can have the greatest intentions when implementing best practices and yet detrimental results can occur. Table 9–2 identifies some possible expectations and the detrimental results that can occur. The poor results could have been the result of poor expectations or not fully understanding the possible ramifications after implementation.

TABLE 9–2. RESULTS OF IMPLEMENTING BEST PRACTICES

Type of Best Practice	Expected Advantage	Potential Disadvantage
Use of traffic light reporting	Speed and simplicity	Poor accuracy of information
Use of a risk management template/form	Forward looking and accurate	Inability to see some potential critical risks
Highly detailed WBS	Control, accuracy, and completeness	Excessive control and cost of reporting
Using EPM on all projects	Standardization and consistency	Too expensive on certain projects
Using specialized software	Better decision making	Too much reliance on tools

There are other reasons why best practices can fail or provide unsatisfactory results. These include:

- Lack of stability, clarity, or understanding of the best practice
- Failure to use best practices correctly
- Identifying a best practice that lacks rigor
- Identifying a best practice based upon erroneous judgment

Best Practices Library

With the premise that project management knowledge and best practices are intellectual property, how does a company retain this information? The solution is usually the creation of a best practices library. Figure 9–3 shows the three levels of best practices that seem most appropriate for storage in a best practices library.

Figure 9–4 shows the process of creating a best practices library. The bottom level is the discovery and understanding of what is or is not a "potential" best practice. The sources for potential best practices can originate anywhere within the organization.

The next level is the evaluation level to confirm that it is a best practice. The evaluation process can be done by the PMO or a committee but should have involvement by the senior levels of management. The evaluation process is very difficult because a one-time positive occurrence may not reflect a best practice. There must exist established criteria for the evaluation of a best practice.

Once it is agreed upon that a best practice exists, it must be classified and stored in some retrieval system such as a company intranet best practices library.

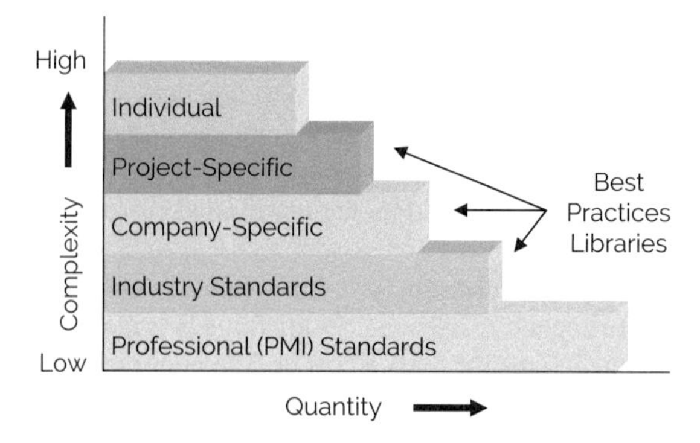

FIGURE 9–3. Levels of best practices.

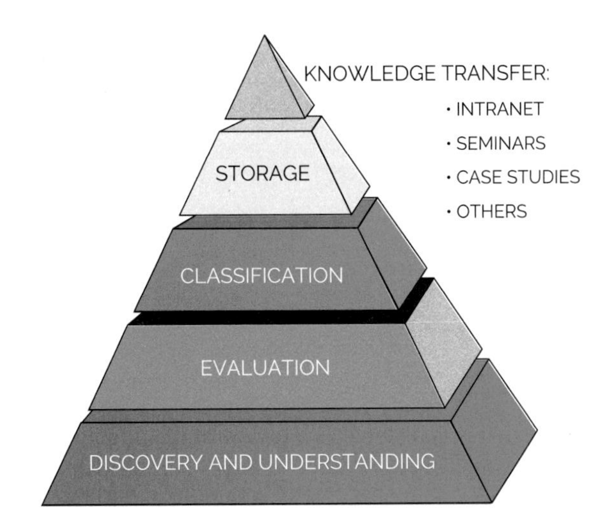

FIGURE 9–4. Creating a best practices library.

Figure 9–1 shows the levels of best practices, but the classification system for storage purposes can be significantly different. Figure 9–5 shows a typical classification system for a best practices library.

Another critical problem is best practices overload. One company started up a best practices library and, after a few years, had amassed hundreds of what were considered to be best practices. Nobody bothered to reevaluate whether all of these were still best practices. After reevaluation had taken place, it was determined that less than one-third of these were still regarded as best practices. Some were no longer best practices, others needed to be updated, and others had to be replaced with newer best practices.

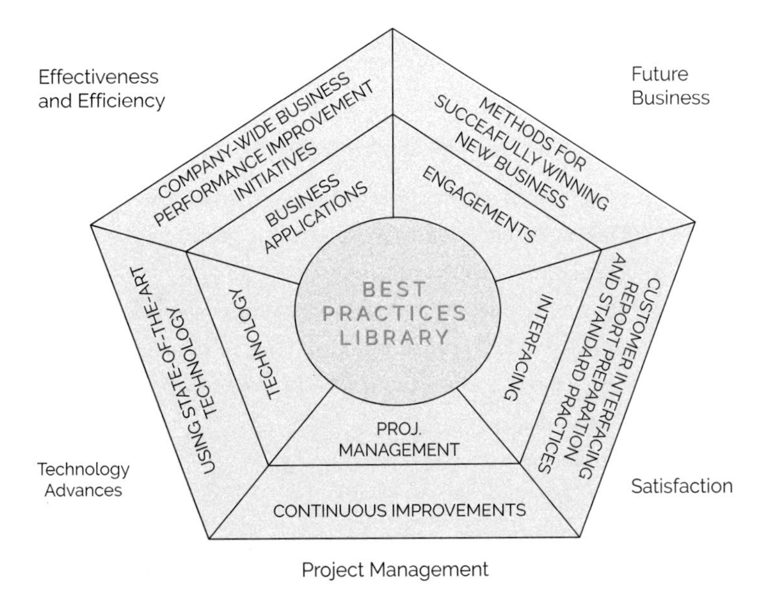

FIGURE 9–5. Best practices library.

Related Case Studies (from Kerzner/*Project Management Case Studies,* 5th ed.)	**Related Workbook Exercises (from Kerzner/*Project Management Workbook and PMP®/ CAPM® Exam Study Guide,* 12th ed.)**	***PMBOK® Guide*, 6th Edition, Reference Section for the PMP® Certification Exam**
• Como Tool and Die (A) • Como Tool and Die (B) • Radiance International*	• Multiple Choice Exam	• All PMBOK® Processes

*Case study appears at end of chapter.

9.6 STUDYING TIPS FOR THE PMI® PROJECT MANAGEMENT CERTIFICATION EXAM

This section is applicable as a review of the principles to support the knowledge areas and domain groups in the *PMBOK® Guide*. This chapter addresses:

- Communications management
- Initiation
- Planning
- Execution
- Monitoring
- Closure

Understanding the following principles is beneficial if the reader is using this text to study for the PMP® Certification Exam:

- Importance of capturing and reporting best practices as part of all project management processes
- Variables for success

The following multiple-choice questions will be helpful in reviewing the principles of this chapter:

1. Lessons learned and best practices are captured:
   A. Only at the end of the project
   B. Only after execution is completed
   C. Only when directed to do so by the project sponsor
   D. At all times but primarily at the closure of each life cycle phase
2. The person responsible for the identification of a best practice is the:
   A. Project manager
   B. Project sponsor
   C. Team member
   D. All of the above

PMP and CAPM are registered marks of the Project Management Institute, Inc.

3. The primary benefit of capturing lessons learned is to:
   A. Appease the customer
   B. Appease the sponsor
   C. Benefit the entire company on a continuous basis
   D. Follow the *PMBOK® Guide* requirements for reporting

ANSWERS

1. D
2. D
3. C

PROBLEMS

9–1 What is an effective working relationship among project managers themselves?

9–2 Must everyone in the organization understand the "rules of the game" for project management to be effective?

9–3 Defend the statement that the first step in making project management work must be a complete definition of the boundaries across which the project manager must interact.

CASE STUDY

RADIANCE INTERNATIONAL

Background

Radiance International (RI) had spent more than half a decade becoming a global leader in managing pollution, hazard, and environmental protection projects for its worldwide clients. It maintained ten offices across the world with approximately 150 people in each office. Its projects ranged from a few hundred thousand dollars to a few million dollars and lasted from six months to two years. When the downturn in corporate spending began in 2008, RI saw its growth stagnate. Line managers who previously spent most of their time interfacing with various project teams were now spending the preponderance of their time writing reports and memos trying to justify their position in case downsizing occurred. Project teams were asked to generate additional information that the line managers needed to justify their existence. This took a toll on the project teams and forced team members to do "busy work" that was sometimes unrelated to their project responsibilities.

Reorganization Plan

Management decided to reorganize the company primarily because of the maturity level of project management. Over the years, project management had matured to the point where senior management explicitly trusted the project managers to make both project-based and business-based decisions without continuous guidance from senior management or line management. The role of line management was simply to staff projects and then "get out of the way." Some line managers remained involved in some of the projects but actually did more harm than good with their interference. Executive sponsorship was also very weak because the project managers were trusted to make the right decisions.

The decision was made to eliminate all line management and go to the concept of pool management. One of the line managers was designated as the pool manager and administratively responsible for the 150 employees that were now assigned to the pool. Some of the previous line managers were let go while others became project managers or subject matter experts within the pool. Line managers that remained with the company were not asked to take a cut in pay.

In the center of the pool were the project managers. Whenever a new project came into the company, senior management and the pool manager would decide which project manager would be assigned to head up the new project. The project manager would then have the authority to talk to anyone in the pool who had the expertise needed on the project. If the person stated that he or she was available to work on the project, the project manager would provide that person with a charge number authorizing budgets and schedules for his or her work packages. If the person overran the budget or elongated the schedule unnecessarily, the project managers would not ask this person to work on his or her project again. Pool workers who ran out of charge numbers or were not being used by the project managers were then terminated from the company. Project managers would fill out a performance review form on each worker at the end of the project and forward it to the pool manager. The pool manager would make the final decision concerning wage and salary administration but relied heavily upon the inputs from the project managers.

The culture fostered effective teamwork, communication, cooperation, and trust. Whenever a problem occurred on a project, the project manager would stand up in the middle of the pool and state his or her crisis, and 150 people would rush to the aid of the project manager asking what they could do to help. The organization prided itself on effective group thinking and group solutions to complex projects. The system worked so well that sponsorship was virtually eliminated. Once a week or even longer, a sponsor would walk into the office of a project manager and ask, "Are there any issues I need to know about?" If the project manager responds "No," then the sponsor would say, "I'll talk to you in a week or two again" and then leave.

Two Years Later

After two years, the concept of pool management was working better than expected. Projects were coming in ahead of schedule and under budget. Teamwork abounded throughout the organization and morale was at an all-time high in every RI location. Everyone embraced the new culture and nobody was terminated from the company after the first year of the reorganization. Business was booming even though the economy was weak. There was no question that RI's approach to pool management had worked, and worked well! By the middle of the third year, RI's success story appeared in business journals around the world. While all of the notoriety was favorable and brought in more business, RI became a takeover target by large construction companies that saw the acquisition of RI as an opportunity. By the end of the third year, RI was acquired by a large construction firm. The construction company believed in strong line management with a span of control of approximately ten employees per supervisor. The pool management concept at RI was eliminated; several line management positions were created in each RI location and staffed with employees from the construction company. Within a year, several employees left the company.

QUESTIONS

1. Is it a good idea to remove all of the line management slots?
2. If pool management does not work, can line management slots then be reinstated?
3. How important is the corporate culture to the pool management concept?
4. Are there project sponsors at RI?

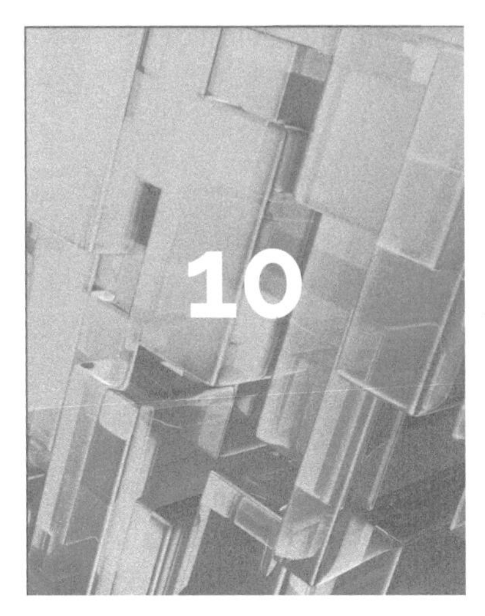

Working with Executives

10.0 INTRODUCTION

PMBOK® Guide, **6th Edition**
Chapter 4 Project Integration Management
Chapter 9 Project Resource Management
Chapter 13 Project Stakeholder Management

In any project management environment, project managers must continually interface with executives during both the planning and execution stages. Unless the project manager understands the executive's role and thought process, a poor working relationship will develop. In order to understand the executive–project interface, two topics are discussed:

- The project sponsor
- The in-house representatives

10.1 THE PROJECT SPONSOR

PMBOK® Guide, **6th Edition**
Chapter 13 Project Stakeholder Management

For more than two decades, the traditional role of senior management, as far as projects were concerned, has been to function as project sponsors.

The project sponsor usually comes from the executive levels and has the primary responsibility of maintaining executive–client contact. The sponsor ensures that the correct information from the contractor's organization is reaching executives in the customer's organization, that there is no filtering of information from the contractor to the customer, and that someone at the executive levels is making sure that the customer's money is being spent wisely. The project sponsor will normally transmit cost and deliverables information to the customer, whereas schedule and performance status data come from the project manager.

PMBOK is a registered mark of the Project Management Institute, Inc.

In addition to executive–client contact, the sponsor also provides guidance on:

- Objective setting
- Priority setting
- Project organizational structure
- Project policies and procedures
- Project master planning
- Up-front planning
- Key staffing
- Monitoring execution
- Conflict resolution

Sponsor-Project Interface

The role of the project sponsor takes on different dimensions based on the life-cycle phase the project is in. During the planning/initiation phase of a project, the sponsor normally functions in an active role, which includes such activities as:

- Assisting the project manager in establishing the correct objectives for the project
- Providing the project manager with information on the environmental/political factors that could influence the project's execution
- Establishing the priority for the project (either individually or through consultation with other executives) and informing the project manager of the established priority and the *reason* for the priority
- Providing guidance for the establishment of policies and procedures by which to govern the project
- Functioning as the executive–client contact point

During the initiation or kickoff phase of a project, the project sponsor must be actively involved in setting objectives and priorities. It is absolutely mandatory that the executives establish the priorities in both business and technical terms.

During the execution phase of the project, the role of the executive sponsor is more passive than active. The sponsor will provide assistance to the project manager on an as-needed basis except for routine status briefings.

During the execution stage of a project, the sponsor must be *selective* in the problems that he or she wishes to help resolve. Trying to get involved in every problem will not only result in severe micromanagement, but will undermine the project manager's ability to get the job done.

The role of the sponsor is similar to that of a referee. Table 10–1 shows the working relationship between the project manager and the line managers in both mature and immature organizations. When conflicts or problems exist in the project–line interface and cannot be resolved at that level, the sponsor might find it necessary to step in and provide assistance. Table 10–2 shows the mature and immature ways that a sponsor interfaces with the project.

TABLE 10–1. THE PROJECT-LINE INTERFACE

Immature Organization	Mature Organization
• Project manager is vested with power/authority over the line managers. • Project manager negotiates for best people. • Project manager works directly with functional employees. • Project manager has no input into employee performance evaluations. • Leadership is project manager-centered.	• Project and line managers share authority and power. • Project manager negotiates for line manager's commitment. • Project manager works through line managers. • Project manager makes recommendations to the line managers. • Leadership is team-centered.

TABLE 10–2. THE EXECUTIVE INTERFACE

Immature Organization	Mature Organization
• Executive is actively involved in projects.	• Executive involvement is passive.
• Executive acts as the project champion.	• Executive acts as the project sponsor.
• Executive questions the project manager's decisions.	• Executive trusts the project manager's decisions.
• Priority shifting occurs frequently.	• Priority shifting is avoided.
• Executive views project management as a necessary evil.	• Executive views project management as beneficial.
• There is very little project management support.	• There is visible, ongoing support.
• Executive discourages bringing problems upstairs.	• Executive encourages bringing problems upstairs.
• Executive is not committed to project sponsorship.	• Executive is committed to sponsorship (and ownership).
• Executive support exists only during project start-up.	• Executive support exists on a continuous basis.
• Executive encourages project decisions to be made.	• Executive encourages business decisions to be made.
• No procedures exist for assigning project sponsors.	• Sponsorship assignment procedures are visible.
• Executives seek perfection.	• Executives seek what is possible.
• Executive discourages use of a project charter.	• Executive recognizes the importance of a charter.
• Executive is not involved in charter preparation.	• Executive takes responsibility for charter preparation.
• Executive does not understand what goes into a charter.	• Executive understands the content of a charter.
• Executives do not believe that the project team is performing.	• Executives trust that performance is taking place.

Sponsor Function

It should be understood that the sponsor exists for everyone on the project, including the line managers and their employees. Project sponsors must maintain open-door policies, even though maintaining an open-door policy can have detrimental effects. First, employees may flood the sponsor with trivial items. Second, employees may feel that they can bypass levels of management and converse directly with the sponsor. The moral here is that employees, including the project manager, must be encouraged to be careful about how many times and under what circumstances they "go to the well."

In addition to his/her normal functional job, the sponsor must be available to provide as-needed assistance to the projects. Sponsorship can become a time-consuming effort, especially if problems occur. Therefore, executives are limited as to how many projects they can sponsor effectively at the same time.

If an executive has to function as a sponsor on several problems at once, problems can occur such as:

- Slow decision making, resulting in problem-solving delays
- Policy issues that remain unresolved and impact decisions
- Inability to prioritize projects when necessary

As an organization matures in project management, executives begin to trust middle- and lower-level management to function as sponsors. There are several reasons for supporting this:

- Executives do not have time to function as sponsors on each and every project.
- Not all projects require sponsorship from the executive levels.
- Middle management is closer to where the work is being performed.
- Middle management is in a better position to provide advice on certain risks.
- Project personnel have easier access to middle management.

Sometimes executives in large diversified corporations are extremely busy with strategic planning activities and simply do not have the time to properly function as a sponsor. In such cases, sponsorship falls one level below senior management.

Figure 10–1 shows the major functions of a project sponsor. At the onset of a project, a senior committee meets to decide whether a given project should be deemed as priority

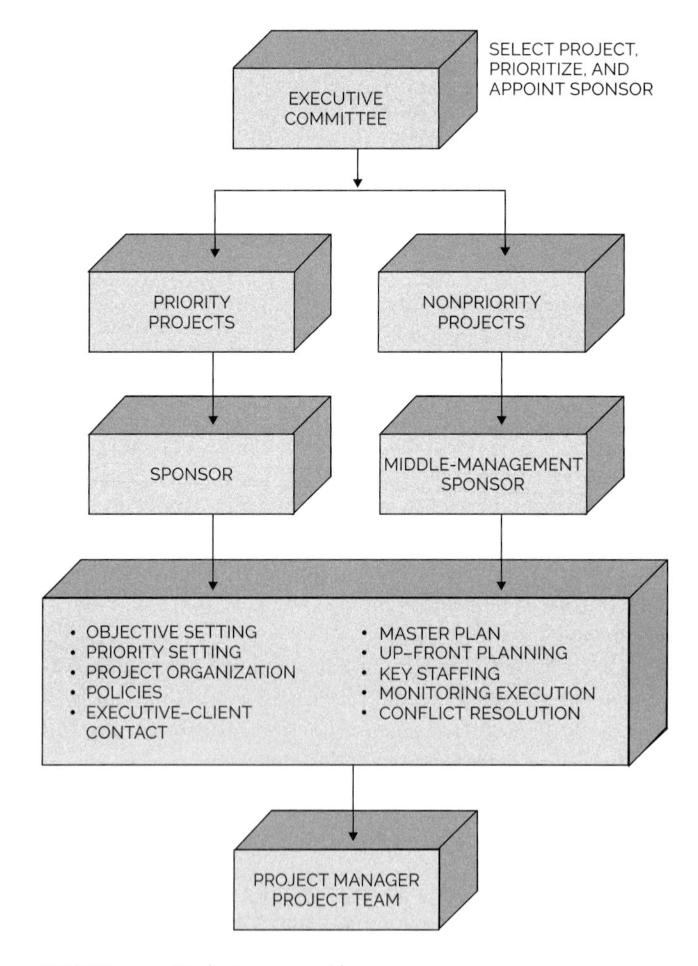

FIGURE 10–1. Project sponsorship.

or nonpriority. If the project is critical or strategic, then the committee may assign a senior manager as the sponsor, perhaps even a member of the committee. It is common practice for steering committee executives to function as sponsors for the projects that the steering committee oversees.

For projects that are routine, maintenance, or noncritical, a sponsor could be assigned from the middle-management levels. This generates an atmosphere of management buy-in at the critical middle levels.

Not all projects need a project sponsor. Sponsorship is generally needed on those projects that require a multitude of resources or a large amount of integration between functional lines or that have the potential for disruptive conflicts or the need for strong customer communications. Quite often customers wish to make sure that the contractor's project manager is spending funds prudently. Customers therefore like it when an executive sponsor supervises the project manager's funding allocation.

It is common practice for companies that are heavily involved in competitive bidding to identify in their proposal not only the resume of the project manager, but the resume of the executive project sponsor as well. This may give the bidder a competitive advantage, all other things being equal, because customers believe they have a direct path of communications to executive management. One such contractor identified the functions of the executive project sponsor as follows:

- Major participation in sales effort and contract negotiations
- Establishes and maintains top-level client relationships
- Assists project manager in getting the project underway (planning, procedures, staffing, etc.)
- Maintains current knowledge of major project activities (receives copies of major correspondence and reports, attends major client and project review meetings, visits project regularly, etc.)
- Handles major contractual matters
- Interprets company policy for the project manager
- Assists project manager in identifying and solving major problems
- Keeps general management and company management advised of major problems

Today's Project Sponsor

Historically, the role of the sponsor was almost exclusively interfacing with just the project manager. This was particularly true for small- to medium-size projects. Today, we are working on projects that are highly complex, requiring the sponsor to interface with the entire project team. As such, new responsibilities are being added in to the role of the sponsor. Some of these are:

- Maintaining close contact with the entire team throughout the project
- Preparing and endorsing the project's charter
- Making sure that the project manager has the appropriate authority for the decisions that must be made throughout the project
- Making sure that the project has the appropriate priority
- Explaining the reasons for the priority to the project team

- Establishing and reinforcing the project's business objectives as well as technical objectives
- Making sure that all deadlines are realistic
- Reaffirming the importance of meeting deadlines
- Explaining the enterprise environmental factors and political factors that can influence the project
- Assisting in the design of the organizational structure for the project
- Developing an emergency backup resource plan for the project
- Providing guidance for the development of project-specific policies and procedures
- Functioning as the focal point for project status for other executives
- Delineating expectations they have of the project manager and the team
- Delineating the skills they expect project managers to possess or develop as part of life-long learning
- Explaining how proprietary information should be handled
- Establishing a structured policy for scope changes
- Providing mentorship for inexperienced project managers
- Providing constructive feedback rather than personal criticism
- Acting as the communications link with the media
- Encouraging both good news and bad news to be brought forth, but not overreacting to bad news
- Recognizing the tell-tale signs that there is excessive pressure being placed upon the project team
- Understanding the cost of paperwork
- Practicing walk-the-halls management
- Placating irate customers
- Establishing a recognition or reward system, even if is a nonmonetary reward system
- Maintaining an open-door policy for the entire project team rather than just the project manager

Multiple Project Sponsors Consider a project that is broken down into two life-cycle phases: planning and execution. For short-duration projects, say two years or less, it is advisable for the project sponsor to be the same individual for the entire project. For long-term projects of five years or so, it is possible to have a different project sponsor for each life-cycle phase, but preferably from the same level of management. The sponsor does not have to come from the same line organization as the one where the majority of the work will be taking place. Some companies even go so far as demanding that the sponsor come from a line organization that has no vested interest in the project.

The project sponsor is actually a "big brother" or advisor for the project manager. Under *no* circumstances should the project sponsor try to function as the project manager. The project sponsor should assist the project manager in solving those problems that the project manager cannot resolve by himself.

Invisible Sponsors

In some industries, such as construction, the project sponsor is identified in the proposal, and thus everyone knows who it is. Unfortunately, there are situations where the project sponsor is "hidden," and the project manager may not realize who it is, or know if the customer realizes who it is. This concept of invisible sponsorship occurs most frequently at the executive level and is referred to as absentee sponsorship.

There are several ways that invisible sponsorship can occur. The first is when the manager who is appointed as a sponsor refuses to act as a sponsor for fear that poor decisions or an unsuccessful project could have a negative impact on his or her career. The second type results when an executive really does not understand either sponsorship or project management and simply provides lip service to the sponsorship function. The third way involves an executive who is already overburdened and simply does not have the time to perform meaningfully as a sponsor. The fourth way occurs when the project manager refuses to keep the sponsor informed and involved. The sponsor may believe that everything is flowing smoothly and that he is not needed.

Some people contend that the best way for the project manager to work with an invisible sponsor is for the project manager to make a decision and then send a memo to the sponsor stating, "This is the decision that I have made and, unless I hear from you in the next 48 hours, I will assume that you agree with my decision."

The opposite extreme is the sponsor who micromanages. One way for the project manager to handle this situation is to bury the sponsor with work in hopes that he will let go.

Unfortunately this could end up reinforcing the sponsor's belief that what he is doing is correct. The better alternative for handling a micromanaging sponsor is to ask for role clarification. The project manager should try working with the sponsor to define the roles of project manager and project sponsor more clearly.

The invisible sponsor and the overbearing sponsor are not as detrimental as the "can't-say-no" sponsor. When a sponsor continuously says "yes" to the customer, everyone in the contractor's organization eventually suffers.

Sometimes the existence of a sponsor can do more harm than good, especially if the sponsor focuses on the wrong objectives around which to make decisions.

It's necessary to have a sponsor who understands project management rather than one who simply assists in decision making. The goals and objectives of the sponsor must be aligned with the goals and objectives of the project, and they must be realistic. If sponsorship is to exist at the executive levels, the sponsor must be visible and constantly informed concerning the project status.

Multinational Sponsors and Stakeholders

When dealing with external stakeholders on global projects, there are significant unknowns that can influence the direction of the project. It may not be easy to identify the stakeholders. To make matters worse, global projects are implemented in complex environments that are highly impacted by sociopolitical factors. Some of the critical factors when dealing with global stakeholders include:

- Not all of the stakeholders are easily identifiable.
- There are significantly more stakeholders than on non-nonglobal projects.
- Global stakeholders are more likely to change over the duration of the project.

- There may be hidden stakeholders higher up in the organizational hierarchy than the project's visible global stakeholders.
- Not all of the issues facing the stakeholders are easy identified.
- Solutions to problems on the project may need to be resolved higher up than the global stakeholders.
- Not all issues with global stakeholders can be resolved the same way we resolve issues with nonglobal stakeholders.
- Global stakeholders may have more hidden agendas than with nonglobal stakeholders.
- Conflicts with global stakeholders may need to be approached with a different set of conflict resolution modes than we use on traditional projects.

Aaltonen and Sivonen have written a paper discussing conflict resolution modes for dealing with stakeholder pressure on global projects. In their paper, they discuss five conflict resolution strategies:[1]

- Adaptation strategy: obeying stakeholder rules, policies and procedures
- Compromising strategy: negotiation and dialogue
- Avoidance strategy: loosening attachments to the stakeholders
- Dismissal strategy: ignoring their demands
- Influence strategy: shaping proactively the demands of the stakeholders

Emerging market global stakeholders can respond differently to issues than global stakeholders in other markets. It is of critical importance to understand the operating and cultural environment in which the global stakeholders reside.

Committee Sponsorship

For years companies have assigned a single individual as the sponsor for a project. The risk was that the sponsor would show favoritism to his line group and suboptimal decision making would occur. Recently, companies have begun looking at sponsorship by committee to correct this.

Committee sponsorship is common in those organizations committed to concurrent engineering and shortening product development time. Committees are composed of middle managers from marketing, R&D, and operations. The idea is that the committee will be able to make decisions in the best interest of the company more easily than a single individual could.

Committee sponsorship also has its limitations. At the executive levels, it is almost impossible to find time when senior managers can convene. For a company with a large number of projects, committee sponsorship may not be a viable approach.

In time of crisis, project managers may need immediate access to their sponsors. If the sponsor is a committee, then how does the project manager get the committee to convene quickly? Also, individual project sponsors may be more dedicated than committees. Committee sponsorship has been shown to work well if one, and only one, member of the committee acts as the prime sponsor for a given project.

---

1. Kirsi Aaltonen and Risto Sivonen, "Response Strategies to Stakeholder Pressure in Global Projects," *International Journal of Project Management*, 27 (2009) 131–141.

When to Seek Help

During status reporting, a project manager can wave either a red, yellow, or green flag. This is known as the "traffic light" reporting system. For each element in the status report, the project manager will illuminate one of three lights according to the following criteria:

- *Green light:* Work is progressing as planned. Sponsor involvement is not necessary.
- *Yellow light:* A potential problem may exist. The sponsor is informed but no action by the sponsor is necessary at this time.
- *Red light:* A problem exists that may affect time, cost, scope, or quality. Sponsor involvement is necessary.

Yellow flags are warnings that should be resolved at the middle levels of management or lower.

If the project manager waves a red flag, then the sponsor will probably wish to be actively involved. Red flag problems can affect the time, cost, or performance constraints of the project and an immediate decision must be made. The main function of the sponsor is to assist in making the best possible decision in a timely fashion.

Both project sponsors and project managers should not encourage employees to come to them with problems unless the employees also bring alternatives and recommendations. Usually, employees will solve most of their own problems once they prepare alternatives and recommendations.

Good corporate cultures encourage people to bring problems to the surface quickly for resolution. The quicker the potential problem is identified, the more opportunities are available for resolution.

Some companies use more than three colors to indicate project status. One company also has an orange light for activities that are still being performed after the target milestone date.

The New Role of the Executive

As project management matures, executives decentralize project sponsorship to middle- and lower-level management. Senior management then takes on new roles such as:

- Establishing a Center for Excellence in project management
- Establishing a project office or centralized project management function
- Creating a project management career path
- Creating a mentorship program for newly appointed project managers
- Creating an organization committed to benchmarking best practices in project management in other organizations
- Providing strategic information for risk management

Because of the pressure placed upon the project manager for schedule compression, risk management could very well become the single most critical skill for project managers. Executives will find it necessary to provide project management with strategic business intelligence, assist in risk identification, and evaluate or prioritize risk-handling options.

TABLE 10–3. VESTED INTEREST OR NOT?

Vested Interest	Impartial
• Finance the fund-starved project	• Provide no funding and limited support
• Keep project alive	• Let project die
• Maximum protection from obstacles	• Limited protection from obstacles
• Fend off internal enemies	• Avoid politics and enemies
• Actively involved	• Go through motions
• Involved in personnel assignments	• Partial involvement in assignment

Active versus Passive Involvement

One of the questions facing senior management in the assigning of a project sponsor is whether or not the sponsor should have a vested interest in the project or be an impartial outsider. Table 10–3 shows the pros and cons of this. Sponsors who do not have a vested interest in the project seem to function more as exit champions than project sponsors.

Managing Scope Creep

***PMBOK® Guide*, 6th Edition**
5.6 Control Scope

Technically oriented team members are motivated not only by meeting specifications, but also by exceeding them. Unfortunately, exceeding specifications can be quite costly. Project managers must monitor scope creep and develop plans for controlling scope changes. But what if it is the project manager who initiates scope creep? The project sponsor must meet periodically with the project manager to review the scope baseline changes or unauthorized changes may occur and significant cost increases will result.

The Executive Champion

Executive champions are needed for those activities that require the implementation of change, such as a new corporate methodology for project management. Executive champions "drive" the implementation of project management down into the organization and accelerate its acceptance because their involvement implies executive-level support and interest.

The Failure of Project Governance

Simply because an executive is placed on a project governance committee is no guarantee that the correct decisions will be made. Executives are prone to making mistakes just like everyone else. Even though it is a thin line between governance failures and project failures, poor decisions by a governance committee can be more costly than poor decisions made by a project manager.

Some of the reasons why projects fail because of governance issues include:

- Lessons learned on projects or previous governance committee assignments are not captured or known by the governance committee
- Refusal to listen to the concerns of the project manager
- Making political rather than project decisions
- Making decisions for a given project without understanding the impact on other projects
- Changing stakeholders in midstream

- Refusing to pull the plug on a failing project
- Providing excessive governance that leads to micromanagement
- Not validating the business case

As the need for governance committees increases, we run the risk that the people that serve on these committees may not fully understand project management. The result may be decisions that appear in the best interest of the project but at the same time can have a detrimental effect on project management. An example is a significant change in scope that could mandate that many of the existing resources be replaced with people that have a different skill set.

10.2 HANDLING DISAGREEMENTS WITH THE SPONSOR

What if the project manager believes that the sponsor has made the wrong decision? Should the project manager have a recourse for action in such a situation?

There are several reasons why disagreements between the project manager and project sponsor occur. First, the project sponsors may not have sufficient technical knowledge or information to evaluate the risks of any potential decision. Second, sponsors may be heavily burdened with other activities and unable to devote sufficient time to sponsorship. Third, some companies prefer to assign sponsors who have no vested interest in the project in hopes of getting impartial decision making. Finally, sponsorship may be pushed down to a middle-management level where the assigned sponsor may not have all of the business knowledge necessary to make the best decisions.

Project managers are expected to challenge the project's assumptions continuously. This could lead to trade-offs. It could also lead to disagreements and conflicts between the project manager and the project sponsor. In such cases, the conflict will be brought to the executive steering committee for resolution. Sponsors must understand that their decisions as a sponsor can and should be challenged by the project manager.

Recognizing that these conflicts can exist, companies are instituting executive steering committees or executive policy board committees to quickly resolve these disputes. Few conflicts ever make it to the executive steering committee, but those that do are usually severe and may expose the company to unwanted risks.

A common conflict that may end up at the executive steering committee level happens when one party wants to cancel the project and the second party wants to continue.

10.3 THE COLLECTIVE BELIEF

Some projects, especially very long-term projects, often mandate that a collective belief exist. The collective belief is a fervent, and perhaps blind, desire to achieve that can permeate the entire team, the project sponsor, and even the most senior levels of management. The collective belief can make a rational organization act in an irrational manner. This is particularly true if the project sponsor spearheads the collective belief.

When a collective belief exists, people are selected based upon their support for the collective belief. Nonbelievers are pressured into supporting the collective belief and team members are not allowed to challenge the results. As the collective belief grows, both advocates and nonbelievers are trampled. The pressure of the collective belief can outweigh the reality of the results.

There are several characteristics of the collective belief, which is why some large, high-technology projects are often difficult to kill:

- Inability or refusal to recognize failure
- Refusing to see the warning signs
- Seeing only what you want to see
- Fearful of exposing mistakes
- Viewing bad news as a personal failure
- Viewing failure as a sign of weakness
- Viewing failure as damage to one's career
- Viewing failure as damage to one's reputation

10.4 THE EXIT CHAMPION

Project sponsors and project champions do everything possible to make their project successful. But what if the project champions, as well as the project team, have blind faith in the success of the project? What happens if the strongly held convictions and the collective belief disregard the early warning signs of imminent danger? What happens if the collective belief drowns out dissent?

In such cases, an exit champion must be assigned. The exit champion sometimes needs to have some direct involvement in the project in order to have credibility, but direct involvement is not always a necessity. Exit champions must be willing to put their reputation on the line and possibly face the likelihood of being cast out from the project team.

The larger the project and the greater the financial risk to the firm, the higher up the exit champion should reside. If the project champion just happens to be the CEO, then someone on the board of directors or even the entire board of directors should assume the role of the exit champion. Unfortunately, there are situations where the collective belief permeates the entire board of directors. In this case, the collective belief can force the board of directors to shirk their responsibility for oversight.

Large projects incur large cost overruns and schedule slippages. Making the decision to cancel such a project, once it has started, is very difficult.

The longer the project, the greater the necessity for the exit champions and project sponsors to make sure that the business plan has "exit ramps" such that the project can be terminated before massive resources are committed and consumed. Unfortunately, when a collective belief exists, exit ramps are purposefully omitted from the project and business plans. Another reason for having exit champions is so that the project closure process can occur as quickly as possible. As projects approach their completion, team members often have apprehension about their next assignment and try to stretch out the existing project until they are ready to leave. In this case, the role of the exit champion is to accelerate the closure process without impacting the integrity of the project.

Some organizations use members of a portfolio review board to function as exit champions. Portfolio review boards have the final say in project selection. They also have the final say as to whether or not a project should be terminated. Usually, one member of the board functions as the exit champion and makes the final presentation to the remainder of the board.

10.5 THE IN-HOUSE REPRESENTATIVES

On high-risk, high-priority projects or during periods of mistrust, customers may wish to place in-house representatives in the contractor's plant. These representatives, if treated properly, are like additional project office personnel who are not supported by your budget. They are invaluable resources for reading rough drafts of reports and making recommendations as to how their company may wish to see the report organized.

In-house representatives are normally not situated in or near the contractor's project office because of the project manager's need for some degree of privacy. The exception would be in the design phase of a construction project, where it is imperative to design what the customer wants and to obtain quick decisions and approvals.

Most in-house representatives know where their authority begins and ends. Some companies demand that in-house representatives have a project office escort when touring the plant, talking to functional employees, or simply observing the testing and manufacturing of components.

It is possible to have a disruptive in-house representative removed from the company. This usually requires strong support from the project sponsor in the contractor's shop. The important point here is that executives and project sponsors must maintain proper contact with and control over the in-house representatives, perhaps more so than the project manager.

10.6 STAKEHOLDER RELATIONS MANAGEMENT

PMBOK® Guide, 6th Edition
Chapter 13 Project Stakeholder Management

As projects become larger and more complex, the role of the sponsor is undertaken by a governance committee where the project sponsor is just one member of the committee. The relationship between the sponsor and the other stakeholders becomes critical.

Stakeholders are, in one way or another, individuals, companies, or organizations that may be affected by the outcome of the project or the way in which the project is managed. Stakeholders may be either directly or indirectly involved throughout the project or may function simply as observers. Stakeholders can shift from a passive role to becoming an

---

Section 10.6 adapted from H. Kerzner and C. Belack, *Managing Complex Projects* (Hoboken, NJ: John Wiley & Sons and IIL, 2010), Chapter 10.

active member of the team and participate in critical decisions. Active stakeholders basically function as sponsors.

On small or traditional projects, project managers generally interface with just the project sponsor as the primary stakeholder, and the sponsor usually is assigned from the organization that funds the project. This is true for both internal and external projects. But the larger the project, the greater the number of stakeholders with which you must interface. The situation becomes even more potentially problematic if you have a large number of stakeholders, geographically dispersed, all at different levels of management in their respective hierarchy, each with a different level of authority, and language and cultural differences. Trying to interface with all of these people on a regular basis and make decisions, especially on a large, complex project is very time-consuming.

Commitments

One of the complexities of stakeholder relations management is figuring out how to do all of this without sacrificing your company's long-term mission or vision. Also, your company may have long-term objectives in mind for this project, and those objectives may not necessarily be aligned to the project's objectives or each stakeholder's objectives. Lining up all the stakeholders in a row and getting them uniformly to agree to all decisions is more wishful thinking than reality. You may discover that it is impossible to get all the stakeholders to agree, and you must simply hope to placate as many as possible at a given point in time.

Stakeholder relations management cannot work effectively without commitments from all of the stakeholders. Obtaining these commitments can be difficult if the stakeholders cannot see what's in it for them at the completion of the project, namely the value that they expect or other personal interests. The problem is that what one stakeholder perceives as value may be perceived to have a different value by another stakeholder. For example, one stakeholder could view the project as a symbol of prestige. Another stakeholder could perceive the value as simply keeping their people employed. A third stakeholder could see value in the final deliverables of the project and the inherent quality in it. And a fourth stakeholder could see the project as an opportunity for future work with particular clients.

Another form of agreement involves developing a consensus on how stakeholders will interact with each other. It may be necessary for certain stakeholders to interact with one another and support one another with regard to sharing resources, providing financial support in a timely manner, and the sharing of intellectual property. While all stakeholders recognize the necessity for these agreements, they can be impacted by politics, economic conditions, and other enterprise environmental factors that may be beyond the control of the project manager. Certain countries may not be willing to work with other countries because of culture, religion, views on human rights, and other such factors.

For the project manager, obtaining these agreements right at the beginning of the project is essential. Some project managers are fortunate in being able to do this while others are not.

It is important to realize that not all of the stakeholders may want the project to be successful. This will happen if stakeholders believe that, at the completion of the project, they may lose power, authority, hierarchical positions in their company, or in a worse case even lose their job. Sometimes these stakeholders will either remain silent or even be supporters of the project until the end date approaches. If the project is regarded as unsuccessful, these stakeholders may respond by saying, "I told you so." If it appears that the project may be a success, these stakeholders may suddenly transform from supporters or the silent majority to adversaries.

It is very difficult to identify stakeholders with hidden agendas. These people can hide their true feelings and be reluctant to share information. There are often no telltale or early warning signs that indicate their true belief in the project. However, if the stakeholders are reluctant to approve scope changes, provide additional investment, or assign highly qualified resources, this could be an indication that they may have lost confidence in the project.

Expectations for the Stakeholder

Not all stakeholders understand project management. Not all stakeholders understand the role of a project sponsor. And not all stakeholders understand how to interface with a project or the project manager even though they may readily accept and support the project and its mission. Simply stated, the majority of the stakeholders are never trained in how to function properly as a stakeholder. Unfortunately, this cannot be detected early on but may become apparent as the project progresses.

Some stakeholders may be under the impression that they are merely observers and need not participate in decision making or authorization of scope changes. For some stakeholders who desire to be just observers, this could be a rude awakening. Some will accept the new role while others will not. Those that do not accept the new role usually are fearful that participating in a decision that turns out to be wrong can be the end of their political career.

Some stakeholders view their role as that of micromanagers, often usurping the authority of the project manager by making decisions that they may not necessarily be authorized to make, at least not alone. Stakeholders that attempt to micromanage can do significantly more harm to the project than stakeholders that remain as observers.

It may be a good idea for the project manager to prepare a list of expectations that he or she has of the stakeholders. This is essential even though stakeholders support the existence of the project. Role clarification for stakeholders should be accomplished early on the same way that the project manager provides role clarification for the team members at the initial kickoff meeting for the project.

Solution Providers

The present view of stakeholder relations management has changed as a result of the implementation of "engagement project management" practices. In the past, whenever a sale was made to the client, the salesperson would then move on to find a new client. Salespeople viewed themselves as providers of products and/or services. Today, salespeople view themselves as the provider of business solutions. In other words, salespeople now tell the client that we can provide you with a solution to all of your business needs and what we want in exchange is to be treated as a strategic business partner. This benefits both the buyer and seller because:

- Not all companies (buyers) have the ability to manage complexity.
- Solution providers must learn while managing the project.
- Solution providers can bring years of history to the table.
- Solution providers have a greater understanding of cultural change, the ability to work within almost any culture, and an understanding of virtual teams.

Therefore, as a solution provider, the project manager focuses heavily on the future and a long-term partnership agreement with the client and the stakeholders. This focus is heavily oriented toward value rather than near-term profitability.

Stakeholder Identification

Stakeholder relations management begins with stakeholder identification. This is easier said than done, especially if the project is multinational. Stakeholders can exist at any level of management. Corporate stakeholders are often easier to identify than political or government stakeholders. Each stakeholder is an essential piece of the project puzzle. Stakeholders must work together and usually interact with the project through the governance process. Therefore it is essential to know which stakeholders will participate in governance and which will not.

As part of stakeholder identification, the project manager must know whether he or she has the authority or perceived status to interface with the stakeholders. Some stakeholders perceive themselves as higher stature than the project manager and, in this case, the project sponsor may be the person to maintain interactions.

There are several ways in which stakeholders can be identified. More than one way can be used on projects:

- Groups: This could include financial institutions, creditors, regulatory agencies, and the like.
- Individuals: This could be by name or title, such as the CIO, COO, CEO, or just the name of the contact person in the stakeholder's organization.
- Contribution: This could be according to financial contributor, resource contributor, or technology contributor.
- Other factors: This could be according to the authority to make decisions or other such factors.

It is important to understand that not all stakeholders have the same expectations of a project. Some stakeholders may want the project to succeed at any cost, whereas other stakeholders may prefer to see the project fail even though they openly seem to support it. Some stakeholders view success as the completion of the project regardless of the cost overruns, whereas others may define success in financial terms only. Some stakeholders are heavily oriented toward the value they expect to see in the project, and this is the only definition of success for them. The true value may not be seen until months after the project has been completed. Some stakeholders may view the project as their opportunity for public notice and increased stature and therefore want to be actively involved. Others may prefer a more passive involvement.

Stakeholder Analysis

On large, complex projects with a multitude of stakeholders, it may be impossible for the project manager to properly cater to all of the stakeholders. Therefore, the project manager must know who the most influential stakeholders are and who can provide the greatest support on the project. Typical questions to ask might include:

- Who are powerful and who are not?
- Who will have or require direct or indirect involvement?
- Who has the power to kill the project?

- What is the urgency of the deliverables?
- Who may require more or less information than others?

Not all stakeholders are equal in influence, power, or authority to make decisions in a timely manner. It is imperative for the project manager to know who sits on the top of the list as having these capabilities.

Finally, it is important to remember that stakeholders can change over the life of a project, especially if it is a long-term project. Also, the importance of certain stakeholders can change over the life of a project and in each life-cycle phase. The stakeholder list is therefore an organic document subject to change.

Stakeholder mapping is most frequently displayed on a grid comparing their power and their level of interest. This is shown in Figure 10–2. The four cells can be defined as:

- Manage closely: These are high-power, interested people who can make or break your project. You must put forth the greatest effort to satisfy them. Be aware that there are factors that can cause them to change quadrants rapidly.
- Keep satisfied: These are high-power, less interested people who can also make or break your project. You must put forth some effort to satisfy them but not with excessive detail that can lead to boredom and total disinterest. They may not get involved until the end of the project approaches.
- Keep informed: These are people with limited power but who are keenly interested in the project. They can function as an early-warning system of approaching problems and may be technically astute to assist with some technical issues. These are the stakeholders who often provide hidden opportunities.
- Monitor only: These are people with limited power who may not be interested in the project unless a disaster occurs. Provide them with some information but not with too much detail such that they will become disinterested or bored.

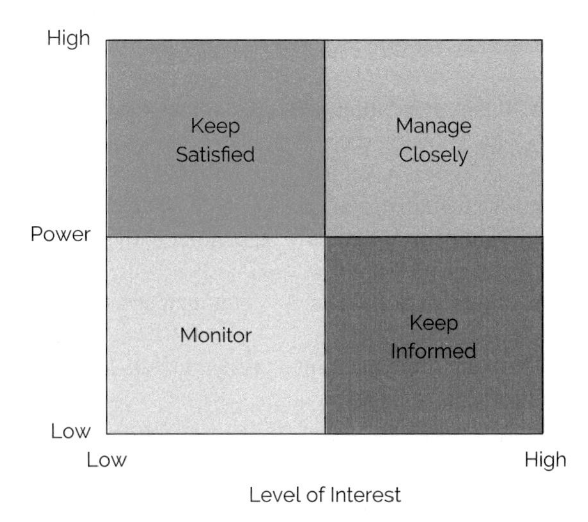

FIGURE 10–2. Stakeholder power-interest grid.

Stakeholder Engagement

Stakeholder engagement is when you physically meet with the stakeholders and determine their needs and expectations. As part of this, you must:

- Understand them and their expectations
- Understand their needs
- Value their opinions
- Find ways to win their support on a continuous basis
- Identify any stakeholder problems early on that can influence the project

Even though stakeholder engagement follows stakeholder identification, it is often through stakeholder engagement that we determine which stakeholders are supporters, advocates, neutral, or opponents. This may also be viewed as the first step in building a trusting relationship between the project manager and the stakeholders.

Information Flow

As part of stakeholder engagements, it is necessary for the project manager to understand each stakeholder's interests. One of the ways to accomplish this is to ask the stakeholders (usually the key stakeholders) what information they would like to see in performance reports. This information will help identify the key performance indicators (KPI) needed to service this stakeholder.

Each stakeholder may have a different set of KPI interests. It then becomes a costly endeavor for the project manager to maintain multiple KPI tracking and reporting flows, but it is a necessity for successful stakeholder relations management. Getting all of the stakeholders to agree on a uniform set of KPI reports and dashboards may be almost impossible.

There must be an agreement on what information is needed for each stakeholder, when the information is needed, and in what format the information will be presented. Some stakeholders may want a daily or weekly information flow whereas others may be happy with monthly data. For the most part, the information will be provided via the Internet.

The need for effective stakeholder communications is clear:

- Communicating with stakeholders on a regular basis is a necessity.
- By knowing the stakeholders, you may be able to anticipate their actions.
- Effective stakeholder communications builds trust.
- Virtual teams thrive on effective stakeholder communications.
- Although we classify stakeholders by groups or organizations, we still communicate with people.
- Ineffective stakeholder communications can cause a supporter to become a blocker.

There are three additional critical factors that must be considered for successful stake holder relations management:

- Effective stakeholder relations management takes time. It may be necessary to share this responsibility with the sponsor, executives, and members of the project team.

- Based upon the number of stakeholders, it may not be possible to address their concerns face-to-face. You must maximize your ability to communicate via the Internet. This is also important when managing virtual teams.
- Regardless of the number of stakeholders, documentation on the working relationships with the stakeholders must be archived. This is critical for success on future projects.

Effective stakeholder relations management can be the difference between an outstanding success and a terrible failure. Successful stakeholder relations management can result in binding agreements. The resulting benefits may be:

- Better decision making and in a more timely manner
- Better control of scope changes; prevention of unnecessary changes
- Follow-on work from stakeholders
- End-user satisfaction and loyalty
- Minimize the impact that politics can have on your project

Sometimes, regardless of how hard we try, we will fail at stakeholder relations management. Typical reasons include:

- Inviting stakeholders to participate too early, thus encouraging scope changes and costly delays
- Inviting stakeholders to participate too late such that their views cannot be considered without costly delays
- Inviting the wrong stakeholders to participate in critical decisions, thus leading to unnecessary changes and criticism by key stakeholders
- Key stakeholders becoming disinterested in the project
- Key stakeholders becoming impatient with the lack of progress
- Allowing the key stakeholders to believe that their contributions are meaningless
- Managing the project with an unethical leadership style or interfacing with the stakeholders in an unethical manner

10.7 PROJECT PORTFOLIO MANAGEMENT

***PMBOK® Guide*, 6th Edition**
1.2.3.3 Portfolio Management

One of the downside risks of even good project management practices is that project decision making is often suboptimized whereby decisions can be made for what is in the best interest of one project rather than what is in the best interest of the portfolio of projects. Some of the shortcomings that companies face include:

- Intense competition for critical resources
- Lack of a prioritization system
- Even if a prioritization system exists, having too many scope changes that result in significant project rework

- Inability to align projects to business strategies
- Unable to recognize the benefits and value of projects as part of the project approval process
- Working on too many non-value-added projects

To alleviate these problems, companies are now embarking upon project portfolio management practices. A portfolio is a grouping of projects and programs, usually those that are considered strategic in nature, and may be necessary for the survival of the firm. The portfolio can be limited to the ten or twelve most critical projects whereas there could be hundreds of other projects being worked on at the same time. There can be several different portfolios of projects in the same company.

Portfolio management is the centralization of the management of the processes, methods, and technologies used by the project managers to make sure that the value of the portfolio is maximized and that the projects are aligned to strategic business objectives. The portfolio governance committee is made up of sponsors and stakeholders whose function is to determine if a business opportunity exists, select the right projects to take advantage of this opportunity, prioritize the projects if necessary, and determine the optimal mix of resources both human and nonhuman based upon existing resource capacity, financial controls, scope change control, and portfolio risk management. Project managers who are managing portfolio projects usually report to the portfolio governance committee as well as any other sponsors who may be needed for this project.

On a macro level, portfolio governance committees are expected to make informed decisions in order to answer the following three critical questions related to the entire portfolio:

- Are we doing the right projects?
- Are we doing the right projects the right way?
- Are we doing enough of the right projects?

On a micro or individual project level, the portfolio governance committee's responsibilities are to:

- Verify that value is being created
- Know the risks and how risks are being mitigated
- Know when to intervene
- Know when to pull the plug on poorly performing projects
- Predict future corporate performance
- Confirm that the projects are still aligned to strategic objectives
- Perform resource reoptimization if necessary
- Know when to reprioritize projects if necessary

Based upon the information provided, the governance committee must then periodically assess:

- Do we have any weak investments that need to be canceled or replaced?
- Must any of the programs and/or projects be consolidated?

- Must any of the projects be accelerated or decelerated?
- How well are we aligned to strategic objectives?
- Does the portfolio have to be rebalanced?

For the governance committee to make informed decisions, they need evidence based upon metrics other than time, cost, and scope. Traditionally, these were the three primary reported metrics. But today, because of competing constraints, project managers may have to report ten to twelve metrics, and many of them are business-oriented metrics rather than traditional metrics. As stated in Chapter 1, today's project managers are becoming more business-oriented.

10.8 POLITICS

PMBOK® Guide, 6th Edition
3.4.4.3 Politics, Power, and Getting Things Done

Every young and/or inexperienced project manager embarks upon their first project with "stars in their eyes" thinking this will be a great achievement for them. Then, often without warning, the reality of project politics, external politics, or both sets in and the delusions of grandeur are replaced by "What have I gotten myself into?" It is natural for people to make decisions based upon what's in it for them. As an example, politics can occur when:

- Trying to define the project's requirements and disagreements exist among all of the stakeholders
- Creating the statement of work
- Negotiating for resources that must service multiple projects
- Having to prepare a schedule around the preferences of the team members
- Having conflicts over priorities
- Trying to explain the reasons for unfavorable variances

The list could go on and on. Politics will occur and project sponsors are limited in how far they can go to insulate the team from politics. Sponsors are often the first line of defense to protect the project from external influences. Politics can exist in any life-cycle phase of a project. For politics to occur, all you need are two or more people or a group of people.

Most people believe that politics on projects are bad. That's not necessarily true. Politics can be good or bad or at least lead to positive or negative project outcomes. Asking your sponsor to exert his or her political influence in helping you get the qualified resources you need or a high priority for your project could certainly lead to a positive outcome.

Unfavorable outcomes occur when people play politics for more power, authority, control, or advancement opportunities. These types of political games can make a project disruptive or dysfunctional. These types of bad politics occur more frequently than good politics. If these bad political games are not controlled, the result can be win–lose positions during conflicts and decision making. The overall result can be low team morale.

Related Case Studies (from Kerzner/*Project Management Case Studies*, 5th ed.)	**Related Workbook Exercises (from Kerzner/*Project Management Workbook and PMP®/CAPM® Exam Study Guide*, 12th ed.)**	***PMBOK® Guide*, 6th ed., Reference Section for the PMP® Certification Exam**
• Greyson Corporation • The Blue Spider Project • Corwin Corporation • The Prioritization of Projects* • The Irresponsible Sponsors* • Selling Executives on Project Management*	• Multiple Choice Exam	• Project Integration Management • Project Scope Management • Project Resource Management

*Case Study also appears at end of chapter.

10.9 STUDYING TIPS FOR THE PMI® PROJECT MANAGEMENT CERTIFICATION EXAM

This section is applicable as a review of the principles to support the knowledge areas and domain groups in the *PMBOK® Guide*. This chapter addresses:

- Integration management
- Scope management
- Human resources management
- Initiation
- Planning
- Execution
- Monitoring
- Closure

Understanding the following principles is beneficial if the reader is using this text to study for the PMP® Certification Exam:

- Role of the executive sponsor or project sponsor
- That the project sponsor need not be at the executive levels
- That some projects have committee sponsorship
- When to bring a problem to the sponsor and what information to bring with you

In Appendix C, the following Dorale Products mini case studies are applicable:

- Dorale Products (G) (Integration and Scope Management)

The following multiple-choice questions will be helpful in reviewing the principles of this chapter:

1. The role of the project sponsor during project initiation is to assist in:
   A. Defining the project's objectives in both business and technical terms
   B. Developing the project plan
   C. Performing the project feasibility study
   D. Performing the project cost-benefit analysis
2. The role of the project sponsor during project execution is to:
   A. Validate the project's objectives
   B. Validate the execution of the plan
   C. Make all project decisions
   D. Resolve problems/conflicts that cannot be resolved elsewhere in the organization
3. The role of the project sponsor during the closure of the project or a life-cycle phase of the project is to:
   A. Validate that the profit margins are correct
   B. Sign off on the acceptance of the deliverables
   C. Administer performance reviews of the project team members
   D. All of the above

ANSWERS

1. A 2. D 3. B

PROBLEMS

10–1 How should a project manager react when he or she finds inefficiency in the functional lines? Should executive management become involved?

10–2 Should project managers be permitted to establish prerequisites for top management regarding standard company procedures?

10–3 Does a project sponsor have the right to have an in-house representative removed from his or her company?

10–4 When does project management turn into overmanagement?

10–5 You are a line manager, and two project managers (each reporting to a divisional vice president) enter your office soliciting resources. Each project manager claims that his project is top priority as assigned by his own vice president. How should you, as the line manager, handle this situation? What are the recommended solutions to keep this situation from recurring repeatedly?

10–6 Should a client have the right to communicate directly to the project staff (i.e., project office) rather than directly to the project manager, or should this be at the discretion of the project manager?

10–7 Your company has assigned one of its vice presidents to function as your project sponsor. Unfortunately, your sponsor refuses to make any critical decisions, always "passing the buck" back to you. What should you do? What are your alternatives and the pros and cons of each? Why might an executive sponsor act in this manner?

CASE STUDIES

THE PRIORITIZATION OF PROJECTS

Background

The directorates of Engineering, Marketing, Manufacturing, and R&D all had projects that they were working on and each directorate established its own priorities for the projects. The problem was that the employees were working on multiple projects and had to deal with competing priorities.

Prioritization Issues

Lynx Manufacturing was a low-cost producer of cables and wires. The industry itself was considered as a low-technology industry and some of its products had been manufactured the same way for decades. There were some projects to improve the manufacturing processes, but they were few and far between.

Each of the four directorates, namely Engineering, Marketing, Manufacturing, and R&D, had projects, but the projects were generally quite small and used resources from only its own directorate.

By the turn of the century, manufacturing technologies began to grow and Lynx had to prepare for the technology revolution that was about to impact its business. Each directorate began preparing lists of projects that it would need to work on, and some of the lists contained as many as 200 projects. These projects were more complex than projects worked on previously and project team members from all directorates were assigned on either a full-time or part-time basis.

Each directorate's chief officer would establish the priorities for the projects originating in his or her directorate even though the projects required resources from other directorates. This created significant staffing issues and numerous conflicts:

- Each directorate would hoard its best project resources even though some projects outside of the directorate were deemed more important to the overall success of the company.
- Each directorate would put out fires by using people that were assigned to projects outside of its directorate rather than using people that were working on internal projects.
- Each directorate seemed to have little concern about any projects done in other directorates.
- Project priorities within each directorate could change on a daily basis because of the personal whims of the chief of that directorate.

The only costs and schedules that were important were those related to projects that originated within the directorate.

Senior management at the corporate level refused to get involved in the resolution of conflicts between directorates.

The working relationships between the directorates deteriorated to the point where senior management reluctantly agreed to step in. The total number of projects that the four directorates wanted to complete over the next few years exceeded 350, most of which required a team with members coming from more than one division.

QUESTIONS

1. Why is it necessary for senior management to step in rather than let the chiefs of the directorates handle the conflicts?
2. What should the senior management team do to resolve the problem?
3. Let's assume that the decision was to create a list that included all of the projects from the four directorates. How many of the projects on the list should have a priority number or priority code?
4. Can the directorate chiefs assign the priority or must it be done with the involvement of senior management?
5. How often should the list of prioritized projects be reviewed and who should be in attendance at the review meetings?
6. Suppose that some of the directorate chiefs refuse to assign resources according to the prioritized list and still remain focused on their own pet projects. How should this issue now be resolved?

THE IRRESPONSIBLE SPONSORS

Background

Two executives in this company each funded a "pet" project that had little chance of success. Despite repeated requests by the project managers to cancel the projects, the sponsors decided to throw away good money after bad money. The sponsors then had to find a way to prevent their embarrassment from such a blunder from becoming apparent to all.

Story Line

Two vice presidents came up with ideas for pet projects and funded the projects internally using money from their functional areas. Both projects had budgets close to $2 million and schedules of approximately one year. These were somewhat high-risk projects because they both required that a similar technical breakthrough be made. There was no guarantee that the technical breakthrough could be made at all. And even if the technical breakthrough could be made, both executives estimated that the shelf life of the products would be about one year before becoming obsolete but that they could easily recover their R&D costs.

These two projects were considered pet projects because they were established at the personal request of two senior managers and without any real business case. Had these two projects been required to go through the formal process of portfolio selection of projects, neither project would have been approved. The budgets for these projects were way out of line for the value that the company would receive, and the return on investment would be below minimum levels even if the technical breakthrough could be made. Personnel from the project management office (PMO), which are actively involved in the portfolio selection of projects, also stated that they would never recommend approval of a project where the end result would have a shelf life of one year or less. Simply stated, these projects existed for the self-satisfaction of the two executives and to get them prestige with their colleagues.

Nevertheless, both executives found money for their projects and were willing to let the projects go forward without the standard approval process. Both executives were able to get an experienced project manager from their group to manage their pet project.

Gate Review Meetings

At the first gate review meeting, both project managers stood up and recommended that their projects be canceled and the resources be assigned to other more promising projects. They both stated that the technical breakthrough needed could not be made in a timely manner. Under normal conditions, both of these project managers should have received medals for bravery in standing up and recommending that their projects be canceled. This certainly appeared as a recommendation in the best interest of the company.

But neither executive was willing to give up that easily. Canceling both projects would be humiliating for the two executives who were sponsoring these projects. Instead, both executives stated that the projects were to continue until the next gate review meeting, at which time a decision would be made for possible cancellation of both projects.

At the second gate review meeting, both project managers once again recommended that their projects be canceled. And, as before, both executives asserted that the projects should continue to the next gate review meeting before a decision would be made.

As luck would have it, the necessary technical breakthrough was finally made, but six months late. That meant that the window of opportunity to sell the products and recover the R&D costs would be six months rather than one year. Unfortunately, the thinking in the marketplace was that these products would be obsolete in six months and no sales occurred of either product.

Both executives had to find a way to save face and avoid the humiliation of having to admit that they squandered a few million dollars on two useless R&D projects.

This could very well impact their year-end bonuses.

QUESTIONS

1. Is it customary for companies to allow executives to have pet or secret projects that do not follow the normal project approval process?
2. Who got promoted and who got fired? In other words, how did the executives save face?

SELLING EXECUTIVES ON PROJECT MANAGEMENT

Background

The executives at Levon Corporation watched as their revenue stream diminished and refused to listen to their own employees who were arguing that project management implementation was necessary for growth. Finally, the executives agreed to listen to a presentation by a project management consultant.

Need for Project Management

Levon Corporation had been reasonably successful for almost twenty years as an electronics component manufacturer. The company was a hybrid between project-driven and non-project-driven businesses. A large portion of its business came from development of customized products for government agencies and private-sector companies around the world.

The customized or project-driven portion of the business was beginning to erode. Even though Levon's reputation was good, the majority of these contracts were awarded through competitive bidding. Every customer's request for proposal asked for a section on the contractor's project management capability. Levon had no real project management capability. Since most of the contracts were awarded on points rather than going to the lowest bidder, Levon was constantly downgraded in the evaluation of the proposals because of no project management capability.

The sales and marketing personnel continuously expressed their concerns to senior management, but the concerns fell upon deaf ears. Management was afraid that their support of project management could result in a shift in the balance of power in the company. Also, whatever executive ended up with control of the project management function could become more powerful than the other executives.

Gap Analysis

Reluctantly, the executives agreed to hire a project management consultant. The consultant was asked to identify the gaps between Levon and the rest of the industry and to show how project management could benefit the company. The consultant was also asked to identify the responsibilities of senior management once project management is implemented.

After a few weeks of research, the consultant was ready to make his presentation before the senior staff. The first slide that the consultant presented was Exhibit 10–1, which showed that Levon's revenue stream was not as good as they thought. Levon was certainly lagging the industry average and distance between Levon and the industry leader was getting larger.

Exhibit 10–1. Levon's gap analysis

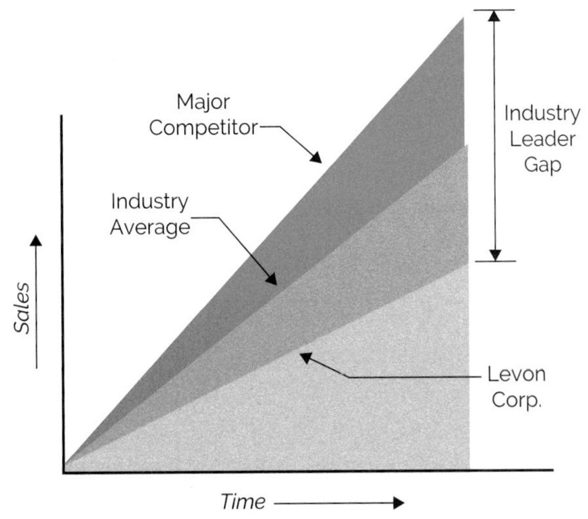

The consultant then showed Exhibit 10–2. The consultant had developed a project management maturity factor based upon such elements as time, cost, meeting scope, ability to handle risks, providing quality products, and customer interfacing and reporting. Using the project management maturity factor, the consultant showed that Levon's understanding and use of project management were lagging behind the industry trend.

Exhibit 10–2. Project management performance trend

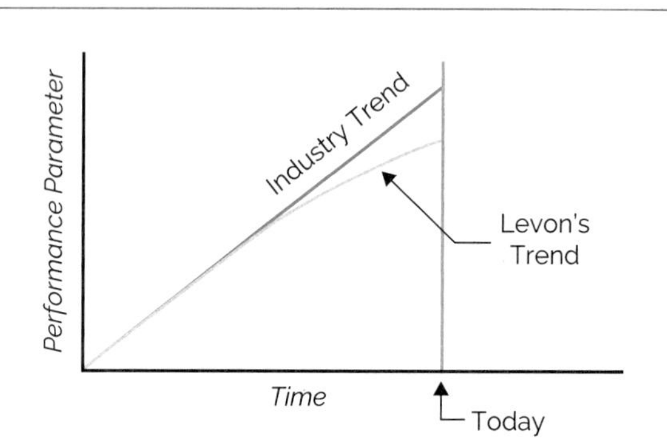

The consultant then showed Exhibit 10–3, which clearly illustrated that, unless Levon takes decisive action to improve its project management capability, the gap will certainly increase. The executives seemed to understand this but the consultant could still see their apprehension in supporting project management.

Exhibit 10–3. Increasing performance gap

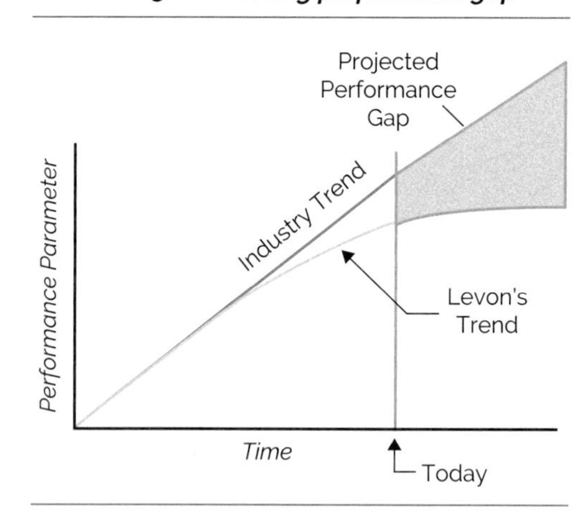

QUESTIONS

1. Why were the executives opposed to listening to their own employees but were willing to listen to a consultant?
2. Was the consultant correct in beginning the presentation by showing the gap between Levon and the rest of the industry?
3. Why did the executives still seem apprehensive even after the consultant's presentation?
4. What should the consultant say next to get the executives to understand and support project management?

11 Planning

11.0 INTRODUCTION

PMBOK® Guide, 6th Edition
Chapter 5 Scope Management
5.3 Define Scope

The most important responsibilities of a project manager are often recognized as planning, integrating, and executing plans. Almost all projects, because of their relatively short duration and often prioritized control of resources, require formal, detailed planning. The integration of the planning activities is necessary because each functional unit may develop its own planning documentation with little regard for other functional units.

Planning, in general, can best be described as the function of selecting the enterprise objectives and establishing the policies, procedures, and programs necessary for achieving them. Planning in a project environment may be described as establishing a predetermined course of action within a forecasted environment. The project's requirements set the major milestones. If line managers cannot commit because the milestones are perceived as unrealistic, the project manager may have to develop alternatives, one of which may be to move the milestones. Upper-level management must become involved in the selection of alternatives.

The project manager is the key to successful project planning. It is desirable that the project manager be involved from project conception through execution. Project planning must be *systematic, flexible* enough to handle unique activities, *disciplined* through reviews and controls, and capable of accepting *multifunctional* inputs. Successful project managers realize that project planning is an iterative process and must be performed throughout the life of the project.

PMBOK® Guide, 6th Edition
5.3 Define Scope

One of the objectives of project planning is to completely define all work required (possibly through the development of a documented project plan) so that it will be readily identifiable to each project participant. This is a necessity in a project environment because:

- If the task is well understood prior to being performed, much of the work can be preplanned.
- If the task is not understood, then during the actual task execution more knowledge is gained, which in turn leads to changes in resource allocations, schedules, and priorities.

PMBOK is a registered mark of the Project Management Institute, Inc.

- The more uncertain the task, the greater the amount of information that must be processed in order to ensure effective performance.

These considerations are important in a project environment because each project can be different from the others, requiring a variety of different resources, but having to be performed under time, cost, and performance constraints with little margin for error.

11.1 BUSINESS CASE

PMBOK® Guide, 6th Edition
1.2.6.1 Project Business Case
1.2.6.2 Project Benefits Management Plan
5.2 Collecting Requirements

Projects usually begin with the development of a business case for the project. This occurs most often well before the actual scope of the project is defined. A business case is a document that provides the reasoning why a project should be initiated. Historically, business cases were small documents or presentations and the decision to initiate the project was based upon the rank of the person making the request. Today, business cases are well-structured written documents that support a specific business need. Each business case should describe the boundaries to the project in sufficient detail such that the decision makers can determine that the expected business value and benefits exceed the cost of performing the project.

The business case must contain both quantifiable and unquantifiable information that justifies the investment in the project. Typical information that is part of a business case includes:

- The business need: This identifies the gap that currently exists and the need for the investment.
- The opportunity options: This identifies how the project is linked to strategic business objectives.
- The benefit realization plan: This identifies the value/benefits (rather than products or deliverables) that can be obtained whether they are cost savings, additional profits, or opportunities.
- Assumptions made: This identifies all of the assumptions that are made to justify the project.
- High-level objectives: This identifies the high-level or strategic objectives for the project.
- Recommendation for evaluation: This identifies what techniques should be used for evaluation such as a benefit-to-cost ratio, cash flow considerations, strategic options, opportunity costs, return on investment, net present value, and risks.
- Project metrics: This identifies the financial and nonfinancial metrics that will be used to track the performance of the project.
- Exit strategies: This identifies the cancellation criteria to be used to cancel the project if necessary.
- Project risks: This helps the decision makers evaluate the project by listing briefly the business, legal, technical, and other risks of the project.

- Project complexity: This identifies how complex the project might be, perhaps even from a risk perspective, if the organization can manage the complexity, and if it can be done with existing technology.
- Resources needed: This identifies the human and nonhuman resources needed.
- Timing: This identifies the major milestones for the project.
- Legal requirements: This identifies and legal requirements that must be followed.

The above information is used not only to approve the project, but also to be able to prioritize it with all of the other projects in the queue.

Templates can be established for most of the items in the business case. Sometimes, the benefits realization is a separate document rather than being included as part of the business case. A template for a benefits realization plan might include the following:

- A description of the benefits
- Identification of each benefit as tangible or intangible
- Identification of the recipient of each benefit
- How the benefits will be realized
- How the benefits will be measured
- The realization date for each benefit
- The handover activities to another group that may be responsible for converting the project's deliverables into benefits realization

Decision makers must understand that, over the life cycle of a project, circumstances can change requiring modification of the requirements, shifting of priorities, and redefinition of the desired outcomes. It is entirely possible that the business case and benefits realization plan can change to a point where the outcome of the project provides detrimental results and the project should be cancelled or backlogged for consideration at a later time. Some of the factors that can induce changes in the business case or benefits realization plan include:

- Changes in business owner or executive leadership: Over the life of a project, there can be a change in leadership. Executives that originally crafted the project may have passed it along to others who either have a tough time understanding the benefits, are unwilling to provide the same level of commitment, or see other projects as providing more important benefits.
- Changes in assumptions: Based upon the length of the project, the assumptions can and most likely will change, especially those related to enterprise environmental factors. Tracking metrics must be established to make sure that the original or changing assumptions are still aligned with the expected benefits.
- Changes in constraints: Changes in market conditions (i.e., markets served and consumer behavior) or risks can induce changes in the constraints. Companies may approve scope changes to take advantage of additional opportunities or reduce funding based upon cash flow restrictions. Metrics must also track for changes in the constraints.

- Changes in resource availability: The availability or loss of resources with the necessary critical skills is always an issue and can impact benefits if a breakthrough in technology is needed to achieve the benefits or to find a better technical approach with less risk.

One of the axioms of project management is that the earlier the project manager is assigned, the better the plan and the greater the commitment to the project. There is a trend today for bringing the project manager on board enough to participate in business case development.

There are valid arguments for assigning the project manager after the business case is developed:

- The project manager may have limited knowledge and not be able to contribute to the project at this time.
- The project might not be approved and/or funded, and it would be an added cost to have the project manager on board this early.
- The project might not be defined well enough at this early stage to determine the best person to be assigned as the project manager.

While these arguments seem to have merit, there is a more serious issue in that the project manager ultimately assigned may not fully understand the assumptions, constraints, and alternatives considered during the business case development. This could lead to a less than optimal development of a project plan. It is wishful thinking to believe that the business case development effort, which may have been prepared by someone completely separated from the execution of the project, contains all of the necessary assumptions, alternatives, and constraints.

Business case development often results in a highly optimistic approach with little regard for the schedule and/or the budget. Pressure is then placed upon the project manager to accept unrealistic arguments and assumptions made during business case development. If the project fails to meet business case expectations, then the blame may be placed upon the project manager.

11.2 VALIDATING THE ASSUMPTIONS

PMBOK® Guide, 6th Edition
2.2 Enterprise Environmental Factors
2.3 Organizational Process Assets
4.1.3.2 Assumption Log
5.2 Collecting Requirements

Planning begins with an understanding of the requirements, constraints, and assumptions. Project planning is based upon the expectation that future results can be extrapolated from past experiences. If experience is lacking or if extrapolation will generate misleading information, then assumptions must be made to predict future outcomes.

Quite often, the assumptions included in the project's business case are made by marketing and sales personnel and then approved by senior management as part of the project selection and approval process. The expectations for the final results are based upon the assumptions made.

Why is it that, more often than not, the final results of a project do not satisfy senior management's expectations? At the beginning of a project, it is impossible to ensure that the benefits expected by senior management will be realized at project completion. While project length is a critical factor, the real culprit is changing assumptions. Sometimes, we have assumption "blind spots" where we fail to recognize that some of our conventional or often used assumptions are no longer valid.

Assumptions can be made for items that are or are not under the direct control of the project team but can influence the outcome of the project. Assumptions made by the project manager are often surrounding the enterprise environment factors and the organizational process assets.

- ***Enterprise Environmental Factors:*** These are assumptions about external environmental conditions that can affect the success of the project, such as interest rates, market conditions, changing customer demands and requirements, customer involvement, changes in technology, political climate, and even government policies.
- ***Organizational Process Assets:*** These are assumptions about present and future company assets that can impact the success of the project, such as the capability of your project management methodology, the project management information system, forms, templates, guidelines, checklists, the ability to capture and use lessons learned data and best practices, resource availability, and skill level.

PMBOK® *Guide*, 6th Edition
2.3 Organizational Process Assets
2.2 Enterprise Environmental Factors

At the onset of the project, all assumptions must be challenged to verify their validity. As the project progresses, the assumptions must be tracked and validated. If the assumptions change or are no longer valid, then perhaps the project should be redirected or even canceled. Unfortunately, many project managers do not track the validity of the assumptions and end up completing a project within time and cost, but the project does not add business value to the company or the client.

It may be impossible to make assumptions with any degree of accuracy. However, this should not prevent risk management from taking place on the assumptions and contingency plans established if the assumptions are proven to be false.

Types of Assumptions

There are several types of assumptions. The two most common categories are explicit and implicit assumptions, and critical and noncritical assumptions. Critical and noncritical assumptions are also referred to as primary and secondary assumptions. These two categories of assumptions are not mutually exclusive.

Explicit assumptions may be quantified and are expressed without any ambiguity. Implicit assumptions may be hidden and may go undetected. Explicit instructions often contain hidden implicit assumptions. As an example, we could make an explicit assumption that five people will be needed full time to complete the project. Hidden is the implicit assumption that the people assigned will be available full time and possess the necessary skills. Serious consequences can occur if the implicit assumptions are proven to be false.

Critical assumptions are those assumptions that can cause significant damage to a project if even small changes take place. Critical assumptions must be tracked closely, whereas noncritical assumptions may not be tracked and may not require any action as long as they do not become critical assumptions. Project managers must develop a plan for how they will measure, track, and report the critical assumptions. Measurement implies that the assumptions can be quantified. Since assumptions predict future outcomes, testing and measurement may not be able to be made until well into the future or unless some risk triggers appear. Sensitivity analysis may be required to determine the risk triggers.

In Agile fixed-price contracts, the project manager and the customer work together to identify the assumptions. An agreement must be reached on the critical assumptions especially with regard to business value, risks and costs. An understanding must also be reached on what changes to the critical assumptions may trigger the need for scope changes. This requires that the project manager and the customer remain in close collaboration throughout the life of the project.

There are some assumptions that project managers may never see or even know about. These are referred to as strategic assumptions and are retained by decision makers when approving a project or selecting a portfolio of projects. These types of assumptions may contain company proprietary information that executives do not want the project team to know about.

Documenting Assumptions

Assumptions must be documented at project initiation using the project charter as a possible means. Throughout the project, the project manager must revalidate and challenge the assumptions. Changing assumptions may mandate that the project be terminated or redirected toward a different set of objectives.

A project management plan is based upon the assumptions described in the project charter. But there may be additional assumptions made by the team that are inputs to the project management plan. One of the primary reasons why companies use a project charter is that project managers were most often brought on board well after the project selection process and approval process were completed. As a result, project managers needed to know what assumptions were considered.

Enterprise Environmental Factors: These are assumptions about the external environmental conditions that can affect the success of the project, such as interest rates, market conditions, changing customer demands and requirements, changes in technology, and even government policies.

Organizational Process Assets: These are assumptions about present or future company assets that can impact the success of the project such as the capability of your enterprise project management methodology, the project management information system, forms, templates, guidelines, checklists, and ability to capture and use lessons learned data and best practices.

Documenting assumptions is necessary in order to track the changes. Examples of assumptions that are likely to change over the duration of a project, especially on a long-term project, might be that:

- The cost of borrowing money and financing the project will remain fixed.
- The procurement costs will not increase.

TABLE 11–1. ASSUMPTION VALIDATION CHECKLIST

Checklist for Validating Assumptions	YES	NO
Assumption is outside of the control of the project team.		
Assumption is outside of the control of the stakeholder(s).		
The assumption can be validated as correct.		
Changes in the assumption can be controlled.		
The assumed condition is not fatal.		
The probability of the assumption holding true is clear.		
The consequences of this assumption pose a serious risk to the project.		
Unfavorable changes in the assumption can be fatal to the project.		

- The breakthrough in technology will take place as scheduled.
- The resources with the necessary skills will be available when needed.
- The marketplace will readily accept the product.
- Our competitors will not catch up to us.
- The risks are low and can be easily mitigated.
- The political environment in the host country will not change.

The problem with having faulty assumptions is that they can lead to faulty conclusions, bad results, and unhappy customers. The best defense against poor assumptions is good preparation at project initiation, including the development of risk mitigation strategies. One possible way to do this is with a validation checklist, as shown in Table 11–1.

11.3 VALIDATING THE OBJECTIVES

***PMBOK® Guide*, 6th Edition**
5.5 Validate Scope

When project managers are assigned to a project and review the business case, they look first at the assumptions and objectives for the project. A project's objectives, which are usually high-level objectives, provide an aim or desired end of action. Project managers must then prepare the interim objectives to satisfy the high-level objectives.

Objectives are described in specific terms, are measureable, and are attainable and action-oriented, realistic, and bound by time. Clearly written and well-understood objectives are essential so that the project team will know when the project is over. Unfortunately the objectives are usually imposed upon the project manager, rather than having the project manager assigned early enough so as to participate in the establishment of the objectives.

Clearly written objectives follow the SMART rule: **S**pecific, **M**easurable, **A**ttainable, **R**ealistic or relevant, **T**angible or time bound.

Project managers may not be able to establish the objectives themselves without some assistance from perhaps the project sponsor. Most project managers may be able to establish the technical components of the objectives but must rely heavily upon the project sponsor for the business components.

If the project manager believes that the requirements are unrealistic, then he may consider scaling back the scope of the objectives. While other techniques are available for scaling back the objectives, scope reduction is often the first attempt. According to Eric Verzuh:

> If the goals of the project will take too long to accomplish or cost too much, the first step is to scale down the objectives—the product scope. The result of this alternative will be to reduce the functionality of the end product. Perhaps an aircraft will carry less weight, a software product will have fewer features, or a building will have fewer square feet or less expensive wood paneling. (Remember the difference between product scope and project scope; *product scope* describes the functionality and performance of the product; *project scope* is all of the work required to deliver the product scope.
>
> *Positive.* This will save the project while saving both time and money.
>
> *Negative.* When a product's functionality is reduced, its value is reduced. If the airplane won't carry as much weight, will the customers still want it? If a software product has fewer features, will it stand up to competition? A smaller office building with less expensive wood paneling may not attract high-enough rents to justify the project.
>
> *Best application.* The key to reducing a product's scope without reducing its value is to reevaluate the true requirements for the business case. Many a product has been over budget because it was overbuilt. Quality has best been defined as "conformance to requirements." Therefore, reducing product scope so that the requirements are met more accurately actually improves the value of the product, because it is produced more quickly and for a lower cost.[1]

11.4 GENERAL PLANNING

PMBOK® Guide, 6th Edition
Chapter 5 Scope Management
Chapter 3 Role of the Project Manager

Planning is determining what needs to be done, by whom, and by when, in order to fulfill one's assigned responsibility. There are nine major components of the planning phase:

1. *Objective:* a goal, target, or quota to be achieved by a certain time
2. *Program:* the strategy to be followed and major actions to be taken in order to achieve or exceed objectives
3. *Schedule:* a plan showing when individual or group activities or accomplishments will be started and/or completed
4. *Budget:* planned expenditures required to achieve or exceed objectives
5. *Forecast:* a projection of what will happen by a certain time
6. *Organization:* design of the number and kinds of positions, along with corresponding duties and responsibilities, required to achieve or exceed objectives
7. *Policy:* a general guide for decision making and individual actions
8. *Procedure:* a detailed method for carrying out a policy
9. *Standard:* a level of individual or group performance defined as adequate or acceptable

1. Eric Verzuh, *The Fast Forward MBA in Project Management* (Hoboken, NJ: John Wiley & Sons, 2016), p. 278.

An item that has become important in recent years is documenting assumptions that go into the objectives or the project/subsidiary plans. Assumptions are items that are believed to be true, real, or certain. They are not grounded in fact and carry some element of risk. As projects progress, even for short-term projects, assumptions can change because of the economy, technological advances, or market conditions. These changes can invalidate original assumptions or require that new assumptions be made. These changes could also mandate that projects be canceled. Companies are now validating assumptions during gate review meetings.

Several of these factors require additional comment. Forecasting what will happen may not be easy, especially if predictions of environmental reactions are required. Forecasting is customarily defined as either strategic, tactical, or operational. Strategic forecasting is generally for five years or more, tactical can be for one to five years, and operational is the here and now of six months to one year. Although most projects are operational, they can be considered as strategic, especially if spin-offs or follow-up work is promising. Forecasting also requires an understanding of strengths and weaknesses in:

- The competitive situation
- Marketing
- Research and development
- Production
- Financing
- Personnel
- The management structure

If project forecasting is strictly operational, then these factors may be clearly definable. However, if strategic or long-range forecasting is necessary, then the future economic outlook can vary, say, from year to year, and forecasting must be revisited at regular intervals because the goals and objectives can change.

The last three factors—policies, procedures, and standards—can vary from project to project because of their uniqueness. Each project manager can establish project policies, provided that they fall within the broad limits set forth by top management.

Project policies must often conform closely to company policies, and are usually similar in nature from project to project. Procedures, on the other hand, can be drastically different from project to project, even if the same activity is performed. For example, the signing off of manufacturing plans may require different signatures on two selected projects even though the same end-item is being produced.

Planning varies at each level of the organization. At the individual level, planning is required so that cognitive simulation can be established before irrevocable actions are taken. At the working group or functional level, planning must include:

- Agreement on purpose
- Assignment and acceptance of individual responsibilities
- Coordination of work activities
- Increased commitment to group goals
- Lateral communications

At the organizational or project level, standards must include:

- Recognition and resolution of group conflict on goals
- Assignment and acceptance of group responsibilities
- Increased motivation and commitment to organizational goals
- Vertical and lateral communications
- Coordination of activities between groups

The logic of planning requires answers to several questions at various stages in order for the alternatives and constraints to be fully understood. A list of questions would include:

- Prepare environmental analysis
  - Where are we? How and why did we get here?
- Set objectives
  - Is this where we want to be? Where would we like to be? In a year? In five years?
- List alternative strategies
  - Where will we go if we continue as before? Is that where we want to go? How could we get to where we want to go?
- List threats and opportunities
  - What might prevent us from getting there? What might help us to get there?
- Prepare forecasts
  - Where are we capable of going? What do we need to take us where we want to go?
- Select strategy portfolio
  - What is the best course for us to take? What are the potential benefits? What are the risks?
- Prepare action programs
  - What do we need to do? When do we need to do it? How will we do it? Who will do it?
- Monitor and control
  - Are we on course? If not, why? What do we need to do to be on course? Can we do it?

One of the most difficult activities in the project environment is to keep the planning on target. The following procedures can assist project managers during planning activities:

- Let functional managers do their own planning. Too often operators are operators, planners are planners, and never the twain shall meet.
- Establish goals before you plan. Otherwise short-term thinking takes over.
- Set goals for the planners. This will guard against the nonessentials and places your effort where there is payoff.
- Stay flexible. Use people-to-people contact, and stress fast response.
- Keep a balanced outlook. Don't overreact; position yourself for an upturn.
- Welcome top-management participation. Top management has the capability to make or break a plan, and may well be the single most important variable.

- Beware of future spending plans. This may eliminate the tendency to underestimate.
- Test the assumptions behind the forecasts. This is necessary because professionals are generally too optimistic. Do not depend solely on one set of data.
- Don't focus on today's problems. Try to get away from crisis management and firefighting.
- Reward those who dispel illusions. Do not "kill the messenger"; in other words, project participants should not be wary of relaying bad news. Reward the first to come forth with bad news.

Finally, effective total program planning cannot be accomplished unless all of the necessary information becomes available at project initiation. These information requirements are:

- The statement of work (SOW) (Section 11.10)
- The project specifications (Section 11.11)
- The milestone schedule (Section 11.6 and 11.12)
- The work breakdown structure (WBS) (Section 11.13)

11.5 LIFE-CYCLE PHASES

***PMBOK® Guide*, 6th Edition**
1.2.4.1 Project and Development Life-Cycle Phases
1.2.4.2 Project Phase

Project planning takes place at two levels. The first level is the corporate cultural approach; the second is the individual's approach. The corporate cultural approach breaks the project down into life-cycle phases. The life-cycle phase approach is *not* an attempt to put handcuffs on the project manager but to provide a methodology for uniformity in project planning. Many companies, including government agencies, prepare checklists of activities that should be considered in each phase. These checklists are for consistency in planning. The project manager can still exercise his own planning initiatives within each phase.

Another benefit of life-cycle phases is control. At the end of each phase there is a meeting of the project manager, sponsor, senior management, and even the customer, to assess the accomplishments of this life-cycle phase and to get approval for the next phase. These meetings are often called critical design reviews, "on-off ramps," and "gates." In some companies, these meetings are used to firm up budgets and schedules for the follow-on phases. In addition to monetary and schedule considerations, life-cycle phases can be used for staffing deployment and equipment/facility utilization. Some companies go so far as to prepare project management policy and procedure manuals where all information is subdivided according to life-cycle phasing. They include:

- Proceed with the next phase based on an approved funding level
- Proceed to the next phase but with a new or modified set of objectives
- Postpone approval to proceed based on a need for additional information
- Terminate project

11.6 LIFE-CYCLE MILESTONES

When we discuss life-cycle phases, it is generally understood that there are "go" or "no go" milestones at the end of each life-cycle phase. These milestones are used to determine if the project should be continued, and if so, should there be any changes to the funding or requirements? But there are also other interim milestones that support the end-of-phase milestones or appear within certain life-cycle phases.

Scope Freeze Milestones

Scope freeze milestones are locations in the project's timeline where the scope is frozen and further scope changes will not be allowed. In traditional project management, we often do not use scope freeze milestones because it is assumed that the scope is well-defined at project initiation. But on other projects, especially in IT, the project may begin based upon just an idea and the project's scope must evolve as the project is being implemented. This is quite common with techniques such as Agile and Scrum.

According to Melik:

> There will be resistance to scope freeze; if it is too early, you may have an unhappy customer, too late and the project will be over budget or late. Striking the right balance between the customer's needs and your ability to deliver on time and on budget is a judgment call; timing varies from project to project, but the scope should always be frozen before a [significant] financial commitment is made.
>
> What if the scope baseline cannot be agreed to? One of the following approaches may be helpful:
>
> - Work more with the customer until scope freeze is achieved.
> - Agree to work in phases; proceed with partial project execution, clarify and agree to the rest of the scope during the course of the project. Pilot, prototype, or fit-to-business-analysis mini-projects are helpful.
> - Structure the project with many predetermined milestones and phases of short duration; such a project is far more likely to remain aligned with its ultimate objectives.[2]

Design Freeze Milestones

In addition to scope freeze milestones, there can also be design freeze milestones. Even if the scope is well defined and agreed to, there may be several different designs to meet the scope. Sometimes, the best design may require that the scope be changed and the scope freeze milestone must be moved out.

A design freeze milestone is a point in a project where no further changes to the design of the product came be made without incurring a financial risk especially if the design must then go to manufacturing. The decision point for the freeze usually occurs at the end of a specific life-cycle phase. There are several types of freezes, and they can occur in just about any type of project. However, they are most common in new product development (NPD) projects.

---

2. Rudolf Melik, *The Rise of the Project Workforce* (Hoboken, NJ: John Wiley & Sons, 2007), pp. 205–206.

In NPD projects, we normally have a both a specification freeze and a design freeze. The specification dictates the set of requirements upon which the final design must be made. Following a specification freeze, we have a design freeze whereby the final design is handed over to manufacturing. The design freeze may be necessary for timely procurement of long lead items such as parts and tooling that are necessary for the final product to be manufactured. The timing of the design freeze is often dictated by the lead times and may be beyond the control of the company. Failing to meet design freeze points has a significantly greater impact on manufacturing than engineering design.

Design freezes have the additional benefit of controlling downstream scope changes. However, even though design changes can be costly, they are often necessary for safety reasons, to protect the firm against possible product liability lawsuits, and to satisfy a customer's specific needs.

Changes to a product after the product's design is handed over to manufacturing can be costly. As a rule, we generally state that the cost of a change in any life-cycle phase after the design freeze is ten times the cost of performing the change in the previous life-cycle phase. If a mistake is made prior to the design freeze point, the correction could have been made for $100. But if the mistake is not detected until the manufacturing stage, then the correction cost could be $1000. The same mistake, if detected after the customer receives the product, could cost $10,000 to correct a $100 planning mistake or a $1000 manufacturing mistake. While the rule of ten may seem a little exaggerated, it does show the trend in correcting costs downstream.

Customer Approval Milestones Project managers often neglect to include in the life-cycle phases timeline milestones and the accompanying time durations for customer approvals of the project's schedule baseline with the mistaken belief that the approval process will happen quickly. The approval process in the project manager's parent company may be known with some reasonable degree of certainly, but the same cannot be said for the client's approval process. Factors that can impact the speed by which the approvals take place can include:

- How many people are involved in the approval process
- Whether any of the people are new to the project
- When all of the necessary participants can find a mutually agreed upon time to meet
- The amount of time the people need to review the data, understand the data, and determine the impact of their decision
- Their knowledge of project management
- A review of the previous project decisions that were made by them or others
- How well they understand the impact of a delayed decision
- Whether they need additional information to make a decision
- Whether the decision can be made verbally or if it must be written in report format

Simply adding a milestone to a schedule that says "customer approval" does not solve the problem. Project managers must find out how long their customers need to make a decision, and it may be better to indicate customer approval as an activity rather than as a milestone.

11.7 KICKOFF MEETINGS

PMBOK® *Guide*, 6th Edition
5.1 Plan Scope Management

The typical launch of a project begins with a kickoff meeting involving the major players responsible for planning, including the project manager, assistant project managers for certain areas of knowledge, subject matter experts (SME), and functional leads. A typical sequence is shown in Figure 11–1.

There can be multiple kickoff meetings based upon the size, complexity, and time requirements for the project. The major players are usually authorized by their functional areas to make decisions concerning timing, costs, and resource requirements.

Some of the items discussed in the initial kickoff meeting include:

- Wage and salary administration, if applicable
- Letting the employees know that their boss will be informed as to how well or how poorly they perform
- Initial discussion of the scope of the project including both the technical objective and the business objective
- The definition of success on this project

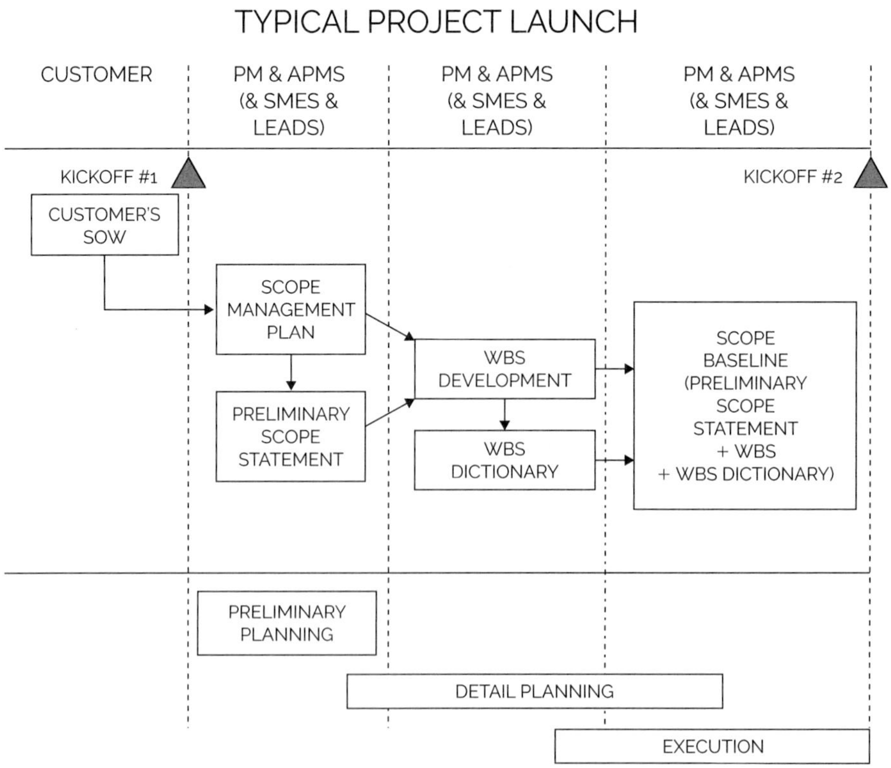

TYPICAL PROJECT LAUNCH

CUSTOMER
PM & APMS (& SMES & LEADS)
PM & APMS (& SMES & LEADS)
PM & APMS (& SMES & LEADS)
KICKOFF #1
KICKOFF #2
CUSTOMER'S SOW
SCOPE MANAGEMENT PLAN
PRELIMINARY SCOPE STATEMENT
WBS DEVELOPMENT
WBS DICTIONARY
SCOPE BASELINE (PRELIMINARY SCOPE STATEMENT + WBS + WBS DICTIONARY)
PRELIMINARY PLANNING
DETAIL PLANNING
EXECUTION

FIGURE 11–1. Typical project launch.

- The assumptions and constraints as identified in the project charter
- The project's organizational chart (if known at that time)
- The participants' roles and responsibilities

For a small or short-term project, estimates on cost and duration may be established in the kickoff meeting. In this case, there may be little need to establish a cost estimating schedule. But where the estimating cycle is expected to take several weeks, and where inputs will be required from various organizations and/or disciplines, an essential tool is an estimating schedule. In this case, there may be a need for a prekickoff meeting simply to determine the estimates. The minimum key milestones in a cost estimating schedule are (1) a "kickoff" meeting, (2) a "review of ground rules" meeting, (3) "resources input and review" meeting, and (4) summary meetings and presentations. Descriptions of these meetings and their approximate places in the estimating cycle follow.[3]

- **The Prekickoff Meeting**: The very first formal milestone in an estimate schedule is the estimate kickoff meeting. This is a meeting of all the individuals who are expected to have an input to the cost estimate. Sufficient time should be allowed in the kickoff meeting to describe all project ground rules, constraints, and assumptions; to hand out technical specifications, drawings, schedules, and work element descriptions and resource estimating forms; and to discuss these items and answer any questions that might arise. This kickoff meeting may be 6 weeks to 3 months prior to the estimate completion date to allow sufficient time for the overall estimating process.
- **The Review of Ground Rules Meeting**: Several days after the estimate kickoff meeting, when the participants have had the opportunity to study the material, a review of ground rules meeting should be conducted. In this meeting the estimate manager answers questions regarding the conduct of the cost estimate, assumptions, ground rules, and estimating assignments.
- **The Resources Input and Review Meeting**: Several weeks after the kickoff and review of ground rules meetings, each team member that has a resources (man-hour and/or materials) input is asked to present his or her input before the entire estimating team. Thus starts one of the most valuable parts of the estimating process: the interaction of team members to reduce duplications, overlaps, and omissions in resource data. The estimator of each discipline area has the opportunity to justify and explain the rationale for his estimates in view of his peers, an activity that tends to iron out inconsistencies, overstatements, and incompatibilities in resources estimates.
- **Summary Meetings and Presentations**: Once the resources inputs have been collected, adjusted, and "priced," the cost estimate is presented to the estimating team as a "dry run" for the final presentation to the company's management or to the requesting organization. This dry run can produce visibility into further inconsistencies or errors that have crept into the estimate during the process of consolidation and reconciliation.

---

3. R. D. Stewart, *Cost Estimating* (New York: John Wiley & Sons, 1982), pp. 56–57.

11.8 UNDERSTANDING PARTICIPANTS' ROLES

Companies that have histories of successful plans also have employees who fully understand their roles in the planning process. Good up-front planning may not eliminate the need for changes, but may reduce the number of changes required. The responsibilities of the major players are as follows:

- Project manager will define:
  - Goals and objectives
  - Major milestones
  - Requirements
  - Ground rules and assumptions
  - Time, cost, and performance constraints
  - Operating procedures
  - Administrative policy
  - Reporting requirements
- Line manager will define:
  - Detailed task descriptions to implement objectives, requirements, and milestones
  - Detailed schedules and manpower allocations to support budget and schedule
  - Identification of areas of risk, uncertainty, and conflict
- Senior management (project sponsor) will:
  - Act as the negotiator for disagreements between project and line management
  - Provide clarification of critical issues
  - Provide communication link with customer's senior management
  - Successful planning requires that project, line, and senior management are in agreement with the plan.

11.9 ESTABLISHING PROJECT OBJECTIVES

PMBOK® Guide, 6th Edition
Chapter 5 Project Scope Management
5.3 Define Scope

Successful project management, whether in response to an in-house project or a customer request, must utilize effective planning techniques. The first step, as stated previously, is understanding the project objectives. These goals may be to develop expertise in a given area, to become competitive, to modify an existing facility for later use, or simply to keep key personnel employed.

The objectives are generally not independent; they are all interrelated, both implicitly and explicitly. Many times it is not possible to satisfy all objectives. At this point, management must prioritize the objectives as to which are strategic and which are not. Typical problems with developing both high level and interim objectives include:

- Project objectives/goals are not agreeable to all parties.
- Project objectives are too rigid to accommodate changing priorities.

- Insufficient time exists to define objectives well.
- Objectives are not adequately quantified.
- Objectives are not documented well enough.
- Efforts of client and project personnel are not coordinated.
- Personnel turnover is high.

Once the objectives are clearly defined, four questions must be considered:

- What are the major elements of the work required to satisfy the objectives, and how are these elements interrelated?
- Which functional divisions will assume responsibility for accomplishment of these objectives and the major-element work requirements?
- Are the required corporate and organizational resources available?
- What are the information flow requirements for the project?

11.10 THE STATEMENT OF WORK

***PMBOK® Guide*, 6th Edition**
5.1 Plan Scope Management
5.1.3.1 Scope Management Plan
5.3.3.1 Project Scope Statement

Planning generally addresses four elements related to scope:

- Scope: Scope is the summation of all deliverables required as part of the project. This includes all products, services, and results.
- Project Scope: This is the work that must be completed to achieve the final scope of the project, namely the products, services, and end results.
- Scope statement: This is a document that provides the basis for making future decisions such as scope changes. The intended use of the document is to make sure that all stakeholders have a common knowledge of the project scope. Included in this document are the objectives, description of the deliverables, end result or product, and justification for the project. The scope statement addresses seven questions: who, what, when, why, where, how, and how many. This document validates the project scope against the statement of work provided by the customer.
- Statement of work: This is a narrative description of the end results to be provided under the contract. For the remainder of this section, we will focus our attention on the statement of work.

The statement of work (SOW) is a narrative description of the work required for the project and the end results to be provided under the contract. The complexity of the SOW is determined by the desires of top management, the customer, and/or the user groups. For projects internal to the company, the SOW is prepared by the project office with input from the user groups because the project office is usually composed of personnel with writing skills.

For projects external to the organization, as in competitive bidding, the contractor may have to prepare the SOW for the customer because the customer may not have people trained in SOW preparation. In this case, as before, the contractor would submit the SOW to the customer for approval. It is also quite common for the project manager to rewrite a customer's SOW so that the contractor's line managers can price out the effort.

In a competitive bidding environment, there are two SOWs—the SOW used in the proposal and a contract statement of work (CSOW). There might also be a proposal WBS and a contract work breakdown structure (CWBS). Special care must be taken by contract and negotiation teams to discover all discrepancies between the SOW/WBS and CSOW/CWBS, or additional costs may be incurred. A good (or winning) proposal is *no guarantee* that the customer or contractor understands the SOW. For large projects, fact-finding is usually required before final negotiations because it is *essential* that both the customer and the contractor understand and agree on the SOW, what work is required, what work is proposed, the factual basis for the costs, and other related elements. In addition, it is imperative that there be agreement between the final CSOW and CWBS.

SOW preparation is not as easy as it sounds. Misinterpretations of the SOW can result in losses of hundreds of millions of dollars. Common causes of misinterpretation are:

- Mixing tasks, specifications, approvals, and special instructions
- Using imprecise language ("nearly," "optimum," "approximately," etc.)
- No pattern, structure, or chronological order
- Wide variation in size of tasks
- Wide variation in how to describe details of the work
- Failing to get third-party review

Misinterpretations of the statement of work can and will occur no matter how careful everyone has been. The best way to control such misinterpretations is with a good definition of the requirements up front, if possible.

SOW preparation manuals typically contain guides for editors and writers:[4]

- Every SOW that exceeds two pages in length should have a table of contents conforming to the CWBS coding structure. There should rarely be items in the SOW that are not shown on the CWBS.
- Clear and precise task descriptions are essential as they will be read and interpreted by persons of varied background. A good SOW states precisely the product or service desired. The clarity of the SOW will affect administration of the contract.
- Every effort must be made to avoid ambiguity.
- Remember that any provision that takes control of the work away from the contractor, even temporarily, may result in relieving the contractor of responsibility.
- In specifying requirements, use active rather than passive terminology.
- Limit abbreviations to those in common usage. Provide a list of all pertinent abbreviations and acronyms at the beginning of the SOW.
- Any division of responsibilities between the contractor, other agencies, etc., should be included and delineated.

4. Adapted from *Statement of Work Handbook* NHB5600.2, National Aeronautics and Space Administration, February 1975.

- Include procedures. When immediate decisions cannot be made, it may be possible to include a procedure for making them.
- Do not overspecify. The ideal situation may be to specify results required or end-items to be delivered and let the contractor propose his best method.
- Describe requirements in sufficient detail to assure clarity, not only for legal reasons, but for practical application. It is easy to overlook many details. It is equally easy to be repetitious. Beware of doing either.
- Avoid incorporating extraneous material and requirements. They may add unnecessary cost.
- Screen out unnecessary data requirements, and specify only what is essential and when.
- Do not repeat detailed requirements or specifications that are already spelled out in applicable documents. Instead, incorporate them by reference.

11.11 PROJECT SPECIFICATIONS

PMBOK® Guide, 6th Edition
5.2 Collect Requirements
5.3 Define Scope

An abbreviated specification list as shown in Table 11–2 is separately identified or called out as part of the statement of work. Specifications are used for man-hour, equipment, and material estimates. Small changes in a specification can cause large cost overruns.

TABLE 11–2. SPECIFICATION FOR STATEMENT OF WORK

Description	Specification No.
Civil	100 (Index)
• Concrete	101
• Field equipment	102
• Piling	121
• Roofing and siding	122
• Soil testing	123
• Structural design	124
Electrical	200 (Index)
• Electrical testing	201
• Heat tracing	201
• Motors	209
• Power systems	225
• Switchgear	226
• Synchronous generators	227
HVAC	300 (Index)
• Hazardous environment	301
• Insulation	302
• Refrigeration piping	318
• Sheet metal ductwork	319
Installation	400 (Index)
• Conveyors and chutes	401
• Fired heaters and boilers	402
• Heat exchangers	403
• Reactors	414
• Towers	415
• Vessels	416

Another reason for identifying the specifications is to make sure that there are no surprises for the customer downstream. The specifications should be the most current revision. It is not uncommon for a customer to hire outside agencies to evaluate the technical proposal and to make sure that the proper specifications are being used.

Specifications are, in fact, standards for pricing out a proposal. If specifications do not exist or are not necessary, then work standards should be included in the proposal. The work standards can also appear in the cost volume of the proposal. Labor justification backup sheets may or may not be included in the proposal, depending on RFP/RFQ (request for quotation) requirements.

11.12 DATA ITEM MILESTONE SCHEDULES

PMBOK® Guide, 6th Edition
Chapter 6 Project Schedule Management

Typical project milestone schedules contain such information as:

- Project start date
- Project end date
- Other major milestones (such as discussed in Section 11.6)
- Data items (deliverables or reports)

Project start and end dates, if known, must be included. Other major milestones, such as review meetings, prototype available, procurement, testing, and so on, should also be identified. The last topic, data items, is often overlooked. There are two good reasons for preparing a separate schedule for data items. First, the separate schedule will indicate to line managers that personnel with writing skills may have to be assigned. Second, data items require direct-labor man-hours for writing, typing, editing, retyping, proofing, graphic arts, and reproduction. Many companies identify on the data item schedules the approximate number of pages per data item, and each data item is priced out at a cost per page, say $1,000/page. Pricing out data items separately often induces customers to require fewer reports.

The steps required to prepare a report, after the initial discovery work or collection of information, include:

- Organizing the report
- Writing
- Typing
- Editing
- Retyping
- Proofing
- Graphic arts
- Submittal for approvals
- Reproduction and distribution

Typically, 6–8 hours of work are required per page. At a burdened hourly rate of $150/hour, it is easy for the cost of documentation to become exorbitant.

11.13 WORK BREAKDOWN STRUCTURE

PMBOK® Guide, 6th Edition
5.4 Create WBS
5.4.2.2 Decomposition

In planning a project, the project manager must structure the work into small elements that are:

- Manageable, in that specific authority and responsibility can be assigned
- Independent, or with minimum interfacing with and dependence on other ongoing elements
- Integratable so that the total package can be seen
- Measurable in terms of progress

A work breakdown structure (WBS) is a product-oriented family tree subdivision of the hardware, services, and data required to produce the end product. The WBS is structured in accordance with the way the work will be performed and reflects the way in which project costs and data will be summarized and eventually reported. Preparation of the WBS also considers other areas that require structured data, such as scheduling, configuration management, contract funding, and technical performance parameters. The WBS is the single most important element because it provides a common framework from which:

- The total program can be described as a summation of subdivided elements.
- Planning can be performed.
- Costs and budgets can be established.
- Time, cost, and performance can be tracked.
- Objectives can be linked to company resources in a logical manner.
- Schedules and status-reporting procedures can be established.
- Network construction and control planning can be initiated.
- The responsibility assignments for each element can be established.

The work breakdown structure acts as a vehicle for breaking the work down into smaller elements, thus providing a greater probability that every major and minor activity will be accounted for. Although a variety of work breakdown structures exist, the most common is the six-level indented structure shown below:

	Level	*Description*
Managerial levels	1	Total program
	2	Project
	3	Task
Technical levels	4	Subtask
	5	Work package
	6	Level of effort

Level 1 is the total program and is composed of a set of projects. The summation of the activities and costs associated with each project must equal the total program. Each project (level 2), however, can be broken down into tasks (level 3), where the summation of all tasks equals the summation of all projects, which, in turn, comprises the total program. The reason for this subdivision of effort is simply ease of control. Program management therefore becomes synonymous with the integration of activities, and the project manager acts as the integrator, using the work breakdown structure as the common framework.

Careful consideration must be given to the design and development of the WBS. From Figure 11–2, the work breakdown structure can be used to provide the basis for:

- The responsibility matrix
- Network scheduling
- Costing
- Risk analysis
- Organizational structure
- Coordination of objectives
- Control (including contract administration)

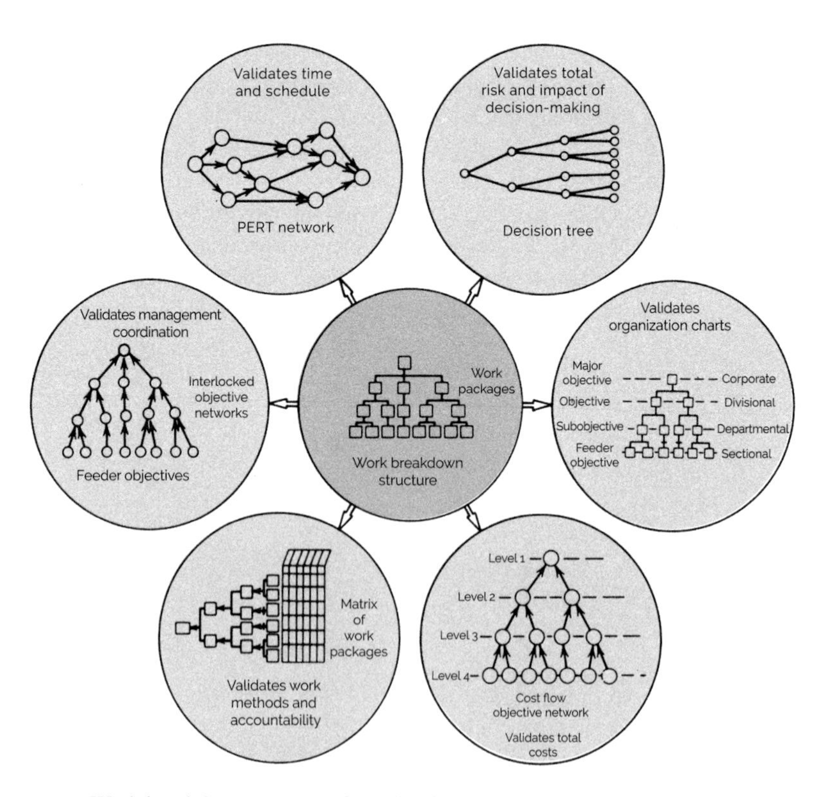

FIGURE 11–2. Work breakdown structure for objective control and evaluation. *Source:* Paul Mali, *Managing by Objectives* (New York: John Wiley & Sons, 1972), p. 163. Copyright © 1972 by John Wiley & Sons. Reprinted by permission of the publisher.

The upper three levels of the WBS are normally specified by the customer (if part of an RFP/RFQ) as the summary levels for reporting purposes. The lower levels are generated by the contractor for in-house control. Each level serves a vital purpose: Level 1 is generally used for the authorization and release of all work, budgets are prepared at level 2, and schedules are prepared at level 3. Certain characteristics can now be generalized for these levels:

- The top three levels of the WBS reflect integrated efforts and should not be related to one specific department. Effort required by departments or sections should be defined in subtasks and work packages.
- The summation of all elements in one level must be the sum of all work in the next lower level.
- Each element of work should be assigned to one and only one level of effort. For example, the construction of the foundation of a house should be included in one project (or task), not extended over two or three. (At level 5, the work packages should be identifiable and homogeneous.)
- The level at which the project is managed is generally called the work package level. Actually, the work package can exist at any level below level one.
- The WBS must be accompanied by a description of the scope of effort required, or else only those individuals who issue the WBS will have a complete understanding of what work has to be accomplished. It is common practice to reproduce the customer's statement of work as the description for the WBS.
- It is often the best policy for the project manager, regardless of his technical expertise, to allow all of the line managers to assess the risks in the SOW. After all, the line managers are usually the recognized experts in the organization.

Project managers normally manage at the top three levels of the WBS and prefer to provide status reports to management at these levels also.

The work package is the critical level for managing a work breakdown structure, as shown in Figure 11–3. However, it is possible that the actual management of the work packages is supervised and performed by the line managers with status reporting provided to the project manager at higher levels of the WBS.

Work packages are natural subdivisions of cost accounts and constitute the basic building blocks used by the contractor in planning, controlling, and measuring contract performance. A work package is simply a low-level task or job assignment. It describes the work to be accomplished by a specific performing organization or a group of cost centers and serves as a vehicle for monitoring and reporting progress of work. Work package descriptions must permit cost account managers and work package supervisors to understand and clearly distinguish one work package effort from another.

In setting up the work breakdown structure, tasks should:

- Have clearly defined start and end dates
- Be usable as a communications tool in which results can be compared with expectations
- Be estimated on a "total" time duration, not when the task must start or end
- Be structured so that a minimum of project office control and documentation (i.e., forms) is necessary

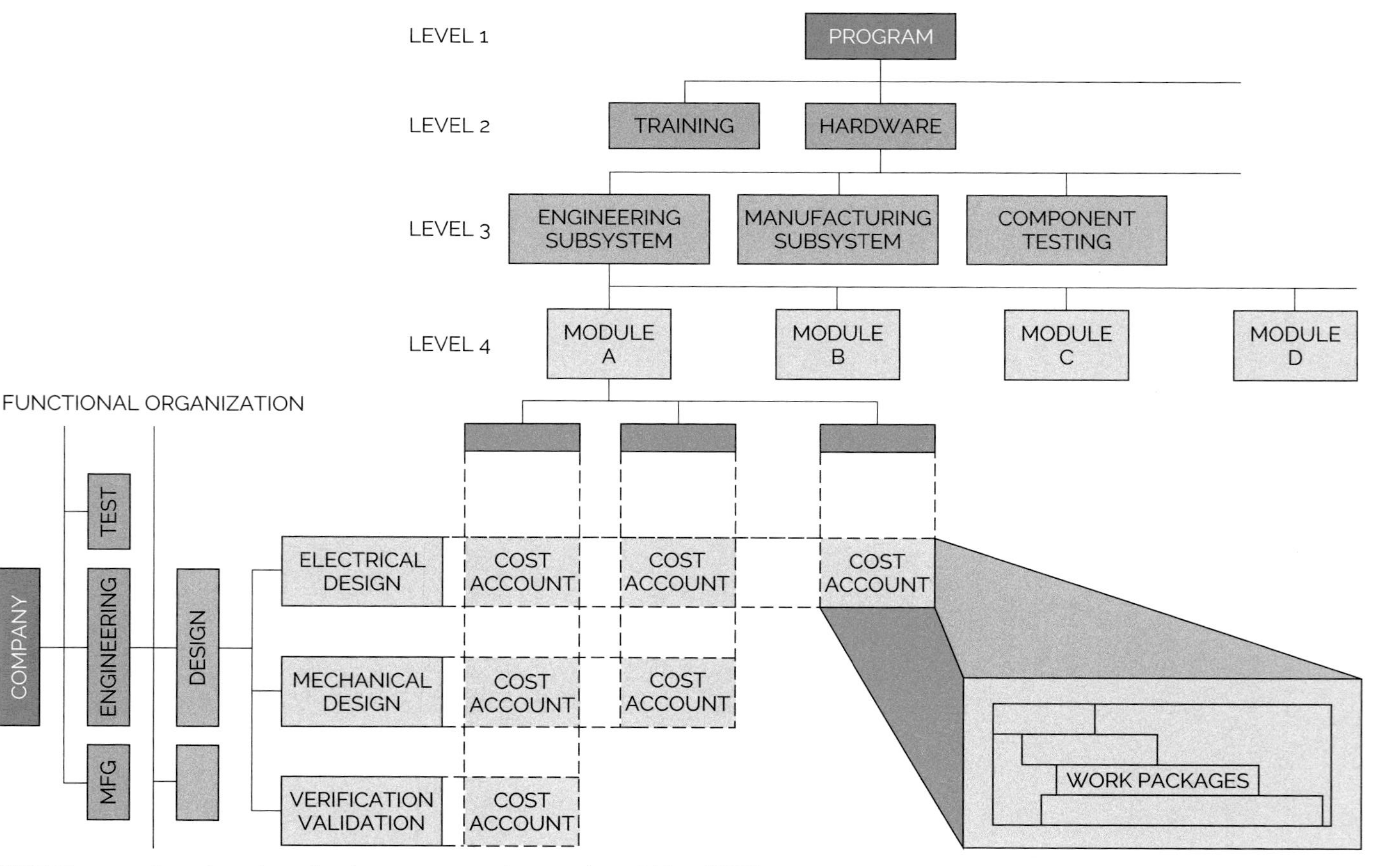

FIGURE 11–3. A work package for the cost account intersection within a WBS.

For large projects, planning will be time phased at the work package level of the WBS. The work package has the following characteristics:

- Represents units of work at the level where the work is performed
- Clearly distinguishes one work package from all others assigned to a single functional group
- Contains clearly defined start and end dates that are representative of physical accomplishment (This is accomplished after scheduling has been completed.)
- Specifies a budget in terms of dollars, man-hours, or other measurable units
- Limits the work to be performed to relatively short periods of time to minimize the work-in-process effort

Table 11-3 shows a simple work breakdown structure with the associated numbering system following the work breakdown. The first number represents the total program (in this case, it is represented by 01), the second number represents the project, and the third number identifies the task. Therefore, number 01–03–00 represents project 3 of program 01, whereas 01–03–02 represents task 2 of project 3. This type of numbering system is not standard; each company may have its own system, depending on how costs are to be controlled.

The preparation of the work breakdown structure is not easy. The WBS is a communications tool, providing detailed information to different levels of management. If it does not contain enough levels, then the integration of activities may prove difficult. If too many levels exist, then unproductive time will be made to have the same number of levels for all projects, tasks, and so on. Each major work element should be considered by itself. Remember, the WBS establishes the number of required networks for cost control.

TABLE 11-3. WORK BREAKDOWN STRUCTURE FOR NEW PLANT CONSTRUCTION AND START-UP

PLANT CONSTRUCTION AND START-UP	
Program: New Plant Construction and Start-up	01–00–00
Project 1: Analytical Study	01–01–00
Task 1: Marketing/Production Study	01–01–01
Task 2: Cost Effectiveness Analysis	01–01–02
Project 2: Design and Layout	01–02–00
Task 1: Product Processing Sketches	01–02–01
Task 2: Product Processing Blueprints	01–02–02
Project 3: Installation	01–03–00
Task 1: Fabrication	01–03–01
Task 2: Setup	01–03–02
Task 3: Testing and Run	01–03–03
Project 4: Program Support	01–04–00
Task 1: Management	01–04–01
Task 2: Purchasing Raw Materials	01–04–02

11.14 WBS DECOMPOSITION PROBLEMS

There is a common misconception that WBS decomposition is an easy task to perform. In the development of the WBS, the top three levels or management levels are usually roll-up levels. Preparing templates at these levels is becoming common practice. However, at levels 4–6 of the WBS, templates may not be appropriate. There are reasons for this.

- Breaking the work down to extremely small and detailed work packages may require the creation of hundreds or even thousands of cost accounts and charge numbers. This could increase the management, control, and reporting costs of these small packages to a point where the costs exceed the benefits.
- Breaking the work down to small work packages can provide accurate cost control if, and only if, the line managers can determine the costs at this level of detail. Line managers must be given the right to tell project managers that costs *cannot* be determined at the requested level of detail.
- The work breakdown structure is the basis for scheduling techniques such as the Arrow Diagramming Method and the Precedence Diagramming Method. At low levels of the WBS, the interdependencies between activities can become so complex that meaningful networks cannot be constructed.

From a cost control point of view, cost analysis down to the fifth level is advantageous. However, it should be noted that the cost required to prepare cost analysis data to each lower level may increase exponentially, especially if the customer requires data to be presented in a specified format that is not part of the company's standard operating procedures. The level-5 work packages are normally for in-house control only. Some companies bill customers separately for each level of cost reporting below level 3.

Once the WBS is established and the program is "kicked off," it becomes a very costly procedure to either add or delete activities, or change levels of reporting because of cost control. Many companies do not give careful forethought to the importance of a properly developed WBS, and ultimately they risk cost control problems downstream. One important use of the WBS is that it serves as a cost control standard for any future activities that may follow on or may just be similar.

Quite often work breakdown structures accompanying customer RFPs contain much more scope of effort, as specified by the statement of work, than the existing funding will support. This is done intentionally by the customer in hopes that a contractor may be willing to "buy in." If the contractor's price exceeds the customer's funding limitations, then the scope of effort must be reduced by eliminating activities from the WBS. By developing a separate project for administrative and indirect support activities, the customer can easily modify his costs by eliminating the direct support activities of the canceled effort.

PMBOK® *Guide*, 6th Edition
5.4 Create WBS

There are also checklists that can be used in the preparation of the WBS:[5]

- Develop a preliminary WBS to not lower than the top three levels for solicitation purposes (or lower if deemed necessary for some special reason).

5. *Handbook for Preparation of Work Breakdown Structures,* NHB5610.1, National Aeronautics and Space Administration, February 1975.

- Ensure that the contractor is required to extend the preliminary WBS in response to the solicitation, to identify and structure all contractor work to be compatible with his organization and management system.
- Following negotiations, the CWBS included in the contract should not normally extend lower than the third level.
- Ensure that the negotiated CWBS structure is compatible with reporting requirements.
- Ensure that the negotiated CWBS is compatible with the contractor's organization and management system.
- Review the CWBS elements to ensure correlation with:
  - The specification tree
  - Contract line items
  - End-items of the contract
  - Data items required
  - Work statement tasks
  - Configuration management requirements
- Define CWBS elements down to the level where such definitions are meaningful and necessary for management purposes (WBS dictionary).
- Specify reporting requirements for selected CWBS elements if variations from standard reporting requirements are desired.
- Ensure that the CWBS covers measurable effort, level of effort, apportioned effort, and subcontracts, if applicable.
- Ensure that the total costs at a particular level will equal the sum of the costs of the constituent elements at the next lower level.

On simple projects, the WBS can be constructed as a "tree diagram" or according to the logic flow. In Figure 11–4, the tree diagram can follow the work or even the organizational

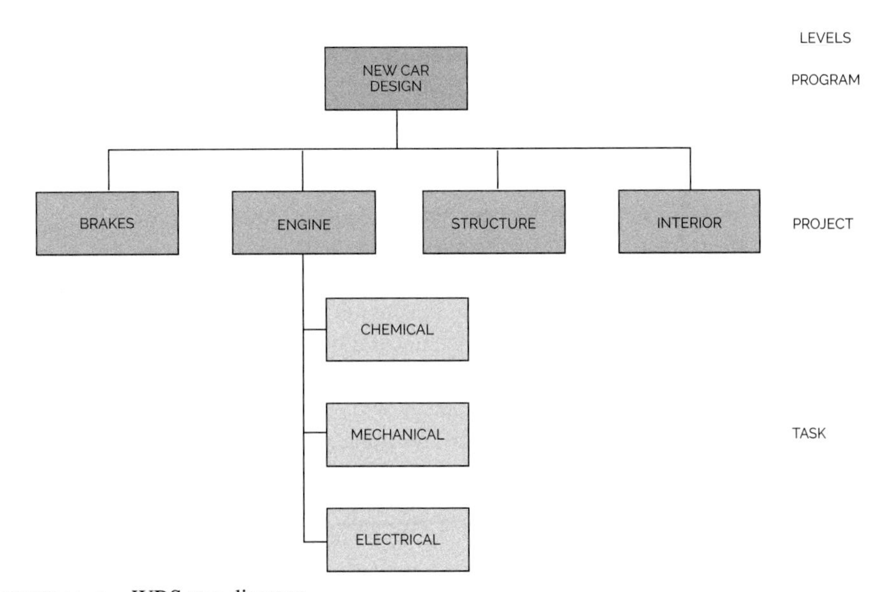

FIGURE 11–4. WBS tree diagram.

structure of the company (i.e., division, department, section, unit). The second method is to create a logic flow (see Figure 12.21) and cluster certain elements to represent tasks and projects. In the tree method, lower-level functional units may be assigned to one, and only one, work element, whereas in the logic flow method the lower-level functional units may serve several WBS elements.

The table below shows the three most common methods for structuring the WBS:

Level	Method		
	Flow	**Life Cycle**	**Organization**
Program	Program	Program	Program
Project	System	Life cycle	Division
Task	Subsystem	System	Department
Subtask	People	Subsystem	Section
Work package	People	People	People
Level of effort	People	People	People

The flow method breaks the work down into systems and major subsystems. This method is well suited for projects less than two years in length. For longer projects, we use the life-cycle method, which is similar to the flow method. The organization method is used for projects that may be repetitive or require very little integration between functional units.

11.15 WORK BREAKDOWN STRUCTURE DICTIONARY

PMBOK® Guide, 6th Edition
5.4.3.1 Scope Baseline (WBS Dictionary)

Work breakdown structures are actually numbering systems such as the last column in Table 11–3. Wording is often added into the WBS to provide clarity. As an example, project management software treats the WBS as a numbering system but may ask you for a description of the work package and the name or initials of the person responsible for that work package.

Perhaps the best way to understand the meaning and intent of each work package is to use a WBS dictionary. For each element in the WBS, the dictionary provides a brief description of each element, the name of the person or cost center responsible for that element such as in a responsibility assignment matrix, the element's milestones, and the final deliverable. The WBS dictionary can also identify the cost associated with that element, the charge number to be used, and the required resources by name or skill level. The dictionary can also provide a detailed technical description of each element and cross-listing to other WBS elements, quality requirements, and contractual documentation. The WBS dictionary is also cross-listed to the project's work authorization form to be discussed in Chapter 15.

The WBS and the WBS dictionary can be used to support the scope verification process. Norman, Brotherton, and Fried state:[6]

6. E. S. Norman, S. A. Brotherton, and R. T. Fried, *Work Breakdown Structures* (Hoboken, NJ: John Wiley & Sons, 2008), pp. 144–145.

During project execution, validation of the deliverables can be accomplished by referencing the deliverables as they have been described in the WBS and WBS Dictionary. Since the WBS and WBS Dictionary both describe project deliverables, including acceptance and completion criteria, these then become the reference points for validation and acceptance of the completed deliverables. The WBS and WBS Dictionary often are used additionally as a baseline for monitoring and measuring "wants" and "needs" versus the agreed upon project scope. This ensures that the project does not attempt to deliver outcomes that are not included in the requirements. The WBS and WBS Dictionary help ensure the project team does attempt to deliver outcomes or quality that exceed the boundaries of the requirements while they also contain and control scope creep.

The WBS and WBS Dictionary help support communications between the project manager, project team, sponsor(s) and stakeholders regarding the content and completion criteria for the project deliverables. Without first developing the WBS, frequently the criteria for deliverable acceptance and completion are ill-defined, leading to misunderstanding and disagreement about the completion of specific project outcomes.

As work proceeds on the project, the WBS can be used as a checklist to determine what deliverables have and have not been completed or accepted. When communicated via the status report and other vehicles in the project's Communications Plan, this helps ensure that all project stakeholders clearly understand the current state of the project.

At the end of the project, Scope Verification supports the transition of the project to ongoing operations as well as closure of any open contracts or subcontracts. Here again the WBS is used as the basis for verification and as a key input to the contract and project closure processes.

11.16 PROJECT SELECTION

A prime responsibility of senior management (and possibly project sponsors) is the selection of projects. Most organizations have an established selection criteria, which can be subjective, objective, quantitative, qualitative, or simply a seat-of-the-pants guess. In any event, there should be a valid reason for selecting the project.

From a financial perspective, project selection is basically a two-part process. First, the organization will conduct a feasibility study to determine whether the project *can* be done. The second part is to perform a benefit-to-cost analysis to see whether the company *should* do it.

The purpose of the feasibility study is to validate that the project meets feasibility of cost, technological, safety, marketability, and ease of execution requirements. The company may use outside consultants or subject matter experts (SMEs) to assist in both feasibility studies and benefit-to-cost analyses. A project manager may not be assigned until after the feasibility study is completed.

As part of the feasibility process during project selection, senior management often solicits input from SMEs and lower-level managers through rating models. The rating models normally identify the business and/or technical criteria against which the ratings will be made. Figure 11–5 shows a scaling model for a single project. Figure 11–6 shows a checklist rating system to evaluate three projects at once. Figure 11–7 shows a scoring model for multiple projects using weighted averages.

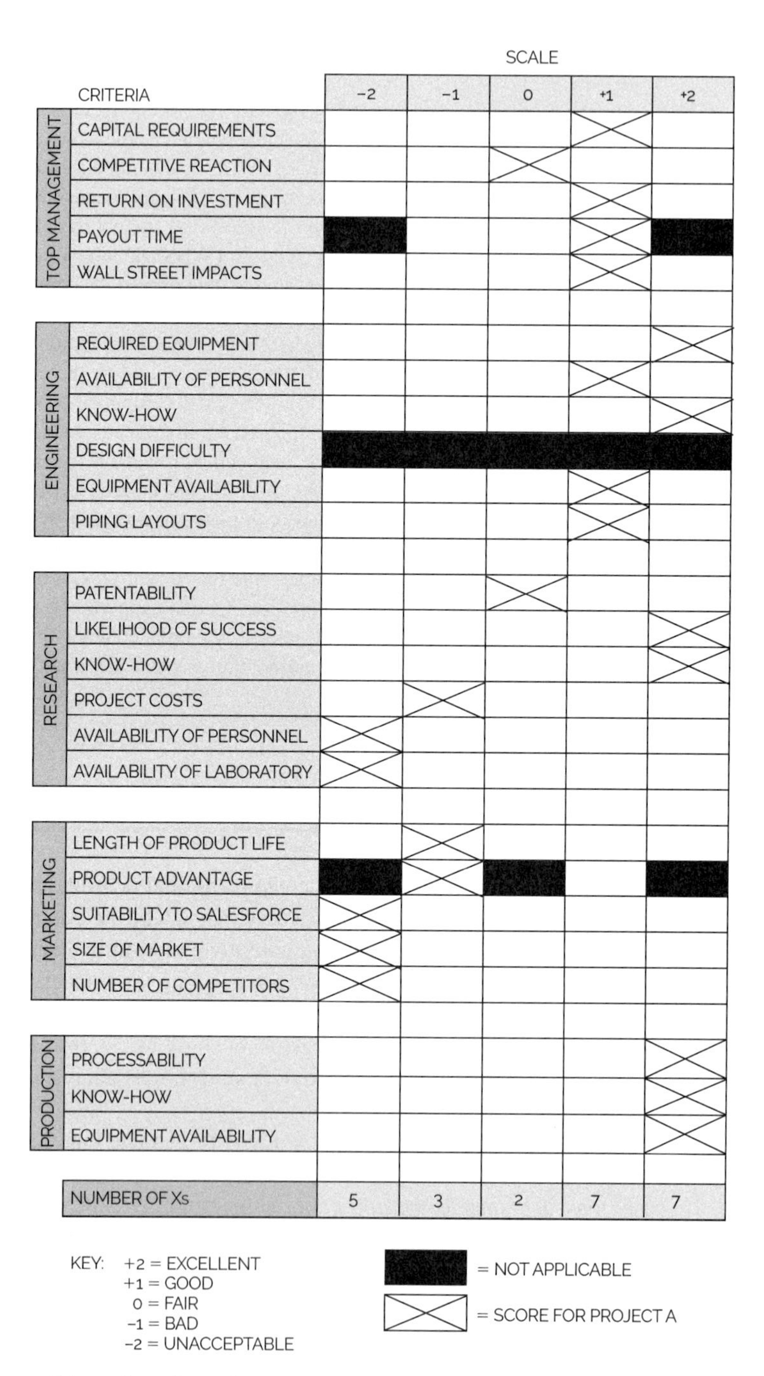

FIGURE 11–5. Illustration of a scaling model for one project, Project A. *Source:* William E. Souder, Project Selection and Economic Appraisal, p. 66.

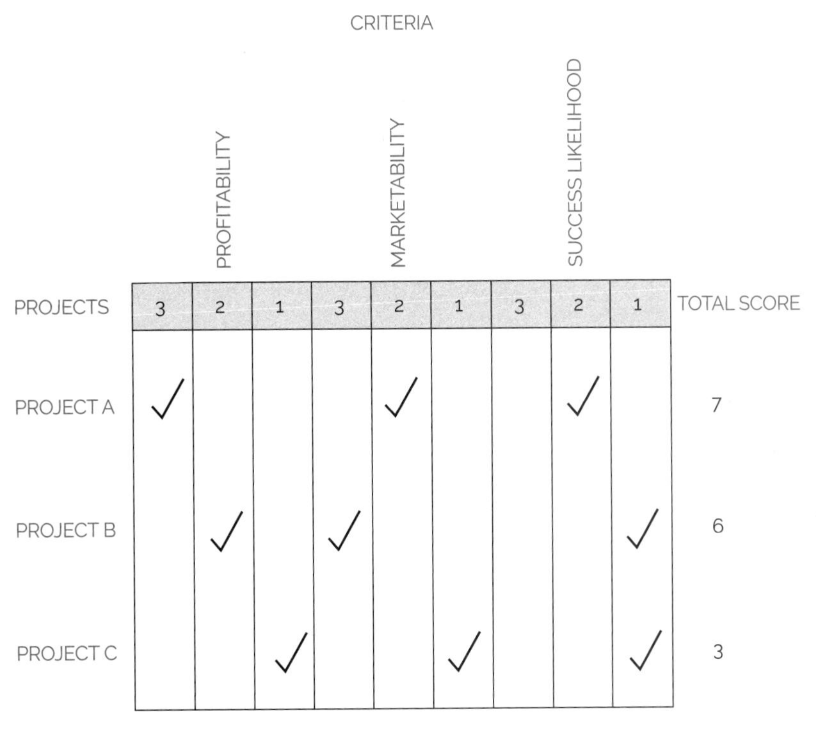

FIGURE 11–6. Illustration of a checklist for three projects. *Source:* William Souder, Project Selection and Economic Appraisal, p. 68.

***PMBOK® Guide*, 6th Edition**
5.3 Scope Definition

If the project is deemed feasible and a good fit with the strategic plan, then the project is prioritized for development along with other projects. Once feasibility is determined, a benefit-to-cost analysis is performed to validate that the project will, if executed correctly, provide the required financial and nonfinancial benefits. Benefit-to-cost analyses require significantly more information to be scrutinized than is usually available during a feasibility study. This can be an expensive proposition.

Estimating benefits and costs in a timely manner is very difficult. Benefits are often defined as:

- Tangible benefits for which dollars may be reasonably quantified and measured.
- Intangible benefits that may be quantified in units other than dollars or may be identified and described subjectively.

Costs are significantly more difficult to quantify. The minimum costs that must be determined are those that specifically are used for comparison to the benefits. These include:

- The current operating costs or the cost of operating in today's circumstances.
- Future period costs that are expected and can be planned for.
- Intangible costs that may be difficult to quantify. These costs are often omitted if quantification would contribute little to the decision-making process.

CRITERIA	PROFITABILITY	PATENTABILITY	MARKETABILITY	PRODUCEABILITY
CRITERION WEIGHTS	4	3	2	1

PROJECTS	CRITERION SCORES*				TOTAL WEIGHTED SCORE
PROJECT D	10	6	4	3	69
PROJECT E	5	10	10	5	75
PROJECT F	3	7	10	10	63

TOTAL WEIGHTED SCORE = Σ (CRITERION SCORE × CRITERION WEIGHT)

* SCALE: 10 = EXCELLENT; 1 = UNACCEPTABLE

FIGURE 11–7. Illustration of a scoring model. *Source:* William Souder, Project Selection *and Economic Appraisal,* p. 69.

There must be careful documentation of all known constraints and assumptions that were made in developing the costs and the benefits. Unrealistic or unrecognized assumptions are often the cause of unrealistic benefits. The go or no-go decision to continue with a project could very well rest upon the validity of the assumptions.

Table 11–4 shows the major differences between feasibility studies and benefit-to-cost analyses.

Today, the project manager may end up participating in the project selection process. In Chapter 1, we discussed the new breed of project manager, namely a person that has excellent business skills as well as project management skills. These business skills now

TABLE 11–4. FEASIBILITY STUDY AND BENEFIT-COST ANALYSIS

	Feasibility Study	**Benefit-Cost Analysis**
Basic Question	**Can We Do It?**	**Should We Do It?**
Life-Cycle Phase	Preconceptual	Conceptual
PM Selected	Usually not yet	Usually identified but partial involvement
Analysis	Qualitative	Quantitative
Critical Factors for Go/No-Go	• Technical • Cost • Quality • Safety • Ease of performance • Economical • Legal	• Net present value • Discounted cash flow • Internal rate of return • Return on investment • Probability of success • Reality of assumptions and constraints
Executive Decision Criteria	Strategic fit	Benefits exceed costs by required margin

allow us to bring the project manager on board the project at the beginning of the initiation phase rather than at the end of the initiation phase because the project manager can now make a valuable contribution to the project selection process. The project manager can be of assistance during project selection by providing business case knowledge, including:

- Opportunity options (sales volume, market share, and follow-on business)
- Resource requirements (team knowledge requirements and skill set)
- Refined project costs
- Refined savings
- Benefits (financial, strategic, payback)
- Project metrics (key performance indicators and critical success factors)
- Benefits realization (consistency with the corporate business plan)
- Risks
- Exit strategies
- Organizational readiness and strengths
- Schedule/milestones
- Overall complexity
- Technology complexity and constraints, if any[7]

11.17 THE ROLE OF THE EXECUTIVE IN PLANNING

Executives are responsible for selecting the project manager, and the person chosen should have planning expertise. Not all technical specialists are good planners. Likewise, some people who are excellent in execution have minimal planning skills. Executives must make

7. For additional factors that can influence project selection decision making, see J. R. Meredith and S. J. Mantel, Jr., *Project Management,* 3rd ed. (New York: John Wiley & Sons, 1995), pp. 44–46.

sure that whoever is assigned as the project manager has both planning and execution skills. In addition, executives must take an active role during project planning activities especially if they also function as project sponsors.[8]

Executives must not arbitrarily set unrealistic milestones and then "force" line managers to fulfill them. Both project and line managers should try to adhere to unrealistic milestones, but if a line manager says he cannot, executives should comply because the line manager is supposedly the expert.

Executives should interface with project and line personnel during the planning stage in order to define the requirements and establish reasonable deadlines. Executives must realize that creating an unreasonable deadline may require the reestablishment of priorities, and, of course, changing priorities can push milestones backward.

11.18 MANAGEMENT COST AND CONTROL SYSTEM

On long-term projects, phasing can be overdone, resulting in extra costs and delays. To prevent this, many project-driven companies resort to other types of systems, such as a management cost and control system (MCCS). No program or project can be efficiently organized and managed without some form of management cost and control system. Figure 11–8 shows the five phases of a management cost and control system. The first phase constitutes the planning cycle, and the next four phases identify the operating cycle.

Figure 11–9 shows the activities included in the planning cycle. The work breakdown structure serves as the initial control from which all planning emanates. The WBS acts as a vital artery for communications and operations in all phases. A comprehensive analysis of management cost and control systems is presented in Chapter 15.

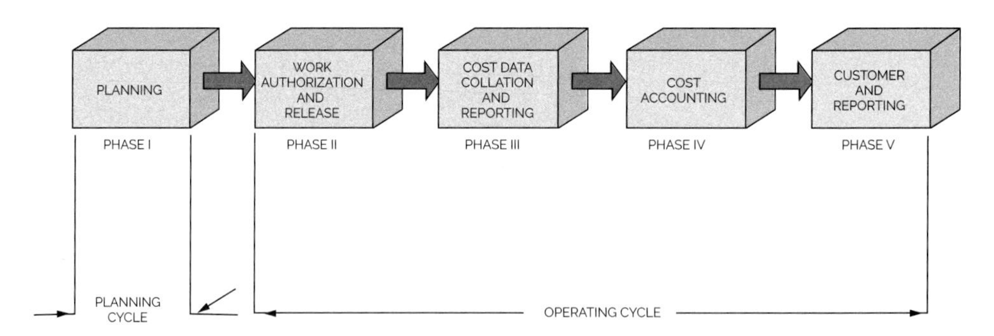

FIGURE 11–8. Phases of a management cost and control system.

8. Although this section is called "The Role of the Executive in Planning," it also applies to line management if project sponsorship is pushed down to the middle-management level or lower. This is quite common in highly mature project management organizations where senior management has sufficient faith in line management's ability to serve as project sponsors.

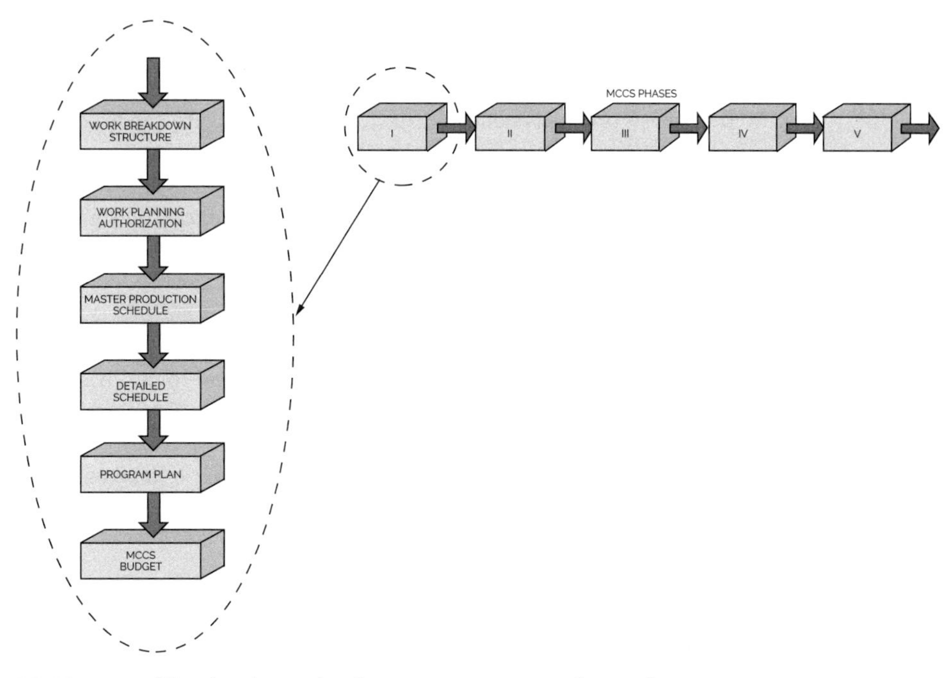

FIGURE 11–9. The planning cycle of a management cost and control system.

11.19 WORK PLANNING AUTHORIZATION

> ***PMBOK® Guide*, 6th Edition**
> 4.3 Direct and Manage Project Work

After receipt of a contract, some form of authorization is needed before work can begin, even in the planning stage. Both work authorization and work planning authorization are used to release funds, but for different purposes. Work planning authorization releases funds (primarily for functional management) so that scheduling, costs, budgets, and all other types of plans can be prepared prior to the release of operational cycle funds, which hereafter shall be referred to simply as work authorization. Both forms of authorization require the same paperwork. In many companies this work authorization is identified as a subdivided work description (SWD), which is a narrative description of the effort to be performed by the cost center (division-level minimum).

The SWD is one of the key elements in the planning of a program, as shown in Figure 11–9. Contract control and administration releases the contract funds by issuing a SWD, which sets forth general contractual requirements and authorizes program management to proceed. Program management issues the SWD to set forth the contractual guidelines and requirements for the functional units. The SWD specifies how the work will be performed, which functional organizations will be involved, and who has what specific responsibilities, and authorizes the utilization of resources within a given time period.

The subdivided work description package is used by the operating organizations to further subdivide the effort defined by the WBS into small segments or work packages.

Many people contend that if the data in the work authorization document are different from what was originally defined in the proposal, the project is in trouble right at the start. This may not be the case, because most projects are priced out assuming "unlimited" resources, whereas the hours and dollars in the work authorization document are based upon "limited" resources. This situation is common for companies that thrive on competitive bidding.

11.20 WHY DO PLANS FAIL?

No matter how hard we try, planning is not perfect, and sometimes plans fail. Typical reasons include:

- Corporate goals are not understood at the lower organizational levels.
- Plans encompass too much in too little time.
- Financial estimates are poor.
- Plans are based on insufficient data.
- No attempt is being made to systematize the planning process.
- Planning is performed by a planning group.
- No one knows the ultimate objective.
- No one knows the staffing requirements.
- No one knows the major milestone dates, including written reports.
- Project estimates are best guesses, and are not based on standards or history.
- Not enough time has been given for proper estimating.
- No one has bothered to see if there will be personnel available with the necessary skills.
- People are not working toward the same specifications.
- People are consistently shuffled in and out of the project with little regard for schedule.

Why do these situations occur? If corporate goals are not understood, it is because corporate executives have been negligent in providing the necessary strategic information and feedback. If a plan fails because of extreme optimism, then the responsibility lies with both the project and line managers for not assessing risk. Project managers should ask the line managers if the estimates are optimistic or pessimistic, and expect an honest answer. Erroneous financial estimates are the responsibility of the line manager. If the project fails because of a poor definition of the requirements, then the project manager is totally at fault.

Sometimes project plans fail because simple details are forgotten or overlooked. Examples of this might be:

- Neglecting to tell a line manager early enough that the prototype is not ready and that rescheduling is necessary.
- Neglecting to see if the line manager can still provide additional employees for the next two weeks because it was possible to do so six months ago.

11.21 STOPPING PROJECTS

PMBOK® *Guide*, 6th Edition
4.7 Close Project or Phase

There are always situations in which projects have to be stopped. Nine reasons for stopping are:

- Final achievement of the objectives
- Poor initial planning and market prognosis
- A better alternative is found
- A change in the company interest and strategy
- Allocated time is exceeded
- Budgeted costs are exceeded
- Key people leave the organization
- Personal whims of management
- Problem too complex for the resources available

Today most of the reasons why projects are not completed on time and within cost are behavioral rather than quantitative. They include:

- Poor morale
- Poor human relations
- Poor labor productivity
- No commitment by those involved in the project

The last item appears to be the cause of the first three items in many situations.

Once the reasons for cancellation are defined, the next problem concerns how to stop the project. Some of the ways are:

- Orderly planned termination
- The "hatchet" (withdrawal of funds and removal of personnel)
- Reassignment of people to higher priority tasks
- Redirection of efforts toward different objectives
- Burying it or letting it die on the vine (i.e., not taking any official action)

11.22 HANDLING PROJECT PHASEOUTS AND TRANSFERS

PMBOK® *Guide*, 6th Edition
4.5 Monitor and Control Project Work

By definition, projects (and even life-cycle phases) have an end point. Closing out is a very important phase in the project life cycle, which should follow particular disciplines and procedures with the objective of:

- Effectively bringing the project to closure according to agreed-on contractual requirements
- Preparing for the transition of the project into the next operational phase, such as from production to field installation, field operation, or training

- Analyzing overall project performance with regard to financial data, schedules, and technical efforts
- Closing the project office, and transferring or selling off all resources originally assigned to the project, including personnel
- Identifying and pursuing follow-on business

Although most project managers are completely cognizant of the necessity for proper planning for project start-up, many project managers neglect planning for project termination. Planning for project termination includes:

- Transferring responsibility
- Completion of project records
  - Historic reports
  - Post-project analysis
- Documenting results to reflect "as built" product or installation
- Acceptance by sponsor/user
- Satisfying contractual requirements
- Releasing resources
  - Reassignment of project office team members
  - Disposition of functional personnel
  - Disposition of materials
- Closing out work orders (financial closeout)
- Preparing for financial payments

Project success or failure often depends on management's ability to handle personnel issues properly during this final phase.

Given business realities, it is difficult to transfer project personnel under ideal conditions. The following suggestions may increase organizational effectiveness and minimize personal stress when closing out a project:

- Carefully plan the project closeout on the part of both project and functional managers. Use a checklist to prepare the plan.
- Establish a simple project closeout procedure that identifies the major steps and responsibilities.
- Treat the closeout phase like any other project, with clearly delineated tasks, agreed-on responsibilities, schedules, budgets, and deliverable items or results.
- Understand the interaction of behavioral and organizational elements in order to build an environment conducive to teamwork during this final project phase.
- Emphasize the overall goals, applications, and utilities of the project as well as its business impact.
- Secure top-management involvement and support.
- Be aware of conflict, fatigue, shifting priorities, and technical or logistic problems. Try to identify and deal with these problems when they start to develop. Communicating progress through regularly scheduled status meetings is the key to managing these problems.

- Keep project personnel informed of upcoming job opportunities. Resource managers should discuss and negotiate new assignments with personnel and involve people already in the next project.
- Be aware of rumors. If a reorganization or layoff is inevitable, the situation should be described in a professional manner or people will assume the worst.
- Assign a contract administrator dedicated to company-oriented projects. He will protect your financial position and business interests by following through on customer sign-offs and final payment.

11.23 DETAILED SCHEDULES AND CHARTS

The scheduling of activities is the first major requirement of the program office after program go-ahead. The program office normally assumes full responsibility for activity scheduling if the activity is not too complex.

Activity scheduling is probably the single most important tool for determining how company resources should be integrated. Activity schedules are invaluable for projecting time-phased resource utilization requirements, providing a basis for visually tracking performance and estimating costs. The schedules serve as master plans from which both the customer and management have an up-to-date picture of operations.

Certain guidelines should be followed in the preparation of schedules, regardless of the projected use or complexity:

- All major events and dates must be clearly identified. If a statement of work is supplied by the customer, those dates shown on the accompanying schedules must be included. If for any reason the customer's milestone dates cannot be met, the customer should be notified immediately.
- The exact sequence of work should be defined through a network in which interrelationships between events can be identified.
- Schedules should be directly relatable to the work breakdown structure. If the WBS is developed according to a specific sequence of work, then it becomes an easy task to identify work sequences in schedules using the same numbering system as in the WBS. The minimum requirement should be to show where and when all tasks start and finish.
- All schedules must identify the time constraints and, if possible, should identify those resources required for each event.

Although these four guidelines relate to schedule preparation, they do not define how complex the schedules should be. Before preparing schedules, three questions should be considered:

- How many events or activities should each network have?
- How much of a detailed technical breakdown should be included?
- Who is the intended audience for this schedule?

Most organizations develop multiple schedules: summary schedules for management and planners and detailed schedules for the doers and lower-level control. The detailed schedules may be strictly for interdepartmental activities. Program management must approve all schedules down through the first three levels of the work breakdown structure. For lower-level schedules (i.e., detailed interdepartmental), program management may or may not request a sign of approval.

One of the most difficult problems to identify in schedules is a hedge position. A hedge position is a situation in which the contractor may not be able to meet a customer's milestone date without incurring a risk, or may not be able to meet activity requirements following a milestone date because of contractual requirements.

Detailed schedules are prepared for almost every activity. It is the responsibility of the program office to marry all of the detailed schedules into one master schedule to verify that all activities can be completed as planned. The preparation sequence for schedules (and also for program plans) is shown in Figure 11–10. The program office submits a request for detailed schedules to the functional managers and the functional managers prepare summary schedules, detailed schedules, and, if time permits, interdepartmental schedules. Each functional manager then reviews his schedules with the program office. The program office, together with the functional program team members, integrates all of the plans and schedules and verifies that all contractual dates can be met.

Before the schedules are finalized, rough drafts of each schedule and plan should be reviewed with the customer. This procedure accomplishes the following:

- Verifies that nothing has fallen through the cracks
- Prevents immediate revisions to a published document and can prevent embarrassing moments
- Minimizes production costs by reducing the number of early revisions
- Shows customers early in the program that you welcome their help and input into the planning phase

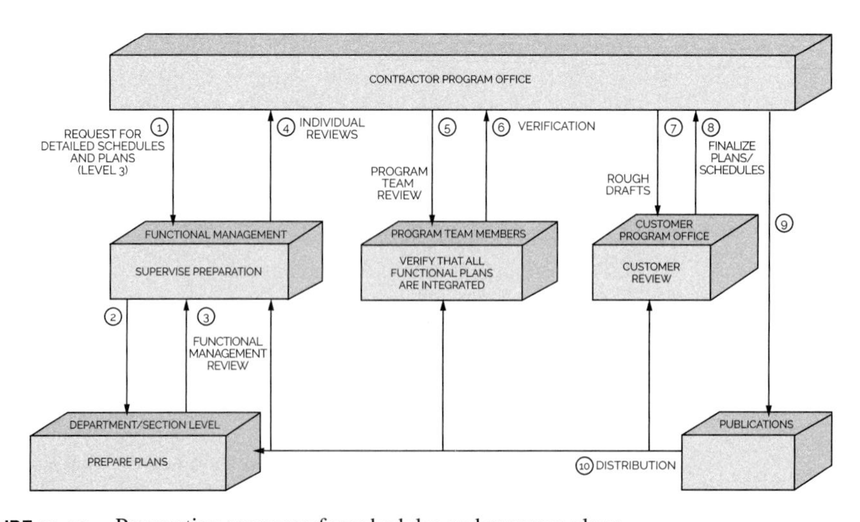

FIGURE 11–10. Preparation sequence for schedules and program plans.

After the document is finalized, it should be distributed to all program office personnel, functional team members, functional management, and the customer. In addition to the detailed schedules, the program office, with input provided by functional management, must develop organization charts. The charts show who has responsibility for each activity and display the formal (and often the informal) lines of communication. The program office may also establish linear responsibility charts (LRCs). In spite of the best attempts by management, many functions in an organization may overlap between functional units. Also, management might wish to have the responsibility for a certain activity given to a functional unit that normally would not have that responsibility. This is a common occurrence on short-duration programs where management desires to cut costs and red tape.

Project personnel should keep in mind why the schedule was developed. The primary objective is usually to coordinate activities to complete the project with the:

- Best time
- Least cost
- Least risk

There are also secondary objectives of scheduling:

- Studying alternatives
- Developing an optimal schedule
- Using resources effectively
- Communicating
- Refining the estimating criteria
- Obtaining good project control
- Providing for easy revisions

Large projects, especially long-term efforts, may require a "war room." All of the walls are covered with large schedules, perhaps printed on blueprint paper, and each wall could have numerous sliding panels. The schedules and charts on each wall could be updated on a daily basis. The room would be used for customer briefings, team meetings, and any other activities related specifically to this project.

11.24 MASTER PRODUCTION SCHEDULING

A *master production schedule* is a statement of what will be made, how many units will be made, and when they will be made.[9] It is a production plan, not a sales plan. The MPS considers the total demand on a plant's resources, including finished product sales, spare (repair) part needs, and inter-plant needs. The MPS must also consider the capacity of the plant and the requirements imposed on vendors. Provisions are made in the overall plan for each manufacturing facility's operation. All planning for materials, manpower, plant, equipment, and financing for the facility is driven by the master production schedule.

9. The master production schedule is being discussed here because of its importance in the planning cycle. The MPS cannot be fully utilized without effective inventory control procedures.

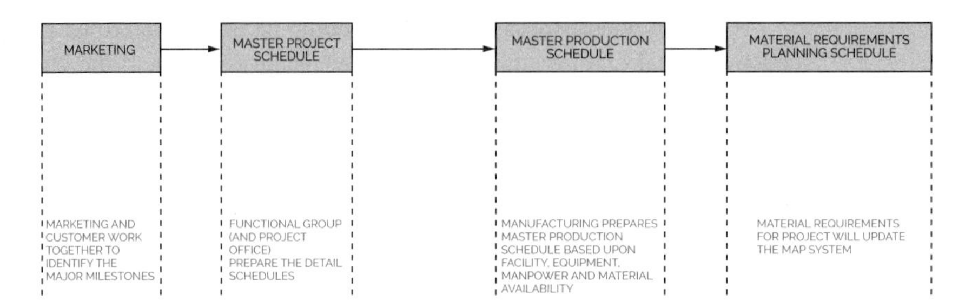

FIGURE 11–11. Material requirements planning interrelationships.

Objectives of master production scheduling are:

- To provide top management with a means to authorize and control manpower levels, inventory investment, and cash flow
- To coordinate marketing, manufacturing, engineering, and finance activities by a common performance objective
- To reconcile marketing and manufacturing needs
- To provide an overall measure of performance
- To provide data for material and capacity planning

The development of a master production schedule is a very important step in a planning cycle. Master production schedules directly tie together personnel, materials, equipment, and facilities, as shown in Figure 11–11. Master production schedules also identify key dates to the customer, should he wish to visit the contractor during specific operational periods.

11.25 PROJECT PLAN

> ***PMBOK® Guide*, 6th Edition**
> Chapter 5 Project Scope Management
> Chapter 4 Project Integration Management

A project plan is fundamental to the success of any project. For large and often complex projects, customers may require a project plan that documents all activities within the program. The project plan then serves as a management guideline for the lifetime of the project and may be revised as often as once a month, depending on the circumstances and the type of project (i.e., research and development projects require more revisions to the project plan than manufacturing or construction projects). The project plan provides the following framework:

- Eliminates conflicts between functional managers
- Eliminates conflicts between functional management and program management
- Provides a standard communications tool throughout the lifetime of the project (It should be geared to the work breakdown structure)
- Provides verification that the contractor understands the customer's objectives and requirements
- Provides a means for identifying inconsistencies in the planning phase
- Provides a means for early identification of problem areas and risks so that no surprises occur downstream
- Contains all of the schedules needed for progress analysis and reporting

Development of a project plan can be time-consuming and costly. All levels of the organization participate. The upper levels provide summary information, and the lower levels provide the details. The project plan, like activity schedules, does not preclude departments from developing their own plans.

The project plan must identify how the company resources will be integrated. The process is similar to the sequence of events for schedule preparation, shown in Figure 11–10. Since the project plan must explain the events in Figure 11–10, additional iterations are required, which can cause changes in a project. This can be seen in Figure 11–12.

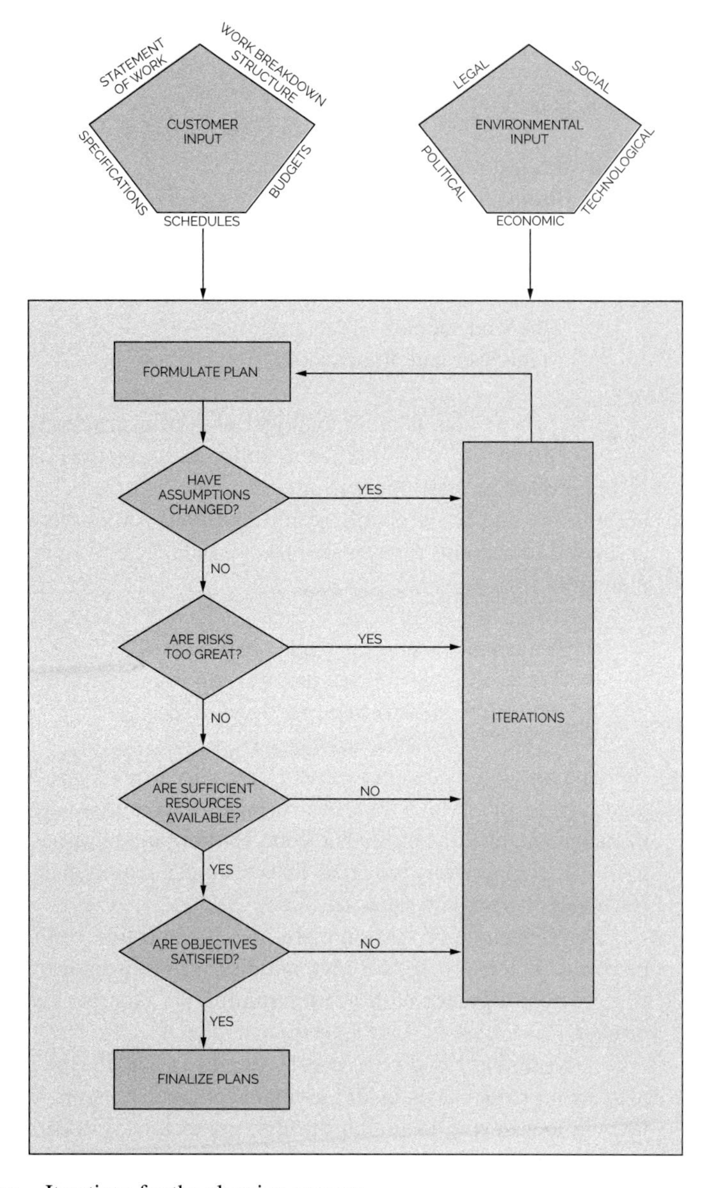

FIGURE 11–12. Iterations for the planning process.

The project plan is a standard from which performance can be measured by the customer and the project and functional managers. The plan serves as a cookbook by answering these questions for all personnel identified with the project:

- What will be accomplished?
- How will it be accomplished?
- Where will it be accomplished?
- When will it be accomplished?
- Why will it be accomplished?

The answers to these questions force both the contractor and the customer to take a hard look at:

- Project requirements
- Project management
- Project schedules
- Facility requirements
- Logistic support
- Financial support
- Manpower and organization

The project plan is more than just a set of instructions. It is an attempt to eliminate crisis by preventing anything from "falling through the cracks." The plan is documented and approved by both the customer and the contractor to determine what data, if any, are missing and the probable resulting effect. As the project matures, the project plan is revised to account for new or missing data. The most common reasons for revising a plan are:

- "Crashing" activities to meet end dates
- Trade-off decisions involving manpower, scheduling, and performance
- Adjusting and leveling manpower requests

The makeup of the project plan may vary from contractor to contractor. Most project plans can be subdivided into four main sections: introduction, summary, and conclusions, management, and technical. The complexity of the information is usually up to the discretion of the contractor, provided that customer requirements, as may be specified in the statement of work, are satisfied.

The introductory section contains the definition of the project and the major parts involved. If the project follows another, or is an outgrowth of similar activities, this is indicated, together with a brief summary of the background and history behind the project.

The summary and conclusion chapter is usually the second section in the project plan so that upper-level customer management can have a complete overview of the project without having to search through the technical information. This section identifies the targets and objectives of the project and includes the necessary "lip service" on how successful the project will be and how all problems can be overcome. This section must also include the project master schedule showing how all projects and activities are related.

The management section of the project plan contains procedures, charts, and schedules as follows:

- The assignment of key personnel to the project is indicated. This usually refers only to the project office personnel and team members, since under normal operations these will be the only individuals interfacing with customers.
- Manpower, planning, and training are discussed to assure customers that qualified people will be available from the functional units.
- A linear responsibility chart might also be included to identify to customers the authority relationships that will exist in the program.

Situations exist in which the management section may be omitted from the proposal. For a follow-up program, the customer may not require this section if management's positions are unchanged. Management sections are also not required if the management information was previously provided in the proposal or if the customer and contractor have continuous business dealings.

The technical section may include as much as 75 to 90 percent of the program plan, especially if the effort includes research and development, and may require constant updating as the project matures. The following items can be included as part of the technical section:

- A detailed breakdown of the charts and schedules used in the project master schedule, possibly including schedule/cost estimates.
- A listing of the testing to be accomplished for each activity. (It is best to include the exact testing matrices.)
- Procedures for accomplishment of the testing. This might also include a description of the key elements in the operations or manufacturing plans, as well as a listing of the facility and logistic requirements.
- Identification of materials and material specifications. (This might also include system specifications.)
- An attempt to identify the risks associated with specific technical requirements (not commonly included). This assessment tends to scare management personnel who are unfamiliar with the technical procedures, so it should be omitted if possible.

The project plan, as used here, contains a description of all phases of the project. For many projects, especially large ones, detailed planning is required for all major events and activities. Table 11–5 identifies the type of individual plans that may be required in place of a (total) project plan. These are often called subsidiary plans.

TABLE 11–5. TYPES OF PLANS

Type of Plan	Description
Budget	How much money is allocated to each event?
Configuration management	How are technical changes made?
Facilities	What facilities resources are available?
Logistics support	How will replacements be handled?
Management	How is the program office organized?
Manufacturing	What are the time-phase manufacturing events?
Procurement	What are my sources? Should I make or buy? If vendors are not qualified, how shall I qualify them?
Quality assurance	How will I guarantee specifications will be met?
Research/development	What are the technical activities?
Scheduling	Are all critical dates accounted for?
Tooling	What are my time-phased tooling requirements?
Training	How will I maintain qualified personnel?
Transportation	How will goods and services be shipped?

The project plan, once agreed on by the contractor and customer, is then used to provide project direction. This is shown in Figure 11–13. If the project plan is written clearly, then any functional manager or supervisor should be able to identify what is expected of him. The project plan should be distributed to each member of the project team, all functional managers and supervisors interfacing with the project, and all key functional personnel.

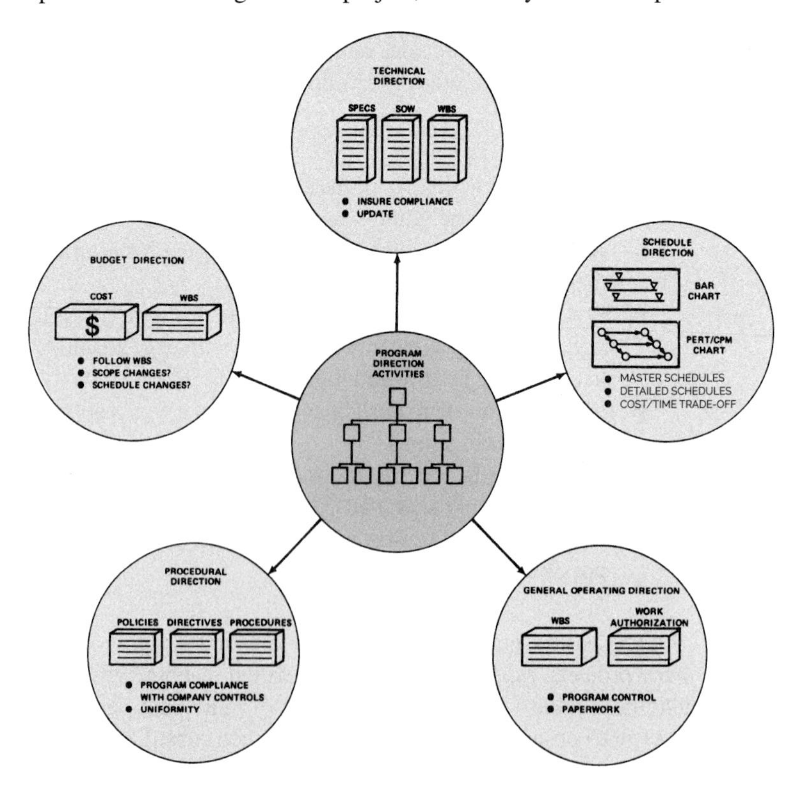

FIGURE 11–13. Project direction activities.

One final note need be mentioned concerning the legality of the project plan. The project plan may be specified contractually to satisfy certain requirements as identified in the customer's statement of work. The contractor retains the right to decide how to accomplish this, unless, of course, this is also identified in the SOW.

11.26 THE PROJECT CHARTER

PMBOK® *Guide*, 6th Edition
4.1 Develop Project Charter

The original concept behind the project charter was to document the project manager's authority and responsibility, especially for projects implemented away from the home office. Today, the project charter is more of an internal legal document identifying to the line managers and their personnel the project manager's authority and responsibility and the management- and/or customer-approved scope of the project.

Theoretically, the sponsor prepares the charter and affixes his/her signature, but in reality, the project manager may prepare it for the sponsor's signature. At a minimum, the charter should include:

- Identification of the project manager and his/her authority to apply resources to the project
- The business purpose that the project was undertaken to address, including all assumptions and constraints
- Summary of the conditions defining the project
- Description of the project
- Objectives and constraints on the project
- Project scope (inclusions and exclusions)
- Key stakeholders and their roles
- Risks
- Involvement by certain stakeholders

The charter is a "legal" agreement between the project manager and the company. Some companies supplement the charter with a "contract" that functions as an agreement between the project and the line organizations.

Some companies have converted the charter into a highly detailed document containing:

- The scope baseline/scope statement
  - Scope and objectives of the project (SOW)
  - Specifications
  - WBS (template levels)
  - Timing
  - Spending plan (S-curve)
- The management plan
  - Resource requirements and manloading (if known)
  - Resumes of key personnel
  - Organizational relationships and structure
  - Responsibility assignment matrix

- Support required from other organizations
- Project policies and procedures
- Change management plan
- Management approval of above

When the project charter contains a scope baseline and management plan, the project charter may function as the project plan. This is not really an effective use of the charter, but it may be acceptable on certain types of projects for internal customers.

11.27 PROJECT BASELINES

PMBOK® Guide, 6th Edition
5.4.3.1 Project Baselines
7.2.1.1 Project Management Plan
6.1.3.1 Schedule Management Plan
8.1.3.1 Quality Management Plan
7.1.3.1 Cost Management Plan

Executives and clients expect project managers to effectively monitor and control projects. As part of monitoring and control, project managers must prepare progress, status, and forecast reports that clearly articulate the performance of the project. But to measure performance, one needs a reference point or baseline from which measurements can be made. The necessity for a baseline is clear:

- Without a baseline, performance cannot be measured.
- If performance cannot be measured, it cannot be managed.
- Performance that can be measured gets watched.
- What gets watched gets done.

For a project to be able to be controlled, it must be organized as a closed system. The closed system is defined as the project management plan. This requires that baselines be established for scope, quality, time, and cost at a minimum. Without such baselines, a project is considered out of control and it may be impossible to track what has changed without knowing where you started.

The project management plan usually contains component plans which describe the baselines:

- Scope management plan: describes the scope baseline which includes the scope statement, WBS and WBS dictionary
- Schedule management plan: describes the steps for planning, monitoring and controlling the schedule
- Quality management plan: describes the activities needed to achieve the quality objectives
- Cost management plan: describes the steps for estimating and controlling the costs

There can be other baselines and plans that are included in the project management plan.

Performance Measurement Baseline

The reference point for measuring performance is the performance measurement baseline (PMB). It serves as the metric benchmark against which performance is measured in terms of time, cost, and scope. It is also used as the basis for business value tracking.

The principal reasons for establishing, approving, controlling, and documenting the PMB are to:

- Ensure achievement of project objectives
- Manage and monitor progress during project execution
- Ensure accurate information on the accomplishment of the deliverables and requirements
- Establish performance measurement criteria

The PMB is finalized at the end of the planning phase once the requirements have been defined, the initial costs have been developed and approved, and the schedule has been set. Once established, the PMB serves as the benchmark from which to measure and gauge the project's progress. The baseline is used to measure how actual progress compares to planned performance. Performance measurement may be meaningless without an accurate baseline as a starting point. Unfortunately, project managers tend to create baselines based upon just those elements of work they feel are important and this may or may not be in full alignment with customer requirements. *The baseline is what the project manager plans to do, not necessarily what the customer has asked for.*

Rebaselining

***PMBOK® Guide*, 6th Edition**
5.6 Control Scope

Projects undergo scope changes for a variety of reasons, including:

- Customer requested changes or add-ons
- Team requested changes or add-ons
- Poor initial understanding or interpretation of customer requirements
- Poorly defined performance or flawed baseline
- Unfavorable variances that cannot be corrected

When changes to the baseline are requested, we go through a change control board (CCB) made up of stakeholders from both the customer's and contractor's organizations. At this change control board meeting, the following questions are addressed:

- The cost of the change
- The impact on the schedule
- The impact on other competing constraints
- The value added for the client
- Any additional risks
- The potential impact on other projects

If the CCB approves the change, then the very first document to be updated is the performance baseline. Once rebaselining occurs, the new baseline is redefined as the original or previous baseline plus the approved changes. Records of all changes are maintained in archives to show how the plan changed over time. It is important to remember that projects seldom run exactly according to plan and traceability of changes is essential.

Developing the PMB

The following steps show a logical approach to creating a PMB:

- *Review the project's business case and accompanying constraints and assumptions*: This is a necessity in order to understand the business boundary of the PMB.
- Establish a requirements baseline: This comes from a review of the customer's requirements and the contractual statement of work (CSOW). The requirements baseline is what the project manager plans to achieve and may contain inclusions and exclusions from the customer's original requirements documentation. This establishes the technical boundary for the PMB and feeds into the project's scope statement.
- *Convert the requirements baseline into a WBS*: Decompose the work into work packages. Each work package should have measurable milestones such that accomplishment of deliverables and performance can be measured. Create a WBS dictionary. The scope baseline can now be defined as the scope statement plus the WBS plus the WBS dictionary.
- *Arrange the work packages into a logical network of activities*: This then becomes the schedule baseline.
- *Price out the time-phased schedule, including both direct and indirect costs:* This then becomes the distributed budget. If the project is multiyear, then work for the following years may not be broken down into work packages yet, even though a budget has been established for each year. This is called an undistributed budget. The summation of the distributed and undistributed budgets make up the cost baseline for the project.
- *The cost baseline does not include the management reserve*: The cost baseline is based upon distributed and undistributed budgets you plan on spending over the life of the project. The management reserve is money you hopefully do not plan on spending.
- *Finalize the PMB*: The PMB is the summation of the scope, schedule, and cost baselines.
- *Prepare a requirements traceability matrix (RTM)*: The RTM links the projects requirements to the WBS and the PMB and is used to verify project requirements are being met.
- *Identify the key metrics or key performance indicators (KPIs)*: These metrics are what will be monitored to determine performance and accomplishment of deliverables and requirements.

These baselines are a necessity for change/version control. Without these baselines, status and the measurement of progress may become meaningless. And if measurement cannot be determined with some degree of accuracy, then no objective information may be found and it may be impossible to determine the true value of what has been accomplished.

While the baseline serves as an excellent reference point, projects can still become derailed, resulting in continuous changes to the PMB. Typical causes include:

- Failing to administer the work orders correctly
- Failing to control the budget

- Having a project management information system that does not provide meaningful data
- Poor understanding and use of the earned-value measurement system
- Improper use of the management reserve
- Constant replanning and baseline fluctuations
- Unnecessary or unwanted changes by management

Types of Baselines In addition to the scope baseline, also known as the technical baseline, the cost baseline, and the schedule baseline, there can be other baselines based upon the firm's business practices and the type of industry. A brief list of these might include:

- *Functional baseline*: System and/or functional requirements such as specifications, contracts, etc.
- *Allocated baseline*: State of the work products once requirements are approved
- *Developmental baseline*: State of work and products during development
- *Product baseline*: Functional and physical characteristics of the project
- Resources baseline: Number and quality of the resources over the project's duration
- *Fixed baseline*: A baseline that remains fixed for the lifetime of the project
- *Revisable baseline*: A baseline that is allowed to vary over the life of the project
- *Project-specific baseline*: A baseline designed for one and only one project
- *Multiproject baseline*: A baseline that can be applied to a number of similar projects

11.28 VERIFICATION AND VALIDATION

PMBOK® Guide, 6th Edition
5.2.3.2 Requirements Traceability Matrix
5.5 Validate Scope

The terms *verification* and *validation* (V&V) are often used in conjunction with the PMB. Verification is sometimes seen as a quality control process that is used to evaluate whether or not a product, service, or system complies with regulations, specifications, or conditions imposed at the start of a development phase. Verification can occur in development, scale-up, or production. This is often an internally performed assessment process.

Verification is actually the acceptance of the deliverables, whereas quality control refers to the correctness of the deliverables. Sometimes, verification and quality control can be done in parallel, but it is more common for quality control to come first.

Validation is the quality assurance process of establishing evidence or assurance that a product, service, or system will accomplish its intended requirements. This often involves meeting acceptance criteria or fitness for purpose. When providing a list of the project's requirements, stakeholders and clients may provide product acceptance criteria, which state the criteria and processes for accepting completed deliverables. The acceptance criteria can include information on:

- Target dates
- Functionality

TABLE 11–6. COMPARISON OF VERIFICATION AND VALIDATION

Verification	Validation
Are we building the product right?	Are we building the right product?
Performed internally, possibly by the project team	Performed internally and by the customer
Measures conformance and compliance to specifications, requirements, regulations, and other imposed conditions	Measures conformance to the customer's acceptance criteria
Use of inspections, audits, reviews, walkthroughs, and analyses	Testing by the client or users on functionality of the deliverable

- Appearance
- Performance levels
- Ease of use
- Capacity
- Availability
- Maintainability
- Reliability
- Operating costs
- Security

It is sometimes said that verification can be expressed by the query "Are you building the thing right?" and validation by "Are you building the right thing?" "Building the right thing" refers back to the user's needs, while "building it right" checks that the specifications are correctly implemented by the system (see Table 11.6). In some contexts, it is required to have written requirements for both as well as formal procedures or protocols for determining compliance.

Verification does not necessarily detect incorrect input specifications. Therefore, verification and validation must be performed to ensure that the system or deliverable is operational. At the completion of verification and validation, we often obtain a certificate or written guarantee that the system, component, or deliverable complies with its specified requirements and is acceptable for operation use.

11.29 MANAGEMENT CONTROL

PMBOK® Guide, 6th Edition
4.6 Perform Integrated Change Control

Because the planning phase provides the fundamental guidelines for the remainder of the project, careful management control must be established.

In addition, since planning is an ongoing activity for a variety of different programs, management guidelines must be established on a company-wide basis in order to achieve unity and coherence.

All functional organizations and individuals working directly or indirectly on a program are responsible for identifying, to the project manager, scheduling and planning problems that require corrective action during both the planning cycle and the operating cycle. The program manager bears the ultimate and final responsibility for identifying requirements for corrective actions. Management policies and directives are written specifically to assist the program manager in defining the requirements. Without clear definitions during the planning phase, many projects run off in a variety of directions.

11.30 CONFIGURATION MANAGEMENT

PMBOK® Guide, 6th Edition
4.6 Perform Integrated Change Control

A critical tool employed by a project manager is configuration management or configuration change control. As projects progress downstream through the various life-cycle phases, the cost of engineering changes can grow boundlessly. It is not uncommon for companies to bid on proposals at 40 percent below their own cost hoping to make up the difference downstream with engineering changes. It is also quite common for executives to "encourage" project managers to seek out engineering changes because of their profitability.

Configuration management is a control technique, through an orderly process, for formal review and approval of configuration changes. If properly implemented, configuration management provides

- Appropriate levels of review and approval for changes
- Focal points for those seeking to make changes
- A single point of input to contracting representatives in the customer's and contractor's office for approved changes

At a minimum, the configuration control committee should include representation from the customer, contractor, and line group initiating the change. Discussions should answer the following questions:

- What is the cost of the change?
- Do the changes improve quality?
- Is the additional cost for this quality justifiable?
- Is the change necessary?
- Is there an impact on the delivery date?

Changes cost money. Therefore, it is imperative that configuration management be implemented correctly. The following steps can enhance the implementation process:

- Define the starting point or "baseline" configuration
- Define the "classes" of changes
- Define the necessary controls or limitations on both the customer and contractor
- Identify policies and procedures, such as board chairman, voters/alternatives, meeting time, agenda, approval forums, step-by-step processes, and expedition processes in case of emergencies

Effective configuration control pleases both customer and contractor. Overall benefits include:

- Better communication among staff
- Better communication with the customer
- Better technical intelligence
- Reduced confusion for changes
- Screening of frivolous changes
- Providing a paper trail

As a final note, it must be understood that configuration control, as used here, is not a replacement for design review meetings or customer interface meetings. These meetings are still an integral part of all projects.

11.31 ENTERPRISE PROJECT MANAGEMENT METHODOLOGIES

Enterprise project management methodologies can enhance the project planning process as well as provide some degree of standardization and consistency.

Companies have come to the realization that enterprise project management methodologies work best if the methodology is based upon templates rather than rigid policies and procedures. The International Institute for Learning has created a Unified Project Management Methodology (UPMM™) with templates categorized according to the *PMBOK® Guide* Areas of Knowledge.[10] Table 11–7 identifies the types of templates based on the different areas of a project.

TABLE 11–7. TYPES OF PROJECT TEMPLATES

Communication	Cost	Human Resources
Project Charter	Project Schedule	Project Charter
Project Procedures Document	Risk Response Plan and Register	Work Breakdown Structure (WBS)
Project Change Requests Log	Work Breakdown Structure (WBS)	Communications Management Plan
Project Status Report	Work Package	Project Organization Chart
PM Quality Assurance Report	Cost Estimates Document	Project Team Directory
Procurement Management Summary	Project Budget	Responsibility Assignment Matrix (RAM)
Project Issues Log	Project Budget Checklist	Project Management Plan
Project Management Plan		Project Procedures Document
Project Performance Report		Kick-off Meeting Checklist
		Project Team Performance Assessment
		Project Manager Performance Assessment
Integration	**Procurement**	**Quality**
Project Procedures Overview	Project Charter	Project Charter

10. Unified Project Management Methodology (UPMM™) a trademark of the International Institute for Learning, Inc., © 2003–2012 by the International Institute for Learning, Inc., all rights reserved.

Communication	**Cost**	**Human Resources**
Project Proposal	Scope Statement	Project Procedures Overview
Communications Management Plan	Work Breakdown Structure (WBS)	Work Quality Plan
Procurement Plan	Procurement Plan	Project Management Plan
Project Budget	Procurement Planning Checklist	Work Breakdown Structure (WBS)
Project Procedures Document	Procurement Statement of Work (SOW)	PM Quality Assurance Report
Project Schedule	Request for Proposal Document Outline	Lessons Learned Document
Responsibility Assignment Matrix (RAM)	Project Change Requests Log	Project Performance Feedback
Risk Response Plan and Register	Contract Formation Checklist	Project Team Performance Assessment
Scope Statement	Procurement Management Summary	PM Process Improvement Document
Work Breakdown Structure (WBS)		
Project Management Plan	**Risk**	**Time**
Project Change Requests Log	Procurement Plan	Activity Duration Estimating Worksheet
Project Issues Log	Project Charter	Cost Estimates Document
Project Management Plan Changes Log	Project Procedures Document	Risk Response Plan and Register Medium
Project Performance Report	Work Breakdown Structure (WBS)	Work Breakdown Structure (WBS)
Lessons Learned Document	Risk Response Plan and Register	Work Package
Project Performance Feedback		Project Schedule
Product Acceptance Document	**Scope**	Project Schedule Review Checklist
Project Charter	Project Scope Statement	
Closing Process Assessment Checklist	Work Breakdown Structure (WBS)	
Project Archives Report	Work Package	
	Project Charter	

11.32 PROJECT AUDITS

The necessity for a structured independent review of various parts of a business, including projects, has taken on a more important role. Part of this can be attributed to the Sarbanes–Oxley law compliance requirements. These independent reviews are audits that focus on either discovery or decision making. The audits can be scheduled or random and can be performed by in-house personnel or external examiners.

Some common audit types include:

- Performance audits: Used to appraise the progress and performance of a given project. The project manager, project sponsor, or an executive steering committee can conduct this audit.
- Compliance audits: Usually performed by the project management office (PMO) to validate that the project is using the project management methodology properly. Usually the PMO has the authority to perform the audit but may not have the authority to enforce compliance.
- Quality audits: Ensure that the planned project quality is being met and that all laws and regulations are being followed. The quality assurance group performs this audit.
- Exit audits: Usually for projects that are in trouble and may need to be terminated. Personnel external to the project, such as an exit champion or an executive steering committee, conduct the audits.

- Best practices audits: Conducted at the end of each life-cycle phase or at the end of the project. Project managers may not be the best individuals to perform the audit. Professional facilitators trained in conducting best practices reviews are used.

Related Case Studies (from Kerzner/*Project Management Case Studies*, 5th ed.)	**Related Workbook Exercises (from Kerzner/*Project Management Workbook and PMP®/CAPM® Exam Study Guide*, 12th ed.)**	***PMBOK® Guide*, 6th Edition, Reference Section for the PMP® Certification Exam**
• Quantum Telecom • Margo Company • Project Overrun • The Two-Boss Problem • Denver International Airport (DIA)	• The Statement of Work • Technology Forecasting • The Noncompliance Project • Multiple Choice Exam • Crossword Puzzle on Scope Management	• Scope Management

11.33 STUDYING TIPS FOR THE PMI® PROJECT MANAGEMENT CERTIFICATION EXAM

This section is applicable as a review of the principles to support the knowledge areas and domain groups in the *PMBOK® Guide*. This chapter addresses:

- Project scope management
- Initiation
- Planning
- Execution
- Monitoring
- Closure

Understanding the following principles is beneficial if the reader is using this text to study for the PMP® Certification Exam:

- Need for effective planning
- Components of a project plan and subsidiary plans
- Need for and components of a statement of work (both proposal and contractual)
- How to develop a work breakdown structure and advantages and disadvantages of highly detailed levels
- Types of work breakdown structures
- Purpose of a work package
- Purpose of configuration management and role of the change control board
- Need for a project charter and components of a project charter
- Need for the project team to be involved in project-planning activities
- That changes to a plan or baseline need to be managed

In Appendix C, the following Dorale Products minicase studies are applicable:

- Dorale Products (C) (Scope Management)

PMP and CAPM are registered marks of the Project Management Institute, Inc.

- Dorale Products (D) [Scope Management]
- Dorale Products (E) [Scope Management]

The following multiple-choice questions will be helpful in reviewing the principles of this chapter:

1. The document that officially sanctions the project is the:
   A. Project charter
   B. Project plan
   C. Feasibility study
   D. Cost-benefit analysis
2. The work breakdown structure "control points" for the management of a project are the:
   A. Milestones
   B. Work packages
   C. Activities
   D. Constraints
3. One of the most common reasons why projects undergo scope changes is:
   A. Poor work breakdown structure
   B. Poorly defined statement of work
   C. Lack of resources
   D. Lack of funding
4. Which of the following generally cannot be validated using a work breakdown structure?
   A. Schedule control
   B. Cost control
   C. Quality control
   D. Risk management

Answer questions 5–8 using the work breakdown structure (WBS) shown below (numbers in parentheses show the dollar value for a particular element):

1.00.00	
1.1.0	($25K)
1.1.1	
1.1.2	($12K)
1.2.0	
1.2.1	($16K)
1.2.2.0	
1.2.2.1	($20K)
1.2.2.2	($30K)

5. The cost of WBS element 1.2.2.0 is:
   A. $20K
   B. $30K
   C. $50K
   D. Cannot be determined
6. The cost of WBS element 1.1.1 is:
   A. $12K
   B. $13K
   C. $25K
   D. Cannot be determined
7. The cost of the entire program (1.00.00) is:
   A. $25K
   B. $66K
   C. $91K
   D. Cannot be determined
8. The work packages in the WBS are at WBS level(s):
   A. 2 only
   B. 3 only
   C. 4 only
   D. 3 and 4
9. The performance measurement baseline is most often composed of three baselines:
   A. Cost, schedule, and risk baselines
   B. Cost, schedule, and scope baselines
   C. Cost, risk, and quality baselines
   D. Schedule, risk, and quality baselines
10. Which of the following is (are) the benefit(s) of developing a WBS to low levels?
    A. Better estimation of costs
    B. Better control
    C. Less likely that something will "fall through the cracks"
    D. All of the above
11. Baselines, once established, identify:
    A. What the customer and contractor agree to
    B. What the sponsor and the customer agree to
    C. What the customer wants done but not necessarily what the project manager plans to do
    D. What the project manager plans on doing but not necessarily what the customer has asked for

12. One of your contractors has sent you an e-mail requesting that they be allowed to conduct only eight tests rather than the ten tests required by the specification. What should the project manager do first?
    A. Change the scope baseline.
    B. Ask the contractor to put forth a change request.
    C. Look at the penalty clauses in the contract.
    D. Ask your sponsor for his or her opinion.
13. One of your contractors sends you an e-mail request to use high quality raw materials in your project stating that this will be value-added and improve quality. What should the project manager do first?
    A. Change the scope baseline
    B. Ask the contractor to put forth a change request
    C. Ask your sponsor for his or her opinion
    D. Change the WBS
14. What are the maximum number of subsidiary plans a program management plan can contain?
    A. 10
    B. 15
    C. 20
    D. Unlimited number
15. The change control board, of which you are a member, approves a significant scope change. The first document that the project manager should updated would be the:
    A. Scope baseline
    B. Schedule
    C. WBS
    D. Budget

ANSWERS

1. A
2. B
3. B
4. C
5. C
6. B
7. C
8. D
9. B
10. D
11. D
12. B
13. B
14. D
15. A

PROBLEMS

11–1 You have been asked to develop a work breakdown structure for a project. How should you go about accomplishing this? Should the WBS be time-phased, department-phased, division-phased, or some combination?

11–2 You have just been instructed to develop a schedule for introducing a new product into the marketplace. Below are the elements that must appear in your schedule. Arrange these elements into a work breakdown structure (down through level 3), and then draw the arrow diagram. You may feel free to add additional topics as necessary.

- Production layout
- Review plant costs
- Market testing
- Select distributors
- Analyze selling cost
- Lay out artwork
- Analyze customer reactions
- Approve artwork
- Storage and shipping costs
- Introduce at trade show
- Select salespeople
- Distribute to salespeople
- Train salespeople
- Establish billing procedure
- Train distributors
- Establish credit procedure
- Literature to salespeople
- Revise cost of production
- Literature to distributors
- Revise selling cost
- Print literature
- Approvals*
- Sales promotion
- Review meetings*
- Sales manual
- Final specifications
- Trade advertising
- Material requisitions

(*Approvals and review meetings can appear several times.)

11–3 The project start-up phase is complete, and you are now ready to finalize the operational plan. Below are six steps that are often part of the finalization procedure. Place them in the appropriate order.

1. Draw diagrams for each individual WBS element.
2. Establish the work breakdown structure and identify the reporting elements and levels.
3. Create a coarse (arrow-diagram) network and decide on the WBS.
4. Refine the diagram by combining all logic into one plan. Then decide on the work assignments.
5. If necessary, try to condense the diagram as much as possible without losing clarity.
6. Integrate diagrams at each level until only one exists. Then begin integration into higher WBS levels until the desired plan is achieved.

11–4 Below are seven factors that must be considered before finalizing a schedule. Explain how a base case schedule can change as a result of each of these:

- Introduction or acceptance of the product in the marketplace
- Present or planned manpower availability
- Economic constraints of the project
- Degree of technical difficulty
- Manpower availability
- Availability of personnel training
- Priority of the project

11–5 Below are twelve instructions. Which are best described as planning, and which are best described as forecasting?

a. Give a complete definition of the work.
b. Lay out a proposed schedule.
c. Establish project milestones.
d. Determine the need for different resources.
e. Determine the skills required for each WBS task or element.
f. Change the scope of the effort and obtain new estimates.
g. Estimate the total time to complete the required work.
h. Consider changing resources.
i. Assign appropriate personnel to each WBS element.
j. Reschedule project resources.
k. Begin scheduling the WBS elements.
l. Change the project priorities.

11–6 "Expecting trouble." Good project managers know what type of trouble can occur at the various stages in the development of a project. The activities in the numbered list below indicate the various stages of a project. The lettered list that follows identifies major problems. For each project stage, select and list all of those problems that are applicable.

Request for proposal ____________________
Submittal to customer ____________________
Contract award ____________________
Design review meetings ____________________
Testing the product____________________
Customer acceptance ____________________

a. Engineering does not request manufacturing input for end-item producibility.
b. The work breakdown structure is poorly defined.
c. Customer does not fully realize the impact that a technical change will have upon cost and schedule.
d. Time and cost constraints are not compatible with the state of the art.
e. The project–functional interface definition is poor.
f. Improper systems integration has created conflicts and a communications breakdown.
g. Several functional managers did not realize that they were responsible for certain risks.
h. The impact of design changes is not systematically evaluated.

11–7 Table 11.8 identifies twenty-six steps in project planning and control. Below is a description of each of the twenty-six steps. Using this information, fill in columns 1 and 2 (column 2 is a group response). After your instructor provides you with column 3, fill in the remainder of the table.

1. *Develop the linear responsibility chart.* This chart identifies the work breakdown structure and assigns specific authority/responsibility to various individuals as groups in order to be sure that all WBS elements are accounted for. The linear responsibility chart can be prepared with either the titles or names of individuals. Assume that this

is prepared after you negotiate for qualified personnel, so that you know either the names or capabilities of those individuals who will be assigned.

2. *Negotiate for qualified functional personnel.* Once the work is decided on, the project manager tries to identify the qualifications for the desired personnel. This then becomes the basis for the negotiation process.
3. *Develop specifications.* This is one of the four documents needed to initially define the requirements of the project. Assume that these are either performance or material specifications, and are provided to you at the initial planning stage by either the customer or the user.
4. *Determine the means for measuring progress.* Before the project plan is finalized and project execution can begin, the project manager must identify the means for measuring progress; specifically, what is meant by an out-of-tolerance condition and what are the tolerances/variances/thresholds for each WBS base case element?
5. *Prepare the final report.* This is the final report to be prepared at the termination of the project.
6. *Authorize departments to begin work.* This step authorizes departments to begin the actual execution of the project, *not* the planning. This step occurs generally after the project plan has been established, finalized, and perhaps even approved by the customer or user group. This is the initiation of the work orders for project implementation.
7. *Develop the work breakdown structure.* This is one of the four documents required for project definition in the early project planning stage. Assume that WBS is constructed using a bottom-up approach. In other words, the WBS is constructed from the logic network (arrow diagram) and checkpoints which will eventually become the basis for the PERT/CPM charts (see Activity 25).
8. *Close out functional work orders.* This is where the project manager tries to prevent excessive charging to his project by closing out the functional work orders (i.e., Activity 6) as work terminates. This includes canceling all work orders except those needed to administer the termination of the project and the preparation of the final report.
9. *Develop scope statement and set objectives.* This is the statement of work and is one of the four documents needed in order to identify the requirements of the project. Usually, the WBS is the structuring of the statement of work.
10. *Develop gross schedule.* This is the summary or milestone schedule needed at project initiation in order to define the four requirements documents for the project. The gross schedule includes start and end dates (if known), other major milestones, and data items.
11. *Develop priorities for each project element.* After the base case is identified and alternative courses of action are considered (i.e., contingency planning), the project team performs a sensitivity analysis for each element of the WBS. This may require assigning priorities for each WBS element, and the highest priorities may *not* necessarily be assigned to elements on the critical path.
12. *Develop alternative courses of action.* Once the base case is known and detailed courses of action (i.e., detailed scheduling) are prepared, project managers conduct "what if" games to develop possible contingency plans.

13. *Develop PERT network.* This is the finalization of the PERT/CPM network and becomes the basis from which detailed scheduling will be performed. The logic for the PERT network can be conducted earlier in the planning cycle (see Activity 25), but the finalization of the network, together with the time durations, are usually based on who has been (or will be) assigned, and the resulting authority/responsibility of the individual. In other words, the activity time duration is a function not only of the performance standard, but also of the individual's expertise and authority/ responsibility.
14. *Develop detailed schedules.* These are the detailed project schedules, and are constructed from the PERT/CPM chart and the capabilities of the assigned individuals.
15. *Establish functional personnel qualifications.* Once senior management reviews the base case costs and approves the project, the project manager begins the task of conversion from rough to detail planning. This includes identification of the required resources, and then the respective qualifications.
16. *Coordinate ongoing activities.* These are the ongoing activities for project execution, not project planning. These are the activities that were authorized to begin in Activity 6.
17. *Determine resource requirements.* After senior management approves the estimated base case costs obtained during rough planning, detailed planning begins by determining the resource requirements, including human resources.
18. *Measure progress.* As the project team coordinates ongoing activities during project execution, the team monitors progress and prepares status reports.
19. *Decide on a basic course of action.* Once the project manager obtains the rough cost estimates for each WBS element, the project manager puts together all of the pieces and determines the basic course of action.
20. *Establish costs for each WBS element.* After deciding on the base case, the project manager establishes the base case cost for each WBS element in order to prepare for the senior management pricing review meeting. These costs are usually the same as those that were provided by the line managers.
21. *Review WBS costs with each functional manager.* Each functional manager is provided with the WBS and told to determine his role and price out his functional involvement. The project manager then reviews the WBS costs to make sure that everything was accounted for and without duplication of effort.
22. *Establish a project plan.* This is the final step in detail planning. Following this step, project execution begins. (Disregard the situation where project plan development can be run concurrently with project execution.)
23. *Establish cost variances for the base case elements.* Once the priorities are known for each base case element, the project manager establishes the allowable cost variances that will be used as a means for measuring progress. Cost reporting is minimum as long as the actual costs remain within these allowable variances.
24. *Price out the WBS.* This is where the project manager provides each functional manager with the WBS for initial activity pricing.
25. *Establish logic network with checkpoints.* This is the bottom-up approach that is often used as the basis for developing both the WBS and later the PERT/CPM network.
26. *Review base case costs with director.* Here the project manager takes the somewhat rough costs obtained during the WBS functional pricing and review and seeks management's approval to begin detail planning.

Activity	Description	Column 1: Your sequence	Column 2: Group sequence	Column 3: Expert's sequence	Column 4: Difference between 1 & 3	Column 5: Difference between 2 & 3
1.	Develop linear responsibility chart					
2.	Negotiate for qualified functional personnel					
3.	Develop specifications					
4.	Determine means for measuring progress					
5.	Prepare final report					
6.	Authorize departments to begin work					
7.	Develop work breakdown structure					
8.	Closeout functional work order					
9.	Develop scope statement and set objectives					
10.	Develop gross schedule					
11.	Develop priorities for each project element					
12.	Develop alternative courses of action					
13.	Develop PERT network					
14.	Develop detailed schedules					
15.	Establish functional personnel qualifications					
16.	Coordinate ongoing activities					
17.	Determine resource requirements					
18.	Measure progress					
19.	Decide upon a basic course of action					
20.	Establish costs for each WBS element					
21.	Review WBS costs with each functional manager					
22.	Establish a project plan					
23.	Establish cost variances for base case elements					
24.	Price out WBS					
25.	Establish logic network with checkpoints					
26.	Review base case costs with director					

12 Network Scheduling Techniques

12.0 INTRODUCTION

PMBOK® Guide, 6th Edition
Chapter 6 Project Schedule Management

PMBOK® Guide, 6th Edition
6.2.3.3 Milestone Lists

PMBOK® Guide, 5th Edition
6.3 Sequencing Activities

Management is continually seeking new and better control techniques to cope with the complexities, masses of data, and tight deadlines that are characteristic of highly competitive industries. Managers also want better methods for presenting technical and cost data to customers.

Scheduling techniques help achieve these goals. The most common techniques are:

- Gantt or bar charts
- Milestone charts
- Line of balance[1]
- Networks
- Program Evaluation and Review Technique (PERT)
- Arrow Diagram Method (ADM) (Sometimes called the Critical Path Method [CPM])[2]
- Precedence Diagram Method (PDM)
- Graphical Evaluation and Review Technique (GERT)

---

1. Line of balance is more applicable to manufacturing operations for production line activities. However, it can be used for project management activities where a finite number of deliverables must be produced in a given time period. The reader need only refer to the multitude of texts on production management for more information on this technique.
2. The text uses the term CPM instead of ADM. The reader should understand that they are interchangeable.

Advantages of network scheduling techniques include:

- They form the basis for all planning and predicting and help management decide how to use its resources to achieve time and cost goals.
- They provide visibility and enable management to control "one-of-a-kind" programs.
- They help management evaluate alternatives by answering such questions as how time delays will influence project completion, where slack exists between elements, and what elements are crucial to meet the completion date.
- They provide a basis for obtaining facts for decision making.
- They utilize a so-called time network analysis as the basic method to determine manpower, material, and capital requirements, as well as to provide a means for checking progress.
- They provide the basic structure for reporting information.
- They reveal interdependencies of activities.
- They facilitate "what if" exercises.
- They identify the longest path or critical paths.
- They aid in scheduling risk analysis.

PERT was originally developed in 1958 and 1959 to meet the needs of the "age of massive engineering." Since that time, PERT has spread rapidly throughout almost all industries. At about the same time, the DuPont Company initiated a similar technique known as the critical path method (CPM), which also has spread widely, and is particularly concentrated in the construction and process industries.

In the early 1960s, the basic requirements of PERT/time were as follows:

- All of the individual tasks to complete a program must be clear enough to be put down in a network, which comprises events and activities; that is, follow the work breakdown structure.
- Events and activities must be sequenced on the network under a highly logical set of ground rules that allow the determination of critical and subcritical paths. Networks may have more than one hundred events, but not fewer than ten.
- Time estimates must be made for each activity on a three-way basis. Optimistic, most likely, and pessimistic elapsed-time figures are estimated by the person(s) most familiar with the activity.
- Critical path and slack times are computed. The critical path is that sequence of activities and events whose accomplishment will require the greatest time.

A big advantage of PERT lies in its extensive planning. Network development and critical path analysis reveal interdependencies and problems that are not obvious with other planning methods. PERT therefore determines where the greatest effort should be made to keep a project on schedule.

The second advantage of PERT is that one can determine the probability of meeting deadlines by development of alternative plans. If the decision maker is statistically sophisticated, he can examine the standard deviations and the probability of accomplishment data. If there exists a minimum of uncertainty, one may use the single-time approach, of course, while retaining the advantage of network analysis.

A third advantage is the ability to evaluate the effect of changes in the program. For example, PERT can evaluate the effect of a contemplated shift of resources from the less critical activities to the activities identified as probable bottlenecks. PERT can also evaluate the effect of a deviation in the actual time required for an activity from what had been predicted.

Finally, PERT allows a large amount of sophisticated data to be presented in a well-organized diagram from which contractors and customers can make joint decisions.

PERT, unfortunately, is not without disadvantages. The complexity of PERT adds to implementation problems. There exist more data requirements for a PERT-organized reporting system than for most others. PERT, therefore, becomes expensive to maintain and is utilized most often on large, complex programs.

Many companies have taken a hard look at the usefulness of PERT on small projects. The result has been the development of PERT/LOB procedures, which can do the following:

- Cut project costs and time
- Coordinate and expedite planning
- Eliminate idle time
- Provide better scheduling and control of subcontractor activities
- Develop better troubleshooting procedures
- Cut the time required for routine decisions, but allow more time for decision making

Even with these advantages, many companies should ask whether they actually need PERT because incorporating it may be difficult and costly, even with canned software packages. Criticism of PERT includes:

- Time and labor intensive
- Decision-making ability reduced
- Lacks functional ownership in estimates
- Lacks historical data for time–cost estimates
- Assumes unlimited resources
- Requires too much detail

12.1 NETWORK FUNDAMENTALS

> ***PMBOK® Guide*, 6th Edition**
> 6.3 Sequencing Activities
> 6.3.2 Sequencing Activities Tools and Techniques

The major discrepancy with Gantt, milestone, or bubble charts is the inability to show the interdependencies between events and activities. These interdependencies must be identified so that a master plan can be developed that provides an up-to-date picture of operations at all times.

Interdependencies are shown through the construction of networks. Network analysis can provide valuable information for planning, integration of plans, time studies, scheduling, and resource management. The primary purpose of network planning is to eliminate the need for crisis management by providing a pictorial representation of the total program. The following management information can be obtained from such a representation:

- Interdependencies of activities
- Project completion time
- Impact of late starts
- Impact of early starts
- Trade-offs between resources and time
- "What if" exercises
- Cost of a crash program
- Slippages in planning/performance
- Evaluation of performance

Networks are composed of events and activities. The following terms are helpful in understanding networks:

- *Event*: Equivalent to a milestone indicating when an activity starts or finishes.
- *Activity*: The element of work that must be accomplished.
- *Duration*: The total time required to complete the activity.
- Effort: The amount of work that is actually performed within the duration. For example, the duration of an activity could be one month but the effort could be just a two-week period within the duration.
- *Critical path*: This is the longest path through the network and determines the duration of the project. It is also the shortest amount of time necessary to accomplish the project.

Figure 12–1 shows the standard nomenclature for PERT networks. The circles represent events, and arrows represent activities. The numbers in the circles signify the specific events or accomplishments. The number over the arrow specifies the time needed (hours, days, months), to go from event 6 to event 3. The events need not be numbered in any specific order. However, event 6 must take place before event 3 can be completed (or begun). In Figure 12–2A, event 26 must take place prior to events 7, 18, and 31. In Figure 12–2B, the opposite holds true, and events 7, 18, and 31 must take place prior to event 26. Figure 12–2B is similar to "and gates" used in logic diagrams.[3]

In this chapter's introduction we have summarized the advantages and disadvantages of Gantt and milestone charts. These charts, however, can be used to develop the PERT network, as shown in Figure 12–3. The bar chart in Figure 12–3A can be converted to the milestone chart in Figure 12–3B. By then defining the relationship between the events on different bars in the milestone chart, we can construct the PERT chart in Figure 12–3C.

PERT is basically a management planning and control tool. It can be considered as a road map for a particular program or project in which all of the major elements (events)

3. PERT diagrams can, in fact, be considered as logic diagrams. Many of the symbols used in PERT have been adapted from logic flow nomenclature.

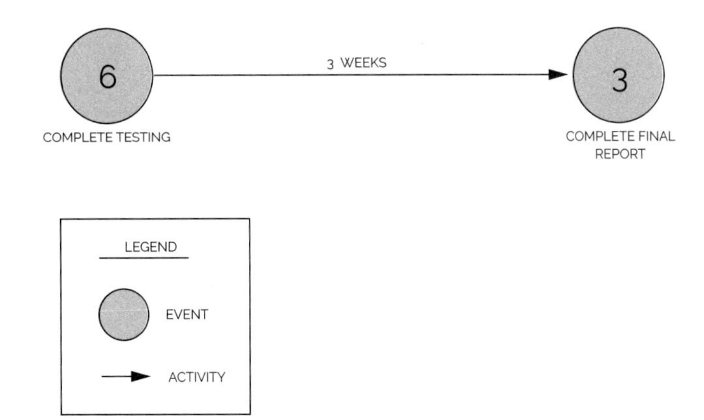

FIGURE 12–1. Standard PERT nomenclature.

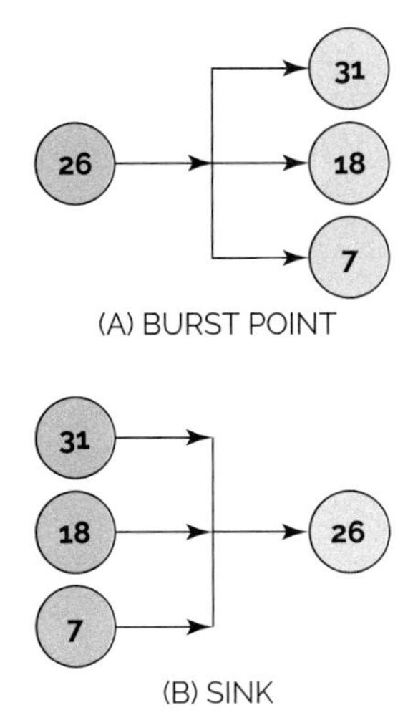

FIGURE 12–2. PERT sources (burst points) and sinks.

have been completely identified, together with their corresponding interrelations.[4] PERT charts are often constructed from back to front because, for many projects, the end date is fixed and the contractor has front-end flexibility.

PMBOK® Guide, 6th Edition
6.3.2 Sequencing Activities Tools and techniques

One of the purposes of constructing the PERT chart is to determine how much time is needed to complete the project. PERT, therefore, uses time as a common denominator to analyze those elements that directly influence the success of the project, namely, time, cost,

4. These events in the PERT charts should be broken down to at least the same reporting levels as defined in the work breakdown structure.

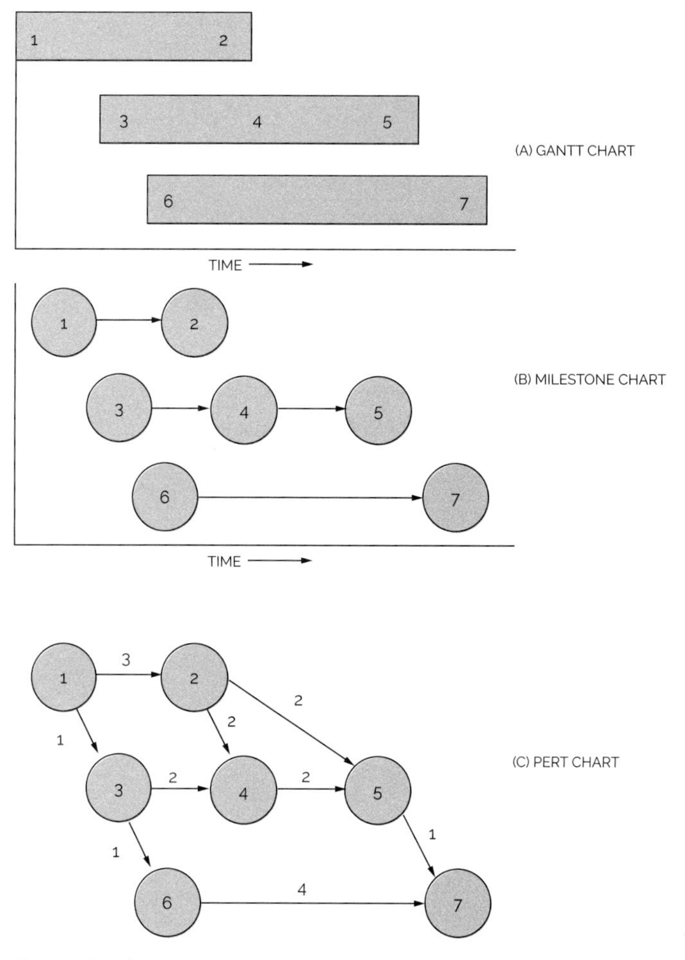

FIGURE 12–3. Conversion from bar chart to PERT chart.

and performance. The construction of the network requires two inputs. First, do events represent the start or the completion of an activity? Event completions are generally preferred. The next step is to define the sequence of events, as shown in Table 12–1, which relates each event to its immediate predecessor. Large projects can easily be converted into PERT networks once the following questions are answered:

- What job immediately precedes this job?
- What job immediately follows this job?
- What jobs can be run concurrently?

Figure 12–4 shows a typical PERT network. The bold line in Figure 12–4 represents the critical path, which is established by the longest time span through the total system of

TABLE 12–1. SEQUENCE OF EVENTS

Activity	Title	Immediate Predecessors	Activity Time, Weeks
1–2	A	—	1
2–3	B	A	5
2–4	C	A	2
3–5	D	B	2
3–7	E	B	2
4–5	F	C	2
4–8	G	C	3
5–6	H	D,F	2
6–7	I	H	3
7–8	J	E,I	3
8–9	K	G,J	2

events. The critical path is composed of events 1–2–3–5–6–7–8–9. The critical path is vital for successful control of the project because it tells management two things:

- Because there is no slack time in any of the events on this path, any slippage will cause a corresponding slippage in the end date of the program unless this slippage can be recovered during any of the downstream events (on the critical path).

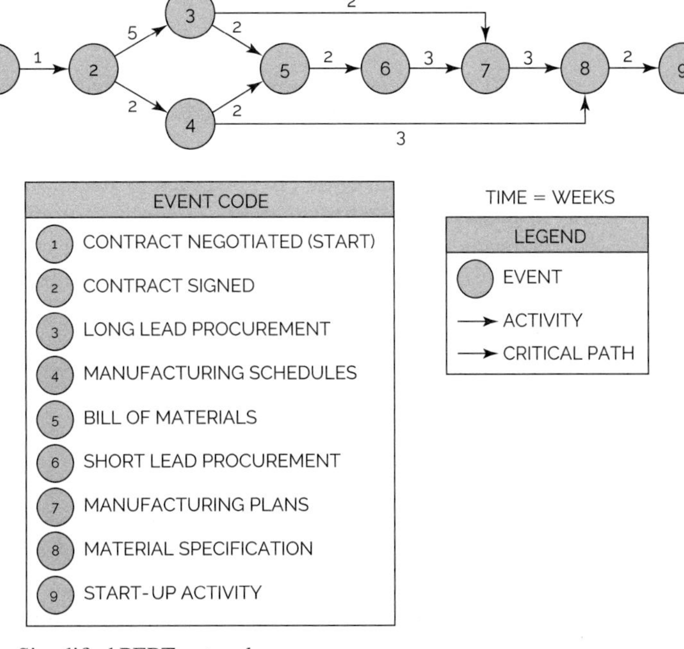

FIGURE 12–4. Simplified PERT network.

- Because the events on this path are the most critical for the success of the project, management must take a hard look at these events in order to improve the total program.

Using PERT we can now identify the earliest possible dates on which we can expect an event to occur, or an activity to start or end. There is nothing overly mysterious about this type of calculation, but without a network analysis the information might be hard to obtain.

PERT charts can be managed from either the events or the activities. For levels 1–3 of the Work Breakdown Structure (WBS), the project manager's prime concerns are the milestones, and therefore, the events are of prime importance. For levels 4–6 of the WBS, the project manager's concerns are the activities.

The principles that we have discussed thus far also apply to CPM. The nomenclature is the same and both techniques are often referred to as arrow diagramming methods, or activity-on-arrow networks. The differences between PERT and CPM are:

PMBOK® Guide, 6th Edition
6.3 Sequencing Activities

- PERT uses three time estimates (optimistic, most likely, and pessimistic as shown in Section 12.7) to derive an expected time. CPM uses one time estimate that represents the normal time (i.e., better estimate accuracy with CPM).
- PERT is probabilistic in nature, based on a beta distribution for each activity time and a normal distribution for expected time duration (see Section 12.7). This allows us to calculate the "risk" in completing a project. CPM is based on a single time estimate and is deterministic in nature.
- Both PERT and CPM permit the use of dummy activities in order to develop the logic.
- PERT is used for R&D projects where the risks in calculating time durations have a high variability. CPM is used for construction projects that are resource dependent and based on accurate time estimates.
- PERT is used on those projects, such as R&D, where percent complete is almost impossible to determine except at completed milestones. CPM is used for those projects, such as construction, where percent complete can be determined with reasonable accuracy and customer billing can be accomplished based on percent complete.

12.2 GRAPHICAL EVALUATION AND REVIEW TECHNIQUE (GERT)

PMBOK® Guide, 6th Edition
6.5.2.1 Schedule Network Analysis

Graphical evaluation and review techniques are similar to PERT but have the distinct advantages of allowing for looping, branching, and multiple project end results. With PERT one cannot easily show that if a test fails, we may have to repeat the test several times. With PERT, we cannot show that, based upon the results of a test, we can select one of several different branches to continue the project. These problems are easily overcome using GERT.

PMBOK® Guide, 6th Edition
6.3 Sequencing Activities

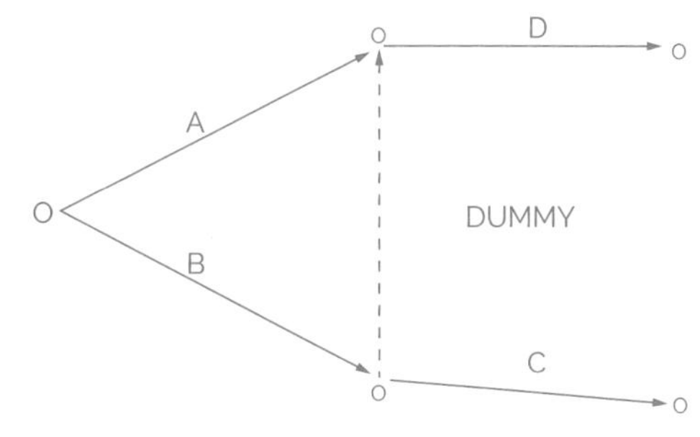

FIGURE 12–5 Dummy activity.

12.3 DEPENDENCIES

PMBOK® Guide, 6th Edition
6.3 Sequencing Activities
6.3.2.2 Dependency Determination and Integration

There are three basic types of interrelationships or dependencies:

- *Mandatory dependencies (i.e., hard logic):* These are dependencies that cannot change, such as erecting the walls of a house before putting up the roof.
- *Discretionary dependencies (i.e., soft logic):* These are dependencies that may be at the discretion of the project manager or may simply change from project to project. As an example, one does not need to complete the entire bill of materials prior to beginning procurement.
- *External dependencies:* These are dependencies that may be beyond the control of the project manager such as having contractors sit on your critical path.

Sometimes, it is impossible to draw network dependencies without including dummy activities. Dummy activities are artificial activities, represented by a dotted line, and do not consume resources or require time. They are added into the network simply to complete the logic.

In Figure 12–5, activity C is preceded by activity B only. Now, let's assume that there exists an activity D that is preceded by both activities A and B. Without drawing a dummy activity (i.e., the dashed line), there is no way to show that activity D is preceded by both activities A and B. Using two dummy activities, one from activity A to activity D and another one from activity B to activity D, could also accomplish this representation. Software programs insert the minimum number of dummy activities, and the direction of the arrowhead is important. In Figure 12–5, the arrowhead must be pointed upward.

12.4 SLACK TIME

PMBOK® Guide, 6th Edition
6.5 Develop Schedule
6.5.2.2 Critical Path Method

Since there exists only one path through the network that is the longest, the other paths must be either equal in length to or shorter than that path.

Therefore, there must exist events and activities that can be completed before the time when they are actually needed. The time differential between the scheduled completion date and the required date to meet critical path is referred to as the slack time.[5] In Figure 12–4, event 4 is not on the critical path. To go from

5. There are special situations where the critical path may include some slack. These cases are not considered here.

event 2 to event 5 on the critical path requires seven weeks taking the route 2–3–5. If route 2–4–5 is taken, only four weeks are required. Therefore, event 4, which requires two weeks for completion, should begin anywhere from zero to three weeks after event 2 is complete. During these three weeks, management might find another use for the resources of people, money, equipment, and facilities required to complete event 4.

The critical path is vital for resource scheduling and allocation because the project manager, with coordination from the functional manager, can reschedule those events not on the critical path for accomplishment during other time periods when maximum utilization of resources can be achieved, provided that the critical path time is not extended. This type of rescheduling through the use of slack times provides for a better balance of resources throughout the company, and may possibly reduce project costs by eliminating idle or waiting time.

Slack can be defined as the difference between the latest allowable date and the earliest expected date based on the nomenclature below:

T_E = the earliest time (date) on which an event can be expected to take place

T_L = the latest date on which an event can take place without extending the completion date of the project

Slack time = $T_L - T_E$

The calculation for slack time is performed for each event in the network, as shown in Figure 12–6, by identifying the earliest expected date and the latest starting date. For event 1, $TL - TE = 0$. Event 1 serves as the reference point for the network and could just as easily have been defined as a calendar date. As before, the critical path is represented as a bold line. The events on the critical path have no slack (i.e., $TL = TE$) and provide the boundaries for the noncritical path events. Since event 2 is critical, $TL = TE = 3 + 7 = 10$ for event =. Event 6 terminates the critical path with a completion time of fifteen weeks.

The earliest time for event 3, which is not on the critical path, would be two weeks ($TE = 0 + 2 = 2$), assuming that it started as early as possible. The latest allowable date is

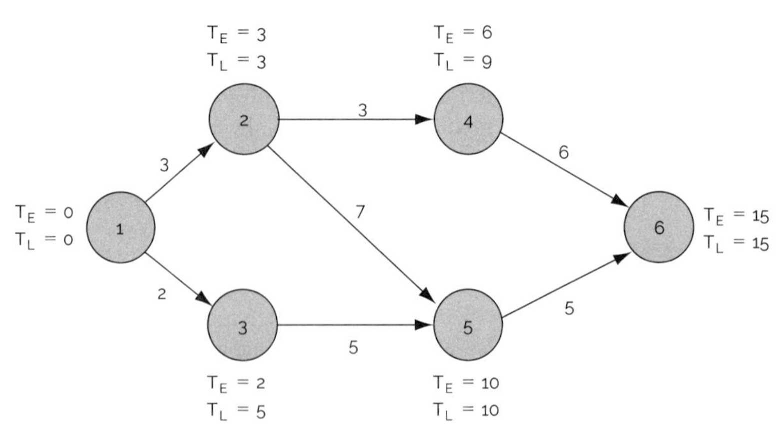

FIGURE 12–6. Network with slack time.

obtained by subtracting the time required to complete the activity from events 3 to 5 from the latest starting date of event 5. Therefore, *TL* (for event 3) = 10 – 5 = 5 weeks. Event 3 can now occur anywhere between weeks 2 and 5 without interfering with the scheduled completion date of the project. This same procedure can be applied to event 4, in which case *TE* = 6 and *TL* = 9.

Figure 12–6 contains a simple PERT network, and therefore the calculation of slack time is not too difficult. For complex networks containing multiple paths, the earliest starting dates must be found by proceeding from start to finish through the network, while the latest allowable starting date must be calculated by working backward from finish to start. The importance of knowing exactly where the slack exists cannot be overstated. Proper use of slack time permits better technical performance.

Because of these slack times, PERT networks are often not plotted with a time scale. Planning requirements, however, can require that PERT charts be reconstructed with time scales, in which case a decision must be made as to whether we wish early or late time requirements for slack variables. This is shown in Figure 12–7 for comparison with total program costs and manpower planning. Early time requirements for slack variables are utilized in this figure.

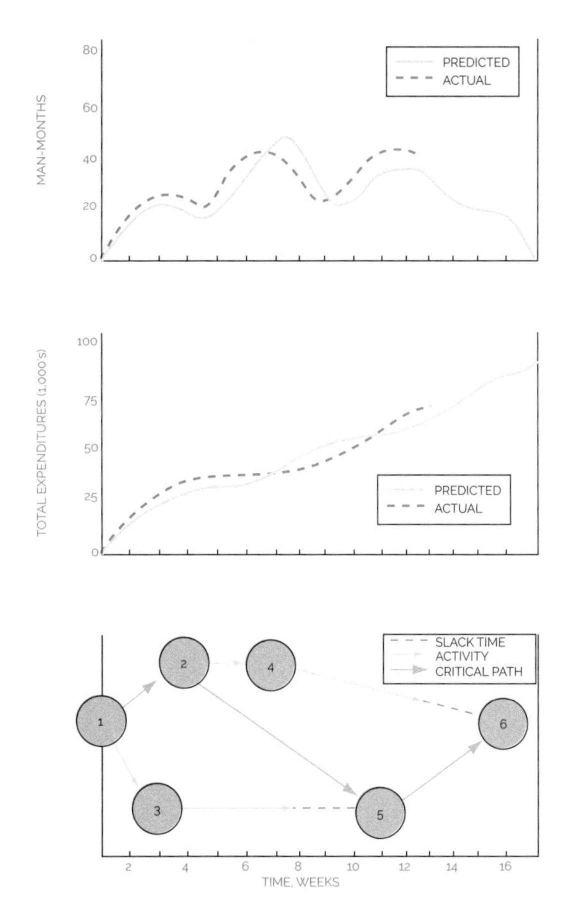

FIGURE 12–7. Comparison models for a time-phase PERT chart.

The earliest times and late times can be combined to determine the probability of successfully meeting the schedule. A sample of the required information is shown in Table 12–2. The earliest and latest times are considered as random variables. The original schedule refers to the schedule for event occurrences that were established at the beginning of the project. The last column in Table 12–2 gives the probability that the earliest time will not be greater than the original schedule time for this event. The exact method for determining this probability, as well as the variances, is described in Section 12.5.

In the example shown in Figure 12–6, the earliest and latest times were calculated for each event. Some people prefer to calculate the earliest and latest times for each activity instead. Also, the earliest and latest times were identified simply as the time or date when

TABLE 12–2. PERT CONTROL OUTPUT INFORMATION

	Earliest Time		Latest Times				
Event Number	Expected	Variance	Expected	Variance	Slack	Original Schedule	Probability of Meeting Schedule

an event can be expected to take place. To make full use of the capabilities of PERT/CPM, we could identify four values:

- The earliest time when an activity can start (ES)
- The earliest time when an activity can finish (EF)
- The latest time when an activity can start (LS)
- The latest time when an activity can finish (LF)

Figure 12–8 shows the earliest and latest times identified on the activity.

To calculate the earliest starting times, we must make a forward pass through the network (i.e., left to right). The earliest starting time of a successor activity is the latest of the earliest finish dates of the predecessors. The earliest finishing time is the total of the earliest starting time and the activity duration.

To calculate the latest times, we must make a *backward* pass through the network by calculating the latest finish time. Since the activity time is known, the latest starting time can be

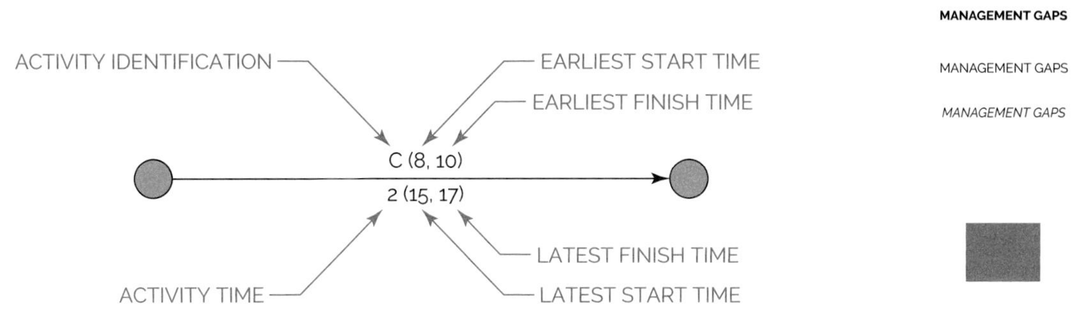

FIGURE 12–8. Slack identification.

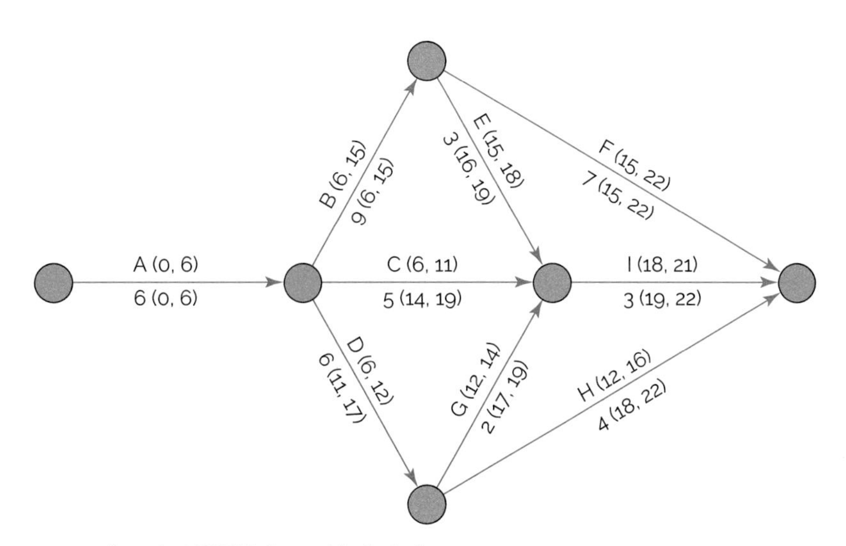

FIGURE 12–9. A typical PERT chart with slack times.

calculated by subtracting the activity time from the latest finishing time. The latest finishing time for an activity entering a node is the earliest starting time of the activities exiting the node. Figure 12–9 shows the earliest and latest starting and finishing times for a typical network.

The identification of slack time can function as an early warning system for the project manager. As an example, if the total slack time available begins to decrease from one reporting period to the next, that could indicate that work is taking longer than anticipated or that more highly skilled labor is needed. A new critical path could be forming.

Looking at the earliest and latest start and finish times can identify slack. In Situation a, the slack is easily identified as four work units, where the work units can be expressed in hours, days, weeks, or even months. In Situation b, the slack is *negative* five units of work. This is referred to as negative slack or negative float.

What can cause the slack to be negative? Look at Figure 12–10. When performing a forward pass through a network, we work from left to right beginning at the customer's starting milestone (position 1). The backward pass, however, begins at the customer's end date milestone (position 2), *not* (as is often taught) where the forward pass ends. If the forward pass ends at position 3, which is before the customer's end date, it is possible to have slack on the critical path. This slack is often called reserve time and may be added to other activities or filled with activities such as report writing so that the forward pass will extend to the customer's completion date.

Negative slack usually occurs when the forward pass extends beyond the customer's end date, as shown by position 4 in the figure. However, the backward pass is still measured from the customer's completion date, thus creating negative slack. This is most likely to result when:

- The original plan was highly optimistic, but unrealistic
- The customer's end date was unrealistic
- One or more activities slipped during project execution
- The assigned resources did not possess the correct skill levels
- The required resources would not be available until a later date

In any event, negative slack is an early warning indicator that corrective action is needed to maintain the customer's end date.

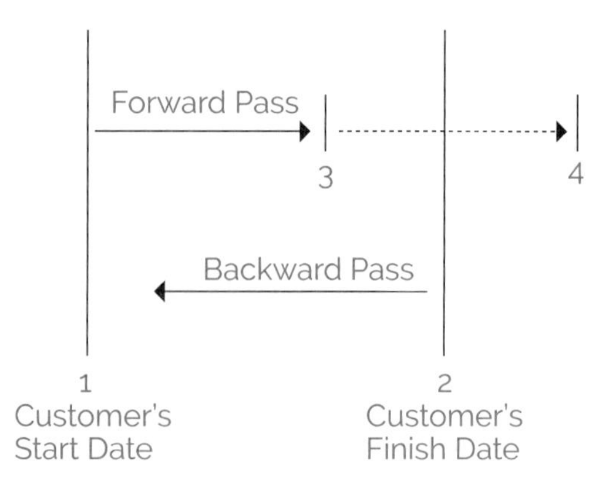

FIGURE 12–10. Slack time.

At this point, it is important to understand the physical meaning of slack. Slack measures how early or how late an event can start or finish. In Figure 12–6, the circles represented events and the slack was measured on the events. Most networks today, however, focus on the activity rather than the event, as shown in Figure 12–9. When slack is calculated on the activity, it is usually referred to as float rather than slack, but most project managers use the terms interchangeably. For activity C in Figure 12–9, the float is eight units. If the float in an activity is zero, then it is a critical path activity, such as seen in activity F. If the slack in an event is zero, then the event is a critical path event.

Another term is maximum float. The equation for maximum float is:

$$\text{Maximum float} = \text{latest finish} - \text{earliest start} - \text{duration}$$

For activity H in Figure 12–9, the maximum float is six units.

12.5 NETWORK REPLANNING

PMBOK® Guide, 6th Edition
6.5 Develop Schedule
6.5.2.5 Schedule Compression

Once constructed, the PERT/CPM charts provide the framework from which detailed planning can be initiated and costs can be controlled and tracked. Many iterations, however, are normally made during the planning phase before the PERT/CPM chart is finished. Figure 12–11 shows this iteration process. The slack times form the basis from which additional iterations, or network replanning, can be performed. Network replanning is performed either at the conception of the program in order to reduce the length of the critical path, or during the program, should the unexpected occur. If all were to go according to schedule, then the original PERT/CPM chart would be unchanged for the duration of the project. But, how many programs or projects follow an exact schedule from start to finish?

Suppose that activities 1–2 and 1–3 in Figure 12–6 require manpower from the same functional unit. Upon inquiry by the project manager, the functional manager asserts that he can reduce activity 1–2 by one week if he shifts resources from activity 1–3 to activity 1–2. Should this happen, however, activity 1–3 will increase in length by one week. Reconstructing the PERT/CPM network as shown in Figure 12–12, the length of the critical path is reduced by one week, and the corresponding slack events are likewise changed.

There are two network replanning techniques based almost entirely upon resources: resource leveling and resource allocation.

PMBOK® Guide, 5th Edition
6.5.2.3 Resource Optimization

- Resource leveling is an attempt to eliminate the manpower peaks and valleys by smoothing out the period-to-period resource requirements. The ideal situation is to do this without changing the end date. However, in reality, the end date moves out and additional costs are incurred.
- Resource allocation (also called resource-limited planning) is an attempt to find the shortest possible critical path based upon the available or fixed resources. The problem with this approach is that the employees may not be qualified technically to perform on more than one activity in a network.

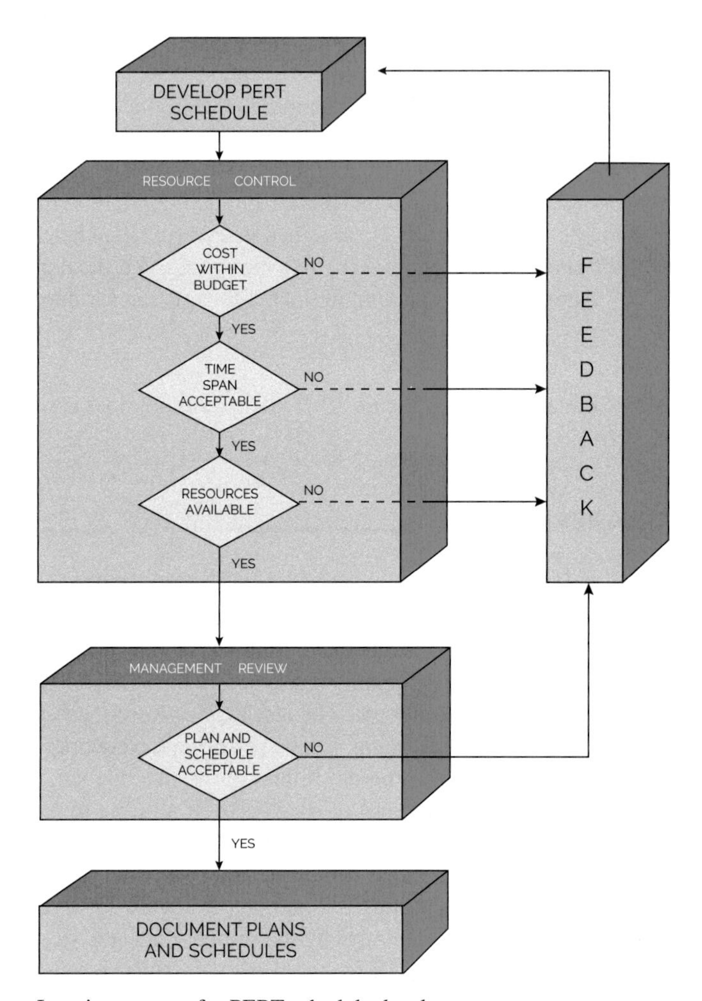

FIGURE 12–11. Iteration process for PERT schedule development.

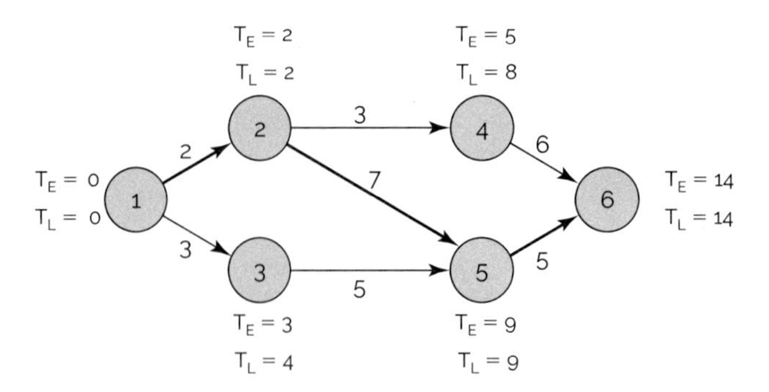

FIGURE 12–12. Network replanning of Figure 12–6.

Resource leveling attempts to minimize the fluctuations in the day-to-day resources throughout the life of the project. Leveling efforts tend to elongate the project and drive up the cost. However, there are some projects, such as in construction, where resource leveling can in fact save money. According to Saleh Mubarak,

> Leveling may also be necessary for an expensive piece of equipment such as a crane (which may cost money not only in rental expenses but also in mobilization, setup, operation, maintenance and demobilization). Say, for example, two activities require a tower crane at the same time. If you can delay the start of one activity (the more critical) until the other (the less critical) has finished, you will re-assign the tower crane, or any other major resource, to the other activity. By doing this, you will have reduced the maximum demand of tower cranes at any time to only one, which will save expenses.[6]
>
> Some resources may be shared among projects. The question is which resources and how much should be shared. For small projects in a relative close vicinity, for example, some staff (project manager, safety manager, quality manager, secretary, etc.) and equipment can be shared. A project manager must make decisions when the situation looks like a borderline case. For instance, whether it would be more efficient to have someone travel between two jobs versus hiring another person, even though that person would not be fully occupied of the time. The same argument holds for equipment. In general, convenience and simple economics are mostly the driving criteria in making such decisions. However, other issues may be considered, such as short- and long-term need; future market expectations; staff morale, fatigue and satisfaction; relationships with stakeholders; the possibly of a need suddenly occurring, and so forth. Keep in mind that transferring resources from one project to another permanently is frequently normally done, and it is not considered sharing.
>
> Staff members who do not have to be present at the job site every day can be spread out, either by dividing their day between two or more jobs or by assigning certain entire days to different jobs. Certain high-paid staff, such as safety officers, schedulers and project control managers, who need to spend only one day every week or every two weeks at the job site, may even fly hundreds of miles between jobs. With the advancement of telecommunications tools (phones, Internet, video-conferencing, etc.) many functions can now be performed from a remote location.[7]

Unfortunately, not all PERT/CPM networks permit such easy rescheduling of resources. Project managers should make every attempt to reallocate resources to reduce the critical path, provided that the slack was not intentionally planned as a safety valve.

Transferring resources from slack paths to more critical paths is only one method for reducing expected project time. Several other methods are available:

- Elimination of some parts of the project
- Addition of more resources (i.e., crashing)
- Substitution of less time-consuming components or activities
- Parallelization of activities
- Shortening critical path activities

---

6. Saleh Mubarak, *Construction Project Scheduling and Control*, 3rd ed. (Hoboken, NJ: John Wiley & Sons, 2015), pp. 9. 125.
7. Mubarak, p. 126–128.

- Shortening early activities
- Shortening longest activities
- Shortening easiest activities
- Shortening activities that are least costly to speed up
- Shortening activities for which you have more resources
- Increasing the number of work hours per day

Under the ideal situation, the project start and end dates are fixed, and performance within this time scale must be completed within the guidelines described by the statement of work. Should the scope of effort have to be reduced in order to meet other requirements, the contractor incurs a serious risk that the project may be canceled, or meeting performance expectations may no longer be possible.

Adding resources is not always possible. If the activities requiring these added resources also call for certain expertise, then the contractor may not have qualified or experienced employees, and may avoid the risk. The contractor might still reject this idea, even if time and money were available for training new employees, because on project termination he might not have any other projects for these additional people. However, if the project is the construction of a new facility, then the labor-union pool may be able to provide additional experienced manpower.

Parallelization of activities can be regarded as accepting a risk by assuming that a certain event can begin in parallel with a second event that would normally be in sequence with it. This is shown in Figure 12–13. One of the biggest headaches at the beginning of

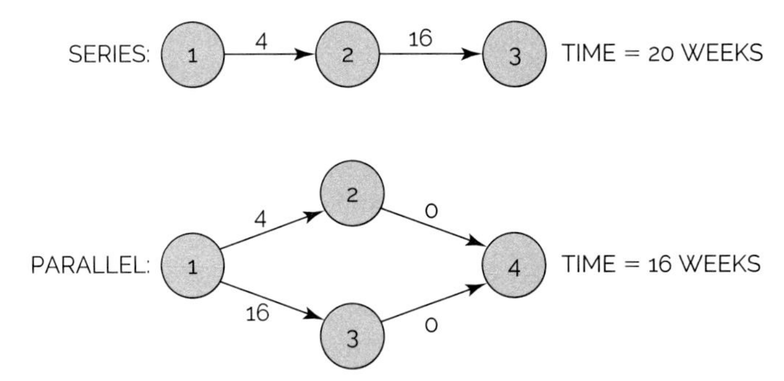

FIGURE 12–13. Parallelization of PERT activities.

any project is the purchasing of tooling and raw materials. As shown in Figure 12–13, four weeks can be saved by sending out purchase orders after contract negotiations are completed, but before the one-month waiting period necessary to sign the contract. Here the contractor incurs a risk. Should the effort be canceled or the statement of work change prior to the signing of the contract, the customer incurs the cost of the termination expenses from the vendors. This risk is normally overcome by the issuance of a long-lead procurement letter immediately following contract negotiations.

There are two other types of risk that are common. In the first situation, engineering has not yet finished the prototype, and manufacturing must order the tooling in order to keep the end date fixed. In this case, engineering may finally design the prototype to fit the tooling. In the second situation, the subcontractor finds it difficult to perform according to the original blueprints. In order to save time, the customer may allow the contractor to work without blueprints, and the blueprints are then changed to represent the as-built end-item.

Because of the complexities of large programs, network replanning becomes an almost impossible task when analyzed on total program activities. It is often better to have each department or division develop its own PERT/CPM networks, on approval by the project office, and based on the work breakdown structure. The individual PERT charts are then integrated into one master chart to identify total program critical paths, as shown in Figure 12–14.

***PMBOK® Guide*, 6th Edition**
1.2.4.1 Project and Development Life Cycles

Segmented PERT charts can also be used when a number of contractors work on the same program. Each contractor (or subcontractor) develops his own PERT chart. It then becomes the responsibility

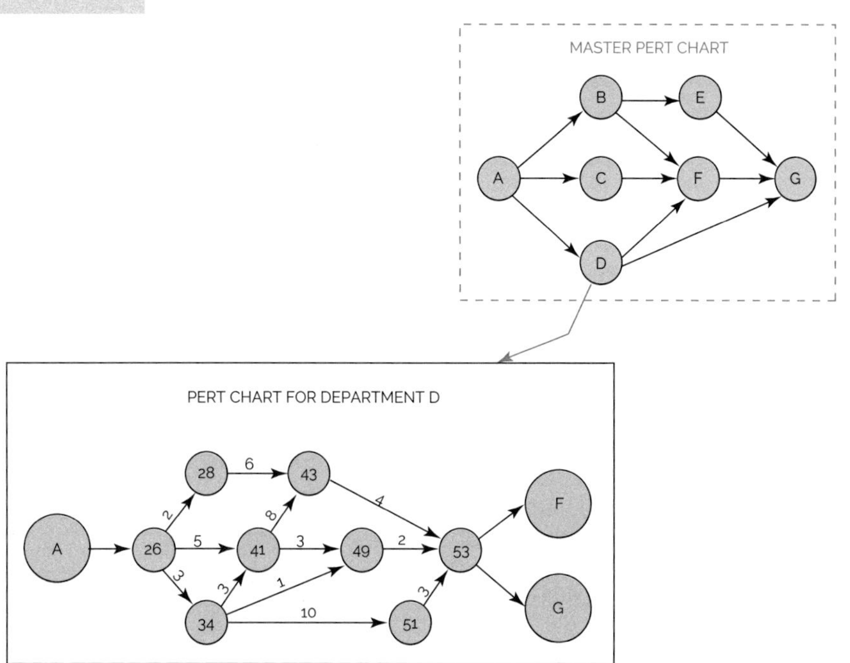

FIGURE 12–14. PERT chart for one department.

of the prime contractor to integrate all of the subcontractors' PERT charts to ensure that total program requirements can be met.

12.6 ESTIMATING ACTIVITY TIME

PMBOK® Guide, 6th Edition
6.4 Estimating Activity Durations

Determining the elapsed time between events requires that responsible functional managers evaluate the situation and submit their best estimates.

The calculations for critical paths and slack times in the previous sections were based on these best estimates.

In this ideal situation, the functional manager would have at his disposal a large volume of historical data from which to make his estimates. Obviously, the more historical data available, the more reliable the estimate. Many programs, however, include events and activities that are nonrepetitive. In this case, the functional managers must submit their estimates using three possible completion assumptions:

PMBOK® Guide, 6th Edition
6.4.2.4 Three-Point Estimates

- *Optimistic completion time*: This time assumes that everything will go according to plan and with minimal difficulties. This should occur approximately 1 percent of the time.
- *Pessimistic completion time*: This time assumes that everything will not go according to plan and maximum difficulties will develop. This should also occur approximately 1 percent of the time.
- *Most likely completion time*: This is the time that, in the mind of the functional manager, would most often occur should this effort be reported over and over again.[8]

Before these three times can be combined into a single expression for expected time, two assumptions must be made. The first assumption is that the standard deviation, s, is one-sixth of the time requirement range. This assumption stems from probability theory, where the end points of a curve are three standard deviations from the mean. The second assumption requires that the probability distribution of time required for an activity be expressible as a beta distribution.

PMBOK® Guide, 6th Edition
6.4.2.4 Three-Point Estimates

The expected time between events can be found from the expression:

$$t_e = \frac{a + 4m + b}{6}$$

Where
t_e = expected time
a = most optimistic time
b = most pessimistic time
m = most likely time.

8. It is assumed that the functional manager performs all of the estimating. The reader should be aware that there are exceptions where the program or project office would do their own estimating.

As an example, if $a = 3$, $b = 7$, and $m = 5$ weeks, then the expected time, t_e, would be 5 weeks. This value for *te* would then be used as the activity time between two events in the construction of a PERT chart. This method for obtaining best estimates contains a large degree of uncertainty. If we change the variable times to $a = 2$, $b = 12$, and $m = 4$ weeks, then t_e will still be 5 weeks. The latter case, however, has a much higher degree of uncertainty because of the wider spread between the optimistic and pessimistic times. Care must be taken in the evaluation of risks in the expected times.

12.7 ESTIMATING TOTAL PROJECT TIME

PMBOK® *Guide*, 6th Edition
6.4 Estimate Activity Durations

In order to calculate the probability of completing the project on time, the standard deviations of each activity must be known. This can be found from the expression:

$$\sigma_{t_e} = \frac{b-a}{6}$$

where σ_{t_e} is the standard deviation of the expected time, t_e. Another useful expression is the variance, υ, which is the square of the standard deviation. The variance is primarily useful for comparison to the expected values. However, the standard deviation can be used just as easily, except that we must identify whether it is a one, two, or three sigma limit deviation. Figure 12–15 shows the critical path of Figure 12–6, together with the corresponding values from which the expected times were calculated, as well as the standard deviations. The total path standard deviation is calculated by the square root of the sum of the squares of the activity standard deviations using the following expression:

$$\begin{aligned}\sigma_{\text{total}} &= \sqrt{\sigma_{1-2}^2 + \sigma_{2-5}^2 + \sigma_{5-6}^2} \\ &= \sqrt{(0.33)^2 + (1.0)^2 + (0.67)^2} \\ &= 1.25\end{aligned}$$

The purpose of calculating s is that it allows us to establish a confidence interval for each activity and the critical path. From statistics, using a normal distribution, we know

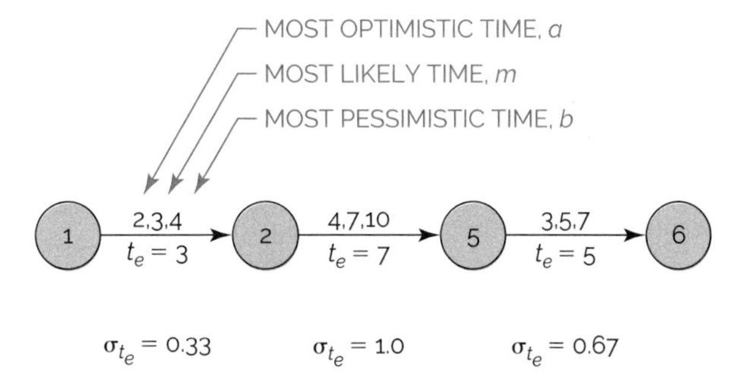

FIGURE 12–15. Expected time analysis for critical path events in Figure 12–6.

that there is a 68 percent chance of completing the project within one standard deviation, a 95 percent chance within two standard deviations, and a 99.73 percent chance within three standard deviations.

This type of analysis can be used to measure the risks in the estimates, the risks in completing each activity, and the risks in completing the entire project. In other words, the standard deviation, σ, serves as a measurement of the risk. This analysis, however, assumes that normal distribution applies, which is not always the case.

As an example of measuring risk, consider a network that has only three activities on the critical path as shown below (all times in weeks):

Activity	Optimistic Time	Most Likely Time	Pessimistic Time	T_{ex}	σ	$σ^2$
A	3	4	5	4	$\frac{2}{6}$	$\frac{4}{36}$
B	4	4.5	8	5	$\frac{4}{6}$	$\frac{16}{36}$
C	4	6	8	6	$\frac{4}{6}$	$\frac{16}{36}$
				15		1.0

From the above table, the length of the critical path is 15 weeks. Since the variance (i.e., $σ^2$) is 1.0, then $σ_{path}$, which is the square root of the variance, must be 1 week.

We can now calculate the probability of completing the project within certain time limits:

- The probability of getting the job done within 16 weeks is 50% + (1/2) × (68%), or 84%.
- Within 17 weeks, we have 50% + (1/2) × (95%), or 97.5%.
- Within 14 weeks, we have 50% – (1/2) × (68%), or 16%.
- Within 13 weeks, we have 50% – (1/2) × (95%), or 2.5%.

12.8 TOTAL PERT/CPM PLANNING

Before we continue, it is necessary to discuss the methodology for preparing PERT schedules. PERT scheduling is a six-step process. Steps one and two begin with the project manager laying out a list of activities to be performed and then placing these activities in order of precedence, thus identifying the interrelationships. These charts drawn by the project manager are called either logic charts, arrow diagrams, work flow, or simply networks. The arrow diagrams will look like Figure 12–6 with two exceptions: The activity time is not identified, and neither is the critical path.

Step three is reviewing the arrow diagrams with the line managers (i.e., the true experts) in order to obtain their assurance that neither too many nor too few activities are identified, and that the interrelationships are correct.

In step four the functional manager converts the arrow diagram to a PERT chart by identifying the time duration for each activity. It should be noted here that the time estimates that the line managers provide are based on the *assumption of unlimited resources* because the calendar dates have not yet been defined.

Step five is the first iteration on the critical path. It is here that the project manager looks at the critical calendar dates in the definition of the project's requirements. If the critical path does not satisfy the calendar requirements, then the project manager must try to shorten the critical path using methods explained in Section 12.3 or by asking the line managers to take the "fat" out of their estimates.

Step six is often the most overlooked step. Here the project manager places calendar dates on each event in the PERT chart, thus converting from planning under unlimited resources to planning with *limited resources.* Even though the line manager has given you a time estimate, there is no guarantee that the correct resources will be available when needed. That is why this step is crucial. If the line manager cannot commit to the calendar dates, then replanning will be necessary. Most companies that survive on competitive bidding lay out proposal schedules based on unlimited resources. After contract award, the schedules are analyzed again because the company now has limited resources. After all, how can a company bid on three contracts simultaneously and put a detailed schedule into each proposal if it is not sure how many contracts, if any, it will win? For this reason customers require that formal project plans and schedules be provided thirty to ninety days after contract award.

Finally, PERT replanning should be an ongoing function during project execution. The best project managers continually try to assess what can go wrong and perform perturbation analysis on the schedule. (This should be obvious because the constraints and objectives of the project can change during execution.) Primary objectives on a schedule are:

- Best time
- Least cost
- Least risk

Secondary objectives include:

- Studying alternatives
- Optimum schedules
- Effective use of resources
- Communications
- Refinement of the estimating process
- Ease of project control
- Ease of time or cost revisions

Obviously, these objectives are limited by such constraints as:

- Calendar completion
- Cash or cash flow restrictions
- Limited resources
- Management approvals

12.9 CRASH TIMES

PMBOK® Guide, 6th Edition
6.5.2.6 Schedule Compression

In the preceding sections, no distinction was made between PERT and CPM. The basic difference between PERT and CPM lies in the ability to calculate percent complete. PERT is used in R&D or just development activities, where a percent-complete determination is almost impossible. Therefore, PERT is event oriented rather than activity oriented. In PERT, funding is normally provided for each milestone (i.e., event) achieved because incremental funding along the activity line has to be based on percent complete. CPM, on the other hand, is activity oriented because, in activities such as construction, percent complete along the activity line can be determined. CPM can be used as an arrow diagram network without PERT. The difference between the two methods lies in the environments in which they evolved and how they are applied.

The CPM (activity-type network) has been widely used in the process industries, in construction, and in single-project industrial activities. Common problems include no place to store early arrivals of raw materials and project delays for late arrivals.

Using strictly the CPM approach, project managers can consider the cost of speeding up, or crashing, certain phases of a project. In order to accomplish this, it is necessary to calculate a crashing cost per unit time as well as the normal expected time for each activity. CPM charts, which are closely related to PERT charts, allow visual representation of the effects of crashing. There are these requirements:

- For a CPM chart, the emphasis is on activities, not events. Therefore, the PERT chart should be redrawn with each circle representing an activity rather than an event.
- In CPM, both time and cost of each activity are considered.[9]
- Only those activities on the critical path are considered, starting with the activities for which the crashing cost per unit time is the lowest.

Figure 12–16 shows a CPM network with the corresponding crash time for all activities on and off the critical path. The activities are represented by circles and include an activity identification number and the estimated time. The costs expressed in the figure are usually direct costs only.

To determine crashing costs we begin with the lowest weekly crashing cost, activity A, at $2,000 per week. Although activity C has a lower crashing cost, it is not on the critical path. Only critical path activities are considered for crashing. Activity A will be the first to be crashed for a maximum of two weeks at $2,000 per week. The next activity to be considered would be F at $3,000 per week for a maximum of three weeks. These crashing costs are additional expenses above the normal estimates.

A word of caution concerning the selection and order of the activities that are to crash: There is a good possibility that as each activity is crashed, a new critical path will be developed. This new path may or may not include those elements that were bypassed because they were not on the original critical path.

Returning to Figure 12–16 (and assuming that no new critical paths are developed), activities A, F, E, and B would be crashed in that order. The crashing cost would then be an

9. Although PERT considers mainly time, modifications through PERT/cost analysis can be made to consider the cost factors.

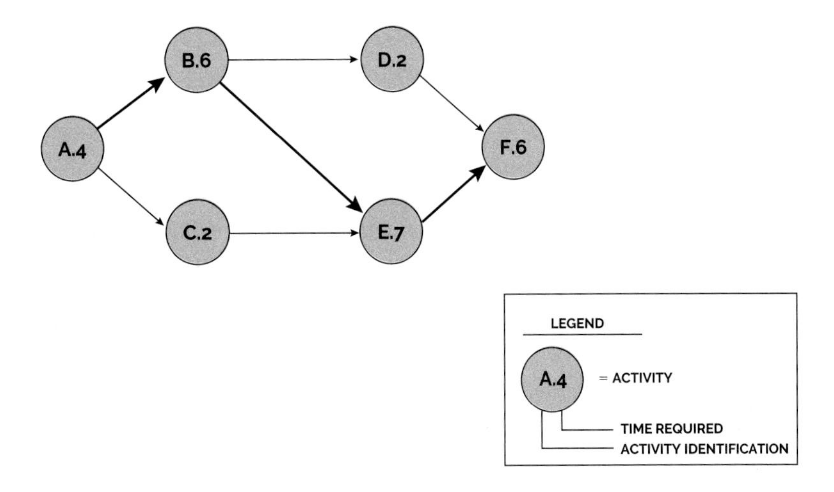

ACTIVITY	TIME REQUIRED, WEEKS		COST $		CRASHING COST PER WEEK, $
	NORMAL	CRASH	NORMAL	CRASH	
A	4	2	10,000	14,000	2,000
B	6	5	30,000	42,500	12,500
C	2	1	8,000	9,500	1,500
D	2	1	12,000	18,000	6,000
E	7	5	40,000	52,000	6,000
F	6	3	20,000	29,000	3,000

FIGURE 12–16. CPM network.

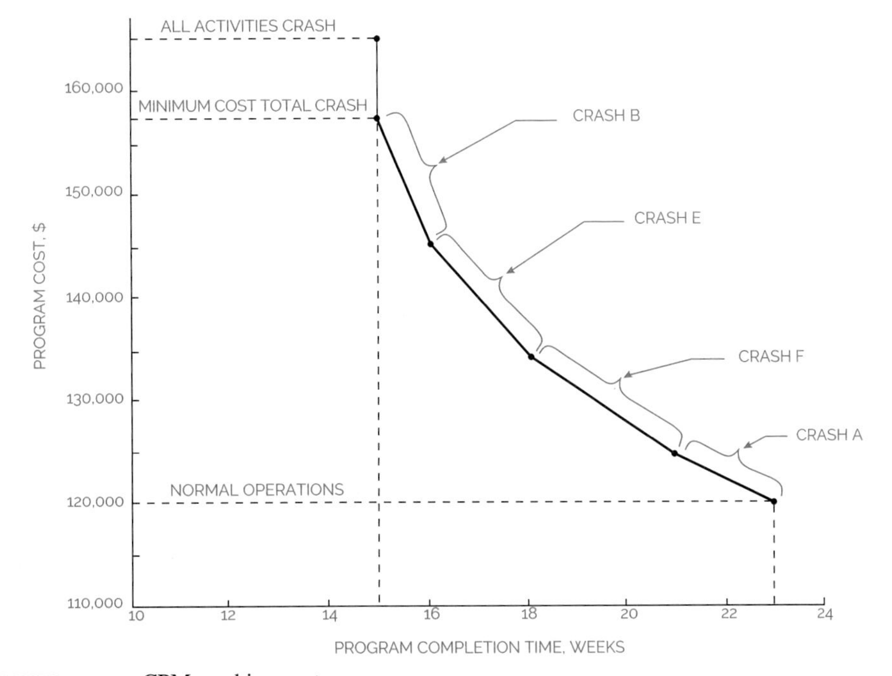

FIGURE 12–17. CPM crashing costs.

increase of $37,500 from the base of $120,000 to $157,500. The corresponding time would then be reduced from twenty-three weeks to fifteen weeks. This is shown in Figure 12–17 to illustrate how a trade-off between time and cost can be obtained. Also shown in Figure 12–17 is the increased cost of crashing elements not on the critical path. Crashing these elements would result in a cost increase of $7,500 without reducing the total project time. There is also the possibility that this figure will represent unrealistic conditions because sufficient resources are not or cannot be made available for the crashing period.

The purpose behind balancing time and cost is to avoid wasting resources. If the direct and indirect costs can be accurately obtained, then a region of feasible budgets can be found, bounded by the early start (crash) and late-start (or normal) activities. This is shown in Figure 12–18.

Since the direct and indirect costs are not necessarily expressible as linear functions, time–cost trade-off relationships are made by searching for the lowest possible total cost (i.e., direct and indirect) that likewise satisfies the region of feasible budgets. This method is shown in Figure 12–19.

Like PERT, CPM also contains the concept of slack time, the maximum amount of time that a job may be delayed beyond its early start without delaying the project completion time. Figure 12–20 shows a typical representation of slack time using a CPM chart. In addition, the figure shows how target activity costs can be identified. Figure 12–20 can be modified to include normal and crash times as well as normal and crash costs. In this case, the cost box in the figure would contain two numbers: The first number would be the normal cost, and the second would be the crash cost. These numbers might also appear as running totals.

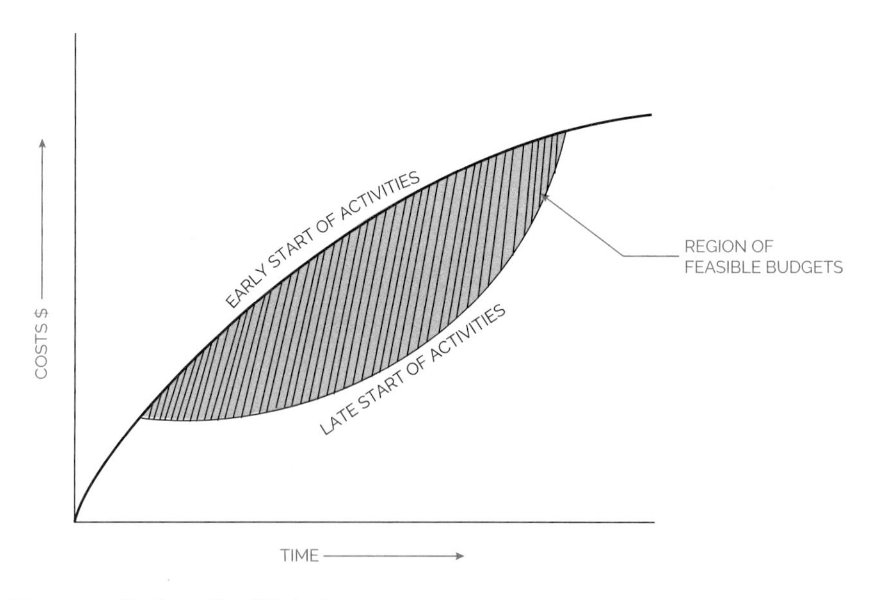

FIGURE 12–18. Region of feasible budgets.

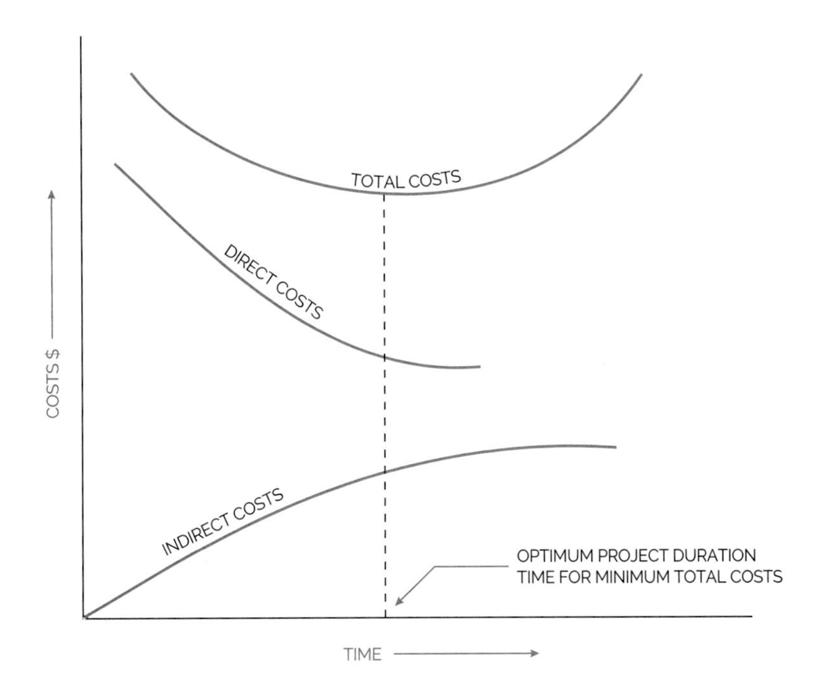

FIGURE 12–19. Determining project duration.

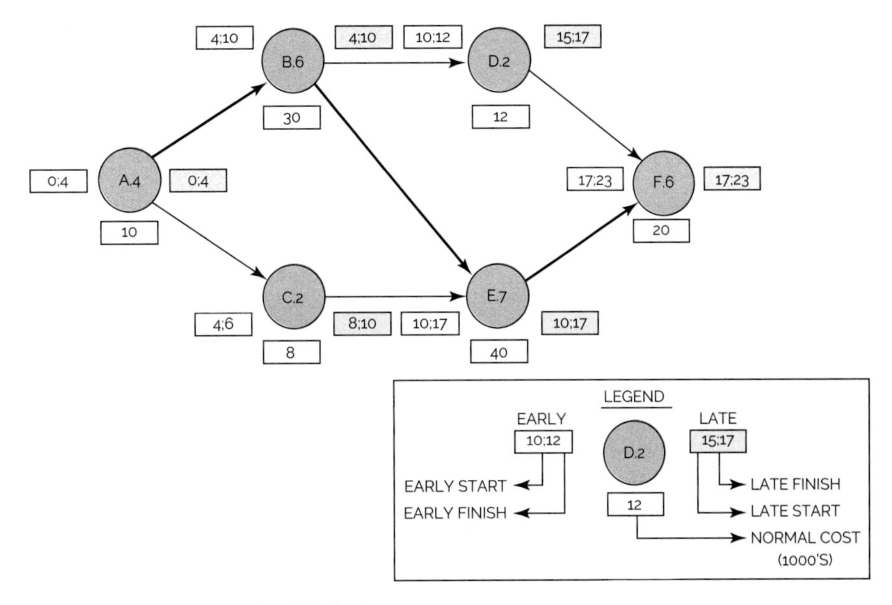

FIGURE 12–20. CPM network with slack.

12.10 PERT/CPM PROBLEM AREAS

PERT/CPM models are not without their disadvantages and problems. Even the largest organizations with years of experience in using PERT and CPM have the same ongoing problems as newer or smaller companies.

Many companies have a difficult time incorporating PERT systems because PERT is end-item oriented. Many upper-level managers feel that the adoption of PERT/CPM removes a good part of their power and ability to make decisions. This is particularly evident in companies that have been forced to accept PERT/CPM as part of contractual requirements.

In PERT systems, there are planners and doers. In most organizations PERT planning is performed by the program office and functional management. Yet once the network is constructed, the planners and managers become observers and rely on the doers to accomplish the job within time and cost limitations. Management must convince the doers that they have an obligation to the successful completion of the established PERT/CPM plans.

PERT networks are based on the assumption that all activities start as soon as possible. This assumes that qualified personnel and equipment are available. Regardless of how well we plan, there are almost always differences in performance times from what would normally be acceptable. For the selected model, time and cost should be well-considered estimates, not spur-of-the-moment decisions.

Cost control problems arise when the project cost and control system is not compatible with company policies. Project-oriented costs may be meshed with non-PERT-controlled jobs in order to develop the annual budget. This becomes a difficult chore for cost reporting, especially when each project may have its own method for analyzing and controlling costs.

12.11 ALTERNATIVE PERT/CPM MODELS

Because of the many advantages of PERT/time, numerous industries have found applications for this form of network. A partial list of these advantages includes capabilities for:

- Trade-off studies for resource control
- Providing contingency planning in the early stages of the project
- Visually tracking up-to-date performance
- Demonstrating integrated planning
- Providing visibility down through the lowest levels of the work breakdown structure
- Providing a regimented structure for control purposes to ensure compliance with the work breakdown structure and the statement of work
- Increasing functional members' ability to relate to the total program, thus providing participants with a sense of belonging

Even with these advantages, in many situations PERT/time has proved ineffective in controlling resources within the parameters of time, cost, and performance. With these factors in mind, companies began reconstructing PERT/time into PERT/cost and PERT/performance models.

PERT/cost is an extension of PERT/time and attempts to overcome the problems associated with the use of the most optimistic and most pessimistic time for estimating completion. PERT/cost can be regarded as a cost accounting network model based on the work breakdown structure and capable of being subdivided down to the lowest elements, or work packages. The advantages of PERT/cost are that it:

- Contains all the features of PERT/time
- Permits cost control at any WBS level

The primary reason for the development of PERT/cost was so that project managers could identify critical schedule slippages and cost overruns in time to correct them.

12.12 PRECEDENCE NETWORKS

PMBOK® Guide, 6th Edition
6.3.2.1 Precedence Diagramming Method

Project management can provide answers to such questions as:

- How will the project be affected by limited resources?
- How will the project be affected by a change in the requirements?
- What is the cash flow for the project (and for each WBS element)?
- What is the impact of overtime?
- What additional resources are needed to meet the constraints of the project?
- How will a change in the priority of a certain WBS element affect the total project?

The more sophisticated packages can provide answers to schedule and cost based on:

- Adverse weather conditions
- Weekend activities
- Unleveled manpower requirements
- Variable crew size
- Splitting of activities
- Assignment of unused resources

PMBOK® Guide, 6th Edition
6.3.2.1 Precedence Diagramming Method

Regardless of the sophistication of computer systems, printers and plotters prefer to draw straight lines rather than circles. Most software systems today use precedence networks, as shown in Figure 12–21, which attempt to show interrelationships on bar charts. In Figure 12–21, task 1 and task 2 are related because of the solid line between them. Task 3 and task 4 can begin when task 2 is half finished. (This cannot be shown easily on PERT without splitting activities.) The dotted lines indicate slack. The critical path can be identified by putting an asterisk (*) beside the critical elements, or by putting the critical connections in a different color or boldface.

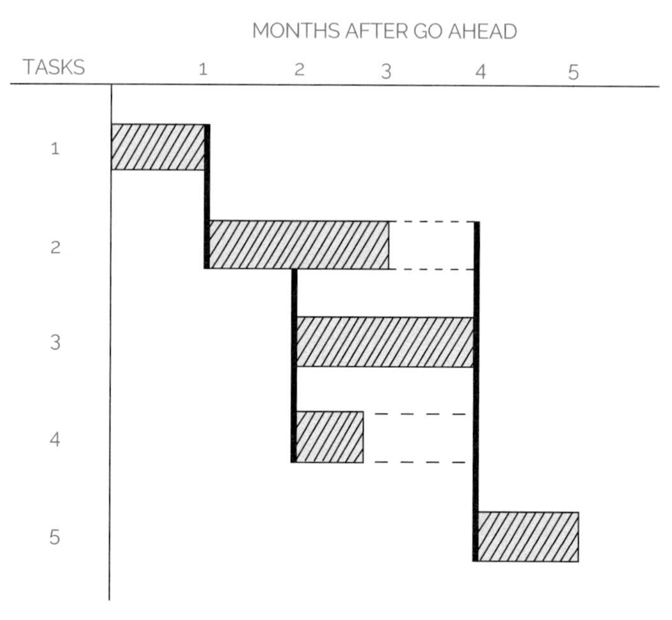

FIGURE 12–21. Precedence network.

The more sophisticated software packages display precedence networks in the format shown in. In each of these figures, work is accomplished during the activity. This is sometimes referred to as the activity-on-node method. The arrow represents the relationship or constraint between activities.

Figure 12–22A illustrates a finish-to-start constraint. In this figure, activity 2 can start no earlier than the completion of activity 1. All PERT charts are finish-to-start constraints. Figure 12–23B illustrates a start-to-start constraint. Activity 2 cannot start prior to the start of activity 1. Figure 12–22C illustrates a finish-to-finish constraint. In this figure, activity 2 cannot finish until activity 1 finishes. Figure 12–22D illustrates a start-to-finish constraint.

An example might be that you must start studying for an exam some time prior to the completion of the exam. This is the least common type of precedence chart. Figure 12–22E illustrates a percent complete constraint. In this figure, the last 20 percent of activity 2 cannot be started until 50 percent of activity 1 has been completed.[10]

PMBOK® *Guide*, 6th Edition
6.2.3.2 Activity Attributes

Figure 12–23 shows the typical information that appears in each of the activity boxes shown in Figure 12–22. The box identified as "responsibility cost center" could also have been identified as the name, initials, or badge number of the person responsible for this activity.

10. Meredith and Mantel categorize precedence relationships in three broad categories: Natural Precedences, Environmental Precedences, and Preferential Precedences. For additional information on these precedence relationships, see Jack R. Meredith and Samuel J. Mantel, Jr., *Project Management*, 3rd ed. (New York: John Wiley & Sons, 1995), pp.385–386.

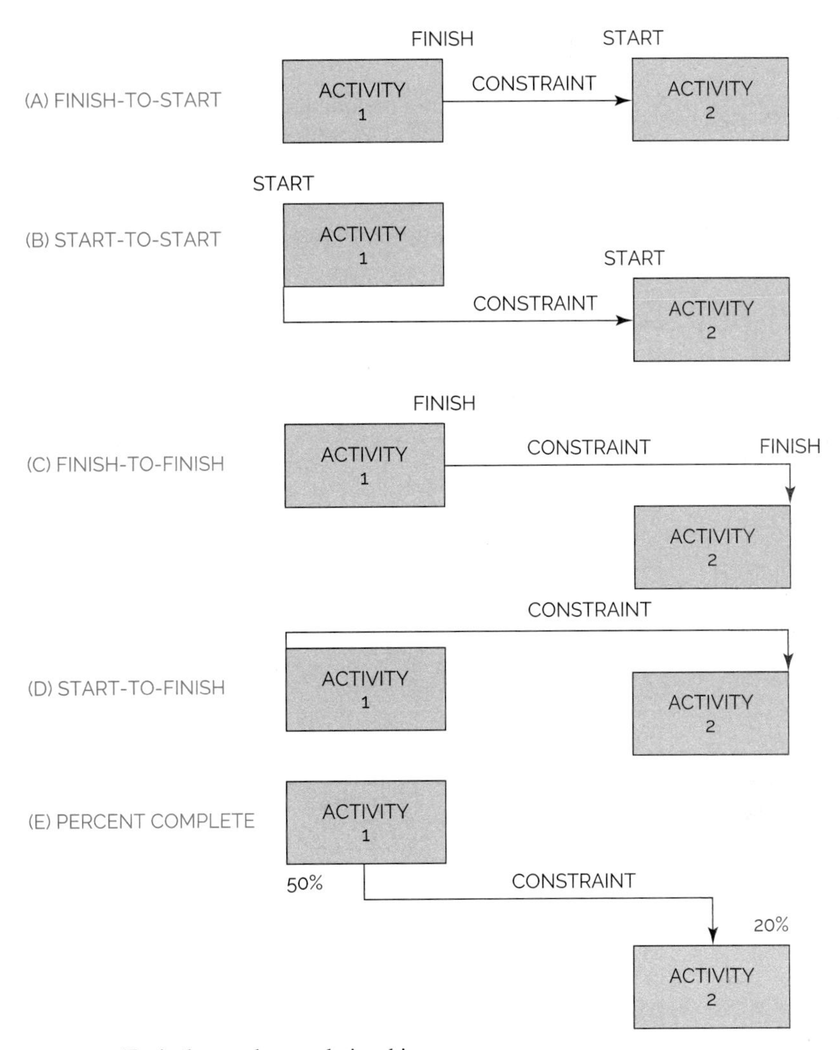

FIGURE 12–22. Typical precedence relationships.

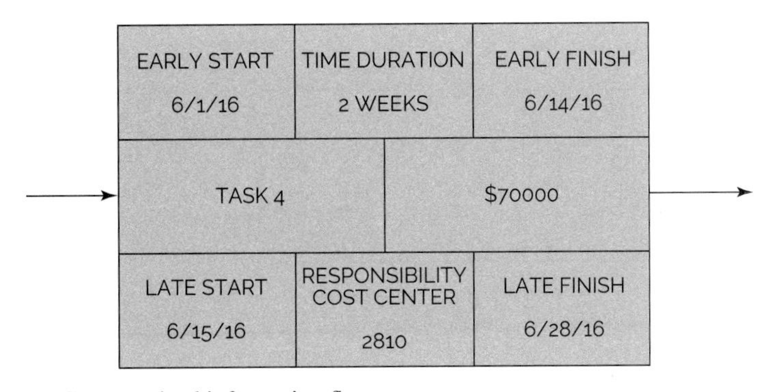

FIGURE 12–23. Computerized information flow.

12.13 LAG

PMBOK® Guide, 6th Edition
6.3.2.3 Leads and Lags
6.3.2.1 Precedence Diagramming Method

The time period between the early start or finish of one activity and the early start or finish of another activity in the sequential chain is called lag. Lag is most commonly used in conjunction with precedence networks. Figure 12–24 shows five different ways to identify lag on the constraints.

(A) FINISH-TO-START (FS) RELATIONSHIP. THE START OF B MUST LAG 6 DAYS AFTER THE FINISH OF A.

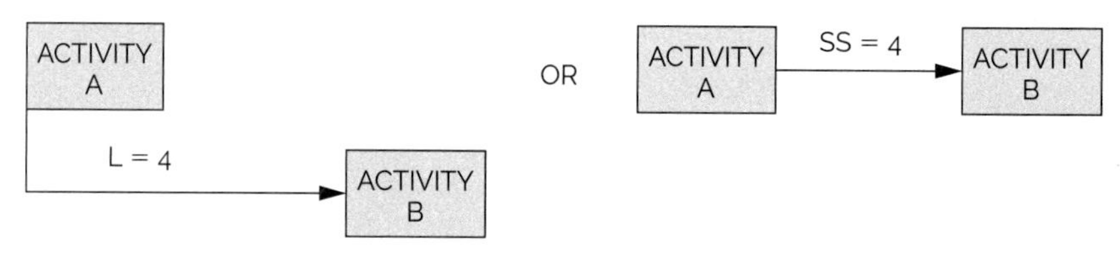

(B) START-TO-START (SS) RELATIONSHIP. THE START OF B MUST LAG 4 DAYS AFTER THE START OF A.

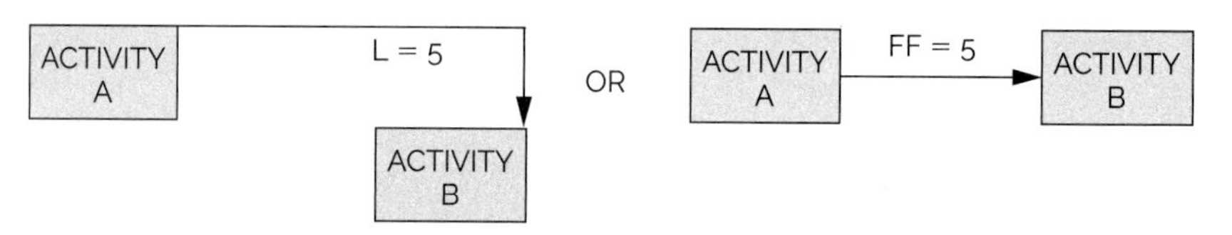

(C) FINISH-TO-FINISH (FF) RELATIONSHIP. THE FINISH OF B MUST LAG 5 DAYS AFTER THE FINISH OF A.

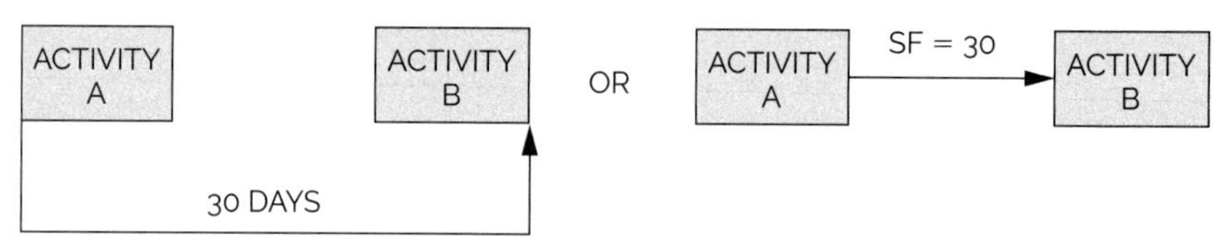

(D) START-TO-FINISH (SF) RELATIONSHIP. THE FINISH OF B MUST LAG 30 DAYS AFTER THE START OF A.

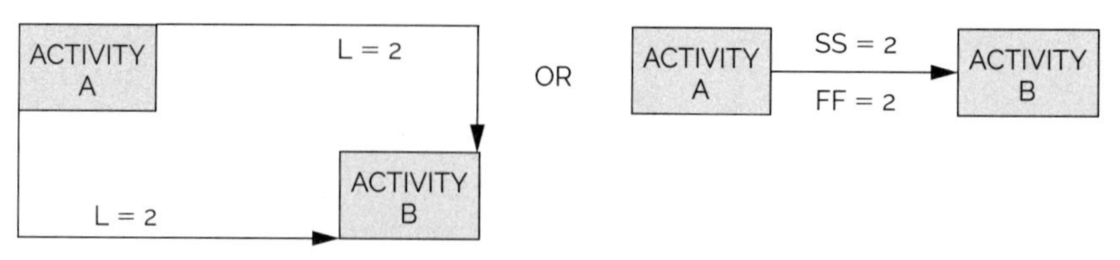

(E) COMPOSITE START-TO-START AND FINISH-TO-FINISH RELATIONSHIP. THE START OF B MUST LAG 2 DAYS AFTER THE START OF A, AND THE FINISH OF B MUST LAG 2 DAYS AFTER THE FINISH OF A.

FIGURE 12–24. Precedence charts with lag.

Slack is measured within activities, whereas lag is measured between activities. As an example, look at Figure 12–24A. Suppose that activity A ends at the end of the first week of March. Since it is a finish-to-start precedence chart, one would expect the start of activity B to be the beginning of the second week in March. But if activity B cannot start until the beginning of the third week of March, that would indicate a week of lag between activity A and activity B even though both activities can have slack within the activity. The lag may be the result of resource constraints.

Another common term is lead. Again looking at Figure 12–24A, suppose that activity A actually finishes on March 15 but the precedence chart shows activity B is scheduled to start on March 8, seven days prior to the completion of activity A. In this case, *L* 527, a negative value, indicating that the start of activity B leads the completion of activity A by seven days. To illustrate how this can happen, consider the following example: The line manager responsible for activity B promised you that his resources would be available on March 16, the day after activity A was scheduled to end. The line manager then informs you that these resources will be available on March 8, and if you do not pick them up on your charge number at that time, they may be assigned elsewhere and not be available on the 16th. Most project managers would take the resources on the 8th and find some work for them to do even though logic says that the work cannot begin until after activity A has finished.

12.14 SCHEDULING PROBLEMS

Every scheduling technique has advantages and disadvantages. Some scheduling problems are the result of organizational indecisiveness, such as having a project sponsor that refuses to provide the project manager guidance on whether the schedule should be based upon a least time, least cost, or least risk scheduling objective. As a result, precious time is wasted in having to redo the schedules.

However, there are some scheduling problems that can impact all scheduling techniques. These include:

- Using unrealistic estimates for effort and duration
- Inability to handle employee workload imbalances
- Having to share critical resources across several projects
- Overcommitted resources
- Continuous readjustments to the WBS primarily from scope changes
- Unforeseen bottlenecks

12.15 THE MYTHS OF SCHEDULE COMPRESSION

Simply because schedule compression techniques may exist does not mean that they will work. There is a tendency for managers to be aggressively positive in their thinking at the onset of a project, believing that compression techniques can be applied effectively.

TABLE 12–3. MYTHS AND REALITIES OF SCHEDULE COMPRESSION

Compression Technique	Myth	Reality
Use of overtime	Work will progress at the same rate on overtime.	The rate of progress is less on overtime; more mistakes may occur; and prolonged overtime may lead to burnout.
Adding more resources (i.e., crashing)	The performance rate will increase due to the added resources.	It takes time to find the resources; it takes time to get them up to speed; the resources used for the training must come from the existing resources.
Reducing scope (i.e., needed, reducing functionality)	The customer always requests more work than actually needed.	The customer needs all of the tasks agreed to in the statement of work.
Outsourcing	Numerous qualified suppliers exist.	The quality of the suppliers' work can damage your reputation; the supplier may go out of business; and the supplier may have limited concern for your scheduled dates.
Doing series work in parallel	An activity can start before the previous activity has finished.	The risks increase and rework becomes expensive because it may involve multiple activities.

There are five common techniques for schedule compression, and each technique has significant limitations that may make this technique more of a myth than reality. This is shown in Table 12–3.

12.16 PROJECT MANAGEMENT SOFTWARE

PMBOK® Guide, **6th Edition**
6.3.2.4 Project Management Information System

While it is clear that even the most sophisticated software package is not a substitute for competent project leadership—and by itself does not identify or correct any task-related problems—it can be a terrific aid to the project manager in tracking the many interrelated variables and tasks that come into play with a project. Specific examples of these capabilities are:

- Project data summary: expenditure, timing, and activity
- Project management and business graphics capabilities
- Data management and reporting capabilities
- Critical path analysis
- Customized and standard reporting formats
- Multiproject tracking
- Subnetworking
- Impact analysis (what if . . .)
- Early warning systems
- On-line analysis of recovering alternatives
- Graphical presentation of cost, time, and activity data
- Resource planning and analysis

- Cost analysis, variance analysis
- Multiple calendars
- Resource leveling

Features

Project management software capabilities and features vary a great deal. However, the variation is more in the depth and sophistication of the features, such as storage, display, analysis, interoperability, and user friendliness, rather than in the type of features offered, which are very similar for most software programs. Most packages offer the following features:

1. *Planning, tracking, and monitoring.* These features provide for planning and tracking the projects' tasks, resources, and costs. The data format for describing the project to the computer is usually based on standard network typologies such as the CPM, PERT, or Precedence Diagram Method (PDM). Task elements, with their estimated start and finish times, their assigned resources, and actual cost data, can be entered and updated as the project progresses. The software provides an analysis of the data and documents the technical and financial status of the project against its schedule and original plan. Usually, the software also provides impact assessments of plan deviations and resource and schedule projections. Many systems also provide resource leveling, a feature that averages out available resources to determine task duration and generates a leveled schedule for comparison.
2. *Reports.* Project reporting is usually achieved via a menu-driven report writer system that allows the user to request several standard reports in a standard format. The user can also modify these reports or create new ones. Reporting capabilities include:

PMBOK® *Guide*, 6th Edition
7.4.2 Cost Control Tools and Techniques

- Budgeted cost for work scheduled (BCWS) or planned value of work (PV)
- Budgeted cost for work performed (BCWP) or earned value of work (EV)
- Actual versus planned expenditure
- Earned value analysis
- Cost and schedule performance indices
- Cash-flow
- Critical path analysis
- Change order
- Standard government reports (DoD, DoE, NASA), formatted for the performance monitoring system (PMS)

In addition, many software packages feature a user-oriented, free-format report writer for styled project reporting.

3. *Project calendar.* This feature allows the user to establish work weeks based on actual workdays. Hence, the user can specify nonwork periods such as weekends, holidays, and vacations. It is available in detail or in a summary format and is automatically the basis for all resource scheduling.

4. *What-if analysis.* Some software is designed to make what-if analyses easy. A separate, duplicate project database is established and the desired changes are entered. A comparative analysis is performed and displays the new against the old project plan in tabular or graphical form for fast and easy management review and analysis.
5. *Multiproject analysis.* Some of the more sophisticated packages feature a single, comprehensive database that facilitates cross-project analysis and reporting. Cost and schedule modules share common files that allow integration among projects and minimize problems of data inconsistencies and redundancies.

Reporting

PMBOK® Guide, 5th Edition
Chapter 6 Project Schedule Management
Chapter 10 Project Communications Management
6.6.2.1 Control Schedule Tools and Techniques

In Chapter 11, we defined the steps involved in establishing a formal program plan with detailed schedules to manage the total program. Any plan, schedule, drawing, or specification that will be read by more than one person must be expressed in a language understood by all recipients.

The ideal situation is to construct charts and schedules in suitable Bar Charts notation that can be used for both in-house control and out-of-house customer status reporting. Unfortunately, this is easier said than done.

All schedules and charts should consider time, cost, and performance and their relationship to corporate resources.

Information to ensure proper project evaluation is usually obtained through four methods:

- Firsthand observation
- Oral and written reports
- Review and technical interchange meetings
- Graphical displays

Firsthand observations are an excellent tool for obtaining unfiltered information, but they may not be possible on large projects. Although oral and written reports are a way of life, they often contain either too much or not enough detail, and significant information may be disguised. Review and technical interchange meetings provide face-to-face communications and can result in immediate agreement on problem definitions or solutions, such as changing a schedule. The difficulty is in the selection of attendees from the customer's and the contractor's organizations. Good graphical displays make the information easy to identify and are the prime means for tracking cost, schedule, and performance. Proper graphical displays can result in:

- Cutting project costs and reducing the time scale
- Coordinating and expediting planning
- Eliminating idle time
- Obtaining better scheduling and control of subcontractor activities
- Developing better troubleshooting procedures
- Cutting time for routine decisions, but allowing more time for decision making

Related Case Studies (from Kerzner/*Project Management Case Studies,* 5th ed.)	**Related Workbook Exercises (from Kerzner/*Project Management Workbook and PMP®/CAPM® Exam Study Guide,* 12th ed.)**	***PMBOK® Guide*, Sixth Edition, Reference Section for the PMP® Certification Exam**
• The Invisible Sponsor*	• Crashing the Effort • Multiple Choice Exam • Crossword Puzzle on Time • (Schedule) Management	• Time Management

*Case Study also appears at end of chapter.

12.17 STUDYING TIPS FOR THE PMI® PROJECT MANAGEMENT CERTIFICATION EXAM

This section is applicable as a review of the principles to support the knowledge areas and domain groups in the *PMBOK® Guide*. This chapter addresses:

- Project Schedule Management
- Planning
- Controlling

Understanding the following principles is beneficial if the reader is using this text to study for the PMP® Certification Exam:

- How to identify the three types of scheduling techniques and their respective advantages and disadvantages
- Difference between activity-on-arrow and activity-on-node networks
- Four types of precedence networks
- Basic network terminology such as activities, events, critical path, and slack (float)
- Difference between positive and negative slack
- Schedule compression techniques and crashing and fast-tracking (concurrent engineering)
- Importance of the work breakdown structure in network development
- The steps, and their order, for the development of a network
- Three types of dependencies
- How to perform a forward and backward pass
- Resources leveling
- Resource-limited planning
- Difference between effort and duration
- Which network technique uses optimistic, most likely, and pessimistic estimates
- Use of dummy activities
- Lag
- Difference between unlimited versus limited resource planning/scheduling

---

PMP and CAPM are registered marks of the Project Management Institute, Inc.

The following multiple-choice questions will be helpful in reviewing the principles of this chapter:

1. The shortest time necessary to complete all of the activities in a network is called the:
   A. Activity duration length
   B. Critical path
   C. Maximum slack path
   D. Compression path
2. Which of the following *cannot* be identified after performing a forward and backward pass?
   A. Dummy activities
   B. Slack time
   C. Critical path activities
   D. How much overtime is scheduled
3. Which of the following is *not* a commonly used technique for schedule compression?
   A. Resource reduction
   B. Reducing scope
   C. Fast-tracking activities
   D. Use of overtime
4. A network-based schedule has four paths, namely 7, 8, 9, and 10 weeks. If the 10-week path is compressed to 8 weeks, then:
   A. We now have two critical paths.
   B. The 9-week path is now the critical path.
   C. Only the 7-week path has slack.
   D. Not enough information is provided to make a determination.
5. The major disadvantage of using bar charts to manage a project is that bar charts:
   A. Do not show dependencies between activities
   B. Are ineffective for projects under one year in length
   C. Are ineffective for projects under $1 million in size
   D. Do not identify start and end dates of a schedule
6. The first step in the development of a schedule is a:
   A. Listing of the activities
   B. Determination of dependencies
   C. Calculation of effort
   D. Calculation of durations
7. Reducing the peaks and valleys in manpower assignments in order to obtain a relatively smooth manpower curve is called:
   A. Manpower allocation
   B. Manpower leveling
   C. Resource allocation
   D. Resource commitment planning

8. Activities with no time duration are called:
   A. Reserve activities
   B. Dummy activities
   C. Zero slack activities
   D. Supervision activities
9. Optimistic, pessimistic, and most likely activity times are associated with:
   A. PERT
   B. GERT
   C. PDM
   D. ADM
10. The most common "constraint" or relationship in a precedence network is:
    A. Start-to-start
    B. Start-to-finish
    C. Finish-to-start
    D. Finish-to-finish
11. A network-based technique that allows for branching and looping is:
    A. PERT
    B. GERT
    C. PDM
    D. ADM
12. If an activity on the critical path takes longer than anticipated, then:
    A. Activities not on the critical path have additional slack.
    B. Activities not on the critical path have less slack.
    C. Additional critical path activities will appear.
    D. None of the above
13. Which of the following is not one of the three types of dependencies?
    A. Mandatory
    B. Discretionary
    C. Internal
    D. External
14. You have an activity where the early start is week 6, the early finish is week 10, the latest start is week 14, and the latest finish is week 18. The slack in this activity is:
    A. 4 weeks
    B. 6 weeks
    C. 8 weeks
    D. 18 weeks

ANSWERS

1. B	2. D	3. A
4. D	5. A	6. A
7. B	8. B	9. A
10. C	11. B	12. A
13. C	14. C	

PROBLEMS

12–1 Should a PERT/CPM network become a means of understanding reports and schedules, or should it be vice versa?

12–2 Should PERT networks follow the work breakdown structure?

12–3 Should key milestones be established at points where trade-offs are most likely to occur?

12–4 Would you agree or disagree that the cost of accelerating a project rises exponentially, especially as the project nears completion?

12–5 What are the major difficulties with PERT, and how can they be overcome?

12–6 Draw the network and identify the critical path. Also calculate the earliest–latest starting and finishing times for each activity:

Activity	Preceding Activity	Time (Weeks)
A	—	4
B	—	6
C	A, B	7
D	B	8
E	B	5
F	C	5
G	D	7
H	D, E	8
I	F, G, H	4

12–7 Identify the critical path for the following network for a small MIS project (all times are in days; network proceeds from node 1 to node 10):

Network			
Job Activity	Initial Node	Final Node	Estimated Time
A	1	2	2
B	1	3	3
C	1	4	3

	Network		
Job Activity	Initial Node	Final Node	Estimated Time
D	2	5	3
E	2	9	3
F	3	5	1
G	3	6	2
H	3	7	3
I	4	7	5
J	4	8	3
K	5	6	3
L	6	9	4
M	7	9	4
N	8	9	3
O	9	10	2

12–8 Can PERT charts have more depth than the WBS?

12–9 For the network shown in Figure P12–9 with all times indicating weeks, answer the following questions:

a. What is the impact on the end date of the project if activity B slips by two weeks?
b. What is the impact on the end date of the project if activity E slips by one week?
c. What is the impact on the end date of the project if activity D slips by two weeks?
d. If the customer offered you a bonus for completing the project in sixteen weeks or less, which activities would you focus on first as part of compression ("crashing") analyses?

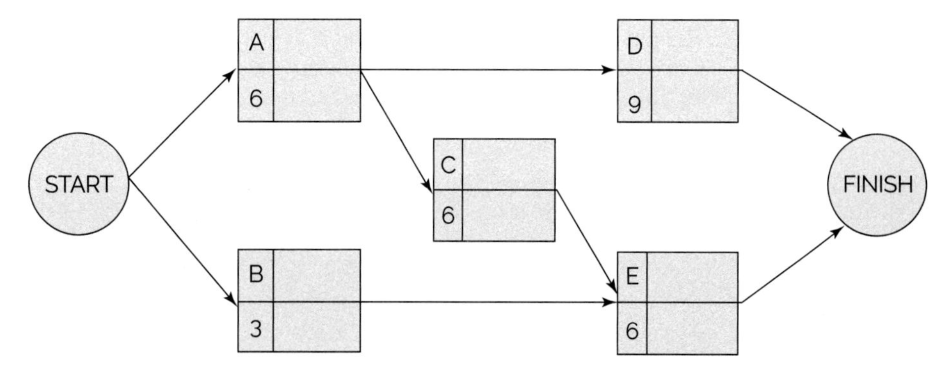

FIGURE P12–9.

12–10 A project manager discovers that his team has neglected to complete the network diagram for the project. The network diagram is shown in Figure P12–10. However, the project manager has some information available, specifically that each activity, labeled A–G, has a different duration between one and seven weeks. Also, the slack time for each of the activities is known as shown below:

Duration (weeks): 1, 2, 3, 4, 5, 6, 7
Slack time (weeks): 0, 0, 0, 3, 6, 8, 8

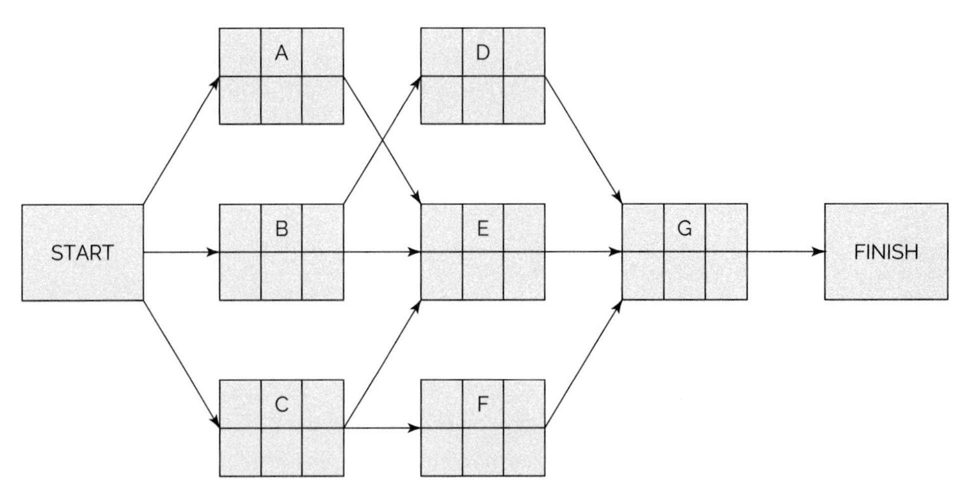

FIGURE P12–10.

Using the clues provided below, determine the duration of each activity as well as the early start, early finish, latest start, and latest finish times for each activity.

Clues

1. There exists only one critical path, and it is the largest possible number given the possible durations shown.
2. Activity E has the smallest amount of slack that is greater than zero.
3. The early finish (EF) time for activity A is four weeks, and this does not equal the latest finish (LF) time. (Note: There is no negative slack in the network.)
4. The slack in activity C is eight weeks.
5. The duration of activity F is greater than the duration of activity C by at least two weeks.

Activity	Duration	Early Start	Early Finish	Latest Start	Latest Finish
A	______	______	______	______	______
B	______	______	______	______	______
C	______	______	______	______	______
D	______	______	______	______	______
E	______	______	______	______	______
F	______	______	______	______	______
G	______	______	______	______	______

CASE STUDY

THE INVISIBLE SPONSOR

Background

Some executives prefer to micromanage projects, whereas other executives are fearful of making a decision because, if they were to make the wrong decision, it could impact their career. In this case study, the president of the company assigned one of the vice presidents to act as the project sponsor on a project designed to build tooling for a client. The sponsor, however, was reluctant to make any decisions.

Assigning the VP

Moreland Company was well-respected as a tooling design-and-build company. Moreland was project-driven because all of its income came from projects. Moreland was also reasonably mature in project management. When the previous VP for engineering retired, Moreland hired an executive from a manufacturing company to replace him. The new VP for engineering, Al Zink, had excellent engineering knowledge about tooling but had worked for companies that were not project-driven. Al had very little knowledge about project management and had never functioned as a project sponsor. Because of Al's lack of experience as a sponsor, the president decided that Al should "get his feet wet" as quickly as possible and assigned him as the project sponsor on a medium-sized project. The project manager on this project was Fred Cutler. Fred was an engineer with more than twenty years of experience in tooling design and manufacturing. Fred reported directly to Al Zink administratively.

Fred's Dilemma: Fred understood the situation; he would have to train Al Zink on how to function as a project sponsor. This was a new experience for Fred because subordinates usually do not train senior personnel on how to do their job. Would Al Zink be receptive? Fred explained the role of the sponsor and how there are certain project documents that require the signatures of both the project manager and the project sponsor. Everything seemed to be going well until Fred informed Al that the project sponsor is the person that the president eventually holds accountable for the success or failure of the project. Fred could tell that Al was quite upset over this statement.

Al realized that the failure of a project where he was the sponsor could damage his reputation and career. Al was now uncomfortable about having to act as a sponsor but knew that he might eventually be assigned as a sponsor on other projects. Al also knew that this project was somewhat of a high risk. If Al could function as an invisible sponsor, he could avoid making any critical decisions.

In the first meeting between Fred and Al where Al was the sponsor, Al asked Fred for a copy of the schedule for the project. Fred responded:

> I'm working on the schedule right now. I cannot finish the schedule until you tell me whether you want me to lay out the schedule based upon best time, least cost, or least risk.

Al stated that he would think about it and get back to Fred as soon as possible.

During the middle of the next week, Fred and Al met in the company's cafeteria. Al asked Fred again, "How is the schedule coming along?" and Fred responded as before:

> I cannot finish the schedule until you tell me whether you want me to lay out the schedule based upon best time, least cost, or least risk.

Al was furious, turned around, and walked away from Fred. Fred was now getting nervous about how upset Al was and began worrying if Al might remove him as the project manager. But Fred decided to hold his ground and get Al to make a decision.

At the weekly sponsor meeting between Fred and Al, once again Al asked the same question, and once again Fred gave the same response as before. Al now became quite angry and yelled out:

> Just give me a least time schedule.

Fred had gotten Al to make his first decision. Fred finalized his schedule and had it on Al's desk two days later awaiting Al's signature. Once again, Al procrastinated and refused to sign off on the schedule. Al believed that, if he delayed making the decision, Fred would take the initiative and begin working on the schedule without Al's signature.

Fred kept sending emails to Al asking when he intended to sign off on the schedule or, if something was not correct, what changes needed to be made. As expected, Al did not respond. Fred then decided that he had to pressure Al one way or another into making timely decisions as the project sponsor. Fred then sent an email to Al that stated:

> I sent you the project schedule last week. If the schedule is not signed by this Friday, there could be an impact on the end date of the project. If I do not hear from you, one way or another, by this Friday, I will assume you approve the schedule and I can begin implementation.

The president's email address was also included in the CC location on the email. The next morning, Fred found the schedule on his desk, signed by Al Zink.

QUESTIONS

1. Why do some executives refuse to function as project sponsors?
2. Can an executive be "forced" to function as a sponsor?
3. Is it right for the sponsor to be the ultimate person responsible for the success or failure of the project?
4. Were Al Zink's actions that of someone trying to be an invisible sponsor?
5. Did Fred Cutler act appropriately in trying to get Al Zink to act as a sponsor?
6. What is your best guess as to what happened to the working relationship between Al Zink and Fred Cutler?

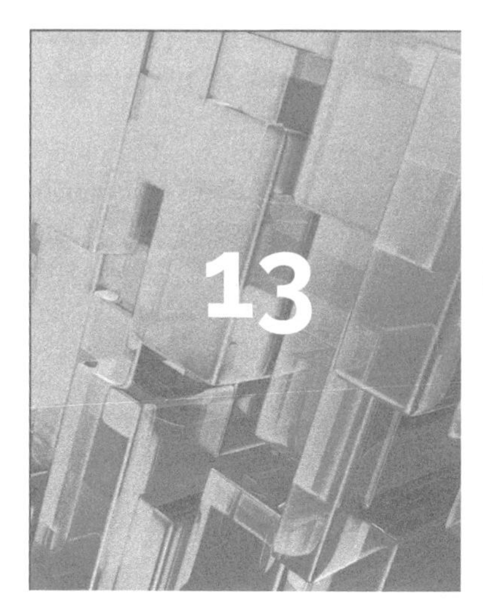

13 Pricing and Estimating

13.0 INTRODUCTION

PMBOK® Guide, 6th Edition
6.4.2.5 Bottom-Up Estimating
6.4 Estimate Activity Durations

With the complexities involved, it is not surprising that many business managers consider pricing an art. Having information on customer cost budgets and competitive pricing would certainly help.

However, the reality is that whatever information is available to one bidder is generally available to the others. A disciplined approach helps in developing all the input for a rational pricing recommendation. A side benefit of using a disciplined management process is that it leads to the documentation of the many factors and assumptions involved at a later time. These can be compared and analyzed, contributing to the learning experiences that make up the managerial skills needed for effective business decisions. Estimates are *not* blind luck. They are well-thought-out decisions based on either the best available information, some type of cost estimating relationship, or some type of cost model. Cost estimating relationships (CERs) are generally the output of cost models. Typical CERs might be:

- Mathematical equations based on regression analysis
- Cost–quantity relationships such as learning curves
- Cost–cost relationships
- Cost–noncost relationships based on physical characteristics, technical parameters, or performance characteristics

13.1 GLOBAL PRICING STRATEGIES

Specific pricing strategies must be developed for each individual situation. Frequently, however, one of two situations prevails when one is pursuing project acquisitions competitively. First, the new business opportunity may be a one-of-a-kind project with little or no

PMBOK is a registered mark of the Project Management Institute, Inc.

follow-on potential, a situation classified as type I acquisition. Second, the new business opportunity may be an entry point to a larger follow-on or repeat business, or may represent a planned penetration into a new market. This acquisition is classified as type II.

Clearly, in each case, we have specific but different business objectives. The objective for type I acquisition is to win the project and execute it profitably and satisfactorily according to contractual agreements. The type II objective is often to win the project and perform well, thereby gaining a foothold in a new market segment or a new customer community in place of making a profit. Accordingly, each acquisition type has its own, unique pricing strategy, as summarized in Table 13–1.

Comparing the two pricing strategies for the two global situations (as shown in Table 13–1) reveals a great deal of similarity for the first five points. The fundamental difference is that for a profitable new business acquisition the bid price is determined according to actual cost, whereas in a "must-win" situation the price is determined by the market forces. It should be emphasized that one of the most crucial inputs in the pricing decision is the cost estimate of the proposed baseline. The design of this baseline to the minimum requirements should be started early, in accordance with well-defined ground rules, cost models, and established cost targets. Too often the baseline design is performed in parallel

TABLE 13–1. TWO GLOBAL PRICING STRATEGIES

Type I Acquisition: One-of-a-Kind Project with Little or No Follow-On Business	**Type II Acquisition: New Project with Potential for Large Follow-On Business or Representing a Desired Penetration into New Markets**
1. Develop cost model and estimating guidelines; design proposed project baseline for minimum cost, to minimum customer requirements.	1. Design proposed project baseline compliant with customer requirements, with innovative features but minimum risks.
2. Estimate cost realistically for minimum requirements.	2. Estimate cost realistically.
3. Scrub the baseline. Squeeze out unnecessary costs.	3. Scrub baseline. Squeeze out unnecessary costs.
4. Determine realistic minimum cost. Obtain commitment from performing organizations.	4. Determine realistic minimum cost. Obtain commitment from performing organizations.
5. Adjust cost estimate for risks.	5. Determine "should-cost" including risk adjustments.
6. Add desired margins. Determine the price.	6. Compare your final cost estimate to customer budget and the "most likely" winning price.
7. Compare price to customer budget and competitive cost information.	7. Determine the gross profit margin necessary for your winning proposal. This margin could be negative!
8. Bid only if price is within competitive range.	8. Decide whether the gross margin is acceptable according to the must-win desire.
	9. Depending on the strength of your desire to win, bid the "most likely" winning price or lower.
	10. If the bid price is below cost, it is often necessary to provide a detailed explanation to the customer of where the additional funding is coming from. The source could be company profits or sharing of related activities. In any case, a clear resource picture should be given to the customer to ensure cost credibility.

with the proposal development. At the proposal stage it is too late to review and fine-tune the baseline for minimum cost. Also, such a late start does not allow much of an option for a final bid decision. Even if the price appears outside the competitive range, it makes little sense to terminate the proposal development. As all the resources have been sent anyway, one might just as well submit a bid in spite of the remote chance of winning.

Clearly, effective pricing begins a long time before proposal development. It starts with preliminary customer requirements, well-understood subtasks, and a top-down estimate with should-cost targets. This allows the functional organization to design a baseline to meet the customer requirements and cost targets, and gives management the time to review and redirect the design before the proposal is submitted. Furthermore, it gives management an early opportunity to assess the chances of winning during the acquisition cycle, at a point when additional resources can be allocated or the acquisition effort can be terminated before too many resources are committed to a hopeless effort.

The final pricing review session should be an integration and review of information already well known in its basic context. The process and management tools outlined here should help to provide the framework and discipline for deriving pricing decisions in an orderly and effective way.

13.2 TYPES OF ESTIMATES

PMBOK® Guide, 6th Edition

6.4 Estimate Activity Durations
7.2.2 Estimate Costs Tools and Techniques
7.2.2.1 Expert Judgment
7.2.2.2 Analogous Estimating
7.2.2.3 Parametric Estimating
7.2.2.4 Bottom-Up Estimating
7.2.3.2 Basis for estimates

Any company or corporation that wants to remain profitable must continuously improve its estimating and pricing methodologies. While it is true that some companies have been successful without good cost estimating pricing, very few remain successful without them.

Good estimating requires that information be collected prior to the initiation of the estimating process. Typical information includes:

- Recent experience in similar work
- Professional and reference material
- Market and industry surveys
- Knowledge of the operations and processes
- Estimating software and databases if available
- Interviews with subject matter experts

Projects can range from a feasibility study, through modification of existing facilities, to complete design, procurement, and construction of a large complex. Whatever the project may be, whether large or small, the estimate and type of information desired may differ radically. To save time, companies usually start with a parametric estimate, an analogy estimate, or expert judgment. A parametric estimate is based upon statistical data. An analogy estimate compares the features of a completed project to the features of the new project and adjusts the costs based upon the change in the degree of difficulty. Expert judgment comes from individuals or groups with specialized knowledge because of previous work on these types of projects.

The first type of estimate is an *order-of-magnitude* analysis, which is made without any detailed engineering data. The order-of-magnitude analysis may have an accuracy of ±35 percent within the scope of the project. This type of estimate may use past experience (not necessarily similar), scale factors, parametric curves, or capacity estimates (i.e., $/# of product or $/kW electricity).

Order-of-magnitude estimates are top-down estimates usually applied to level 1 of the WBS, and in some industries, use of parametric estimates is included. A parametric estimate is based upon statistical data. For example, assume that you live in a Chicago suburb and wish to build the home of your dreams. You contact a construction contractor who informs you that the parametric or statistical cost for a home in this suburb is $120 per square foot. In Los Angeles, the cost may be $350 per square foot.

Next, there is the *approximate estimate* (or top-down estimate), which is also made without detailed engineering data, and may be accurate to ±15 percent. This type of estimate is prorated from previous projects that are similar in scope and capacity, and may be titled as estimating by analogy, parametric curves, rule of thumb, and indexed cost of similar activities adjusted for capacity and technology. In such a case, the estimator may say that this activity is 50 percent more difficult than a previous (i.e., reference) activity and requires 50 percent more time, man-hours, dollars, materials, and so on.

The *definitive estimate,* or grassroots buildup estimate, or bottom-up estimate is prepared from well-defined engineering data including (as a minimum) vendor quotes, fairly complete plans, specifications, unit prices, and estimate to complete. The definitive estimate, also referred to as detailed estimating, has an accuracy of ±5 percent.

Another method for estimating is the use of *learning curves.* Learning curves are graphical representations of repetitive functions in which continuous operations will lead to a reduction in time, resources, and money. The theory behind learning curves is usually applied to manufacturing operations.

Each company may have a unique approach to estimating. However, for normal project management practices, Table 13–2 would suffice as a starting point.

Many companies try to standardize their estimating procedures by developing an *estimating manual.* The estimating manual is then used to price out the effort, perhaps as much as 90 percent. Estimating manuals usually give better estimates than industrial engineering standards because they include groups of tasks and take into consideration such items as downtime, cleanup time, lunch, and breaks. Table 13–3 shows the table of contents for a construction estimating manual.

TABLE 13–2. STANDARD PROJECT ESTIMATING

Estimating Method	Generic Type	WBS Relationship	Accuracy	Time to Prepare
Parametric	ROM*	Top down	225% to 175%	Days
Analogy	Budget	Top down	210% to 125%	Weeks
Engineering (grass roots)	Definitive	Bottom up	25% to 110%	Months

*ROM 5 Rough order of magnitude.

TABLE 13–3. ESTIMATING MANUAL TABLE OF CONTENTS

Introduction
Purpose and types of estimates
Major Estimating Tools
Cataloged equipment costs
Automated investment data system
Automated estimate system
Computerized methods and procedures
Classes of Estimates
Definitive estimate
Capital cost estimate
Appropriation estimate
Feasibility estimate
Order of magnitude
Charts—estimate specifications quantity and pricing guidelines
Data Required
Chart—comparing data required for preparation of classes of estimates
Presentation Specifications
Estimate procedure—general
Estimate procedure for definitive estimate
Estimate procedure for capital cost estimate
Estimate procedure for appropriation estimate
Estimate procedure for feasibility estimate

Estimating manuals, as the name implies, provide estimates. The question, of course, is "How good are the estimates?" Most estimating manuals provide accuracy limitations by defining the type of estimates. Using Table 13–3, we can create Tables 13–4, 13–5, and 13–6, which illustrate the use of the estimating manual.

Not all companies can use estimating manuals. Estimating manuals work best for repetitive tasks or similar tasks that can use a previous estimate adjusted by a degree-of-difficulty factor. Activities such as R&D do not lend themselves to the use of estimating manuals other than for benchmark, repetitive laboratory tests. Proposal managers must carefully consider whether the estimating manual is a viable approach. The literature abounds with examples of companies that have spent millions trying to develop estimating manuals for situations that just do not lend themselves to the approach.

TABLE 13–4. CLASSES OF ESTIMATES

Class	Types	Accuracy
I	Definitive	±5%
II	Capital cost	±10–15%
III	Appropriation (with some capital cost)	±15–20%
IV	Appropriation	±20–25%
V	Feasibility	±25–35%
VI	Order of magnitude	> 635%

TABLE 13–5. CHECKLIST FOR WORK NORMALLY REQUIRED FOR THE VARIOUS CLASSES (I-IV) OF ESTIMATES

Item	I	II	III	IV	V	VI
1. Inquiry	X	X	X	X	X	X
2. Legibility	X	X	X			
3. Copies	X	X				
4. Schedule	X	X	X	X		
5. Vendor inquiries	X	X	X			
6. Subcontract packages	X	X				
7. Listing	X	X	X	X	X	
8. Site visit	X	X	X	X		
9. Estimate bulks	X	X	X	X	X	
10. Labor rates	X	X	X	X	X	
11. Equipment and subcontract selection	X	X	X	X	X	
12. Taxes, insurance, and royalties	X	X	X	X	X	
13. Home office costs	X	X	X	X	X	
14. Construction indirects	X	X	X	X	X	
15. Basis of estimate	X	X	X	X	X	X
16. Equipment list	X					
17. Summary sheet	X	X	X	X	X	
18. Management review	X	X	X	X	X	X
19. Final cost	X	X	X	X	X	X
20. Management approval	X	X	X	X	X	X
21. Computer estimate	X	X	X	X		

During competitive bidding, it is important that the type of estimate be consistent with the customer's requirements. For in-house projects, the type of estimate can vary over the life cycle of a project:

- *Conceptual stage:* Venture guidance or feasibility studies for the evaluation of future work. This estimating is often based on minimum-scope information.
- *Planning stage:* Estimating for authorization of partial or full funds. These estimates are based on preliminary design and scope.
- *Main stage:* Estimating for detailed work.
- *Termination stage:* Reestimation for major scope changes or variances beyond the authorization range.

13.3 PRICING PROCESS

This activity schedules the development of the work breakdown structure and provides management with two of the three operational tools necessary for the control of a system or project. The development of these two tools is normally the responsibility of the project office with input from the functional units.

The integration of the functional unit into the project environment or system occurs through the pricing-out of the work breakdown structure. The total project costs obtained by pricing out the activities over the scheduled period of performance provide management with the third tool necessary to successfully manage the project. During the pricing

TABLE 13–6. DATA REQUIRED FOR PREPARATION OF ESTIMATES

	Classes of Estimates					
	I	II	III	IV	V	VI
General						
Product	X	X	X	X	X	X
Process description	X	X	X	X	X	X
Capacity	X	X	X	X	X	X
Location—general					X	X
Location—specific	X	X	X	X		
Basic design criteria	X	X	X	X		
General design specifications	X	X	X	X		
Process						
Process block flow diagram						X
Process flow diagram (with equipment size and material)				X	X	
Mechanical P&Is	X	X	X			
Equipment list	X	X	X	X	X	
Catalyst/chemical specifications	X	X	X	X	X	
Site						
Soil conditions	X	X	X	X		
Site clearance	X	X	X			
Geological and meteorological data	X	X	X			
Roads, paving, and landscaping	X	X	X			
Property protection	X	X	X			
Accessibility to site	X	X	X			
Shipping and delivery conditions	X	X	X			
Major cost is factored					X	X
Major Equipment						
Preliminary sizes and materials			X	X	X	
Finalized sizes, materials, and appurtenances	X	X				
Bulk Material Quantities						
Finalized design quantity take-off		X				
Preliminary design quantity take-off	X	X	X	X		
Engineering						
Plot plan and elevations	X	X	X	X		
Routing diagrams	X	X	X			
Piping line index	X	X				
Electrical single line	X	X	X	X		
Fire protection	X	X	X			
Sewer systems	X	X	X			
Pro-services—detailed estimate	X	X				
Pro-services—ratioed estimate			X	X	X	
Catalyst/chemicals quantities	X	X	X	X	X	
Construction						
Labor wage, F/B, travel rates	X	X	X	X	X	
Labor productivity and area practices	X	X				
Detailed construction execution plan	X	X				
Field indirects—detailed estimate	X	X				
Field indirects—ratioed estimate			X	X	X	
Schedule						
Overall timing of execution				X	X	
Detailed schedule of execution	X	X	X			
Estimating preparation schedule	X	X	X			

activities, the functional units have the option of consulting project management about possible changes in the activity schedules and work breakdown structure.

The work breakdown structure and activity schedules are priced out through the lowest pricing units of the company. It is the responsibility of these pricing units, whether they be sections, departments, or divisions, to provide accurate and meaningful cost data (based on historical standards, if possible). All information is priced out at the lowest level of performance required, which, from the assumption of Chapter 11, will be the task level. Costing information is rolled up to the project level and then one step further to the total program level.

Under ideal conditions, the work required (i.e., man-hours) to complete a given task can be based on historical standards. Unfortunately, for many industries, projects and programs are so diversified that realistic comparison between previous activities may not be possible. The costing information obtained from each pricing unit, whether or not it is based on historical standards, should be regarded only as an estimate. How can a company predict the salary structure three years from now? What will be the cost of raw materials two years from now? Will the business base (and therefore overhead rates) change over the duration of the project? The final response to these questions shows that costing data are explicitly related to an environment that cannot be predicted with any high degree of certainty. The systems approach to management, however, provides for a more rapid response to the environment than less structured approaches permit.

Once the cost data are assembled, they must be analyzed for their potential impact on the company resources of people, money, equipment, and facilities. It is only through a total project cost analysis that resource allocations can be analyzed. The resource allocation analysis is performed at all levels of management, ranging from the section supervisor to the vice president and general manager. For most projects, the chief executive must approve final cost data and the allocation of resources.

Proper analysis of the total project costs can provide management (both project and corporate) with a strategic planning model for integration of the current project with other projects in order to obtain a total corporate strategy. Meaningful planning and pricing models include analyses for monthly manloading schedules per department, monthly costs per department, monthly and yearly total project costs, monthly material expenditures, and total project cash-flow and man-hour requirements per month.

Previously we identified several of the problems that occur at the nodes where the horizontal hierarchy of project management interfaces with the vertical hierarchy of functional management. The pricing-out of the work breakdown structure provides the basis for effective and open communication between functional and project management where both parties have one common goal. This is shown in Figure 13–1. After the pricing effort is completed, and the project is initiated, the work breakdown structure still forms the basis of a communications tool by documenting the performance agreed on in the pricing effort, as well as establishing the criteria against which performance costs will be measured.

13.4 ORGANIZATIONAL INPUT REQUIREMENTS

Once the work breakdown structure and activity schedules are established, the project manager calls a meeting for all organizations that will submit pricing information. It is imperative that all pricing or labor-costing representatives be present for the first meeting. During

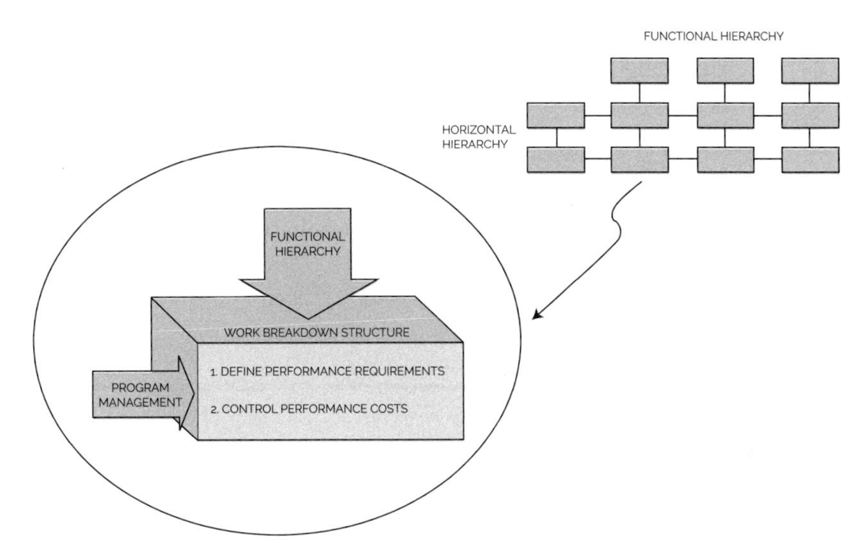

FIGURE 13–1. The vertical–horizontal interface.

this "kickoff" meeting, the work breakdown structure is described in depth so that each pricing unit manager will know exactly what his responsibilities are during the project. The kickoff meeting also resolves the struggle for power among functional managers whose responsibilities may be similar. Unfortunately, one meeting is not always sufficient to clarify all problems. Follow-up or status meetings are held, normally with only those parties concerned with the problems that have arisen. Some companies prefer to have all members attend the status meetings so that all personnel will be familiar with the total effort and the associated problems. The advantage of not having all project-related personnel attend is that time is of the essence when pricing out activities. Many functional divisions carry this policy one step further by having a divisional representative together with possibly key department managers or section supervisors as the only attendees at the kickoff meeting. The divisional representative then assumes all responsibility for assuring that all costing data are submitted on time. This arrangement may be beneficial in that the project office need contact only one individual in the division to learn of the activity status, but it may become a bottleneck if the representative fails to maintain proper communication between the functional units and the project office, or if the individual simply is unfamiliar with the pricing requirements of the work breakdown structure.

During proposal activities, time may be extremely important. There are many situations in which a request for proposal (RFP) requires that all responders submit their bids by a specific date. Under a proposal environment, the activities of the project office, as well as those of the functional units, are under a schedule set forth by the proposal manager. The proposal manager's schedule has very little, if any, flexibility and is normally under tight time constraints so that the proposal may be typed, edited, and published prior to the date of submittal. In this case, the RFP will indirectly define how much time the pricing units have to identify and justify labor costs.

The justification of the labor costs may take longer than the original cost estimates, especially if historical standards are not available. Many proposals often require that comprehensive labor justification be submitted. Other proposals, especially those that request an almost immediate response, may permit vendors to submit labor justification at a later date.

In the final analysis, it is the responsibility of the lowest pricing unit supervisors to maintain adequate standards, so that an almost immediate response can be given to a pricing request from a project office.

13.5 LABOR DISTRIBUTIONS

The functional units supply their input to the project office in the form of man-hours, as shown in Figure 13–2. The input may be accompanied by labor justification, if required. The man-hours are submitted for each task, assuming that the task is the lowest pricing element, and are time-phased per month. The man-hours per month per task are converted to dollars after multiplication by the appropriate labor rates. The labor rates are generally known with certainty over a twelve-month period, but from then on are only estimates. How can a company predict salary structures five years hence? If the company underestimates the salary structure, increased costs and decreased profits will occur. If the salary structure is overestimated, the company may not be competitive; if the project is government funded, then the salary structure becomes an item under contract negotiations.

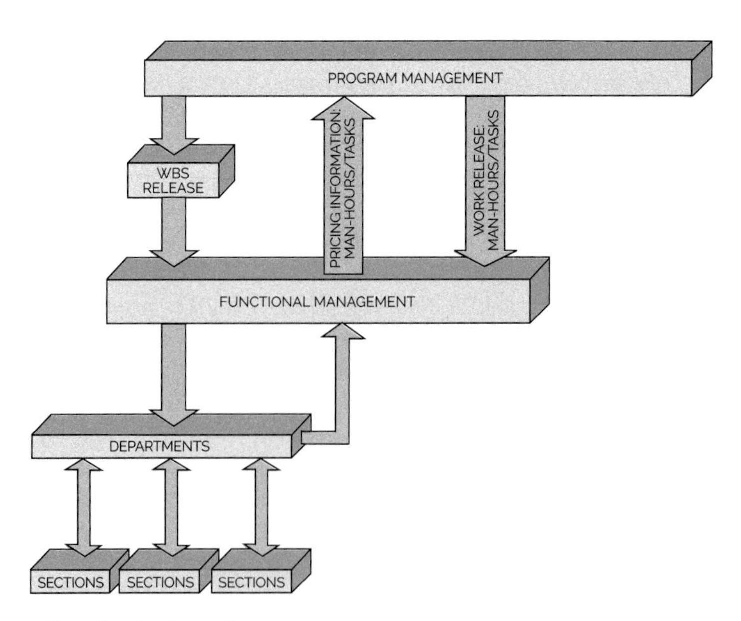

FIGURE 13–2. Functional pricing flow.

The development of the labor rates to be used in the projection is based on historical costs in business base hours and dollars for the most recent month or quarter. Average hourly rates are determined for each labor unit by direct effort within the operations at the department level. The rates are only averages, and include both the highest-paid employees and lowest-paid employees, together with the department manager and the clerical support.[1] These base rates are then escalated as a percentage factor based on past experience, budget as approved by management, and the local outlook and similar industries. If the company has a predominant aerospace or defense industry business base, then these salaries are negotiated with local government agencies prior to submittal for proposals.

The labor hours submitted by the functional units are often overestimated for fear that management will "massage" and reduce the labor hours while attempting to maintain the same scope of effort. Many times management is forced to reduce man-hours either because of insufficient funding or just to remain competitive in the environment.

The reduction of man-hours often causes heated discussions between the functional and project managers. Project managers tend to think in terms of the best interests of the project, whereas functional managers lean toward maintaining their present staff.

The most common solution to this conflict rests with the project manager. If the project manager selects members for the project team who are knowledgeable in man-hour standards for each of the departments, then an atmosphere of trust can develop between the project office and the functional department so that man-hours can be reduced in a manner that represents the best interests of the company. This is one of the reasons why project team members are often promoted from within the functional ranks.

The man-hours submitted by the functional units provide the basis for total project cost analysis and project cost control.

13.6 OVERHEAD RATES

PMBOK® Guide, 6th Edition
7.2.1 Estimate Costs Inputs

The ability to control project costs involves more than tracking labor dollars and labor hours; overhead dollars, one of the biggest headaches, must also be tracked. Although most projects have an assistant project manager for cost whose responsibilities include monthly overhead rate analysis, the project manager can drastically increase the success of his program by insisting that each program team member understand overhead rates. For example, if overhead rates apply only to the first forty hours of work, then, depending on the overhead rate, program dollars can be saved by performing work on overtime where the increased salary is at a lower burden.

1. Problems can occur if the salaries of the people assigned to the project exceed the department averages. Methods to alleviate this problem are discussed later. Also, in many companies department managers are included in the overhead rate structure, not in direct labor, and therefore their salaries are not included as part of the department average.

Regardless of whether one analyzes a project or a system, all costs must have associated overhead rates. Unfortunately, many project managers and systems managers consider overhead rates as a magic number pulled out of the air. The preparation and assignment of overheads to each of the functional divisions is a science. Although the *total dollar pool* for overhead rates is relatively constant, management retains the option of deciding how to distribute the overhead among the functional divisions. A company that supports its R&D staff through competitive bidding projects may wish to keep the R&D overhead rate as low as possible. Care must be taken, however, that other divisions do not absorb additional costs so that the company no longer remains competitive on those manufactured products that may be its bread and butter.

The development of the overhead rates is a function of three separate elements: direct labor rates, direct business base projections, and projection of overhead expenses. Direct labor rates have already been discussed. The direct business base projection involves the determination of the anticipated direct labor hours and dollars along with the necessary direct materials and other direct costs required to perform and complete the project efforts included in the business base. Those items utilized in the business base projection include all contracted projects as well as the proposed or anticipated efforts. The foundation for determination of the business base required for each project can be one or more of the following:

- Actual costs to date and estimates to completion
- Proposal data
- Marketing intelligence
- Management goals
- Past performance and trends

The projection of the overhead expenses is made by an analysis of each of the elements that constitute the overhead expense. A partial listing of those items is shown in Table 13–7. Projection of expenses within the individual elements is then made based on one or more of the following:

- Historical direct/indirect labor ratios
- Regression and correlation analysis
- Manpower requirements and turnover rates
- Changes in public laws
- Anticipated changes in company benefits
- Fixed costs in relation to capital asset requirements
- Changes in business base
- Bid and proposal (B&P) budgets to win more business
- Internal research and development (IR&D) budgets

For many industries, such as aerospace and defense, the federal government funds a large percentage of the B&P and IR&D activities. This federal funding is a necessity since many companies could not otherwise be competitive within the industry. The federal

TABLE 13–7. ELEMENTS OF OVERHEAD RATES

Building maintenance	New business directors
Building rent	Office supplies
Cafeteria	Payroll taxes
Clerical	Personnel recruitment
Clubs/associations	Postage
Consulting services	Professional meetings
Corporate auditing expenses	Reproduction facilities
Corporate salaries	Retirement plans
Depreciation of equipment	Sick leave
Executive salaries	Supplies/hand tools
Fringe benefits	Supervision
General ledger expenses	Telephone/telegraph facilities
Group insurance	Transportation
Holiday	Utilities
Moving/storage expenses	Vacation

government employs this technique to stimulate research and competition. Therefore, B&P and IR&D are included in the above list.

The prime factor in the control of overhead costs is the annual budget. This budget, which is the result of goals and objectives established by the chief executive officer, is reviewed and approved at all levels of management. It is established at department level, and the department manager has direct responsibility for identifying and controlling costs against the approved plan.

The departmental budgets are summarized, in detail, for higher levels of management. This summarization permits management, at these higher organizational levels, to be aware of the authorized indirect budget in their area of responsibility.

Reports are published monthly indicating current month and year-to-date budget, actuals, and variances. These reports are published for each level of management, and an analysis is made by the budget department through coordination and review with management. Each directorate's total organization is then reviewed with the budget analyst who is assigned the overhead cost responsibility. A joint meeting is held with the directors and the vice president and general manager, at which time overhead performance is reviewed.

13.7 MATERIALS/SUPPORT COSTS

***PMBOK® Guide*, 6th Edition**
7.2.1 Estimate Costs Inputs

The salary structure, overhead structure, and labor hours fulfill three of four major pricing input requirements. The fourth major input is the cost for materials and support. Six subtopics are included under materials/support: materials, purchased parts, subcontracts, freight, travel, and other. Freight and travel can be handled in one of two ways, both normally dependent on the size of the project. For small-dollar-volume projects, estimates are made for travel and freight. For large-dollar-volume projects, travel is normally expressed as between 3 and 5 percent of the direct labor costs, and freight is likewise between 3 and 5 percent of all costs for material, purchased parts, and subcontracts. The category labeled "other support costs" may include such topics as special consultants.

Determination of the material costs is very time-consuming, more so than cost determination for labor hours. Material costs are submitted via a bill of materials that includes all vendors from whom purchases will be made, projected costs throughout the project, scrap factors, and shelf lifetime for those products that may be perishable.

Upon release of the work statement, work breakdown structure, and subdivided work description, the end-item bill of materials and manufacturing plans are prepared as shown in Figure 13–3. End-item materials are those items identified as an integral part of the production end-item. Support materials consist of those materials required by engineering and operations to support the manufacture of end-items, and are identified on the manufacturing plan.

A procurement plan/purchase requisition is prepared as soon as possible after contract negotiations (using a methodology as shown in Figure 13–4). This plan is used to monitor material acquisitions, forecast inventory levels, and identify material price variances.

Manufacturing plans prepared upon release of the subdivided work descriptions are used to prepare tool lists for manufacturing, quality assurance, and engineering. From these plans a special tooling breakdown is prepared by tool engineering, which defines those tools to be procured and the material requirements of tools to be fabricated in-house. These items are priced by cost element for input on the planning charts.

The materials/support costs are submitted by month for each month of the project. If long-lead funding of materials is anticipated, then they should be assigned to the first month of the project. In addition, an escalation factor for costs of materials/support items must be applied to all materials/support costs. Some vendors may provide fixed prices over time periods in excess of a twelve-month period. As an example, vendor Z may quote a firm-fixed price of $130.50 per unit for 650 units to be delivered over the next eighteen months if the order is placed within sixty days. There are additional factors that influence the cost of materials.

13.8 PRICING OUT THE WORK

Using logical pricing techniques will help in obtaining detailed estimates. The following thirteen steps provide a logical sequence to help a company control its limited resources. These steps may vary from company to company.

Step 1: Provide a complete definition of the work requirements.

Step 2: Establish a logic network with checkpoints.

Step 3: Develop the work breakdown structure.

Step 4: Price out the work breakdown structure.

Step 5: Review WBS costs with each functional manager.

Step 6: Decide on the basic course of action.

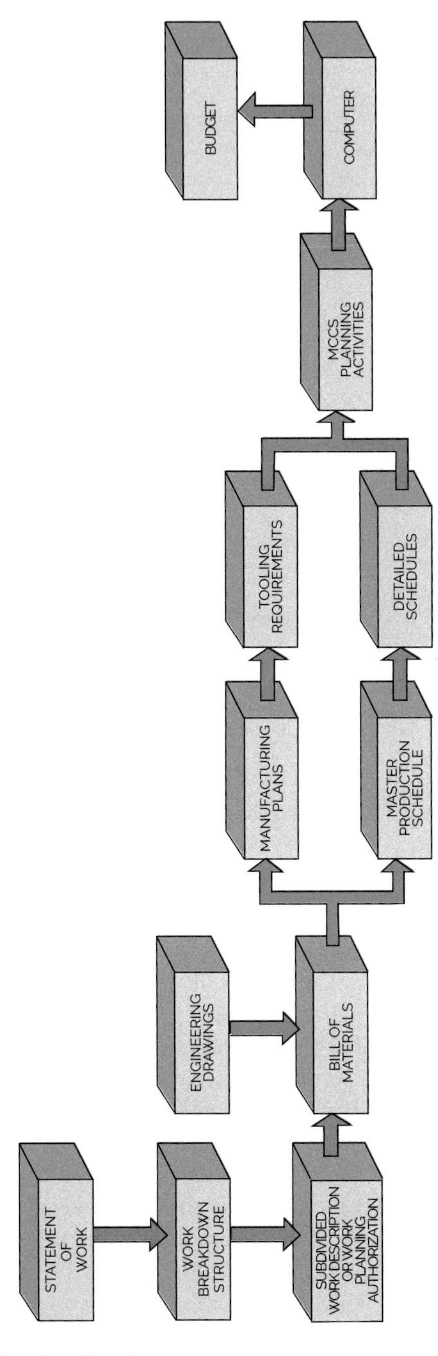

FIGURE 13–3. Material planning flowchart.

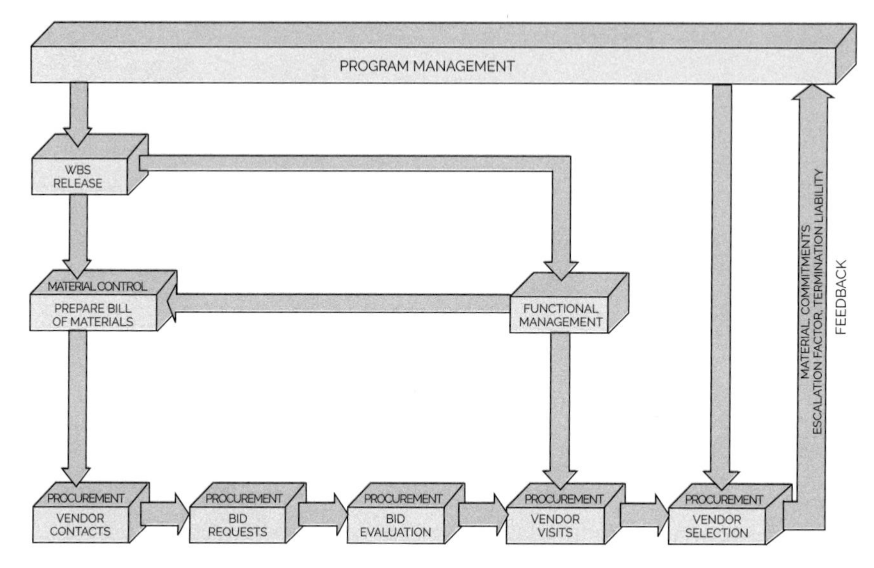

FIGURE 13–4. Procurement activity.

Step 7: Establish reasonable costs for each WBS element.

Step 8: Review the base case costs with upper-level management.

Step 9: Negotiate with functional managers for qualified personnel.

Step 10: Develop the linear responsibility chart.

Step 11: Develop the final detailed schedules.

Step 12: Establish pricing cost summary reports.

Step 13: Document the result in a project plan.

Although the pricing of a project is an iterative process, the project manager must still develop cost summary reports at each iteration point so that key project decisions can be made during the planning. Detailed pricing summaries are needed at least twice: in preparation for the pricing review meeting with management and at pricing termination. At all other times it is possible that "simple cosmetic surgery" can be performed on previous cost summaries, such as perturbations in escalation factors and procurement cost of raw materials. The list below shows the typical pricing reports:

- *A detailed cost breakdown for each WBS element.* If the work is priced out at the task level, then there should be a cost summary sheet for each task, as well as rollup sheets to the top level of the WBS.
- *A total project manpower curve for each department.* These manpower curves show how each department has contracted with the project office to supply functional resources. If the departmental manpower curves contain several "peaks and valleys," then the project manager may have to alter some of his schedules to obtain some degree of manpower smoothing. Functional managers always prefer manpower-smoothed resource allocations.

- *A monthly equivalent manpower cost summary.* This table normally shows the fully burdened cost for the average departmental employee carried out over the entire period of project performance. If project costs have to be reduced, the project manager performs a parametric study between this table and the manpower curve tables.
- *A yearly cost distribution table.* This table is broken down by WBS element and shows the yearly (or quarterly) costs that will be required. This table, in essence, is a project cash-flow summary per activity.
- *A functional cost and hour summary.* This table provides top management with an overall description of how many hours and dollars will be spent by each major functional unit, such as a division. Top management would use this as part of the forward planning process to make sure that there are sufficient resources available for all projects. This also includes indirect hours and dollars.
- *A monthly labor hour and dollar expenditure forecast.* This table can be combined with the yearly cost distribution, except that it is broken down by month, not activity or department. In addition, this table normally includes manpower termination liability information for premature cancellation of the project by outside customers.
- *A raw material and expenditure forecast.* This shows the cash flow for raw materials based on vendor lead times, payment schedules, commitments, and termination liability.
- *Total project termination liability per month.* This table shows the customer the monthly costs for the entire project. This is the customer's cash flow, not the contractor's. The difference is that each monthly cost contains the termination liability for man-hours and dollars, on labor and raw materials. This table is actually the monthly costs attributed to premature project termination.

These tables are used by project managers as the basis for project cost control and by upper-level executives for selecting, approving, and prioritizing projects.

13.9 SMOOTHING OUT DEPARTMENT MAN-HOURS

The dotted curve in Figure 13–5 indicates projected manpower requirements for a given department as a result of a typical project manloading schedule. Department managers, however, attempt to smooth out the manpower curve as shown by the solid line in Figure 13–5. Smoothing out the manpower requirements benefits department managers by eliminating fractional man-hours per day. The project manager must understand that if departments are permitted to eliminate peaks, valleys, and small-step functions in manpower planning, small project and task man-hour (and cost) variances can occur, but should not, in general, affect the total project cost significantly.

Two important questions to ask are whether the department has sufficient personnel available to fulfill manpower requirements and what is the rate at which the functional departments can staff the project? Figure 13–6 indicates the types of problems that can occur. Curve A shows the manpower requirements for a given department after

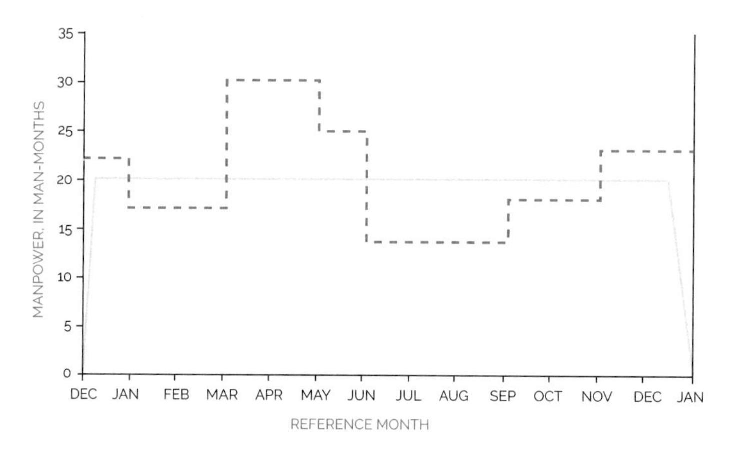

FIGURE 13–5. Typical manpower loading.

time-smoothing. Curve B represents the modification to the time-phase curve to account for reasonable project manning and demanning rates. The difference between these two curves (i.e., the shaded area) therefore reflects the amount of money the contractor may have to forfeit owing to manning and demanning activities. This problem can be partially overcome by increasing the manpower levels after time-smoothing (see curve C) such that the difference between curves B and C equals the amount of money that would be forfeited from curves A and B. Of course, project management would have to be able to justify this increase in average manpower requirements, especially if the adjustments are made in a period of higher salaries and overhead rates.

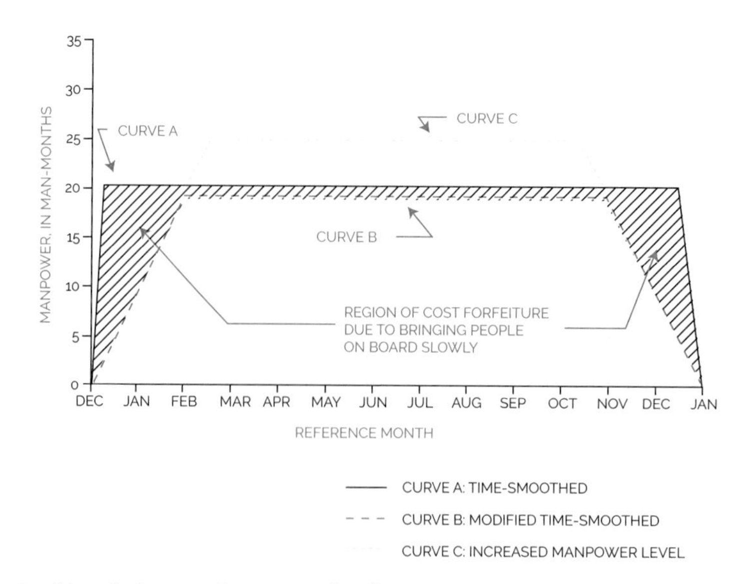

FIGURE 13–6. Linearly increased manpower loading.

13.10 THE PRICING REVIEW PROCEDURE

The ability to project, analyze, and control problem costs requires coordination of pricing information and cooperation between the functional units and upper-level management. A typical company policy for cost analysis and review is shown in Figure 13–7. Corporate management may be required to initiate or authorize activities, if corporate/company resources are or may be strained by the project, if capital expenditures are required for new facilities or equipment, or simply if corporate approval is required for all projects in excess of a certain dollar amount.

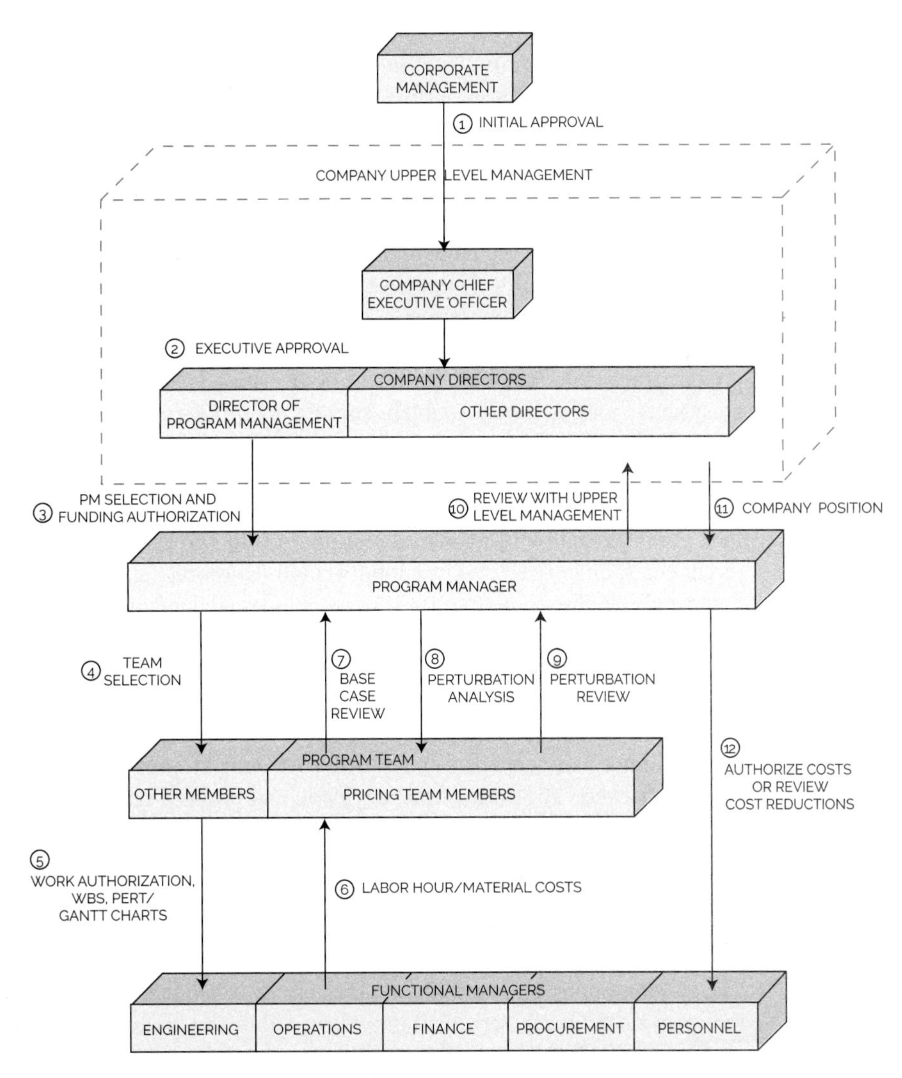

FIGURE 13–7. The pricing review procedure.

Upper-level management, upon approval by the chief executive officer of the company, approves and authorizes the initiation of the project or program. The actual performance activities, however, do not begin until the director of project management selects a project manager and authorizes either the bid and proposal budget (if the project is competitive) or project planning funds.

The newly appointed project manager then selects this project's team. Team members, who are also members of the project office, may come from other projects, in which case the project manager may have to negotiate with other project managers and upper-level management to obtain these individuals. The members of the project office are normally support-type individuals. In order to obtain team members representing the functional departments, the project manager must negotiate directly with the functional managers. Functional team members may not be selected or assigned to the project until the actual work is contracted for. Many proposals, however, require that all functional team members be identified, in which case selection must be made during the proposal stage of a project.

The first responsibility of the project office (not necessarily including functional team members) is the development of the activity schedules and the work breakdown structure. The project office then provides work authorization for the functional units to price out the activities. The functional units then submit the labor hours, material costs, and justification, if required, to the pricing team member. The pricing team member is normally attached to the project office until the final costs are established, and becomes part of the negotiating team if the project is competitive.

Once the base case is formulated, the pricing team member, together with the other project office team members, performs perturbation analyses. These analyses are designed as systems approaches to problem-solving where alternatives are developed in order to respond to management's questions during the final review.

The base case, with the perturbation analysis costs, is then reviewed with upper-level management in order to formulate a company position for the project and to take a hard look at the allocation of resources required for the project. The company position may be to cut costs, authorize work, or submit a bid. Corporate approval may be required if the company's chief executive officer has a ceiling on the amount he can authorize.

If labor costs must be cut, the project manager must negotiate with the functional managers as to the size and method for the cost reductions. Otherwise, this step would simply entail authorization for the functional managers to begin the activities.

Figure 13–7 represents the system approach to determining total program costs. This procedure normally creates a synergistic environment, provides open channels of communication between all levels of management, and ensures agreement among all individuals as to program costs.

13.11 SYSTEMS PRICING

The systems approach to pricing out the activity schedules and the work breakdown structure provide a means for obtaining unity within the company. The flow of information readily admits the participation of all members of the organization in the project, even if on a part-time basis. Functional managers obtain a better understanding of how their labor fits into the total project and how their activities interface with those of other departments. For the first time, functional managers can accurately foresee how their activity can lead to corporate profits.

The project pricing model (sometimes called a strategic project planning model) acts as a management information system, forming the basis for the systems approach to resource control, as shown in Figure 13–8. The summary sheets from the strategic pricing model help management select programs that will best utilize resources. The strategic pricing model also provides management with an invaluable tool for performing perturbation analysis on the base case costs and an opportunity for design and evaluation of contingency plans, if necessary.

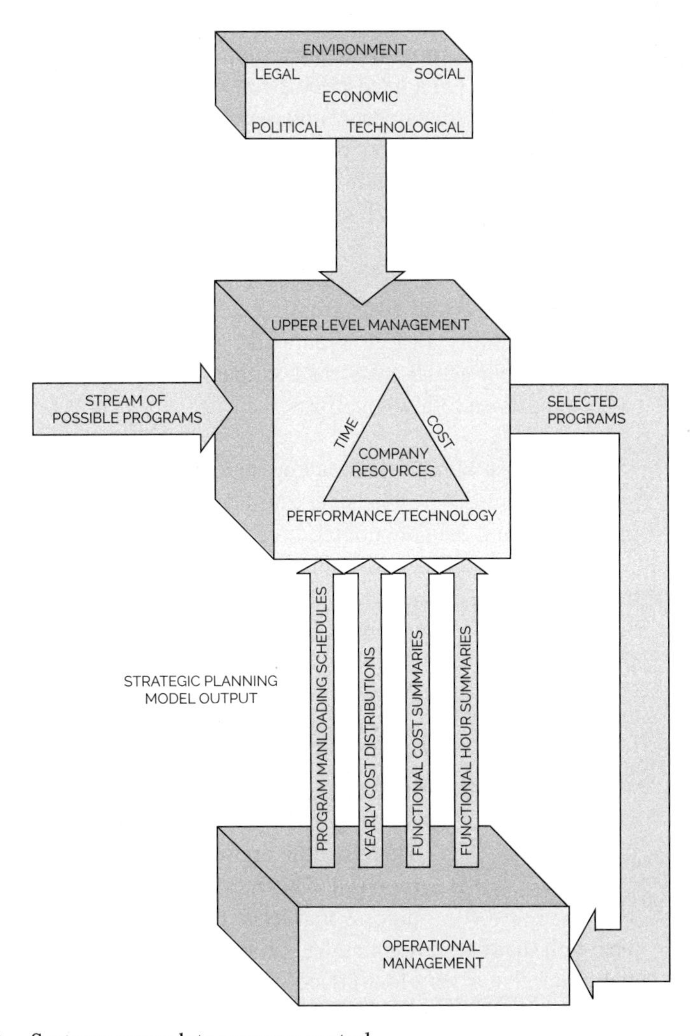

FIGURE 13–8. System approach to resource control.

13.12 DEVELOPING THE SUPPORTING/BACKUP COSTS

PMBOK® Guide, 6th Edition
7.2.2 Estimate Costs Tools and Techniques

Not all cost proposals require backup support, but for those that do, the backup support should be developed along with the pricing. The itemized prices should be compatible with the supporting data. Government pricing requirements are a special case.

Most supporting data come from external (subcontractor or outside vendor) quotes. Internal data must be based on historical data, and these historical data must be updated continually as each new project is completed. The supporting data should be traceable by itemized charge numbers.

Customers may wish to audit the cost proposal. In this case, the starting point might be the supporting data. It is not uncommon on sole-source proposals to have the supporting data audited before the final cost proposal is submitted to the customer.

Not all cost proposals require supporting data; the determining factor is usually the type of contract. On a fixed-price effort, the customer may not have the right to audit your books. However, for a cost-reimbursable package, your costs are an open book, and the customer usually compares your exact costs to those of the backup support.

Most companies usually have a choice of more than one estimate to be used for backup support. In deciding which estimate to use, consideration must be given to the possibility of follow-on work:

- If your actual costs grossly exceed your backup support estimates, you may lose credibility for follow-on work.
- If your actual costs are less than the backup costs, you must use the new actual costs on follow-on efforts.

The moral here is that backup support costs provide future credibility. If you have well documented, "livable" cost estimates, then you may wish to include them in the cost proposal even if they are not required.

Since both direct and indirect costs may be negotiated separately as part of a contract, supporting data, such as those in Tables 13–8 through 13–10 and Figure 13–9, may be necessary to justify any costs that may differ from company (or customer-approved) standards.

13.13 THE LOW-BIDDER DILEMMA

PMBOK® Guide, 6th Edition
12.3.2.1 Select Sellers

There is little argument about the importance of the price tag to the proposal. The question is, what price will win the job? The decision process that leads to the final price of your proposal is highly complex with many uncertainties. Yet proposal managers, driven by the desire to win the job, may think that a very low-priced proposal will help. But winning is only the beginning.

TABLE 13–8. CONTRACTOR'S MANPOWER AVAILABILITY

	Number of Personnel			
	Total Current Staff		Available for This Project and Other New York 1/15 Permanent + Agency	Anticipated Growth by 1/15 Permanent + Agency
	Permanent Employees	Agency Personnel		
Process engineers	93	—	70	4
Project managers/engineers	79	—	51	4
Cost estimating	42	—	21	2
Cost control	73	—	20	2
Scheduling/scheduling control	14	—	8	1
Procurement/purchasing	42	—	20	1
Inspection	40	—	20	2
Expediting	33	—	18	1
Home office construction management	9	—	6	0
Piping	90	13	67	6
Electrical	31	—	14	2
Instrumentation	19	—	3	1
Vessels/exchangers	24	—	19	1
Civil/structural	30	—	23	2
Other	13	—	8	0

TABLE 13–9. STAFF TURNOVER DATA

	For Twelve-Month Period 1/1/14 to 1/1/15	
Hired	Number Terminated	Number
Process engineers	5	2
Project managers/engineers	1	1
Cost estimating	1	2
Cost control	12	16
Scheduling/scheduling control	2	5
Procurement/purchasing	13	7
Inspection	18	6
Expediting	4	5
Home office construction management	0	0
Design and drafting—total	37	29
Engineering specialists—total	26	45
Total	119	118

TABLE 13–10. STAFF EXPERIENCE PROFILE

	Number of Years' Employment with Contractor				
	0–1	1–2	2–3	3–5	5 or more
Process engineers	2	4	15	11	18
Project managers/engineers	1	2	5	11	8
Cost estimating	0	4	1	5	7
Cost control	5	9	4	7	12
Scheduling and scheduling control	2	2	1	3	6
Procurement/purchasing	4	12	13	2	8
Inspection	1	2	6	14	8
Expediting	6	9	4	2	3
Piping	9	6	46	31	22
Electrical	17	6	18	12	17
Instrumentation	8	8	12	13	12
Mechanical	2	5	13	27	19
Civil/structural	4	8	19	23	16
Environmental control	0	1	1	3	7
Engineering specialists	3	3	3	16	21
Total	64	81	161	180	184

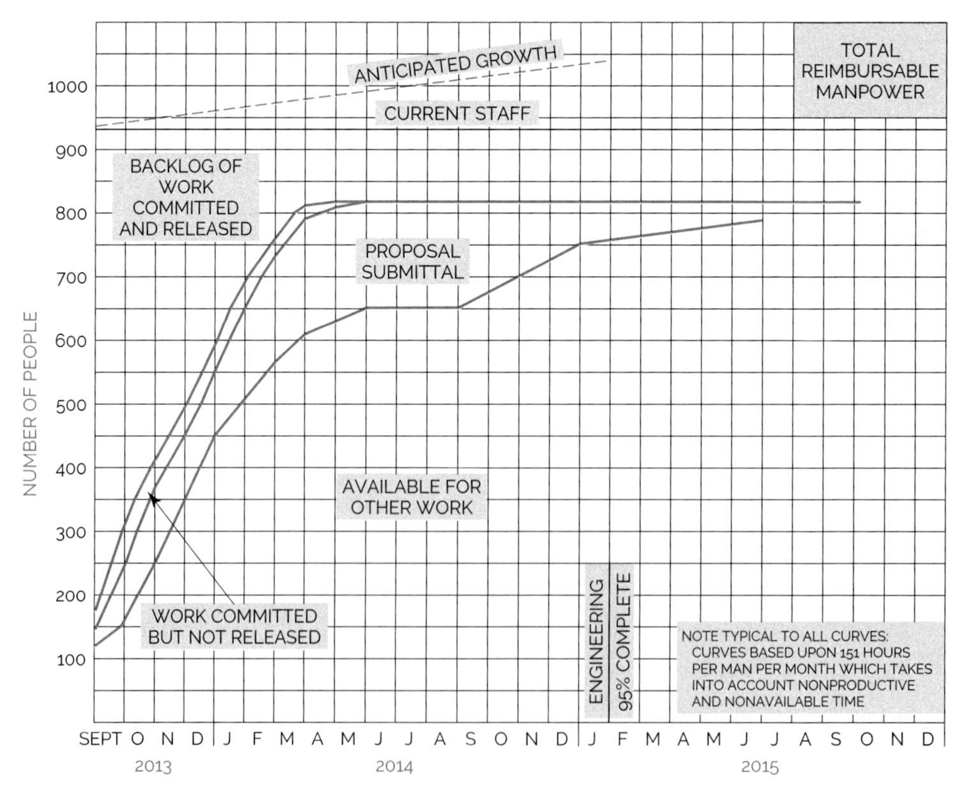

FIGURE 13–9. Total reimbursable manpower.

Companies have short- and long-range objectives on profit, market penetration, new product development, and so on. These objectives may be incompatible with or irrelevant to a low-price strategy. For example:

- A suspiciously low price, particularly on cost-plus type proposals, might be perceived by the customer as unrealistic, thus affecting the bidder's cost credibility or even the technical ability to perform.
- The bid price may be unnecessarily low, relative to the competition and customer budget, thus eroding profits.
- The price may be irrelevant to the bid objective, such as entering a new market. Therefore, the contractor has to sell the proposal in a credible way, e.g., using cost sharing.
- Low pricing without market information is meaningless. The price level is always relative to (1) the competitive prices, (2) the customer budget, and (3) the bidder's cost estimate.
- The bid proposal and its price may cover only part of the total project. The ability to win phase II or follow-on business depends on phase I performance and phase II price.
- The financial objectives of the customer may be more complex than just finding the lowest bidder. They may include cost objectives for total system life-cycle cost (LCC), for design to unit production cost (DTUPC), or for specific logistic support items. Presenting sound approaches for attaining these system cost–performance parameters and targets may be just as important as, if not more important than, a low bid for the system's development.

The lowest bidder is certainly not an automatic winner. Both commercial and governmental customers are increasingly concerned about cost realism and the ability to perform under contract. A compliant, sound, technical and management proposal, based on past experience with realistic, well-documented cost figures, is often chosen over the lowest bidder, who may project a risky image regarding technical performance, cost, or schedule.

13.14 SPECIAL PROBLEMS

There are always special problems that, if overlooked, can have a severe impact on the pricing effort. As an example, pricing must include an understanding of cost control—specifically, how costs are billed back to the project. There are three possible situations:

- Work is priced out at the department average, and all work performed is charged to the project at the department average salary, regardless of who accomplished the work. This technique is obviously the easiest, but encourages project managers to fight for the highest salary resources, since only average wages are billed to the project.

- Work is priced out at the department average, but all work performed is billed back to the project at the actual salary of those employees who perform the work. This method can create a severe headache for the project manager if he tries to use only the best employees on his project. If these employees are earning substantially more money than the department average, then a cost overrun will occur unless the employees can perform the work in less time. Some companies are forced to use this method by government agencies and have estimating problems when the project that has to be priced out is of a short duration where only the higher-salaried employees can be used. In such a situation it is common to "inflate" the direct labor hours to compensate for the added costs.
- The work is priced out at the actual salary of those employees who will perform the work, and the cost is billed back the same way. This method is the ideal situation as long as the people can be identified during the pricing effort.

Some companies use a combination of all three methods. In this case, the project office is priced out using the third method (because these people are identified early), whereas the functional employees are priced out using the first or second method.

13.15 ESTIMATING PITFALLS

PMBOK® Guide, 6th Edition
7.2.1 Estimate Costs Inputs

Several pitfalls can impede the pricing function. Probably the most serious pitfall, and the one that is usually beyond the control of the project manager, is the "buy-in" decision, which is based on the assumption that there will be "bail-out" changes or follow-on contracts later. These changes and/or contracts may be for spare parts, maintenance, maintenance manuals, equipment surveillance, optional equipment, optional services, and scrap factors. Other types of estimating pitfalls include:

- Misinterpretation of the statement of work
- Omissions or improperly defined scope
- Poorly defined or overly optimistic schedule
- Inaccurate work breakdown structure
- Applying improper skill levels to tasks
- Failure to account for risks
- Failure to understand or account for cost escalation and inflation
- Failure to use the correct estimating technique
- Failure to use forward pricing rates for overhead, general and administrative, and indirect costs

Unfortunately, many of these pitfalls do not become evident until detected by the cost control system, well into the project.

13.16 ESTIMATING HIGH-RISK PROJECTS

PMBOK® Guide, 6th Edition
6.5 Develop Schedule
7.2 Estimate Costs
Chapter 11 Project Risk Management

Whether a project is high-risk or low-risk depends on the validity of the historical estimate. Construction companies have well-defined historical standards, which lowers their risk, whereas many R&D and MIS projects are high risk. Typical accuracies for each level of the WBS are shown in Table 13–11.

A common technique used to estimate high-risk projects is the "rolling wave" or "moving window" approach. This is shown in Figure 13–10 for a high-risk R&D project. The project lasts for twelve months. In part A, the R&D effort to be accomplished for the first six months is well defined and can be estimated to level 5 of the WBS. However, the effort for the latter six months is based on the results of the first six months and can be estimated at level 2 only, thus incurring a high risk. Now consider part B, which shows a six-month moving window. At the end of the first month, in order to maintain a six-month moving window (at level 5 of the WBS), the estimate for month seven must be improved from a level-2 to a level-5 estimate. Likewise, in parts C and D, we see the effects of completing the second and third months.

There are two key points to be considered in utilizing this technique. First, the length of the moving window can vary from project to project, and usually increases in length as you approach downstream life-cycle phases. Second, this technique works best when upper-level management understands how the technique works. All too often senior management hears only one budget and schedule number during project approval and might not realize that at least half of the project might be time/cost accurate to only 50–60 percent. Simply stated, when using this technique, the word "rough" is not synonymous with the word "detailed."

Methodologies can be developed for assessing risk. Figures 13–11 and 13–12 show such methodologies.

TABLE 13–11. LOW- VERSUS HIGH-RISK ACCURACIES

WBS		Accuracy	
Level	Description	Low-Risk Projects	High-Risk Projects
1	Program	±35	±75–100
2	Project	20	50–60
3	Task	10	20–30
4	Subtask	5	10–15
5	Work package	2	5–10

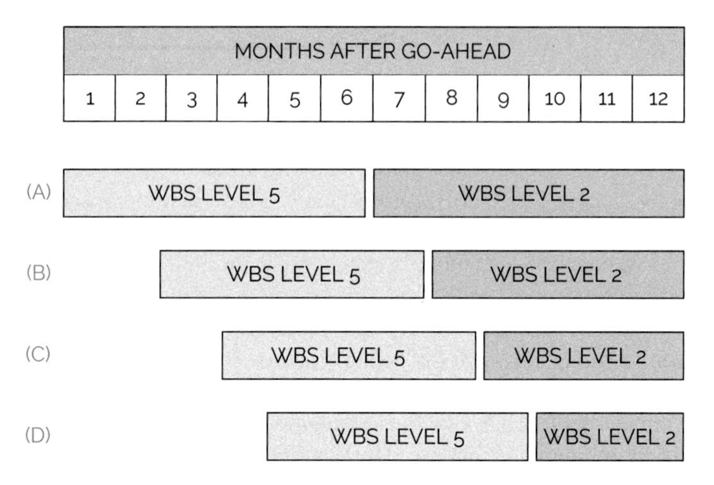

FIGURE 13–10. The moving window/rolling wave concept.

13.17 PROJECT RISKS

> ***PMBOK® Guide*, 6th Edition**
> 11.2 Identify Risks

Project plans are "living documents" and are therefore subject to change. Changes are needed in order to prevent or rectify unfortunate situations. These unfortunate situations can be called project risks.

Risk refers to those dangerous activities or factors that, if they occur, will *increase* the probability that the project's goals of time, cost, and performance will not be met. Many risks can be anticipated and controlled. Furthermore, risk management must be an integral part of project management throughout the entire life cycle of the project.

Some common risks include:

- Poorly defined requirements
- Lack of qualified resources
- Lack of management support
- Poor estimating
- Inexperienced project manager

Risk identification is an art. It requires the project manager to probe, penetrate, and analyze all data. Tools that can be used by the project manager include:

- Decision support systems
- Expected value measures
- Trend analysis/projections
- Independent reviews and audits

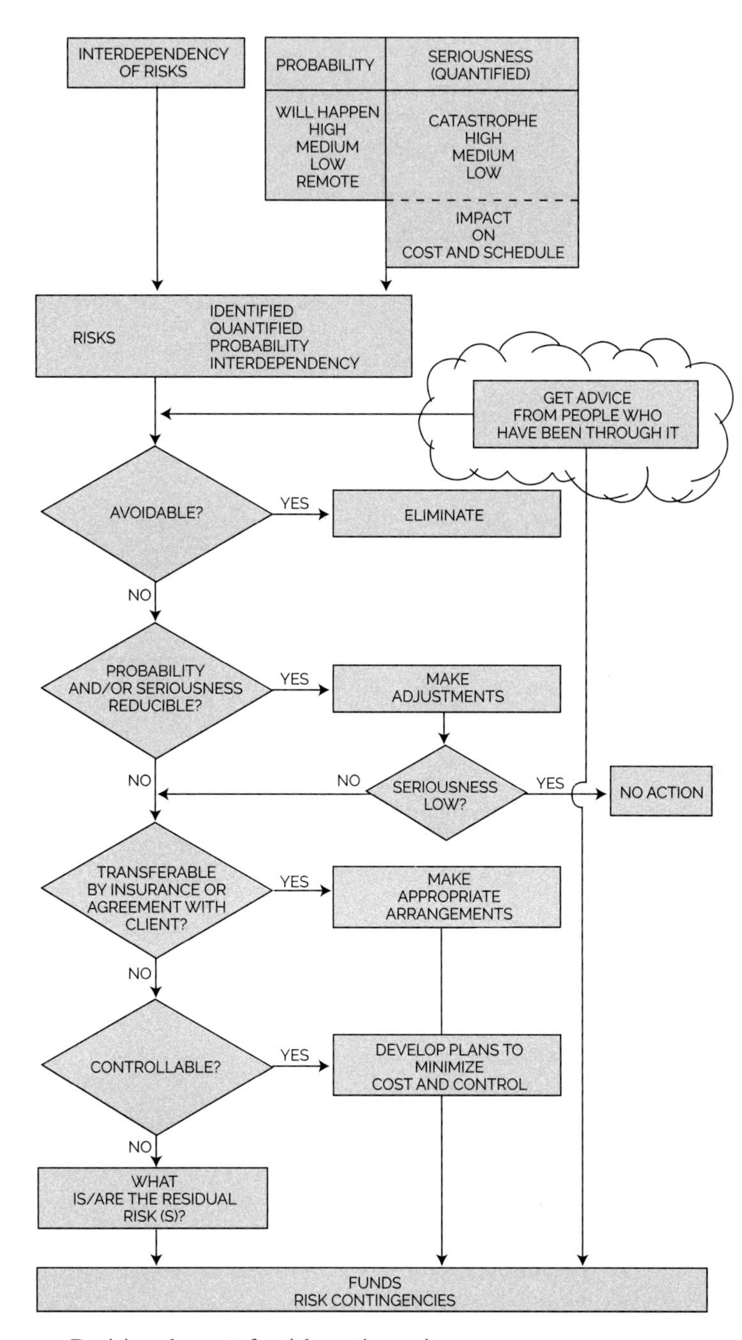

FIGURE 13–11. Decision elements for risk contingencies.

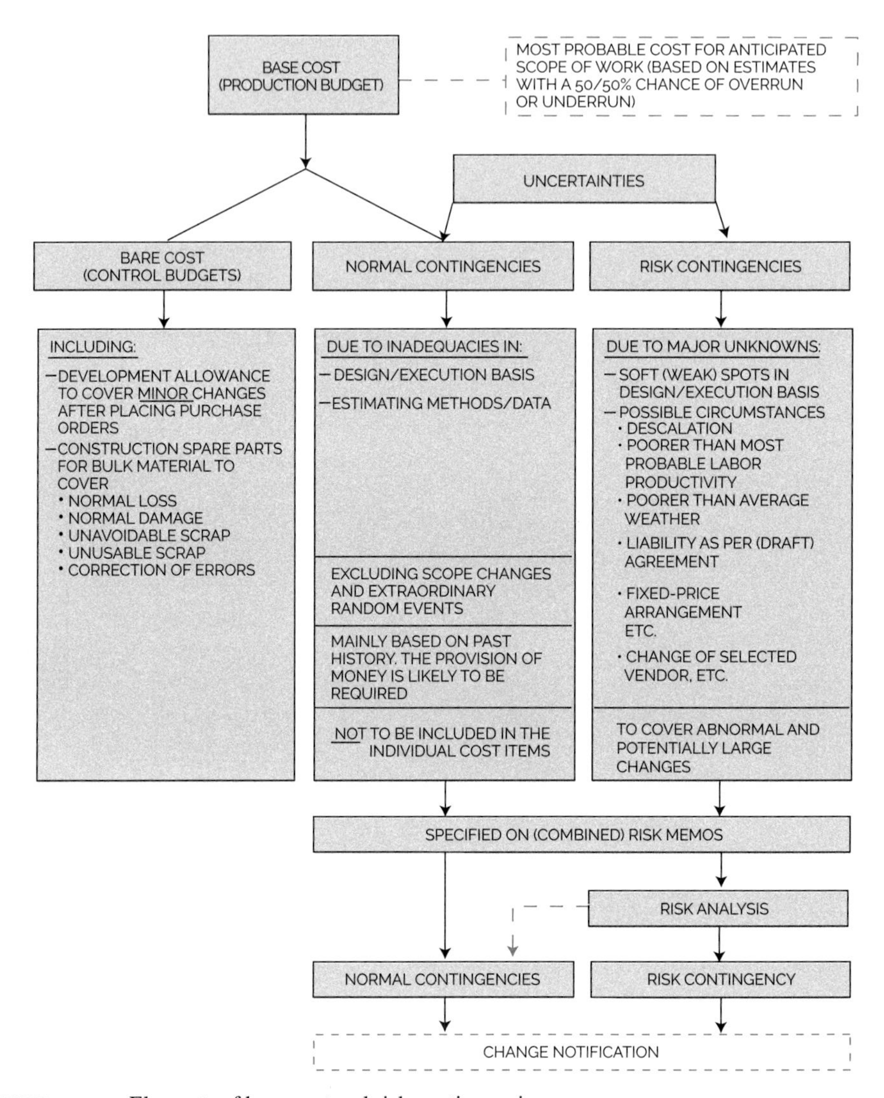

FIGURE 13–12. Elements of base cost and risk contingencies.

Managing project risks is not as difficult as it may seem. There are six steps in the risk management process:

- Identification of the risk
- Quantifying the risk
- Prioritizing the risk
- Developing a strategy for managing the risk
- Project sponsor/executive review
- Taking action

Figures 13–11 and 13–12 identify the process of risk evaluation on capital projects. In all three exhibits, it is easily seen that the attempt is to quantify the risks, possibly by developing a contingency fund.

13.18 THE DISASTER OF APPLYING THE 10 PERCENT SOLUTION TO PROJECT ESTIMATES

Economic crunches can and do create chaos in all organizations. For the project manager, the worst situation is when senior management arbitrarily employs "the 10 percent solution," which is a budgetary reduction of 10 percent for each and every project, especially those that have already begun. The 10 percent solution is used to "create" funds for additional activities for which budgets are nonexistent. The 10 percent solution very rarely succeeds. For the most part, the result is simply havoc, resulting in schedule slippages, a degradation of quality and performance, and eventual budgetary increases rather than the expected decreases.

Most projects are initiated through an executive committee, governing committee, or screening committee. The two main functions of these committees are to select the projects to be undertaken and to prioritize the efforts. Budgetary considerations may also be included, as they pertain to project selection. The real budgets, however, are established from the middle-management levels and sent upstairs for approvals.

Although the role of executive committee is often ill-defined with regard to budgeting, the real problem is that the committee does not realize the impact of adopting the 10 percent solution. If the project budget is an honest one, then a reduction in budget *must* be accompanied by a trade-off in either time or performance. It is often said that 90 percent of the budget generates the first 10 percent of the desired service or quality levels, and that the remaining 10 percent of the budget will produce the remaining 90 percent of the target requirements. If this is true, then a 10 percent reduction in budget must be accompanied by a loss of performance much greater than the target reduction in cost.

It is true that some projects have "padded" estimates, and the budgetary reduction will force out the padding. Most project managers, however, provide realistic estimates and schedules with marginal padding. Likewise, a trade-off between time and cost is unlikely to help, since increasing the duration of the project will increase the cost.

However, there are two viable alternatives. The first is to use the 10 percent solution, but only on selected projects and *after* an "impact study" has been conducted, so that the executive committee understands the impact on the time, cost, and performance constraints. The second choice, which is by far the better one, is for the executive committee to cancel or descope selected projects. Since it is impossible to reduce budget without reducing scope, canceling a project or simply delaying it until the next fiscal year is a viable choice. After all, why should all projects have to suffer?

Terminating one or two projects within the queue allows existing resources to be used more effectively, more productively, and with higher organizational morale. However, it does require strong leadership at the executive committee level for the participants to terminate a project rather than to "pass the buck" to the bottom of the organization with the

10 percent solution. Executive committees often function best if the committee is responsible for project selection, prioritization, and tracking, with the middle managers responsible for budgeting.

13.19 LIFE-CYCLE COSTING (LCC)

***PMBOK® Guide*, 6th Edition**
7.2.1 Estimate Costs Inputs

For years, many R&D organizations have operated in a vacuum where technical decisions made during R&D were based entirely on the R&D portion of the plan, with little regard for what happens after production begins. Today, industrial firms are adopting the life-cycle costing approach that has been developed and used by military organizations. Simply stated, LCC requires that decisions made during the R&D process be evaluated against the total life-cycle cost of the system. As an example, the R&D group has two possible design configurations for a new product. Both design configurations will require the same budget for R&D and the same costs for manufacturing. However, the maintenance and support costs may be substantially greater for one of the products. If these downstream costs are not considered in the R&D phase, large unanticipated expenses may result at a point where no alternatives exist.

Life-cycle costs are the total cost to the organization for the ownership and acquisition of the product over its full life. This includes the cost of R&D, production, operation, support, and, where applicable, disposal. A typical breakdown description might include:

- *R&D costs:* The cost of feasibility studies; cost-benefit analyses; system analyses; detail design and development; fabrication, assembly, and test of engineering models; initial product evaluation; and associated documentation.
- *Production cost:* The cost of fabrication, assembly, and testing of production models; operation and maintenance of the production capability; and associated internal logistic support requirements, including test and support equipment development, spare/repair parts provisioning, technical data development, training, and entry of items into inventory.
- *Construction cost:* The cost of new manufacturing facilities or upgrading existing structures to accommodate production and operation of support requirements.
- *Operation and maintenance cost:* The cost of sustaining operational personnel and maintenance support; spare/repair parts and related inventories; test and support equipment maintenance; transportation and handling; facilities, modifications, and technical data changes; and so on.
- *Product retirement and phaseout cost (also called disposal cost):* The cost of phasing the product out of inventory due to obsolescence or wearout, and subsequent equipment item recycling and reclamation as appropriate.

Life-cycle cost analysis is the systematic analytical process of evaluating various alternative courses of action early on in a project, with the objective of choosing the best way to employ scarce resources. Life-cycle cost is employed in the evaluation of alternative design configurations, alternative manufacturing methods, alternative support schemes, and so on. This process includes defining the problem (what information is needed), defining the requirements of the cost model being used, collecting historical data–cost relationships, and developing estimate and test results.

Successful application of LCC will provide downstream resource impact visibility, provide life-cycle cost management, influence R&D decision-making, and support downstream strategic budgeting.

There are also several limitations to life-cycle cost analyses. They include:

- The assumption that the product, as known, has a finite life-cycle
- A high cost to perform, which may not be appropriate for low-cost/low-volume production
- A high sensitivity to changing requirements

Life-cycle costing requires that early estimates be made. The estimating method selected is based on the problem context (i.e., decisions to be made, required accuracy, complexity of the product, and the development status of the product) and the operational considerations (i.e., market introduction date, time available for analysis, and available resources).

The estimating methods available can be classified as follows:

- Informal estimating methods
  - Judgment based on experience
  - Analogy
  - SWAG method
  - ROM method
  - Rule-of-thumb method
- Formal estimating methods
  - Detailed (from industrial engineering standards)
  - Parametric

Table 13–12 shows the advantages/disadvantages of each method.

Life-cycle cost analysis is an integral part of strategic planning since today's decision will affect tomorrow's actions. Yet there are common errors made during life-cycle cost analyses:

- Loss or omission of data
- Lack of systematic structure
- Misinterpretation of data
- Wrong or misused techniques
- A concentration on insignificant facts
- Failure to assess uncertainty
- Failure to check work
- Estimating the wrong items

PMBOK® Guide, 66th Edition
7.2.2 Estimate Costs Tools and Techniques

TABLE 13–12 ESTIMATING METHODS

Estimating Technique	Application	Advantages	Disadvantages
Engineering estimates empirical)	Reprocurement Production Development	• Most detailed technique • Best inherent accuracy • Provides best estimating base for future project change estimates	• Requires detailed project and product definition • Time-consuming and may be expensive • Subject to engineering bias • May overlook system integration costs
Parametric estimates and scaling statistical)	Production Development	• Application is simple and low cost • Statistical database can provide expected values and prediction intervals • Can be used for equipment or systems prior to detail design or project planning	• Requires parametric cost relationships to be established • Limited frequently to specific subsystems or functional hardware of systems • Depends on quantity and quality of the data • Limited by data and number of independent variables
Equipment/subsystem analogy estimates (comparative)	Reprocurement Production Development Project planning	• Relatively simple • Low cost • Emphasizes incremental project and product changes • Good accuracy for similar systems	• Requires analogous product and project data • Limited to stable technology • Narrow range of electronic applications • May be limited to systems and equipment built by the same firm
Expert opinion	All project phases	• Available when there are insufficient data, parametric cost relationships, or project/product definition	• Subject to bias • Increased product or project complexity can degrade estimates • Estimate substantiation is not quantifiable

13.20 LOGISTICS SUPPORT

There is a class of projects called "material" projects where the deliverable may require maintenance, service, and support after development. This support will continue throughout the life cycle of the deliverable. Providing service to these deliverables is referred to as logistics support.

The two key parameters used to evaluate the performance of material systems are supportability and readiness. Supportability is the ability to maintain or acquire the necessary human and nonhuman resources to support the system. Readiness is a measure of how good we are at keeping the system performing as planned and how quickly we can make repairs during a shutdown. Clearly, proper planning during the design stage of a project can reduce supportability requirements, increase operational readiness, and minimize or lower logistics support costs.

The ten elements of logistics support are:

1. *Maintenance planning:* The process conducted to evolve and establish maintenance concepts and requirements for the lifetime of a materiel system.
2. *Manpower and personnel:* The identification and acquisition of personnel with the skills and grades required to operate and support a material system over its lifetime.
3. *Supply support:* All management actions, procedures, and techniques used to determine requirements to acquire, catalog, receive, store, transfer, issue, and dispose of secondary items. This includes provisioning for initial support as well as replenishment supply support.
4. *Support equipment: All equipment (mobile or fixed) required to support the operation and maintenance of a materiel system. This includes associated multiuse end-items; ground-handling and maintenance equipment; tools, metrology, and calibration equipment; and test and au*tomatic test equipment. It includes the acquisition of logistics support for the support and test equipment itself.
5. *Technical data:* Recorded information regardless of form or character (such as manuals and drawings) of a sc*ientific or technical nature. Computer programs and related software are not technical data; documentation of computer programs and related software are. Also other information related t*o contract administration.
6. *Training and training support:* The processes, procedures, techniques, training devices, and equipment used to train *personnel to operate and support a materiel system. This includes individual and crew training; new equipment training; initial, formal, and on-the-job training; and logistic support p*lanning for training equipment and training device acquisitions and installations.
7. *Computer resource support: The facilities, hardware, software, documentation, manpower, and personnel needed t*o operate and support embedded computer systems.
8. *Facilities:* The permane*nt or semipermanent real property assets required to support the materiel system. Facilitie*s management includes conducting studies to define types of facilities or facility improvement, locations, space needs, environment requirements, and equipment.
9. *Packaging, handling, storage, and transportation:* The resources, processes, procedures, design considerations, a*nd methods to ensure that all system, equipment, and support items are preserved, packaged,* handled, and transported properly. This includes environmental considerations and equipment preservation requirements for short- and long-term storage and transportability.
10. *Design interface:* The relationship of logistics-related design parameters to readiness and support resource requirements. These logistics-related design parameters are expressed in operational terms rather than as inherent values and specifically relate to system readiness objectives and support costs of the material system.

13.21 ECONOMIC PROJECT SELECTION CRITERIA: CAPITAL BUDGETING

PMBOK® Guide, 6th Edition
1.2.6.1 Business Case

Project managers are often called upon to be active participants during the benefit-to-cost analysis of project selection. It is highly unlikely that companies will approve a project where the costs exceed the benefits. Benefits can be measured in either financial or nonfinancial terms.

The process of identifying the financial benefits is called capital budgeting, which may be defined as the *decision-making process* by which organizations evaluate projects that include the purchase of major fixed assets such as buildings, machinery, and equipment. Sophisticated capital budgeting techniques take into consideration depreciation schedules, tax information, and cash flow. Since only the principles of capital budgeting will be discussed in this text, we will restrict ourselves to the following topics:

- Payback Period
- Discounted Cash Flow (DCF)
- Net Present Value (NPV)
- Internal Rate of Return (IRR)

13.22 PAYBACK PERIOD

PMBOK® Guide, 6th Edition
1.2.6.1 Business Case

The payback period is the exact length of time needed for a firm to recover its initial investment as calculated from cash inflows. Payback period is the *least* precise of all capital budgeting methods because the calculations are in dollars and not adjusted for the time value of money. Table 13–13 shows the cash flow stream for Project A.

From Table 13–13, Project A will last for exactly five years with the cash inflows shown. The payback period will be exactly four years. If the cash inflow in Year 4 were $6,000 instead of $5,000, then the payback period would be three years and 10 months.

The problem with the payback method is that $5,000 received in Year 4 is not worth $5,000 today. This unsophisticated approach mandates that the payback method be used as a supplemental tool to accompany other methods.

TABLE 13–13. CAPITAL EXPENDITURE DATA FOR PROJECT A

Initial Investment	**Expected Cash Inflows**				
$10,000	*Year 1*	*Year 2*	*Year 3*	*Year 4*	*Year 5*
	$1,000	$2,000	$2,000	$5,000	$2,000

13.23 THE TIME VALUE OF MONEY AND DISCOUNTED CASH FLOW (DCF)

***PMBOK*® *Guide*, 6th Edition**
1.2.6.1 Business Case

Everyone knows that a dollar today is worth more than a dollar a year from now. The reason for this is because of the time value of money. To illustrate the time value of money, let us look at the following equation:

$$FV = PV(1+k)^n$$

$$\begin{aligned} \text{where FV} &= \text{Future value of an investment} \\ \text{PV} &= \text{Present value} \\ k &= \text{Investment interest rate (or cost of capital)} \\ n &= \text{Number of years} \end{aligned}$$

Using this formula, we can see that an investment of $1,000 today (PV) invested at 10% (k) for one year (n) will give us a future value of $1,100. If the investment is for two years, then the future value would be worth $1,210.

Now, let us look at the formula from a different perspective. If an investment yields $1,000 a year from now, then how much is it worth *today* if the cost of money is 10%? To solve the problem, we must discount future values to the present for comparison purposes. This is referred to as discounted cash flows (DCF).

The previous equation can be written as: FV

$$PV = \frac{FV}{(1+k)^n}$$

Using the data given: $1,000

$$PV = \frac{\$1,000}{(1+0.1)^1} = \$909$$

Therefore, $1,000 a year from now is worth only $909 today. If the interest rate, k, is known to be 10%, then you should *not* invest more than $909 to get the $1,000 return a year from now. However, if you could purchase this investment for $875, your interest rate would be more than 10%.

Discounting cash flows to the present for comparison purposes is a viable way to assess the value of an investment. As an example, you have a choice between two investments. Investment A will generate $100,000 two years from now and investment B will generate $110,000 three years from now. If the cost of capital is 15%, which investment is better?

Using the formula for discounted cash flow, we find that:

$$PV_A = \$75,614$$
$$PV_B = \$72,327$$

This implies that a return of $100,000 in two years is worth more to the firm than a $110,000 return three years from now.

13.24 NET PRESENT VALUE (NPV)

PMBOK® Guide, 6th Edition
1.2.6.1 Business Case

The net present value (NPV) method is a sophisticated capital budgeting technique that equates the discounted cash flows against the initial investment. Mathematically, FV_t

$$NPV=\sum_{t=1}^{n}\left[\frac{FV_t}{(1+k)^t}\right]-II$$

where FV is the future value of the cash inflows, II represents the initial investment, and k is the discount rate equal to the firm's cost of capital. Table 13–14 calculates the NPV for the data provided previously in Table 13–13 using a discount rate of 10%.

This indicates that the cash inflows discounted to the present will *not* recover the initial investment. This, in fact, is a bad investment to consider. Previously, we stated that the cash flow stream yielded a payback period of four years. However, using discounted cash flow, the actual payback is greater than five years, assuming that there will be cash inflow in years 6 and 7.

If in Table 13–14 the initial investment was \$5,000, then the net present value would be \$3,722. The decision-making criteria using NPV are as follows:

- If the NPV is greater than or equal to zero dollars, accept the project.
- If the NPV is less than zero dollars, reject the project.

A positive value of NPV indicates that the firm will earn a return equal to or greater than its cost of capital.

13.25 INTERNAL RATE OF RETURN (IRR)

PMBOK® Guide, 6th Edition
1.2.6.1 Business Case

The internal rate of return (IRR) is perhaps the most sophisticated capital budgeting technique and also more difficult to calculate than NPV. The internal rate of return is the discount rate where the present

TABLE 13–14, NPV CALCULATION FOR PROJECT A

Year	Cash Inflows	Present Value
1	$1,000	$ 909
2	2,000	1,653
3	2,000	1,503
4	5,000	3,415
5	2,000	1,242
	Present value of cash inflows	$ 8,722
	Less investment	10,000
	Net Present Value	<1,278>

TABLE 13–15. IRR CALCULATION FOR PROJECT A

IRR	NPV
10%	$3,722
20%	1,593
25%	807
30%	152
31%	34
32%	<78>

value of the cash inflows exactly equals the initial investment. In other words, IRR is the discount rate when NPV = 0. Mathematically FV*t*

$$\sum_{t=1}^{n}\left[\frac{FV_t}{(1+IRR)^t}\right]-II=0$$

The solution to problems involving IRR is basically a trial-and-error solution. Table 13–15 shows that with the cash inflows provided, and with a $5,000 initial investment, an IRR of 10% yielded a value of $3,722 for NPV. Therefore, as a second guess, we should try a value greater than 10% for IRR to generate a zero value for NPV. Table 13–15 shows the final calculation.

The table implies that the cash inflows are equivalent to a 31% return on investment. Therefore, if the cost of capital were 10%, this would be an excellent investment. Also, this project is "probably" superior to other projects with a lower value for IRR.

13.26 COMPARING IRR, NPV, AND PAYBACK

***PMBOK*® *Guide*, 6th Edition**
1.2.6.1 Business Case

For most projects, both IRR and NPV will generate the same accept-reject decision. However, there are differences that can exist in the underlying assumptions that can cause the projects to be ranked differently. The major problem is the differences in the magnitude and timing of the cash inflows. NPV assumes that the cash inflows are reinvested at the cost of capital, whereas IRR assumes reinvestment at the project's IRR. NPV tends to be a more conservative approach.

The timing of the cash flows is also important. Early year cash inflows tend to be at a lower cost of capital and are more predictable than later year cash inflows. Because of the downstream uncertainty, companies prefer larger cash inflows in the early years rather than the later years.

Magnitude and timing are extremely important in the selection of capital projects. Consider Table 13–16.

TABLE 13–16. CAPITAL PROJECTS

Project	IRR	Payback Period with DCF
A	10%	1 year
B	15%	2 years
C	25%	3 years
D	35%	5 years

If the company has sufficient funds for one and only one project, the natural assumption would be to select Project D with a 35% IRR. Unfortunately, companies shy away from long-term payback periods because of the relative uncertainties of the cash inflows after Year 1.

13.27 RISK ANALYSIS

PMBOK® Guide, 6th Edition
11.4.2. Perform Quantitative Risk Analysis

Suppose you have a choice between two projects, both of which require the same initial investment, have identical net present values, and require the same yearly cash inflows to break even. If the cash inflow of the first investment has a probability of occurrence of 95% and that of the second investment is 70%, then risk analysis would indicate that the first investment is better.

Risk analysis refers to the chance that the selection of this project will prove to be unacceptable. In capital budgeting, risk analysis is almost entirely based upon how well we can predict cash inflows since the initial investment is usually known with some degree of certainty. The inflows, of course, are based upon sales projections, taxes, cost of raw materials, labor rates, and general economic conditions.

Sensitivity analysis is a simple way of assessing risk. A common approach is to estimate NPV based upon an optimistic (best case) approach, most likely (expected) approach, and pessimistic (worst case) approach. This can be illustrated using Table 13–17. Both Projects A and B require the same initial investment of $10,000, with a cost of capital of 10%, and with expected five-year annual cash inflows of $5,000/year. The range for Project A's NPV is substantially less than that of Project B, thus implying that Project A is less risky. A risk lover might select Project B because of the potential reward of $27,908, whereas a risk avoider would select Project A, which offers perhaps no chance for loss.

13.28 CAPITAL RATIONING

PMBOK® Guide, 6th Edition
11.4.2 Perform Quantitative Risk Analysis

Capital rationing is the process of selecting the best group of projects such that the highest overall net present value will result without exceeding the total budget available. An assumption with capital

TABLE 13–17. SENSITIVITY ANALYSIS

Initial Investment	Project A $10,000	Project B $10,000
	Annual Cash Inflows	
Optimistic	$ 8,000	$10,000
Most likely	5,000	5,000
Pessimistic	3,000	1,000
Range	$ 5,000	$ 9,000
	Net Present Values	
Optimistic	$20,326	$27,908
Most likely	8,954	8,954
Pessimistic	1,342	<6,209>
Range	$18,984	$34,117

rationing is that the projects under consideration are mutually exclusive. There are two approaches often considered for capital rationing.

The internal rate of return approach plots the IRRs in descending order against the cumulative dollar investment. The resulting figure is often called an investment opportunity schedule. As an example, suppose a company has $300,000 committed for projects and must select from the projects identified in Table 13–18. Furthermore, assume that the cost of capital is 10%.

Figure 13–13 shows the investment opportunity schedule. Project G should not be considered because the IRR is less than the firm's cost of capital, but we should select Projects, A, B, and C, which will consume $280,000 out of a total budget of $300,000. This allows us to have the three largest IRRs.

The problem with the IRR approach is that it does not guarantee that the projects with the largest IRRs will maximize the total dollar returns. The reason is that not all of the funds have been consumed.

TABLE 13–18. PROJECTS UNDER CONSIDERATION

Project	Investment	IRR	Discounted Cash Flows at 10%
A	$ 50,000	20%	$116,000
B	120,000	18%	183,000
C	110,000	16%	147,000
D	130,000	15%	171,000
E	90,000	12%	103,000
F	180,000	11%	206,000
G	80,000	8%	66,000

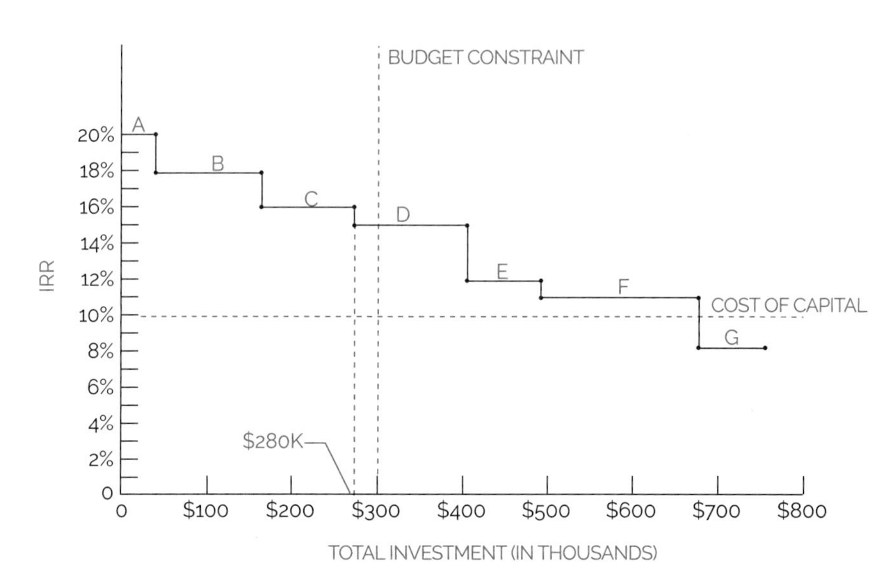

FIGURE 13–13. Investment opportunity schedule (IOS) for Table 13–18.

A better approach is the net present value method. In this method, the projects are again ranked according to their IRRs, but the combination of projects selected will be based upon the highest net present value. As an example, the selection of Projects A, B, and C from Table 13–18 requires an initial investment of $280,000 with resulting discounted cash flows of $446,000. The net present value of Projects A, B, and C is, therefore, $166,000. This assumes that unused portions of the original budget of $300,000 do not gain or lose money. However, if we now select Projects A, B, and D, we will invest $300,000 with a net present value of $170,000 ($470,000 less $300,000). Selection of Projects A, B, and D will, therefore, maximize net present value.

13.29 PROJECT FINANCING

Project financing involves the establishment of a legally independent project company, usually for large-scale investments (LSI) and long term where the providers of funds are repaid out of cash flow and earnings, and where the assets of the unit (and only the unit) are used as collateral for the loans. Debt repayment would come from the project company only rather than from any other entity. A risk with project financing is that the capital assets may have a limited life. The potential limited life constraint often makes it difficult to get lenders to agree to long-term financial arrangements.

Another critical issue with project financing especially for high-technology projects is that the projects are generally long term. It may be nearly eight to ten years before service will begin, and in terms of technology, eight years can be an eternity. Project financing is often considered a "bet on the future." And if the project were to fail, the company could be worth nothing after liquidation.

There are several risks that must be considered to understand project financing. The risks commonly considered are

Financial Risks

- Use of project versus corporate financing
- Use of corporate bonds, stock, zero coupon bonds, and bank notes
- Use of secured versus unsecured debt
- The best sequence or timing for raising capital
- Bond rating changes
- Determination of the refinancing risk, if necessary

Development Risks

- Reality of the assumptions
- Reality of the technology
- Reality of development of the technology
- Risks of obsolescence

Political Risks

- Sovereignty risks
- Political instability
- Terrorism and war
- Labor availability
- Trade restrictions
- Macroeconomics such as inflation, currency conversion, and transferability of funding and technology

Organizational Risks

- Members of the board of directors
- Incentives for the officers
- Incentives for the board members
- Bonuses as a percentage of base compensation
- Process for the resolution of disputes

Execution Risks

- Timing when execution will begin
- Life expectancy of execution
- Ability to service debt during execution

Related Case Studies (from Kerzner/*Project Management Case Studies*, 5th ed.)	**Related Workbook Exercises (from Kerzner/*Project Management Workbook and PMP®/CAPM® Exam Study Guide*, 12th ed.)**	***PMBOK® Guide*, 6th Edition, Reference Section for the PMP® Certification Exam**
• Capital Industries • Small Project Cost Estimating at Percy Company • Cory Electric • Camden Construction Corporation • Payton Corporation • The Estimating Problem*	• The Automobile Problem • Life-Cycle Costing • Multiple Choice Exam	• Project Integration Management • Project Scope Management • Project Cost Management

*Case study also appears at end of chapter.

13.30 STUDYING TIPS FOR THE PMI® PROJECT MANAGEMENT CERTIFICATION EXAM

This section is applicable as a review of the principles to support the knowledge areas and domain groups in the *PMBOK® Guide*. This chapter addresses:

- Project Integration Management
- Project Scope Management
- Project Schedule Management
- Project Management
- Initiating
- Planning

Understanding the following principles is beneficial if the reader is using this text to study for the PMP® Certification Exam:

- What is meant by cost-estimating relationships (CER)
- Three basic types of estimates
- Relative accuracy of each type of estimate and the approximate time to prepare the estimate
- Information that is needed to prepare the estimates (i.e., labor, material, overhead rates, etc.)
- Importance of backup data for costs
- Estimating pitfalls
- Concept of rolling wave planning
- What is meant by life cycle costing
- Different ways of evaluating a project's financial feasibility or benefits (i.e., ROI, payback period, net present value, internal rate of return, depreciation, scoring models)

The following multiple-choice questions will be helpful in reviewing the principles of this chapter:

1. Which of the following is a valid way of evaluating the financial feasibility of a project?
   A. Return on investment
   B. Net present value
   C. Internal rate of return
   D. All of the above
2. The three common classification systems for estimates includes all of the following except:
   A. Parametric estimates
   B. Quick-and-dirty estimates
   C. Analogy estimates
   D. Engineering estimates

3. The most accurate estimates are:
   A. Parametric estimates
   B. Quick-and-dirty estimates
   C. Analogy estimates
   D. Engineering estimates
4. Which of the following is considered to be a bottom-up estimate rather than a top-down estimate?
   A. Parametric estimates
   B. Analogy estimates
   C. Engineering estimates
   D. None of the above
5. Which of the following would be considered as a cost-estimating relationship (CER)?
   A. Mathematical equations based upon regression analysis
   B. Learning curves
   C. Cost–cost or cost–quantity relationships
   D. All of the above
6. If a worker earns $30 per hour in salary but the project is charged $75 per hour for each hour the individual works, then the overhead rate is:
   A. 100%
   B. 150%
   C. 250%
   D. None of the above
7. Information supplied to a customer to support the financial data provided in a proposal is commonly called:
   A. Backup data
   B. Engineering support data
   C. Labor justification estimates
   D. Legal rights estimates
8. Estimating pitfalls can result from:
   A. Poorly defined statement of work
   B. Failure to account for risks in the estimates
   C. Using the wrong estimating techniques
   D. All of the above
9. The source of many estimating risks is:
   A. Poorly defined requirements
   B. An inexperienced project manager
   C. Lack of management support during estimating
   D. All of the above

10. A project where the scope evolves as the work takes place is called either progressive planning or:
    A. Synchronous planning
    B. Continuous planning
    C. Rolling wave planning
    D. Continuous reestimation planning
11. The calculation of the total cost of a product, from R&D to operational support and disposal, is called:
    A. Birth-to-death costing
    B. Life-cycle costing
    C. Summary costing
    D. Depreciation costing

ANSWERS

1. D
2. B
3. D
4. C
5. D
6. B
7. A
8. D
9. D
10. C
11. B

PROBLEMS

13–1 How does a project manager price out a job in which the specifications are not prepared until the job is half over?

13–2 With reference to Figure 13.7, under what conditions could *each* of the following situations occur:

a. Project manager and project office determine labor hours by pricing out the work breakdown structure without coordination with functional management.
b. Upper-level management determines the price of a bid without forming a project office or consulting functional management.
c. Perturbations on the base case are not performed.
d. The chief executive officer selects the project manager without consulting his directors.
e. Upper-level management does not wish to have a cost review meeting prior to submittal of a bid.

13–3 How can upper-level management use the functional cost and hour summary to determine manpower planning for the entire company? How would you expect management to react if the functional cost and hour summary indicated a shortage or an abundance of trained personnel?

13–4 Two contractors decide to enter into a joint venture on a project. What difficulties can occur if the contractors have decided on who does what work, but changes may take place if problems occur? What happens if one contractor has higher salary levels and overhead rates?

13–5 The Jones Manufacturing Company is competing for a production contract that requires that work begin in January 2016. The cost package for the proposal must be submitted by July 2015. The business base, and therefore the overhead rates, are uncertain because Jones has the possibility of winning another contract, to be announced in September 2015. How can the impact of the announcement be included in the proposal? How would you handle a situation where another contract may not be renewed after January 2016 (i.e., assume that the announcement would not be made until March)?

13–6 During initial pricing activities, one of the functional managers discovers that the work breakdown structure requires costing data at a level that is not normally made, and will undoubtedly incur additional costs. How should you, as a project manager, respond to this situation? What are your alternatives?

13–7 Should the project manager give the final manpower loading curves to the functional managers? If so, at what point in time?

13–8 Should a project manager be appointed in the bidding stage of a project? If so, what authority should he have, and who is responsible for winning the contract?

13–9 Explain how useful each of the following can be during the estimating of project costs:

a. Contingency planning and estimating
b. Using historical databases
c. Usefulness of computer estimating
d. Usefulness of performance factors to account for inefficiencies and uncertainties.

CASE STUDY

THE ESTIMATING PROBLEM

Barbara just received the good news: She was assigned as the project manager for a project that her company won as part of competitive bidding. Whenever a request for proposal (RFP) comes into Barbara's company, a committee composed mainly of senior managers reviews the RFP. If the decision is made to bid on the job, the RFP is turned over to the Proposal Department. Part of the Proposal Department is an estimating group that is responsible for estimating all work. If the estimating group has no previous history concerning some of the deliverables or work packages and is unsure about the time and cost for the work, the estimating team will then ask the functional managers for assistance with estimating.

Project managers like Barbara do not often participate in the bidding process. Usually, their first knowledge about the project comes after the contract is awarded to their company and they are assigned as the project manager. Some project managers are highly optimistic and trust the estimates that were submitted in the bid implicitly unless, of course, a significant span of time has elapsed between the date of submittal of the proposal and the final contract award date. Barbara, however, is somewhat pessimistic. She believes that accepting the estimates as they were submitted in the proposal is like playing Russian roulette. As such, Barbara prefers to review the estimates.

One of the most critical work packages in the project was estimated at twelve weeks using one grade 7 employee full time. Barbara had performed this task on previous projects and it required one person full time for fourteen weeks. Barbara asked the estimating group how they arrived at this estimate. The estimating group responded that they used the three-point estimate where the optimistic time was four weeks, the most likely time was thirteen weeks, and the pessimistic time was sixteen weeks.

Barbara believed that the three-point estimate was way off of the mark. The only way that this work package could ever be completed in four weeks would be for a very small project nowhere near the complexity of Barbara's project. Therefore, the estimating group was not considering any complexity factors when using the three-point estimate. Had the estimating group used the triangular distribution where each of the three estimates had an equal likelihood of occurrence, the final estimate would have been thirteen weeks. This was closer to the fourteen weeks that Barbara thought the work package would take. While a difference of 1 week seems small, it could have a serious impact on Barbara's project and incur penalties for late delivery.

Barbara was now still confused and decided to talk to Peter, the employee who was assigned to do this task. Barbara had worked with Peter on previous projects. Peter was a grade 9 employee and considered to be an expert in this work package. As part of the discussions with Barbara, Peter made the following comments:

> I have seen estimating data bases that include this type of work package and they all estimate the work package at about 14 weeks. I do not understand why our estimating group prefers to use the three point estimate.

"Does the typical data base account for project complexity when considering the estimates?" asked Barbara. Peter responded:

> Some data bases have techniques for considering complexity, but mostly they just assume an average complexity level. When complexity is important, as it is in our project, analogy estimating would be better. Using analogy estimating and comparing the complexity of the work package on this project to the similar works packages I have completed, I would say that 16–17 weeks is closer to reality, and let's hope I do not get removed from the project to put out a fire somewhere else in the company. That would be terrible. It is impossible for me to get it done in 12 weeks. And adding more people to this work package will not shorten the schedule. It may even make it worse.
>
> Barbara then asked Peter one more question:
>
> Peter, you are a grade 9 and considered as the subject matter expert. If a grade 7 had been assigned, as the estimating group had said, how long would it have taken the grade 7 to do the job?
>
> "Probably about 20 weeks or so," responded Peter.

QUESTIONS

1. How many different estimating techniques were discussed in the case?
2. If each estimate is different, how does a project manager decide that one estimate is better than another?
3. If you were the project manager, which estimate would you use?

14 Cost Control

14.0 INTRODUCTION

PMBOK® Guide 6th Edition
7.0 Project Cost Management
7.1.3.1 Cost Management Plan
7.4 Control Costs

Cost control is important to all companies, regardless of size. Small companies generally have tighter monetary controls because the failure of even one project can put the company at risk, but they have less sophisticated control techniques. Large companies may have the luxury to spread project losses over several projects, whereas the small company may have few projects.

Many people have a poor understanding of cost control yet it is crucial to long-term success. Cost control is not only "monitoring" costs and recording data, but also analyzing the data in order to take corrective action before it is too late. It should be performed by all personnel who incur costs, not merely the project office.

Cost control implies good cost management, which must include:

- Cost estimating
- Cost accounting
- Project cash flow
- Company cash flow
- Direct labor costing
- Overhead rate costing
- Other tactics, such as incentives, penalties, and profit-sharing

Cost control is actually a subsystem of the management cost and control system (MCCS) rather than a complete system per se. This is shown in Figure 14–1, where the MCCS is represented as a two-cycle process: a planning cycle and an operating cycle. The operating cycle is what is commonly referred to as the

PMBOK is a registered mark of the Project Management Institute, Inc.

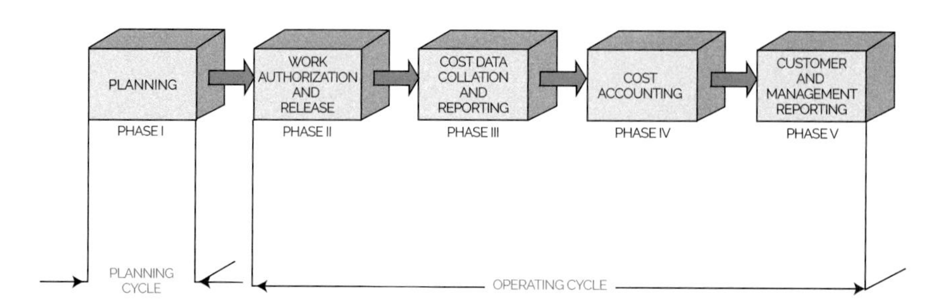

FIGURE 14–1. Phases of a management cost and control system.

cost control system. Failure of a cost control system to accurately describe the true status of a project does not necessarily imply that the cost control system is at fault. Any cost control system is only as good as the original plan against which performance will be measured. Therefore, the designing of a planning system must take into account the cost control system. For this reason, it is common for the planning cycle to be referred to as planning and control, whereas the operating cycle is referred to as cost (or cost monitoring and collection) and control.

The planning and control system must help management project the status toward objective completion. Its purpose is to establish policies, procedures, and techniques that can be used in the day-to-day management and control of projects and programs. It must, therefore, provide information that:

- Gives a picture of true work progress
- Will relate cost and schedule performance
- Identifies potential problems with respect to their sources.
- Provides information to project managers with a practical level of summarization
- Demonstrates that the milestones are valid, timely, and auditable

The planning and control system, in addition to being a tool by which objectives can be defined (i.e., hierarchy of objectives and organization accountability), exists as a tool to develop planning, measure progress, and control change. As a tool for planning, the system must be able to:

- Plan and schedule work
- Identify those indicators that will be used for measurement
- Establish direct labor budgets
- Establish overhead budgets
- Identify management reserve

The project budget that results from the planning cycle of the MCCS must be reasonable, attainable, and based on contractually negotiated costs and the statement of work. The basis for the budget is either historical cost, best estimates, or industrial engineering standards. The budget must be based upon an acceptable level of accuracy in the estimates. The budget must identify planned manpower requirements, contract-allocated funds, and management reserve. All of this information becomes the basis for the cost management plan which describes how cost are planned, structured, and controlled over the life of the project.

Establishing budgets requires that the planner fully understand the meaning of standards. There are two categories of standards. Performance results standards are quantitative measurements and include such items as quality of work, quantity of work, cost of work, and time-to-complete. Process standards are qualitative, including personnel, functional, and physical factors relationships. Standards are advantageous in that they provide a means for unity, a basis for effective control, and an incentive for others. The disadvantage of standards is that performance is often frozen, and employees are quite often unable to adjust to the differences.

As a tool for measuring progress and controlling change, the systems must be able to:

- Measure resources consumed
- Measure status and accomplishments
- Compare measurements to projections and standards
- Provide the basis for diagnosis and replanning

For MCCS to be effective, both the scheduling and budgeting systems must be disciplined and formal in order to prevent inadvertent or arbitrary budget or schedule changes. This does *not* mean that the baseline budget and schedule, once established, is static or inflexible. Rather, it means that changes must be controlled and result only from deliberate management actions.

Disciplined use of MCCS is designed to put pressure on the project manager to perform exceptionally good project planning so that changes will be minimized. As an example, government subcontractors may not:

- Make retroactive changes to budgets or costs for work that has been completed
- Rebudget work-in-progress activities
- Transfer work or budget independently of each other
- Reopen closed work packages

14.1 UNDERSTANDING CONTROL

PMBOK® Guide 6th Edition
1.2.4.5 Project Management Process Groups: Monitoring and Control

Effective management of a program during the operating cycle requires that a well-organized cost and control system be designed, developed, and implemented so that immediate feedback can be obtained, whereby the up-to-date usage or resources can be compared to target objectives established during the planning cycle. The requirements for an effective control system (for both cost and schedule/performance) should include[1]:

- Thorough planning of the work to be performed to complete the project
- Good estimating of time, labor, and costs
- Clear communication of the scope of required tasks
- A disciplined budget and authorization of expenditures
- Timely accounting of physical progress and cost expenditures

1. Russell D. Archibald, *Managing High-Technology Programs and Projects* (New York: John Wiley & Sons, 1976), p. 191.

- Periodic reestimation of time and cost to complete remaining work
- Frequent, periodic comparison of actual progress and expenditures to schedules and budgets, both at the time of comparison and at project completion

Management must compare the time, cost, and performance of the program to the budgeted time, cost, and performance, not independently but in an integrated manner. Being within one's budget at the proper time serves no useful purpose if performance is only 75 percent. Likewise, having a production line turn out exactly 200 items, as planned, loses its significance if a 50 percent cost overrun is incurred. All three resource parameters (time, cost, and performance) must be analyzed as a group, or else we might "win the battle but lose the war." The use of the expression "management cost and control system" is vague in that the implication is that only costs are controlled. This is not true—an effective control system monitors schedule and performance as well as costs by setting budgets, measuring expenditures against budgets and identifying variances, assuring that the expenditures are proper, and taking corrective action when required.

The WBS is the total project broken down into successively lower levels until the desired control levels are established. The work breakdown structure therefore serves as the tool from which performance can be subdivided into objectives and subobjectives. As work progresses, the WBS provides the framework on which costs, time, and schedule/performance can be compared against the budget for each level of the WBS.

The first purpose of control therefore becomes a verification process accomplished by the comparison of actual performance to date with the predetermined plans and standards set forth in the planning phase. The comparison serves to verify that:

- The objectives have been successfully translated into performance standards.
- The performance standards are, in fact, a reliable representation of program activities and events.
- Meaningful budgets have been established such that actual versus planned comparisons can be made.

In other words, the comparison verifies that the correct standards were selected, and that they are properly used.

The second purpose of control is decision-making. Three useful reports are required by management in order to make effective and timely decisions:

- The project plan, schedule, and budget prepared during the planning phase
- A detailed comparison between resources expended to date and those predetermined. This includes an estimate of the work remaining and the impact on activity completion.
- A projection of resources to be expended through program completion

These reports, supplied to the managers and the doers, provide three useful results:

- Feedback to management, the planners, and the doers
- Identification of any major deviations from the current program plan, schedule, or budget

- The opportunity to initiate contingency planning early enough that cost, performance, and time requirements can undergo corrected action without loss of resources

These reports provide management with the opportunity to minimize downstream changes by making proper corrections here and now. As shown in Figures 14–2 and 14–3, cost reductions are more available in the early project phases, but are reduced as we go further into the project life-cycle phases. Figure 14–3 identifies the people that most likely have the greatest influence on possibly initiating changes to a project. Downstream the cost of changes could easily exceed the original cost of the project. This is an example of the "iceberg" syndrome, where problems become evident too late in the project to be solved easily, resulting in a very high cost to correct them.

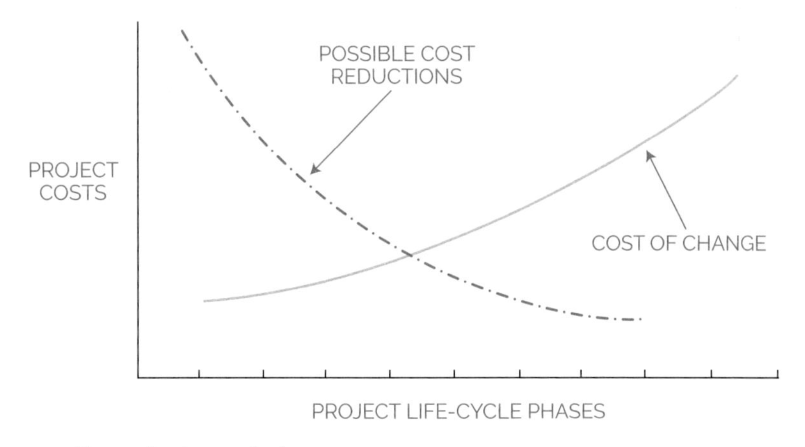

FIGURE 14–2. Cost reduction analysis.

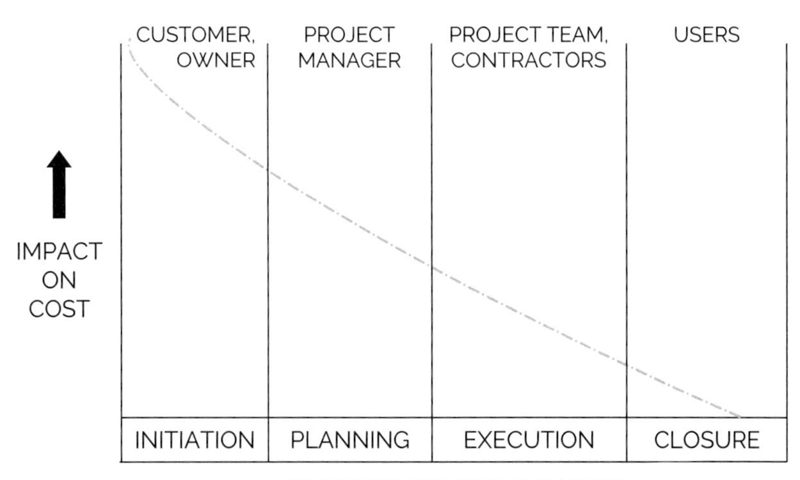

FIGURE 14–3. Cost influencers.

14.2 THE OPERATING CYCLE

The MCCS takes on paramount importance during the operating cycle of the project. The operating cycle is composed of four phases:

- Work authorization and release (phase II)
- Cost data collection and reporting (phase III)
- Cost analysis (phase IV)
- Reporting: customer and management (phase V)

These four phases, when combined with the planning cycle (phase I), constitute a closed system network that forms the basis for the management cost and control system.

After planning is completed and a contract is received, phase II is authorized via a work description document. The work description, or project work authorization form, is a contract that contains the narrative description, organization, and time frame for *each* WBS level. This multipurpose form is used to release the contract, authorize planning, record detail description of the work outlined in the work breakdown structure, and release work to the functional departments.

Contract services may require a work description form to release the contract. The contractual work description form sets forth general contractual requirements and authorizes program management to proceed.

Program management may then issue a subdivided work description form to the functional units so that work can begin. The subdivided work description specifies how contractual requirements are to be accomplished, the functional organizations involved, and their specific responsibilities, and authorizes the expenditure of resources within a particular time frame.

The work control center assigns a work order number to the subdivided work description form, if no additional instructions are required, and releases the document to the performing organizations. If additional instructions are required, the work control center can prepare a more detailed work-release document (shop traveler, tool order, work order release), assign the applicable work order number, and release it to the performing organization.

A work order number is required for all in-house direct and indirect charging. The work order number also serves as a cross-reference number for automatic assignment of the indentured work breakdown structure number to labor and material data records in the computer.

Small companies can avoid this additional paperwork cost by going directly from an awarded contract to a single work order, which may be the only work order needed for the entire contract.

14.3 COST ACCOUNT CODES

PMBOK® *Guide* 6th Edition
7.3.2.2 Cost Aggregation
7.4 Control Costs Tools and Techniques

Since project managers control resources through the line managers rather than directly, project managers end up controlling direct labor costs by opening and closing work orders. Work orders define the charge numbers for each cost account. By definition, a cost account is an identified level at a natural intersection point of the work breakdown structure and the organizational breakdown structure (OBS) at which functional responsibility for the work is assigned, and actual direct labor, material, and other direct costs are compared with actual work performed for management control purposes.

Cost accounts are the focal point of the MCCS and may comprise several work packages, as shown in Figure 14–4. Work packages are detailed short-span job or material items identified for the accomplishment of required work. Costs need not be reported at the work package level but can be rolled up to higher levels in the WBS. This referred to as cost aggregation.

To illustrate how the work packages are used, consider the cost account code breakdown shown in Figure 14–5 and the work authorization form shown in Figure 14–6. The work authorization form specifically identifies the cost centers that are "open" for this charge number, the man-hours available for each cost center, and the operational time period for the charge number. Because the exact dates of operation are completely defined, the charge number can be assigned perhaps as much as a year in advance of the work-begin date. This can be shown pictorially, as in Figure 14–7.

PMBOK® Guide 6th Edition
7.4 Control Costs Tools and Techniques

PMBOK® Guide 6th Edition
4.5 Monitor and Control Project Work

If a cost center needs additional time or additional man-hours, then a cost account change notice form must be initiated, usually by the requesting cost center, and approved by the project office. Figure 14–8 shows a typical cost account change notice form.

Cost data collection and reporting constitute the second phase of the operating cycle of the MCCS (and phase III in the overall MCCS). Actual cost (ACWP) and the budgeted cost for work performed (BCWP) for each contract or in-house project are accumulated in detailed cost accounts by cost center and cost element, and reported in accordance with the flow charts shown in Figure 14–9. These detailed elements, for both actual costs incurred and the budgeted cost for work performed, are usually made available monthly for all levels of the work breakdown structure. In addition, weekly supplemental direct labor reports can be printed showing the actual labor charge incurred, and can be compared to the predicted efforts.

Most weekly labor reports provide current month subtotals and previous month totals. Although these also appear on the detailed monthly report, they are included in the weekly report for a quick-and-dirty comparison. Year-to-date totals are usually not on the weekly report unless the users request them for an immediate comparison to the estimate at completion (EAC) and the work order release.

Weekly labor output is a vital tool for members of the program office in that these reports can indicate trends in cost and performance in sufficient time for contingency plans to be established and implemented. If these reports are not available, then cost and labor overruns would not be apparent until the following month when the detailed monthly labor, cost, and materials output was obtained.

Work order releases are used to authorize certain cost centers to begin charging their time to a specific cost reporting element. Work orders specify hours, not dollars. The hours indicate the "targets" that the program office would like to have the department shoot for. If the program office wished to be more specific and "compel" the departments to live within these hours, then the budgeted cost for work scheduled (BCWS) should be changed to reflect the reduced hours.

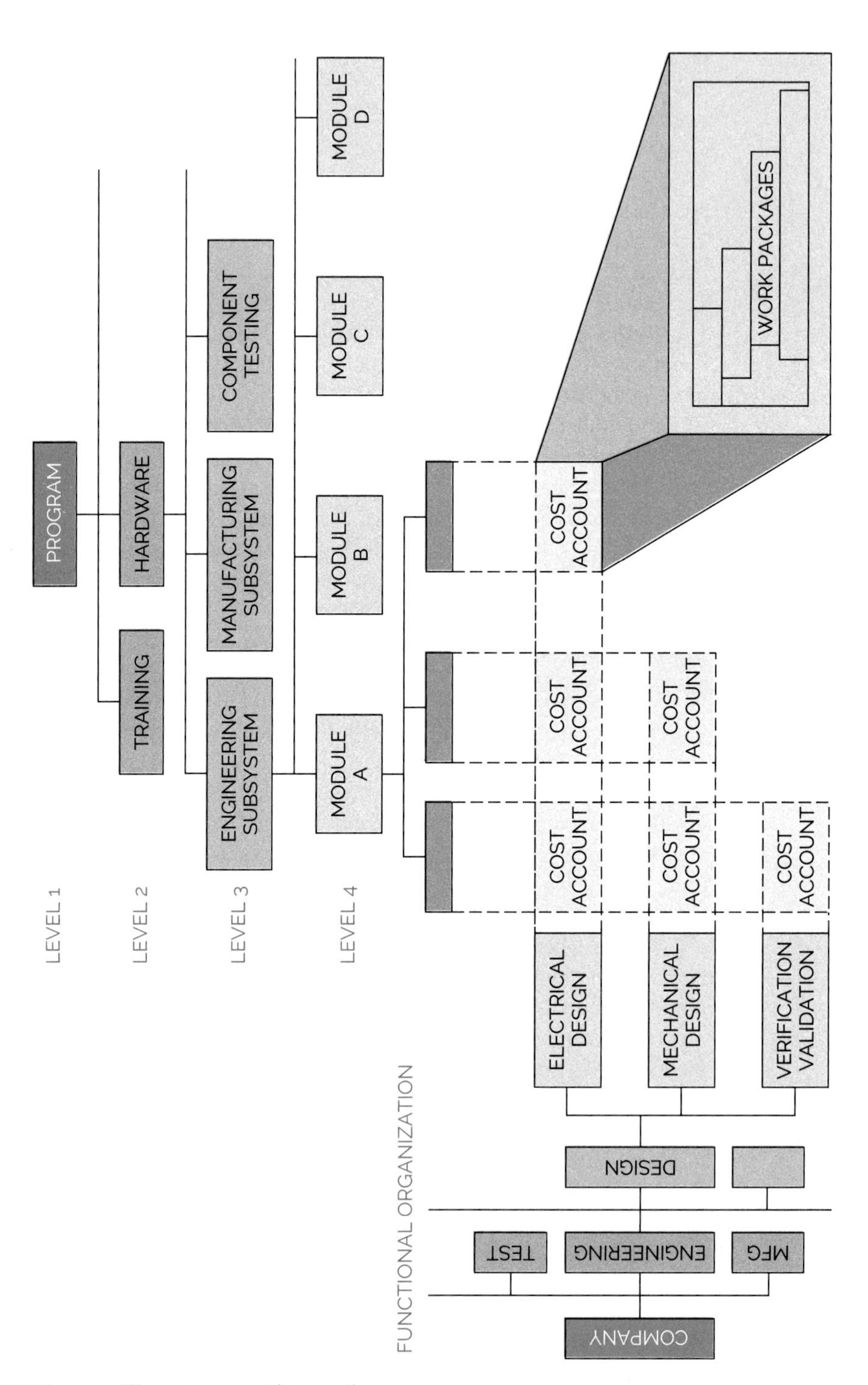

FIGURE 14–4. The cost account intersection.

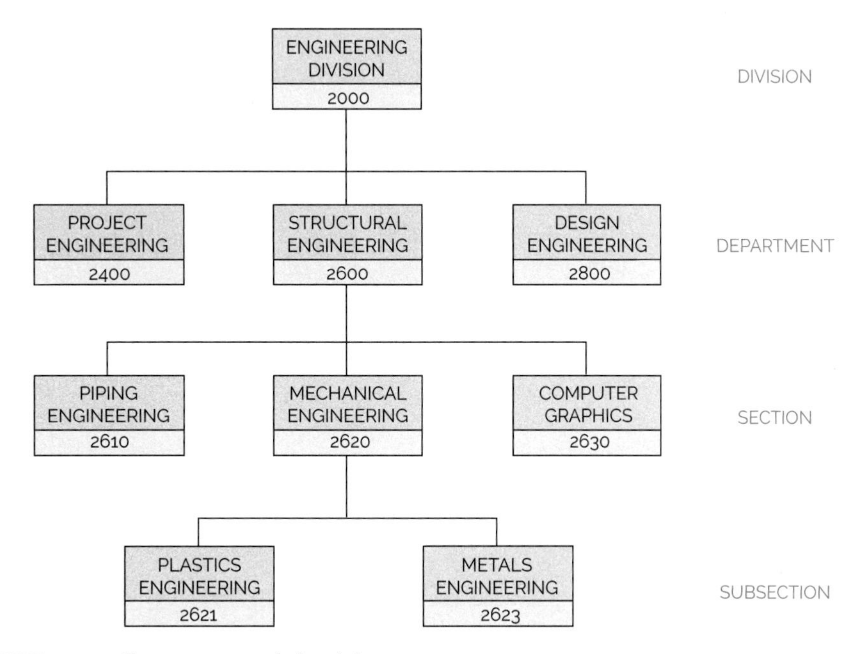

FIGURE 14–5. Cost account code breakdown.

WORK AUTHORIZATION FORM

WBS NO: 31-03-02 WORK ORDER NO: D1385

DATE OF ORIGINAL RELEASE: 3 FEB 16

DATE OF REVISION: : 18 MAR 16

REVISION NUMBER: : C

DESCRIPTION	COST CENTERS	HOURS	WORK BEGINS	WORK ENDS
TEST MATERIAL VB-2 IN ACCORDANCE WITH THE PROGRAM PLAN AND MIL STANDARD G1483-52. THIS TASK INCLUDES A WRITTEN REPORT.	2400 2610 2621 2623 5000*	150 160 140 46 600	1 AUG 16	15 SEPT 16

PROJECT OFFICE AUTHORIZATION SIGNATURE ____________________

*NOTE: SOME COMPANIES DO NOT PERMIT DIVISION COST CENTERS TO CHARGE AT LEVEL 3 OF THE WBS

FIGURE 14–6. Work authorization form.

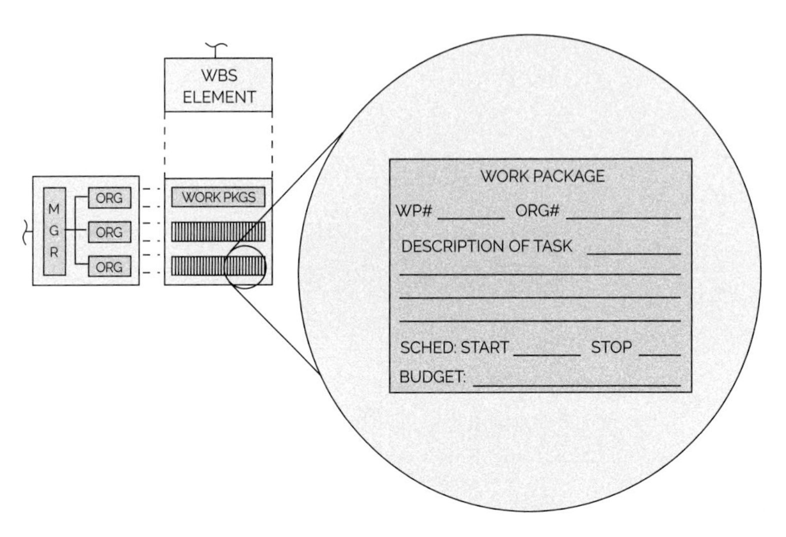

FIGURE 14–7. Planning and budgeting describe, plan, and schedule the work.

CACN No. ____________ Revision to Cost Account No. ________ Date ________

DESCRIPTION OF CHANGE:

____

____

____

____

____

____

REASON FOR CHANGE:

____

____

____

____

____

____

	Requested Budget	Authorized Budget	
Labor Hours	____________	____________	Period of Performance:
Material $	____________	____________	From ________
Indirect $	____________	____________	To ________

BUDGET SOURCE:

- ☐ Funded Contract Change
- ☐ Management Reserve
- ☐ Undistributed Budget
- ☐ Other ____________

APPROVALS: Program Mgr. ________
Prog. Control ________

INITIATED BY: ____________

FIGURE 14–8. Cost account change notice (CACN).

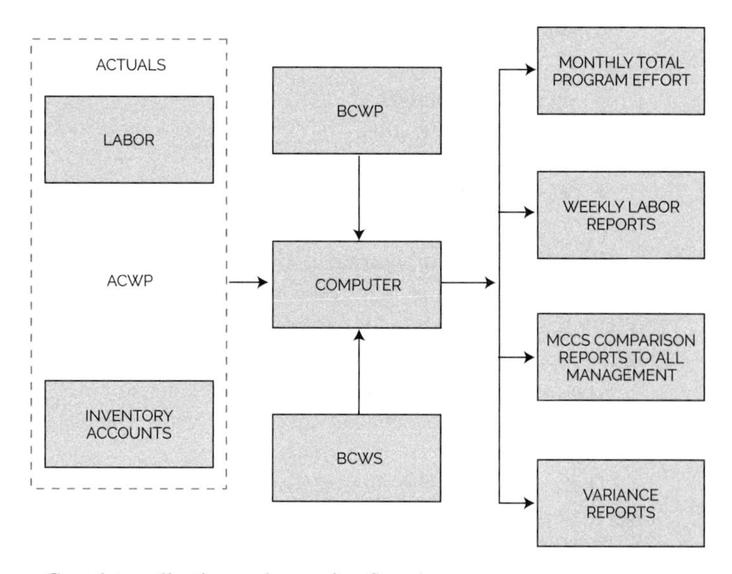

FIGURE 14–9. Cost data collection and reporting flowchart.

Four categories of cost data are normally accumulated:

- Labor
- Material
- Other direct charges
- Overhead

***PMBOK*® *Guide* 6th Edition**
7.4 Control Costs
7.2.2.6 Data Analysis
7.2.2.7 Project Management Information System

Project managers can maintain reasonable control over labor, material, and other direct charges. Overhead costs, on the other hand, are calculated yearly or monthly and applied retroactively to all applicable programs. Management reserves are often used to counterbalance the effects of adverse changes in overhead rates. Generally, management reserves are used for known-unknowns such as uncertainties in estimating.

14.4 BUDGETS

The project budget, which is the final result of the planning cycle of the MCCS, must be reasonable, attainable, and based on contractually negotiated costs and the statement of work. The basis for the budget is either historical cost, best estimates, or industrial engineering standards. The budget must identify planned manpower requirements, contract allocated funds, and management reserve.

All budgets must be traceable through the budget "log," which includes:

- Distributed budget
- Management reserve
- Undistributed budget
- Contract changes

The distributed or normal performance budget is the time-phased budget that is released through cost accounts and work packages. Management reserve is generally the dollar amount established for categories of unforeseen problems and contingencies resulting in special out-of-scope work to the performers.

The management reserve should be established based upon the project's risks. Some project may require no management reserve at all, whereas others may necessitate a reserve of 15 percent.

There is always the question of who should get to keep any unused management reserve at the end of the project. If the project is under a firm-fixed price contract, then the management reserve becomes extra profit for the performing organization. If the contract is a cost reimbursable type, all or part of the unused management reserve may have to be returned to the customer.

Although the management reserve may appear as a line item in the work breakdown structure, it is neither part of the distributed budget nor part of the cost baseline. Budgets are established on the assumption that they will be spent, whereas management reserve is money that you try not to spend. It would be inappropriate to consider the management reserve as an undistributed budget.

In addition to the "normal" performance budget and the management reserve budget, there are two other budgets:

- Undistributed budget, which is that budget associated with contract changes where time constraints prevent the necessary planning to incorporate the change into the performance budget. (This effort may be time-constrained.)
- Unallocated budget, which represents a logical grouping of contract tasks that have not yet been identified and/or authorized.

14.5 THE EARNED VALUE MEASUREMENT SYSTEM (EVMS)

In the early years of project management, it became evident that project managers were having difficulty determining project status. Some people believed that status could be determined only by a mystical approach, such as a fortune teller.

The critical question was whether project managers were managing costs or just monitoring costs. The government wanted costs to be managed rather than just monitored, accounted for, or reported. This need resulted in the creation of the EVMS.

The basis for the EVMS, which some consider to be a component of the MCCS, is the determination of earned value. Earned value is a management technique that relates resource planning to schedules and technical performance requirements. Earned value management (EVM) is a systematic process that uses earned value as the primary tool for integrating cost, schedule, technical performance management, and risk management.

Without using the EVMS, determining status can be difficult. Consider the following:

- The project
  - A total budget of $1.2 million
  - A 12-month effort
  - Produce 10 deliverables
- Reported status
  - Time elapsed: 6 months
  - Money spent to date: $700,000
  - Deliverables produced: 4 complete, 2 partial

What is the real status of the project? How far along is the project: 40, 50, 60 percent, etc.? Another problem was how to accurately relate cost to performance. If you spent 20 percent of the budget, does that imply that you are 20 percent complete? If you are 30 percent complete, then have you spent 30 percent of the budget?

The EVMS provides the following benefits:

- Accurate display of project status
- Early and accurate identification of trends
- Early and accurate identification of problems
- Basis for course corrections

The EVMS can answer the following questions:

- What is the true status of the project?
- What are the problems?
- What can be done to fix the problems?
- What is the impact of each problem?
- What are the present and future risks?

The EVMS emphasizes prevention over cures by identifying and resolving problems early. The EVMS is an early warning system allowing for early identification of trends and variances from the plan. The EVMS provides an early warning system, thus allowing the project manager sufficient time to make course corrections in small increments. It is usually easier to correct small variances as opposed to large variances. Therefore, the EVMS should be used continuously throughout the project in order to detect the variances while they are small and possibly easy to correct. Large variances are more difficult to correct and run the risk that the cost to correct the large variance may displease management to the point where the project may be canceled.

14.6 VARIANCE AND EARNED VALUE

A variance is defined as any schedule, technical performance, or cost deviation from a specific plan. Variances must be tracked and reported. They should be mitigated through corrective actions and not eliminated through a baseline change unless there is a good reason. Variances are used by all levels of management to verify the budgeting system and the scheduling system. The budgeting and scheduling system variance must be compared because:

- The cost variance compares deviations only from the budget and does not provide a measure of comparison between work scheduled and work accomplished.
- The scheduling variance provides a comparison between planned and actual performance but does not include costs.

There are two primary methods of measurement:

- *Measurable efforts:* Discrete increments of work with a definable schedule for accomplishment, whose completion produces tangible results.
- *Level of effort:* Work that does not lend itself to subdivision into discrete scheduled increments of work, such as project support and project control.

Calculating Variances

Variances are used on both types of measurement. In order to calculate variances, we must define the three basic variances for budgeting and actual costs for work scheduled and performed:

> ***PMBOK® Guide* 6th Edition**
> 7.4.2 Control Costs Tools and Techniques
> 7.4.3.5 Project Documents Updates

- Budgeted cost for work scheduled (BCWS) is the budgeted amount of cost for work scheduled to be accomplished plus the amount or level of effort or apportioned effort scheduled to be accomplished in a given time period.
- Budget cost for work performed (BCWP) is the budgeted amount of cost for completed work, plus budgeted for level of effort or apportioned effort activity completed within a given time period. This is sometimes referred to as "earned value."
- Actual cost for work performed (ACWP) is the amount reported as actually expended in completing the work accomplished within a given time period.

NOTE: The Project Management Institute has changed the nomenclature in their new version of the *PMBOK® Guide*, whereby BCWS is now PV, BCWP is now EV, and ACWP is now AC. However, the majority of heavy users of these acronyms, specifically government contractors, still use the old acronyms. Until the PMI acronyms are accepted across all industries, we will continue to focus on the most commonly used acronyms.

BCWS represents the time-phased budget plan against which performance is measured. For the total contract, BCWS is normally the negotiated contract plus the estimated cost of authorized but unpriced work (less any management reserve). It is time-phased by the assignment of budgets to scheduled increments of work. For any given time period, BCWS is determined at the cost account level by totaling budgets for all work packages, plus the budget for the portion of in-process work (open work packages), plus the budget for level of effort and apportioned effort.

A contractor must utilize anticipated learning when developing the time-phased BCWS. Any recognized method used to apply learning is usually acceptable as long as the BCWS is established to represent as closely as possible the expected actual cost (ACWP) that will be charged to the cost account/work package.

These costs can then be applied to any level of the work breakdown structure (i.e., program, project, task, subtask, work package) for work that is completed, in-program, or anticipated. Using these definitions, the following variance definitions are obtained:

- Cost variance (CV) calculation:

$$CV = BCWP - ACWP$$

A negative variance indicates a cost-overrun condition.

- Schedule variance (SV) calculation:

$$SV = BCWP - BCWS$$

A negative variance indicates a behind-schedule condition.

In the analysis of both cost and schedule, costs are used as the lowest common denominator. In other words, the schedule variance is given as a function of cost. To alleviate this problem, the variances are usually converted to percentages:

$$\text{Cost variance \% (CVP)} = \frac{CV}{BCWP}$$

$$\text{Schedule variance \% (SVP)} = \frac{SV}{BCWS}$$

The schedule variance may be represented by hours, days, weeks, or even dollars.

As an example, consider a project that is scheduled to spend \$100K for each of the first four weeks of the project. The actual expenditures at the end of week four are \$325K. Therefore, BCWS 5 \$400K and ACWP 5 \$325K. From these two parameters alone, there are several possible explanations as to project status. However, if BCWP is now known, say \$300K, then the project is behind schedule and overrunning costs.

It is important to understand the physical meaning of CV and SV. Consider the following example:

- BCWS = \$1000
- BCWP = \$800
- ACWP = \$700

In this example, CV = \$800 – \$700 = +\$100. Because CV is a positive value, it indicates that physical progress was accomplished at a lower cost than the forecasted cost. This is a favorable situation. Had CV been negative, it would have indicated that physical progress was accomplished at a greater cost than what was forecasted. If CV 50, then the physical accomplishment was as budgeted.

Although CV is measured in hours or dollars, it is actually a measurement of the efficiency with which physical progress was accomplished compared with the plan. To correct a negative cost variance, emphasis should be placed upon the productivity rate (i.e., burn rate) at which work is being performed.

Returning to the above example, SV = $800 –$1000 = –$200. In this example, the schedule variance is a negative value, indicating that physical progress is being accomplished at a slower rate than planned. This is an unfavorable condition. If the schedule variance were positive, this would indicate physical progress being accomplished at a faster rate than planned. If SV = 0, physical progress is being accomplished as planned.

The schedule variance, SV, measures the timeliness of the physical progress compared to the plan whereas the cost variance, CV, measures the efficiency. To correct a negative schedule variance, emphasis should be placed upon improving the speed by which work is being performed.

The CV relates to the real cost. However, the problem with SV is how it relates to the real schedule. The schedule variance is determined from cost account or work package financial numbers and does not necessarily relate to the real schedule. The schedule variance does not distinguish between critical path and non–critical path work packages. The schedule variance by itself does not measure time. A negative schedule variance indicates a behind-schedule condition but does not mean that the critical path has slipped. On the contrary, the real schedule (i.e., precedence networks or the arrow diagramming networks) could indicate that the project will be ahead of schedule. A detailed analysis of the real schedule is still required irrespective of the value for the schedule variance.

Variance Controls

Variances are almost always identified as critical items and are reported to all organizational levels. Critical variances are established for each level of the organization in accordance with management policies.

Not all companies have a uniform methodology for variance thresholds. Permitted variances may be dependent on such factors as:

- Life-cycle phase
- Length of life-cycle phase
- Length of project
- Type of estimate
- Accuracy of estimate

Variance controls may be different from program to program. Table 14–1 identifies sample variance criteria for program X.

For many programs and projects, variances are permitted to change over the duration of the program. For strict manufacturing programs (product management), variances may be fixed over the program time span using criteria like those in Table 14–1. For programs that include research and development, larger deviations may be permitted during the earlier phases than during the later phases. Figure 14–10 shows time-phased cost variances for a program requiring research and development, qualification, and production phases. Since the risk should decrease as time goes on, the variance boundaries are reduced. The variance envelope in such a case may be dependent on the type of estimate.

TABLE 14–1. VARIANCE CONTROL FOR PROGRAM X

Organizational Level	Variance Thresholds*
Section	Variances greater than \$20,000 that exceed 25% of costs
Section	Variances greater than \$50,000 that exceed 10% of costs
Section	Variances greater than \$100,000 0
Department	Variances greater than \$100,000 that exceed 25% of costs
Department	Variances greater than \$250,000 that exceed 10% of costs
Department	Variances greater than \$400,000 0
Division	Variances greater than \$1,000,000 that exceed 10% of costs

*Thresholds are usually tighter within company reporting system than required external to government. Thresholds for external reporting are usually adjusted during various phases of program (% lower at end).

By using both cost and schedule variance, we can develop an integrated cost/schedule reporting system that provides the basis for variance analysis by measuring cost performance in relation to work accomplished. This system ensures that both cost budgeting and performance scheduling are constructed on the same database.

Performance Index

PMBOK® Guide 6th Edition
7.4.2.2 Data Analysis

In addition to calculating the cost and schedule variances in terms of dollars or percentages, we also want to know how efficiently the work has been accomplished. The formulas used to calculate the performance efficiency as a percentage of EV are:

$$\text{Cost performance index (CPI)} = \frac{\text{BCWP}}{\text{ACWP}}$$

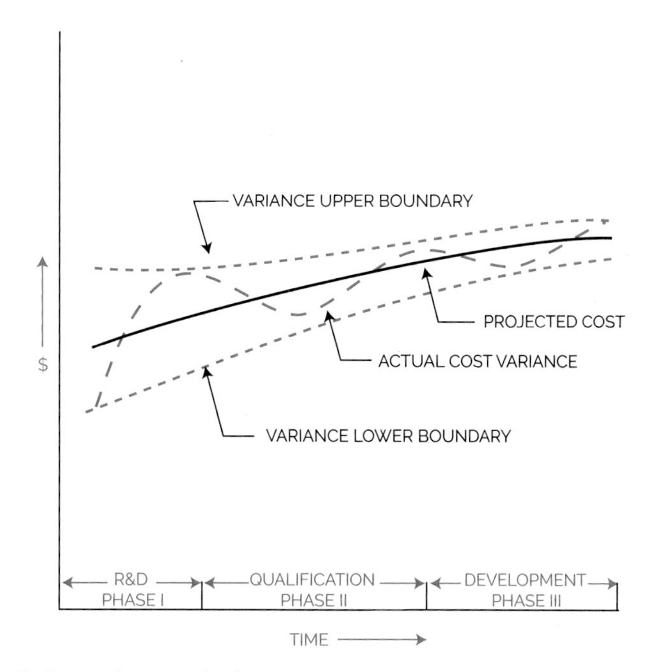

FIGURE 14–10. Project variance projection.

$$\text{Schedule performance index (SPI)} = \frac{\text{BCWP}}{\text{BCWS}}$$

If the CPI and SPI = 1.0, we have perfect cost and schedule performance. If CPI and SPI are less than 1.0, physical progress is being accomplished at a greater cost or slower rate than forecasted. This is unfavorable. If CPI and SPI are greater than 1.0, physical progress is being accomplished at less cost or a faster rate than planned, which is favorable. Similar to CV, CPI measures the efficiency by which the physical progress was accomplished compared to the plan or baseline. For an unfavorable value of CPI or SPI, emphasis should be placed upon improving the productivity by which work was being performed or the timeliness of the physical progress.

SPI and CPI are expressed as ratios compared to the performance factor of 1.0 whereas CV and SV are expressed in dollars or hours. One historic reason for this is that SPI and CPI can be used to show performance for a specified time period or trends over a long time horizon without disclosing actual company sensitive numbers. This makes SPI and CPI valuable tools for customer status reporting without disclosing hard numbers.

Trend Analysis and Reporting The cost and schedule performance index is most often used for trend analysis as shown in Figure 14–11. Companies use either three-month, four-month, or six-month moving averages to predict trends. Trend analysis provides an early warning system and allows managers to take corrective action. Unfortunately, its use may be restricted to long-term projects because of the time needed to correct the situation.

Figure 14–12 shows an integrated cost/schedule system. The figure identifies a performance slippage to date because the actual performance is less than the scheduled performance. This might not be a bad situation if the costs are proportionately underrun. However, from the upper portion of Figure 14–12, we find that costs are overrun (actual costs are greater than the target cost), thus adding to the severity of the situation.

Also shown in Figure 14–12 is the management reserve. Management reserves cover unforeseen events *within* a defined project scope, but are not used for unlikely major events or changes in scope. These changes are funded separately, perhaps through management-established contingency funds. Actually, there is a difference between management reserves (which come from project budgets) and contingency funds (which come from external sources) although most people do not differentiate. It is a natural tendency for a functional manager (and some project managers) to substantially inflate estimates to protect the particular organization and provide a certain amount of cushion. Furthermore, if the inflated budget is approved, managers will undoubtedly use all of the allocated funds, including reserves.

The line indicated as actual cost in Figure 14–12 shows a cost overrun compared to the budget. However, costs are still within the contractual requirement if we consider the management reserve. Therefore, things may not be as bad as they seem.

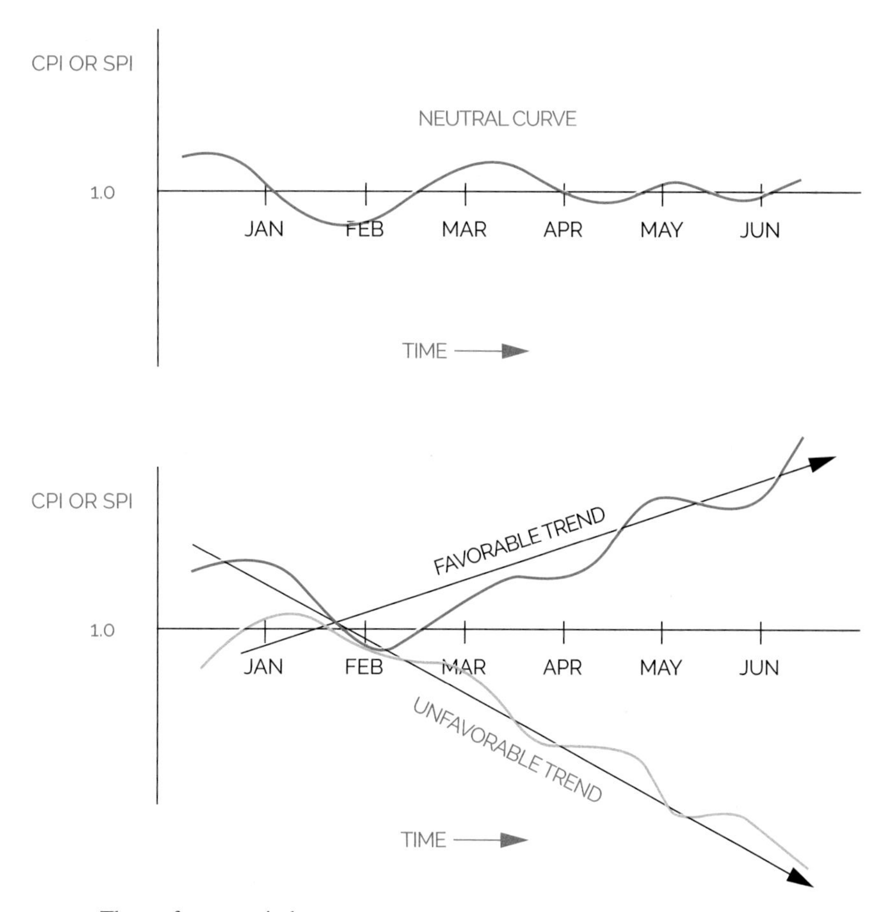

FIGURE 14–11. The performance index.

Government subcontractors are required to have a government-approved cost/schedule control system. The information requirements that must be demonstrated by such a system include:

- Budgeted cost for work scheduled (BCWS)
- Budgeted cost for work performed (BCWP)
- Actual cost for work performed (ACWP)
- Estimated cost at completion
- Budgeted cost at completion
- Cost and schedule variances/explanations
- Traceability

The last two items imply that standardized policies and procedures should exist for reporting and controlling variances. When permitted variances are exceeded, cost account

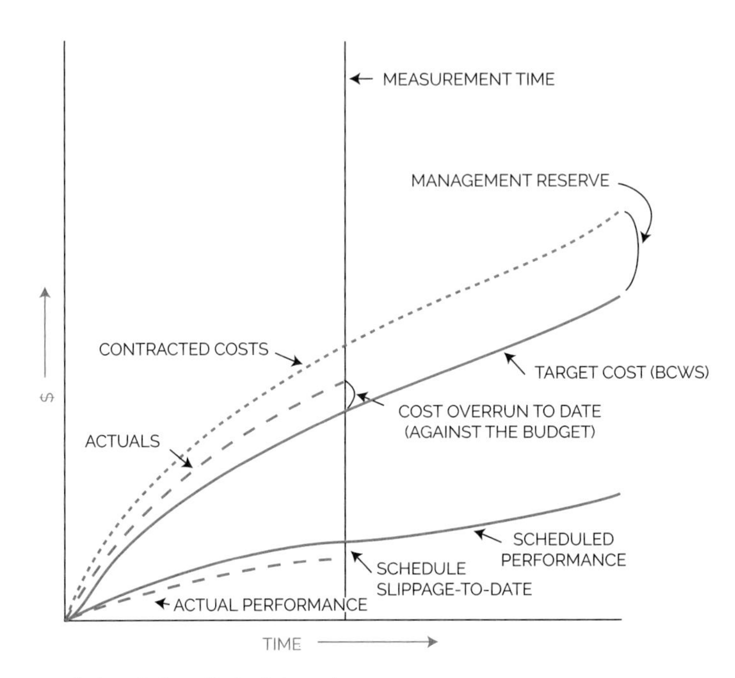

FIGURE 14–12. Integrated cost/schedule system.

variance analysis reports, as shown in Figure 14–13, are required. Required signatures may include:

- The functional employees responsible for the work
- The functional managers responsible for the work
- The cost accountant and/or the assistant project manager for cost control
- The project manager, work breakdown structure element manager, or someone with signature authority from the project office

For variance analysis, the goal of the cost account manager (whether project officer or functional employee) is to take action that will correct the problem within the original budget or justify a new estimate.

Five questions must be addressed during variance analysis:

- What is the problem causing the variance?
- What is the impact on time, cost, and performance?
- What is the impact on other efforts, if any?
- What corrective action is planned or under way?
- What are the expected results of the corrective action?

One of the key parameters used in variance analysis is the "earned value" concept, which is the same as BCWP. Earned value is a forecasting variable used to predict whether the project will finish over or under the budget.

COST ACCOUNT NO/CAM						REPORTING LEVEL		
WBS/DESCRIPTION						AS OF		
COST PERF. DATA				VARIANCE		AT COMPLETION		
	BCWS	BCWP	ACWP	SCH	COST	BUDGET	EAC	VAR.
MONTH TO DATE ($)								
CONTRACT TO DATE ($K)								
PROBLEM CAUSE AND IMPACT								
CORRECTIVE ACTION (INCLUDE EXPECTED RECOVERY DATE)								
COST ACCOUNT MANAGER DATE		COST CENTER MGR. DATE		WBS ELEMENT MANAGER DATE		DATE		

FIGURE 14–13. Cost account variance analysis report.

The major difficulty encountered in the determination of BCWP is the evaluation of in-process work (work packages that have been started but have not been completed at the time of cutoff for the report). The use of short-span work packages or establishment of discrete value milestones within work packages will significantly reduce the work-in-process evaluation problem, and procedures used will vary depending on work package length. For example, some contractors prefer to take no BCWP credit for a short-term work package until it is completed, while others take credit for 50 percent of the work package budget when it starts and the remaining 50 percent at completion. Some contractors use formulas that approximate the time-phasing of the effort, others use earned standards, while still others prefer to make physical assessments of the work completed to determine the applicable budget earned. For longer work packages, many contractors use discrete milestones with preestablished budget or progress values to measure work performed.

The difficulty in performing variance analysis is the calculation of BCWP because one must predict the percent complete. The simplest formula for calculating BCWP is:

$$\text{BCWP} = (\%\ \text{complete}) \times \text{BAC}$$

Most people calculate "percent complete" based upon task durations. However, a more accurate representation would be to calculate "percent work complete." However, this requires a schedule that is resource loaded. To eliminate this problem, many companies use standard dollar expenditures for the project, regardless of percent complete. For example, we could say that 10 percent of the costs are to be "booked" for each 10 percent of the time interval. Another technique, and perhaps the most common, is the 50/50 rule:

> Half of the budget for each element is recorded at the time that the work is scheduled to begin, and the other half at the time that the work is scheduled to be completed. For a project with a large number of elements, the amount of distortion from such a procedure is minimal. (Figures 14–14 and 14–15 illustrate this technique.)

One advantage of using the 50/50 rule is that it eliminates the necessity for the continuous determination of the percent complete. However, if percent complete can be determined, then percent complete can be plotted against time expended, as shown in Figure 14–16.

There are techniques available other than the 50/50 rule:

- *0/100:* Usually limited to work packages (activities) of small duration (i.e., less than one month). No value is earned until the activity is complete.
- *Milestone:* This is used for long work packages with associated interim milestones, or a functional group of activities with a milestone established at identified

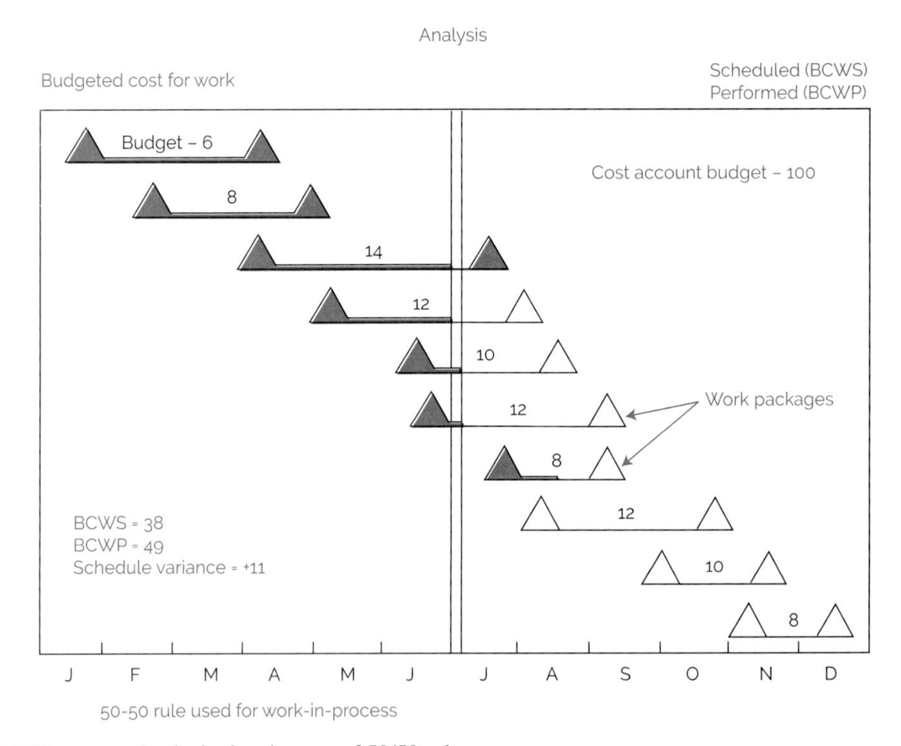

FIGURE 14–14. Analysis showing use of 50/50 rule.

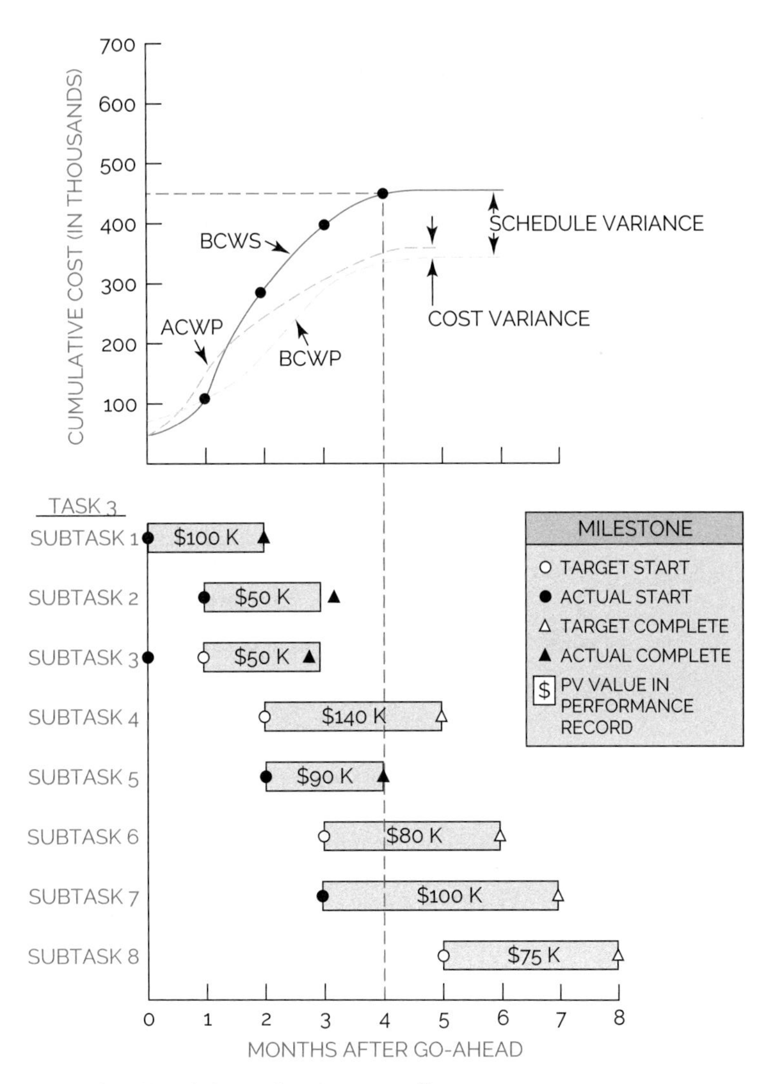

FIGURE 14–15. Project Z, task 3 cost data (contractual).

control points. Value is earned when the milestone is completed. In these cases, a budget is assigned to the milestone rather than the work packages.

- *Percent complete:* Usually invoked for long-duration work packages (i.e., three months or more) where milestones cannot be identified. The value earned would be the reported percent of the budget.
- *Equivalent units:* Used for multiple similar-unit work packages, where earnings are on completed units, rather than labor.
- *Cost formula (80/20):* A variation of percent complete for long-duration work packages.

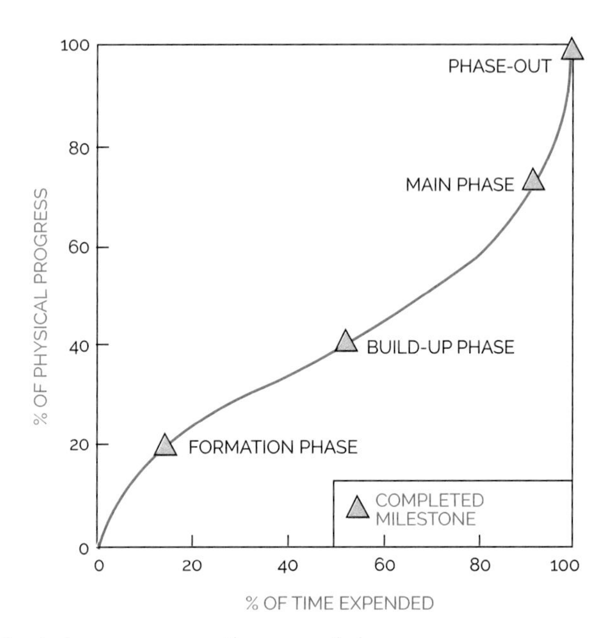

FIGURE 14–16. Physical progress versus time expended.

- *Level of effort:* This method is based on the passage of time, often used for supervision and management work packages. The value earned is based on time expended over total scheduled time. It is measured in terms of resources consumed over a given period of time and does not result in a final product.
- *Apportioned effort:* A rarely used technique, for special related work packages. As an example, a production work package might have an apportioned inspection work package of 20 percent. There are only a few applications of this technique. Many people will try to use this for supervision, which is not a valid application. This technique is used for effort that is not readily divisible into short-span work packages but that is in proportion to some other measured effort.

Estimate at Completion (EAC)

PMBOK® Guide 6th Edition
7.4.2.2 Data Analysis
7.4.2.3 To-Complete Performance Index

The estimate at completion (EAC) is the best estimate of the total cost at the completion of the project. The EAC is a periodic evaluation of the project status, usually on a monthly basis or until a significant change has been identified. It is usually the responsibility of the performing organization to prepare the EAC.

The calculation of a new EAC and subsequent revision does not imply that corrective action has been taken. Consider a three-month task that is 99 percent complete and was budgeted to spend $400K (BCWS). The actual costs to date (ACWP) are $395K. Using the 50/50 rule, BCWP is $200K. The estimated cost-to-complete (EAC) ratio is $395K/$200K, which implies that we are heading for a 100 percent cost overrun. Obviously, this is not the case.

TABLE 14–2. PROJECT Z, TASK 3 COST SUMMARY FOR WORK COMPLETED OR IN PROGRESS (COST IN THOUSANDS)

		Cumulative to Date				
	Contractual	BCWS	BCWP	ACWP	Cost Variance	Schedule Variance
Direct labor hours	8650	6712	5061	4652	409	
Direct labor dollars	241	187	141	150	(9)	(46)
Labor overhead (140%)	338	263	199	210	(11)	(64)
Subtotal	579	450	340	360	(20)	
Material dollars	70	66	26	30	(4)	
Subtotal	649					
G&A (10%)	65					
Subtotal	714					
Fee (12%)	86					
Total	800					

Note: This table assumes a 50/50 ratio for planned and earned values of budget.

Using the data in Table 14–2, we can calculate the estimate at completion (EAC) by the expression where BAC is the value of BCWS at completion.

$$\begin{aligned}\text{EAC} &= (\text{ACWP/BCWP}) \times \text{BAC} = \text{BAC/CIP} \\ &= (360/340) \times 579{,}000 \\ &= \$613{,}059\end{aligned}$$

The discussion of what value to use for BAC is argumentative. In the above calculation, we used burdened direct labor dollars. Some people prefer to use nonburdened labor with the argument that the project manager controls only direct labor hours and dollars. Also, the calculation for EAC did not include material costs or general and administrative costs.

The above calculation of EAC implies that we are overrunning labor costs by 6.38 percent and that the final burdened labor cost will exceed the budgeted burdened labor cost by $34,059. For a more precise calculation of EAC we would need to include material cost (assumed at $70,000) and G&A. This would give us a final cost, excluding profit, of $751,365, which is an overrun of $37,365. The resulting profit would be $86,000 less $37,365, or $48,635. The final analysis is that work is being accomplished almost on schedule except for subtask 4 and subtask 6, but costs are being overrun.

Cost Overruns

The question that remains is, "Where is the cost overrun occurring?" To answer this question, we must analyze the cost summary sheet for project Z, task 3. Table 14–2 represents a hypothetical case for the cost elements of project

Z, task 3. From Table 14–2 we see that negative (overrun) variances exist for labor dollars, overhead dollars, and material costs. Because labor overhead is measured as a percentage of direct labor dollars, the problem appears to be in the direct labor dollars.

From the contractual column in Table 14–2 the project was estimated at $27.86 per hour direct labor ($241,000/8650 hours), but actuals to date are $150,000/4652 hours, or $32.24 per hour. Therefore, higher-salaried people than anticipated are being employed. This salary increase is partially offset by the fact that there exists a positive variance of 409 direct labor hours, indicating that these higher-salaried employees are performing at a more favorable position than expected on the learning curve. Since the milestones (from Figure 14–15) appear to be on target, work is progressing as planned, except for subtask 4.

The labor overhead rate has not changed. The contractual, BCWS, and BCWP overhead rates were estimated at 140 percent. The actuals, obtained from month-end reports, indicate that the true overhead rate is as predicted.

The following conclusions can be drawn:

- Work is being performed as planned (almost on schedule, although at a more favorable position on the learning curve), except for subtask 4, which is giving us a schedule delay.
- Direct labor costs are increasing through the use of higher-salaried employees.
- Overhead rates are as anticipated.
- Direct labor hours must be reduced even further to compensate for increased costs, or profits will be drastically reduced.

This type of analysis could have been carried out to one more level by identifying exactly which departments were using the more expensive employees. This step should probably be completed anyway to see if lower-paid employees are available and can work at the required position on the learning curve. Had the labor costs been a result of increased labor hours, this step would have definitely been necessary to identify the reason for the overrun in-house. Perhaps poor estimating was the cause.

In Table 14–2, there also appears a positive variance in materials. This likewise should undergo further analysis. The cause may be the result of improperly identified hardware, material escalation costs increasing beyond what was planned, increased scrap factors, or a change in subcontractors.

It should be obvious from this analysis that a detailed investigation into the cause of variances appears to be the best method for identifying causes. The concept of earned value, although a crude estimate, identifies trends concerning the status of specific WBS elements.

EAC Formulas

There are several formulas that can be used to calculate EAC. Using the data shown below, we can illustrate how each of three different formulas can give a different result. Assume that your project consists of these three activities only.

Activity	%Complete	BCWS	BCWP	ACWP
A	100	1000	1000	1200
B	50	1000	500	700
C	0	1000	0	0

$$\text{Formula I. EAC} = \frac{\text{ACWP}}{\text{BCWP}} \times \text{BAC}$$

$$= \frac{1900}{1500}(3000) = \$3800$$

$$\text{Formula II. EAC} = \frac{\text{ACWP}}{\text{BCWP}} \times \begin{bmatrix}\text{Work completed} \\ \text{and in progress}\end{bmatrix} + \begin{bmatrix}\text{Actual (or revised) cost} \\ \text{of work packages not} \\ \text{yet begun}\end{bmatrix}$$

$$= \frac{1900}{1500}(2000) + \$1000 = \$3533$$

$$\text{Formula III. EAC} = [\text{Actual to date}] \times \begin{bmatrix}\text{All remaining work to be at planned} \\ \text{cost including remaining work in} \\ \text{progress}\end{bmatrix}$$

$$= 1900 + [\underset{\text{B}}{\underset{\downarrow}{500}} + \underset{\text{C}}{\underset{\downarrow}{1000}}] = \$3400$$

Advantages and disadvantages exist for each formula. Formula I assumes that the burn rate (i.e., ACWP/BCWP) will be the same for the remainder of the project. This is the easiest formula to use. The burn rate is updated each reporting period.

Formula II assumes that all work packages not yet opened will be completed at the planned cost. However, it is possible for planned cost to be revised based upon history from completed work packages.

Formula III assumes that all remaining work is independent of the burn rate incurred thus far. This may be unrealistic unless all remaining work can be reestimated if necessary.

Organization-Level Analysis Each critical variance identified on the organizational MCCS reports may require the completion of MCCS variance analysis procedures by the supervisor of the cost center involved. Analyzing both the work breakdown and organizational structure, the supervisor systematically concentrates his efforts on cost and schedule problems appearing within his organization.

Analysis begins at the lowest organizational level by the supervisor involved. Critical variances are noted at the cost account on the MCCS report. If a schedule variance is involved and the subtask consists of a number of work packages, the supervisor may refer to a separate report that breaks down each cost account into the various work packages that are ahead or behind schedule. The supervisor can then analyze the variance on the basis of the work package involved and determine with the aid of supporting organizations the cause of the variance, the corrective action that can be taken, or the possible effect on associated or future planned effort.

Cost variances involving labor are analyzed by the supervisor on the basis of the performance of his organization in accomplishing the work assigned, within the budgeted man-hours and planned labor rate. The cause of any variance to this performance is determined, and corrective action is then implemented.

Cost variances on non-labor efforts are analyzed by the supervisor with the aid of the program team member and other supporting organizations.

All material variance analyses are normally initiated by cost accounting as a service to the using organization. These variance analyses are completed, including cause and corrective action, to the extent that can be explained by cost accounting. They are then sent to the using organization, which reviews the analyses and completes those resulting from schedule performance or usage. If a variance is recognized as a change in the material acquisition price, this information is supplied by cost accounting to the responsible organization and a change to the estimate-to-complete is initiated by the using organization.

The supervisor should forward copies of each completed MCCS variance analysis/EAC change form to his higher-level manager and the program team member.

Program Team Analysis

The program team member may receive a team critical variance report that lists variances in his organization at the lowest level of the work breakdown structure at the division cost center level by cost element. Upon request of the program manager, analyses of variances contributing to the variances on the team critical variance report are summarized by the responsible program team member and reviewed with the program manager. The preparation of status reports, whether they be for internal management or for the customer, should, at a minimum, answer two fundamental questions:

- Where are we today (with respect to time and cost)?
- Where will we end up (with respect to time and cost)?

The information necessary to answer these questions can be obtained from the following formulas:

- Where are we today?
  - Cost variances (in dollars/hours and percent complete)
  - Schedule variances (in dollars/hours and percent complete)
  - Percent complete
  - Percent money spent
- Where will we end up?
  - Estimate at completion (EAC)
  - The remaining critical path
  - SPI (trend analysis)
  - CPI (trend analysis)

Since SPI and CPI are used for trend analyses, we can use CPI and SPI to forecast the expected final cost and the expected end date of the project. We can express the cost at completion, EAC, as:

$$\text{EAC} = \frac{\text{BAC}}{\text{CPI}}$$

The time at completion uses SPI for the forecast and can be expressed as:

$$\text{New project length} = \frac{\text{Original project length}}{\text{SPI}}$$

Care must be taken with the use of SPI to calculate the new project length because a favorable value for SPI (i.e., .1.0) could be the result of work packages that are not on the critical path.

Once EAC and the new project length are calculated, we can calculate the variance at completion (VAC) and the estimated cost to complete (ETC) using the following two formulas:

$$\text{VAC} = \text{BAC} - \text{EAC} \quad \text{and} \quad \text{ETC} = \text{EAC} - \text{ACWP}$$

Percent complete and percent money spent can be obtained from the following formulas:

$$\text{Percent complete} = \frac{\text{BCWP}}{\text{BAC}}$$

$$\text{Percent money spent} = \frac{\text{ACWP}}{\text{BAC}}$$

Another formula that appears to be important today is the to-complete performance index. This shows the cost performance that is necessary to complete a component of work. The formula is:

$$\text{TCPI} = (\text{BAC} - \text{EV})/(\text{BAC} - \text{AC})$$

The program manager uses this information to review the program status with upper-level management. This review is normally on a monthly basis on large projects. In addition, the results of these analyses are used to explain variances in the contractually required reports to the customer.

After the analyses of the variances have been made, reports must be developed for both the customer and in-house (upper-level) management.

14.7 THE COST BASELINE

PMBOK® Guide, 6th Edition
7.3.3.1 Cost Baseline
7.4.2.2 Data Analysis

Once the project is initiated, the project team establishes the cost or financial baseline against which status will be reported and variances will be measured. Figure 14–17 represents a cost baseline. Each block represents a cost account or work package element. The summation of all of the cost accounts or work packages would then equal the time-phased

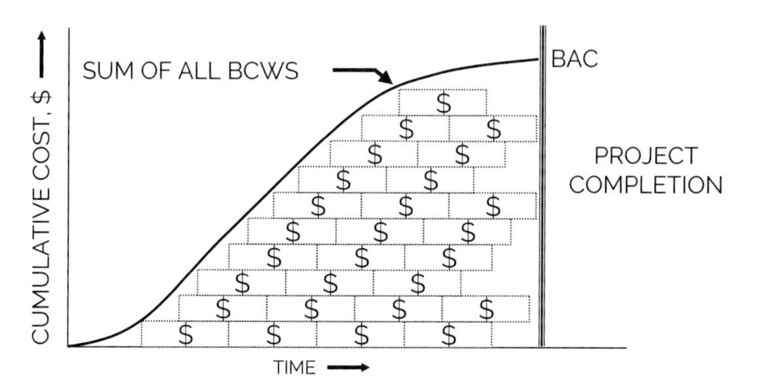

FIGURE 14–17. The cost baseline.

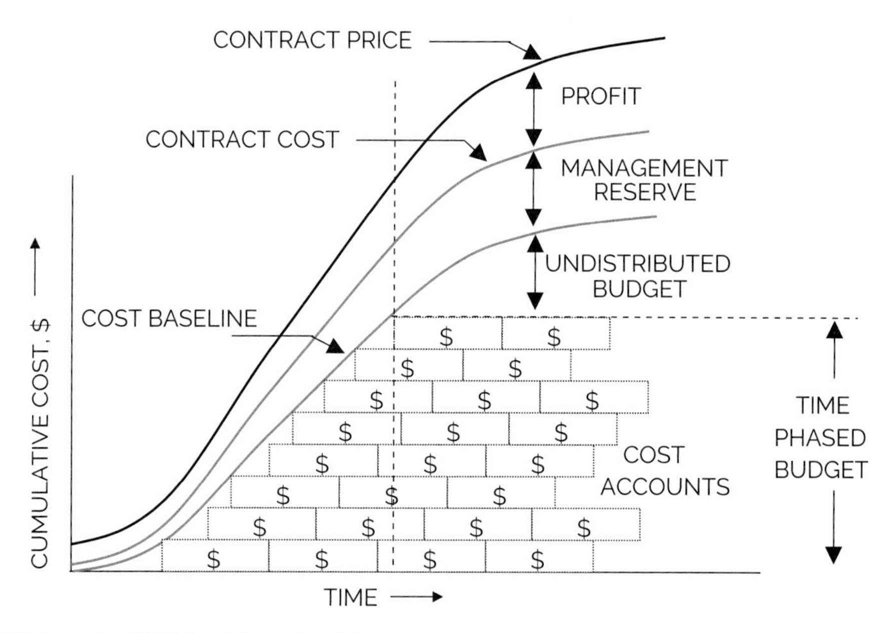

FIGURE 14–18. WBS level 1 cost breakdown.

budget. Each work package would then be described through the work authorization form for that work package.

The cost baseline in Figure 14–17 is just part of the cost breakdown. An illustration of a cost breakdown appears in Figure 14–18.

There are certain distinguishing features of Figure 14–18:

- The time-phased budget, which is the released budget, is the summation of all BCWS elements.

- The cost baseline is the summation of the time-phased budget (i.e., the distributed budget) and the undistributed budget. This will equal the released, planned budget at completion (BAC).
- The contractual cost to complete the project is the summation of the cost baseline and the management reserve, assuming that a management reserve exists.
- The contract price is the contract cost plus the profit, if any.

14.8 JUSTIFYING THE COSTS

Project pricing is often based upon best guesses rather than concrete estimates. This is particularly true for companies that survive on competitive bidding and where the preparation cost of a bid may vary between $50,000 and $500,000. If the probability of winning a bid is low, then the company may spend the minimum amount of time and cost during bid preparation.

Table 14–3 shows a typical project pricing summary where each functional area or division can have its own overhead rate. In this summary, the overhead rate for engineering is 110 percent, whereas the manufacturing overhead rate is 200 percent. If this company is a subsidiary of a larger company, then a corporate general and administrative (G&A) cost may be included. If the project is for an external customer, then a profit margin will be included.

Once the project pricing summary is completed, the costs must be justified before executive committee. Every company has its own evaluation criteria cost summary approval process.

Typical elements that must be justified or supported by hard data include:

Labor Rates: For estimating purposes, department averages or skill set weighted averages can be used. This is sometimes called the blended rate. The best-case scenario would

TABLE 14–3. TYPICAL PROJECT PRICING SUMMARY

	Direct Labor			Overhead		
Department	**Hours**	**Rate**	**Dollars**	**%**	**Dollars**	**Total**
Engineering	1000	$42.00	42,000	110	46,200	$ 88,200
Manufacturing	500	$35.00	17,500	200	35,000	$ 52,500
					Total Labor	$140,700
				Other: Subcontracts	$10,000	
				Consultants	$ 2,000	$ 12,000
			Total labor and material			$152,700
			Corporate G&A: 10%			$ 15,270
						$167,970
			Profit: 15%			$ 25,196
						$193,166

TABLE 14–4. FORWARD PRICING RATES: SALARY (DEPARTMENT PAY STRUCTURE)

		Salary (per hour)		
Pay Grade	**Title**	**2015**	**2016***	**2017***
9	Engineering Consultant	$73	$81	$85
8	Senior Engineer	60	63	67
7	Engineer	51	55	58
6	Junior Engineer	46	49	52
5	Apprentice Engineer	41	43	45

*Projected rates.

be estimating from the actual salary or skill set of the workers to be assigned. This may be impossible during competitive bidding because we do not know who will be available or who will be assigned assuming the contract is received. Also, if the project is a multiyear effort, we may need forward pricing rates, which are the predicted, full burdened salaries anticipated in the next few years. This is illustrated in Table 14–4.

- Overtime: If resources are scarce and the company has no intention of hiring additional resources, then some of the work must be accomplished on overtime. This could increase the cost of the project and an allowance must be made for possible mistakes made during this period of excessive overtime.
- Scrap factors: If the project includes procurement of raw materials, then some scrap factor allowance may be necessary. This calculation may be impacted by the skill set of the resources assigned and using the materials, previous experience using these materials, and experience on these types of projects.
- Risks: Risk analysis may be based upon the quality of the estimates and experience of those who made the estimates. Other risks considered include the company's ability to achieve the anticipated benefits or the designated profits and, if a disaster occurs, the company's exposure and liability for lawsuits.
- Hidden costs: These costs, such as travel, shipping and postage, capital costs, and meeting attendance, can erode all of the profitability expected on a project. Another potentially hidden cost is the yearly or monthly workload availability. A typical calculation appears in Table 14–5. If we use Table 14–5 and all of the workers are long-term employees, then there may be less than 1840 hours available per year to account for vacations, sick leave, etc.

14.9 THE COST OVERRUN DILEMMA

The lifeblood of most organizations is a continuous stream of new products or services. Because of the word "new," historical data may be at a minimum and cost overruns are expected. Figure 14–19 shows a typical range of overruns.

TABLE 14–5. HOURS AVAILABLE FOR WORK

Hours available per year (52 × 40):	2080 hours
Vacation (3 weeks):	–120 hours
Sick leave (3 days):	–24 hours
Paid holidays (11 days):	–88 hours
Jury duty (1 day):	–8 hours
	1840 hours
(1840 hours/year) ÷ 12 months = 153 hours/month	

Rough order-of-magnitude (ROM) estimates are often made from "soft" data, which can result in a wide range of overruns, and are used in the initiation phase of a project. As we go from soft data to hard data and enter the planning phase of a project, the accuracy of the estimates improves and the range of the overruns narrows.

When overruns occur, the project manager looks for ways of reducing costs. The simplest way is to reduce scope. This begins with a search for items that are easy to cut. The items that are easiest to cut are those items that were poorly understood during the estimating process and were therefore underestimated. Typical items that are cut or reduced in magnitude include project management supervision, line management supervision, process controls, quality assurance, and testing.

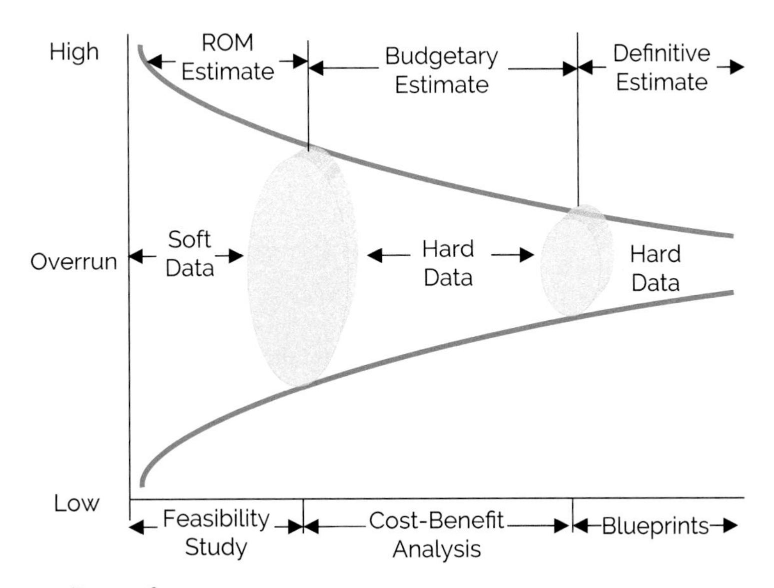

FIGURE 14–19. Range of overruns.

If the easy-to-cut items do not provide sufficient cost reductions, then a desperate search begins among the hard-to-cut items. Hard-to-cut items include direct labor hours, materials, equipment, facilities, and other items.

If the cost reductions are unacceptable to management, then management must decide whether or not to pull the plug and cancel the project.

14.10 RECORDING MATERIAL COSTS USING EARNED VALUE MEASUREMENT

Recording direct labor costs usually presents no problem since labor costs are normally recorded as the labor is accomplished. Therefore, recorded and reported labor will be the same. Material costs, on the other hand, may be recorded at various times. Material costs can be recorded as commitments, expenditures, accruals, and applied costs. All provide useful information and are important for control purposes.

Because of the choices available for material cost analysis, material costs should be reported *separately* from the standard labor hour/labor dollar earned value report. For example, cost variances associated with the procurement of material may be determined at the time that the purchase orders are negotiated and placed with the vendors since this information provides the *earliest* visibility of potential cost variance problems. Significant variances in the anticipated and actual costs of materials can have a serious effect on the total contract cost and should be reflected promptly in the estimated cost at completion (EAC) and explained in the narrative part of the project status report.

Separating labor from material costs is essential. Consider the following example: You are budgeted to spend $1,000,000 in burdened labor and $600,000 in material. At the end of the first month of your project, the following information is available to you:

Labor:	ACQP = $90,000
	BCWP = $100,000
	BAC = $1,000,000
Material:	AWCP = $450,000
	BCWP = $400,000
	BAC = $600,000

For simplicity's sake, let us use the following formula for EAC:

$$\text{EAC} = (\text{ACWP}/\text{BCWP}) \times \text{BAC}$$

Therefore,

EAC (labor) = $900,000

EAC (material = $675,000

If we add together both EACs, the estimated cost at completion will be $1,575,000, which is $25,000 *below* the planned budget. If the costs are combined before we calculate EAC, then

$$\text{EAC} = [(\$450{,}000 + \$90{,}000)/\$500{,}000] \times (\$1{,}600{,}000) = \$1{,}728{,}000$$

which is a $128,000 *overrun*. Therefore, it is usually best to separate material from labor in status reporting. Another major problem is how to account for the costs of material placed on order, which does *not* reflect the cost of work completed and is not normally used in status reporting. For performance measurement purposes, it is desirable that material costs be recorded at the time that the materials are received, paid for, or used rather than as of the time that they are ordered. Therefore, the actual costs reported for materials should be derived in accordance with established procedures, and normally will be recorded for earned value measurement purposes at or after time of material receipt. In addition, costs should always be recorded on the same basis as budgets are prepared in order to make comparisons between budgeted and actual costs meaningful. For example, material should not be budgeted on the basis of when it is used and then have its costs collected/reported on the basis of when it is received.

14.11 MATERIAL VARIANCES: PRICE AND USAGE

When the actual material costs exceed a material budget, there are normally two causes:

- The articles purchased cost more than was planned, called a "price variance."
- More articles were consumed than were planned, called a "usage variance."

Price variances (PV) occur when the budgeted price value (BCWS) of the material was different than what was actually experienced (ACWP). This condition can arise for a host of reasons: poor initial estimates, inflation, different materials used than were planned, too little money available to budget, and so on.

The formula for price variance (PV) is:

$$\text{PV} = (\text{Budgeted price} - \text{Actual price}) \times (\text{Actual quantity})$$

Price variance is the difference between the budgeted cost for the bill of materials and the price paid for the bill of materials. By contrast, usage variances (UV) occur when a greater quantity of materials is consumed than were planned. The formula for usage variance (UV) is:

$$\text{UV} = (\text{Budgeted quantity} - \text{Actual quantity}) \times (\text{Budgeted price})$$

Section 14.11 is adapted from Quentin W. Fleming, *Cost/Schedule Control Systems Criteria* (Chicago: Probus Publishers, 1992), pp. 151–152.

Normally, usage variances are the resulting costs of materials used over and above the quantity called for in the bill of materials.

Consider the following example: The project manager establishes a material budget of 100 units (which includes 10 units for scrap factor) at a price of $150 per unit. Therefore, the material budget was set at $15,000. At the end of the short project, material actuals (ACWP) came in at $15,950, which was $950 over budget. What happened?

Applying the formulas defined previously,

$$\begin{aligned}\text{Price variance (PV)} &= (\text{BCWS price} - \text{ACWP price}) \times \text{Actual quantity}\\ &= (\$150 \text{ per unit} - \$145 \text{ per unit}) \times 110 \text{ units}\\ &= \$550 \text{ favorable}\end{aligned}$$

$$\begin{aligned}\text{Usage variance (UV)} &= (\text{BCWP qty} - \text{ACWP qty}) \times \text{BCWS price}\\ &= (100 \text{ units} - 110 \text{ units}) \times \$150 \text{ per unit}\\ &= \$1{,}500 \text{ unfavorable}\end{aligned}$$

The analysis indicates that your purchase price was less than you anticipated, thus generating a cost savings. However, you used 10 units more than planned for, thus generating an unfavorable usage variance. Further investigation indicated that your line manager had increased the scrap factor from 10 to 20 units.

14.12 SUMMARY VARIANCES

Summary variances can be calculated for both labor and material. Consider the information shown below:

	Direct Material	Direct Labor
Planned price/unit	$ 30.00	$ 24.30
Actual units	17,853	9,000
Actual price/unit	$ 31.07	$ 26.24
Actual cost	$554,630	$236,200

We can now calculate the total price variance for direct material and the rate cost variance:

- *Total* price variance for direct material

$$\begin{aligned}&= \text{Actual units} \times (\text{BCWP} - \text{ACWP})\\ &= 17{,}853 \times (\$30.00 - \$31.07)\\ &= \$19{,}102.71 \text{ (unfavorable)}\end{aligned}$$

- Labor *rate* cost variance

$$\begin{aligned}&= \text{Budgeted rate} - \text{Actual rate}\\ &= \$24.30 - \$26.24\\ &= \$1.94 \text{ (unfavorable)}\end{aligned}$$

14.13 STATUS REPORTING

PMBOK® Guide, 6th Edition
7.4.3.5 Project Documents updates

There are four types of performance reports that are generally printed out from the earned value measurement system:

- Progress reports: These reports indicate the physical progress to date, namely, BCWS, BCWP, and ACWP. The report might also include information on material procurement, delivery, and usage, but most companies have separate reports on materials.
- Status reports: These reports identify where we are today and use the information from the performance reports to calculate SV and CV.
- Projection reports: These reports calculate EAC, ETC, SPI, and CPI as well as any other forward-looking projections. These reports emphasize where we will end up.
- Exception reports: These reports identify exceptions, problems, or situations that exceed the threshold limits on such items as variances, cash flow, resources assigned, and other such topics.

Reporting procedures for variance analysis should be as brief as possible. The reason for this is simple: the shorter and more concise the report, the faster that feedback can be generated and responses developed. Time is critical if rescheduling must be accomplished with limited resources.

This by no means implies that all variances require corrective action. There are four major responses to a variance report:

- Ignoring it
- Functional modification
- Replanning
- System redesign

Permissible variances exist for all levels of the organization. If the variance is within these permitted deviations, then there will be no response, and the variance may be ignored. In some situations where the variance is marginal (or even within limits), corrective action may be required. This would normally occur at the functional level and might simply involve using another test procedure or possibly considering some alternative not delineated in the program plan.

If major variances occur, then either replanning or system redesign must take place.

14.14 COST CONTROL PROBLEMS

PMBOK® Guide, 6th Edition
7.4 Control Costs

No matter how good the cost and control system is, problems can occur. Common causes of cost problems include:

- Poor estimating techniques and/or standards, resulting in unrealistic budgets
- Out-of-sequence starting and completion of activities and events
- Inadequate work breakdown structure
- No management policy on reporting and control practices
- Poor work definition at the lower levels of the organization
- Management reducing budgets or bids to be competitive or to eliminate "fat"
- Inadequate formal planning that results in unnoticed, or often uncontrolled, increases in scope of effort
- Poor comparison of actual and planned costs
- Comparison of actual and planned costs at the wrong level of management
- Unforeseen technical problems
- Schedule delays that require overtime or idle time costing
- Material escalation factors that are unrealistic

Cost overruns can occur in any phase of project development. The most common causes for cost overruns are:

- Proposal phase
  - Failure to understand customer requirements
  - Unrealistic appraisal of in-house capabilities
  - Underestimating time requirements
- Planning phase
  - Omissions
  - Inaccuracy of the work breakdown structure
  - Misinterpretation of information
  - Use of wrong estimating techniques
  - Failure to identify and concentrate on major cost elements
  - Failure to assess and provide for risks
- Negotiation phase
  - Forcing a speedy compromise
  - Procurement ceiling costs
  - Negotiation team that must "win this one"
- Contractual phase
  - Contractual discrepancies
  - SOW different from RFP requirements
  - Proposal team different from project team
- Design phase
  - Accepting customer requests without management approval
  - Problems in customer communications channels and data items
  - Problems in design review meetings
- Production phase
  - Excessive material costs
  - Specifications that are not acceptable
  - Manufacturing and engineering disagreement

Related Case Studies (from Kerzner/*Project Management Case Studies*, 4th ed.)	**Related Workbook Exercises (from Kerzner/*Project Management Workbook and PMP®/CAPM® Exam Study Guide*, 11th ed.)**	***PMBOK® Guide*, 6th Edition, Reference Section for the PMP® Certification Exam**
• The Bathtub Period* • Trouble in Paradise • Franklin Electronics*	• Using the 50/50 Rule • Multiple Choice Exam • Crossword Puzzle on Cost Management	• Project Cost Management • Project Scope Management

*Case Study also appears at end of chapter.

14.15 STUDYING TIPS FOR THE PMI® PROJECT MANAGEMENT CERTIFICATION EXAM

This section is applicable as a review of the principles to support the knowledge areas and domain groups in the *PMBOK® Guide*. This chapter addresses:

- Project Scope Management
- Project Cost Management
- Initiating
- Planning
- Controlling

Understanding the following principles is beneficial if the reader is using this text to study for the PMP® Certification Exam:

- What is meant by a management cost and control system
- What is meant by earned value measurement
- The meaning of control
- Code of cost accounts
- Work authorization for and its relationship to the code of accounts
- Sources of funds for a project or changes to a project
- Four primary elements of cost monitoring and control: BCWS, BCWP, ACWP, and BAC
- How to calculate the cost and schedule variances, in hours, dollars, and percentages
- Importance of SPI and CPI in trend analysis
- Ways to forecast the time and cost to completion as well as variances at completion
- Different types of reports: performance, status, forecasting, and exception
- Use of the management reserve
- Escalation factors and how they affect a project
- What is a cost or financial baseline for a project
- Different ways to calculate either BCWP or percent complete

The following multiple-choice questions will be helpful in reviewing the principles of this chapter:

1. In earned value measurement, earned value is represented by:
   A. BCWS
   B. BCWP
   C. ACWP
   D. None of the above
2. If BCWS = 1000, BCWP = 1200, and ACWP = 1300, the project is:
   A. Ahead of schedule and under budget
   B. Ahead of schedule and over budget
   C. Behind schedule and over budget
   D. Behind schedule and under budget
3. If BAC = $20,000 and the project is 40 percent complete, then the earned value is:
   A. $5,000
   B. $8,000
   C. $20,000
   D. Cannot be determined
4. If BAC = $12,000 and CPI = 1.2, then the variance at completion is:
   A. –$2,000
   B. +2,000
   C. –$3,000
   D. +$3000
5. If BAC = $12,000 and CPI = 0.8, then the variance at completion is:
   A. –$3,000
   B. +$2,000
   C. –$3,000
   D. $3,000
6. If BAC for a work package is $10,000 and BCWP = $4,000, then the work package is:
   A. 40 percent complete
   B. 80 percent complete
   C. 100 percent complete
   D. 120 percent complete
7. If CPI = 1.1 and SPI = 0.95, then the trend for the project is:
   A. Running over budget but ahead of schedule
   B. Running over budget but behind schedule
   C. Running under budget but ahead of schedule
   D. Running under budget but behind schedule

8. The document that describes a work package, identifies the cost centers allowed to charge against this work package, and establishes the charge number for this work package is the:
   A. Code of accounts
   B. Work breakdown structure
   C. Work authorization form
   D. None of the above
9. Unknown problems such as escalation factors are often budgeted for using the:
   A. Project manager's charge number
   B. Project sponsor's charge number
   C. Management reserve
   D. Configuration management cost account
10. EAC, ETC, SPI, and CPI most often appear in which type of report?
    A. Performance
    B. Status
    C. Forecast
    D. Exception
11. If BAC 5 $24,000, BCWP 5 12,000, ACWP 5 $10,000, and CPI 5 1.2, then the cost that remains to finish the project is:
    A. $10,000
    B. $12,000
    C. $14,000
    D. Cannot be determined
12. There are several purposes for the 50–50 rule, but the *primary* purpose of the 50–50 rule is to calculate:
    A. BCWS
    B. BCWP
    C. ACWP
    D. BAC
13. When a project is completed, which of the following *must* be true?
    A. BAC = ACWP
    B. ACWP = BCWP
    C. SV = 0
    D. BAC = ETC
14. In March CV = –$20,000, and in April CV = –$30,000. In order to determine whether or not the situation has really deteriorated because of a larger unfavorable cost variance, we would need to calculate:
    A. CV in percent
    B. SV in dollars
    C. SV in percent
    D. All of the above

15. If a project manager is looking for revenue for a value-added scope change, the project manager's first choice would be:
    A. Management reserve
    B. Customer-funded scope change
    C. Undistributed budget
    D. Retained profits
16. A project was originally scheduled for 20 months. If CPI is 1.25, then the new schedule date is:
    A. 16 months
    B. 20 months
    C. 25 months
    D. Cannot be determined
17. The cost or financial baseline of a project is composed of:
    A. Distributed budget only
    B. Distributed and undistributed budgets only
    C. Distributed budget, undistributed budget, and the management reserve only
    D. Distributed budget, undistributed budget, management reserve, and profit only

ANSWERS

1. B
2. B
3. B
4. B
5. C
6. A
7. D
8. C
9. C
10. C
11. A
12. B
13. C
14. A
15. B
16. D
17. B

PROBLEMS

14–1 Do cost overruns just happen, or are they caused?

14–2 What impact would there be on BCWS, BCWP, ACWP, and cost and schedule variances as a result of the:

A. Early start of an activity on a PERT chart?
B. Late start of an activity on a PERT chart?

14–3 Alpha Company has implemented a plan whereby functional managers will be held totally responsible for all cost overruns against their (the functional managers') original estimates. Furthermore, all cost overruns must come out of the functional managers' budgets, whether they be overhead or otherwise, not the project budget. What are the advantages and disadvantages of this approach?

14–4 What would be the result if all project managers decided to withhold a management reserve? What criteria should be used for determining when a management reserve is necessary?

14–5 Consider a situation in which several tasks may be for one to two years rather than the 200 hours normally used in the work-package level of the WBS.

A. How will this affect cost control?

B. Can we still use the 50-50 rule?

C. How frequently should costs be updated?

14–6 Complete the table below and plot the EAC as a function of time. What are your conclusions?

	Cumulative Cost, in Thousands			Variance $		
Week	**BCWS**	**BCWP**	**ACWP**	**Schedule**	**Cost**	**EAC**
1	50	50	25			
2	70	60	40			
3	90	80	67			
4	120	105	90			
5	130	120	115			
6	140	135	130			
7	165	150	155			
8	200	175	190			
9	250	220	230			
10	270	260	270			
11	300	295	305			
12	350	340	340			
13	380	360	370			
14	420	395	400			
15	460	460	450			

14–7 Calculate the total price variance for direct labor and the labor rate cost variance from the following data:

	Direct Material	Direct Labor
Planned price/unit	$ 10.00	$ 22.00
Actual units	9,300	12,000
Actual price/unit	$ 9.25	$ 22.50
Actual cost	$86,025,00	$270,000

14–8 Companies usually estimate work based upon man-months. If the work must be estimated in man-weeks, the man-month is then converted to man-weeks. The problem is in the determination of how many man-hours per month are actually available for actual direct labor work.

Your company has received a request for proposal (RFP) from one of your customers and management has decided to submit a bid. Only one department in your company will be required to perform the work and the department manager estimates that 3,000 hours of *direct* labor will be required.

Your first step is to calculate the number of hours available in a typical man-month. The Human Resources Department provides you with the following *yearly* history for the average employee in the company:

- Vacation (3 weeks)
- Sick days (4 days)
- Paid holidays (10 days)
- Jury duty (1 day)

A. How many direct labor hours are available per month per person?

B. If only one employee can be assigned to the project, what will be the duration of the effort, in months?

C. If the customer wants the job completed within one year, how many employees should be assigned?

CASE STUDIES

THE BATHTUB PERIOD

The award of the Scott contract on January 3, 2010, left Park Industries elated. The Scott Project, if managed correctly, offered tremendous opportunities for follow-on work over the next several years. Park's management considered the Scott Project as strategic in nature.

The Scott Project was a ten-month endeavor to develop a new product for Scott Corporation. Scott informed Park Industries that sole-source production contracts would follow, for at least five years, assuming that the initial R&D effort proved satisfactory. All follow-on contracts were to be negotiated on a year-to-year basis.

Jerry Dunlap was selected as project manager. Although he was young and eager, he understood the importance of the effort for future growth of the company. Dunlap was given some of the best employees to fill out his project office as part of Park's matrix organization. The Scott Project maintained a project office of seven full-time people, including Dunlap, throughout the duration of the project. In addition, eight people from the functional department were selected for representation as functional project team members, four full-time and four half-time.

Although the workload fluctuated, the manpower level for the project office and team members was constant for the duration of the project at 2,080 hours per month. The company assumed that each hour worked incurred a cost of $120.00 per person, fully burdened.

At the end of June, with four months remaining on the project, Scott Corporation informed Park Industries that, owing to a projected cash flow problem, follow-on work would not be awarded until the first week in March (2011). This posed a tremendous problem for Jerry Dunlap because he did not wish to break up the project office. If he permitted his key people to be assigned to other projects, there would be no guarantee that he could get them back at the beginning of the follow-on work. Good project office personnel are always in demand.

Jerry estimated that he needed $240,000 per month during the "bathtub" period to support and maintain his key people. Fortunately, the bathtub period fell over Christmas and New Year's, a time when the plant would be shut down for seventeen days. Between the vacation days that his key employees would be taking, and the small special projects that his people could be temporarily assigned to on other programs, Jerry revised his estimate to $200,000 for the entire bathtub period.

At the weekly team meeting, Jerry told the program team members that they would have to "tighten their belts" in order to establish a management reserve of $200,000. The project team understood the necessity for this action and began rescheduling and replanning until a management reserve of this size could be realized. Because the contract was firm-fixed-price, all schedules for administrative support (i.e., project office and project team members) were extended through February 28 on the supposition that this additional time was needed for final cost data accountability and program report documentation.

Jerry informed his boss, Frank Howard, the division head for project management, as to the problems with the bathtub period. Frank was the intermediary between Jerry and the general manager. Frank agreed with Jerry's approach to the problem and requested to be kept informed.

On September 15, Frank told Jerry that he wanted to "book" the management reserve of $200,000 as excess profit since it would influence his (Frank's) Christmas bonus. Frank and Jerry argued for a while, with Frank constantly saying, "Don't worry! You'll get your key people back. I'll see to that. But I want those uncommitted funds recorded as profit and the program closed out by November 1."

Jerry was furious with Frank's lack of interest in maintaining the current organizational membership.

QUESTIONS

a. Should Jerry go to the general manager?
b. Should the key people be supported on overhead?
c. If this were a cost-plus program, would you consider approaching the customer with your problem in hopes of relief?
d. If you were the customer of this cost-plus program, what would your response be for additional funds for the bathtub period, assuming cost overrun?
e. Would your previous answer change if the program had the money available as a result of an underrun?
f. How do you prevent this situation from recurring on all yearly follow-on contracts?

FRANKLIN ELECTRONICS

In October 2013 Franklin Electronics won an 18-month labor-intensive product development contract awarded by Spokane Industries. The award was a cost reimbursable contract with a cost target of $2.66 million and a fixed fee of 6.75 percent of the target. This contract would be Franklin's first attempt at using formal project management, including a newly developed project management methodology.

Franklin had won several previous contracts from Spokane Industries, but they were all fixed-price contracts with no requirement to use formal project management with earned value reporting. The terms and conditions of this contract included the following key points:

- Project management (formalized) was to be used.
- Earned value cost schedule reporting was a requirement.

- The first earned value report was due at the end of the second month's effort and monthly thereafter.
- There would be two technical interchange meetings, one at the end of the sixth month and another at the end of the twelfth month.

Earned value reporting was new to Franklin Electronics. In order to respond to the original request for proposal (RFP), a consultant was hired to conduct a four-hour seminar on earned value management. In attendance were the project manager who was assigned to the Spokane RFP and would manage the contract after contract award, the entire cost accounting department, and two line managers. The cost accounting group was not happy about having to learn earned value management techniques, but they reluctantly agreed in order to bid on the Spokane RFP. On previous projects with Spokane Industries, monthly interchange meetings were held. On this contract, it seemed that Spokane Industries believed that fewer interchange meeting would be necessary because the information necessary could just as easily be obtained through the earned value status reports. Spokane appeared to have tremendous faith in the ability of the earned value measurement system to provide meaningful information. In the past, Spokane had never mentioned that it was considering the possible implementation of an earned value measurement system as a requirement on all future contracts.

Franklin Electronics won the contact by being the lowest bidder. During the planning phase, a work breakdown structure was developed containing forty-five work packages of which only four work packages would be occurring during the first four months of the project.

Franklin Electronics designed a very simple status report for the project. The table below contains the financial data provided to Spokane at the end of the third month.

	Totals at End of Month 2					Totals at End of Month 3				
Work Packages	**PV**	**EV**	**AC**	**CV**	**SV**	**PV**	**EV**	**AC**	**CV**	**SV**
A	38K	30K	36K	<6K>	<8K>	86K	74K	81K	<7K>	<12K>
B	17K	16K	18K	<2K>	<1K>	55K	52K	55K	<3K>	<3K>
C	26K	24K	27K	<3K>	<2K>	72K	68K	73K	<5K>	<4K>
D	40K	20K	23K	<3K>	<20K>	86K	60K	70K	<10K>	<26K>

Note: BCWS = PV, BCWP = EV, and ACWP = AC.

A week after sending the status report to Spokane Industries, Franklin's project manager was asked to attend an emergency meeting requested by Spokane's vice president for engineering, who was functioning as the project sponsor. The vice president was threatening to cancel the project because of poor performance. At the meeting, the vice president commented, "Over the past month the cost variance overrun has increased by 78 percent from $14,000 to $25,000, and the schedule variance slippage has increased by 45 percent from $31,000 to $45,000. At these rates, we are easily looking at a 500 percent cost overrun and a schedule slippage of at least one year. We cannot afford to let this project continue at this lackluster performance rate. If we cannot develop a plan to control time and cost any better than we have in the past three months, then I will just cancel the contract now, and we will find another contractor who can perform."

QUESTIONS

1. Are the vice president's comments about cost and schedule variance correct? What information did the vice president fail to analyze?

2. What additional information should have been included in the status report?
3. Does Franklin Electronics understand earned value measurement? If not, then what went wrong?
4. Does Spokane Industries understand project management?
5. Does proper earned value measurement serve as a replacement for interchange meetings?
6. What should the project manager from Franklin say in his defense?

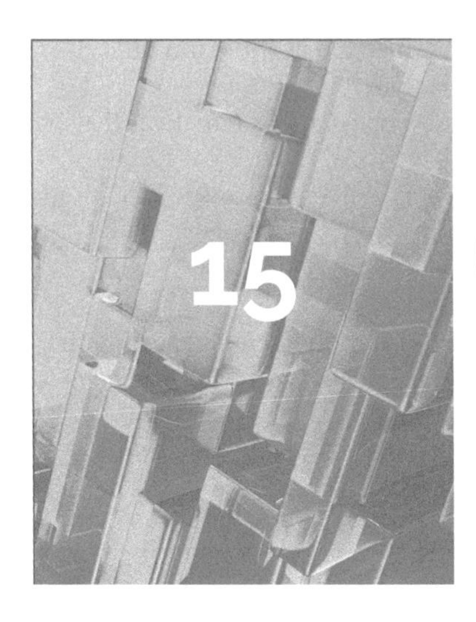

Metrics

15.0 INTRODUCTION

For more than 50 years, relatively project management has been in use but perhaps not on a worldwide or company-wide basis. What differentiated companies in the early years was whether or not they used project management, not how well they used it. Today, almost every company uses project management and the differentiation is whether they are simply good at project management or whether they truly excel at project management. The difference between using project management and being good at project management is relatively small and most companies can become good at project management in a relatively short time period, especially if they have executive-level support and good metrics.

Time and cost has traditionally been the only two primary metrics used to measure and report project performance. Everyone knew decades ago that time and cost alone could not accurately show project performance, but these were to two easiest metrics to track and report. Today, advances in measurement techniques and available performance tracking software have given companies the opportunity to look at new metrics such as benefits expected, value created, safety, risk, customer satisfaction, image, reputation and many others. Executives, sponsors and project managers are now able to make informed decisions based upon evidence rather than merely guesses about the health of the projects.

15.1 PROJECT MANAGEMENT INFORMATION SYSTEMS

A project management information system (PMIS) contains all of the essential and supporting information for project approval, initiation, planning, scheduling, execution, monitoring and control, and closure. While an earned-value measurement system (EVMS) is a critical component of the PMIS, today's PMIS contains significantly more metrics than just time and cost. The PMIS can provide significant benefits if designed properly, such as:

- Satisfying the information needs for the various stakeholders in a timely manner
- Providing the correct Information for informed decision making

- Having the correct amount of information, rather than too much or too little
- Lowering the cost of collecting the right information
- Providing information on how the project interacts with various initiatives that are part of the ongoing business
- Providing information on how one project interacts with other projects being supported by line managers
- Providing value to the company

A good PMIS can prevent projects from failing because of the derailment in project communications. PMIS also makes it easy for team members and functional managers to input the information necessary for effective status reporting.

15.2 ENTERPRISE RESOURCE PLANNING

For several decades, PMISs were seen as report generators providing information on time, cost, and what work was left to do on the project. Time and cost were the two primary metrics that were tracked. Today, that has changed.

Companies have come to the realization that everything they do in their company can be considered as a project. We are managing our business by projects. Therefore, decision-makers have a need for information on both business as well as project management processes, and the two are now related. Also, we are now looking at significantly more metrics than just time and cost.

Information is the key to effective decision making. Companies have developed enterprise resource planning (ERP) systems which are enterprise-wide information systems designed to coordinate all the resources, information, and tasks needed to complete various business and project processes. ERP supports supply chain management, finance and accounting, human resource management, and project management. PMIS is now part of ERP systems.

One of the most important reasons for integrating PMIS and ERP is capacity planning. Functional managers must supply resources to both projects and ongoing business activities. ERP systems are invaluable in this regard. Capacity planning is an essential activity in the portfolio selection of projects. As an example, the ERP system states that a given functional department has fifteen employees available for work assignments and ten of these workers are committed to ongoing work. The ERP system then relays information to the PMIS stating that the remaining five employees are available for project work assignments. The information can also contain the pay grade of the available workers in case specific skill levels must be available for the selection of certain projects.

15.3 PROJECT METRICS

Metrics keep stakeholders informed as to the status of the project. Stakeholders must be confident that the correct metrics are being used and that the measurement portrays a clear and truthful representation of the status. At the beginning of a project, the project manager and the appropriate stakeholders must come to an agreement on which metrics to use and how measurements will be made. We are now using more metrics than just time and cost. This is partially due to the growth in PMIS and ERP technology as well as stakeholders now possessing a greater understanding about project management.

Today, part of the project manager's new role is to understand what the critical metrics are that need to be identified and managed for the project to be viewed as a success by all of the stakeholders. Project managers have come to the realization that defining project-specific metrics and key performance indicators are joint ventures between the project manager, client, and stakeholders. Getting stakeholders to agree on the metrics is difficult, but it must be done as early as possible in the project.

Unlike financial metrics, project-based metrics can change during each life-cycle phase as well as from project to project. Therefore, the establishment and measurement of metrics may be an expensive necessity to validate the critical success factors (CSFs) and maintain customer satisfaction. Many people believe that the future will be metric-driven project management.

Understanding Metrics

Although most companies use some type of metrics for measurement, they seem to have a poor understanding of what constitutes a metric, at least for use in project management. You cannot effectively manage a project without having metrics and accompanying measurement capable of providing you with complete or almost complete information. Therefore, the simplest definition of a metric is something that is measured. Consider the following:

- If it cannot be measured, then it cannot be managed.
- What gets measured gets done.
- You never really understand anything fully unless it can be measured.
- Without good metrics, you make seat-of-the-pants decisions that are just guesses rather than decisions based upon evidence and facts.

Metrics can be measured and recorded as numbers, percentages, dollars, counts, ratings (good, bad, or neutral), or qualitatively versus quantitatively. If you cannot offer a stakeholder something that can be measured, then how can you promise that their expectations will be met? You cannot control what you cannot measure. Good metrics lead to proactive project management rather than reactive project management if the metrics are timely and informative.

For years, measurement itself was not well understood. We avoided metrics management because we did not understand it. But authors such as Douglas Hubbard have helped to resolve the problem:[1]

- It has been measured before.
- You have far more data than you think.

1. D. W. Hubbard, *How to Measure Anything; Finding the Value of Intangibles in Business*, 3rd ed. (Wiley, Hoboken, NJ, 2014), p. 59.

- You need far less data than you think.
- Useful, new observations are more accessible than you think.

Over the years, numerous benefits have surfaced from the use of metrics management. Some benefits of using metrics are:

- Metrics tell us if we are hitting the targets/milestones, getting better, or getting worse.
- Metrics allow you to catch mistakes before they lead to other mistakes; early identification of issues.
- Good metrics lead to informed decision making, whereas poor or inaccurate metrics lead to bad management decisions.
- Good metrics can assess performance accurately.
- Metrics allow for proactive management in a timely manner.
- Metrics improve future estimating.
- Metrics improve performance for the future.
- Metrics make it easier to validate baselines and maintain the baselines with minimal disruptions.
- Metrics can more accurately assess success and failure.
- Metrics can improve client satisfaction.
- Metrics are a means of assessing the project's health.
- Metrics track the ability to meet the project's critical success factors.
- Good metrics allow the definition of project success to be made in terms of factors other than the traditional triple constraints.

While metrics are most frequently used to validate the health of a project, they can also be used to discover best practices in the processes. Capturing best practices and lessons learned are necessities for long-term continuous improvement. Without effective use of metrics, companies could spend years trying to achieve sustained improvements. In this regard, metrics are a necessity because:

- Project approvals are often based upon insufficient information and poor estimating.
- Project approvals are based upon unrealistic return on investment (ROI), net present value (NPV), and payback period calculations.
- Project approvals are often based upon a best-case scenario.
- The true time and cost requirements may be either hidden or not fully understood during the project approval process.

Metrics require a need or purpose, a target, baseline, or reference point, a means of measurement, a means of interpretation, and a reporting structure. Even with good metrics, metrics management can fail. The most common causes of failure are:

- Poor governance, especially by stakeholders
- Slow decision-making processes
- Overly optimistic project plans

- Trying to accomplish too much in too little time
- Poor project management practices and/or methodology
- Poor understanding of how the metrics will be used

During the past few years, one of the drivers for effective metrics management has been the growth in complex projects. The larger and more complex the project is, the greater the difficulty in measuring and determining success. Therefore, the larger and more complex the project is, the greater the need for metrics.

Identifying Metrics

Unlike business environments, which are long term, project environments are much shorter and therefore more susceptible to changing metrics. In a project environment, metrics can change from project to project, during each life-cycle phase, and at any time because of:

- The way the company defines value internally
- The way the customer and the contractor jointly define success and value at project initiation
- The way the customer and contractor come to an agreement at project initiation as to what metrics should be used on a given project
- New or updated versions of tracking software
- Improvements to the enterprise project management methodology and accompanying project management information system
- Changes in the enterprise environmental factors
- Changes in the project's business case assumptions

Metrics can be classified. As an example, below are seven types of metrics or metric indicators that could appear in a metrics library:

- Quantitative metrics (planning dollars or hours as a percentage of total labor)
- Practical metrics (improved efficiencies)
- Directional metrics (risk ratings getting better or worse)
- Actionable metrics (affect change as the number of unstaffed hours)
- Financial metrics (profit margins, ROI, etc.)
- Milestone metrics (number of work packages on time)
- End result or success metrics (customer satisfaction)

Finding a compromise on the correct number of metrics is not easy, but we must determine how many metrics are needed for a particular project.

- With too many metrics:
  - Metric management steals time from other work.
  - We end up providing too much information to stakeholders such that they cannot determine what information is critical.
  - We end up providing information that has limited value.

- With too few metrics:
  - Not enough critical information is provided.
  - Informed decision making becomes difficult.

There are certain ground rules we can establish as part of the metric selection process:

- Make sure that the metrics are worth collecting.
- Make sure that we use what we collect.
- Make sure that the metrics are informative.
- Train the team in the use and value of metrics.

Selecting metrics is a lot easier when you have competent baselines from which to make measurements. It is very difficult or even impossible to use metrics management effectively when the baselines undergo continuous transformation. For work that has not been planned yet, benchmarks and standards can be used instead of baselines.

Metrics by themselves are just numbers or trends resulting from measurements. Metrics have no real value unless they can be properly interpreted by the stakeholders or subject matter experts and a corrective plan, if necessary, can be developed. It is important to know who will benefit from each metric. The level of importance can vary from stakeholder to stakeholder.

There are several questions that can be addressed during metric selection:

- How knowledgeable are the stakeholders in project management?
- How knowledgeable are the stakeholders in metrics management?
- Do we have the necessary organizational process assets for metric measurements?
- Will the baselines and standards undergo transformations during the project?

There are two additional factors that must be considered when selecting metrics. First, there is a cost involved in performing the measurements and, based upon the frequency of the measurements, the costs can be quite large. Second, we must recognize that metrics needed to be updated. Metrics are like best practices; they age and may no longer provide the value or information that was expected. There are several reasons therefore for periodically reviewing the metrics:

- Customers may desire real-time reporting rather than periodic reporting, thus making some metrics inappropriate.
- The cost and complexity of the measurement may make a metric inappropriate for use.
- The metric does not fit well with the organizational process assets available for an accurate measurement.
- Project funding limits may restrict the number of metrics that can be used.

In reviewing the metrics, there are three possible outcomes:

- Update the metric.
- Leave the metric as is but possibly put it on hold.
- Retire the metric from use.

Finally, metrics should be determined after the project is selected and approval is obtained. Selecting a project based upon available or easy-to-use metrics often results in either the selection of the wrong project or metrics that provide useless data.

15.4 KEY PERFORMANCE INDICATORS (KPIS)

Part of the project manager's role is to understand what the critical metrics are that need to be identified, measured, reported, and managed such that the project will be viewed as a success by all of the stakeholders, if possible. The term "metric" is generic whereas a "KPI" is specific. KPIs serve as early warning signs that, if an unfavorable condition exists and is not addressed, the results could be poor. KPIs and metrics can be displayed in dashboards, scorecards, and reports.

Defining the correct metrics or KPIs is a joint venture between the project manager, client, and stakeholders and is necessary in order to get stakeholder agreement. One of the keys to a successful project is the effective and timely management of information. This includes the KPIs. KPIs give us information for making informed decisions by reducing uncertainty.

Getting stakeholder agreement on the KPIs is difficult. If you provide the stakeholders with fifty metrics to select from, they will somehow justify the need for all fifty of them. If you show them one hundred metrics, they will find a reason why all one hundred should be reported. The hard part is to select from the metrics library those critical metrics which can function as KPIs.

For years, metrics and KPIs were used primarily as part of business intelligence techniques. When applied to projects, KPIs answer the question, "What is really important for different stakeholders to monitor on the project?" In business, once a KPI is established, it becomes difficult to change as enterprise environmental factors change for fear that historical comparison data will be lost. But benchmarking industry KPIs is still possible because the KPIs are long term. In project management, because of the uniqueness of projects, benchmarking is more complex because of the relatively short life span of the KPIs.

Need for KPIs

KPIs are high-level snapshots of how a project is progressing toward predefined targets. Some people confuse a KPI with leading indicators. A leading indicator is actually a KPI that measures how the work you are doing now will affect the future. KPIs can be treated as indicators but not necessarily leading indicators.

While some metrics may appear as leading indicators, care must be taken as to how they are interpreted. The misinterpretation of a metric or the mistaken belief that a metric is a leading indicator can lead to faulty conclusions.

KPIs are critical components of all earned-value measurement systems. Terms such as cost variance, schedule variance, schedule performance index, cost performance index, and time/cost at completion are actually KPIs if used correctly but not always referred to as such. The need for these KPIs is simple: What gets measured gets done! If the goal of a performance measurement system is to improve efficiency and effectiveness, then the KPI must reflect controllable factors. There is no point in measuring an activity if the users cannot change the outcome.

Typical KPIs that project managers may use include:

- Percent of work packages adhering to the schedule
- Percent of work packages adhering to the budget
- Number of assigned resources versus planned resources
- Percent of actual versus planned baselines completed to date
- Percent of actual versus planned best practices used
- Project complexity factor
- Time to achieve value
- Customer satisfaction ratings
- Number of critical assumptions made
- Percent of critical assumptions that have changed
- Number of cost revisions
- Number of schedule revisions
- Number of scope change review meetings
- Number of critical constraints
- Percent of work packages with a critical risk designation
- Net operating margins
- Grade levels of assigned resources versus planned resources

Project managers must explain to the stakeholders the differences between metrics and KPIs and why only the KPIs should be reported on dashboards. As an example, metrics focus on the completion of work packages, achievement of milestones, and accomplishment of performance objectives. KPIs focus on future outcomes and this is the information stakeholders need for decision making. Neither metrics nor KPIs can truly predict that the project will be successful, but KPIs provide more accurate information on what might happen in the future if the existing trends continue. Both metrics and KPIs provide useful information, but neither can tell you what action to take or whether a distressed project can be recovered.

Once the stakeholders understand the need for correct KPIs, other questions must be discussed, including:

- How many KPIs are needed?
- How often should they be measured?
- What should be measured?
- How complex will the KPI become?
- Who will be accountable for the KPI (i.e., the KPI owner)?
- Will the KPI serve as a benchmark?

We stated previously that what gets measured gets done, and it is through measurement that a true understanding of the information is obtained. If the goal of a metric measurement system is to improve efficiency and effectiveness, then the KPI must reflect controllable factors. There is no point in measuring an activity or a KPI if the users cannot change the outcome. Such KPIs would not be acceptable to stakeholders.

Using the KPIs

Although most companies use metrics and perform measurement, they seem to have a poor understanding of what constitutes a KPI for projects and how they should be used. Some general principles include:

- KPIs are agreed to beforehand and reflect the CSFs on the project.
- KPIs indicate how much progress has been made toward the achievement of the project's targets, goals, and objectives.
- KPIs are not performance targets.
- The ultimate purpose of a KPI is the measurement of items directly relevant to performance and to provide information on controllable factors appropriate for decision making such that it will lead to positive outcomes.
- Good KPIs drive change but do not prescribe a course of action. They indicate how close you are to a target but do not tell you what must be done to correct deviations from the target.
- KPIs assist in the establishment of objectives to be targeted with the ultimate purpose of either adding value to the project or achieving the prescribed value.

Some people argue that the high-level purposes of a KPI are to encourage effective measurement. In this regard, the three high-level purposes are:

- Measurements that lead to motivation of the team
- Measurements that lead to compliance with use of organizational process assets and alignment to business objectives
- Measurements that lead to performance improvements and the capturing of lessons learned and best practices

Anatomy of a KPI

Some metrics, such as project profitability, can tell us if things look good or bad but do not necessarily provide meaningful information on what we must do to improve performance. Therefore, a typical KPI must do more than just function as a metric. If we dissect the KPIs we will see the following:

- KEY = a major contributor to the success or failure of the project. A KPI metric is therefore only "key" when it can make or break the project.
- PERFORMANCE = a metric that can be measured, quantified, adjusted, and controlled. The metric must be controllable to improve performance.
- INDICATOR = reasonable representation of present and future performance.

A KPI is part of a measurable objective. Defining and selecting the KPIs are much easier if you define the CSFs first. KPIs should not be confused with CSFs. CSFs are things that must be in place to achieve an objective. A KPI is not a CSF but may provide a leading indication that the CSF can be met.

Selecting the right KPIs and the right number of KPIs will:

- Allow for better decision making
- Improve performance on the project
- Help identify problem areas faster
- Improve customer–contractor–stakeholder relations

David Parmenter[2] defines three categories of metrics:

- Results Indicators (RIs): what have we accomplished?
- Performance Indicators (PIs): what must we do to increase or meet performance?
- Key Performance Indicators (KPIs): what are the critical performance indicators that can drastically increase performance or accomplishment of the objectives?

The number of KPIs can vary from project to project and may be impacted by the number of stakeholders. Some people select the number of KPIs based upon the Pareto principle, which states that 20 percent of the total indicators will impact 80 percent of the project. David Parmenter states that the 10/80/10 rule is usually applied when selecting the number of KPIs[3]:

- RIs: 10
- PIs: 80
- KPIs: 10

Typically, between six and ten KPIs are standard. Factors influencing the number of KPIs include:

- The number of information systems that the project manager uses (i.e., one, two, or three)
- The number of stakeholders and their reporting requirements
- The ability to measure the information
- The organizational process assets available to collect the information
- The cost of measurement and collection
- Dashboard reporting limitations

KPI Characteristics The literature abounds with articles defining the characteristics of metrics and KPIs. All too often, authors use the "SMART" rule as a means of identifying the characteristics:

- S = Specific: clear and focused toward performance targets or a business purpose
- M = Measurable: can be expressed quantitatively
- A = Attainable: the targets are reasonable and achievable
- R = Realistic or relevant: the KPI is directly pertinent to the work done on the project
- T = Time-based: the KPI is measurable within a given time period

The SMART rule was originally developed for establishing meaningful objectives for projects and later adapted to the identification of metrics and KPIs. While the use of the SMART rule does have some merit, its applicability to KPIs is questionable.

---

2. David Parmenter, *Key Performance Indicators* (Hoboken, NJ: John Wiley & Sons, 2007), p. 1.
3. Parmenter, *Key Performance Indicators*, p. 9.

The most important attribute of a KPI may be that it is actionable. If the trend of the metric is unfavorable, then the users should know what action is necessary to correct the unfavorable trend. The user must be able to control the outcome. This is a weakness when using the SMART rule to select KPIs.

Wayne Eckerson has developed a more sophisticated set of characteristics for KPIs. The list is more appropriate for business-oriented KPIs than project-oriented KPIs but can be adapted for project management usage. Table 15–1 shows Eckerson's twelve characteristics.[4]

Business or financial metrics are usually the results of many factors, and it therefore may be difficult to isolate what must be done to implement change. For project-oriented KPIs, the following six characteristics may very well be sufficient:

- Predictive: able to predict the future of this trend
- Measurable: can be expressed quantitatively
- Actionable: triggers changes that may be necessary for corrective action
- Relevant: the KPI is directly related to the success or failure of the project

TABLE 15–1. TWELVE CHARACTERISTIC OF EFFECTIVE KPIs

Aligned	KPIs are always aligned with corporate strategy and objectives.
Owned	Every KPI is "owned" by an individual or group on the business side who is accountable for its outcome.
Predictive	KPIs measure drivers of business value. Thus, they are "leading" indicators of performance desired by the organization.
Actionable	PIs are populated with timely, actionable data so users can intervene to improve performance before it is too late.
Few in number	KPIs should focus users on a few high-value tasks, not scatter their attention and energy on too many things.
Easy to understand	KPIs should be straightforward and easy to understand, not based on complex indexes that users do not know how to influence directly.
Balanced and linked	KPIs should balance and reinforce each other, not undermine each other and suboptimize processes.
Trigger changes	The act of measuring a KPI should trigger a chain reaction of positive changes in the organization, especially when it is monitored by the CEO.
Standardized	KPIs are based on standard definitions, rules, and calculations so they can be integrated across dashboards throughout the organization.
Context driven	KPIs put performance in context by applying targets and thresholds to performance so users can gauge their progress over time.
Reinforced with incentives	Organizations can magnify the impact of KPIs by attaching compensation or incentives to them. However, they should do this cautiously, applying incentives only to well-understood and stable KPIs.
Relevant	KPIs gradually lose their impact over time, so they must be periodically reviewed and refreshed.

4. W. W. Eckerson, *Performance Dashboards: Measuring, Monitoring and Managing Your Business* (Hoboken, NJ: John Wiley & Sons, 2006), p. 201.

- Automated: reporting minimizes the chance of human error
- Few in number: only what is necessary

Sometimes KPIs are categorized according to what they are intended to indicate, similar to the metrics categories discussed in the previous section:

- Quantitative KPIs: numerical values
- Practical KPIs: interfacing with company processes
- Directional KPIs: getting better or worse
- Actionable KPIs: effect change
- Financial KPIs: performance measurements

Another means of classification might be leading or lagging indicators or KPIs:

- Lagging KPIs measure past performance.
- Leading KPIs measure drivers for future performance.

Most dashboards have a compromise of both leading and lagging metrics.

KPI Failures

There are several reasons why the use of KPIs often fails on projects. Some of the reasons are:

- People believe that the tracking of a KPI ends at the first line manager level.
- The actions needed to regulate unfavorable indications are beyond the control of the employees doing the monitoring or tracking.
- The KPIs are not related to the actions or work of the employees doing the monitoring.
- The rate of change of the KPIs is too slow, thus making them unsuitable for managing the daily work of the employees.
- Actions needed to correct unfavorable KPIs take too long.
- Measurement of the KPIs does not provide enough meaning or data to make them useful.
- The company identifies too many KPIs to the point where confusion reigns among the people doing the measurements.

Years ago, the only metrics that some companies used were those identified as part of the earned-value measurement system. The metrics generally focused only on time and cost and neglected metrics related to business success as opposed to project success. Thus, the measurement metrics were the same on each project and the same for each life-cycle phase. Today, metrics can change from phase to phase and from project to project. The hard part is obviously deciding upon which metrics to use. Care must be taken that whatever metrics are established does not end up comparing apples and oranges. Fortunately, there are several good books in the marketplace that can assist in identifying proper or meaningful metrics.[5]

5. Three books that provide examples of metric identification are P. F. Rad and G. Levin, *Metrics for Project Management* (Vienna, VA: Management Concepts, 2006); M. Schnapper and S. Rollins, *Value-Based Metric for Improving Results* (Fort Lauderdale, FL: J. Ross Publishing, 2006); and D. W. Hubbard, *How to Measure Anything,* 3rd ed. (Hoboken, NJ: John Wiley & Sons, 2014).

15.5 VALUE-BASED METRICS

For years, customers and contractors have been working toward different definitions of project success. The project manager's definition of success was profitability and tracked through financial metrics. The customer's definition of success was usually the quality of the deliverables. Unfortunately, quality was measured at the closure of the project because it was difficult to track throughout the project. Yet quality was often considered the only measurement of success.

Today, clients and stakeholders appear to be more interested in the value they will receive at the end of the project. If you were to ask ten people, including project personnel, the meaning of value, you would probably get ten different answers. Likewise, if you were to ask which CSF has the greatest impact on value, you would get different answers. Each answer would be related to the individual's work environment and industry. Today, companies seem to have more of an interest in value than quality. This does not mean that we are giving up on quality. Quality is part of value. Some people believe that value is simply quality divided by the cost of obtaining that quality. In other words, the less you pay for obtaining the customer's desired level of quality, the greater the value to the customer.

The problem with this argument is that we assume that quality is the only attribute of value that is important to the client and therefore we need to determine better ways of measuring and predicting just quality. Unfortunately, there are other attributes of value and many of these other attributes are equally as difficult to measure and predict. Customers can have many attributes that they consider as value, but not all of the value attributes are equal in importance.

Unlike the use of quality as the solitary parameter, value allows a company to better measure the degree to which the project will satisfy its objectives. Quality can be regarded as an attribute of value along with other attributes. Today, everyone has quality and produces quality in some form. This is necessary for survival. But what differentiates one company from another are the other attributes, components, or factors used to define value. Some of these attributes might include price, timing, image, reputation, customer service, and sustainability.

In today's world, customers make decisions to hire a contractor based upon the value they expect to receive and the price they must pay to receive this value. Actually it is more of a "perceived" value that may be based upon trade-offs on the attributes of the client's definition of value. *The client may perceive the value of your project to be used internally in their company or pass it on to their customers through their customer value management program.* If your organization does not or cannot offer recognized value to your clients and stakeholders, then you will not be able to extract value (i.e., loyalty) from them in return. Over time, they will defect to other contractors.

As a project manager, you must establish metrics so that the client and the stakeholders can track the value that you will be creating. Measuring and reporting customer value throughout the project is now a competitive necessity. If it is done correctly, it will build emotional bonds with your clients.

For years, the principles of value management have been applied to engineering and manufacturing activities, but only recently have the same principles been applied to project management. According to Venkataraman and Pinto:[6]

---

6. Adapted from R. R. Venkataraman and J. K. Pinto, *Cost and Value Management in Projects* (Hoboken, NJ: John Wiley & Sons, 2008), pp.164–165.

> Value can be added to projects in several ways. These include providing greater levels of client satisfaction, maintaining acceptable levels of satisfaction while lowering resource expenditures, or some combination of the two. It is also possible to improve value by simultaneously increasing satisfaction and resources, provided that satisfaction increases more than the resources used to achieve it.
>
> When managing projects for value, five fundamental concepts must be embraced:
>
> Concept #1: Projects derive their value from the benefits the organization accrues by achieving its stated goals
>
> Concept #2: Project can be viewed as investments made by management
>
> Concept #3: Project investors and sponsors tolerate risk
>
> Concept #4: Project value is related to investment and risks
>
> Concept #5: Value is a balance among the three key project elements: performance, resource usage, and risk

Traditionally, business plans have identified the benefits and resulting value expected from the project. The business plans were usually prepared by a business analyst (BA), and all of this was done prior to the project manager being assigned and brought on board the project. Unfortunately, once the project kicked off, the metrics being monitored and reported generally focused on time and cost rather than the value that the customer was or would be receiving. Value-based metrics were not reported because we simply did not know how to perform the measurements.

Today, we can define a project as a collection of value scheduled for realization. The role of the BA and the PM are now coming together. As stated by Robert Wisocki:[7]

> Meeting time and cost constraints has very little to do with project success. Project success is measured in terms of business value expected compared to business value delivered. Both the PM and the BA should be making every effort to maximize business value for the time and cost invested. This puts the goals of the PM and the BA in alignment.

We can now define project success as the ability to achieve the desired value within the competing constraints imposed upon the project.

Today, with the growth of measurement management techniques, value-based metrics are a necessity for determining project success and are being considered as critical KPIs to be monitored and reported to the client. One contractor reports to their client on a monthly basis the amount of time left before the client will achieve the value that is expected, and the date may be beyond the end date of the project. However, many of the value-based metrics are still considered as a measurement challenge, as shown in Table 15–2.

Not all metrics are value-based metrics, and the metrics can change from project to project as well as in each life-cycle phase. Time and cost can be treated as value-based metrics if coming in under budget can increase profitability and coming in ahead of schedule allows us to enter the marketplace sooner to generate revenue.

---

7. R. K. Wysocki, *The Business Analyst/Project Manager* (Hoboken, NJ: John Wiley & Sons, 2011), p. 2. The author provided an excellent discussion of the relationship between the PM and the BA.

TABLE 15–2. METRIC MEASUREMENT COMPLEXITY

Metric or KPI	Measurement Complexity
Profitability	Easy
Customer satisfaction	Hard
Goodwill	Hard
Penetrate new markets	Easy
Develop new technology	Medium
Technology transfer	Medium
Reputation	Hard
Stabilize work force	Easy
Utilize unused capacity	Easy

Even with the best possible metrics, measuring value can be difficult. Some values are easy to measure while others are more difficult. The easy values to measure are often called soft or tangible values whereas the hard values are often considered as intangible values. Table 15–3 illustrates some of the easy and hard value metrics to measure. Table 15–4 shows some of the problems associated with measuring both hard and soft value metrics.

TABLE 15–3. MEASURING VALUES

Easy (Soft/Tangible) Values	Hard (Intangible) Values
ROI calculators	Stockholder satisfaction
Net present value	Stakeholder satisfaction
Internal rate of return	Customer satisfaction
Cash flow	Employee retention
Payback period	Brand loyalty
Profitability	Time-to-market
Market share	Business relationships
	Safety Reliability
	Reputation Goodwill
	Image

TABLE 15–4. PROBLEMS WITH MEASURING VALUES

Easy (Soft/Tangible) Values	Hard (Intangible) Values
Assumptions are often not disclosed and can affect decision making.	Value is almost always based upon subjective-type attributes of the person doing the measurement.
Measurement is very generic.	It is more of an art than a science.
Measurement never meaningfully captures the correct data.	Limited models are available to perform the measurement.

Value Measurement

The intangible elements or metrics are now considered by some to be more important than the tangible elements. This appears to be happening on IT projects where executives are giving significantly more attention to intangible value metrics. The critical issue with intangible value metrics is not necessarily in the end result, but in the way that the intangibles were calculated.[8] Tangible values are usually expressed quantitatively, whereas intangible values are expressed through a qualitative assessment.

There are three schools of thought for value measurement:

- School 1: The only thing that is important is ROI.
- School 2: ROI can never be calculated effectively; only the intangibles are important.
- School 3: If you cannot measure it, then it does not matter.

The three schools of thought appear to be an all-or-nothing approach, where value is either 100 percent quantitative or 100 percent qualitative. The best approach is most likely a compromise between a quantitative and qualitative assessment of value.

The timing of value measurement is absolutely critical. During the life cycle of a project, it may be necessary to switch back and forth from qualitative to quantitative assessment and, as stated previously, the actual metrics or KPIs can change as well. Certain critical questions must be addressed:

- When or how far along the project life cycle can we establish concrete metrics, assuming it can be done at all?
- Can value be simply perceived and therefore no value metrics are required?
- Even if we have value metrics, are they concrete enough to reasonably predict actual value?
- Will we be forced to use value-driven project management on all projects or are there some projects where this approach is not necessary?
  - Well-defined versus ill-defined
  - Strategic versus tactical
  - Internal versus external
- Can we develop a criterion for when to use value-driven project management, or should we use it on all projects but at a lower intensity level?

For some projects, assessing value at project closure may be difficult. We must establish a time frame for how long we are willing to wait to measure the value or benefits from a project. This is particularly important if the actual value cannot be identified until sometime after the project has been completed. Therefore, it may not be possible to appraise the success of a project at closure if the true economic values cannot be realized until sometime in the future.

---

8. For additional information on the complexities of measuring intangibles, see J. J. Phillips, T. W. Bothell, and G. L. Snead, *The Project Management Scorecard* (Butterworth Heinemann, An Imprint of Elsevier, Oxford, UK, 2002), Chapter 10. The authors emphasize that the true impact on a business must be measured in business units.

Some practitioners of value measurement question whether value measurement is better using boundary boxes instead of life-cycle phases. For value-driven projects, the potential problems with life-cycle phases include:

- Metrics can change between phases and even during a phase.
- Inability to account for changes in the enterprise environmental factors.
- Focus may be on the value at the end of the phase rather than the value at the end of the project.
- Team members may get frustrated not being able to quantitatively calculate value.

Boundary boxes, as show in Figure 15–1, have some degree of similarity to statistical process control charts. We establish a performance target for the value metric. The goal is to stay between 6 and 10 percent of the optimal value. If we are greater than 10 percent above the optimal value, we are exceeding the expected value. If we are below 10 percent of the target value, performance is poor. If we are greater than 20 percent below the target value, urgent attention is needed.

Projects that focus heavily upon value-based metrics must undergo value health checks to confirm that the project will make a contribution of value to the company. Value metrics, such as KPIs, indicate the current value. What is also needed is an extrapolation of the present into the future. Using traditional project management combined with the traditional enterprise project management methodology we can calculate the time at completion and the cost at completion. These are common terms that are part of earned-value measurement systems. But as stated previously, being on time and within budget is no guarantee that the perceived value will be there at project completion.

Therefore, instead of using an enterprise project management methodology which focuses on earned-value measurement, we may need to create a value management methodology (VMM) which stresses the value variables. With VMM, time to complete and cost to complete are still used, but we introduce a new term entitled value (or benefits) at completion. Determination of value at completion must be done periodically throughout

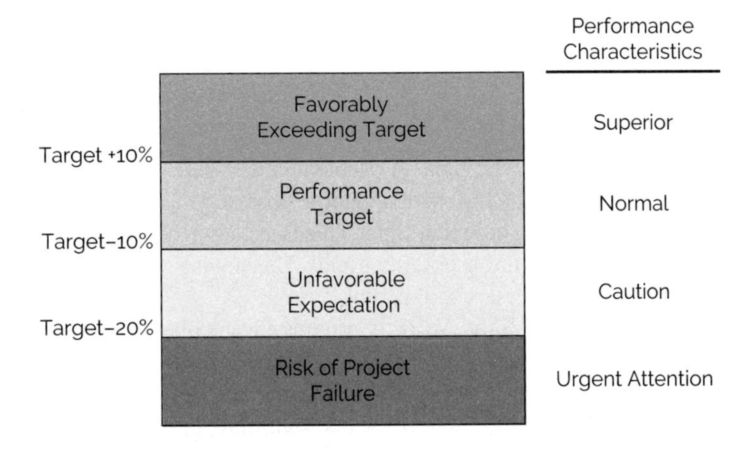

FIGURE 15–1. The boundary box

the project. However, periodic reevaluation of benefits and value at completion may be difficult because:

- There may be no reexamination process.
- Management is not committed and believes that the reexamination process is unreal.
- Management is overoptimistic and complacent with existing performance.
- Management is blinded by unusually high profits on other projects (misinterpretation).
- Management believes that the past is an indication of the future.

An assessment of value at completion can tell us if value trade-offs are necessary. Reasons for value trade-offs include:

- Changes in the enterprise environmental factors
- Changes in the assumptions
- Better approaches have been found, possibly with less risk
- Availability of highly skilled labor
- A breakthrough in technology

Traditional tools and techniques may not work well on value-driven projects. The creation of a VMM may be necessary to achieve the desired results. A VMM can include the features of EVMSs and enterprise project management systems (EPMs), but additional variables must be included for the capturing, measurement, and reporting of value.

15.6 DASHBOARDS AND SCORECARDS

In our attempt to go to paperless project management, emphasis is being given to visual displays such as dashboards and scorecards. Executives and customers desire a visual display of the most critical project performance information in the least amount of space. Simple dashboard techniques, such as traffic light reporting, can convey critical performance information. As an example,

- Red traffic light: A problem exists which may affect time, cost, quality, or scope. Sponsorship involvement is necessary.
- Yellow or amber light: This is a caution. A potential problem may exist, perhaps in the future if not monitored. The sponsor is informed but no action by the sponsor is necessary at this time.
- Green light: Work is progressing as planned. No involvement by the sponsor is necessary.

While a traffic light dashboard with just three colors is most common, some companies use many more colors. The IT group of a retailer had an eight-color dashboard for IT projects. Amber meant that the targeted end date had passed and the project was still not complete. Purple meant that this work package was undergoing a scope change that could have an impact on the triple constraint.

Some people confuse dashboards with scorecards, but there is a difference. According to Eckerson:[9]

- Dashboards are visual display mechanisms used in an operationally oriented performance measurement system that measure performance against targets and thresholds using right-time data.
- Scorecards are visual displays used in a strategically oriented performance measurement system that chart progress towards achieving strategic goals and objectives by comparing performance against targets and thresholds.

Both dashboards and scorecards are visual display mechanisms within a performance measurement system that convey critical information. The primary difference between dashboards and scorecards is that dashboards monitor operational processes such as those used in project management, whereas scorecards chart the progress of tactical goals. Table 15–5 and the description following it show how Eckerson compares the features of dashboards and scorecards.[10]

> Dashboards. Dashboards are more like automobile dashboards. They let operational specialists and their supervisors monitor events generated by key business processes. But unlike automobiles, most business dashboards do not display events in "real time" as they occur; they display them in "right time" as users need to view them. This could be every second, minute, hour, day, week, or month depending on the business process, its volatility, and how critical it is to the business. However, most elements on a dashboard are updated on an intra-day basis, with latency measured in either minutes or hours.
>
> Dashboards often display performance visually, using charts or simple graphs, such as gauges and meters. However, dashboard graphs are often updated in place, causing the graph to "flicker" or change dynamically. Ironically, people who monitor operational processes often find the visual glitz distracting and prefer to view the data in its original form, as numbers or text, perhaps accompanied by visual graphs.

TABLE 15–5. COMPARING FEATURES

Feature	Dashboard	Scorecard
Purpose	Measures performance	Charts progress
Users	Supervisors, specialists	Executives, managers, and staff
Updates	Right-time feeds	Periodic snapshots
Data	Events	Summaries
Display	Visual graphs, raw data	Visual graphs, comments

9. See Eckerson, *Performance Dashboards*, pp. 293, 295. Chapter 12 provides an excellent approach to designing dashboard screens.
10. Eckerson, *Performance Dashboards*, p.13.

Scorecards. Scorecards, on the other hand, look more like performance charts used to track progress toward achieving goals. Scorecards usually display monthly snapshots of summarized data for business executives who track strategic and long-term objectives, or daily and weekly snapshots of data for managers who need to chart the progress of their group of project toward achieving goals. In both cases, the data are fairly summarized so users can view their performance status at a glance.

Like dashboards, scorecards also make use of charts and visual graphs to indicate performance state, trends, and variance against goals. The higher up the users are in the organization, the more they prefer to see performance encoded visually. However, most scorecards also contain (or should contain) a great deal of textual commentary that interprets performance results, describes action taken, and forecasts future results.

Summary. In the end, it does not really matter whether you use the term dashboard or scorecard as long as the tool helps to focus users and organizations on what really matters. Both dashboards and scorecards need to display critical performance information on a single screen so users can monitor results at a glance.

Although the terms are used interchangeably, most project managers prefer to use dashboards and/or dashboard reporting. Eckerson defines three types of dashboards as shown in Table 15–6 and the description that follows:[11]

Operational dashboards monitor core operational processes and are used primarily by frontline workers and their supervisors who deal directly with customers or manage the creation or delivery of organizational products and services. Operational dashboards primarily deliver detailed information that is only lightly summarized. For example, an online Web merchant may track transactions at the product level rather than the customer level. In addition, most metrics in an operational dashboard are updated on an intra-day basis, ranging from minutes to hours depending on the application. As a result, operational dashboards emphasize monitoring more than analysis and management.

Tactical dashboards track departmental processes and projects that are of interest to a segment of the organization or a limited group of people. Managers and business analysts use tactical dashboards to compare performance of their areas or projects, to budget plans, forecasts, or last period's results. For example, a project to reduce the number of errors in a customer database might use a tactical dashboard to display, monitor and analyze progress

TABLE 15–6. THREE TYPES OF PERFORMANCE DASHBOARDS

	Operational	**Tactical**	**Strategic**
Purpose	Monitor operations	Measure progress	Execute strategy
Users	Supervisors, specialists	Managers, analysts	Executives, managers, staff
Scope	Operational	Departmental	Enterprise
Information	Detailed	Detailed/summary	Detailed/summary
Updates	Intra-day	Daily/weekly	Monthly/quarterly
Emphasis	Monitoring	Analysis	Management

11. Eckerson, *Performance Dashboards*, pp. 17–18.

> during the previous 12 months toward achieving 99.9 percent defect-free customer data by 2007.
>
> Strategic dashboards monitor the execution of strategic objectives and are frequently implemented using a Balanced Scorecard approach, although Total Quality Management, Six Sigma, and other methodologies are used as well. The goal of a strategic dashboard is to align the organization around strategic objectives and get every group marching in the same direction. To do this, organizations roll out customized scorecards to every group in the organization and sometimes to every individual as well. These "cascading" scorecards, which are usually updated weekly or monthly, give executives a powerful tool to communicate strategy, gain visibility into operations, and identify the key drivers of performance and business value. Strategic dashboards emphasize management more than monitoring and analysis.

There are three critical steps that must be considered when using dashboards: (1) the target audience for the dashboard, (2) the type of dashboard to be used, and (3) the frequency in which the data will be updated. In some companies, there is a fourth step that includes security of the intellectual property on the dashboard. Some project dashboards focus on the key performance indicators that are part of earned-value measurement. These dashboards may need to be updated daily or weekly. Dashboards related to the financial health of the company may be updated weekly or quarterly.

15.7 BUSINESS INTELLIGENCE

Corporations have been using the concept of business intelligence (BI) for more than two decades. In recent years, business intelligence applications have been replaced by strategic intelligence (SI) applications. Both applications are designed around the monitoring and surveillance of business metrics. According to Corine Cohen:[12]

> The general surveillance field covers notions of watch, scanning, intelligence, competitive intelligence, vigilance, business intelligence, economic intelligence, economic and strategic intelligence, etc. . . .
>
> SI is defined here as a formalized process of research, collection, information processing and distribution of knowledge useful to strategic management. Besides its information function, the main goals of SI are to anticipate environmental threats and opportunities (anticipatory function), help in strategic decision making and improve competitiveness and performance of the organization. It requires an organizational network structure, and human technical and financial resources.
>
> A distinction must therefore be made between Strategic Watch and SI. SI goes beyond Strategic Watch with its proactivity and its deeper involvement in the strategic decision process. Watch can (must) indicate the impacts of a detected event for example. However, it becomes intelligence when it produces recommendations and provides instructions to the recipient (all the more so when it implements them).

12. C. Cohen, *Business Intelligence* (Hoboken, NJ: John Wiley & Sons and ISTE Ltd Publishers, 2009), p. xiii.

BI and SI applications have taught us that the way we try to monitor and control projects must change. In a project management environment, BI would be represented by metrics and SI would be represented by key performance indicators (KPIs). Key performance indicators are the "strategic" metrics that provide us with the critical information for informed decision making. BI metrics are simply monitoring metrics whereas SI metrics, or KPIs, provide information on the future rather than just the present and indicate changes that may be necessary. Since project managers today and in the future will become more business-oriented managers, the relationship between metrics and BI and SI will become more important.

Related Case Studies (from Kerzner/*Project Management Case Studies*, 4th ed.)	**Related Workbook Exercises (from Kerzner/*Project Management Workbook and PMP®/CAPM® Exam Study Guide*, 11th ed.)**	***PMBOK® Guide*, 6th Edition, Reference Section for the PMP® Certification Exam**
• Trouble in Paradise	• Using the 50/50 Rule • Multiple Choice Exam • Crossword Puzzle on Cost Management	• Project Cost Management • Project Scope Management

15.8 STUDYING TIPS FOR THE PMI® PROJECT MANAGEMENT CERTIFICATION EXAM

This section is applicable as a review of the principles to support the knowledge areas and domain groups in the *PMBOK® Guide*. This chapter addresses:

- Project Scope Management
- Project Cost Management
- Initiating
- Planning
- Controlling

Understanding the following principles is beneficial if the reader is using this text to study for the PMP® Certification Exam:

- What is meant by a management cost and control system
- How important is it to have good metrics for each area of knowledge in the *PMBOK® Guide*

The following multiple-choice questions will be helpful in reviewing the principles of this chapter:

1. In the future, we must better understand how to:
   A. Identify metrics
   B. Measure metrics
   C. Report metrics
   D. All of the above

PMBOK, CAPM, and PMP are registered marks of the Project Management Institute, Inc.

2. Metrics must be established for those activities that:
   A. Are the most difficult to manage
   B. The most difficult to measure
   C. The most difficult to report
   D. Have a direct bearing on the success or failure of the project
3. Completing a project within the triple constraint is not necessarily success if the stakeholders cannot recognize the value at the completion of the project.
   A. True
   B. False
4. Metrics are design to:
   A. Keep stakeholders informed
   B. Portray a clear and truthful representation of the project
   C. Provide information for informed decision making
   D. All of the above
5. In the early years of project management, how many core metrics were looked at on each project other than time and cost?
   A. None
   B. 2
   C. 3
   D. 5
6. In the early years of project management, we discovered that:
   A. Measurement of time and cost was usually quite accurate.
   B. The project was a success if we met the time and cost constraints.
   C. Unfavorable metrics did not often provide us with sufficient information for corrective action.
   D. Customers and stakeholders quickly understood the meaning of these metrics.
7. Which of the following is true?
   A. Metrics should not be allowed to change as the project progresses.
   B. Good metrics can provide precise predictions of performance rather than a close estimate.
   C. Metrics do not provide any real value unless they can be measured.
   D. All of above are true.
8. Metrics focus on the future whereas KPIs focus more so on the here and now.
   A. True
   B. False
9. The Rule of Inversion states that:
   A. Only metrics that are difficult to measure should be selected as KPIs.
   B. Risk KPIs are essential for tracking performance.
   C. Only core metrics should be tracked.
   D. Only the easy metrics, such as time and cost, should be selected.

10. Some metrics and KPIs may not be measured until well into the future because the measurement is based upon the beneficial use of the deliverable.
    A. True
    B. False
11. Which of the following is not one of the six characteristics of a KPI?
    A. Predicts the present and the future
    B. Can be displayed with more than one icon
    C. Relevant to the success and failure of the project
    D. Few in number
12. Which of the following can influence the selection of KPIs?
    A. Size and number of dashboards
    B. Type of audience and audience requirements
    C. Audience project management maturity level
    D. All of the above
13. KPIs provide no real value if they cannot be measured with any "reasonable" degree of accuracy.
    A. True
    B. False
14. Sophisticated measurement techniques will eliminate uncertainties and provide us with complete information for project decision making.
    A. True
    B. False
15. The true health of a project can be determined from a single metric or KPI.
    A. True
    B. False
16. During a project kick-off meeting, team members should be briefed on what KPIs will be used as well as what KPIs will not be used.
    A. True
    B. False

ANSWERS

1. D
2. D
3. A
4. D
5. A
6. C
7. C
8. B
9. D
10. A
11. B
12. D
13. A
14. B
15. B
16. A

PROBLEMS

15–1 What are the differences between a traditional metric and a KPI?

15–2 Is there a standard way that metric information will be reported?

15–3 What are the requirements for a metric to exist?

15–4 What causes a lack of support for the use of metrics?

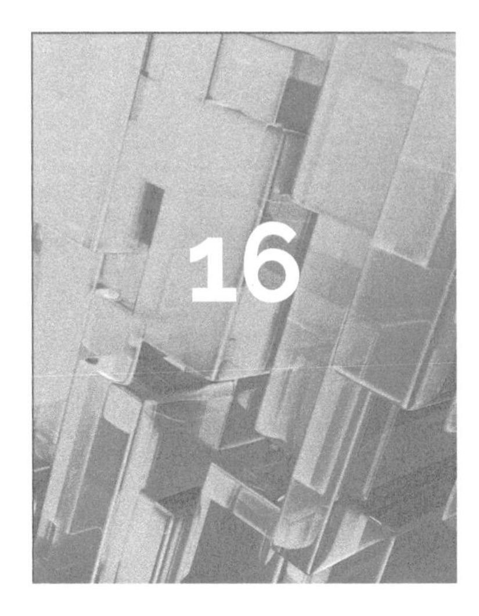

16 Trade-Off Analysis in a Project Environment

"When we try to pick out anything by itself, we find it hitched to everything else in the universe." —MUIR'S LAW

16.0 INTRODUCTION

Successful project management is both an art and a science and attempts to control corporate resources within the constraints of time, cost, and performance. Most projects are unique, one-of-kind activities for which there may not have been reasonable standards for forward planning. As a result, the project manager may find it extremely difficult to stay within the time–cost–performance triangle of Figure 16–1. Even though today we focus on competing constraints, the trade-offs in this section will focus heavily upon just time, cost, and performance.

The time–cost–performance triangle is the "magic combination" that is continuously pursued by the project manager throughout the life cycle of the project. If the project were to flow smoothly, according to plan, there might not be a need for trade-off analysis. Unfortunately, this rarely happens. The situation will become even more complex as we add in more constraints and metrics. But for simplicity's sake, only time, cost, and performance will be considered in this chapter.

Trade-offs are illustrated in Figure 16–2, where the Ds represent deviations from the original estimates. The time and cost deviations are normally overruns, whereas the performance error will be an underrun. No two projects are exactly alike, and trade-off analysis will be an ongoing effort throughout the life of the project, continuously influenced by both the internal and the external environment. Experienced project managers have predetermined trade-offs in reserve, recognizing that trade-offs are part of a continuous thought process.

Trade-offs are always based on the constraints of the project. Table 16–1 illustrates the types of constraints commonly imposed. Situations A and B are the typical trade-offs encountered in project management. For example, situation A-3 portrays most research and development projects. The performance of an R&D project is usually well defined, and it is cost and time that may be allowed to go beyond budget and schedule. The determination of what to sacrifice is based on the available alternatives. If there are

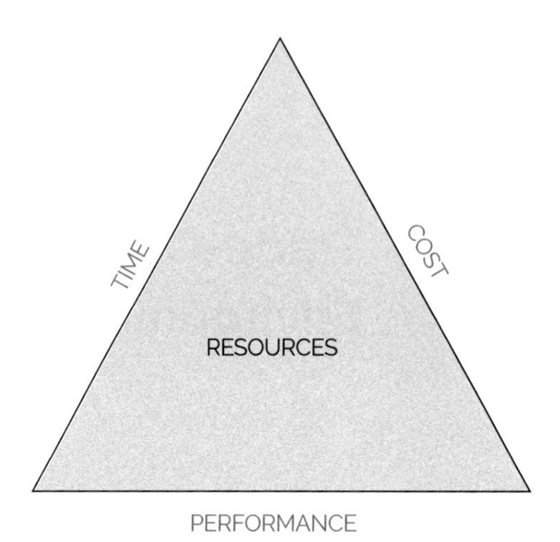

FIGURE 16–1. Overview of project management.

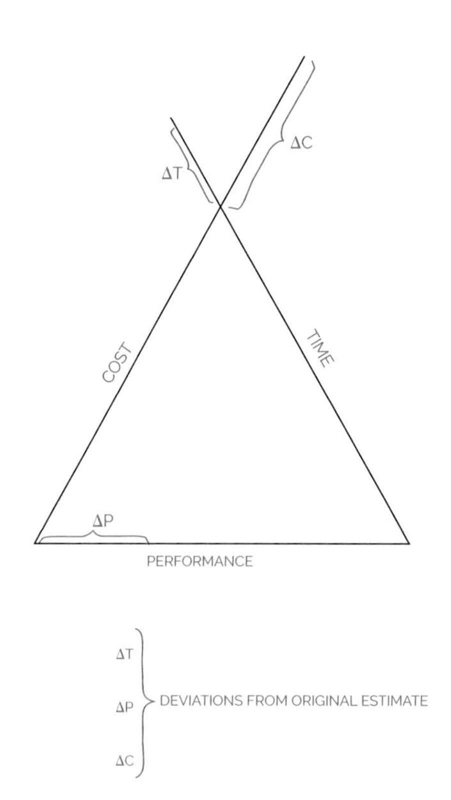

FIGURE 16–2. Project management with trade-offs.

TABLE 16–1. CATEGORIES OF CONSTRAINTS

	Time	Cost	Performance
A. One Element Fixed at a Time			
A-1	Fixed	Variable	Variable
A-2	Variable	Fixed	Variable
A-3	Variable	Variable	Fixed
B. Two Elements Fixed at a Time			
B-1	Fixed	Fixed	Variable
B-2	Fixed	Variable	Fixed
B-3	Variable	Fixed	Fixed
C. Three Elements Fixed or Variable			
C-1	Fixed	Fixed	Fixed
C-2	Variable	Variable	Variable

no alternatives to the product being developed and the potential usage is great, then cost and time are the trade-offs.

Most capital equipment projects would fall into situation A-1 or B-2, where time is of the essence. The sooner the piece of equipment gets into production, the sooner the return of investment can be realized. Often there are performance constraints that determine the profit potential of the project. If the project potential is determined to be great, cost will be the slippage factor, as in situation B-2.

Non–process-type equipment, such as air pollution control equipment, usually develops a scenario around situation B-3. Performance is fixed by the Environmental Protection Agency. The deadline for compliance can be delayed through litigation, but if the lawsuits fail, most firms then try to comply with the least expensive equipment that will meet the minimum requirements.

The professional consulting firm operates primarily under situation B-1. In situation C, the trade-off analysis will be completed based on the selection criteria and constraints. If everything is fixed (C-1), there is no room for any outcome other than total success, and if everything is variable (C-2), there are no constraints and thus no trade-off.

Many factors go into the decision to sacrifice either time, cost, or performance. It should be noted, however, that it is not always possible to sacrifice one of these items without affecting the others. For example, reducing the time could have a serious impact on performance and cost (especially if overtime is required).

There are several factors, such as those shown in Figure 16–3, that tend to "force" trade-offs. Poorly written documents (e.g., statements of work, contracts, and specifications) are almost always inward forces for conflict in which the project manager tends to look for performance relief. In many projects, the initial sale and negotiation, as well as the specification writing, are done by highly technical people who are driven to create a monument rather than meet the operational needs of the customer. When the operating forces dominate outward from the project to the customer, project managers may tend to seek cost relief.

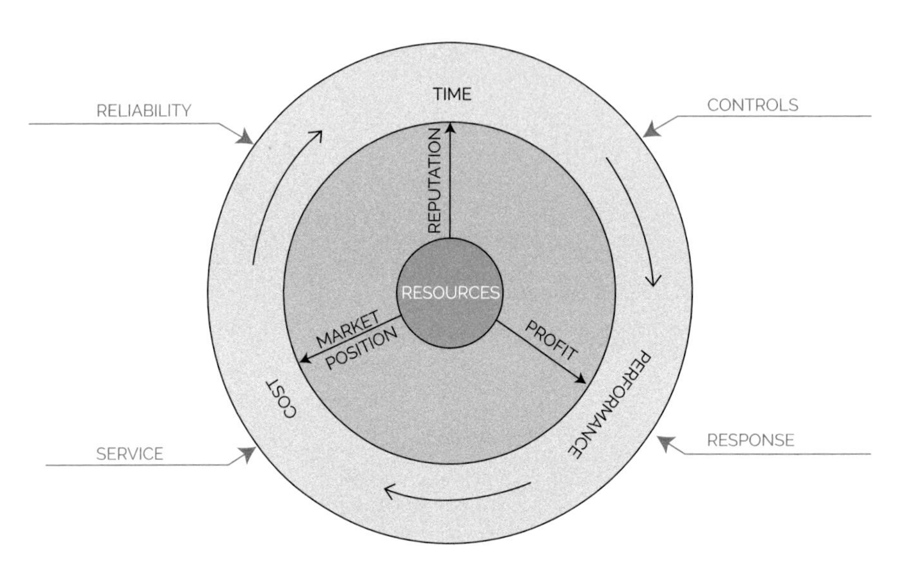

FIGURE 16–3. Trade-off forcing factors.

16.1 METHODOLOGY FOR TRADE-OFF ANALYSIS

> ***PMBOK® Guide*, 6th Edition**
> Chapter 4 Project Integration Management
> Chapter 5 Project Scope Management

Any process for managing time, cost, and performance trade-offs should emphasize the systems approach to management by recognizing that even the smallest change in a project or system could easily affect all of the Planning Process Group Triangle organization's systems. A typical systems model is shown in Figure 16–4. Because of this, it is often better to develop a process for decision-making/trade-off analysis rather than to maintain hard-and-fast rules on trade-offs. The following six steps may help:

1. Recognizing and understanding the basis for project conflicts
2. Reviewing the project objectives
3. Analyzing the project environment and status
4. Identifying the alternative courses of action
5. Analyzing and selecting the best alternative
6. Revising the project plan

Step 1: Recognizing Project Conflicts The first step in any decision-making process must be recognition and understanding of the conflict. Most projects have management cost and control systems that compare actual versus planned results, scrutinize the results through variance analyses, and provide status reports so that corrective action can be taken to resolve the problems. Project managers must carefully evaluate

PMBOK is a registered mark of the Project Management Institute, Inc.

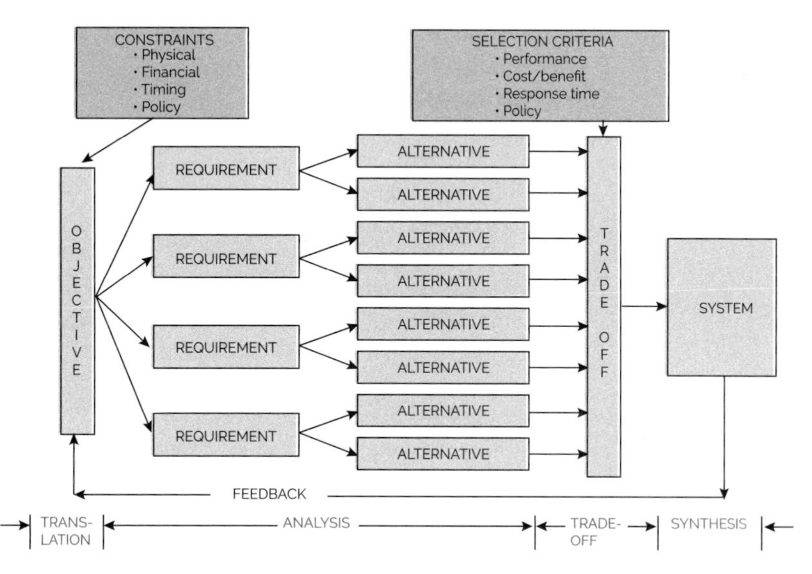

FIGURE 16–4. The typical systems approach.

information about project problems because it may not always be what it appears to be. Typical questions to ask are:

- Is the information pertinent?
- Is the information current?
- Are the data complete?
- Who has determined that this situation exists?
- How does he know this information is correct?
- If this information is true, what are the implications for the project?

The reason for this first step is to understand the cause of the conflict and the need for trade-offs. Most causes can be categorized as human errors or failures, uncertain problems, and totally unexpected problems, as shown below:

- Human errors/failures
  - Impossible schedule commitments
  - Poor control of design changes
  - Poor project cost accounting
  - Machine failures
  - Test failures
  - Failure to receive a critical input
  - Failure to receive anticipated approvals
- Uncertain problems
  - Too many concurrent projects
  - Labor contract expiration

  - Change in project leadership
  - Possibility of project cancellation
- Unexpected problems
  - Overcommitted company resources
  - Conflicting project priorities
  - Cash flow problems
  - Labor contract disputes
  - Delay in material shipment
  - "Fast-track" people having been promoted off the project
  - "Temporary" employees having to be returned to their home base
  - Inaccurate original forecast
  - Change in market conditions
  - New standards having been developed
  - Customers are requesting scope changes

Step 2: Reviewing Objectives The second step in the decision-making process is a complete review of the project objectives as seen by the various participants in the projects, ranging from top management to project team members. These objectives and/or priorities were originally set after considering many environmental factors, some of which may have changed over the lifetime of the project.

The nature of these objectives will usually determine the degree of rigidity that has been established between time, cost, and performance. This may require reviewing project documentation, including:

- Project objectives
- Project integration into sponsor's objectives and strategic plan
- Statement of work
- Schedule, cost, and performance specifications
- Resources consumed and projected

Step 3: Analyzing Project Status The third step is the analysis of the project environment and status, including a detailed measurement of the actual time, cost, and performance results with the original or revised project plan. This step should not turn into a "witch hunt" but should focus on project results, problems, and roadblocks. Factors such as financial risk, potential follow-up contracts, the status of other projects, and relative competitive positions are just a few of the environmental factors that should be reviewed. Some companies have established policies toward trade-off analysis, such as "never compromise performance." Even these policies, however, have been known to change when environmental factors add to the financial risk of the company. The following topics may be applicable under step 3:

- Discuss the project with the project management office to:
  - Determine relative priorities for time, cost, and performance
  - Determine impact on firm's profitability and strategic plan
  - Get a management assessment (even a hunch as to what the problems are)

- If the project is a contract with an outside customer, meet with the customer's project manager to assess his views relative to project status and assess the customer's priorities for time, cost, and performance.
- Meet with the functional managers to determine their views on the problem and to gain an insight regarding their commitment to a successful project. Where does this project sit in their priority list?
- Review in detail the status of each project work package. Obtain a clear and detailed appraisal by the responsible project office personnel as to:
  - Time to complete
  - Cost to complete
  - Work to complete
- Review past data to assess credibility of cost and schedule information in the previous step.

The project manager may have sufficient background to quickly assess the significance of a particular variance and the probable impact of that variance on project team performance. Knowledge of the project requirements (possibly gained with the assistance of the project sponsor) will usually help a project manager determine whether corrective action must be taken at all, or whether the project should simply be permitted to continue as originally conceived.

Whether or not immediate action is required, a quick analysis of why a potential problem has developed is in order. Obviously, it will not help to "cure the symptoms" if the "disease" itself is not remedied. The project manager must remain objective in such problem identification, since he himself is a key member of the project team and may be personally responsible for problems that are occurring. Suspect areas typically include:

- *Inadequate planning:* Either planning was not done in sufficient detail or controls were not established to determine that the project is proceeding according to the approved plan.
- *Scope changes:* Cost and schedule overruns are the normal result of scope changes that are permitted without formal incorporation in the project plan or increase in the resources authorized for the project.
- *Poor performance:* Because of the high level of interdependencies that exist within any project team structure, unacceptable performance by one individual may quickly undermine the performance of the entire team.
- *Excess performance:* Frequently an overzealous team member will unintentionally distort the planned balance between cost, schedule, and performance on the project.
- *Environmental restraints—particularly on projects involving "third-party approvals" or dependent on outside resources:* Changes, delays, or nonperformance by parties outside the project team may have an adverse impact on the team performance.

Some projects appear to be out of tolerance when, in fact, they are not. For example, some construction projects are so front-loaded with costs that there appears to be a major discrepancy when one actually does not exist. The front-end loading of cost was planned for.

Step 4: Analyzing Alternatives The fourth step in the project trade-off process is to list alternative courses of action. This step usually means brainstorming the possible methods of completing the project by compromising some combination of time, cost, or performance. Hopefully, this step will refine these possible alternatives into the three or four most likely scenarios for project completion. At this point, some intuitive decision making may be required to keep the list of alternatives at a manageable level.

In order fully to identify the alternatives, the project manager must have specific answers to key questions involving time, cost, and performance:

***PMBOK® Guide*, 6th Edition**
1.2.4.5 Monitoring and Controlling Process Group

- Time
  - Is a time delay acceptable to the customer?
  - Will the time delay change the completion date for other projects and other customers?
  - What is the cause for the time delay?
  - Can resources be recommitted to meet the new schedule?
  - What will be the cost for the new schedule?
  - Will the increased time give us added improvement?
  - Will an extension of this project cause delays on other projects in the customer's house?
  - What will the customer's response be?
  - Will the increased time change our learning curve?
  - Will this hurt our company's ability to procure future contracts?
- Cost
  - What is causing the cost overrun?
  - What can be done to reduce the remaining costs?
  - Will the customer accept an additional charge?
  - Should we absorb the extra cost?
  - Can we renegotiate the time or performance standards to stay within cost?
  - Are the budgeted costs for the remainder of the project accurate?
- Will there be any net value gains for the increased funding?
  - Is this the only way to satisfy performance?
  - Will this hurt our company's ability to procure future contracts?
  - Is this the only way to maintain the schedule?
- Performance
  - Can the original specifications be met?
  - If not, at what cost can we guarantee compliance?
  - Are the specifications negotiable?
  - What are the advantages to the company and customer for specification changes?
  - What are the disadvantages to the company and customer for performance changes?
  - Are we increasing or decreasing performance?
  - Will the customer accept a change?
  - Will there be a product or employee liability incurred?
  - Will the change in specifications cause a redistribution of project resources?
  - Will this change hurt our company's ability to procure future contracts?

Once the answers to these questions are obtained, it is often best to plot the results graphically. Graphical methods have been used during the past two decades to determine crashing costs for shortening the length of a project. To use the graphical techniques, we must decide on which of the three parameters to hold fixed.

Situation 1: Performance Is Held Constant (to Specifications): With performance fixed, cost can be expressed as a function of time. Sample curves appear in Figures 16–5 and 16–6. In Figure 16–5, the circled X indicates the target cost and target time. Unfortunately, the cost to complete the project at the target time is higher than the budgeted cost. It may be possible to add resources and work overtime so that the time target can be met. Depending upon the way that overtime is burdened, it may be possible to find a minimum point in the curve where further delays will cause the total cost to escalate.

Curve A in Figure 16–6 shows the case where "time is money," and any additional time will increase the cost to complete. Factors such as management support time will always increase the cost to complete. There are, however, some situations where the increased costs occur in plateaus. This is shown in curve B of Figure 16–6. This could result from having to wait for temperature conditioning of a component before additional work can be completed, or simply waiting for nonscheduled resources to be available. In the latter case, the trade-off decision points may be at the end of each plateau.

With performance fixed, there are four methods available for constructing and analyzing the time–cost curves:

- Additional resources may be required. This will usually drive up the cost very fast. Assuming that the resources are available, cost control problems can occur as a result of adding resources after initial project budgeting.
- The scope of work may be redefined and some work deleted without changing the project performance requirements. Performance standards may have been set

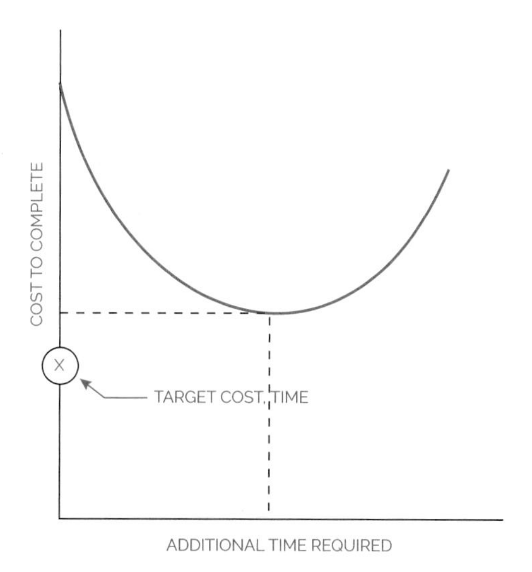

FIGURE 16–5. Trade-offs with fixed performance.

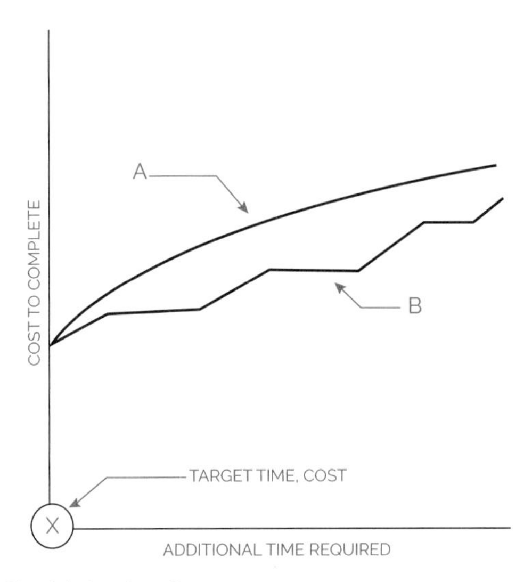

FIGURE 16–6. Trade-offs with fixed performance.

too high, or the probability of success demanded of the project team may have been simply unrealistic. Reductions in cost and improvements in schedules would typically result from relaxing performance specifications, provided that the lower quality level will still meet the requirements of the customer.

- Available resources may be shifted in order to balance project costs or to speed up activities that are on the "critical" path work element that is trailing. This process of replanning shifts elements from noncritical to critical activities.
- Given a schedule problem, a change in the logic diagram may be needed to move from the current position to the desired position. Such a change could easily result in the replanning and reallocation of resources. An example of this would be to convert from "serial" to "parallel" work efforts. This is often risky.

Trade-offs with fixed performance levels must take into account the dependence of the firm on the customer, priority of the project within the firm, and potential for future business. A basic assumption here is that the firm may never sacrifice its reputation by delivering a product that doesn't perform to the specifications. The exception might be a change that would enhance performance and pull the project back on schedule. This is always worth investigating before entering into time–cost trade-offs.

Time and cost are interrelated in a labor-intensive project. As delivery slips, costs usually rise. Slipping delivery schedules and minimizing cost growth are usually the recommended alternative for projects in which the dependence of the firm on the customer, the priority of the project within the firm's stream of projects, and the future business potential in terms of sales represent a low- to medium-risk. Even in some high-risk situations, the contractor may have to absorb the additional cost. This decision is often based on estimating the future projects from this customer so that the loss is amortized against future business. Not all projects are financial successes.

Sometimes projects may have fixed time and costs, leaving only the performance variable for trade-offs. However, as shown in the following scenario, the eventual outcome may be to modify the "fixed" cost constraint.

Situation 2: Cost Is Fixed: With cost fixed, performance will vary as a function of time, as shown in Figure 16–7. The decision of whether to adhere to the target schedule data is usually determined by the level of performance. In curve A, performance may increase rapidly to the 90 percent level at the beginning of the project. A 10 percent increase in time may give a 20 percent increase in performance. After a certain point, a 10 percent increase in time may give only a 1 percent increase in performance. The company may not wish to risk the additional time necessary to attain the 100 percent performance level if it is possible to do so. In curve C, the additional time must be sacrificed because it is unlikely that the customer will be happy with a 30 to 40 percent performance level. Curve B is the most difficult curve to analyze unless the customer has specified exactly which level of performance will be acceptable.

If cost is fixed, then it is imperative that the project have a carefully worded and understood contract with clear specifications as to the required level of performance and very clear statements of inclusion and exclusion. Careful attention to costs incurred because of customer changes or additional requirements can help reduce the possibility of a cost overrun. Experience in contracting ensures that costs that may be overlooked by the inexperienced project manager are included, thus minimizing the need for such trade-offs downstream. Common, overlooked items that can drive up costs include:

- Excessive detailed reporting
- Unnecessary documentation
- Excessive tracking documentation for time, cost, and performance

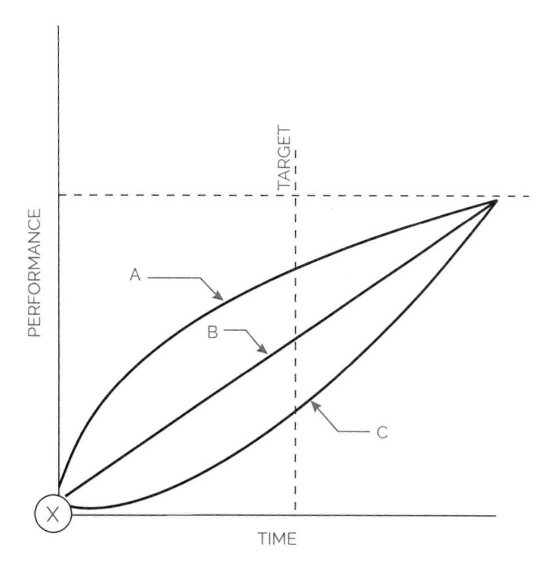

FIGURE 16–7. Trade-offs with fixed cost.

- Detailed specification development for equipment that could be purchased externally for less cost
- Wrong type of contract for this type of project

Often with a fixed-cost constraint, the first item that is sacrificed is performance. But such an approach can contain hidden disasters over the life of a project if the sacrificed performance turns out to have been essential to meeting some unspecified requirement such as long-term maintenance. In the long run, a degraded performance can actually increase costs rather than decrease them. Therefore, the project manager should be sure he has a good understanding of the real costs associated with trade-offs in performance.

Situation 3: Time Is Fixed: Figure 16–8 identifies the situation in which time is fixed and cost varies with performance. Figure 16–8 is similar to Figure 16–7 in that the rate of change of performance with cost is the controlling factor. If performance is at the 90 percent level with the target cost, then the contractor may request performance relief. This is shown in curve A. However, if the actual situation reflects curve B or C, additional costs must be incurred with the same considerations of situation 1—namely, how important is the customer and what emphasis should be placed on his follow-on business? Completing the project on schedule can be extremely important in certain cases. For example, if an aircraft pump is not delivered when the engine is ready for shipment, it can hold up the engine manufacturer, the airframe manufacturer, and ultimately the customer. All three can incur substantial losses due to the delay of a single component. Moreover, customers who are unable to perform and who incur large unanticipated costs tend to have long memories. An irate vice president in the customer's shop can kill further contracts out of all proportion to

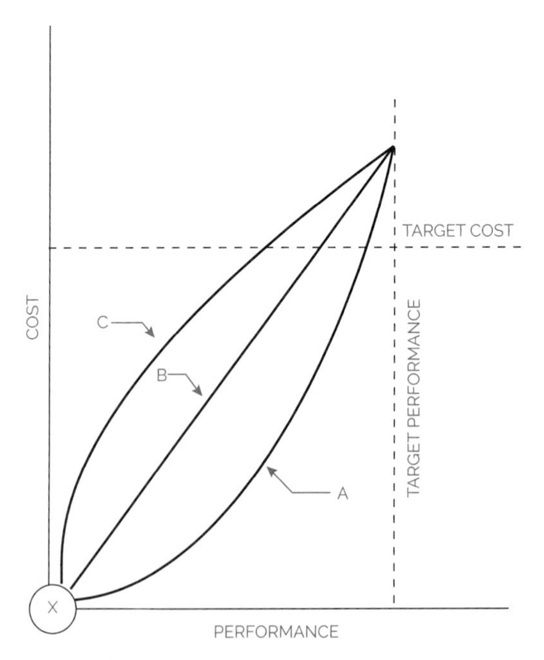

FIGURE 16–8. Trade-offs with fixed time.

the real failure to deliver on time. Sometimes, even though time is supposedly fixed, there may be latitude without inconvenience to the customer. This could come about because the entire program (of which your project is just one subcontract) is behind schedule, and the customer is not ready for your particular project. Another aspect of the time factor is that "early warning" of a time overrun can often mitigate the damage to the customer and greatly increase his favorable response. Careful planning and tracking, close coordination with all functions involved, and realistic dealing with time schedules before and during the project can ensure early notification to the customer and the possible negotiation of a trade-off of time and dollars or even technical performance. The last thing that a customer wants is to have a favorable progress report right up to the end of scheduled time and then to be surprised with a serious schedule overrun.

When time is fixed, the customer may find that he has some flexibility in determining how to arrive at the desired performance level. As shown in Figure 16–9, the contractor may be willing to accept additional costs to maximize employee safety.

Situation 4: No Constraints Are Fixed: Another common situation is that in which neither time, cost, nor performance is fixed. The best method for graphically showing the trade-off relationships is to develop parametric curves as in Figure 16–10. Cost and time trade-offs can now be analyzed for various levels of performance. The curves can also be redrawn for various cost levels (i.e., 100, 120, 150 percent of target cost) and schedule levels.

Another method for showing a family of curves is illustrated in Figure 16–11. Here, the contractor may have several different cost paths for achieving the desired time and performance constraints. The final path selected depends on the size of the risk that the contractor wishes to take.

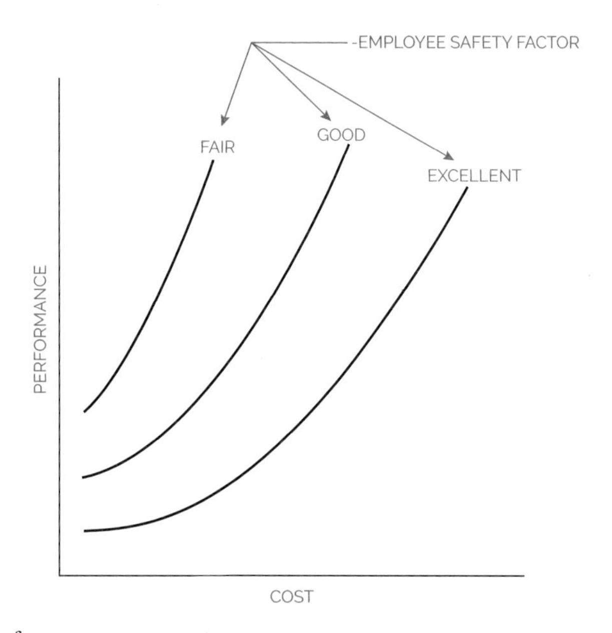

FIGURE 16–9. Performance versus cost.

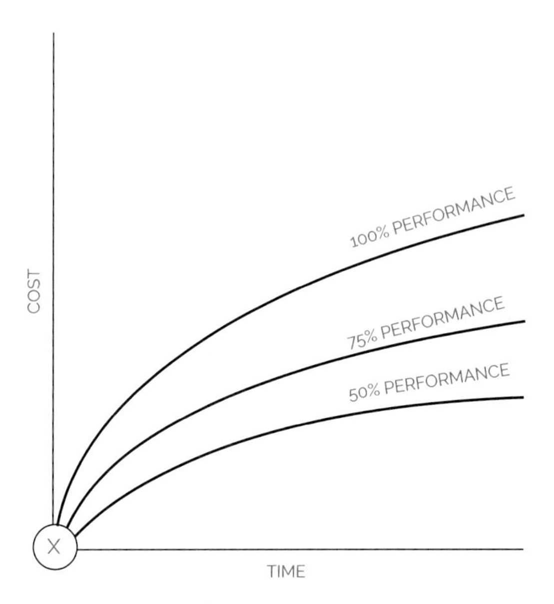

FIGURE 16–10. Trade-off analysis with family of curves.

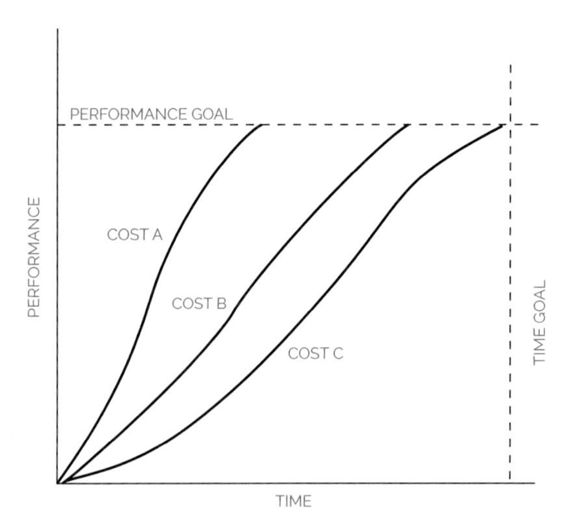

FIGURE 16–11. Cost–time–performance family of curves.

Trade-offs can also be necessary at any point during the life cycle of a project. Figure 16–12 identifies how the relative importance of the constraints of time, cost, and performance can change over the life cycle of the project. At project initiation, costs may not have accrued to a point where they are important. On the other hand, project performance may become even more important than the schedule. At this point, additional performance can be "bought." As the project nears termination, the relative importance of the cost

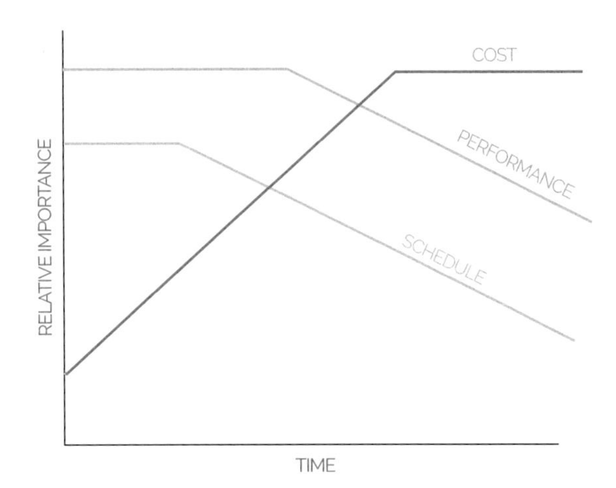

FIGURE 16–12. Life-cycle trade-offs. (Schedule not necessarily typical.)

constraint may increase drastically, especially if project profits are the company's major source of revenue. Likewise, it is probable that the impact of performance and schedule will be lower.

Step 5: Selecting an Alternative Once the alternative courses of action are determined, step 5 in the methodology is employed in order to analyze and select the feasible alternatives. Analyzing the alternatives should include the preparation of the revised project objectives for cost, performance, and time, along with an analysis of the required resources, general schedules, and revised project plans necessary to support each scenario. It is then the function of top management in conjunction with the project and functional managers to choose the solution that minimizes the overall impact to the company. This impact need not be measured just in short-term financial results, but should include long-term strategic and market considerations.

The following tasks can be included in this step:

- Prepare a formal project update report including alternative work scopes, schedules, and costs to achieve.
  - Minimum cost overrun
  - Conformance to project objectives
  - Minimum schedule overrun
- Construct a decision tree including costs, work objectives, and schedules, and an estimate of the probability of success for each condition leading to the decision point.
- Present to internal and external project management the several alternatives along with an estimate of success probability.
- With management's agreement, select the appropriate completion strategy, and begin implementation. This assumes that management does not insist on an impossible task.

- The last item requires further clarification. Many companies use a checklist to establish the criteria for alternative evaluation as well as for assessment of potential future problems. The following questions may be part of such a checklist:
- Will other projects be affected?
- Will rework be required in previous tasks?
- Are repair and/or maintenance made more difficult?
- Will additional tasks be required in the future?
- How will project personnel react?
- What is the effect on the project life cycle?
- Will project flexibility be reduced?
- What is the effect on key employees?
- What is the effect on the customer(s)?

The probability of occurrence and severity should be assessed for all potential future problems. If there is a high probability that the problem will recur and be severe, a plan should be developed to reduce this probability. Internal restrictions, such as manpower, materials, machines, money, management, time, policies, quality, and changing requirements, can cause problems throughout the life cycle of a project. External restrictions of capital, completion dates, and liability also limit project flexibility.

One of the best methods for comparing the alternatives is to list them and then rank them in order of perceived importance relative to certain factors such as customer, potential follow-on business, cost deficit, and loss of goodwill. This is shown in Table 16–2. In the table each of the objectives is weighted according to some method established by management. The percentages represent the degree of satisfactory completion for each alternative. This type of analysis, often referred to as decision making under risk, is commonly taught in operations research and management science coursework. Weighting factors are often used to assist in the decision-making process. Unfortunately, this can add mass confusion to the already confused process.

Table 16–3 shows that some companies perform trade-off analysis by equating all alternatives to a lowest common denominator—dollars. Although this conversion can be very difficult, it does ensure that we are comparing "apples to apples." All resources such as capital equipment can be expressed in terms of dollars. Difficulties arise in assigning

TABLE 16–2. WEIGHING THE ALTERNATIVES

Objectives / Weights / Alternatives	Increase Future Business	Ready on Time	Meet Current Cost	Meet Current Specs	Maximize Profits
Alternatives	0.4	0.25	0.10	0.20	.05
Add resources	100%	90%	20%	90%	10%
Reduce scope of work	60%	90%	90%	30%	95%
Reduce specification change	90%	80%	95%	5%	80%
Complete project late	80%	0%	20%	95%	0%
Bill customer for added cost	30%	85%	0%	60%	95%

TABLE 16–3. TRADE-OFF ANALYSIS FOR IMPROVING PERFORMANCE CAPABILITY

Assumption	Description	Capital Expenditure, $	Time to Complete, Months	Project Profit, $	Ranking in Profit, $
1	No change	0	6	100,000	5
2	Hire higher-salaried people	0	5	105,000	3
3	Refurbish equipment	10,000	7	110,000	2
4	Purchase new equipment	85,000	9	94,000	6
5	Change specifications	0	6	125,000	1
6	Subcontract	0	6	103,000	4

dollar values to such items as environmental pollution, safety standards, or the possible loss of life.

There are often several types of corrective action that can be utilized, including:

- Overtime
- Double shifts
- Expediting
- Additional manpower
- More money
- Change of vendors
- Change of specifications
- Shift of project resources
- Waiving equipment inspections
- Change in statement of work
- Change in work breakdown structure
- Substitution of equipment
- Substitution of materials
- Use of outside contractors
- Providing bonus payments to contractors
- Single-sourcing
- Waiving drawing approvals

The corrective actions defined above can be used for time, cost, and performance. However, there are specific alternatives for each area. Assuming that a PERT/CPM analysis was done initially to schedule the project, then the following options are available for schedule manipulation:

- Prioritize all tasks and see the effect on the critical path of eliminating low-priority efforts.
- Use resource leveling.
- Carry the work breakdown structure to one more level, and reassess the time estimates for each task.

Performance trade-offs can be obtained as follows:

- Excessive or tight specifications that are not critical to the project may be eased. (Many times standard specifications such as mil-specs are used without regard for their necessity.)
- Requirements for testing can be altered to accommodate automation (such as accelerated life testing) to minimize costs.

- Set an absolute minimum acceptable performance requirement below which you will not pursue the project. This gives a bound at the low end of performance that can't be crossed in choosing between trade-off alternatives.
- Give up only those performance requirements that have little or no bearing on the overall project goals (including implied goals) and their achievement. This may require the project manager to itemize and prioritize major and minor objectives.
- Consider absorbing tasks with dedicated project office personnel. This is a resource trade-off that can be effective when the tasks to be performed require in-depth knowledge of the project. An example would be the use of dedicated project personnel to perform information gathering on rehabilitation-type projects. The improved performance of these people in the design and testing phases due to their strong background can save considerable time and effort.

The most promising areas for cost analysis include:

- Incremental costing (using sensitivity analysis)
- Reallocation of resources
- Material substitution where lower-cost materials are utilized without changing project specifications

Depending on the magnitude of the problem, the timeliness of its identification, and the potential impact on the project results, it may be that no actions exist that will bring the project in on time, within budget, and at an acceptable level of performance. The following viable alternatives usually remain:

- A renegotiation of project performance criteria could be attempted with the project sponsor. Such action would be based on a pragmatic view of the acceptability of the probable outcome. Personal convenience of the project manager is not a factor. Professional and legal liability for the project manager, project team, or parent organization may be very real concerns.
- If renegotiation is not considered a viable alternative, or if it is rejected, the only remaining option is to "stop loss" in completing the project. Such planning should involve both line and project management, since the parent organization is at this point seeking to defend itself. Options include:
  - Completing the project on schedule, to the minimum quality level required by the project sponsor. This results in cost overruns (financial loss) but should produce a reasonably satisfied project sponsor. (Project sponsors are not really comfortable when they know a project team is operating in a "stop-loss" mode!)
  - Controlling costs and performance, but permitting the schedule to slide. The degree of unhappiness this generates with the project sponsor will be determined by the specific situation. Risks include loss of future work or consequential damages.
  - Maintaining schedule and cost performance by allowing quality to slip. The high-risk approach has a low probability of achieving total success and a high probability of achieving total failure. Quality work done on the project will be lost if the final results are below minimum standards.

- Seeking to achieve desired costs, schedule, and performance results in the light of impossible circumstances. This approach "hopes" that the inevitable won't happen, and offers the opportunity to fail simultaneously in all areas. Criminal liability could become an issue.
- Project cancellation, in an effort to limit exposure beyond that already encountered. This approach might terminate the career of a project manager but could enhance the career of the staff counsel!

Step 6: Replanning the Project The sixth and final step in the methodology of the management of project trade-offs is to obtain management approval and replan the project. The project manager usually identifies the alternatives and prepares his recommendation. He then submits his recommendation to top management for approval. Top-management involvement is necessary because the project manager may try to make corrective action in a vacuum. Top management normally makes decisions based on the following:

- The firm's policies on quality, integrity, and image
- The ability to develop a long-term client relationship
- Type of project (R&D, modernization, new product)
- Size and complexity of the project
- Other projects underway or planned
- Company's cash flow
- Bottom line—ROI
- Competitive risks
- Technical risks
- Impact on affiliated organizations

After choosing a new course of action from the list of alternatives, management and especially the project team must focus on achieving the revised objectives. This may require a detailed replanning of the project, including new schedules, PERT charts, work breakdown structures, and other key benchmarks. The entire management team (i.e., top management, functional managers, and project managers) must all be committed to achieving the revised project plan.

16.2 CONTRACTS: THEIR INFLUENCE ON PROJECTS

> ***PMBOK® Guide*, 6th Edition**
> Chapter 12 Project Procurement Management
> 12.3 Control Procurements

The final decision on whether to trade off cost, time, or performance can vary depending on the type of contract. Table 16–4 identifies seven common types of contracts and the order in which trade-offs will be made.

The firm-fixed-price (FFP) contract. Time, cost, and performance are all specified within the contract, and are the contractor's responsibility. Because all constraints are equally important with respect to this type of contract, the sequence of

TABLE 16–4. SEQUENCE OF RESOURCES SACRIFICED BASED ON TYPE OF CONTRACT

	Firm-Fixed-Price (FFP)	Fixed-Price-Incentive-Fee (FPIF)	Cost Contract	Cost Sharing	Cost-Plus-Incentive-Fee (CPIF)	Cost-Plus-Award-Fee (CPAF)	Cost-Plus-Fixed-Fee (CPFF)
Time	2	1	2	2	1	2	2
Cost	1	3	3	3	3	1	1
Performance	3	2	1	1	2	3	3

1 = first to be sacrificed.
2 = second to be sacrificed.
3 = third to be sacrificed.

resources sacrificed is the same as for the project-driven organization shown previously in Table 16–1.

The fixed-price-incentive-fee (FPIF) contract: Cost is measured to determine the incentive fee, and thus is the last constraint to be considered for trade-off. Because performance is usually more important than schedule for project completion, time is considered the first constraint for trade-off, and performance is the second.

The cost-plus-incentive-fee (CPIF) contract: The costs are reimbursed and measured for determination of the incentive fee. Thus cost is the last constraint to be considered for trade-off. As with the FPIF contract, performance is usually more important than schedule for project completion, and so the sequence is the same as for the FPIF contract.

The cost-plus-award-fee (CPAF) contract: The costs are reimbursed to the contractor, but the award fee is based on performance by the contractor. Thus cost would be the first constraint to be considered for trade-off, and performance would be the last constraint to be considered.

The cost-plus-fixed-fee (CPFF) contract: Costs are reimbursed to the contractor. Thus, cost would be the first constraint to be considered for trade-off. Although there are no incentives for efficiency in time or performance, there may be penalties for bad performance. Thus time is the second constraint to be considered for trade-off, and performance is the third.

16.3 INDUSTRY TRADE-OFF PREFERENCES

Table 16–5 identifies twenty-one industries that were surveyed on their preferential process for trade-offs. Obviously, there are variables that affect each decision. The data in the table reflect the interviewees' general responses, neglecting external considerations, which might have altered the order of preference.

Table 16–6 shows the relative grouping of Table 16–5 into four categories: project-driven, non–project-driven, nonprofit, and banks.

In all projects in the banking industry, whether regulated or nonregulated, cost is the first resource to be sacrificed. The major reason for this trade-off is that banks in general do not have a quantitative estimation of what actual costs they incur in providing a given service.

TABLE 16–5. INDUSTRY GENERAL PREFERENCE FOR TRADE-OFFS

Industry	Time	Cost	Performance
Construction	1	3	2
Chemical	2	1	3
Electronics	2	3	1
Automotive manu.	2	1	3
IT	2	1	3
Government	2	1	3
Health (nonprofit)	2	3	1
Medicine (profit)	1	3	2
Nuclear	2	1	3
Manu. (plastics)	2	3	1
Manu. (metals)	1	2	3
Consulting (mgt.)	2	1	3
Consulting (eng.)	3	1	2
Office products	2	1	3
Machine tool	2	1	3
Oil	2	1	3
Primary batteries	1	3	2
Utilities	1	3	2
Aerospace	2	1	3
Retailing	3	2	1
Banking	2	1	3

Note: Numbers in table indicate the order (first, second, third) in which the three parameters are sacrificed.

One example of this phenomenon is that a number of commercial banks heavily emphasize the use of *Functional Cost Analysis,* a publication of the Federal Reserve, for pricing their services. This publication is a summary of data received from member banks, of which the user is one. This results in questionable output because of inaccuracies of the input.

TABLE 16–6. SPECIAL CASES

	Type of Organization					
	Project-Driven Organization				Banks	
	Early Life-Cycle Phase	Late-Life-Cycle Phase	Non-Project-Driven Organizations	Nonprofit Organizations	Leader	Follower
Time	2	1	1	2	3	2
Cost	1	3	3	3	1	1
Performance	3	2	2	1	2	3

In cases where federal regulations prescribe time constraints, cost is the only resource of consideration, since performance standards are also delineated by regulatory bodies.

In nonregulated banking projects, the next resource to be sacrificed depends on the competitive environment. When other competitors have developed a new service or product that a particular bank does not yet offer, then the resource of time will be less critical than the performance criteria.

In some banking projects, the time factor is extremely important. A number of projects depend on federal laws. The date that a specific law goes into effect sets the deadline for the project.

Generally, in a nonprofit organization, performance is the first resource that will be compromised. The United Way, free clinics, March of Dimes, American Cancer Society, and Goodwill are among the many nonprofit agencies that serve community needs. They derive their income from donations and/or federal grants, and this funding mechanism places a major constraint on their operations. Cost overruns are prohibited by the very nature of the organization. Inexperienced staff and time constraints result in poor customer service.

The non–project-driven organization is structured along the lines of the traditional vertical hierarchy. Functional managers in areas such as marketing, engineering, accounting, and sales are involved in planning, organizing, staffing, and controlling their functional areas. Many projects that materialize, specifically in a manufacturing concern, are a result of a need to improve a product or process and can be initiated by customer request, competitive climate, or internal operations. The first resource to be sacrificed in the non–project-driven organization is time, followed by performance and cost, respectively. In most manufacturing concerns, budgetary constraints outweigh performance criteria.

In a non–project-driven organization, new projects will take a back seat to the day-to-day operations of the functional departments. The organizational funds are allocated to individual departments rather than to the project itself. When functional managers are required to maintain a certain productivity level in addition to supporting projects, their main emphasis will be on operations at the expense of project development. When it becomes necessary for the firm to curtail costs, special projects will be deleted in order to maintain corporate profit margins.

Resource trade-offs in a project-driven organization depend on the life-cycle phase of a given project. During the conceptual, definition, and production phases and into the operational phase of the project, the trade-off priorities are cost first, then time, and finally performance. In these early planning phases the project is being designed to meet certain performance and time standards. At this point the cost estimates are based on the figures supplied to the project manager by the functional managers.

During the operational phase the cost factor increases in importance over time and performance, both of which begin to decrease. In this phase the organization attempts to recover its investment in the project and therefore emphasizes cost control. The performance standards may have been compromised, and the project may be behind schedule, but management will analyze the cost figures to judge the success of the project.

The project-driven organization is unique in that the resource trade-offs may vary in priority, depending on the specific project. Research and development projects may have a fixed performance level, whereas construction projects normally are constrained by a date of completion.

16.4 PROJECT MANAGER'S CONTROL OF TRADE-OFFS

It is obvious from the above discussion that a project manager does have options to control a project during its execution. Project managers must be willing to control minor trade-offs as well as major ones. However, the availability of specific options is a function of the particular project environment.

In this chapter we discussed trade-offs solely on the triple constraint. In reality, the trade-off problem is much more complicated because today we have competing constraints, including such topics as quality, image, risk, reputation, goodwill, and legal liability.

Probably the greatest contribution a project manager makes to a project team organization is stability in adverse conditions. Interpersonal relationships have a great deal to do with the alternatives available and their probability of success since team performance will be required. Through a combination of management skill and sensitivity, project managers can make the trade-offs, encourage the team members, and reassure the project sponsor in order to produce a satisfactory project.

Related Case Studies (from Kerzner/*Project Management Case Studies*, 5th ed.)	**Related Workbook Exercises (from Kerzner/*Project Management Workbook and PMP®/CAPM® Exam Study Guide*, 12th ed.)**	***PMBOK® Guide*, 6th Edition, Reference Section for the PMP® Certification Exam**
• The Trade off Decision (A), (B)	• Multiple Choice Exam	• Project Integration Management • Project Procurement Management • Project Scope Management

16.5 STUDYING TIPS FOR THE PMI® PROJECT MANAGEMENT CERTIFICATION EXAM

This section is applicable as a review of the principles to support the knowledge areas and domain groups in the *PMBOK® Guide*. This chapter addresses:

- Project Integration Management
- Project Scope Management
- Project Procurement Management
- Initiating
- Planning
- Execution
- Controlling

Understanding the following principles is beneficial if the reader is using this text to study for the PMP® Certification Exam:

- What is meant by a trade-off
- Who are the major players in performing trade-offs
- That assumptions and circumstances can change mandating that trade-offs take place

The following multiple-choice questions will be helpful in reviewing the principles of this chapter:

1. Trade-offs are almost always necessary because:
   A. Project managers are incapable of planning correctly.
   B. Line managers are unable to provide accurate estimates.
   C. Executives are unable to properly define project objectives.
   D. Circumstances can change, thus mandating trade-offs to take place.
2. The person who may be ultimately responsible for approving the trade-off is the:
   A. Project manager
   B. Line manager
   C. Project sponsor
   D. Customer
3. The most common trade-offs occur on:
   A. Time, cost, and quality
   B. Risk, cost, and quality
   C. Risk, time, and quality
   D. Scope, quality, and risk
4. If the start date of a project is delayed but the budget and specifications remain fixed, what would the project manager most likely trade off first?
   A. Scope
   B. Time
   C. Quality
   D. Risk

ANSWERS

1. D
2. D
3. A
4. C

PROBLEMS

16–1 Will the addition of more constraints and metrics make it easier to perform trade-offs on projects?

16–2 Can customer-requested scope changes impact the order in which trade-offs will happen?

16–3 What should be the role of a project sponsor with regard to trade-offs?

17 Risk Management

17.0 INTRODUCTION

PMBOK® Guide 6th Edition
Chapter 11 Project Risk Management

In the early days of project management on many commercial programs, the majority of project decisions heavily favored cost and schedule. This favoritism occurred because we knew more about cost and scheduling than we did about technical risks. Technology forecasting was very rarely performed other than by extrapolating past technical knowledge into the present.

Today, the state of the art of technology forecasting is being pushed to the limits on many projects. For projects with a duration of less than one year, we normally assume that the environment is known and stable, particularly the technological environment. For projects over a year or so in length, technology forecasting must be considered. Computer technology doubles in performance about every two years. Engineering technology is said to double every three or so years. Given such rapid change, plus the inherent need to balance cost, technical performance, and schedule, how can a project manager accurately define and plan the scope of a three- or four-year project without expecting somewhat uncertain engineering changes resulting from technology improvements? With likely changing and uncertain engineering, technology, and production environments, what are the risks?

We read in the newspaper about cost overruns and schedule slips on a wide variety of medium- to large-scale development projects. Several concerns within the control of the buyer, seller, and/or major

---

This chapter was updated by Dr. E. H. Conrow, CMC, CRM, PMP. Dr. Conrow has extensive experience in developing and implementing risk management on a wide variety of projects. He is a management and technical consultant who is the author of *Effective Risk Management: Some Keys to Success,* 2nd ed. (Washington, DC: American Institute of Aeronautics and Astronautics, 2003). He can be reached at (310) 374–7975 and www.risk-services.com.

PMBOK is a registered mark of the Project Management Institute, Inc.

stakeholders can lead to cost growth and schedule slippage on development projects. These causes include but are not limited to:[1]

- Starting a project with a budget and/or schedule that is inadequate for the desired level of technical performance (or proxies such as integration complexity)
- Starting a project before adequate requirements flowdown and verification have occurred and/or before adequate resources have been committed
- Having an overall development process (or key parts of that process) that favors one or more variables over others (e.g., technical performance over cost and schedule)
- Establishing a design that is near the feasible limit of achievable technical performance at a given point in time
- Making major project design decisions before the relationship between cost, technical performance, schedule, and risk is understood

These five causes will contribute to uncertainty in forecasting technology and the associated design needed to meet technical performance requirements. And the inability to accurately forecast technology and the associated design will contribute to a project's technical risk and can also lead to cost and/or schedule risk.

Today, the competition for technical achievement has become fierce. Companies have gone through life-cycle phases of centralizing all activities, especially management functions, but are decentralizing technical expertise. By the mid-1980s, many companies recognized the need to integrate technical risks with cost and schedule risks and other activities (e.g., quality). Risk management processes were developed and implemented where risk information was made available to key decision makers.

The risk management process, however, should be designed to do more than just identify potential risks. The process must also include a formal *planning* activity, *analysis* to estimate the probability and predict the impact on the project of identified risks, a *risk response* strategy for selected risks, and the ability to *monitor and control* the progress in reducing these selected risks to the desired level.

A project, by definition, is a temporary endeavor used to create something that we have not done previously and will not do again in the future. Because of this uniqueness, we have developed a "live-with-it" attitude on risk and attribute it as part of doing business. If risk management is set up as a continuous, disciplined process of planning, identifying, analyzing, developing risk responses, and monitoring and controlling, then the system will easily supplement other processes such as planning, budgeting, cost control, quality, and scheduling. Surprises that become problems will be diminished because the emphasis will now be on proactive rather than reactive management.

Risk management can be justified on almost all projects. The level of implementation can vary from project to project, depending on such factors as size, type of project, who the customer is, contractual requirements, relationship to the corporate strategic plan, and corporate culture. Risk management is particularly important when the overall stakes are high and/or a great deal of uncertainty exists. In the past, we treated risk as "let's live with it." Today, risk management is a key part of overall project management, and should be a conscious input to decision making. It forces us to focus on the future where uncertainty exists and develop suitable plans of action to prevent potential issues from becoming problems and adversely impacting the project.

1. E. H. Conrow, "Some Long-Term Issues and Impediments Affecting Military Systems Acquisition Reform," *Acquisition Review Quarterly*, Defense Acquisition University, vol. 2, no. 2 (Summer 1995): 199–212.

17.1 DEFINITION OF RISK

PMBOK® Guide 6th Edition
11.1 Identify Risks

Risk is a measure of the probability and consequence of not achieving a defined project goal. Most people agree that risk involves the notion of uncertainty. Can the specified aircraft range be achieved? Can the computer be produced within budgeted cost? Can the new product launch date be met? A probability measure can be used for such questions; for example, the probability of not meeting the new product introduction date is 0.15. However, when risk is considered, the consequences (impact) or damage associated with the event occurring must also be included.

Goal A, with a probability of occurrence of only 0.05, may present a much more serious (risky) situation than goal B, with a probability of occurrence of 0.20, if the consequences of not meeting goal A are, in this case, more than four times more severe than the inability to meet goal B. Risk is not always easy to evaluate, since the probability of occurrence and the consequence of occurrence are usually not directly measurable parameters and must be estimated by judgment, statistical, or other procedures.

Risk has two primary components for a given event:

- A probability of occurrence of that event
- Impact (or consequence) of the event occurring (amount at stake)

Figure 17–1 shows the components of risk.

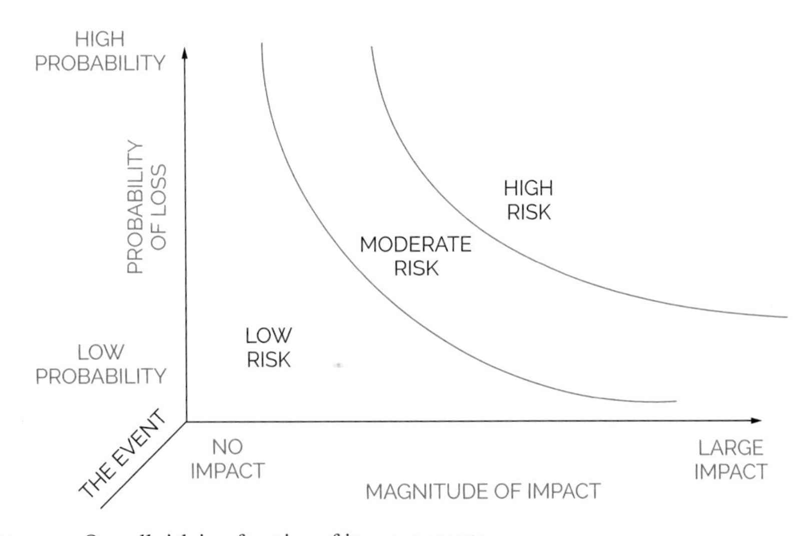

FIGURE 17–1. Overall risk is a function of its components.

Conceptually, the risk for each event can be defined as a function of probability and consequence (impact); that is,

$$\text{Risk} = f(\text{probability, consequence})$$

In general, as either the probability or consequence increases, so does the risk. Both the probability and consequence must be considered in risk management. Risk constitutes a lack of knowledge of future events. Typically, future events (or outcomes) that are favorable are called opportunities, whereas unfavorable events are called risks.

Another aspect of risk is its cause or, more specifically, the root cause(s). Ideally the root cause is known when examining a risk. However, since risks are related to future events, the root cause(s) may not be known and in some cases will never be known.

Something, or the lack of something, can induce a risky situation. We denote this source of danger as the hazard. Certain hazards can be overcome to a great extent by knowing them and taking action to overcome them. For example, a large hole in a road is a much greater danger to a driver who is unaware of it than to one who travels the road frequently and knows enough to slow down and go around the hole. This leads to the second representation of risk:

$$\text{Risk} = f(\text{hazard, safeguard})$$

Risk increases with hazard but decreases with safeguard. The implication of this equation is that good project management should be structured to identify hazards and to allow safeguards to be developed to overcome them. If suitable safeguards are available, then the risk can be reduced to an acceptable level.

Finally, there is often confusion regarding the nature of risks, issues, and problems within the context of project management. All three items are partially related through the consequence (C) dimension but different in either the probability (P) dimension or time frame. A summary of risk, issue, problem, and opportunity with regards to probability, consequence, and time frame is given in Table 17–1.

Both issues and problems have a probability of occurrence equal to one—they will occur, while a risk may not occur ($P < 1$) However, an issue will occur in the future while

TABLE 17–1. CONCISE DEFINITIONS OF RISK, ISSUE, AND PROBLEM

Item	Probability	Consequence	Time Frame
Risk	$0 < P < 1$	$C > 0$	Future
Issue	$P = 1$	$C > 0$	Future
Problem	$P = 1$	$C > 0$	Now
Opportunity	Unclear ($0 < P \leqslant 1$?)	Unclear ($C > 0$ or $C < 0$?)	Unclear (now, future?)

Source: E. H. Conrow, "Risk Analysis for Space Systems," in *Proceedings of the Space Systems Engineering and Risk Management Symposium 2008,* Los Angeles, February 28, 2008. Copyright © 2008, E. H. Conrow. Used with permission of the author.

a problem occurs in the present. The probability dimension for an opportunity is unclear because there is no equivalent differentiation as in the probability dimension for risk, issue, and problem. Moreover, it is not possible to define the consequence dimension in a unique manner since an opportunity may represent, according to three simple definitions, a positive outcome, a less negative outcome, and an outcome that is better than expected. Finally, the time frame associated with an opportunity is also unclear as it may be now or in the future. As evident from the above discussion, while precise definitions have been developed for risk, issue, and problem, precise definitions cannot be defined for opportunities that have universal applicability. Hence, risk and opportunity are not the mirror images or dual of each other in terms of either definitions or the utility associated with potential gains and losses.

17.2 TOLERANCE FOR RISK

***PMBOK*® *Guide* 6th Edition**
11.5 Plan Risk Responses
11.6 Implement Risk Responses

There is no single textbook answer on how to manage risk—one size does not fit all projects or circumstances. The project manager must rely upon sound judgment and the use of the appropriate tools in dealing with risk. The ultimate decision on how to deal with risk is based in part upon the project manager's tolerance for risk, along with contractual requirements, stakeholder preferences, and so on.

The three commonly used classifications of tolerance for risk appear in Figure 17–2: the risk averter or avoider, the neutral risk taker, and the risk taker or seeker. The *Y* axis in Figure 17–2 represents utility (U), which can be defined as the amount of satisfaction or pleasure that the individual receives from a payoff. The *X* axis in this case is the amount of money ($) at stake (but can also potentially represent technical performance or schedule). Curves of this type can represent the project manager or other key decision makers' tolerance for risk.

With the risk averter, utility rises at a *decreasing* rate. In other words, when more money is at stake, the project manager's satisfaction diminishes. With a risk-neutral position, utility rises at a *constant* rate. (*Note*: A risk-neutral position is a specific course of action, and not the average of risk averter and risk taker positions, as is sometimes

The shape of a given decision-maker's curve is derived from comparing response to alternative decision acts.

FIGURE 17–2. Risk preference and the utility function.

erroneously claimed.) With the risk taker, the project manager's satisfaction increases at an increasing rate when more money is at stake. A risk averter prefers a more certain outcome and will demand a premium to accept risk. A risk taker prefers the more uncertain outcome and may be willing to pay a penalty to take a risk. While the project manager's or other key decision makers' tolerance for risk may vary with time, different representations of this tolerance (e.g., risk averter and risk taker) should not exist at the same time else inconsistent decisions may be made.

17.3 DEFINITION OF RISK MANAGEMENT

***PMBOK® Guide* 6th Edition**
11.2 Identify Risks

Risk management is the act or practice of dealing with risk. It includes *planning* for risk, *identifying* risks, *analyzing* risks, developing *risk response* strategies, and *monitoring and controlling* risks to determine how they have changed.

Risk management is not a separate project office activity assigned to a risk management department, but rather is one aspect of sound project management. Risk management should be closely coupled with key project processes, including but not limited to overall project management, systems engineering, configuration management, cost, design/engineering, earned value, manufacturing, quality, schedule, scope, and test. (Project management and systems engineering are typically the two top-level project processes. While risk management can be linked to either of these processes, it is typically associated with project management.)

Proper risk management is proactive rather than reactive and positive rather than negative and seeks to increase the probability of project success. As an example, an item in a network (e.g., router) requires that a new technology be developed. The schedule indicates six months for this development, but project engineers think that nine months is much more likely. If the project manager is proactive, he or she might develop a risk response plan to address potential risks right *now.* If the project manager is reactive (e.g., a "problem solver"), then he or she may do nothing until the problem actually occurs. At that time the project manager must react rapidly to the crisis and may have lost valuable time during which contingencies could have been developed and at least some possible solutions may have been foreclosed. (The resulting cost, technical performance, schedule, and risk design solution space will also have likely shrunk considerably versus when the project was initiated.) Hence, proper risk management will attempt to reduce the probability of an event occurring and/or the magnitude of its impact as well as increase the probability of project success.

17.4 CERTAINTY, RISK, AND UNCERTAINTY

***PMBOK® Guide* 6th Edition**
11.2 Identify Risks

Decision making falls into three categories: certainty, risk, and uncertainty. (Decision making, including but not limited to payoff matrices, expected [monetary] value, and decision trees, can be loosely linked with quantitative risk analysis, discussed in Section 17–10.) Decision making under certainty is the easiest case to work with. With certainty, we assume that all of the necessary information is available to assist us in making the right decision, and we can predict the outcome with a high level of confidence.

Decision Making under Certainty

Decision making under certainty implies that we know with 100 percent accuracy what the states of nature will be and what the expected payoffs will be for each state of nature. Mathematically, this can be shown with payoff matrices. To construct a payoff matrix, we must identify (or select) the states of nature over which *we have no control.* We then select our own action to be taken for each of the states of nature. Our actions are called strategies. The elements in the payoff table are the outcomes for each strategy. A payoff matrix based on decision making under certainty has two controlling features:

- Regardless of which state of nature exists, there will be one strategy that will produce larger gains or smaller losses than any other strategy for all the states of nature.
- There are no probabilities assigned to each state of nature. (It could also be stated that each state of nature has an equal likelihood of occurring.)

Consider a company wishing to invest $50 million to develop a new product. The company decides that the states of nature will be either a strong market demand, an even market demand, or a low market demand. The states of nature shall be represented as *N*1 5 a strong (up) market, *N*2 5 an even market, and *N*3 5 a low market demand. The company also has narrowed its choices to one of three ways to develop the product: either A, B, or C. There also exists a strategy, *S*4, not to develop the product at all, in which case there would be neither profit nor loss. We shall assume that the decision is made to develop the product. The payoff matrix for this example is shown in Table 17–2. Looking for the controlling features in Table 17–2, we see that regardless of how the market reacts, strategy *S*3 will always yield larger profits than the other two strategies. The project manager will therefore always select strategy *S*3 in developing the new product. Strategy *S*3 is the best option to take.

Table 17–2 can also be represented in subscript notation. Let *Pi,j* be the elements of the matrix, where *P* represents profit. The subscript *i* is the row (strategy), and *j is* the column (state of nature). For example, $P_{2,3}$ = profit from choosing strategy 2 with N_3 state of nature occurring. It should be noted that there is no restriction that the matrix be square, but at a minimum it will be a rectangle (i.e., the number of states of nature need not equal the number of possible strategies).

TABLE 17–2. PAYOFF MATRIX (PROFIT IN MILLIONS)

	States of Nature		
Strategy	N^1 **= Up**	N^2 **= Even**	N^3 **= Low**
S1 = A	$50	$40	–$50
S2 = B	$50	$50	$60
S3 = C	$100	$80	$90

Decision Making under Risk

In most cases, there usually does not exist one strategy that dominates for all states of nature. In a realistic situation, higher profits are usually accompanied by higher risks and therefore higher probable losses. When there does not exist a dominant strategy, a probability must be assigned to the occurrence of each state of nature. Risk can be viewed as outcomes (i.e., states of nature) that can be described within established confidence limits (i.e., probability distributions). These probability distributions should ideally be either estimated or defined from experimental data. Consider Table 17–3, in which the payoffs for strategies 1 and 3 of Table 17–2 are interchanged for the state of nature N_3.

From Table 17–3, it is obvious that there does not exist one dominant strategy. When this occurs, probabilities must be assigned to the possibility of each state of nature occurring. The best choice of strategy is therefore the strategy with the largest expected value, where the *expected value* is the summation of the payoff times and the probability of occurrence of the payoff for each state of nature. In mathematical formulation,

$$E_i = \sum_{j=1}^{N} P_{i,j} P_j$$

where Ei is the expected payoff for strategy i, $P_{i,j}$ is the payoff element, and p_j is the probability of each state of nature occurring. The expected value for strategy $S1$ is therefore

Repeating the procedure for strategies 2 and 3, we find that $E_2 = 55$ and $E_3 = 20$. Therefore, based on the expected value, the project manager should always select strategy $S1$. If two strategies of equal value occur, the decision should include other potential considerations (e.g., frequency of occurrence, resource availability, time to impact). (*Note*: Expected value calculations require that a risk-neutral utility relationship exists. If the decision maker is not risk-neutral, such calculations may or may not be useful, and the results should be evaluated to see how they are affected by differences in risk tolerance.)

To quantify potential payoffs, we must identify the strategy we are willing to take, the expected outcome (element of the payoff table), and the probability that the outcome will occur. In the previous example, we should accept the risk associated with strategy $S1$, since it gives us the greatest expected value (all else held constant). If the expected value is

TABLE 17–3. PAYOFF TABLE (PROFIT IN MILLIONS)

Strategy	States of Nature*		
	N^1	N^2	N^3
Probability	0.25	0.25	0.50
S^1	50	40	90
S^2	50	50	60
S^3	100	80	–50

*Numbers are assigned probabilities of occurrence for each state of nature.

positive, then this strategy should be considered. If the expected value is negative, then this strategy should be proactively managed if it is chosen.

An important factor in decision making under risk is the assigning of the probabilities for each of the states of nature. If the probabilities are erroneously assigned, different expected values will result, thus giving us a different perception of the best strategy to take. Suppose in Table 17–3 that the assigned probabilities of the three states of nature are 0.6, 0.2, and 0.2. The respective expected values are

$$E_1 = 56$$
$$E_2 = 52$$
$$E_3 = 66$$

In this case, the project manager would always choose strategy *S*3 (all else held constant).

Decision Making under Uncertainty

The difference between risk and uncertainty is that under risk there are assigned specific probabilities and under uncertainty meaningful assignments of specific probabilities are not possible. As with decision making under risk, uncertainty also implies that there may exist no single dominant strategy. The decision maker, however, does have four basic criteria at his or her disposal from which to make a management decision. The use of each criterion will depend on the type of project as well as the project manager's tolerance to risk.

The first criterion is the Hurwicz criterion, often referred to as the maximax criterion. Under the Hurwicz criterion, the decision maker is always optimistic and attempts to maximize profits by a go-for-broke strategy. This result can be seen from the example in Table 17–3. The maximax criterion says that the decision maker will always choose strategy *S*3 because the maximum profit is 100. However, if the state of nature were *N*3, then strategy *S*3 would result in a maximum loss instead of a maximum gain. The use of the maximax criterion must then be based on how big a risk can be undertaken and how much one can afford to lose. A large corporation with strong assets may use the Hurwicz criterion, whereas the small private company might be more interested in minimizing the possible losses.

A small company may be more apt to use the Wald, or maximin, criterion, where the decision maker is concerned with how much he or she can afford to lose. In this criterion, a pessimistic rather than optimistic position is taken with the viewpoint of minimizing the maximum loss.

For the Wald criterion, we consider only the minimum payoffs. The minimum payoffs in Table 17–3 are 40, 50, and –50 for strategies *S*1, *S*2, *and S*3, respectively. The project manager who wishes to minimize his or her maximum loss will always select strategy *S*2 in this case. If all three minimum payoffs were negative, the project manager would select the smallest loss if these were the only options available. Depending on a company's financial position, there are situations where the project would not be undertaken if all three minimum payoffs were negative.

The third criterion is the Savage, or minimax, criterion. Under this criterion, we assume that the project manager is a sore loser. To minimize the regrets of the sore loser, the project manager attempts to minimize the maximum regret, that is, the minimax criterion.

TABLE 17–4. REGRET TABLE

Strategy	States of Nature			Maximum Regrets
	N^1	N^2	N^3	
S^1	50	40	0	50
S^2	50	30	30	50
S^3	0	0	140	140

The first step in the Savage criterion is to set up a regret table by subtracting all elements in each column from the largest element. Applying this approach to Table 17–3, we obtain Table 17–4.

The regrets are obtained for each column by subtracting each element in a given column from the largest column element. The maximum regret is the largest regret for each strategy, that is, in each row. In other words, if the project manager selects strategy S_1 *or* S_2, he or she will only be sorry for a loss of 50. However, depending on the state of nature, a selection of strategy S_3 may result in a regret of 140. The Savage criterion would select either strategy S_1 or S_2 in this example.

The fourth criterion is the Laplace criterion. The Laplace criterion is an attempt to transform decision making under uncertainty to decision making under risk. Recall that the difference between risk and uncertainty is knowledge of the probability of occurrence of each state of nature. The Laplace criterion makes an a priori assumption based on Bayesian statistics, that if the probabilities of each state of nature are not known, then we can assume that each state of nature has an equal likelihood of occurrence. The procedure then follows decision making under risk, where the strategy with the maximum expected value is selected. Using the Laplace criterion applied to Table 17–3, and thus assuming that $P_1 = P_2 = P_3 = 1/3$ we obtain Table 17–5. The Laplace criterion would select strategy $S1$ in this example.

The important conclusion to be drawn from decision making under uncertainty is the risk that the project manager wishes to incur. For the four criteria presented, we have shown that any strategy can be chosen depending on how much money we can afford to lose and what risks we are willing to take.

Decision Trees

The concept of expected value can also be combined with "probability" or "decision" trees to identify and quantify the potential risks. Another common term is the impact analysis diagram. Decision trees are used when a decision can be viewed not as a single, isolated occurrence, but rather as a sequence of several

TABLE 17–5. LAPLACE CRITERION

Strategy	Expected Value
S^1	60
S^2	53.3
S^3	43.3

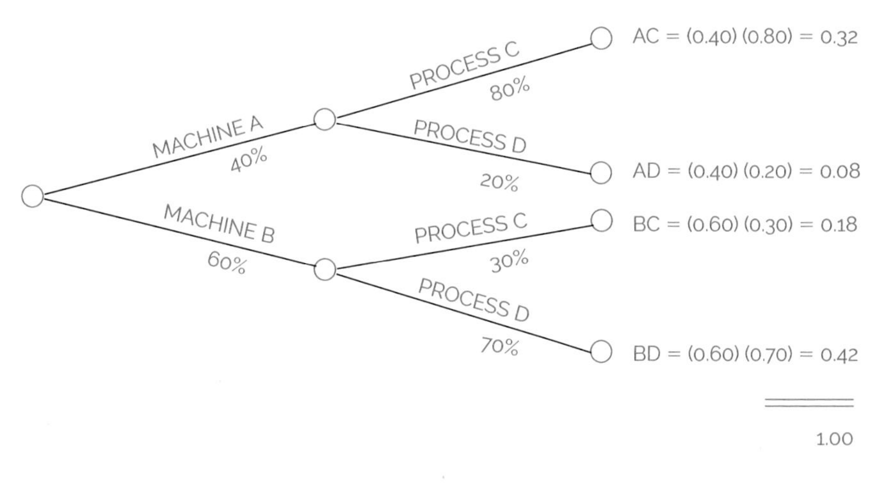

FIGURE 17–3. Decision tree.

interrelated decisions. In this case, the decision maker makes an entire series of decisions simultaneously. (Again, a risk-neutral utility relationship is assumed.)

Figure 17–3 shows a decision tree. The probability at the end of each branch (furthest to the right) is obtained by multiplying the branch probabilities together.

For more sophisticated problems, the process of constructing a decision tree can be complicated. Decision trees contain decision points, usually represented by a box or square, where the decision maker must select one of several available alternatives. Chance points, designated by a circle, indicate that a chance event is expected at this point. (A key assumption required for decision trees is a risk-neutral position, discussed in Section 17–2. Note that the expected value computed in decision trees is not the average outcome, it is the risk-neutral outcome.)

The following three steps are needed to construct a tree diagram:

- Build a logic tree, usually from left to right, including all decision points and chance points.
- Put the probabilities of the states of nature on the branches, thus forming a probability tree.
- Finally, add the conditional payoffs, thus completing the decision tree.

Consider the following problem. You have the chance to make or buy certain widgets for resale. If you make the widgets yourself, you must purchase a new machine for $35,000. If demand is good, which is expected 70 percent of the time, an $80,000 profit will occur on the sale of the widgets. With poor market conditions, $30,000 in profits will occur, not including the cost of the machine. If we subcontract out the work, our contract administration costs will be $5,000. If the market is good, profits will be $50,000; for a

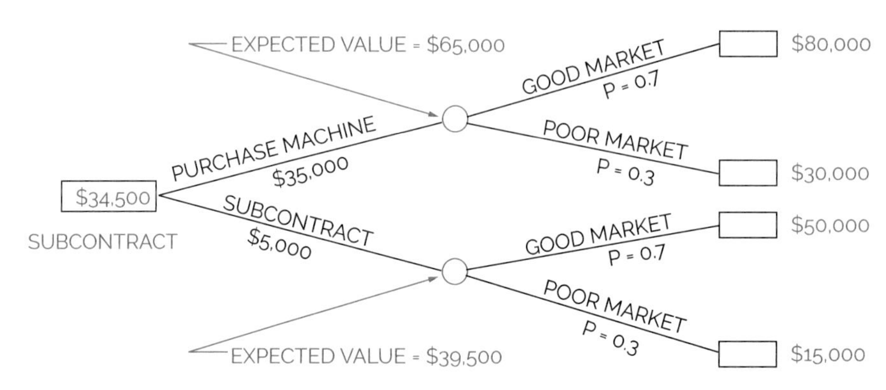

FIGURE 17–4. Expanded decision tree.

poor market, profits will be $15,000. Figure 17–4 shows the tree diagram for this problem. In this case, the expected value of the strategy that subcontracts out the work is $4,500 greater than the expected value for the strategy that manufactures the widgets. Hence, we should select the strategy that subcontracts out the widgets.

17.5 RISK MANAGEMENT PROCESS

***PMBOK® Guide* 6th Edition**
Chapter 11 Project Risk Management

It is important that a risk management strategy be established early in a project and that risk be continually addressed throughout the project life cycle. Risk management includes several related actions, including risk: planning, identification, analysis, response (handling), and monitoring and control.

- Plan risk management is the process of developing and documenting an organized, comprehensive, and interactive strategy and methods for identifying and analyzing risks, developing risk response plans, and monitoring and controlling how risks have changed.
- Identify risks is the process of examining the program areas and each critical technical process to identify and document the associated risk.
- Perform risk analysis is the process of examining each identified risk to estimate the probability and the impact(s) on the project. It includes both qualitative risk analysis and quantitative risk analysis.
- Plan risk response is the process that identifies, evaluates, selects, and implements one or more strategies in order to reduce risk to an acceptable level given program constraints and objectives. This includes the specifics on what should be done, when it should be accomplished, who is responsible, and associated cost and schedule. A risk or opportunity response strategy is composed of an option and implementation approach. Response options for risks include acceptance, avoidance, mitigation (also known as control), and transfer. Response options for

opportunities include acceptance, enhance, exploit, and share. The most desirable response option is selected, and a specific implementation approach is then developed for this option. Finally, resources are allocated to the risk response plan (e.g., budget, personnel, equipment, facilities) and the response plan is implemented.

- Monitor and control risks is the process that systematically tracks and evaluates the performance of risk response actions against established metrics throughout the acquisition process and provides inputs to updating risk response strategies, as appropriate.

17.6 PLAN RISK MANAGEMENT

PMBOK® Guide 6th Edition
11.1 Plan Risk Management
11.2.3.1 Risk Register

Plan for risk management (risk planning) is the detailed formulation of a program of action for the management of risk. It is the process to:

- Develop and document an organized, comprehensive, and interactive risk management strategy
- Determine the methods to be used to execute a program's risk management strategy
- Plan for adequate resources

Risk planning is iterative and includes the entire risk management process, with activities to identify, analyze, respond to, and monitor and control risks. Important outputs of the risk planning process are the risk management plan (RMP) and risk management training. (*Note*: The RMP is an output of risk planning, and *not* the risk planning process itself.)

Risk planning develops a risk management strategy, which includes both the process and implementation approach for the project. Each of these two considerations is of primary importance for achieving effective risk management. However, it is generally far easier to improve a deficient process than remedy a problematic project environment that is unsupportive or hostile towards risk management. Early efforts should establish the purpose and objective, assign responsibilities for specific areas, identify additional technical expertise needed, describe the assessment process and areas to consider, define a risk rating approach, delineate procedures for consideration of response strategies, establish monitoring and control metrics (where possible), and define the reporting, documentation, and communication needs.

The RMP is the risk-related roadmap that tells the project team how to get from where the program is today to where the project manager wants it to be in the future. The key to writing a good RMP is to provide the necessary information so the program team knows the objectives; goals; tools and techniques; reporting, documentation, and communication; organizational roles and responsibilities; and behavioral climate to achieving effective risk management. The RMP should include appropriate definitions, ground rules, and assumptions associated with performing risk management on the project, candidate risk categories, suitable risk identification and analysis methodologies, a suitable risk management organizational implementation, and suitable documentation (e.g., templates or links to an online tool/database) for risk management activities. The RMP should never include results (e.g., risk analysis scores) because these results may frequently change, thus necessitating updates to the RMP. Instead, risk-related results should be included in separate risk documents (e.g., risk register, which contains the details of the individual risks that have been identified and its updates) to avoid unnecessary updates to the RMP.

Since the RMP is a roadmap, it may be specific in some areas, such as the assignment of responsibilities for project personnel and definitions, and general in other areas to allow users to choose the most efficient way to proceed. For example, a description of techniques that suggests several methods to perform a risk identification is appropriate, since every technique will have advantages and disadvantages depending on the situation.

Another important aspect of risk planning is providing risk management training to project personnel. This training should be tailored to various groups within the project as necessary, and a different emphasis may exist for decision makers versus working-level personnel and technical versus nontechnical personnel.

17.7 RISK IDENTIFICATION

PMBOK® Guide 6th Edition
11.2 Identify Risks

The second step in risk management is to identify risks. This may result from a survey of the project, customer, and users for potential concerns.

Some degree of risk always exists in the project, such as in technical, test, logistics, production, engineering, and other areas. Project risks include business, contract relationship, cost, funding, management, political, and schedule risks. (Cost and schedule risks are often so fundamental to a project that they may be treated as standalone risk categories.) The understanding of risks evolves over time, and new risks may appear as the program transitions from examining concepts to evaluating designs, to building prototypes and/or models, and finally production of the selected design. Consequently, risk identification must continue through all project phases.

Project risks should be examined and dissected to a level of detail that permits an evaluator to understand the significance of the risk and its causes and when possible to potentially examine the root cause(s).

Some risks can be categorized according to life-cycle phases, as shown in Figure 17–5. In the early life-cycle phases, the total project risk is high in part because of the lack of information which may preclude comprehensive and accurate risk identification and because risk response plans have yet to be developed and implemented. In the later life-cycle phases, financial risk is generally substantial both because of investments made (such as cost) and because of foreclosed options (opportunity cost).

It is important that all project personnel are involved with risk identification. Designating a small subset of people to perform risk identification almost always diminishes the results from both a technical (number of valid identified risks) and behavioral perspective (sends the "wrong message" to other project personnel) and can lead to decreased risk management effectiveness. This defective risk identification practice should be avoided whenever possible. (*Note*: This is different than occasionally having outside personnel brought into the project to assist in independently identifying candidate risks, which may prove beneficial in challenging projects.)

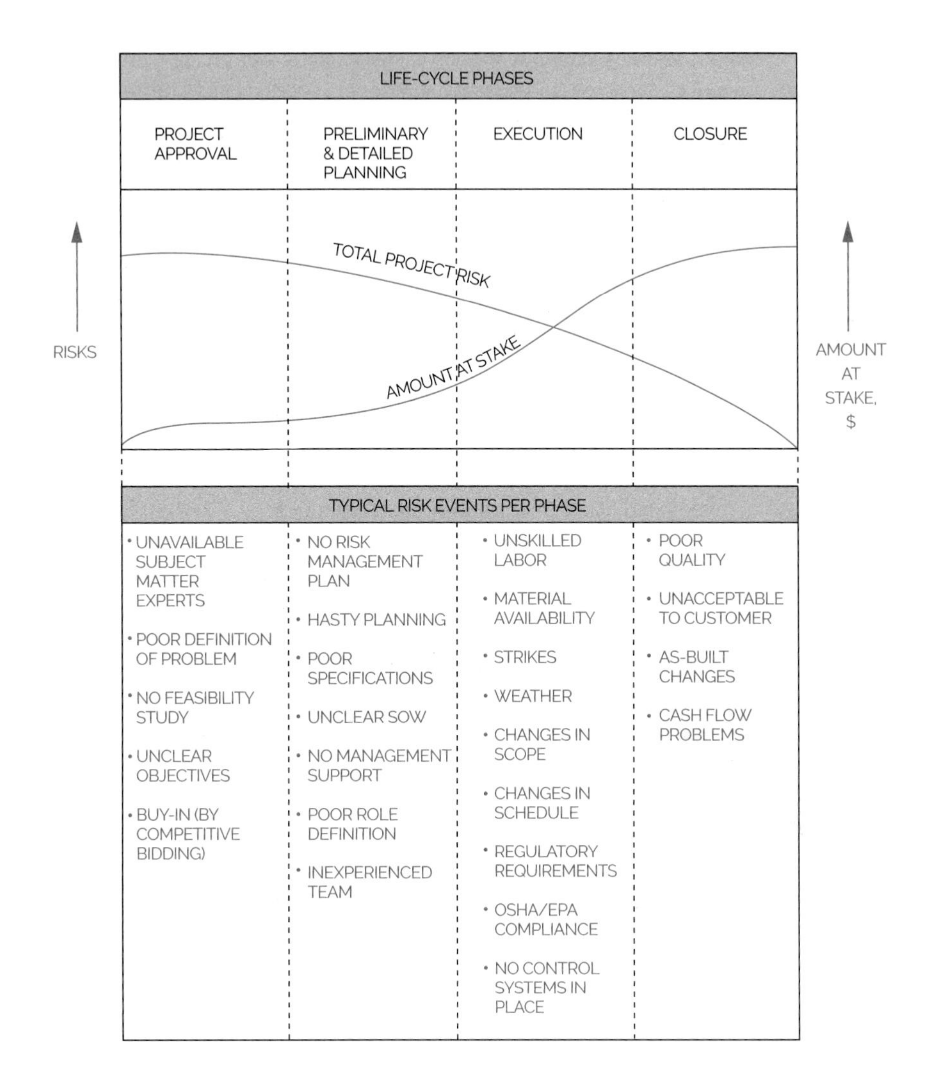

FIGURE 17–5. Life-cycle risk analysis.

17.8 RISK ANALYSIS

PMBOK® Guide **6th Edition**
11.3 Perform Qualitative Risk Analysis
11.4 Perform Quantitative Risk Analysis

Risk analysis is a systematic process to estimate the level of risk for identified and approved risks. This involves estimating the probability of occurrence and consequence of occurrence and converting the results to a corresponding risk level. The approach used depends upon the data available and requirements levied on the project. The most common form of qualitative approach is the use of probability-of-occurrence and consequence-of-occurrence scales together with a risk (mapping) matrix to convert the values to risk levels. Quantitative approaches include but are not limited to expected value [also known as expected (monetary) value for cost-based calculations], decision tree analysis (with branches specified by specific probabilities and/or distributions), payoff matrices, and modeling and simulation. Of key importance is the use of an approved, structured, repeatable methodology rather than a subjective approach that may yield uncertain and/or inaccurate results.

Risk analysis begins with a detailed evaluation of the risks that have been identified and approved by decision makers for further evaluation. The objective is to gather enough information about the risks to estimate the probability of occurrence and consequence of occurrence and convert the resulting values to a corresponding risk level. (*Note*: It is important that only approved risks be analyzed to prevent resources from being expended on items that may not actually be risks.)

Risk analyses are often based on detailed information that may come from a variety of techniques including but not limited to:

- Analysis of plans and related documents
- Comparisons with similar systems
- Data from engineering or other models
- Experience and interviewing
- Modeling and simulation
- Relevant lessons-learned studies
- Results from tests and prototype development
- Sensitivity analysis of alternatives and inputs
- Specialist and expert judgments

After performing a risk analysis it is often necessary to convert the results into risk levels. When a quantitative risk analysis methodology is used, the results can be grouped by existing cost risk, schedule risk, or technical risk boundaries that have specifically been tailored to the program or by performing a (statistical) cluster analysis on the results.

When a qualitative risk analysis is performed, risk ratings can be used as an indication of the potential importance of risks on a program. They are typically a measure of the probability of occurrence and the consequences of occurrence and are often expressed as low, medium, and high (or possibly low, medium low, medium, medium high, and high). A representative ("strawman") set of risk rating definitions follows:

- High risk: Substantial impact on cost, technical performance, or schedule. Substantial action required to alleviate the item. High-priority management attention is required.
- Medium risk: Some impact on cost, technical performance, or schedule. Special action may be required to alleviate the item. Additional management attention may be needed.
- Low risk: Minimal impact on cost, technical performance, or schedule. Normal management oversight is sufficient.

A number of different outputs are possible for both qualitative and/or quantitative risk analyses. These include but are not limited to (1) an overall project risk ranking, (2) a list of prioritized risks, (3) probability of exceeding project cost and/or schedule, (4) probability of not achieving project performance requirements, (5) decision analysis results, (6) failure modes and effects (reliability), (7) fault paths (reliability), and (8) probability of failure (reliability).

17.9 QUALITATIVE RISK ANALYSIS

PMBOK® Guide 6th Edition
11.3 Perform Qualitative Risk Analysis

A commonly used qualitative risk analysis methodology involves risk scales (templates) for estimating probability of occurrence and consequence of occurrence, coupled with a risk mapping matrix. The risk is evaluated using expert opinion against all relevant probability-of-occurrence scales as well as the three consequence-of-occurrence scales (cost, technical performance, and schedule) and the results are then transferred onto a risk mapping matrix to convert these values to a corresponding risk level. The risk is included in a prioritized list based upon the risk level as well as other considerations (e.g., frequency of occurrence, the time to impact, and interrelationships with other risks).

Several different classes of risk scales exist.[2] The first type of scale is a nominal scale. Nominal scales have coefficients with no mathematical meaning, and the values are generally placeholders (e.g., freeway numbers). Nominal scales are not used in risk analyses.

The second type of scale is an interval scale. Interval scales, such as Fahrenheit and Celsius temperature, are cardinal in nature. However, the scales have no meaningful zero point, and ratios between similar scales are not equivalent. Interval scales are not commonly used in risk analyses.

The third type of scale is an ordinal scale. Ordinal scales have levels that are only rank-ordered—they have no cardinal meaning *because the true scale interval values are unknown*. There is no probabilistic or mathematical justification to perform math operations (e.g., addition, multiplication, averaging) on results obtained from ordinal-scale values, and any such results may have large errors. For example, relatively simple examples have been developed that contain errors of 600 percent or more when comparing actual versus assumed ordinal-scale coefficients.[3] These scales may be used to represent different aspects of the probability of occurrence (e.g., technology, design, manufacturing) and consequence of occurrence (e.g., cost, schedule, and technical). Ordinal scales are commonly used in risk analyses. Such scales and a corresponding risk mapping matrix can be a helpful methodology for estimating risk. However, for the reasons discussed above great care must be taken in using this approach.

The fourth type of scale is a calibrated ordinal scale. Calibrated ordinal scales are ordinal scales whose scale-level coefficients are estimated by evaluating an additive utility function (or similar approach). These estimated cardinal coefficients replace the ordinal placeholder values. Limited mathematical operations are possible that yield valid results.

2. This section is derived from Conrow, 2003, note 1, pp. 237–245. Copyright © 2003, Edmund H. Conrow. Used with permission of the author.
3. See Conrow, 2003, note 1, pp. 258–268.

However, the values are often relative rather than absolute and the zero point may not be meaningful. Calibrated ordinal scales are not commonly used in risk analysis, in part because of the difficulty in accurately estimating the associated coefficients.

The fifth type of scale is a ratio scale. Ratio scales, such as Kelvin and Rankine temperature, have cardinal coefficients, indicate absolute position and importance, and the zero point is meaningful. In addition, intervals between scales are consistent, and ratio values between scales are meaningful. Mathematical operations can be performed on ratio scales and yield valid results. Although ratio scales are the ideal scales for use in risk analyses, they rarely exist or are used.

The sixth type of scale is based on subjective estimates assigned to different probability statements (e.g., high), termed here "estimative probability." Estimative probability scales can either be ordinal (more common) or cardinal (less common) in nature depending upon the source of the underlying data and the structure of the scale. In the worst case the probability estimates are point estimates or ranges developed by the scale's author with no rigorous basis to substantiate the values. In the best case the probability estimates are derived from a statistical analysis of survey data from a substantial number of respondents and include point estimates and ranges around the estimate for each probability statement. Estimative probability scales are sometimes used in risk analyses. However, this type of scale should never be the first choice when performing a risk analysis because different people may assign different probability values to the same subjective word (e.g., high). This can lead to nontrivial errors in both the estimated probability value and the resulting risk level.

A risk mapping matrix is typically used to convert ordinal probability of occurrence and consequence of occurrence scale values to a corresponding risk level. While there is no preset size for such a matrix, its dimensions must be less than or equal to the number of scale levels used in both the probability and consequence dimensions. With five-level probability-of-occurrence and consequence-of- occurrence scales this corresponds to a 5 × 5 or smaller matrix. As previously mentioned, risk and opportunity are not the mirror images or dual of each other. Hence, it is incorrect to simply assume that a decision maker will have a neutral position towards both risks and opportunities, then claim that a mirror image mapping matrix (sometimes called a "butterfly matrix") can be used to illustrate the level of risks and opportunities.

17.10 QUANTITATIVE RISK ANALYSIS

***PMBOK® Guide* 6th Edition**
11.4 Perform Quantitative Risk Analysis

Several methodologies are commonly used in quantitative risk analyses. These include but are not limited to payoff matrices, decision analysis (typically decision trees), expected value, and a Monte Carlo process. If the potential probabilities of the states of nature can be represented by a point value, as in Figures 17–3 and 17–4, then the decision tree approach (which relies on expected value calculations assuming risk-neutral participants) is often appropriate. On the other hand, if the states of nature cannot be represented by one or more point values, then probability distributions should be used instead. A common methodology that incorporates a model structure and probability distributions is a Monte Carlo process (commonly called a Monte Carlo simulation).

Two keys to producing accurate quantitative risk analysis results include developing an accurate model structure and incorporating accurate probability information. In project risk management there is often insufficient attention paid to each of these items, and the outcome can be inaccurate results. The model structure should be carefully developed and validated before any output is used for decision-making purposes. While this is easy to do for simple decision trees (e.g., those in Figures 17–3 and 17–4), it can be much more complex when scores or hundreds of branches and potential outcomes are involved.

Quantitative risk analysis outputs can be used in a variety of ways, including but not limited to developing (1) prioritized risk lists (similar to that for calibrated ordinal scales), (2) probabilistic cost estimates at completion per project phase and probabilistic schedule estimates for key milestones to help the project manager allocate reserves accordingly, (3) probabilistic estimates of meeting desired technical performance parameters (e.g., missile accuracy) and validating technical performance of key components (e.g., real-time integrated circuit operation), and (4) estimates of the probability of meeting cost, technical performance, and schedule objectives (e.g., determining the probability of achieving the planned estimate at completion, a key schedule milestone, or key technical performance characteristics.) Trends versus time can also be developed from the above outputs by repeating the quantitative risk analyses during the course of the project phase. (Note, however, that the actual trend information will often be masked by uncertainties in the analysis that should reduce as a function of time [holding all else constant]).

Monte Carlo Process

The Monte Carlo process, as applied to risk management, is an attempt to create a series of probability distributions for potential risks, randomly sample these distributions, and then transform these numbers into useful information that reflects quantification of the associated cost, technical performance, or schedule risks. While often used in technical applications (e.g., integrated circuit performance, structural response to an earthquake), Monte Carlo simulations have also been used to estimate risk in the design of service centers; time to complete key milestones in a project; the cost of developing, fabricating, and maintaining an item; inventory management; and thousands of other applications.

A summary of the steps used in performing a Monte Carlo simulation for cost and schedule follows. (Technical performance simulations can have a widely varying model structure and hence may not fit into the outline below.) Although the details of implementing the Monte Carlo simulation will vary between applications, many cases use a procedure similar to this:

1. Develop and validate a suitable cost or schedule deterministic model without risk and/or uncertainty.
2. Develop the reference point estimate (e.g., cost or schedule duration) for each WBS element or activity contained within the model.
3. Check and recheck the model logic (cost and schedule) and constraints (schedule), as incorrect model logic and constraints are surprisingly common and will lead to erroneous simulation results. For example, the percentage of time spent validating the

deterministic schedule logic and constraints should increase with the number of tasks present. For schedules with several thousand tasks more than half of the time should be spent validating the schedule and less than half the time spent obtaining the probability distributions and interpreting the output.

4. Identify the lowest WBS or activity level for which probability distributions will be constructed. The level selected will depend on the program phase—often lower levels as the project matures.
5. Identify which WBS elements or activities contain estimating uncertainty and/or risk. (For example, technical risk can be present in some cost estimate WBS elements and schedule activities.)
6. Develop suitable probability distributions for each WBS element or activity with estimating uncertainty and/or risk. For cost risk analyses cost estimating uncertainty, schedule risk, and technical risk should be considered as separate distributions. For schedule risk analyses schedule estimating uncertainty, technical risk, and possibly cost risk should be considered as separate distributions. With some tools (e.g., some project scheduling software) only a single probability distribution can be used for a given WBS element or activity. In such cases expert judgement will be needed to develop a single probability distribution from multiple data samples.
7. Aggregate the WBS element or activity probability distributions functions using a Monte Carlo simulation. When performed for cost, the results of this step will typically be a WBS level 1 probabilistic cost estimate at completion at the desired probability level and a cumulative distribution function (CDF) of cost versus probability. These outputs are then analyzed to determine the level of cost risk and to identify the specific cost drivers. When performed for schedule, the results of this step will be CDFs of schedule finish dates (and possibly durations) for selected tasks. These outputs are then analyzed to determine the level of schedule risk at the desired probability level and to identify the specific schedule drivers.
8. Sensitivity and scenario analyses should also be considered for cost and schedule risk analyses. However, they should be performed on the probabilistic (simulation) model, not the deterministic model. If the deterministic model is used, probabilistic considerations will not be taken into consideration. For cost risk analysis the sensitivity analysis identifies which elements with probability distributions affect the results (e.g., total program cost, cost by program phase) the most. For schedule risk analyses the percent of time a task is on the probabilistic critical path (e.g., criticality, critical index) coupled with the influence the probability distribution associated with that task has on the specified output (sensitivity, usually derived from correlation or regression) is of considerable value since neither type of information can be obtained from a deterministic analysis. Furthermore, the product of the criticality times sensitivity yields cruciality, which is a measure of the sensitivity times the percent of time the task is on the probabilistic critical path.

Note: It should be recognized that the quality of Monte Carlo simulation results are only as good as the structure of the model, the quality of the reference point estimates, the selection of probability distributions used in the simulation, and how the simulation is implemented.

17.11 PLAN RISK RESPONSE

PMBOK® Guide 6th Edition
11.5 Plan Risk Response
11.6 Implementing Risk Response

Planning risk responses (risk handling) includes specific methods and techniques to deal with known risks and opportunities, identifies who is responsible for the risk or opportunity, and provides an estimate of the resources associated with handling the risk or opportunity, if any. It involves planning and execution with the objective of reducing risks to an acceptable level and exploiting potential opportunities. There are several factors that can influence our response to a risk or opportunity, including but not limited to:

- Amount and quality of information on the event or situation that caused the risk (descriptive uncertainty)
- Amount and quality of information on the magnitude of the damage (measurement uncertainty)
- Personal benefit to the project manager for accepting the risk or opportunity (voluntary risk or opportunity)
- Risk or opportunity forced upon the project manager (involuntary risk or opportunity)
- The existence of cost-effective alternatives (equitable risks or opportunities)
- The existence of high-cost alternatives or possibly lack of options (inequitable risks or opportunities)
- Length of exposure to the risk or time available for the opportunity

Risk response planning must be compatible with the RMP and any additional guidance the project manager provides. A critical part of risk response planning involves refining and selecting the most appropriate response option(s) and specific implementation approach(es) for selected risks (often those with medium or higher risk levels) and opportunities. The selected risk response option coupled with the specific implementation approach is known as the risk response (handling) strategy, which is documented in the risk response (handling) plan. The procedure to develop a risk response strategy is straightforward. First, the most desirable risk response option (of acceptance, avoidance, control [mitigation], and transfer for risks, and acceptance, enhance, exploit, and share for opportunities) is selected based upon cost, performance, schedule, and risk trade studies, then the best implementation approach is chosen for the selected option. In cases where one or more backup strategies may be warranted (e.g., high risks), the above procedure is repeated. (While the selected option for a backup strategy may be the same as for the primary strategy, the implementation approach will always be different; otherwise, the primary and backup strategy would be identical.) Similarly, contingent responses can be developed for risks and opportunities where action is taken only if certain predefined conditions occur. Finally, handling strategies can be developed using a hybrid of up to all four risk or opportunity options coupled with a suitable implementation approach.

Personnel that evaluate candidate risk response strategies may use the following criteria as a starting point for evaluation:

- Can the strategy be feasibly implemented and still meet the user's needs?
- What is the expected effectiveness of the response strategy in reducing program risk to an acceptable level?

- Is the strategy affordable in terms of dollars and other resources (e.g., use of critical materials, personnel, and test facilities and equipment)?
- Is time available to develop and implement the strategy, and what effect does that have on the overall program schedule?
- What effect does the strategy have on the system's technical performance?

A summary of risk response options for risks and opportunities is given in Table 17–6. For risks this includes acceptance, avoidance, mitigation (control), and transfer while for opportunities this includes acceptance, enhance, exploit, and share. In addition, contingent responses are possible for both risks and opportunities.

A brief discussion of the four response options for risks follows:

- *Acceptance* (i.e., retention): The project manager says, "I know the risk exists and am aware of the possible consequences. I am willing to wait and see what happens. I accept the risk should it occur and have allocated sufficient budget, schedule, and other resources to deal with it."
- *Avoidance*: The project manager says, "I will not accept this design because of the potentially unfavorable results. I will change either the design to preclude the risk or the requirements that lead to the risk."
- *Control* (e.g., mitigation): The project manager says, "I will take the necessary measures required to actively mitigate this risk. I will do what is expected."
- *Transfer*: The project manager says, "I will share this risk with others through insurance or a warranty or transfer the entire risk to them. I may also consider

TABLE 17–6. SUMMARY OF RESPONSE OPTIONS FOR RISKS AND OPPORTUNITIES

Type of Response	Use for Risk or Opportunity	Description
Avoidance	Risk	Eliminate risk by accepting another alternative, changing the design, or changing a requirement. Can affect the probability and/or impact.
Mitigation (control)	Risk	Reduce probability and/or impact through active measures.
Transfer	Risk	Reduce probability and/or impact by transferring ownership of all or part of the risk to another party, use of insurance and warranties, by redesign across hardware/software or other interfaces, etc.
Exploit	Opportunity	Take advantage of opportunities.
Share	Opportunity	Share with another party who can increase the probability and/or impact of opportunities.
Enhance	Opportunity	Increase probability and/or impact of opportunity.
Acceptance	Risk and opportunity	Assume the associated level of risk or opportunity without engaging in any special efforts to control it. Budget, schedule, and other resources must be held in reserve in case the risk occurs or opportunity is selected.

partitioning the risk across hardware and/or software interfaces or using other approaches that share the risk."

A brief discussion of the four response options for opportunities follows:

- *Acceptance* (i.e., retention): The project manager says, "I know an opportunity exists and am aware of the possible benefits. I am willing to wait and see what happens. I will accept the opportunity should it occur."
- *Enhance:* The project manager says, "This is an opportunity. What can we do to increase the probability of occurrence of the opportunity, such as by using more aggressive advertising?"
- *Exploit:* The project manager says, "This is an opportunity. How can we make the most of it? Will assigning more talented resources allow us to get to the market-place quicker?"
- *Share:* The project manager says, "This is an opportunity but we cannot maximize the benefits alone. We should consider sharing the opportunity with a partner."

Finally, while risks and the responses developed to address them may identify potential opportunities, pursuing opportunities will often lead to unanticipated risks. This outcome is rarely if ever discussed by opportunity proponents, yet it can lead to adverse program impacts if the resulting unexpected risks occur.

17.12 MONITOR AND CONTROL RISKS

***PMBOK*® *Guide* 6th Edition**
11.7 Monitor Risks

The monitoring and control process systematically tracks and evaluates the effectiveness of risk response actions against established metrics.[4] Monitoring results should also be fed back to the prior risk management process steps and may also provide a basis for developing additional risk response strategies or updating existing risk response strategies and reanalyzing known risks. In some cases monitoring results may also be used to identify new facets of an existing risk (or new risks) and revise some aspects of risk planning. The key to the risk monitoring and control process is to establish a cost, technical performance, and schedule management indicator system that the project manager and other key personnel use to evaluate the status of the project. The indicator system should be designed to provide early warning of potential problems to allow management actions.

Risk monitoring and control is not a problem-solving technique, but rather a proactive technique to obtain objective information on the progress to date in reducing risks to

4. Material discussing risk monitoring and control was derived in part from *Risk Management Guide for DoD Acquisition*, note 20, pp. 23–24.

acceptable levels. Some techniques suitable for risk monitoring and control that can be used in a program-wide indicator system include:

- Earned Value (EV). This uses standard cost/schedule data to evaluate a program's cost performance (and provide an indicator of schedule performance) in an integrated fashion. As such, it provides a basis to determine if risk response actions are achieving their forecasted results.
- Program Metrics. These are formal, periodic performance assessments of the selected development processes, evaluating how well the development process is achieving its objective. This technique can be used to monitor corrective actions that emerged from an assessment of critical program processes.
- Schedule Performance Monitoring. This is the use of program schedule data to evaluate how well the program is progressing to completion.
- Technical Performance Measurement (TPM). TPM is a product design assessment which estimates, through engineering analysis and tests, the values of essential technical performance parameters of the current design as effected by risk response actions.

The indicator system and periodic reassessments of program risk should provide the program with the means to incorporate risk management into the overall project management structure. Finally, a well-defined test and evaluation program is often a key element in monitoring the performance of selected risk response strategies and updating risk analyses, and identifying candidate risks.

17.13 SOME IMPLEMENTATION CONSIDERATIONS

While it is important to emphasize a comprehensive, structured risk management process, it is equally important that that suitable organizational and behavioral considerations exist so that the process will be properly implemented. While no single set of guidelines will suffice, because implementation considerations vary on a project-by-project basis, it is important that risk management roles and responsibilities be defined in the RMP and carried out in the program. For example, you need to decide (in advance) within the project:

- Which group of managers have responsibility for risk management decision making?
- Which group "owns" and maintains the risk management process?
- Which group or individual is responsible for risk management training and assisting others in risk management implementation?
- Who identifies candidate risks? (Everyone should!)
- How are focal points (sometimes termed "owners") assigned for a particular approved risk?
- How are risk analyses performed and approved?
- How are risk response plans developed and approved, including the necessary resources for their implementation?

- How are data for risk monitoring metrics collected?
- How are independent risk reviews performed to ensure that project risks are properly identified, analyzed, handled, and monitored?

This is but a brief list of some organizational considerations for implementing risk management, which will vary depending upon the size of the project, organizational culture, degree that effective risk management is already practiced within the organization, contractual requirements, and so on. Likewise, while behavioral considerations for effective risk management will also vary on a project-by-project basis, a few key characteristics should apply for all projects.

Risk management must be implemented in both a "top-down" and "bottom-up" manner within the project. The project manager and other decision makers should both use risk management principles in decision making and support and encourage all others within the project to perform risk management. The project manager should generally not be the risk manager (except perhaps on very small projects). However, top-level management must both encourage and foster a positive risk management atmosphere within the project. In addition, they must actively participate in risk management activities and use risk management principles in decision making. Without such active support other project personnel will often view risk management as unimportant, and there may be insufficient encouragement to create or maintain a culture within the project to embrace risk management. Similarly, while it is important for key decision makers within the project to, for example, not "shoot the messenger" for reporting risks, eliminating this behavior does not in and of itself create a positive environment for performing effective risk management because, as mentioned above, a positive atmosphere that is conducive to performing risk management (e.g., identifying risks) needs to be in place.

Working-level personnel are generally quick to decide whether or not decision makers are committed to risk management, and if the appearance is perceived as lip service, then ineffective risk management will almost certainly exist. But working-level personnel must also be actively engaged for risk management to be effective, whereby risk management principles are assimilated as part of their job function.

A key behavioral goal associated with risk management is not to turn every person on the project into a "risk manager" but instead make them sensitive to risk management principles and to apply these principles as part of their job. Accomplishing this is often difficult but nevertheless important in order to achieve effective risk management.

17.14 THE USE OF LESSONS LEARNED

Risks that are analyzed to be medium or higher must be handled to the extent assets allow to reduce their potential to adversely affect the program. All levels of management must be sensitive to hidden "traps" that may induce a false sense of security. If properly interpreted, these signals really indicate a developing problem in a known area of risk. Each trap is usually accompanied by several "warning signs" that show an approaching problem and the probability of failing to treat the problem at its inception.

The ability to turn traps into advantages suggests that much of the technical risk in a program can be actively handled via the risk response avoidance, control, and/or transfer option, not merely watched and resolved after a problem occurs. In some instances it may pay to watch and wait. If the probability that a certain problem will arise is low or if the cost exceeds the benefits of "fixing" the problem before it happens, risk assumption may be advisable. Effective risk management makes selection of the risk assumption option a conscious decision rather than an oversight and may trigger an appropriate addition to the risk "watch list."

"Best practices" acknowledges that all of the traps have not been identified for each risk. The traps are intended to be suggestive, and other potential concerns should be examined as they arise. It is also important to recognize that sources and types of risk evolve over time. Risks may take a long time to occur on a given project. Attention must be properly focused to examine risks and lessons learned.

Lessons learned should be documented so that future project managers can learn from past mistakes. Experience is an excellent teacher in risk management. Yet, no matter how hard we try, risks will occur and projects may suffer.

17.15 DEPENDENCIES BETWEEN RISKS

PMBOK® *Guide*, 6th Edition
11.2.3.1 Risk Registers (updates)

If project managers had unlimited funding, they could generally identify a multitude of risk events, both significant and insignificant. With a large number of possible risk events, it is impossible to address each and every situation, and thus it may be necessary to prioritize risks.

Assume that the project manager categorizes the risks according to the project's time, cost, and performance constraints as illustrated in Figure 17–11. According to the figure, the project manager should focus his efforts on reducing the schedule-related risks. However, it must be recognized that even if schedule has the highest priority, you may also have to start work on cost and technical performance-related issues at the same time, but the schedule-related issues may have the greatest resources applied.

The prioritization of risks could be established by either the project manager or the project sponsor, or even by the customer. The prioritization of risks can also be industry specific, or even country specific, as shown in Figure 17–12. It is highly unlikely that any project management methodology would dictate the prioritization of risks. A well-thought-out risk analysis methodology *does* dictate, or at least reveal, the priority of risks, but then project management input may change the resulting priority. It is simply impossible to develop standardization in this area such that the application could be uniformly applied to each and every project.

The prioritization of risks for an individual project is a good starting point and could work well if most risks were not interrelated. We know from trade-off analysis that changes to a schedule may induce changes in cost and/or performance. The changes may not occur in both dimensions because this depends on the objective functions and market constraints of the buyer and seller. Therefore, even though schedules have the highest priority in Figure 17–6, risk response to the schedule risk events may cause immediate evaluation of the technical performance risk events. Yes, risks are interrelated.

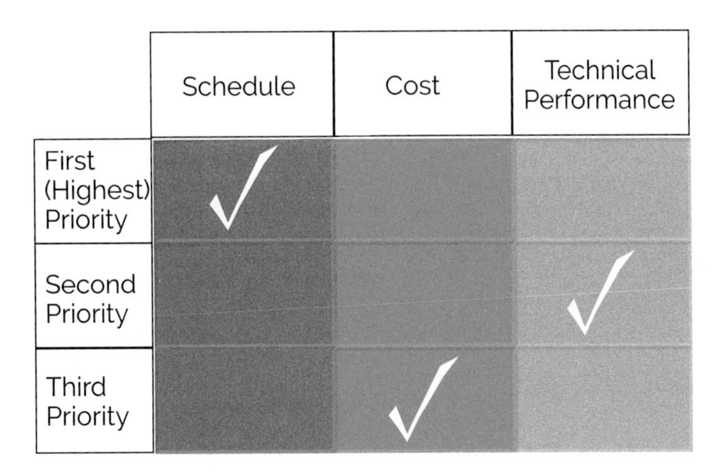

FIGURE 17–6. Prioritization of risks.

The interdependencies between risks can also be seen from Table 17–7. The first column identifies certain actions that the project manager can take in pursuit of the possible benefits listed in column 2. Each of these possible benefits, in turn, can cause additional risks, as shown in column 3. In other words, risk mitigation strategies that are designed to take advantage of a possible benefit could create another risk event that is more severe. As an example, working overtime could save you $15,000 by compressing the schedule. But if the employees make more mistakes on overtime, retesting may be required, additional materials may need to be purchased, and a schedule slippage could

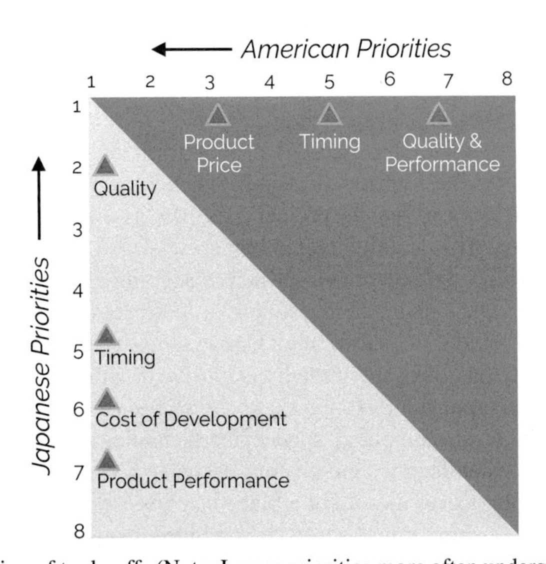

FIGURE 17–7. Ordering of trade-offs (Note: Lower priorities more often undergo trade-offs.).

TABLE 17–7. RISK INTERDEPENDENCIES

Action	Possible Benefit	Risk
• Work overtime	• Schedule compression	• More mistakes; higher cost and longer schedule
• Add resources	• Schedule compression	• Higher cost and learning curve shift
• Parallel work	• Schedule compression	• Rework and higher costs
• Reduce scope	• Schedule compression and lower cost	• Unhappy customer and no follow-on work
• Hire low-cost resources	• Lower cost	• More mistakes and longer time period
• Outsource critical work	• Lower cost and schedule compression	• Contractor possesses critical knowledge at your expense

occur, thus causing a loss of $100,000. Therefore, is it worth risking a loss of $100,000 to save $15,000?

To answer this question, we can use the concept of expected value, assuming we can determine the probabilities associated with mistakes being made and the cost of the mistakes. Without any knowledge of these probabilities, the actions taken to achieve the possible benefits would be dependent upon the project manager's tolerance for risk.

Most project management professionals seem to agree that the most serious risks, and the ones about which we seem to know the least, are the technical risks. The worst situation is to have multiple technical risks that interact in an unpredictable or unknown manner.

Although project management methodologies provide a framework for risk management and the development of a risk management plan, it is highly unlikely that any methodology would be sophisticated enough to account for the identification of technical dependency risks. The time and cost associated with the identification, analysis, and handling of technical risk dependencies could severely tax the project financially.

As companies become successful in project management, risk management evolves into a structured process that is performed continuously throughout the life cycle of the project. The four most common factors supporting the need for continuous risk management are how long the project lasts, how much money is at stake, the degree of developmental maturity, and the interdependencies between the different risks. For example, consider Boeing's aircraft projects where designing and delivering a new plane might require ten years and a financial investment of more than $5 billion.

Table 17–8 shows the characteristics of risks at Boeing. The table does not mean to imply that risks are mutually exclusive of one another. New technologies can appease customers, but production risks increase because the learning curve is lengthened with new technology compared to accepted technology. The learning curve can be lengthened further when features are custom-designed for individual customers. In addition, the loss of suppliers over the life of a plane can affect the level of technical and production risk. The relationships among these risks require the use of a risk management matrix and continuous risk assessment.

TABLE 17–8. RICK CATEGORIES AT BOEING

Type of Risk	Risk Description	Risk Handling Strategy
Financial	Up-front funding and payback period based upon number of planes sold	• Funding by life-cycle phases • Continuous financial risk management • Sharing risks with subcontractors • Risk reevaluation based upon sales commitments
Market	Forecasting customers' expectations on cost, configuration, and amenities based upon a 30–40 year life of a plane	• Close customer contact and input • Willingness to custom-design per customer • Develop a baseline design that allows for customization
Technical	Because of the long lifetime for a plane, must forecast technology and its impact on cost, safety, reliability, and maintainability	• A structured change management process • Using proven designs and technology rather than unproven designs and high risk technology • Parallel product improvement and new product development processes
Production	Coordination of manufacturing and assembly of a large number of subcontractors without impacting cost, schedule, quality, or safety	• Close working relationships with subcontractors • A structured change management process • Lessons learned from other new airplane programs • Use of learning curves

Another critical interdependency is the relationship between change management and risk management, both of which are part of the singular project management methodology. Each risk management strategy can result in changes that generate additional risks. Risks and changes go hand in hand, which is one of the reasons companies usually integrate risk management and change management into a singular methodology. Table 17–9 shows the relationship between managed and unmanaged changes. If changes are unmanaged, then more time and money are needed to perform risk management, which often takes on the appearance and behavior of crisis management. And what makes the situation even worse is that higher-salaried employees and additional time are required to assess the additional risks resulting from unmanaged changes. Managed changes, on the other hand, allow for a lower-cost risk management plan to be developed.

PMBOK® Guide, **6th Edition**
11.2 Identify Risks

TABLE 17–9. UNMANAGED VERSUS MANAGED CHANGED

	Where Time Is Invested	How Energy Is Invested	Which Resources Are Used
Unmanaged Change	Back end	• Rework • Enforcement • Compliance • Supervision	• Senior management and key players only
Managed Change	Front end	• Education • Communication • Planning • Improvements • Value added	• Stakeholders (internal) • Suppliers • Customers

Project management methodologies, no matter how good, cannot accurately define the dependencies between risks. It is usually the responsibility of the project team to make these determinations.

17.16 THE IMPACT OF RISK HANDLING MEASURES

PMBOK® Guide 6th Edition
11.6 Implementing Risk Responses
11.7 Monitoring Risks

Most project management methodologies include risk management, which can be used to:

- Create an understanding of the potential risks and their effects
- Provide an early warning system when the risk event is imminent
- Provide clear guidance on how to manage and contain the risk event, if possible
- Restore the system/process after the risk event occurs
- Provide a means for escape and rescue should all attempts fail

Some guidance in risk management is necessary because each stakeholder could have a different tolerance for risk. Risk and safety system policies, procedures, and guidelines exist primarily for the lower three levels in Figure 17–8. The customer's tolerance for risk could be significantly greater or less than the company's tolerance. Also, based upon the project's requirements, any given project could be willing to accept significantly more or less risk than the organizational procedures normally allow.

The project management methodology may very well dictate the magnitude of the risk handling measures to be undertaken. The risk handling measures for risk assumption may be significantly more complex than measures for avoidance. Figure 17–15 shows the extent of the risk handling strategy versus the magnitude of the risks. As the magnitude of the risk increases, an overreaction may occur that places undue pressure on the risk management process and the project management methodology. The cost of maintaining these risk handling measures should not overly burden the project. Excessive risk management procedures may require that the project manager spend more time and money than appropriate.

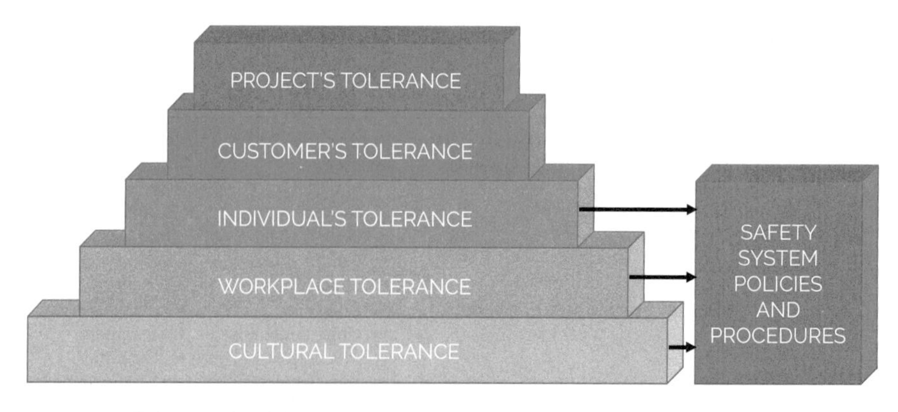

FIGURE 17–8. Tolerance for risk.

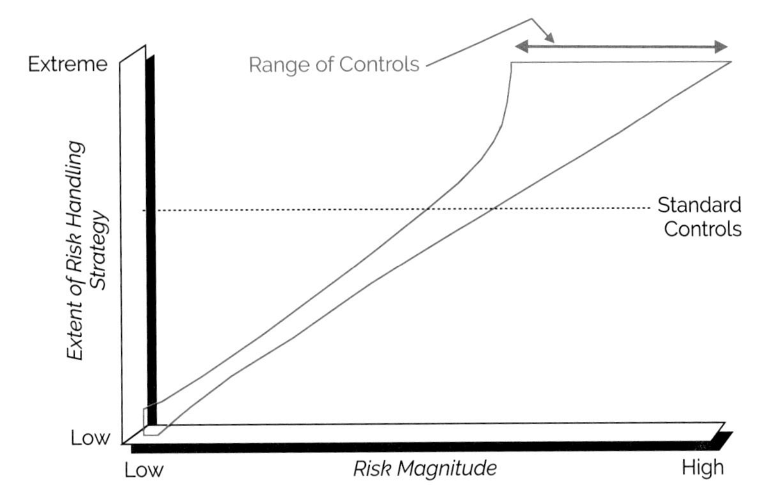

FIGURE 17–9. Risk handling measures.

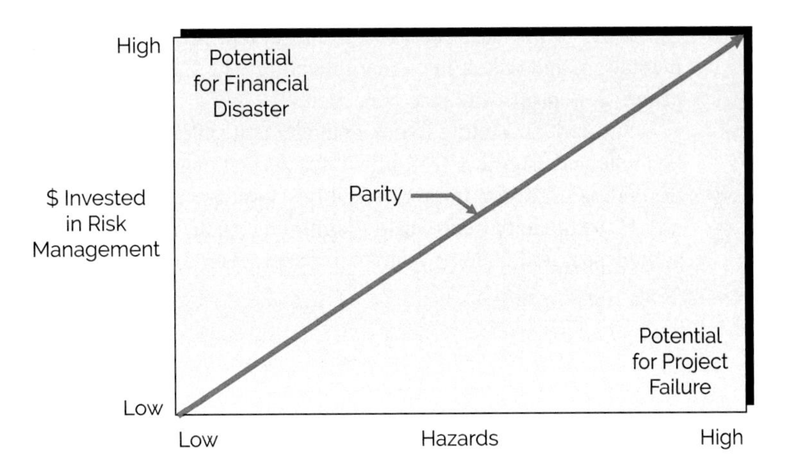

FIGURE 17–10. Investment in risk management.

If an organization goes overboard in its investment in risk management, the results can be devastating, as shown in Figure 17–10. Overinvestment in risk management could lead to financial disaster if the project's risk events do not call for substantial measures or expenses. However, underinvestment in risk management for a project with numerous and complex risk events could lead to heavy losses and damages, possibly leading to project failure. Some sort of parity position is needed.

Determining the proper amount of risk control measures is not easy. This can be seen from Figure 17–11, which illustrates the impact on the schedule constraint. If too few risk handling measures are in place, or if there simply is no risk handling plan, the result may be an elongated schedule due to ineffective risk handling measures. If excessive risk handling measures are in place, such as too many filters and gates, the schedule can likewise

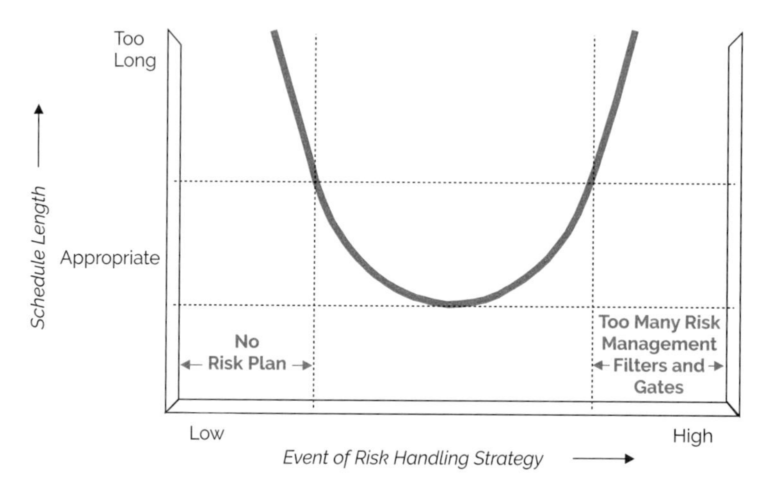

FIGURE 17–11. Risk handling versus schedule length.

be elongated because the workers are spending too much time on contingency planning. The same can be said for a risk management process with excessive risk reporting, documentation, and risk management meetings (i.e., too many gates). This results in very slow progress. A proper balance is needed.

Similarly, investing in risk management is not a guarantee that losses and damages will be prevented. Figure 17–12 illustrates perfect planning for risk management. The organization prepares a primary and possibly secondary risk handling plan for each potential hazard. Unfortunately, real-world planning is often imperfect, as shown in Figure 17–13, and some losses and damages may still occur, even for known risk issues.

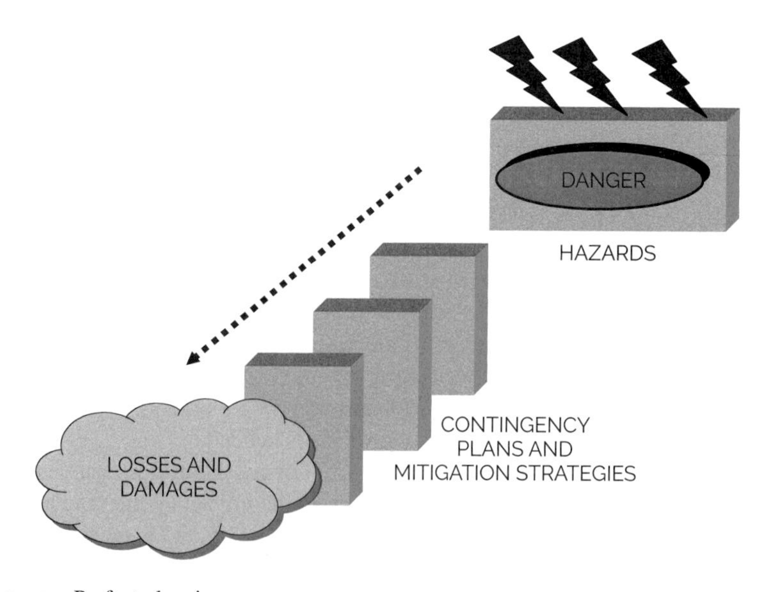

FIGURE 17–12. Perfect planning.

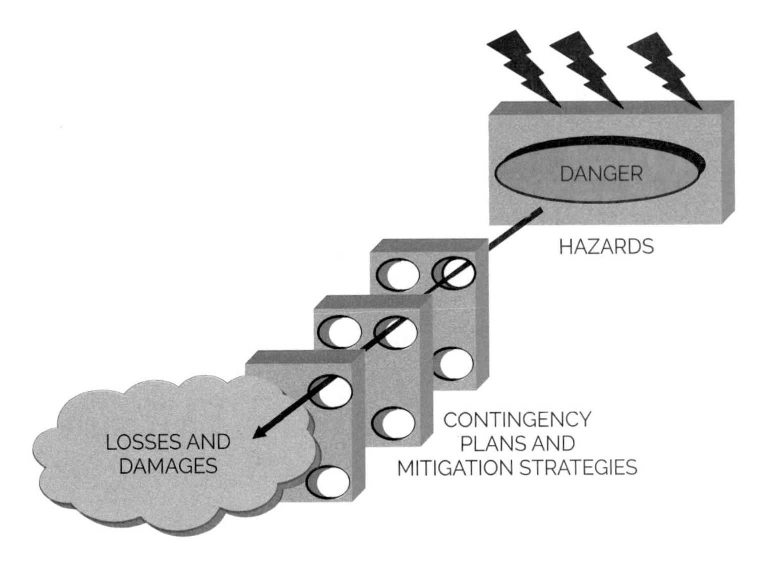

FIGURE 17–13. Imperfect planning.

17.17 RISK AND CONCURRENT ENGINEERING

***PMBOK*® *Guide*, 6th Edition**
11.7 Monitoring Risks

Most companies desire to get to the marketplace in a timely manner because the rewards for being the first-to-market can be huge in both profitability and market share. Getting to the marketplace quickly often entails using concurrent engineering, or overlapping activities. The critical question is, "How much overlapping can we incur before we get diminishing returns?"

The risks involved with overlapping activities are shown in Figure 17–14. Overlapping activities can lead to schedule compression and lower costs. However, too much overlapping can lead to excessive rework and unanticipated problems that can generate significant schedule slippages and cost overruns. Finding the optimal overlapping point that increases benefits while decreasing rework is difficult.

Although there may exist numerous reasons for the rework, two common problems are:

- Combining new technology development and product development technology
- An insufficient test and evaluation program

To illustrate why these problems occur, consider a situation where the sales and marketing force promises the marketplace a new product with advanced technology that hasn't yet been developed. To compress the schedule, the product development team begins designing the product without knowing whether or not (and when) the technology can be developed. Production teams are asked to develop manufacturing plans without having any drawings. This results in massive changes when the product final reaches production.

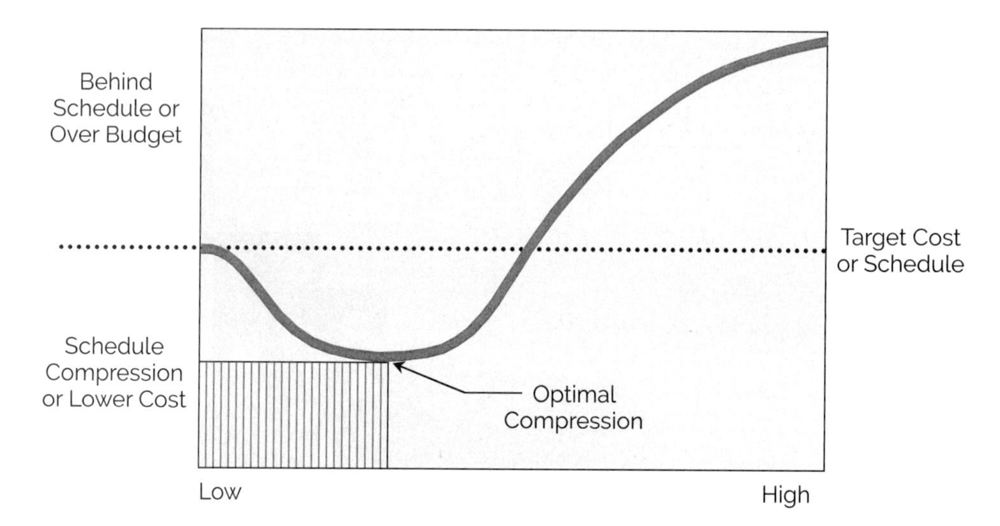

FIGURE 17–14. Overlapping risks.

There are three questions that need to be continuously addressed:

- Can the new technology be developed?
- Can we demonstrate the new technology within the product?
- Can the product then be manufactured within the time, cost, and performance (i.e., reliability) constraints?

Simultaneous development of technologies and products has become commonplace. To decrease the risks of rework, there should be a demonstration that the technology can work as expected. Leading firms that use concurrent engineering do not include a new technology in a product until the technology reaches a prescribed level of maturity. They have disciplined processes that match requirements with technological capability before product development is launched. These companies have learned the hard lesson of not committing to new products that outstrip their technological know-how. These practices stem from their recognition that resolving technology problems after product development begins can result in a tenfold cost increase; resolving these problems in production could increase costs by a hundredfold.

Some commonly accepted practices to reduce risks include:

- Flexibility in both the resources provided and the product's performance requirements to allow for uncertainties of technical progress
- Disciplined paths for technology to be included in products, with strong gatekeepers to decide when to allow it into a product development program
- High standards for judging the maturity of the technology

- The imposition of strict product development cycle times
- Rules concerning how much innovation can be accepted on a product before the next generation must be launched (these rules are sometimes referred to as technology readiness levels)

Collectively, these factors create a healthy environment for developing technology and making good decisions on what to include in a product.

Overlapping activities can be very risky if problems are discovered late in the cycle. One common mistake is to begin manufacturing before a sufficient quantity of engineering drawings is available for review. This normally is the responsibility of systems integration personnel. Systems integration should conclude with a critical design review of engineering drawings and confirmation that the system's design will meet requirements—a key knowledge point. It should also result in firm cost and schedule targets and a final set of requirements for the current version of the product. Decision makers should insist on a mature design, supported with complete engineering drawings, before proceeding to even limited production. Having such knowledge at this point greatly contributes to product success and decreases costly rework.

Related Case Studies (from Kerzner/*Project Management Case Studies*, 5th ed.)	**Related Workbook Exercises (from Kerzner/*Project Management Workbook and PMP®/CAPM® Exam Study Guide*, 12th ed.)**	***PMBOK® Guide*, 6th Edition, Reference Section for the PMP® Certification Exam**
• Teloxy Engineering (A)* • Teloxy Engineering (B)* • The Space Shuttle Challenger Disaster • Packer Telecom • Luxor Technologies • Altex Corporation • Acme Corporation • The Risk Management Department*	• Multiple Choice Exam • Crossword Puzzle on Risk Management	• Project Risk Management • Professional Responsibility

*Case study also appears at end of chapter.

17.18 STUDYING TIPS FOR THE PMI® PROJECT MANAGEMENT CERTIFICATION EXAM

This section is applicable as a review of the principles to support the knowledge areas and domain groups in the *PMBOK® Guide*. This chapter addresses:

- Project Risk Management
- Planning
- Execution
- Controlling
- Professional Responsibility

Understanding the following principles is beneficial if the reader is using this text to study for the PMP® Certification Exam:

- What is meant by a risk
- Components of a risk
- That risk management is performed throughout the project and involves possibly the entire team
- Types of risks
- What is meant by one's tolerance for risk
- Sources of a risk
- What is meant by a risk event
- Components of a risk management plan
- Risk gathering techniques such as the Delphi technique and brainstorming
- Quantitative risk analysis such as expected value and Monte Carlo simulation
- Qualitative risk assessment
- What is meant by decision trees
- Risk response modes

The following multiple-choice questions will be helpful in reviewing the principles of this chapter:

1. The two major components of a risk are:
   A. Time and cost
   B. Uncertainty and impact
   C. Quality and time
   D. Cost and decision-making circumstances
2. Risk management is normally performed by:
   A. Developing contingency plans
   B. Asking the customer for help
   C. Asking the sponsor for help
   D. Developing work-around situations
3. Future outcomes that provide favorable opportunities are called:
   A. Favorable risks
   B. Opportunities
   C. Contingencies
   D. Surprises
4. The cause of a risk event is usually referred to as:
   A. An opportunity
   B. A hazard
   C. An outcome
   D. An unwanted surprise

5. If there is a 40 percent chance of making $100,000 and a 60 percent chance of losing $150,000, then the expected monetary outcome is:
   A. $50,000
   B. –$50,000
   C. $90,000
   D. –$90,000
6. Assumption, mitigation, and transfer are examples of risk:
   A. Contingencies
   B. Uncertainties
   C. Expectations
   D. Responses
7. In which life-cycle phase would project uncertainty be the greatest?
   A. Initiation
   B. Planning
   C. Execution
   D. Closure
8. In which life-cycle phase would the financial risks of a project be the greatest?
   A. Initiation
   B. Planning
   C. Execution
   D. Closure
9. Identifying a risk as high, moderate, or low would be an example of which risk assessment?
   A. Go-for-broke
   B. Adverse
   C. Qualitative
   D. Quantitative
10. Monte Carlo simulation is an example of which risk assessment?
    A. Go-for-broke
    B. Adverse
    C. Qualitative
    D. Quantitative
11. Which of the following is *not* a valid reason for managing a risk?
    A. Minimizing the risk's likelihood
    B. Minimizing the risk's unfavorable consequences
    C. Maximizing the probability of the risk's favorable consequences
    D. Providing a late-as-possible warning system

12. Which of the following is generally *not* part of overall risk management?
    A. Defining the roles and responsibilities of the team members
    B. Establishing a risk reporting format
    C. Selection of the project manager
    D. Risk scoring and interpretation
13. A technique for risk evaluation that uses a questionnaire, a series of rounds, and reports submitted in confidence and then circulated with the source unidentified is called:
    A. The Delphi technique
    B. The work group
    C. Unsolicited team responses
    D. A risk management team
14. Risk symptoms or early warning signs are called:
    A. Vectors
    B. Triggers
    C. Pre-events
    D. Contingency events
15. Which of the following is *not* a risk quantification tool or technique?
    A. Interviewing
    B. Decision tree analysis
    C. Objective setting
    D. Simulation
16. A technique that depicts interactions among decisions and associated events is called:
    A. Decision tree analysis
    B. Earned value measurement system
    C. Network scheduling system
    D. Payoff matrix
17. Varying one risk driver at a time, either in small increments or from optimistic to pessimistic estimates while keeping all other drivers fixed, is called:
    A. Decision tree analysis
    B. Sensitivity analysis
    C. Network analysis
    D. Earned value analysis
18. A risk response strategy that generally reduces the probability or impact of the event without altering the project's objectives is called:
    A. Avoidance
    B. Acceptance
    C. Mitigation
    D. Transfer

19. Earned value measurement is an example of:
    - A. Risk communication planning
    - B. Risk identification planning
    - C. Risk response
    - D. Risk monitoring and control
20. The difference between being proactive and reactive is the development of a:
    - A. Payoff table
    - B. Range of probabilities
    - C. Range of payoffs
    - D. Contingency plan

ANSWERS

1. B
2. A
3. B
4. B
5. B
6. D
7. A
8. D
9. C
10. D
11. D
12. C
13. B
14. C
15. A
16. B
17. C
18. D
19. D

PROBLEMS

17–1 You have been asked to use the expected-value model to assess the risk in developing a new product. Each strategy requires a different sum of money to be invested and produces a different profit payoff as shown below:

	States of Nature		
Strategy	**Complete Failure**	**Partial Success**	**Total Success**
S^1	<$50K>	<30K>	70K
S^2	<80K>	20K	40K
S^3	<70K>	0	50K
S^4	<200K>	<50K>	150K
S^5	0	0	0

Assume that the probabilities for each state are 30 percent, 50 percent, and 20 percent, respectively.

A. Using the concept of expected value, what risk (i.e., strategy) should be taken?

B. If the project manager adopts a go-for-broke attitude, what strategy should be selected?

C. If the project manager is a pessimist and does not have the option of strategy *S*5, what risk would be taken?

D. Would your answer to part c change if strategy *S*5 were an option?

17–2 A telecommunications firm believes that the majority of its income over the next ten years will come from organizations outside of the United States. More specifically, the income will come from third world nations that may have very little understanding or experience in project management. The company prepared Figure 17–15. What causes the increasing risks in Figure 17–15?

17–3 Figure 17–16 shows a probability-impact (or risk mapping) matrix that is frequently used as part of the risk analysis prioritization process. Here, ordinal probability-of-occurrence and consequence-of-occurrence risk scales are used (5, 4, 3, 2, 1 correspond to E, D, C, B, A since the actual scale coefficients are unknown). In this figure, L represents a low risk, which would generally be a risk acceptable to the project manager. The letter M represents a moderate risk, which will likely require a risk response. The letter H represents a high risk, which will need one or more risk responses. What are the advantages of using a high–moderate–low (or red– yellow–green) risk designation as opposed to assigning quantitative numbers to each cell and risk level (e.g., 5 x 5 = 25 to 1 x 1 = 1)?

17–4 Figure 17–17 shows a probability impact matrix for a project involving military operations. The figure has been taken from Y. Y. Haimes, *Risk Modeling, Assessment and Management*, 3rd ed. (Hoboken, NJ: John Wiley & Sons, 2009), p. 312. Can this same figure be used in the pharmaceutical industry for the development of new drugs where the word "mission" is replaced by the word "health"?

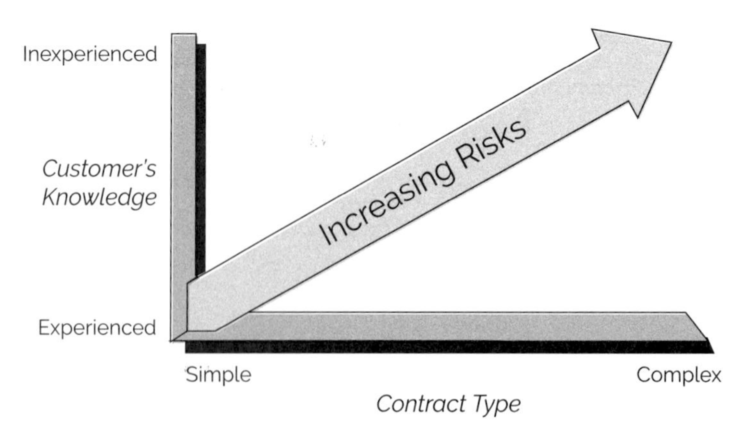

FIGURE 17–15. Future risks.

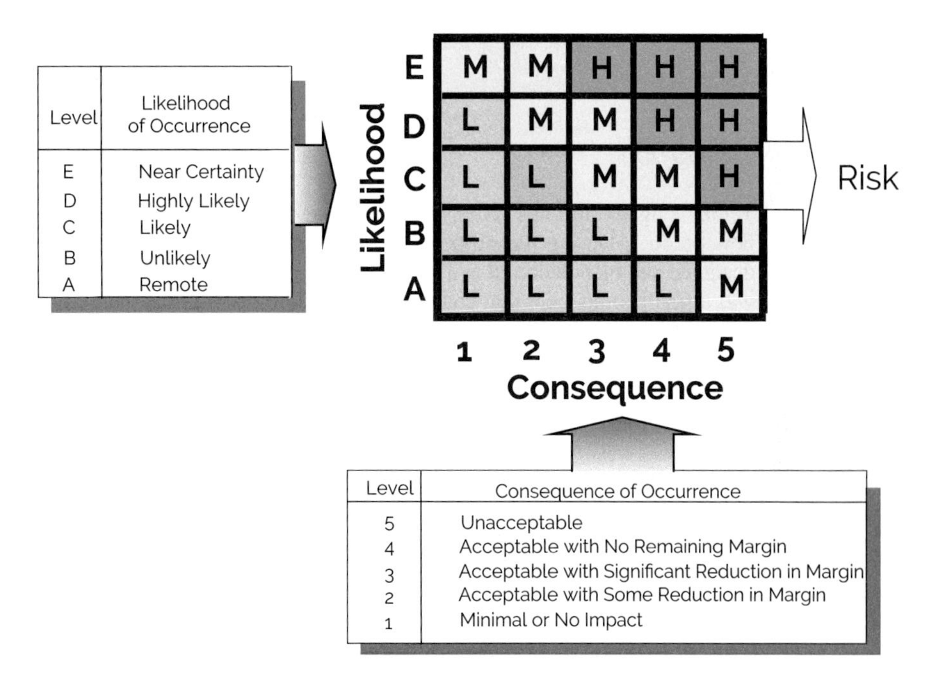

FIGURE 17–16. Risk analysis prioritization.

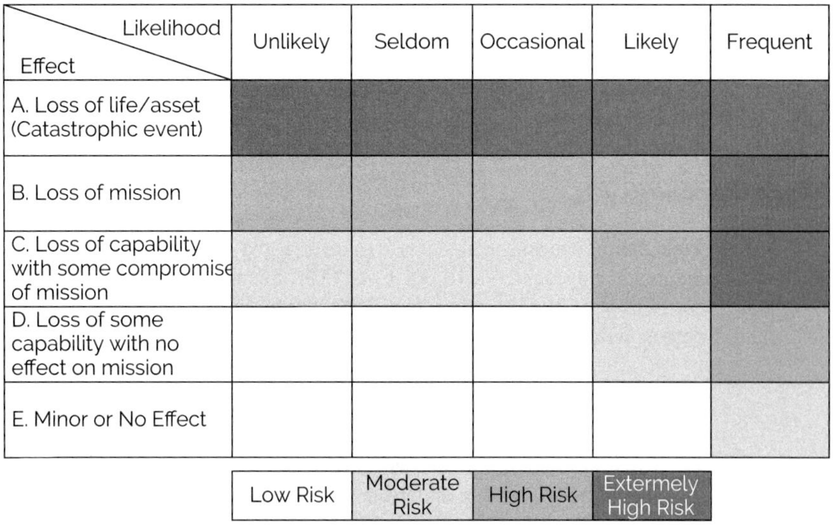

Effect \ Likelihood	Unlikely	Seldom	Occasional	Likely	Frequent
A. Loss of life/asset (Catastrophic event)					
B. Loss of mission					
C. Loss of capability with some compromise of mission					
D. Loss of some capability with no effect on mission					
E. Minor or No Effect					

Low Risk	Moderate Risk	High Risk	Extermely High Risk

FIGURE 17–17. Military operation probability impact matrix.

CASE STUDIES

TELOXY ENGINEERING (A)

Teloxy Engineering has received a one-time contract to design and build 10,000 units of a new product. During the proposal process, management felt that the new product could be designed and manufactured at a low cost. One of the ingredients necessary to build the product was a small component that could be purchased for $60 in the marketplace, including quantity discounts. Accordingly, management budgeted $650,000 for the purchasing and handling of 10,000 components plus scrap.

During the design stage, your engineering team informs you that the final design will require a somewhat higher-grade component that sells for $72 with quantity discounts. The new price is substantially higher than you had budgeted for. This will create a cost overrun.

You meet with your manufacturing team to see if they can manufacture the component at a cheaper price than buying it from the outside. Your manufacturing team informs you that they can produce a maximum of 10,000 units, just enough to fulfill your contract. The setup cost will be $100,000 and the raw material cost is $40 per component. Since Teloxy has never manufactured this product before, manufacturing expects the following defects:

Percent defective	0	10	20	30	40
Probability of occurrence	10	20	30	25	15

All defective parts must be removed and repaired at a cost of $120 per part.

QUESTIONS

1. Using expected value, is it economically better to make or buy the component?
2. Strategically thinking, why might management opt for other than the most economical choice?

TELOXY ENGINEERING (B)

Your manufacturing team informs you that they have found a way to increase the size of the manufacturing run from 10,000 to 18,000 units in increments of 2000 units. However, the setup cost will be $150,000 rather than $100,000 for all production runs greater than 10,000 units and defects will cost the same $120 for removal and repair.

QUESTIONS

1. Calculate the economic feasibility of make or buy.
2. Should the probability of defects change if we produce 18,000 units as opposed to 10,000 units?
3. Would your answer to question 1 change if Teloxy management believes that follow-on contracts will be forthcoming? What would happen if the probability of defects changes to 15 percent, 25 percent, 40 percent, 15 percent, and 5 percent due to learning-curve efficiencies?

THE RISK MANAGEMENT DEPARTMENT

Background

In 1946, shortly after the end of World War II, Cooper Manufacturing Company was created. The company manufactured small appliances for the home. By 2010, Cooper Manufacturing had more than thirty manufacturing plants, all located in the United States. The business now included both small and large household appliances. Almost all of its growth came from acquisitions that were paid for out of cash flow and borrowing from the financial markets.

Cooper's strategic plan called for global expansion beginning in 2003. With this in mind and with large financial reserves, Cooper planned on acquiring five to six companies a year. This would be in addition to whatever domestic acquisitions were also available. Almost all of the acquisitions were manufacturing companies that produced products related to the household marketplace. However, some of the acquisitions included air conditioning and furnace companies as well as home security systems.

Risk Management Department

During the 1980s, when Cooper Manufacturing began its rapid acquisition approach, it established a Risk Management Department. The Risk Management Department reported to the chief financial officer (CFO) and was considered to be part of the financial discipline of the company. The overall objective of the Risk Management Department was to coordinate the protection of the company's assets. The primary means by which this was done was through the implementation of loss prevention programs. The department worked very closely with other internal departments such as Environmental Health and Safety. Outside consultants were brought in as necessary to support these activities.

One method employed by the company to ensure the entire company's cooperation and involvement in the risk management process was to hold each manufacturing division responsible for any specific losses up to a designated self-insured retention level. If there was a significant loss, the division must absorb the loss and its impact on the division's bottom-line profit margin. This directly involved the division in both loss prevention and claims management. When a claim did occur, the Risk Management Department maintained regular contact with the division's personnel to establish protocol on the claim and cash reserves and ultimate disposition.

As part of risk management, the company purchased insurance above the designated retention levels. The insurance premiums were allocated to each division. The premiums were calculated based upon sales volume and claims loss history, with the most significant percentage being allocated against claims loss history.

Risk management was considered an integral part of the due diligence process for acquisitions and divestitures. It began at the onset of the process rather than at the end and resulted in a written report and presentation to the senior levels of management.

A New Risk Materializes

The original intent of the Risk Management Department was to protect the company's assets, especially from claims and lawsuits. The department focused heavily upon financial and business risks with often little regard for human assets. All of this was about to change.

The majority of Cooper's manufacturing processes were labor-intensive assembly line processes. Although Cooper modernized the plants with new equipment to support the assembly lines with hope of speeding up the work, the processes were still heavily labor intensive.

Ergonomics in the Workplace

Ergonomics includes the fundamentals for the flexible workplace variability and compatibility with desk components that flex from individual work activities to team settings. Workstations provide supportive ergonomics for task-intensive environments. Outside the discipline, the term "ergonomics" is generally used to refer to physical ergonomics as it relates to the workplace (as in, e.g., ergonomic chairs and *keyboards*). Ergonomics in the workplace has to do largely with the safety of employees, both long and short term. Ergonomics can help reduce costs by improving safety. This would decrease the money paid out in workers' compensation. For example, over five million workers sustain overextension injuries per year. Through ergonomics, workplaces can be designed so that workers do not have to overextend themselves and the manufacturing industry could save billions in workers' compensation. Workplaces may either take the reactive or proactive approach when applying ergonomics practices. Reactive ergonomics is when something needs to be fixed and corrective action is taken. Proactive ergonomics is the process of seeking areas that could be improved and fixing the issues before they become a large problem. Problems may be fixed through equipment design, task design, or environmental design. Equipment design changes the actual, physical devices used by people. Task design changes what people do with the equipment. Environmental design changes the environment in which people work but not the physical equipment they use.

QUESTIONS

1. Was the original intent of creating the Risk Management Department correct in that it was designed to protect corporate assets? In other words, was this really risk management?
2. Are the new responsibilities of the department, specifically ergonomics, a valid interpretation of risk management?
3. Can the lowering of health care costs and workers' compensation costs be considered as a project?
4. How successful do you think Cooper was in lowering costs?

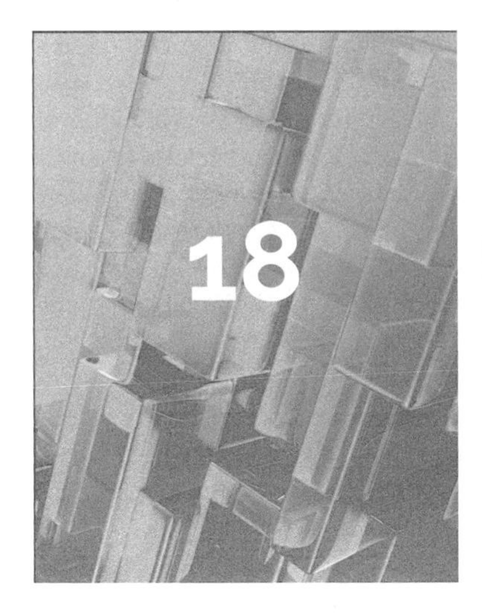

18 Learning Curves

18.0 INTRODUCTION

PMBOK® Guide, 6th Edition
6.4 Estimate Activity Durations

Competitive bidding has become an integral part of the project management responsibility in many industries. A multitude of estimating techniques are available in such fields as construction, aerospace, and defense to assist project managers in arriving at a competitive bid. If the final bid is too high, the company may not be competitive. If the bid is too low, the company may have to incur the cost of the overrun out of its own pocket. For a small firm, this overrun could lead to financial disaster.

Perhaps the most difficult projects to estimate are those that involve the development and manufacturing of a large quantity of units. As an example, a company is asked to bid on the development and manufacture of 15,000 components. The company is able to develop a cost for the manufacture of its first unit, but what will be the cost for the 10th, 100th, 1,000th, or 10,000th unit? Obviously, the production cost of each successive unit should be less than the previous unit, but by how much? Fortunately there exist highly accurate estimating techniques referred to as "learning" or "experience" curves.

18.1 GENERAL THEORY

Experience curves are based on the old adage that practice makes perfect. A product can always be manufactured better and in a shorter time period not only the second time, but each succeeding time. This concept is highly applicable to labor-intensive projects, such as those in manufacturing where labor forecasting has been a tedious and time-consuming effort.

It wasn't until the 1960s that the true implications of experience curves became evident. Personnel from the Boston Consulting Group showed that each time cumulative production doubled, the total manufacturing time and cost fell by a *constant* and *predictable* amount. Furthermore, the Boston Consulting Group showed that this effect extended to a variety of industries such as chemicals, metals, and electronic components.

Today's executives often measure the profitability of a corporation as a function of market share. As market share increases, profitability will increase, more because of lower production costs than increased margins. This is the experience curve effect. Large market shares allow companies to build large manufacturing plants so that the fixed capital costs are spread over more units, thus lowering the unit cost. This increase in efficiency is referred to as *economies of scale* and may be the main reason why large manufacturing organizations may be more efficient than smaller ones.

Capital equipment costs follow the rule of six-tenths power of capacity. As an example, consider a plant that has the capacity of producing 35,000 units each year. The plant's construction cost was \$10 million. If the company wishes to build a new plant with a capacity of 70,000 units, what will the construction cost be?

$$\frac{\$\text{ new}}{\$\text{ old}} = \left(\frac{70,000}{35,000}\right)^{0.6}$$

Solving for \$ new, we find that the new plant will cost approximately \$15 million, or one and one-half times the cost of the old plant. (For a more accurate determination, the costs must be adjusted for inflation.)

18.2 THE LEARNING CURVE CONCEPT

Learning curves stipulate that manufacturing man-hours (specifically direct labor) will decline each time a company doubles its output. Typically, learning curves produce a cost and time savings of 10 to 30 percent each time a company's experience at producing a product doubles. As an example, consider the data shown in Table 18–1, which represents a company operating on a 75 percent learning curve. The time for the second unit is 75 percent of the time of the first unit. The time for the fortieth unit is 75 percent of the time for the twentieth unit. The time for the 800th unit is 75 percent of the time for the 400th unit. Likewise, we can *forecast* the time for the 1,000th unit as being 75 percent of the time for the 500th unit. In this example, the time decreased by a fixed amount of 25 percent. Theoretically, this decrease could occur indefinitely.

In Table 18–1, we could have replaced the man-hours per production unit with the cost per production unit. It is more common to use man-hours because exact costs are either not always known or not publicly disclosed by the firm. Also, the use of costs implies the added complexity of considering escalation factors on salary, cost of living adjustments, and possibly the time value of money. For projects under a year or two, costs are often used instead of man-hours.

These types of costs are often referred to as value-added costs, and can also appear in the form of lower freight and procurement costs through bulk quantities. The value-added costs are actually cost savings for both the customer and contractor.

TABLE 18–1. CUMULATIVE PRODUCTION AND LABOR-HOUR DATA

Cumulative Production	Hours This Unit	Cumulative Total Hours
1	812	812
2	609	1,421
10	312	4,538
12	289	5,127
15	264	5,943
20	234	7,169
40	176	11,142
60	148	14,343
75	135	16,459
100	120	19,631
150	101	25,116
200	90	29,880
250	82	34,170
300	76	38,117
400	68	45,267
500	62	51,704
600	57	57,622
700	54	63,147
800	51	68,349
840	50	70,354

The learning curve was adapted from the historical observation that individuals performing repetitive tasks exhibit an improvement in performance as the task is repeated a number of times. Empirical studies of this phenomenon yielded three conclusions on which the current theory and practice are based:

- The time required to perform a task decreases as the task is repeated.
- The amount of improvement decreases as more units are produced.
- The rate of improvement has sufficient consistency to allow its use as a prediction tool.

The consistency in improvement has been found to exist in the form of a constant percentage reduction in time required over successive doubled quantities of units produced.

It's important to recognize the significance of using the learning curve for manufacturing projects. Consider a project where 75 percent of the total direct labor is in assembly (such as aircraft assembly) and the remaining 25 percent is machine work. With direct labor, learning improvements are possible, whereas with machine work, output may be restricted due to the performance of the machine. In the above example, with 75 percent direct labor and 25 percent machine work, a company may find itself performing on an 80 percent learning curve. But, if the direct labor were 25 percent and the machine work were 75 percent, then the company may find itself on a 90 percent learning curve.

18.3 GRAPHIC REPRESENTATION

Figure 18–1 shows the learning curve plotted from the data in Table 18–1. The horizontal axis represents the total number of units produced. The vertical axis represents the total labor hours (or cost) for each unit. The labor-hour graph in Figure 18–1 represents a hyperbola when drawn on an ordinary rectangular coordinates graph. The curve shows that the difference or amount of labor-hour reduction is *not* consistent. Rather, it declines by a continuously diminishing amount as the quantities are doubled. But the rate of change or decline has been found to be a constant percentage of the prior cost, because the decline in the base figure is proportionate to the decline in the amount of change. To illustrate this, we can use the data in Table 18–1, which was used to construct Figure 18–1. In doubling production from the first to the second unit, a reduction of 203 hours occurs. In doubling from 100 to 200 units, a reduction of 30 hours occurs. However in both cases, the percentage decrease was 25 percent. Again, in going from 400 to 800 units, a 25 percent reduction of 17 hours results. We can therefore conclude that, as more units are produced, the rate of change remains constant but the magnitude of the change diminishes.

When the data from Figure 18–1 are plotted on logarithmic graph, the result is a straight line, which represents the learning curve as shown in Figure 18–2.

There are two fundamental models of the learning curve in general use; the unit curve and the cumulative average curve. Both are shown in Figure 18–2. The unit curve focuses on the hours or cost involved in specific units of production. The theory can be stated as follows: As the total quantity of units produced doubles, the cost per unit decreases by some constant percentage. The constant percentage by which the costs of doubled quantities decrease is called the rate of learning.

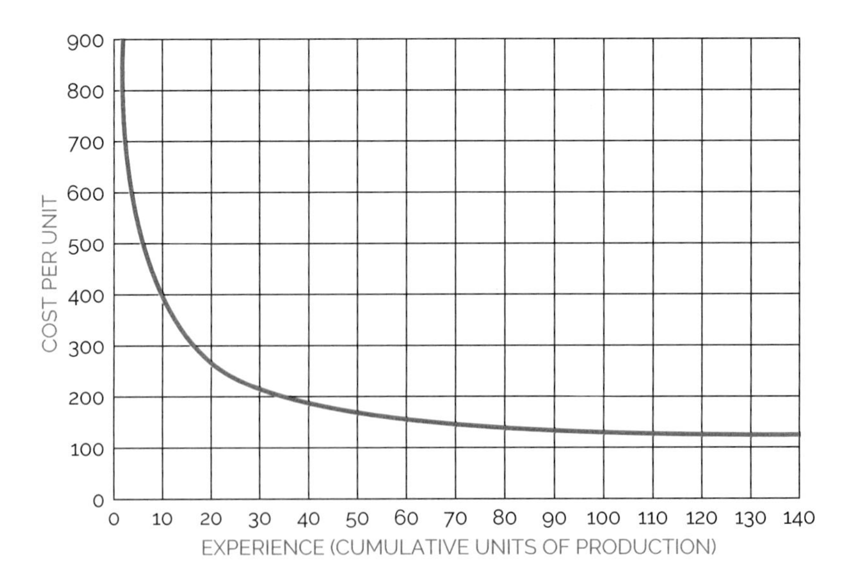

FIGURE 18–1. A 75 percent learning curve.

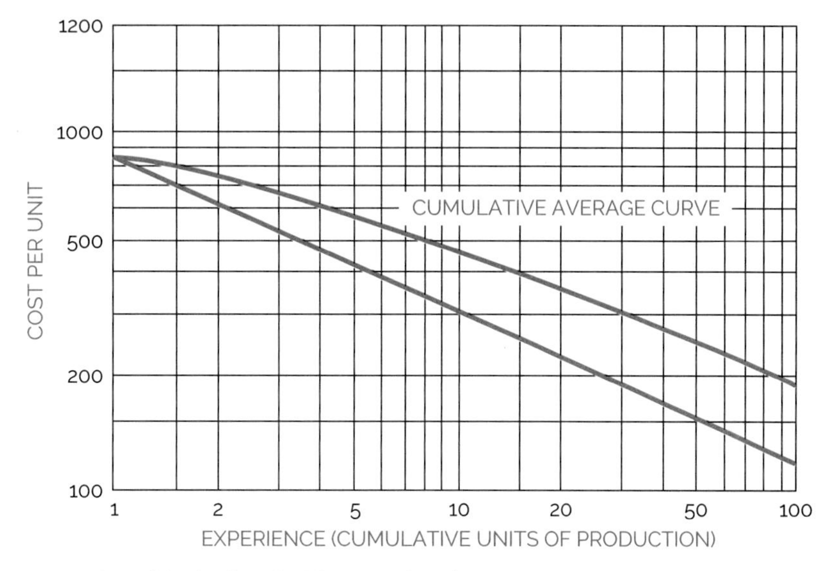

FIGURE 18–2. Logarithmic plot of a 75 percent learning curve.

The "slope" of the learning curve is related to the rate of learning. It is the difference between 100 percent and the rate of learning. For example, if the hours between doubled quantities are reduced by 20 percent (rate of learning), it would be described as a curve with an 80 percent slope.

To plot a straight line, one must know either two points or one point and the slope of the line. Generally speaking, the latter is more common. The question is whether the company knows the man-hours for the first unit or uses a projected number of man-hours for a target or standard unit to be used for pricing purposes.

The cumulative average curve in Figure 18–2 can be obtained from columns 1 and 3 in Table 18–1. Dividing column 3 by column 1, we find that the average hours for the first 100 units is 196 hours. For 200 units, the average is 149 hours. This becomes important in determining the cost for a manufacturing project.

18.4 KEY WORDS ASSOCIATED WITH LEARNING CURVES

Understanding a few key phrases will help in utilizing learning curve theory:

- *Slope of the curve:* A percentage figure that represents the steepness (constant rate of improvement) of the curve. Using the unit curve theory, this percentage represents the value (e.g., hours or cost) at a doubled production quantity in relation to the previous quantity. For example, with an experience curve having 80 percent slope, the value of unit two is 80 percent of the value of unit one, the value of unit four is 80 percent of the value at unit two, the value at unit 1,000 is 80 percent of the value of unit 500, and so on.

- *Unit one:* The first unit of product actually completed during a production run. This is not to be confused with a unit produced in any reproduction phase of the overall acquisition program.
- *Cumulative average hours:* The average hours expended per unit for all units produced through any given unit. When illustrated on a graph by a line drawn through each successive unit, the values form a cumulative average curve.
- *Unit hours:* The total direct labor hours expended to complete any specific unit. When a line is drawn on a graph through the values for each successive unit, the values form a unit curve.
- *Cumulative total hours:* The total hours expended for all units produced through any given unit. The data plotted on a graph with each point connected by a line form a cumulative total curve.

18.5 THE CUMULATIVE AVERAGE CURVE

It is common practice to plot the learning curve on logarithmic graph but to calculate the cumulative average from the following formula:

$$T_x = T_1 X^{-K}$$

where

T_x = the direct labor hours for unit n

T_1 = the direct labor hours for the first unit (unit one)

X = the cumulative unit produced

$-K$ = a factor derived from the slope of the experience curve

Learning curve %	K
100	0.0
95	0.074
90	0.152
85	0.235
80	0.322
75	0.415
70	0.515

As an example, consider a situation where the first unit requires 812 hours and the company is performing on a 75 percent learning curve. The man-hours required for the 250th unit would be:

$$\begin{aligned} T_{250} &= (812)(250)^{-0.415} \\ &= 82 \text{ hours} \end{aligned}$$

This agrees with the data in Table 18–1.

Sometimes companies do not know the time for the first unit. Instead, they assume a target unit and accompanying target man-hours. As an example, consider a company that assumes that the standard for performance will be the 100th unit, which is targeted for 120 man-hours, and performs on a 75 percent learning curve. Solving for $T1$ we have:

$$\begin{aligned} T_1 &= T_x\ X^{-K} \\ &= (120)(100)^{0.415} \\ &= 811 \text{ hours} \end{aligned}$$

This is in approximate agreement with the data in Table 18–1. The cumulative average number of labor hours can be *approximated* from the expression

$$T_c = \frac{T_1\ X^{-K}}{1-K}$$

Where

Tc = cumulative average labor hours for the Xth unit

X = cumulative units produced

$T1$ = direct labor hours for first unit

For the 250th unit,

$$\begin{aligned} T_c &= \frac{(812)(250)^{-0.415}}{1-0.415} \\ &= 135 \text{ hours} \end{aligned}$$

From Table 18–1, the cumulative average for the 250th unit is 34,170 man-hours divided by 250, or 137 hours. We must remember that the above expression is merely an approximation. Significant errors can occur using this expression for fewer than 100 units. For large values of X, the error becomes insignificant.

It is possible to use the learning curve equation to develop Table 18–2, which shows typical cost reductions due to increased experience. Suppose that the production level is quadrupled and you are performing on an 80 percent learning curve. Using Table 18–2, the costs will be reduced by 36 percent.

18.6 SOURCES OF EXPERIENCE

There are several factors that contribute to the learning curve phenomenon. None of the factors performs entirely independently, but are interrelated through a complex network. However, for simplicity's sake, these factors will be sorted out for discussion purposes.

TABLE 18–2. SAMPLE COST REDUCTIONS DUE TO INCREASED EXPERIENCE

Ratio of Old Experience to New Experience	Experience Curve					
	70%	75%	80%	85%	90%	95%
1.1	5%	4%	3%	2%	1%	1%
1.25	11	9	7	5	4	2
1.5	19	15	12	9	6	3
1.75	25	21	16	12	8	4
2.0	30	25	20	15	10	5
2.5	38	32	26	19	13	7
3.0	43	37	30	23	15	8
4.0	51	44	36	28	19	10
6.0	60	52	44	34	24	12
8.0	66	58	49	39	27	14
16.0	76	68	59	48	34	19

Source: Derek F. Abell and John S. Hammond, *Strategic Market Planning* (Upper Saddle River, NJ: Pearson Education, 1979), p. 109. Reprinted by permission.

- *Labor efficiency:* This is the most common factor, which says that we learn more each time we repeat a task. As we learn, the time and cost of performing the task should diminish. As the employee learns the task, less managerial supervision is required, waste and inefficiency can be reduced or even eliminated, and productivity will increase.

Unfortunately, labor efficiency does not occur automatically. Personnel management policies in the area of *workforce stability* and *worker compensation* are of vital importance. As workers mature and become more efficient, it becomes increasingly important to maintain this pool of skilled labor. Loss of a contract or interruption between contracts could force employees to seek employment elsewhere. In certain industries, like aerospace and defense, engineers are often regarded as migratory workers moving from contract to contract and company to company.

Upturns and downturns in the economy can have a serious impact on maintaining experience curves. During downturns in the economy, people work more slowly, trying to preserve their jobs. Eventually the company is forced into a position of having to reassign people to other activities or to lay people off. During upturns in the economy, massive training programs may be needed in order to accelerate the rate of learning.

If an employee is expected to get the job done in a shorter period of time, then the employee expects to be adequately compensated. Wage incentives can produce either a positive or negative effect based on how they are applied. Learning curves and productivity can become a bargaining tool by labor as it negotiates for greater pay.

Fixed compensation plans generally do not motivate workers to produce more. If an employee is expected to produce more at a lower cost, then the employee expects to receive part of the cost savings as either added compensation or fringe benefits.

The learning effect goes beyond the labor directly involved in manufacturing. Maintenance personnel, supervisors, and persons in other line and staff manufacturing positions also increase their productivity, as do people in marketing, sales, administration, and other functions.

- *Work specialization and methods improvements:*[1] Specialization increases worker proficiency at a given task. Consider what happens when two workers, who formerly did both parts of a two-stage operation, each specialize in a single stage. Each worker now handles twice as many items and accumulates experience twice as fast on the more specialized task. Redesign of work operations (methods) can also result in greater efficiency.
- *New production processes:* Process innovations and improvements can be an important source of cost reductions, especially in capital-intensive industries. The low-labor-content semiconductor industry, for instance, achieves experience curves at 70 percent to 80 percent from improved production technology by devoting a large percentage of its research and development to process improvements. Similar process improvements have been observed in refineries, nuclear power plants, and steel mills, to mention a few.
- *Getting better performance from production equipment:* When first designed, a piece of production equipment may have a conservatively rated output. Experience may reveal innovative ways of increasing its output. For instance, capacity of a fluid catalytic cracking unit typically "grows" by about 50 percent over a ten-year period.[2]
- *Changes in the resource mix:* As experience accumulates, a producer can often incorporate different or less expensive resources in the operation. For instance, less skilled workers can replace skilled workers or automation can replace labor.
- *Product standardization:* Standardization allows the replication of tasks necessary for worker learning. Even when flexibility and/or a wider product line are important marketing considerations, standardization can be achieved by modularization. For example, by making just a few types of engines, transmissions, chassis, seats, body styles, and so on, an auto manufacturer can achieve experience effects due to specialization in each part. These in turn can be assembled into a wide variety of models.
- *Product redesign:* As experience is gained with a product, both the manufacturer and customers gain a clear understanding of its performance requirements. This understanding allows the product to be redesigned to conserve material, allow greater efficiency in manufacture, and substitute less costly materials and resources, while at the same time improving performance on relevant dimensions.
- *Incentives and disincentives:* Compensation plans and other sources of experience can be both incentives and disincentives. Incentives can change the slope of the learning curve, as shown in Figure 18–3. This is referred to as a "toe-down" learning curve where a more favorable learning process can occur. In Figure 18–4, we have a "toe-up," or "scallop," learning curve, which is the result of disincentives. After the toe-up occurs, the learning curve may have a new slope that was not as favorable as the original slope.

---

1. The next six elements are from Derek F. Abell and John S. Hammond, *Strategic Market Planning* (Upper Saddle River, NJ: Pearson Education, 1979), pp. 112–113). Reprinted by permission.
2. Winfred B. Hirschmann, "Profit from the Learning Curve," *Harvard Business Review,* 42, no. 1 (January–February 1964), p. 125. Copyright © 1964 by the Harvard Business School Publishing Corporation; all rights reserved. Reprinted by permission.

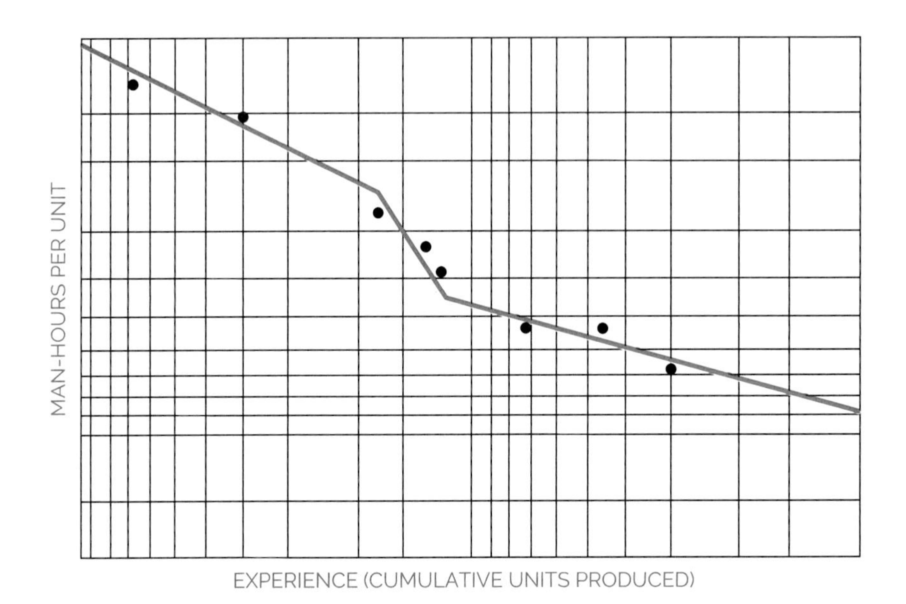

FIGURE 18–3. A "toe-down" learning curve.

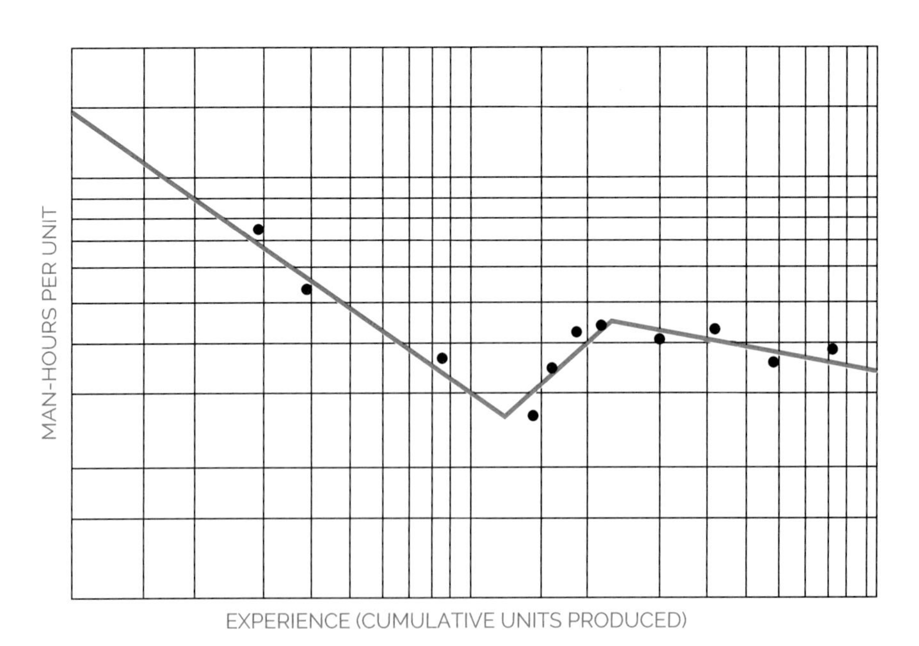

FIGURE 18–4. A "toe-up" learning curve.

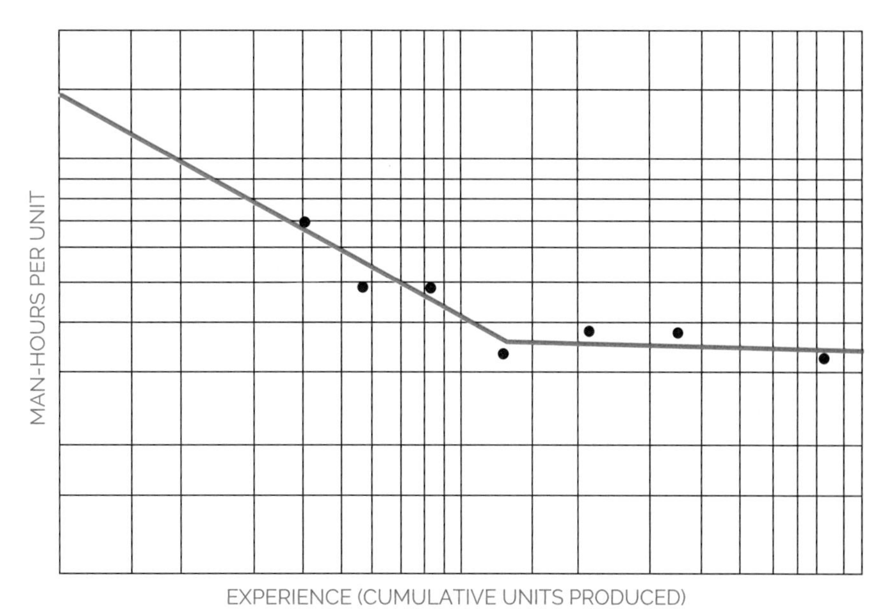

FIGURE 18–5. A leveling off of the learning curve.

Worker dissatisfaction can also create a leveling off of the learning curve, as shown in Figure 18–5. This leveling off can also occur as a result of inefficiencies due to closing out of a production line or transferring workers to other activities at the end of a contract.

18.7 DEVELOPING SLOPE MEASURES

Research by the Stanford Research Institute revealed that many different slopes were experienced by different manufacturers, sometimes on similar manufacturing programs. In fact, manufacturing data collected from the World War II aircraft manufacturing industry had slopes ranging from 69.7 percent to almost 100 percent. These slopes averaged 80 percent, giving rise to an industry average curve of 80 percent. Other research has developed measures for other industries, such as 95.6 percent for a sample of 162 electronics programs. Unfortunately, this industry average curve is frequently misapplied by practitioners, who use it as a standard or norm. When estimating slopes without the benefit of data from the plant of the manufacturer, it is better to use learning curve slopes from similar items at the manufacturer's plant rather than the industry average.

The analyst needs to know the slope of the learning curve for a number of reasons. One is to facilitate communication, because it is part of the language of the learning curve theory. The steeper the slope (lower the percent), the more rapidly the resource requirements (hours) will decline as production increases. Accordingly, the slope of the learning curve is usually an issue in production contract negotiation. The slope of the learning curve is also needed to project follow-on costs, using either learning tables or a computer. Also, a given slope may be established as a standard based on reliable historical experience. Learning curves developed from actual experience on current production can then be compared against this standard slope to determine whether the improvement on a particular contract is or is not reasonable.

18.8 UNIT COSTS AND USE OF MIDPOINTS

The use of the learning curve is dependent on the methods of recording costs that companies employ. An accounting or statistical record system must be devised by a company so that data are available for learning curve purposes. Otherwise, it may be impossible to construct a learning curve. Costs, such as labor hours per unit or dollars per unit, must be identified with the unit of product. It is preferable to use labor hours rather than dollars, because the latter contain an additional variable—the effect of inflation or deflation (both wage-rate and material cost changes)—that the former does not contain. In any event, the record system must have definite cutoff points for such costs permitting identification of the costs with the units involved. Most companies use a lot-release system, whereby costs are accumulated on a job order in which the number of units completed are specified and the costs are cut off at the completion of the number of units. In this case, however, the costs are usually equated with equivalent units rather than actual units. Because the job order system is commonly used, the unit cost is not the actual cost per unit in the lot. This means that when lots are plotted on graph paper, the unit value corresponding with the average cost value must be found.

18.9 SELECTION OF LEARNING CURVES

Existing experience curves, by definition, reflect past experience. Trend lines are developed from accumulated data plotted on logarithmic paper (preferably) and "smoothed out" to portray the curve. The type of curve may represent one of several concepts. The data may have been accumulated by product, process, department, or by other functional or organizational segregations, depending on the needs of the user. But whichever experience curve concept or method of data accumulation is selected for use, based on suitability to the experience pattern, the data should be applied consistently in order to render meaningful information to management. Consistency in curve concept and data accumulation cannot be overemphasized, because existing experience curves play a major role in determining the project experience curve for a new item or product.

When selecting the proper curve for a new production item when only one point of data is available and the slope is unknown, the following, in decreasing order of magnitude, should be considered.

- Similarity between the new item and an item or items previously produced.
- Physical comparisons
    - Addition or deletion of processes and components

    - Differences in material, if any
    - Effect of engineering changes in items previously produced
- Duration of time since a similar item was produced
    - Condition of tooling and equipment
    - Personnel turnover
    - Changes in working conditions or morale
- Other comparable factors between similar items
    - Delivery schedules
    - Availability of material and components
    - Personnel turnover during production cycle of item previously produced
    - Comparison of actual production data with previously extrapolated or theoretical curves to identify deviations

It is feasible to assign weights to these factors as well as to any other factors that are of a comparable nature in an attempt to quantify differences between items. These factors are again historical in nature and only comparison of several existing curves and their actuals would reveal the importance of these factors.

If at least two points of data are available, the slope of the curve may be determined. Naturally the distance between these two points must be considered when evaluating the reliability of the slope. The availability of additional points of data will enhance the reliability of the curve. Regardless of the number of points and the assumed reliability of the slope, comparisons with similar items are considered the most desirable approach and should be made whenever possible.

A value for unit one may be arrived at either by accumulation of data or statistical derivation. When production is underway, available data can be readily plotted, and the curve may be extrapolated to a desired unit. However, if production has yet to be started, actual unit-one data would not be available, and a theoretical unit-one value would have to be developed. This may be accomplished in one of three ways:

- A statistically derived relationship between the preproduction unit hours and first unit hours can be applied to the actual hours from the preproduction phase.
- A cost estimating relationship (CER) for first-unit cost based on physical or performance parameters can be used to develop a first-unit cost estimate.
- The slope and the point at which the curve and the labor standard value converge are known. In this case, a unit-one value can be determined. This is accomplished by dividing the labor standard by the appropriate unit value.

18.10 FOLLOW-ON ORDERS

Once the initial experience curve has been developed for either the initial order or production run, the values through the last unit on the cumulative average and unit curves can be determined. Follow-on orders and continuations of production runs, which are considered extensions of the original orders or runs, are plotted as extensions on the appropriate curve.

However, the cumulative average value through the final point of the extended curve is not the cumulative average for the follow-on portion of that curve. It is the cumulative average for both portions of the curve, assuming no break in production. Thus estimating the cost for the follow-on effort only requires evaluation of the differences between cumulative average costs for the initial run and the follow-on. Likewise, the last-unit value for both portions of the unit curve would represent the last-unit value for the combined curve.

18.11 MANUFACTURING BREAKS

The manufacturing break is the time lapse between the completion of an order or manufacturing run of certain units of equipment and the commencement of a follow-on order or restart of a manufacturing run for identical units. This time lapse disrupts the continuous flow of manufacturing and constitutes a definite cost impact. The time lapse under discussion here pertains to significant periods of time (weeks and months), as opposed to the minutes or hours for personnel allowances, machine delays, power failures, and the like.

It is logical to assume that because the experience curve has a time-cost relationship, a break will affect both time and cost. Therefore, the length of the break becomes as significant as the length of the initial order or manufacturing run. Because the break is quantifiable, the remaining factor to be determined is the cost of this lapse in manufacturing (that is, the additional cost incurred over and above that which would have been incurred had either the initial order or the run continued through the duration of the follow-on order or the restarted run).

When a manufacturer relies on experience curves as management information tools, it can be assumed that the necessary, accurate data for determining the initial curves have been accumulated, recorded, and properly validated. Therefore, if the manufacturer has experienced breaks, the experience curve data for the orders (lots) or runs involved should be available in such form that appropriate curves can be developed.

18.12 LEARNING CURVE LIMITATIONS

There are limitations to the use of learning curves, and care must be taken to avoid erroneous conclusions. Typical limitations include:

- The learning curve does not continue forever. The percentage decline in hours/dollars diminishes over time.
- The learning curve knowledge gained on one product may not be extendable to other products unless there exist shared experiences.
- Cost data may not be readily available in order to construct a meaningful learning curve. Other problems can occur if overhead costs are included with the direct labor cost, or if the accounting codes cannot separate work packages sufficiently in order to identify those elements that truly demonstrate experience effects.

- Quantity discounts can distort the costs and the perceived benefits of learning curves.
- Inflation must be expressed in constant dollars. Otherwise, the gains realized from experience may be neutralized.
- Learning curves are most useful on long-term horizons (i.e., years). On short-term horizons, benefits perceived may not be the result of learning curves.
- External influences, such as limitations on materials, patents, or even government regulations, can restrict the benefits of learning curves.
- Constant annual production (i.e., no growth) may have a limiting experience effect after a few years.

18.13 COMPETITIVE WEAPON

Learning curves are a strong competitive weapon, especially in developing a pricing strategy. The actual pricing strategy depends on the product life-cycle stage, the firm's market position, the competitor's available resources and market position, the time horizon, and the firm's financial position.

From a project management perspective, learning curve pricing can be a competitive weapon. As an example, consider a company that is burdened at $60/hour and is bidding on a job to produce 500 units. Let us assume that the data in Table 18–1 apply. For 500 units of production, the cumulative total hours are 51,704, giving us an average rate of 103.4 hours per unit. The cost for the job would be 51,704 hours 3 $60/hour, or $3,102,240. If the target profit is 10 percent, then the final bid should be $3,412,464. This includes a profit of $310,224.

Even though a 10 percent profit is projected, the *actual* profit may be substantially less. Each product is priced out an average of 103.4 hours/unit. The first unit, however, will require 812 hours. The company will *lose* 708.6 hours 3 $60/hour, or $42,516, on the first unit produced. The 100th unit will require 120 hours, giving us a loss of $996 (i.e., [120 hours 2 103.4 hours] 3 $60/hour). Profit will begin when the 150th unit is produced, because the hours required to produce the 150th unit are less than the average hour per unit of 103.4.

Simply stated, the first 150 units are a drain on cash flow. The cash-flow drain may require the company to "borrow" money to finance operations until the 150th unit is produced, thus lowering the target profit.

Related Case Studies (from Kerzner/*Project Management Case Studies,* 5th ed.)	**Related Workbook Exercises (from Kerzner/*Project Management Workbook and PMP®/CAPM® Exam Study Guide,* 12th ed.)**	***PMBOK® Guide*, 6th Edition, Reference Section for the PMP® Certification Exam**
None	• Multiple Choice Exam	• Project Schedule Management • Project Cost Management

18.14 STUDYING TIPS FOR THE PMI® PROJECT MANAGEMENT CERTIFICATION EXAM

This section is applicable as a review of the principles to support the knowledge areas and domain groups in the *PMBOK® Guide*. This chapter addresses:

- Project Schedule Management
- Project Cost Management

Understanding the following principles is beneficial if the reader is using this text to study for the PMP® Certification Exam:

- What is meant by a learning curve
- Uses of a learning curve
- How learning curves can be used for estimating

The following multiple-choice questions will be helpful in reviewing the principles of this chapter:

1. According to learning curve theory, learning takes place at a fixed rate whenever the production levels:
   A. Increase higher than normal
   B. Increase, but at a lower than normal rate
   C. Double
   D. Quadruple
2. Learning curve theory is most appropriate for estimating which costs?
   A. R&D
   B. Engineering
   C. Marketing
   D. Manufacturing
3. On a 90 percent learning curve, the 100th unit required 80 hours. How many hours would the 200th unit require?
   A. 200
   B. 180
   C. 100
   D. 90
4. Which of the following can be a source of improvement to a learning curve?
   A. New, more efficient production processes
   B. Product redesigns
   C. Higher-quality raw materials
   D. All of the above

ANSWERS

1. C
2. D
3. D
4. D

PROBLEMS

18–1 When a learning curve is plotted on a rectangular coordinates graph, the curve appears to level off. But when the curve is plotted on a logarithmic graph, it appears that the improvements can go on forever. How do you account for the difference? Can the improvements occur indefinitely? If not, what factors could limit continuous improvement?

18–2 A company is performing on an 85 percent learning curve. If the first unit requires 620 hours, how much time will be required for the 300th unit?

18–3 A company working on a 75 percent learning curve has decided that the production standard should be 85 hours of production for the 100th unit. How much time should be required for the first unit? If the first unit requires more hours than you anticipated, does this mean that the learning curve is wrong?

18–4 A company has just received a contract for 700 units of a certain product. The pricing department has predicted that the first unit should require 2,250 hours. The pricing department believes that a 75 percent learning curve is justified. If the actual learning curve is 77 percent, how much money has the company lost? Assume that a fully burdened hour is $65. What percentage error in total hours results from a 2 percent increase in learning curve percentage?

18–5 If the first unit of production requires 1,200 hours and the 150th unit requires 315 hours, then what learning curve is the company performing at?

18–6 A company has decided to bid on a follow-on contract for 500 units of a product. The company has already produced 2,000 units on a 75 percent learning curve. The 2000th unit requires 80 hours of production time. If a fully burdened hour is $80 and the company wishes to generate a 12 percent profit, how much should be bid?

18–7 Referring to question 18–6, how many units of the follow-on contract must be produced before a profit is realized?

18–8 A manufacturing company wishes to enter a new market. By the end of next year, the market leader will have produced 16,000 units on an 80 percent learning curve, and the year-end price is expected to be $475/unit. Your manufacturing personnel tell you that the first unit will require $7,150 to produce and, with the new technology you have developed, you should be able to perform at a 75 percent learning curve. How many units must you produce and sell over the next year in order to compete with the leader at $475/unit at year end? Is your answer realistic, and what assumptions have you made?

18–9 Rylon Corporation is an assembler of electrical components. The company estimates that for the next year, the demand will be 800 units. The company is performing on an 80 percent learning curve. The company is considering purchasing some assembly machinery to accelerate the assembly time. Most assembly activities are 85–90 percent labor intensive.

However, with the new machinery, the assembly activities will be only 25–45 percent labor intensive. If the company purchases and installs the new equipment, it will occur after the 200th unit is produced. Therefore, the remaining 600 units will be produced with the new equipment. The 200th unit will require 620 hours of assembly. However, the 201st unit will require only 400 hours of assembly but on a 90 percent learning curve.

A. Will the new machine shorten product assembly time for all 800 units and, if so, by how many hours?

B. If the company is burdened by $70 per hour, and the new equipment is depreciated over five years, what is the most money that the company should pay for the new equipment? What assumptions have you made?

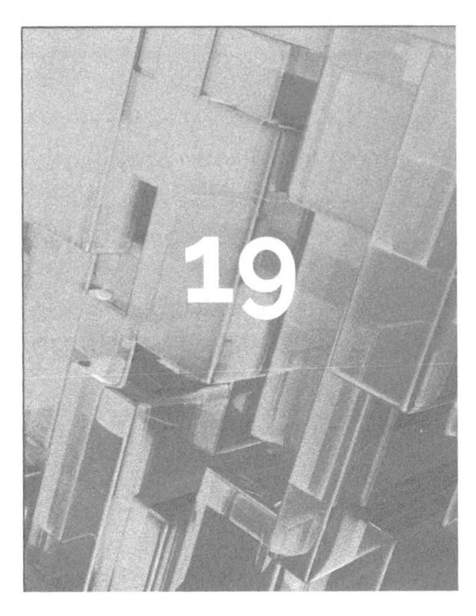

Contract Management

19.0 INTRODUCTION

PMBOK® Guide, 6th Edition
Chapter 12 Project Procurement Management

In general, companies provide services or products based on the requirements set forth in invitations for competitive bids issued by the client or the results of direct contract negotiations with the client. One of the most important factors in preparing a proposal and estimating the cost and profit of a project is the type of contract expected. The confidence by which a bid is prepared is usually dependent on how much of a risk the contractor will incur through the contract. Certain types of contracts provide relief for the contractor since onerous risks[1] exist. The cost must therefore consider how well the contract type covers certain high- and low-risk areas.

Prospective clients are always concerned when, during a competitive bidding process, one bid is much lower than the others. The client may question the validity of the bid and whether the contract can be achieved for the low bid. In cases such as this, the client usually imposes incentive and penalty clauses in the contract for self-protection.

Because of the risk factor, competitors must negotiate not only for the target cost figures but also for the type of contract involved since risk protection is the predominant influential factor. The size and experience of the client's own staff, urgency of completion, availability of qualified contractors, and other factors must be carefully evaluated. The advantages and disadvantages of all basic contractual arrangements must be recognized to select the optimum arrangement for a particular project.

1. *Onerous risks* are unfair risks that the contractor may have to bear. Quite often, the contract negotiations may not reach agreement on what is or is not an onerous risk.

19.1 PROCUREMENT

PMBOK® Guide, 6th Edition
Chapter 12 Project Procurement Management (Introduction)
12.1 Plan Procurement Management
12.1.3.2 Procurement Strategy

Procurement can be defined as the acquisition of goods or services. Procurement (and contracting) is a process that involves two parties with different objectives who interact on a given market segment. Good procurement practices can increase corporate profitability by taking advantage of quantity discounts, minimizing cash flow problems, and seeking out high-quality suppliers. Because procurement contributes to profitability, procurement is often centralized, which results in standardized practices and lower paperwork costs.

All procurement strategies are frameworks by which an organization attains its objectives. There are two basic procurement strategies:

- **Corporate Procurement Strategy:** The relationship of specific procurement actions to the corporate strategy. An example of this would be centralized procurement.
- **Project Procurement Strategy:** The relationship of specific procurement actions to the operating environment of the project. An example of this would be when the project manager is allowed to perform sole source procurement without necessarily involving the centralized procurement group, such as purchasing one ounce of a special chemical for an R&D project.

Project procurement strategies can differ from corporate procurement strategies because of constraints, availability of critical resources, and specific customer requirements. Corporate strategies might promote purchasing small quantities from several qualified vendors, whereas project strategies may dictate sole source procurement.

Procurement planning usually involves the selection of one of the following as the primary objective:

- Procure all goods/services from a single source.
- Procure all goods/services from multiple sources.
- Procure only a small portion of the goods/services.
- Procure none of the goods/services.

PMBOK® Guide, 6th Edition
12.1.5 Enterprise Environmental Factors

Another critical factor is the environment in which procurement must take place. There are two environments: macro and micro. The macro environment includes the general external variables that can influence how and when we do procurement. The *PMBOK® Guide* refers to this as "Enterprise Environmental Factors." These include recessions, inflation, cost of borrowing money, whether a buyer or seller's market exists, and unemployment. As an example, a foreign corporation had undertaken a large project that involved the hiring of several contractors. Because of the country's high unemployment rate, the decision was made to use only domestic suppliers/contractors and to give first preference to contractors in cities where unemployment was the greatest, even though there were other more qualified suppliers/contractors.

The microenvironment is the internal procurement processes of the firm, especially the policies and procedures imposed by the firm, project, or client in the way that procurement will take place. This includes the procurement/contracting system, which contains four processes according to the *PMBOK® Guide*, Sixth Edition:

- Plan Procurements
- Conduct Procurements
- Administer Procurements
- Close Procurements

It is important to understand that, in certain environments such as major projects for the Department of Defense (DoD), the contracting process is used as the vehicle for transitioning the project from one life-cycle phase to the next. For example, a contract can be awarded for the design, development, and testing of an advanced jet aircraft engine. The contract is completed when the aircraft engine testing is completed. If the decision is made at the phase gate review to proceed to aircraft engine production, the contracting process will be reinitiated for the new effort. Thus, the above four *PMBOK® Guide* processes would be repeated for each life-cycle phase. As the project progresses from one phase to the next, and additional project knowledge is acquired through each completed phase, the level of uncertainty (and risk) is reduced. The reduction in project risk allows the use of lower-risk contracts throughout the project life cycle. During higher-risk project phases such as conceptual, development, and testing, cost-type contracts are traditionally used. During the lower-risk project phases such as production and sustainment, fixed-priced contracts are typically used.

It is also important to note that the above four *PMBOK® Guide* processes focus only on the buyer's side of contract management.

Contract management is defined as "art and science of managing a contractual agreement throughout the contracting process."[2] Since contracts involve at least two parties—the buyer and the seller (contractor)—contract management processes are performed by *both* the buyer and seller. The seller's contract management processes, which correspond to the buyer's processes, consist of the following activities[3]:

- **Presales Activity:** The process of identifying prospective and current customers, determining customer's needs and plans, and evaluating the competitive environment.
- **Bid/No Bid Decision Making:** The process of evaluating the buyer's solicitation, assessing the competitive environment and risks against the opportunities of a potential business deal, and then deciding whether to proceed.
- **Bid/Proposal Preparation:** The process of developing offers in response to a buyer's solicitation or based on perceived buyer needs, for the purpose of persuading the buyer to enter into a contract.

---

2 Gregory A. Garrett and Rene G. Rendon, *Contract Management: Organizational Assessment Tools* (Ashburn, VA: National Contract Management Association, 2005), p. 270.
3 See note 2.

- **Contract Negotiation and Formation:** The process of reaching a common understanding of the nature of the project and negotiating the contract terms and conditions for the purpose of developing a set of shared expectations and understandings.
- **Contract Administration:** The process of ensuring that each party's performance meets contractual requirements.
- **Contract Closeout:** The process of verifying that all administrative matters are concluded on a contract that is otherwise physically complete. This involves completing and settling the contract, including resolving any open items.

As can be seen from the previous discussion, the last two phases of the seller's contract management processes are identical to the buyer's contract management processes. This is because the buyer and seller are both performing the same contract management activities and working off of the same contract document.

19.2 PLAN PROCUREMENTS

PMBOK® Guide, 6th Edition
12.1 Plan Procurement Management

The first step in the procurement process is the planning for purchases and acquisitions, specifically the development of a procurement plan that states what to procure, when, and how. This process includes the following:

- Defining the need for the project
- Development of the procurement statement of work, specifications, and work breakdown structure
- Preparing a WBS dictionary, if necessary
- Performing a make or buy analysis
- Laying out the major milestones and the timing/schedule
- Determining if long lead procurement is necessary
- Cost estimating, including life-cycle costing
- Determining whether qualified sellers exist
- Identifying the source selection criteria
- Preparing a listing of possible project/procurement risks (i.e., a risk register)
- Developing a procurement plan
- Obtaining authorization and approval to proceed

PMBOK® Guide, 6th Edition
12.1.3.4 Procurement Statement of Work

There could be separate and different statements of work for each product to be procured. The statement of work (SOW) is a *narrative* description of the work to be accomplished and/or the resources to be supplied. The identification of resources to be supplied has taken on paramount importance during the last 10 years or so. During the 1970s and 1980s, small companies were bidding on megajobs only to subcontract out more than 99 percent of all of the work. Lawsuits were abundant and the solution was to put clauses in the SOW requiring that the contractor identify the names and resumes of the talented *internal* resources that would be committed to the project, including the percentage of their time on the project. In addition to SOWs, organizations also use statements of objectives (SOOs) for projects that are designed as "performance-based" effort. Performance-based projects are now the preference in the federal government. SOOs are used when the procuring organization wants to leverage the advanced technologies, capabilities, and expertise of the potential contractors in the marketplace. Instead of using a SOW, which describes in specific detail to the contractor *what* work needs to be performed and *how* it should be performed, the SOO only describes the end objectives of the project (*what* are the project's end objectives). In response to a solicitation containing a SOO, the potential contractors develop and propose their own SOW that provides the detailed specifics on *how* they intend to perform the work. The source selection process entails comparing the various contractor-developed SOWs, each contractor applying its own unique technologies, capabilities, and expertise to the project effort. The proposal evaluation process includes making trade-offs between differing levels of proposed performance (as reflected in the contractor SOW), as well as proposed price.

Specifications are written, pictorial, or graphic information that describe, define, or specify the services or items to be procured. There are three types of specifications:

- *Design Specifications:* These detail what is to be done in terms of physical characteristics. The risk of performance is on the buyer.
- *Performance Specifications:* These specify measurable capabilities the end product must achieve in terms of operational characteristics. The risk of performance is on the contractor.
- *Functional Specifications:* This is when the seller describes the end use of the item to stimulate competition among commercial items, at a lower overall cost. This is a subset of the performance specification, and the risk of performance is on the contractor.

PMBOK® *Guide*, 6th Edition
12.1.3.6 Make or Buy Analysis Decisions

There are always options in the way the end item can be obtained. Feasible procurement alternatives include make or buy, lease or buy, buy or rent, and lease or rent. Buying domestic or international is also of critical importance, especially to the United Auto Workers Union. Factors involving the make or buy analysis are:

- The make decision
  - Less costly (but not always!)
  - Easy integration of operations
  - Utilize existing capacity that is idle
  - Maintain direct control
  - Maintain design/production secrecy
  - Avoid unreliable supplier base
  - Stabilize existing workforce
- The buy decision
  - Less costly (but not always!)
  - Utilize skills of suppliers

- Small volume requirement (not cost effective to produce)
- Having limited capacity or capability
- Augment existing labor force
- Maintain multiple sources (qualified vendor list)
- Indirect control

The lease or rent decision is usually a financial endeavor. Leases are usually longer term than renting. Consider the following example. A company is willing to rent you a piece of equipment at a cost of $100 per day. You can lease the equipment for $60 per day plus a one-time cost of $5,000. What is the breakeven point, in days, where leasing and renting are the same?

Let X be the number of days;

$$\$100X = \$5{,}000 + \$60X$$

$$\uparrow \qquad\qquad \uparrow$$

Renting Leasing

$$\text{Solving, } X = 125 \text{ days}$$

Therefore, if the firm wishes to use this equipment for more than 125 days, it would be more cost effective to sign a lease agreement rather than a rental agreement.

Procurement planning must address the risks on the contract as well as the risks with procurement. Some companies have project management manuals with sections that specifically address procurement risks using templates. As an example, the following is a partial list of procurement risks as identified in the *ABB Project Management Manual*:[4]

- Contract and agreements (penalty/liquidated damages, specifications open to misinterpretation, vague wording, permits/licenses, paperwork requirements)
- Responsibility and liability (force majeure, liability limits for each party, unclear scope limitations)
- Financial (letters of credit, payment plans, inflation, currency exchange, bonds)
- Political (political stability, changes in legislation, import/export restrictions, arbitration laws)
- Warranty (nonstandard requirements, repairs)
- Schedule (unrealistic delivery time, work by others not finished on time, approval process, limitations on available resources)
- Technical and technology (nonstandard solutions, quality assurance regulations, inspections, customer acceptance criteria)
- Resources (availability, skill levels, local versus external)

---

4 Adapted from Harold Kerzner, *Advanced Project Management: Best Practices on Implementation,* 2nd ed. (Hoboken, NJ: John Wiley & Sons, 2004), pp. 346–348.

The procurement plan will address the following questions:

- How much procurement will be necessary?
- Will they be standard or specialized procurement activities?
- Will we make some of the products or purchase all of them?
- Will there be qualified suppliers?
- Will we need to prequalify some of the suppliers?
- Will we use open bidding or bidding from a preferred supplier list?
- How will we manage multiple suppliers?
- Are there items that require long lead procurement?
- What type of contract will be used, considering the contractual risks?
- Will we need different contract types for multiple suppliers?
- What evaluation criteria will be used to score the proposals?

19.3 CONDUCTING THE PROCUREMENTS

PMBOK® Guide, 6th Edition
12.2 Conduct Procurements

Once the requirements are identified and a procurement plan has been prepared, a requisition form for each item to be procured is sent to procurement to begin the procurement or requisition process. The process of conducting the procurements includes:

- Evaluating/confirming specifications (are they current?)
- Confirming qualified sources
- Reviewing past performance of sources
- Reviewing of team or partnership agreements
- Producing the solicitation package

The solicitation package is prepared during the procurements planning process but utilized during the next process, conduct procurements. In most situations, the same solicitation package for each deliverable must be sent to each possible supplier so that the playing field is level. A typical solicitation package would include:

- Bid documents (usually standardized)
- Listing of qualified vendors (expected to bid)
- Proposal evaluation criteria (source selection criteria)
- Bidder conferences
- How change requests will be managed
- Supplier payment plan

Standardized bid documents usually include standard forms for compliance with EEO, affirmative action, OSHA/EPA, minority hiring, and so on. A listing of qualified vendors appears in order to drive down the cost. Quite often, one vendor will not bid on the job because it knows that it cannot submit a lower bid than one of the other vendors. The cost of bidding on a job is an expensive process.

PMBOK® Guide, 6th Edition
12.2.2.3 Bidder Conferences

The solicitation package also describes the manner in which solicitation questions will be addressed, namely bidder conferences. Bidder conferences are used so that no single bidder has more knowledge than others. If a potential bidder has a question concerning the solicitation package, then it *must* wait for the bidders' conference to ask the question so that all bidders will be privileged to the same information. This is particularly important in government contracting. There may be several bidders' conferences between solicitation and award. Project management may or may not be involved in the bidders' conferences, either from the customer's side or the contractor's side. Some companies do not use bidder conferences and allow bidders to send in questions. However, the answer to each question is provided to all bidders.

The solicitation package usually provides bidders with information on how the bids will be evaluated. Contracts are not necessarily awarded to the lowest bidders. Some proposal evaluation scoring models assign points in regard to each of the following, and the company with the greatest number of points may be awarded the contract:

- Understanding of the requirements
- Overall bid price
- Technical superiority
- Management capability
- Previous performance (or references)
- Financial strength (ability to stay in business)
- Intellectual property rights
- Production capacity (based upon existing contracts and potential new contracts)

Bidder conferences are also held as part of debriefing sessions whereby the bidders are informed as to why they did not win the contract. Under some circumstances, bidders who feel that their bid or proposal was not evaluated correctly can submit a "bid protest," which may require a detailed reappraisal of their bid. The bid protest is not necessarily a complaint that the wrong company won the contract, but rather a complaint that their proposal was not evaluated correctly.

19.4 CONDUCT PROCUREMENTS: REQUEST SELLER RESPONSES

PMBOK® Guide, 6th Edition
12.2.2.2 Advertising
12.2.2.5 Negotiations

Selection of the acquisition method is the critical element to request seller responses. There are three common methods for acquisition:

- Advertising
- Negotiation
- Small purchases (i.e., office supplies)

Advertising is when a company goes out for sealed bids. There are no negotiations.

Competitive market forces determine the price and the award goes to the lowest bidder. Negotiation is when the price is determined through a bargaining process. In such a situation, the customer may go out for one of the following:

PMBOK® Guide, 6th Edition
12.2.1.4 Seller Proposals

- Request for information (RFI)
- Request for quotation (RFQ)

- Request for proposal (RFP)
- Invitation for bids (IFB)

The RFP is the most costly endeavor for the seller. Large proposals may contain separate volumes for cost, technical performance, management history, quality, facilities, subcontractor management, and others. Bidders may be hesitant to spend large sums of money bidding on a contract unless the bidder believes that they have a high probability of winning the contract or will be reimbursed by the buyer for all bidding costs.

Using the IFB process, only selected companies are allowed to bid. Either all or part of the companies on the buyer's preferred contractor list may be allowed to bid.

In government agencies, IFBs are used in sealed bidding procurements. In government sealed bid procurements, the competing offerors submit priced bids in response to IFBs. These IFBs contain all of the necessary technical documents, specifications, and drawings needed for a bidder to develop a priced offer. Thus, in sealed bid procurement, there are no discussions or negotiations, and the contract is always awarded to the lowest acceptable offer using a firm fixed-priced contract.

19.5 CONDUCT PROCUREMENTS: SELECT SELLERS

***PMBOK® Guide*, 6th Edition**
12.2.3.1 Select Sellers

Part of source selection process includes the application of the evaluation criteria to the contractor's proposals. The evaluation criteria reflect the selected contract award strategy, which is typically either a price-based award strategy or best-value award strategy. The priced-based award strategy is used when the contract will be awarded to the lowest-priced, technically acceptable proposal. The best-value award strategy is used when the contract may be awarded to either the lowest-priced, technically acceptable offer or a higher-priced proposal offering a higher level of performance. During a best-value source selection, the procuring organization conducts trade-offs among price, performance, and other nonprice factors to select the proposal that offers the overall best value to the buyer.

While several criteria can be used, the most common are time, cost, expected management team of the project (i.e., quality of assigned resources), and previous performance history. As an example, assume that 100 points maximum can be given to each of the four criteria. The seller that is selected would have the greatest number of total points out of 400 points. Weighing factors can also be applied to each of the four criteria. As an example, previous performance may be worth 200 points, thus giving 500 points as a maximum. Therefore, the lowest-price supplier may be downgraded significantly because of past performance and not receive the contract.

Selecting the appropriate seller is not necessarily left exclusively to the evaluation criteria. A negotiation process can be part of the selection process because the buyer may like several of the ideas among the many bidders and then may try to have the preferred seller take on added work at no additional cost to the buyer. The negotiation process also includes inclusion and exclusions. The negotiation process can be competitive or noncompetitive. Noncompetitive processes are called sole-source procurement.

On large contracts, the negotiation process goes well beyond negotiation of the bottom line. Separate negotiations can be made on:

- Final price of the contract
- Profit margins
- Type of contract
- Length of the contract
- Timing for each of the deliverables
- Quantity of deliverables
- Quality of the deliverables
- Payment schedule
- Assignment of critical personnel
- Ownership of the intellectual property
- Warrantees
- Cancellation and termination liability fees and conditions
- Number and frequency of reports
- Number, frequency, and location of customer-contractor interface meetings

Vendor relations are critical during contract negotiations. The integrity of the relationship and previous history can shorten the negotiation process. The three major factors of negotiations are:

- Compromise ability
- Adaptability
- Good faith

Negotiations should be planned for. A typical list of activities would include:

- Develop objectives (i.e., min-max positions)
- Evaluate your opponent
- Define your strategy and tactics
- Gather the facts
- Perform a complete price/cost analysis
- Arrange "hygiene" factors

If you are the buyer, what is the *maximum* you will be willing to pay? If you are the seller, what is the *minimum* you are willing to accept? You must determine what motivates your opponent. Is your opponent interested in profitability, keeping people employed, developing a new technology, or using your name as a reference? This knowledge could certainly affect your strategy and tactics.

Hygiene factors include where the negotiations will take place. In a restaurant? Hotel? Office? Square table or round table? Morning or afternoon? Who faces the windows and who faces the walls?

There should be a postnegotiation critique in order to review what was learned. The first type of postnegotiation critique is internal to your firm. The second type of postnegotiation critique is with all of the losing bidders to explain why they did not win the contract.

Once negotiations are completed, each selected seller will receive a signed contract. Unfortunately there are several types of contracts. The negotiation process also includes the selection of the type of contract, and the final type of contract may be different than what was identified in the solicitation package.

> ***Conclusion:*** The objective of the conduct procurements process is to negotiate a contract type and price that will result in reasonable contractor risk and provide the contractor with the greatest incentive for efficient and economic performance.

There are some basic contractual terms that should be understood before looking at the various contracts. These include:

- **Agent:** The person or group of people officially authorized to make decisions and represent their firm. This includes signing the contract.
- **Arbitration:** The settling of a dispute by a third party who renders a decision. The third party is not a court of law, and the decision may or may not be legally binding.
- **Breach of Contract:** To violate a law by an act or omission or to break a legal obligation.
- **Contract:** An agreement entered into by two or more parties and the agreement can be enforced in a court of law.
- **Executed Contract:** A contract that has been completed by all concerned parties.
- **Force Majeure Clause:** A provision in a contract that excuses the parties involved from any liability or contractual obligations because of acts of God, wars, terrorism, or other such events.
- **Good Faith:** Honesty and fair dealings between all parties involved in the contract.
- **Infringement:** A violation of someone's legally recognized right.
- **Liquidated Damages:** An amount specified in a contract that stipulates the reasonable estimation of damages that will occur as a result of a breach of contract.
- **Negligence:** The failure to exercise one's activity in such a manner that a reasonable person would do in a similar situation.
- **Noncompete Clause:** A covenant providing restrictions on starting up a competing business or working for a competitor within a specified time period.
- **Nondisclosure Clause:** A covenant providing restrictions on certain proprietary information such that it cannot be disclosed without written permission.
- **Nonconformance:** Performance of work in such a manner that it does not conform to contractual specifications or requirements.
- **Penalty Clause:** An agreement or covenant, identified in financial terms, for failure to perform.
- **Privity of Contract:** The relationship that exists between the buyer and seller of a contract.
- **Termination or Termination Liability:** An agreement between the buyer and seller as to how much money the seller will receive should the project be terminated prior to the scheduled completion date and without all of the contractual deliverable being completed.

- **Truth in Negotiations:** This clause in the contract states that both the buyer and seller have been truthful in the information provided during contract negotiations.
- **Waiver:** An intentional relinquishment of a legal right.
- **Warranty:** A promise, either verbal or written, that certain facts are true as represented.

There are certain basic elements of most contracts:

- **Mutual Agreement:** There must be an offer and acceptance.
- **Consideration:** There must be a down payment.
- **Contract Capability:** The contract is binding only if the contractor has the capability to perform the work.
- **Legal Purpose:** The contract must be for a legal purpose.
- **Form Provided by Law:** The contract must reflect the contractor's legal obligation, or lack of obligation, to deliver end products.

PMBOK® Guide, 6th Edition
12.2.3.1 Select Sellers
12.2.3.2 Agreements

The two most common contract forms are completion contracts and term contracts.

- **Completion Contract:** The contractor is required to deliver a definitive end product. Upon delivery and formal acceptance by the customer, the contract is considered complete, and final payment can be made.
- **Term Contract:** The contract is required to deliver a specific "level of effort," not an end product. The effort is expressed in woman/man-days (months or years) over a specific period of time using specified personnel skill levels and facilities. When the contracted effort is performed, the contractor is under no further obligation. Final payment is made, irrespective of what is actually accomplished technically.

The final contract is usually referred to as a *definitive* contract, which follows normal contracting procedures such as the negotiation of all contractual terms, condition cost, and schedule prior to initiation of performance. Unfortunately, negotiating the contract and preparing it for signatures may require months of preparation. If the customer needs the work to begin immediately or if long-lead procurement is necessary, then the customer may provide the contractor with a *letter contract* or *letter of intent.* The letter contract is a preliminary written instrument authorizing the contractor to begin immediately the manufacture of supplies or the performance of services. The final contract price *may* be negotiated after performance begins, but the contractor may not exceed the "not to exceed" face value of the contract. The definitive contract must still be negotiated.

The type of contract selected is based upon the following:

- Overall degree of cost and schedule risk
- Type and complexity of requirement (technical risk)
- Extent of price competition
- Cost/price analysis

- Urgency of the requirements
- Performance period
- Contractor's responsibility (and risk)
- Contractor's accounting system (is it capable of earned value reporting?)
- Concurrent contracts (will my contract take a back seat to existing work?)
- Extent of subcontracting (how much work will the contractor outsource?)

19.6 TYPES OF CONTRACTS

PMBOK® Guide, 6th Edition
12.2.3.2 Agreements

Before analyzing the various types of contracts, one should be familiar with the terminology found in them.

- The *target cost* or *estimated cost* is the level of cost that the contractor will most likely obtain under normal performance conditions. The target cost serves as a basis for measuring the true cost at the end of production or development. The target cost may vary for different types of contracts even though the contract objectives are the same. The target cost is the most important variable affecting research and development.
- *Target* or *expected profit* is the profit value that is negotiated for, and set forth, in the contract. The expected profit is usually the largest portion of the total profit.
- *Profit ceiling* and *profit floor* are the maximum and minimum values, respectively, of the total profit. These quantities are often included in contract negotiations.
- *Price ceiling* or *ceiling price* is the amount of money for which the government is responsible. It is usually measured as a given percentage of the target cost, and is generally greater than the target cost.
- *Maximum* and *minimum fees* are percentages of the target cost and establish the outside limits of the contractor's profit.
- The *sharing arrangement* or *formula* gives the cost responsibility of the customer to the cost responsibility of the contractor for each dollar spent. Whether that dollar is an overrun or an underrun dollar, the sharing arrangement has the same impact on the contractor. This sharing arrangement may vary depending on whether the contractor is operating above or below target costs. The *production point* is usually that level of production above which the sharing arrangement commences.
- *Point of total assumption* is the point (cost or price) where the contractor assumes all liability for additional costs.

Because no single form of contract agreement fits every situation or project, companies normally perform work in the United States under a wide variety of contractual arrangements, such as:

- Cost-plus percentage fee
- Cost-plus fixed fee
- Cost-plus guaranteed maximum
- Cost-plus guaranteed maximum and shared savings

- Cost-plus incentive (award fee)
- Cost and cost sharing
- Fixed price or lump sum
- Fixed price with redetermination
- Fixed price incentive fee
- Fixed price with economic price adjustment
- Fixed price incentive with successive targets
- Fixed price for services, material, and labor at cost (purchase orders, blanket agreements)
- Time and material/labor hours only
- Bonus-penalty
- Combinations
- Joint venture

At one end of the range is the *cost-plus,* a fixed-fee type of contract where the company's profit, rather than price, is fixed and the company's responsibility, except for its own negligence, is minimal. At the other end of the range is the *lump sum* or *turnkey* type of contract under which the company has assumed full responsibility, in the form of profit or losses, for timely performance and for all costs under or over the fixed contract price. In between are various types of contracts, such as the guaranteed maximum, incentive types of contracts, and the bonus-penalty type of contract. These contracts provide for varying degrees of cost responsibility and profit depending on the level of performance. Contracts that cover the furnishing of consulting services are generally on a per diem basis at one end of the range and on a fixed-price basis at the other end of the range.

There are generally five types of contracts to consider: fixed-price (FP), cost-plus-fixed-fee (CPFF) or cost-plus-percentage-fee (CPPF), guaranteed maximum-shared savings (GMSS), fixed-price-incentive-fee (FPIF), and cost-plus-incentive-fee (CPIF) contracts. Each type is discussed separately.

- Under a *fixed-price* or *lump-sum contract,* the contractor must carefully estimate the target cost. The contractor is required to perform the work at the negotiated contract value. If the estimated target cost was low, the total profit is reduced and may even vanish. The contractor may not be able to underbid the competitors if the expected cost is overestimated. Thus the contractor assumes a large risk.

  This contract provides maximum protection to the owner for the ultimate cost of the project, but has the disadvantage of requiring a long period for preparation and adjudications of bids. Also, there is the possibility that, because of a lack of knowledge of local conditions, all contractors may necessarily include an excessive amount of contingency. This form of contract should never be considered by the owner unless, at the time bid invitations are issued, the building requirements are known exactly. Changes requested by the owner after award of a contract on a lump sum basis lead to troublesome and sometimes costly extras.
- Traditionally, the *cost-plus-fixed-fee* contract has been employed when it was believed that accurate pricing could not be achieved any other way. In the CPFF contract, the cost may vary but the fee remains firm. Because, in a cost-plus contract, the contractor agrees only to use his best efforts to perform the work, good

performance and poor performance are, in effect, rewarded equally. The total dollar profit tends to produce low rates of return, reflecting the small amount of risk that the contractor assumes. The fixed fee is usually a small percentage of the total or true cost. The cost-plus contract requires that the company books be audited.

With this form of contract the engineering-construction contractor bids a fixed dollar fee or profit for the services to be supplied by the contractor, with engineering, materials, and field labor costs to be reimbursed at actual cost. This form of bid can be prepared quickly at a minimal expense to contractor and is a simple bid for the owner to evaluate. Additionally, it has the advantage of establishing incentive to the contractor for quick completion of the job.

If it is a *cost-plus-percentage-fee* contract, it provides maximum flexibility to the owner and permits owner and contractor to work together cooperatively on all technical, commercial, and financial problems. However, it does not provide financial assurance of ultimate cost. Higher building cost may result, although not necessarily so, because of lack of financial incentive to the contractor compared with other forms. The only meaningful incentive that is evident today is the increased competition and prospects for follow-on contracts.

- Under the *guaranteed maximum-share savings* contract, the contractor is paid a fixed fee for his profit and reimbursed for the actual cost of engineering, materials, construction labor, and all other job costs, but only up to the ceiling figure established as the "guaranteed maximum." Savings below the guaranteed maximum are shared between owner and contractor, whereas contractor assumes the responsibility for any overrun beyond the guaranteed maximum price.

  This contract form essentially combines the advantages as well as a few of the disadvantages of both lump sum and cost-plus contracts. This is the best form for a negotiated contract because it establishes a maximum price at the earliest possible date and protects the owner against being overcharged, even though the contract is awarded without competitive tenders. The guaranteed maximum-share savings contract is unique in that the owner and contractor share the financial risk and both have a real incentive to complete the project at lowest possible cost.
- *Fixed-price-incentive-fee* contracts are the same as fixed-price contracts except that they have a provision for adjustment of the total profit by a formula that depends on the final total cost at completion of the project and that has been agreed to in advance by both the owner and the contractor. To use this type of contract, the project or contract requirements must be firmly established. This contract provides an incentive to the contractor to reduce costs and therefore increase profit. Both the owner and contractor share in the risk and savings.
- *Cost-plus-incentive-fee* contracts are the same as cost-plus contracts except that they have a provision for adjustment of the fee as determined by a formula that compares the total project costs to the target cost. This formula is agreed to in advance by both the owner and contractor. This contract is usually used for long-duration or R&D-type projects. The company places more risk on the contractor and forces him to plan ahead carefully and strive to keep costs down. Incentive contracts are covered in greater detail in Section 19.7.
- Another type of contract incentives are *award fees.* Whereas incentive fees are *objectively* determined, that is, based on objective calculations comparing actual

cost to target costs, actual delivery to target delivery, or actual performance to target performance, award fees are more *subjectively* determined. Award fees are used when it is not feasible or effective to determine objective contract incentives. Award fees are earned when the contractor meets higher (over and above the basic requirements of the contract) levels of performance, quality, timeliness, or responsiveness in performing the contract effort. Award fee contracts include an award fee plan that explains the award fee evaluation criteria for any given time period (typically one year), as well as the total dollar amount of the award fee pool. Typically a contract award fee evaluation board convenes at the end of each award fee period to evaluate the contractor's performance in relation to the award fee criteria established in the award fee plan. The award fee determination official, either the project manager or a level above the project manager, makes the actual determination on the amount of award fee earned by the contractor for that specific period. Award fee provisions can be part of cost or fixed-priced contracts.

For major services contracts, award term incentives are used as incentives for the contractor to achieve higher levels of performance, quality, timeliness, or responsiveness in performing the services contract effort. Award term is similar to award fee, but instead of awarding the successful contractor additional dollars (fee), the contractor earns additional time (contract performance periods) on the service contract. Thus, instead of ending the final contract period of performance, and then having to recompete for the follow-on contract, the successful contractor is awarded with time extensions (additional periods of performance) to the contract performance period.

Other types of contracts that are not used frequently include:

- The *fixed-price incentive successive targets* contract is an infrequently used contract type. It has been used in the past in acquiring systems with very long lead time requirements where follow-on production contracts must be awarded before design or even production confirmation costs have been confirmed. Pricing data for the follow-on contract is inconclusive. This type of contract can be used in lieu of a letter contract or cost-plus arrangement.
- The *fixed-price with redetermination* contract can be either prospective or retroactive. The prospective type allows for future negotiations of two or more firm, fixed-price contracts at prearranged times. This is often used when future costs and pricing are expected to change significantly. The retroactive FPR contract allows for adjusting contract price after performance has been completed.
- *Cost* (CR) and *cost-sharing* (CS) contracts have limited use. Cost contracts have a "no fee" feature that has limited use except for nonprofit educational institutions conducting research. Cost-sharing contracts are used for basic and applied research where the contractor is expected to benefit from the R&D by transferring knowledge to other parts of the business for commercial gain and to improve the contractor's competitive position.

Table 19–1 identifies the advantages and disadvantages of various contracting methods that are commonly used.

TABLE 19–1. CONTRACT COMPARISON

Contract Type	Advantages	Disadvantages
Cost-plus-fee	• Provides maximum flexibility to owner • Minimizes contractor profits • Minimizes negotiations and preliminary specification costs • Permits quicker start, earlier completion • Permits choice of best-qualified, not lowest-bidding, contractor • Permits use of same contractor from consultation to completion, usually increasing quality and efficiency	• No assurance of actual final cost • No financial incentive to minimize time and cost • Permits specification of high-cost features by owner's staff • Permits excessive design changes by owner's staff increasing time and costs
Guaranteed maximum-share savings	• Provides firm assurance of ultimate cost at earliest possible date • Insures prompt advice to owner of delays and extra costs resulting from changes • Provides incentive for quickest completion • Owner and contractor share financial risk and have mutual incentive for possible savings • Ideal contract to establish owner–contractor cooperation throughout execution of project	• Requires complete auditing by owner's staff • Requires completion of definitive engineering before negotiation of contract
Fixed price/lump sum	• Provides firm assurance of ultimate cost • Insures prompt advice to owner of delays and extra costs resulting from changes • Requires minimum owner follow-up on work • Provides maximum incentive for quickest completion at lowest cost • Involves minimal auditing by owner's staff	• Requires exact knowledge of what is wanted before contract award • Requires substantial time and cost to develop inquiry specs, solicit, and evaluate bids; delays completion 3–4 months • High bidding costs and risks may reduce qualified bidders • Cost may be increased by excessive contingencies in bids to cover high-risk work
Fixed price for services, material, and labor	• Essentially same as cost-plus-fee contract • Fixes slightly higher percentage of total cost • Eliminates checking and verifying contractor's services	• May encourage reduction of economic studies and detailing of drawings: produce higher costs for operation, construction, maintenance • Other disadvantages same as cost-plus-fee contract
Fixed price for imported goods and services, local costs reimbursable	• Maximum price assured for high percentage of plant costs • Avoids excessive contingencies in bids for unpredictable and highly variable local costs • Permits selection of local suppliers and subcontractors by owner	• Same extended time required for inquiry specs, quotations, and evaluation as fixed lump-sum for complete project • Requires careful definition of items supplied locally to insure comparable bids • No financial incentive to minimize field and local costs

The type of contract that is acceptable to the client and the company is determined by the circumstances of each individual project and the prevailing economic and competitive conditions. Generally, when work is hard to find, clients insist on fixed-price bids. This type of proposal is usually a burden to the contractor because of the proposal costs involved (about 1 percent of the total installed cost of the project), and the higher risk involved in the execution of the project on such a basis.

When there is an upsurge in business, clients are unable to insist on fixed-price bids and more work is awarded on a cost-plus basis. In fact, where a special capacity position exists, or where time is a factor, the client occasionally negotiates a cost-plus contract with only one contractor. Another technique is to award a project on a cost-plus basis with the understanding that the contract will be converted at a later date, when the scope has been better defined and unknowns identified, to another form, such as a lump sum for services. This approach is appealing to both the client and the contractor.

19.7 INCENTIVE CONTRACTS

PMBOK® Guide, 6th Edition
12.2.3.2 Agreements

To alleviate some of the previously mentioned problem areas, clients, especially the government, have been placing incentive objectives into their contracts. The fixed-price-incentive-fee (FPIF) contract is an example of this. The essence of the incentive contract is that it offers a contractor more profit if costs are reduced or performance is improved and less profit if costs are raised or if performance goals are not met. Cost incentives take the form of a sharing formula generally expressed as a ratio. For example, if a 90/10 formula were negotiated, the government would pay for 90 cents and the contractor 10 cents for every dollar above the target cost. Thus it benefits both the contractor and the government to reduce costs, because the contractor must consider that 10 percent of every dollar must be spent by the company. Expected profits can thus be increased by making maximum use of the contractor's managerial skills.

In the FPIF contract, the contractor agrees to perform a service at a given fixed cost. If the total cost is less than the target cost, then the contractor has made a profit according to the incentive-fee formula. If the total cost exceeds the target cost, then the contractor loses money.

Consider the following example, which appears in Figure 19–1. The contractor has a target cost and target profit. However, there is a price ceiling of $11,500, which is the maximum price that the contractor will be paid. If the contractor performs the work below the target cost of $10,000, then additional profit will be made. For example, by performing the work for $9,000, the contractor will receive a profit of $1,150, which is the target profit of $850 plus $300 for 30% of the underrun. The contractor will receive a total price of $10,150.

If the cost exceeds the target cost, then the contractor must pay 30 percent of the overrun out of the contractor's profits. However, the fixed-price-incentive-fee (FPIF) contract

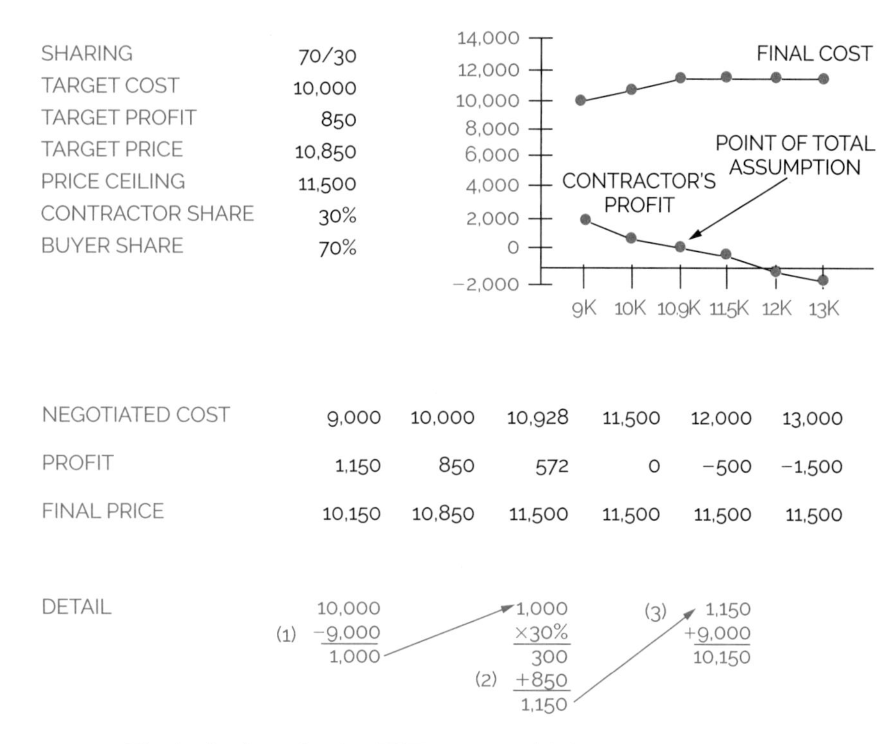

FIGURE 19–1. Fixed-price-incentive-fee (FPIF) contract with firm target.

has a point of total assumption. In this example, the point of total assumption is the point where all additional costs are burdened by the contractor. From Figure 19–1, the point of total assumption is when the cost reaches $10,928. At this point, the final price of $11,500 is reached. If the cost continues to increase, then all profits may disappear and the contractor may be forced to pay the majority of the overrun.

When the contract is completed, the contractor submits a statement of costs incurred in the performance of the contract. The costs are audited to determine allowability and questionable charges are removed. This determines the negotiated cost. The negotiated cost is then subtracted from the target cost. This number is then multiplied by the sharing ratio. If the number is positive, it is added to the target profit. If it is negative, it is subtracted. The new number, the final profit, is then added to the negotiated cost to determine the final price. The final price never exceeds the price ceiling.

Figure 19–2 shows a typical cost-plus-incentive-fee (CPIF) contract. In this contract, the contractor is reimbursed 100% of the costs. However, there is a maximum fee (i.e., profit) of $1,350 and a minimum fee of $300. The final allowable profit will vary between the minimum and maximum fee. Because there appears more financial risk for the customer in a CPIF contract, the target fee is usually less than in an FPIF contract, and the contractor's portion of the sharing ratio is smaller.

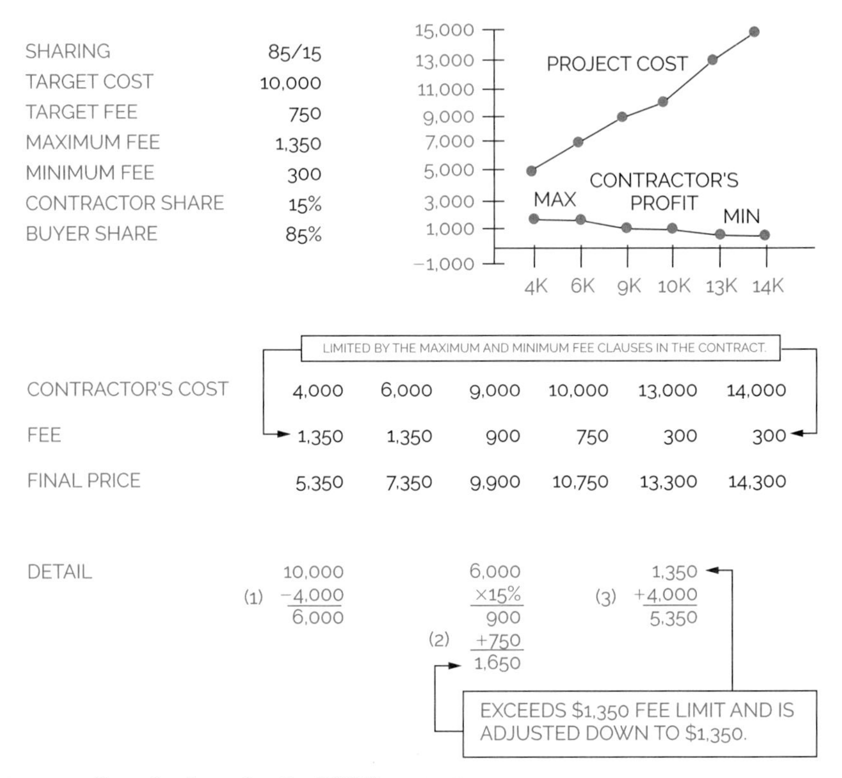

FIGURE 19–2. Cost-plus-incentive-fee (CPIF) contract.

19.8 CONTRACT TYPE VERSUS RISK

***PMBOK® Guide*, 6th Edition**
12.2.3.2 Agreements
12.2.2.4 Data Analysis

The amount of profit on a contract is most frequently based upon how the risks are to be shared between the contractor and the customer. For example, on a firm-fixed-price contract, the contractor absorbs 100 percent of the risks (especially financial) and expects to receive a larger profit than on other types of contracts. On cost, cost-plus, and cost-sharing contracts, the customer absorbs up to 100 percent of the risks and expects the contractor to work for a lower than expected profit margin or perhaps no profit at all.

All other types of contracts may have a risk sharing formula between the customer and the contractor. Figure 19–3 shows the relative degree of risk between the customer and the contractor for a variety of contracts.

19.9 CONTRACT ADMINISTRATION

***PMBOK® Guide*, 6th Edition**
12.3 Conduct Procurements
12.3.3.4 Change Requests

The contract administrator is responsible for compliance by the seller to the buyer's contractual terms and conditions and to make sure that the final product is fit for use. Contract administrators can shut down a manufacturing plant by allowing the seller to make late deliveries.

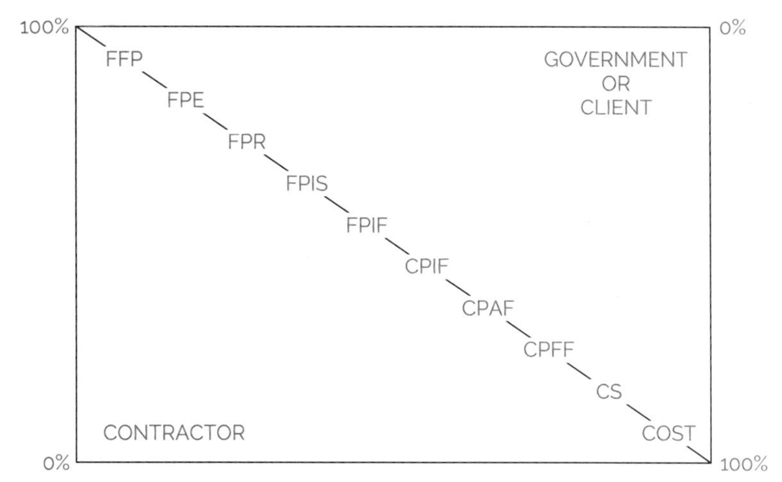

FIGURE 19–3. Contract types and risk types.

Although a contract administrator is a member of the project team for project reporting purposes (dotted line reporting), the contract administrator can report to a line function such as corporate legal and may even be an attorney. The functions of the corporate administrator include:

- Change management
- Specification interpretation
- Adherence to quality requirements
- Inspections and audits
- Warranties
- Performance reporting
- Records management
- Contractor (seller) management
- Contractor (seller) performance report card
- Documenting seller's performance (for future source selection teams)
- Production surveillance
- Approval of waivers
- Breach of contract
- Claims administration
- Resolution of disputes
- Payment schedules
- Project termination
- Project closure

The larger the contract, the greater the need for the contract administrator to resolve ambiguity in the contract. Sometimes, large contracts that are prepared by teams of attorneys contain an *order of precedence* clause. The order of precedence specifies that any inconsistency in the solicitation of the contract shall be resolved in a given order of procedure.

Perhaps the majority of the contract administrator's time is spent handling changes. The following definitions describe the types of changes:

- *Administrative change:* A unilateral contractual change, in writing, that does not affect the substantive rights of the parties (i.e., a change in the paying office or the appropriation funding).
- *Change order:* A written order, signed by the contracting officer, directing the contractor to make a change.
- *Contract modification:* Any written change in the terms of the contract.
- *Undefinitized contractual action:* Any contractual action that authorizes the commencement of work prior to the establishment of a final definitive price.
- *Supplemental agreement:* A contract modification that is accompanied by the mutual action of both parties.
- *Constructive change:* Any effective change to the contract caused by the actions or inaction of personnel in authority, or by circumstances that cause a contractor to perform work differently than required by written contract. The contractor may file a claim for equitable adjustment in the contract.

Based on the type of contract, terms, and conditions, the customer may have the right to terminate a contract for convenience at any time. However, the customer must compensate the contractor for his preparations and for any completed and accepted work relating to the terminated part of the contract.

The following are reasons for termination for convenience of the customer:

- Elimination of the requirement
- Technological advances in the state-of-the-art
- Budgetary changes
- Related requirements and/or procurements
- Anticipating profits not allowed

The following are reasons for termination for default due to contractor's actions:

- Contractor fails to make delivery on scheduled date.
- Contractor fails to make progress so as to endanger performance of the contract and its terms.
- Contractor fails to perform any other provisions of the contract.

If a contract is terminated due to default, then the contractor may not be entitled to compensation of work in progress but not yet accepted by the customer. The customer may even be entitled to repayment from the contractor of any advances or progress payments applicable to such work. Also, the contractor may be liable for any excess reprocurement costs. However, contractors can seek relief through negotiations, a Board of Contracts Appeals, or Claims Court.

The contract administrator is responsible for performance control. This includes inspection, acceptance, and breach of contract/default. If the goods/services do not comply with the contract, then the contract administrator has the right to:

- Reject the entire shipment
- Accept the entire shipment (barring latent defects)
- Accept part of the shipment

In government contracts, the government has the right to have the goods repaired with the costs charged back to the supplier or fix the goods themselves and charge the cost of repairs to the supplier. If the goods are then acceptable to the government, then the government may reduce the contract amount by an appropriate amount to reflect the reduced value of the contract.

Project managers often do financial closeout once the goods are shipped to the customer. This poses a problem if the goods must be repaired. Billing the cost of repairs against a financially closed out project is called *back-charging.* Most companies do not perform financial closeout until at least 90 days after delivery of goods.

19.10 CONTRACT CLOSURE

PMBOK® Guide, 6th Edition
12.3.3.1 Close Procurement

The contract administrator is responsible for verification that all of the work performed and deliverables produced are acceptable to the buyer. Contractual closure is then followed up with administrative closure, which includes:

- Documented verification that the output was accepted by the buyer
- Debriefing the seller on their overall performance
- Documenting seller's performance (documentation will be used in future source selections when evaluating contractor's past performance)
- Identifying room for improvement on future contracts
- Archiving all necessary project documentation
- Performing a lessons-learned review
- Identifying best practices

The seller also performs administrative closure once contractual closure is recognized. For the seller, an important subset of administrative closure is financial closure, which is the closing out of all open charge numbers. If financial closure occurs before contractual closure, then the project manager runs the risk that the charge numbers may have to be reopened to account for the cost of repairs or defects. Back-charging can be a monumental headache for the project manager, especially if the accounting group identified the unused money in the code of accounts as excess profits.

19.11 USING A CHECKLIST

To assist a company in evaluating inquiries and preparing proposals and contracts, a checklist of contract considerations and provisions can be helpful in the evaluation of each proposal and form of contract to ensure that appropriate safeguards are incorporated. This checklist is also used for sales letters and brochures that may promise or represent a commercial commitment. Its primary purpose is to remind users of the legal and commercial factors that should be considered in preparing proposals and contracts. Table 19–2 shows the typical major headings that would be considered in a checklist. A key word concept also provides an excellent checklist of the key issues to be considered. It will be useful as a reminder in preparation for contractor-client agreement discussions.

19.12 PROPOSAL-CONTRACTUAL INTERACTION

PMBOK® Guide, 6th Edition
12.1.3.1 Procurement Management Plan

It is critical during the proposal preparation stage that contract terms and conditions be reviewed and approved before submission of a proposal to the client. The contracts (legal) representative is responsible for the preparation of the contract portion of the proposal. Generally, contracts with the legal department are handled through or in coordination with the proposal group. The contract representative determines or assists with the following:

- Type of contract
- Required terms and conditions
- Any special requirements

TABLE 19–2. TYPICAL MAIN HEADINGS FOR A CONTRACT PROVISIONS CHECKLIST

I.	Definitions of contract terms
II.	Definition of project scope
III.	Scope of services and work to be performed
IV.	Facilities to be furnished by client (for service company use)
V.	Changes and extras
VI.	Warranties and guarantees
VII.	Compensation to service company
VIII.	Terms of payment
IX.	Definition of fee base (cost of the project)
X.	State sales and/or use taxes
XI.	Taxes (other than sales use taxes)
XII.	Insurance coverages
XIII.	Other contractual provisions (including certain general provisions)
XIV.	Miscellaneous general provisions

- Cash-flow requirements
- Patent and proprietary data
- Insurance and tax considerations
- Finance and accounting

The sales department, through the proposal group, has the final responsibility for the content and outcome of all proposals and contracts that it handles. However, there are certain aspects that should be reviewed with others who can offer guidance, advice, and assistance to facilitate the effort. In general, contract agreements should be reviewed by the following departments:

- Proposal
- Legal
- Insurance
- Tax
- Project management
- Engineering
- Estimating
- Construction (if required)
- Purchasing (if required)

Responsibility for collecting and editing contract comments rests with the proposal manager. In preparing contract comments, consideration should be given to comments previously submitted to the client for the same form of agreement, and also previous agreements signed with the client.

Contract comments should be reviewed for their substance and ultimate risk to the company. It must be recognized that in most instances, the client is not willing to make a large number of revisions to his proposed form of agreement. The burden of proof that a contract change is required rests with the company; therefore, each comment submitted must have a good case behind it.

Proposals should send all bid documents, including the client's form of contract, or equivalent information, along with the proposal outline or instructions to the legal department upon receipt of documents from the client. The instructions or outline should indicate the assignment of responsibility and include background information on matters that are pertinent to sales strategy or specific problems such as guarantees, previous experience with client, and so on.

Proposals should discuss briefly with the legal department what is planned by way of the project, the sales effort, and commercial considerations. If there is a kickoff meeting, a representative of the legal department should attend if it is appropriate. The legal department should make a preliminary review of the documents before any such discussion or meeting.

Normal practice is to validate proposals for a period of thirty to sixty days following date of submission. Validation of proposals for periods in excess of this period may be required by special circumstances and should be done only with management's concurrence. Occasionally, it is desirable to validate a bid for fewer than thirty days. The validity period is especially important on lump sum bids. On such bids, the validity period must be consistent with validity times of quotations received for major equipment items. If these are not consistent, additional escalation on equipment and materials may have to be included in the lump sum price, and the company's competitive position could thereby be jeopardized.

One area that is critical to the development of a good contract is the definition of the scope of work covered by the contract. This is of particular importance to the proposal manager, who is responsible for having the proper people prepared for the scope of work description. What is prepared during proposal production most likely governs the contract preparation and eventually becomes part of that contract. The degree to which the project scope of work must be described in a contract depends on the pricing mechanism and contract form used.

An inadequate or unrealistic description of the work to be undertaken or evaluation of the project requirements marks the beginning of an unhappy contract experience.

Related Case Studies (from Kerzner/*Project Management Case Studies*, 5th ed.)	**Related Workbook Exercises (from Kerzner/*Project Management Workbook and PMP®/CAPM® Exam Study Guide*, 12th ed.)**	***PMBOK Guide*, 6th Edition, Reference Section for the PMP® Certification Exam**
• The Scheduling Dilemma • To Bid or Not to Bid* • The Management Reserve*	• Multiple Choice Exam • Crossword Puzzle on Procurement Management	• Project Procurement Management

*Case study also appears at end of chapter.

19.13 STUDYING TIPS FOR THE PMI® PROJECT MANAGEMENT CERTIFICATION EXAM

This section is applicable as a review of the principles to support the knowledge areas and domain groups in the *PMBOK® Guide*. This chapter addresses:

- Project Procurement Management

Understanding the following principles is beneficial if the reader is using this text to study for the PMP® Certification Exam:

- What is meant by procurement planning
- What is meant by solicitation and a solicitation package
- Different types of contracts and relative degree of risk associated with each one
- Role of the contract administrator
- What is meant by contractual closure or closeout

PMP and CAPM are registered marks of the Project Management Institute, Inc.

The following multiple-choice questions will be helpful in reviewing the principles of this chapter:

1. The contractual statement-of-work document is:
   A. A nonbinding legal document used to identify the responsibilities of the contractor
   B. A definition of the contracted work for government contracts only
   C. A narrative description of the work/deliverables to be accomplished and/or the resource skills required
   D. A form of specification
2. A written or pictorial document that describes, defines, or specifies the services or items to be procured is:
   A. A specification document
   B. A Gantt chart
   C. A blueprint
   D. A risk management plan
3. The "order of precedence" is:
   A. The document that specifies the order (priority) in which project documents will be used when it becomes necessary to resolve inconsistencies between project documents
   B. The order in which project tasks should be completed
   C. The relationship that project tasks have to one another
   D. The ordered list (by quality) of the screened vendors for a project deliverable
4. In which type of contract arrangement is the contractor *least likely* to want to control costs?
   A. Cost plus percentage of cost
   B. Firm-fixed price
   C. Time and materials
   D. Purchase order
5. In which type of contract arrangement is the contractor *most likely* to want to control costs?
   A. Cost plus percentage of cost
   B. Firm-fixed price
   C. Time and materials
   D. Fixed-price-incentive-fee
6. In which type of contract arrangement is the *contractor* at the most risk of *absorbing all cost overruns?*
   A. Cost plus percentage of cost
   B. Firm-fixed price
   C. Time and materials
   D. Cost-plus-incentive-fee

7. In which type of contract arrangement is the *customer* at the most risk of absorbing excessive cost overruns?
   A. Cost plus percentage of cost
   B. Firm-fixed price
   C. Time and materials
   D. Fixed-price-incentive-fee
8. What is the primary objective the customer's project manager focuses on when selecting a contract type?
   A. Transferring all risk to the contractor
   B. Creating reasonable contractor risk with provisions for efficient and economical performance incentives for the contractor
   C. Retaining all project risk, thus reducing project contract costs
   D. None of the above
9. Which type of contract arrangement is specifically designed to give a contractor relief for inflation or material/labor cost increases on a long-term contract?
   A. Cost plus percentage of cost
   B. Firm-fixed price
   C. Time and materials
   D. Firm-fixed price with economic price adjustment
10. Which of the following is not a factor to consider when selecting a contract type?
   A. Type/complexity of the requirement
   B. Urgency of the requirement
   C. Extent of price competition
   D. All are factors to consider.
11. In a fixed-price-incentive-fee contract, the "point of total assumption" refers to the point in the project cost curve where:
   A. The customer assumes responsibility for every additional dollar that is spent in fulfillment of the contract.
   B. The contractor assumes responsibility for every additional dollar that is spent in fulfillment of the contract.
   C. The price ceiling is reached after the contractor recovers the target profit.
   D. None of the above
12. A written *preliminary* contractual instrument prepared prior to the issuance of a definitive contract that authorizes the contractor to begin work immediately, within certain limitations, is known as a:
   A. Definitive contract
   B. Preliminary contract
   C. Letter contract/letter of intent
   D. Purchase order

13. A contract entered into after following normal procedures (i.e., negotiation of terms, conditions, cost, and schedule) but prior to initiation of performance is known as a:
    A. Definitive contract
    B. Completed contract
    C. Letter contract/letter of intent
    D. Pricing arrangement
14. Which of the following is not a function of the contract administration activity?
    A. Contract change management
    B. Specification interpretation
    C. Determination of contract breach
    D. Selection of the project manager
15. A fixed-price contract is typically sought by the project manager from the customer's organization when:
    A. The risk and consequences associated with the contracted task are large and the customer wishes to transfer the risk.
    B. The project manager's company is proficient at dealing with the contracted activities.
    C. Neither the contractor nor the project manager understand the scope of the task.
    D. The project manager's company has excess production capacity.
16. Which of the following are typical actions a customer would take if the customer received nonconforming materials or products and the customer did not have the ability to bring the goods into conformance?
    A. Reject the entire shipment but pay the full cost of the contract
    B. Accept the entire shipment, no questions asked
    C. Accept the shipment on condition that the nonconforming products will be brought into conformance by the vendor at the vendor's expense.
    D. Accept the shipment and resell it to a competitor
17. If a project manager requires the use of a piece of equipment, what is the breakeven point where leasing and renting are the same?

Cost Categories	Renting Costs	Leasing Costs
Annual maintenance	$ 0.00	$3,000.00
Daily operation	$ 0.00	$ 70.00
Daily rental	$100.00	$ 0.00

    A. 300 days
    B. 30 days
    C. 100 days
    D. 700 days

18. In which type of incentive contract is there a maximum or minimum value established on the profits allowed for the contract?
    A. Cost-plus-incentive-fee contract
    B. Fixed-price-incentive-fee contract
    C. Time-and-material-incentive-fee contract
    D. Split-pricing-incentive-fee contract
19. In which type of incentive contract is there a maximum or minimum value established on the final price of the contract?
    A. Cost-plus-incentive-fee contract
    B. Fixed-price-incentive-fee contract
    C. Time-and-material-incentive-fee contract
    D. Split-pricing-incentive-fee contract
20. A cost-plus-incentive-fee contract has the following characteristics:

    Sharing ratio: 80/20

    Target cost: $100,000

    Target fee: $12,000

    Maximum fee: $14,000

    Minimum fee: $9,000

    How much will the contractor be reimbursed if the cost of performing the work is $95,000?
    A. $98,000
    B. $100,000
    C. $108,000
    D. $114,000
21. Using the same data from Problem 20, and the same contract type, how much will the contractor be reimbursed if the cost of performing the work is $85,000?
    A. $97,000
    B. B. $99,000
    C. C. $112,000
    D. $114,000
22. Using the same data from Problem 20, and the same contract type, how much will the contractor be reimbursed if the cost of performing the work is $120,000?
    A. $112,000
    B. B. $119,000
    C. C. $126,000
    D. $129,000
23. A fixed-price-incentive-fee contract has the following characteristics:

    Sharing ratio: 70/30

    Target cost: $100,000

    Target fee: $8,000

    Price ceiling: $110,000

How much will the contractor be reimbursed if the cost of performing the work is $90,000?

A. $91,000
B. $101,000
C. $103,000
D. $110,000

24. Using the same data from Problem 23, and the same contract type, how much will the contractor be reimbursed if the cost of performing the work is $102,000?

A. $104,000
B. B. $107,400
C. C. $109,400
D. $110,000

25. Using the same data from Problem 23, and the same contract type, how much will the contractor be reimbursed if the cost of performing the work is $105,000?

A. $105,000
B. B. $106,500
C. C. $110,000
D. $111,500

ANSWERS

1. C
2. A
3. A
4. A
5. B
6. B
7. A
8. B
9. D
10. D
11. B
12. C
13. A
14. D
15. A
16. C
17. C
18. A
19. B
20. C
21. B
22. D
23. B
24. C
25. C

PROBLEMS

19–1 What type of contract would be most suitable for an R&D contract when you expect the customer to request significant scope changes during the execution of the project?

19–2 During competitive bidding activities, why aren't the contractors allowed to have personal conversations with the seller's organization without informing other bidders about the discussion?

19–3 Why are companies reluctant to allow project managers to conduct their own procurement activities without going through a centralized procurement group?

19–4 How do companies handle competitive bidding activities when the statement of work in the proposal is vague?

19–5 A company is bidding on a contract that requires long lead procurement activities. Unfortunately, the buyer needs quite a bit of time to have the final contract ready for signatures. This will create a serious problem for the seller with because of the necessity to start long lead procurement quickly. How should the seller handle this situation?

CASE STUDIES

TO BID OR NOT TO BID

Background

Marvin was the president and chief executive officer (CEO) of his company. The decision of whether or not to bid on a job above a certain dollar value rested entirely upon his shoulders. In the past, his company would bid on all jobs that were a good fit with his company's strategic objectives and the company's win-to-loss ratio was excellent. But to bid on this job would be difficult. The client was requesting certain information in the request for proposal (RFP) that Marvin did not want to release. If Marvin did not comply with the requirements of the RFP, his company's bid would be considered nonresponsive.

Bidding Process

Marvin's company was highly successful at winning contracts through competitive bidding. The company was project-driven and all of the revenue that came into the company came through winning contracts. Almost all of the clients provided the company with long-term contracts as well as follow-on contracts. Almost all of the contracts were firm-fixed-price contracts. Business was certainly good, at least up until now. Marvin established a policy whereby 5 percent of sales would be used for responding to RFPs. This was referred to as a bid-and-proposal (B&P) budget. The cost for bidding on contracts was quite high and clients knew that requiring the company to spend a great deal of money bidding on a job might force a no-bid on the job. That could eventually hurt the industry by reducing the number of bidders in the marketplace. Marvin's company used parametric and analogy estimating on all contracts. This allowed Marvin's people to estimate the work at level 1 or level 2 of the work breakdown structure (WBS). From a financial perspective, this was the most cost-effective way to bid on a project knowing full well that there were risks with the accuracy of the estimates at these levels of the WBS. But over the years continuous improvements to the company's estimating process reduced much of the uncertainty in the estimates.

New RFP

One of Marvin's most important clients announced it would be going out for bids for a potential ten-year contract. This contract was larger than any other contract that Marvin's company had ever received and could provide an excellent cash flow stream for ten years or even longer. Winning the contract was essential. Because most of the previous contracts were firm-fixed-price, only summary-level pricing at the top two levels of the WBS was provided in the proposal. That was usually sufficient for the company's clients to evaluate the cost portion of the bid. The RFP was finally released. For this project, the contract type would be cost-reimbursable. A WBS created by the client was included in the RFP, and the WBS was broken down into five levels. Each bidder had to provide pricing information for each work package in the WBS. By doing this, the client could compare the cost of each work package from each bidder. The client would then be comparing apples and apples from each bidder rather than apples and oranges. To make matters worse, each bidder had to agree to use the WBS created by the client during project execution and to report costs according to the WBS.

Marvin saw the risks right away. If Marvin decided to bid on the job, the company would be releasing its detailed cost structure to the client. All costs would then be clearly exposed to the client. If Marvin were to bid on this project, releasing the detailed cost information could have a serious impact on future bids even if the contracts in the future were firm-fixed-price.

Marvin convened a team composed of his senior officers. During the discussions that followed, the team identified the pros and cons of bidding on the job:

Pros:

- A lucrative ten-year (or longer) contract
- The ability to have the client treat Marvin's company as a strategic partner rather than just a supplier
- Possibly lower profit margins on this and other future contracts but greater overall profits and earnings per share because of the larger business base
- Establishment of a workable standard for winning more large contracts

Cons:

- Release of the company's cost structure
- Risk that competitors will see the cost structure and hire away some of the company's talented people by offering them more pay
- Inability to compete on price and having entire cost structure exposed could be a limiting factor on future bids
- If the company does not bid on this job, the company could be removed from the client's bidder list
- Clients must force Marvin's company to accept lower profit margins

Marvin then asked the team, "Should we bid on the job?"

QUESTIONS

1. What other factors should Marvin and his team consider?
2. Should they bid on the job?

THE MANAGEMENT RESERVE

Background

A project sponsor forces the project management to include a management reserve in the cost of a project. However, the project sponsor intends to use the management reserve for his own "pet" project and this creates problems for the project manager.

Sole-Source Contract

The Structural Engineering Department at Avcon, Inc. made a breakthrough in the development of a high-quality, low-weight composite material. Avcon believed that the new material could be manufactured inexpensively and

Avcon's clients would benefit by lowering their manufacturing and shipping costs.

News of the breakthrough spread through the industry. Avcon was asked by one of its most important clients to submit an unsolicited proposal for design, development, and testing of products for the client using the new material. Jane would be the project manager. She had worked with the client previously as the project manager on several other projects that were considered successes.

Meeting with Tim

Because of the relative newness of the technology, both Avcon and the client understood that this could not be a firm-fixed-price contract. They ultimately agreed to a cost-plus-incentive-fee contract type. However, the target costs still had to be determined. Jane worked with all of the functional managers to determine what their efforts would be on this contract. The only unknown was the time and cost needed for structural testing. Structural testing would be done by the Structural Engineering Department, which was responsible for making the technical breakthrough. Tim was head of the Structural Engineering Department. Jane set up a meeting to discuss the cost of testing on this project. During the meeting, Tim replied:

> A full test matrix will cost about $100,000. I believe that we should price out the full test matrix and also include a management reserve of at least $100,000 should anything go wrong.

Jane was a little perplexed about adding in a management reserve. Tim was usually right on the money on his estimates and Jane knew from previous experience that a full test matrix might not be needed. But Tim was the subject matter expert and Jane reluctantly agreed to include in the contract a management reserve of $100,000. As Jane was about to exit Tim's office, Tim remarked:

> Jane, I had requested to be your project sponsor on this effort and management has given me the okay. You and I will be working together on this effort. As such, I would like to see all of the cost figures before submitting the final bid to the client.

Reviewing Cost Figures

Jane had worked with Tim before but not in a situation where Tim would be the project sponsor. However, it was common on some contracts that lower and middle levels of management would assume the sponsorship role rather than having all sponsorship at the top of the organization. Jane met with Tim and showed him the following information, which would appear in the proposal:

- Sharing ratio: 90–10%
- Contract cost target: $800,000
- Contract profit target: $50,000
- Management reserve: $100,000
- Profit ceiling: $70,000
- Profit floor: $35,000

Tim looked at the numbers and Jane could see that he was somewhat unhappy. Tim then stated:

> Jane, I do not want to identify to the client that we have a management reserve. Let's place the management reserve in with the $800,000 and change the target cost to $900,000. I

know that the cost baseline should not include the management reserve, but in this case I believe it is necessary to do so.

Jane knew that the cost baseline of a project does not include the management reserve, but there was nothing she could do; Tim was the sponsor and had the final say. Jane simply could not understand why Tim was trying to hide the management reserve.

Execution Begins

Tim instructed Jane to include in the structural test matrix work package the entire management reserve of $100,000. Jane knew from previous experience that a full test matrix was not required and that the typical cost of this work package should be between $75,000 and $90,000. Establishing a work package of $200,000 meant that Tim had complete control over the management reserve and how it would be used.

Jane was now convinced that Tim had a hidden agenda. Unsure what to do next, Jane contacted a colleague in the Project Management Office. The colleague informed Jane that Tim had tried unsuccessfully to get some of his pet projects included in the portfolio of projects, but management refused to include any of Tim's projects in the budget for the portfolio.

It was now clear what Tim was asking Jane to be part of and why Tim had requested to be the project sponsor. Tim was forcing Jane to violate PMI's Code of Ethics and Professional Conduct.

QUESTIONS

1. Why did Tim want to add in a management reserve?
2. Why did Tim want to become the project sponsor?
3. Are Tim's actions a violation of the Code of Ethics and Professional Conduct?
4. If Jane follows Tim instructions, is Jane also in violation of the Code of Ethics and Professional Conduct?
5. What are Jane's options if she decides not to follow Tim's instructions?

Quality Management

20.0 INTRODUCTION

> ***PMBOK® Guide*, 6th Edition**
> Chapter 8 Project Quality Management
> 8.1.1 Plan Quality Management Inputs

During the past twenty years, there has been a revolution in quality. Improvements have occurred not only in product quality, but also in leadership quality and project management quality. The changing views of quality appear in Table 20–1.

Unfortunately, it often takes an economic disaster, recession, or downturn in a firm's business base to get management to recognize the need for improved quality. Economic disasters provide some companies with the opportunity to become aggressive competitors in new markets. As an example, many high-tech engineering companies never fully recognized the need for shortening product development time and the relationship between project management, total quality management, and concurrent engineering until they saw their market share diminish.

The push for higher levels of quality appears to be customer driven. Customers are now demanding:

- Higher performance requirements
- Faster product development
- Higher technology levels
- Materials and processes pushed to the limit
- Lower contractor profit margins
- Fewer defects/rejects

One of the critical factors that can affect quality is market expectations. The variables that affect market expectations include:

- Salability: the balance between quality and cost
- Produceability: the ability to produce the product with available technology and workers, and at an acceptable cost

TABLE 20-1. CHANGING VIEWS OF QUALITY

Past	Present
• Quality is the responsibility of blue-collar workers and direct labor employees working on the floor. • Quality defects should be hidden from the customers (and possibly management). • Quality problems lead to blame, faulty justification, and excuses. • Corrections-to-quality problems should be accomplished with minimum documentation. • Increased quality will increase project costs. • Quality is internally focused. • Quality will not occur without close supervision of people. • Quality occurs during project execution	• Quality is everyone's responsibility, including white-collar workers, the indirect labor force, and the overhead staff. • Defects should be highlighted and brought to the surface for corrective action. • Quality problems lead to cooperative solutions. • Documentation is essential for "lessons learned" so that mistakes are not repeated. • Improved quality saves money and increases business. • Quality is customer focused. • People want to produce quality products. • Quality occurs at project initiation and must be planned for within the project

- Social acceptability: the degree of conflict between the product or process and the values of society (i.e., safety, environment)
- Operability: the degree to which a product can be operated safely
- Availability: the probability that the product, when used under given conditions, will perform satisfactorily when called upon
- Reliability: the probability of the product performing without failure under given conditions and for a set period of time
- Maintainability: the ability of the product to be retained in or restored to a performance level when prescribed maintenance is performed

Customer demands are now being handled using total quality management (TQM). Total quality management is an ever-improving system for integrating various organizational elements into the design, development, and manufacturing efforts, providing cost-effective products or services that are fully acceptable to the ultimate customer. Externally, TQM is customer oriented and provides for more meaningful customer satisfaction. Internally, TQM reduces production line bottlenecks and operating costs, thus enhancing product quality while improving organizational morale.

20.1 DEFINITION OF QUALITY

PMBOK® Guide, **6th Edition**
Chapter 8 Plan Quality Management
Introduction

Mature organizations readily admit that they cannot accurately define quality. The reason is that quality is defined by the customer. The Kodak definition of quality is those products and services that are perceived to meet or exceed the needs and expectations of the customer at a cost that represents out-standing value. The ISO 9000 definition is "the totality of feature and characteristics of a product or service that bears on its ability to satisfy stated or implied needs." Terms such as fitness for use, customer satisfaction, and zero defects are goals rather than definitions.

PMBOK® Guide, 6th Edition
8.1 Plan Quality Management

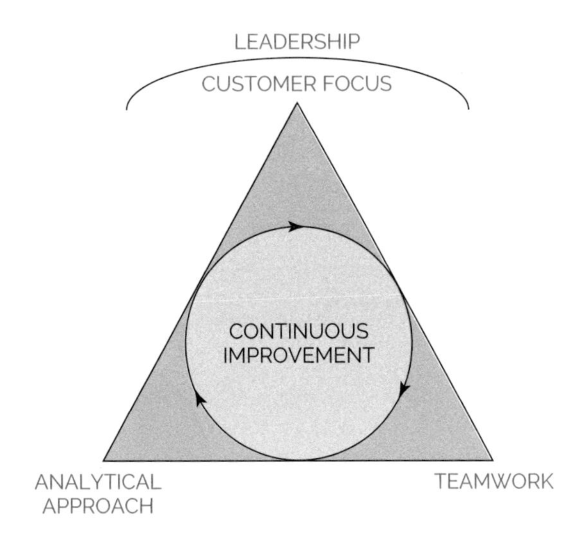

FIGURE 20–1 Kodak's five quality principles.

Most organizations view quality more as a process than a product. To be more specific, it is a continuously improving process where lessons learned are used to enhance future products and services in order to:

- Retain existing customers
- Win back lost customers
- Win new customers

Therefore, companies are developing quality improvement processes. Figure 20–1 shows the five quality principles that support Kodak's quality policy. Figure 20–2 shows a more detailed quality improvement process. These two figures seem to illustrate that organizations are placing more emphasis on the quality process than on the quality product and, therefore, are actively pursuing quality improvements through a continuous cycle.

20.2 THE QUALITY MOVEMENT

PMBOK® Guide, 6th Edition
Chapter 8 Project Quality Management Introduction

During the past hundred years, the views of quality have changed dramatically. Prior to World War I, quality was viewed predominantly as inspection, sorting out the good items from the bad. Emphasis was on problem identification. Following World War I and up to the early 1950s, emphasis was still on sorting good items from bad. However, *quality control* principles were now emerging in the form of:

- Statistical and mathematical techniques
- Sampling tables
- Process control charts

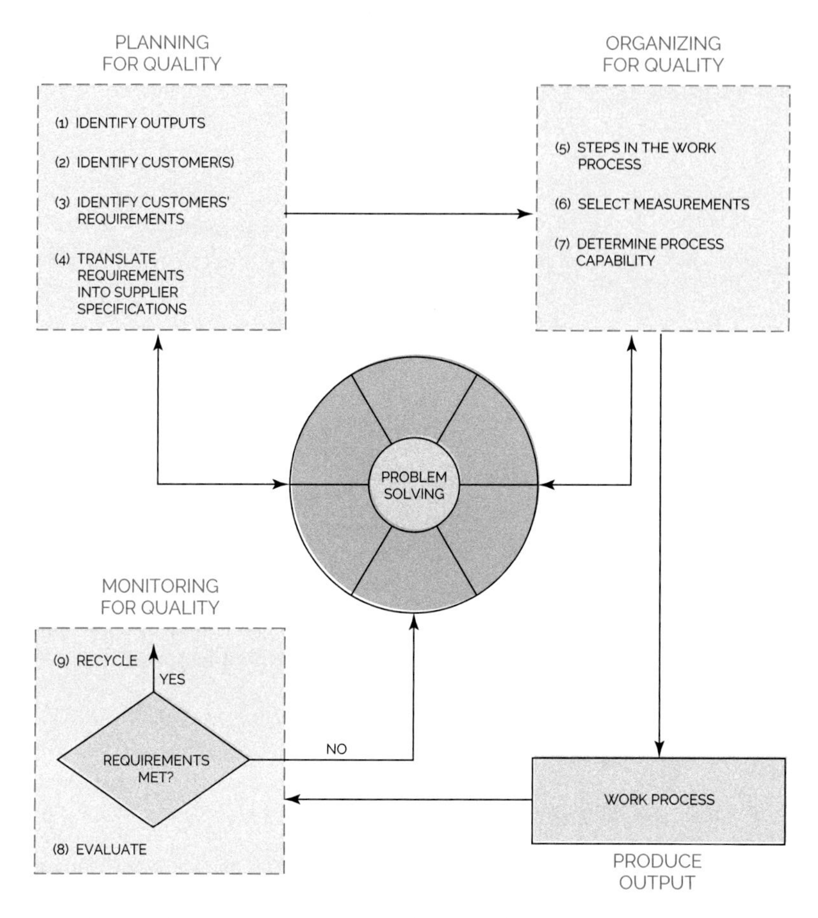

FIGURE 20–2 The quality improvement process. (Source unknown.)

From the early 1950s to the late 1960s, quality control evolved into quality assurance, with its emphasis on problem avoidance rather than problem detection. Additional quality assurance principles emerged, such as:

- The cost of quality
- Zero-defect programs
- Reliability engineering
- Total quality control

Today, emphasis is being placed on strategic quality management, including such topics as:

- Quality is defined by the customer.
- Quality is linked with profitability on both the market and cost sides.

PMBOK® Guide, 6th Edition
8.3.2 Quality Control Tools and Techniques

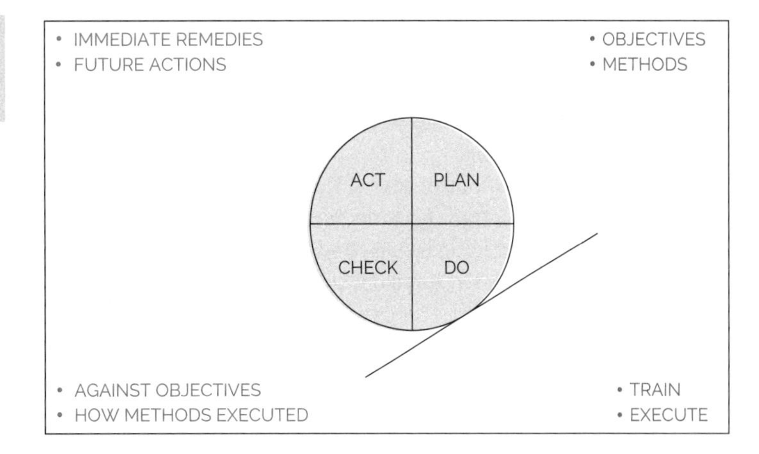

FIGURE 20-3. The Deming Cycle for Improvement.

- Quality has become a competitive weapon.
- Quality is now an integral part of the strategic planning process.
- Quality requires an organization-wide commitment.

Although many experts have contributed to the success of the quality movement, the three most influential contributors are W. Edwards Deming, Joseph M. Juran, and Phillip B. Crosby. Dr. Deming pioneered the use of statistics and sampling methods from 1927 to 1940 at the U.S. Department of Agriculture. During these early years, Dr. Deming was influenced by Dr. Shewhart, and later applied Shewhart's Plan/Do/Check/Act cycle to clerical tasks. Figure 20–3 shows the Deming Cycle for Improvement.

Deming believed that the reason companies were not producing quality products was that management was preoccupied with "today" rather than the future. Deming postulated that 85 percent of all quality problems required management to take the initiative and change the process. Only 15 percent of the quality problems could be controlled by the workers on the floor. As an example, the workers on the floor were not at fault because of the poor quality of raw materials that resulted from management's decision to seek out the lowest-cost suppliers. Management needed to change the purchasing policies and procedures and develop long-term relationships with vendors.

Processes had to be placed under statistical analysis and control to demonstrate the repeatability of quality. Furthermore, the ultimate goals should be a continuous refinement of the processes rather than quotas. Statistical process control charts (SPCs) allowed for the identification of common cause and special (assignable) cause variations. Common cause variations are inherent in any process. They include poor lots of raw material, poor product design, unsuitable work conditions, and equipment that cannot meet the design tolerances. These common causes are *beyond* the control of the workers on the floor and therefore, for improvement to occur, actions by management are necessary.

Special or assignable causes include lack of knowledge by workers, worker mistakes, or workers not paying attention during production. Special causes can be identified by workers on the shop floor and corrected, but management still needs to change the manufacturing process to reduce common cause variability.

Deming contended that workers simply cannot do their best. They had to be shown what constitutes acceptable quality and that continuous improvement is not only possible, but necessary. For this to be accomplished, workers had to be trained in the use of statistical process control charts. Realizing that even training required management's approval, Deming's lectures became more and more focused toward management and what they must do.

Dr. Juran began conducting quality control courses in Japan in 1954, four years after Dr. Deming. Dr. Juran developed his 10 Steps to Quality Improvement (see Table 20–2), as well as the Juran Trilogy: Quality Improvement, Quality Planning, and Quality Control. Juran stressed that the manufacturer's view of quality is adherence to specifications but the customer's view of quality is "fitness for use." Juran defined five attributes of "fitness for use."

- Quality of design: There may be many grades of quality
- Quality of conformance: Provide the proper training; products that maintain specification tolerances; motivation

TABLE 20–2 VARIOUS APPROACHES TO QUALITY IMPROVEMENT

Deming's 14 Points for Management	Juran's 10 Steps to Quality Improvement	Crosby's 14 Steps to Quality Improvement
1. Create constancy of purpose for improvement of product and service.	1. Build awareness of the need and opportunity for improvement.	1. Make it clear that management is committed to quality.
2. Adopt the new philosophy.	2. Set goals for improvement.	2. Form quality improvement teams with representatives from each department.
3. Cease dependence on inspection to achieve quality.	3. Organize to reach the goals (establish a quality council, identify problems, select projects, appoint teams, designate facilitators).	3. Determine where current and potential quality problems lie.
4. End the practice of awarding business on the basis of price tag alone. Instead, minimize total cost by working with a single supplier.	4. Provide training.	4. Evaluate the cost of quality and explain its use as a management tool.
5. Improve constantly and forever every process for planning, production, and service.	5. Carry out projects to solve problems.	5. Raise the quality awareness and personal concern of all employees.
6. Institute training on the job.	6. Report progress.	6. Take actions to correct problems identified through previous steps.
7. Adopt and institute leadership.	7. Give recognition.	7. Establish a committee for the zero-defects program.
8. Drive out fear.	8. Communicate results.	8. Train supervisors to actively carry out their part of the quality improvement program.
9. Break down barriers between staff areas.	9. Keep score.	9. Hold a "zero-defects day" to let all employees realize that there has been a change.
10. Eliminate slogans, exhortations, and targets for the work force.	10. Maintain momentum by making annual improvement part of the regular systems and processes of the company.	10. Encourage individuals to establish improvement goals for themselves and their groups.
11. Eliminate numerical quotas for the workforce and numerical goals for management.		11. Encourage employees to communicate to management the obstacles they face in attaining their improvement goals.
12. Remove barriers that rob people of workmanship. Eliminate the annual rating or merit system.		12. Recognize and appreciate those who participate.
13. Institute a vigorous program of education and self-improvement for everyone.		13. Establish quality councils to communicate on a regular basis.
14. 14. Put everybody in the company to work to accomplish the transformation.		14. Do it all over again to emphasize that the quality improvement program never ends.

- Availability: reliability (i.e., frequency of repairs) and maintainability (i.e., speed or ease of repair)
- Safety: The potential hazards of product use
- Field use: This refers to the way the product will be used by the customer

Dr. Juran also stressed the cost of quality (Section 20.3) and the legal implications of quality. The legal aspects of quality include:

- Criminal liability
- Civil liability
- Appropriate corporate actions
- Warranties

Juran believes that the contractor's view of quality is conformance to specification, whereas the customer's view of quality is fitness for use when delivered and value. Juran also admits that there can exist many grades of quality. The characteristics of quality can be defined as:

- Structural (length, frequency)
- Sensory (taste, beauty, appeal)
- Time-oriented (reliability, maintainability)
- Commercial (warrantee)
- Ethical (courtesy, honesty)

The third major contributor to quality was Phillip B. Crosby. Crosby developed his 14 Steps to Quality Improvement (see Table 20–2) and his Four Absolutes of Quality:

- Quality means conformance to requirements.
- Quality comes from prevention.
- Quality means that the performance standard is "zero defects."
- Quality is measured by the cost of nonconformance.

Crosby found that the cost of not doing things right the first time could be appreciable. In manufacturing, the price of nonconformance averages 40 percent of operating costs.

20.3 QUALITY MANAGEMENT CONCEPTS

***PMBOK® Guide*, 6th Edition**
Chapter 8 Project Procurement Management Introduction
8.1.1 Plan Quality Management Inputs

The project manager has the ultimate responsibility for quality management on the project. Quality management has equal priority with cost and Chapter 8 Introduction schedule management. However, the direct measurement of quality may be the responsibility of the quality assurance department or the assistant project manager for quality. For a labor-intensive project, management support (i.e., the project office) is typically 12–15 percent of the total labor dollars of the project. Approximately 3–5 percent can be attributed to quality management. Therefore, as much as 20–30 percent of all the labor in the project office could easily be attributed to quality management.

From a project manager's perspective, there are six quality management concepts that should exist to support each and every project:

1. Quality policy
2. Quality objectives
3. Quality assurance
4. Quality control
5. Quality audit
6. Quality program plan

Ideally, these six concepts should be embedded within the corporate culture.

Quality Policy

The quality policy is a document that is typically created by quality experts and fully supported by top management. The policy should state the quality objectives, the level of quality acceptable to the organization, and the responsibility of the organization's members for executing the policy and ensuring quality. A quality policy would also include statements by top management pledging its support to the policy. The quality policy is instrumental in creating the organization's reputation and quality image.

Many organizations successfully complete a good quality policy but immediately submarine the good intentions of the policy by delegating the implementation of the policy to lower-level managers. The implementation of the quality policy is the responsibility of top management. Top management must "walk the walk" as well as "talk the talk." Employees will soon see through the ruse of a quality policy that is delegated to middle managers while top executives move onto "more crucial matters that really impact the bottom line."

A good quality policy will:

- Be a statement of principles stating what, not how
- Promote consistency throughout the organization and across projects
- Provide an explanation to outsiders of how the organization views quality
- Provide specific guidelines for important quality matters
- Provide provisions for changing/updating the policy

Quality Objectives

Quality objectives are a part of an organization's quality policy and consist of specific objectives and the time frame for completing them. The quality objectives must be selected carefully. Selecting objectives that are not naturally possible can cause frustration and disillusionment. Examples of acceptable quality objectives might be: to train all members of the organization on the quality policy and objectives before the end of the current fiscal year, to set up baseline measurements of specific processes by the end of the current quarter, to define the responsibility and authority for meeting the organization's quality objectives down to each member of the organization by the end of the current fiscal year, and the like. Good quality objectives should:

- Be obtainable
- Define specific goals
- Be understandable
- State specific deadlines

Quality Assurance

PMBOK® Guide, 6th Edition
8.2 Manage Quality

Quality assurance is the collective term for the formal activities and managerial processes that attempt to ensure that products and services meet the required quality level. Quality assurance also includes efforts external to these processes that provide information for improving the internal processes. It is the quality assurance function that attempts to ensure that the project scope, cost, and time functions are fully integrated.

The Project Management Institute Guide to the Body of Knowledge (*PMBOK® Guide*) refers to quality assurance as the management section of quality management. This is the area where the project manager can have the greatest impact on the quality of his project. The project manager needs to establish the administrative processes and procedures necessary to ensure and, often, prove that the scope statement conforms to the actual requirements of the customer. The project manager must work with his team to determine which processes they will use to ensure that all stakeholders have confidence that the quality activities will be properly performed. All relevant legal and regulatory requirements must also be met.

- A good quality assurance system will:
- Identify objectives and standards
- Be multifunctional and prevention oriented
- Plan for collection and use of data in a cycle of continuous improvement
- Plan for the establishment and maintenance of performance measures
- Include quality audits

Quality Control

Quality control is a collective term for activities and techniques, within the process, that are intended to create specific quality characteristics. Such activities include continually monitoring processes, identifying and eliminating problem causes, use of statistical process control to reduce the variability and to increase the efficiency of processes. Quality control certifies that the organization's quality objectives are being met.

The *PMBOK® Guide* refers to quality control as the technical aspect of quality management. Project team members who have specific technical expertise on the various aspects of the project play an active role in quality control. They set up the technical processes and procedures that ensure that each step of the project provides a quality output from design and development through implementation and maintenance. Each step's output must conform to the overall quality standards and quality plans, thus ensuring that quality is achieved.

A good quality control system will:

- Select what to control
- Set standards that provide the basis for decisions regarding possible corrective action
- Establish the measurement methods used
- Compare the actual results to the quality standards
- Act to bring nonconforming processes and material back to the standard based on the information collected
- Monitor and calibrate measuring devices
- Include detailed documentation for all processes

Quality Audit

PMBOK® Guide, 6th Edition
8.2.2.5 Quality Audit

A quality audit is an independent evaluation performed by qualified personnel that ensures that the project is conforming to the project's quality requirements and is following the established quality procedures and policies.

A good quality audit will ensure that:

- The planned quality for the project will be met.
- The products are safe and fit for use.
- All pertinent laws and regulations are followed.
- Data collection and distribution systems are accurate and adequate.
- Proper corrective action is taken when required.
- Improvement opportunities are identified.

Quality Plan

PMBOK® Guide, 6th Edition
8.1.3.1 Quality Management Plan

The quality plan is created by the project manager and project team members by breaking down the project objectives into a work breakdown structure. Using a treelike diagramming technique, the project activities are broken down into lower-level activities until specific quality actions can be identified. The project manager then ensures that these actions are documented and implemented in the sequence that will meet the customer's requirements and expectations. This enables the project manager to assure the customer that he has a road map to delivering a quality product or service and therefore will satisfy the customer's needs.

A good quality plan will:

- Identify all of the organization's external and internal customers
- Cause the design of a process that produces the features desired by the customer
- Bring in suppliers early in the process
- Cause the organization to be responsive to changing customer needs
- Prove that the process is working and that quality goals are being met

20.4 THE COST OF QUALITY

PMBOK® Guide, 6th Edition
8.1.2.2 Data Gathering
8.1.2.3 Cost of Quality

To verify that a product or service meets the customer's requirements requires the measurement of the costs of quality. For simplicity's sake, the costs can be classified as "the cost of conformance" and "the cost of nonconformance." Conformance costs include items such as training, indoctrination, verification, validation, testing, maintenance, calibration, and audits. Nonconforming costs include items such as scrap, rework, warranty repairs, product recalls, and complaint handling.

Trying to save a few project dollars by reducing conformance costs could prove disastrous. For example, an American company won a contract as a supplier of Japanese parts. The initial contract called for the delivery of 10,000 parts. During inspection and testing at the customer's (i.e., Japanese) facility, two rejects were discovered. The Japanese returned *all* 10,000 components to the American supplier, stating that this batch was not acceptable. In this example, the nonconformance cost could easily be an order of magnitude greater than the conformance cost. The moral is clear: *Build it right the first time.*

Another common method to classify costs includes the following:

- *Prevention costs* are the up-front costs oriented toward the satisfaction of customer's requirements with the first and all succeeding units of product produced without defects. Included in this are typically such costs as design review, training, quality planning, surveys of vendors, suppliers, and subcontractors, process studies, and related preventive activities.
- *Appraisal costs* are costs associated with evaluation of product or process to ascertain how well all of the requirements of the customer have been met. Included in this are typically such costs as inspection of product, lab test, vendor control, in-process testing, and internal–external design reviews.
- *Internal failure costs* are those costs associated with the failure of the processes to make products acceptable to the customer, before leaving the control of the organization. Included in this area are scrap, rework, repair, downtime, defect evaluation, evaluation of scrap, and corrective actions for these internal failures.
- *External failure costs* are those costs associated with the determination by the customer that his requirements have not been satisfied. Included are customer returns and allowances, evaluation of customer complaints, inspection at the customer, and customer visits to resolve quality complaints and necessary corrective action.

Figure 20–4 shows the expected results of the total quality management system on quality costs. Prevention costs are expected to actually rise as more time is spent in prevention activities throughout the organization. As processes improve over the long run, appraisal costs will go down as the need to inspect in quality decreases. The biggest savings will come from the internal failure areas of rework, scrap, reengineering, redo, and so on. The additional time spent in up-front design and development will really pay off here. And, finally, the external costs will also come down as processes yield first-time quality on a regular basis. The improvements will continue to affect the company on a long-term basis in both improved quality and lower costs. Also, as project management matures, there should be further decreases in the cost of both maintaining quality and developing products.

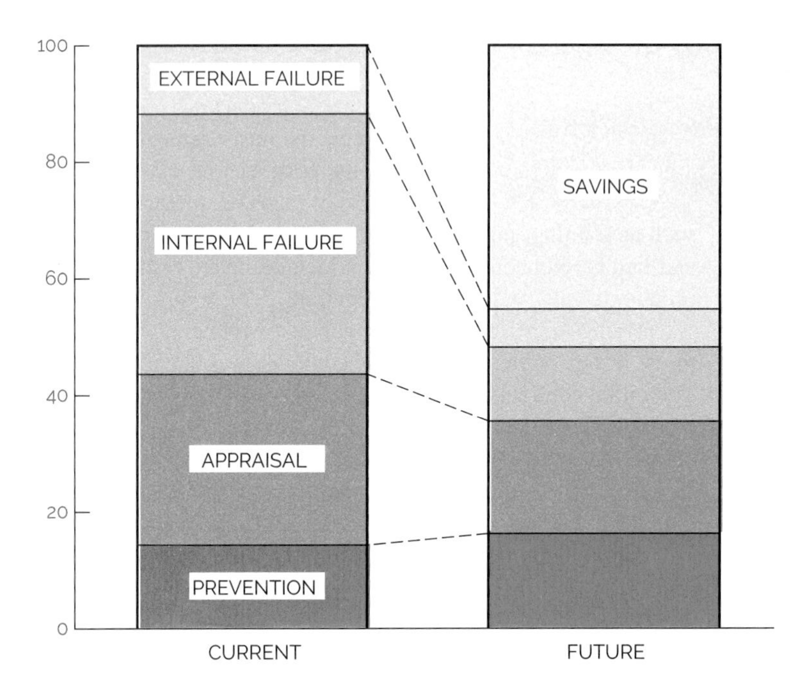

FIGURE 20-4. Total quality cost.

Figure 20–4 shows that prevention costs can increase. This is not always the case. Prevention costs actually decrease without sacrificing the purpose of prevention if we can identify and eliminate the costs associated with waste, such as waste due to

- Rejects of completed work
- Design flaws
- Work in progress
- Improperly instructed manpower
- Excess or noncontributing management (who still charge time to the project)
- Improperly assigned manpower
- Improper utilization of facilities
- Excessive expenses that do not necessarily contribute to the project (i.e., unnecessary meetings, travel, lodgings, etc.)

Another important aspect of Figure 20–4 is that 50 percent or more of the total cost of quality can be attributed to the internal and external failure costs. Complete elimination of failures may seem like an ideal solution but may not be cost-effective. As an example, see Figure 20–5. There are assumptions in the development of this figure. First, the cost of failure (i.e., nonconformance) approaches zero as defects become fewer and fewer. Second, the conformance costs of appraisal and prevention approach infinity as defects become fewer and fewer.

If the ultimate goal of a quality program is to continuously improve quality, then from a financial standpoint, quality improvement may not be advisable if the positive economic

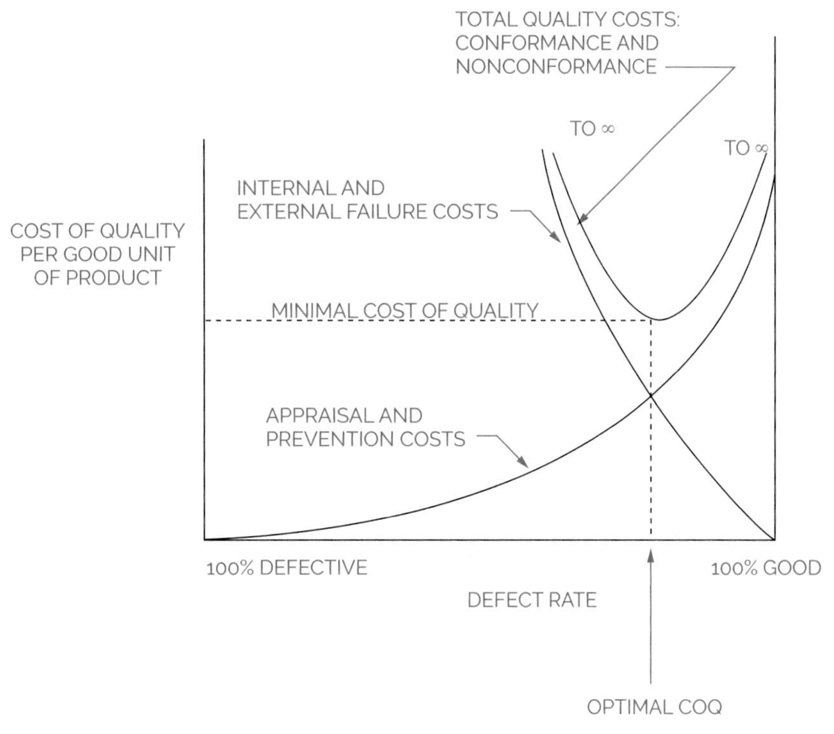

FIGURE 20-5. Minimizing the costs of quality (COQ).

return becomes negative. Juran argued that as long as the per-unit cost for prevention and appraisal were less expensive than nonconformance costs, resources should be assigned to prevention and appraisal. But when prevention and appraisal costs begin to increase the per-unit cost of quality, then the policy should be to maintain quality. The implication here is that zero defects may not be a practical solution since the total cost of quality would not be minimized.

Figure 20–4 shows that the external failure costs are much lower than the internal failure costs. This indicates that most of the failures are being discovered *before* they leave the functional areas or plants.

20.5 THE SEVEN QUALITY CONTROL TOOLS

> ***PMBOK® Guide*, 6th Edition**
> 8.3 Control Quality
> 8.3.2 Control Quality Tools and Techniques

Over the years, statistical methods have become prevalent throughout business, industry, and science. With the availability of advanced, automated systems that collect, tabulate, and analyze data, the practical application of these quantitative methods continues to grow.

This section is taken from H. K. Jackson and N. L. Frigon, *Achieving the Competitive Edge* (New York: John Wiley & Sons, 1996), Chapters 6 and 7.

More important than the quantitative methods themselves is their impact on the basic philosophy of business. The statistical point of view takes decision making out of the subjective autocratic decision-making arena by providing the basis for objective decisions based on quantifiable facts. This change provides some very specific benefits:

- Improved process information
- Better communication
- Discussion based on facts
- Consensus for action
- Information for process changes

Statistical process control (SPC) takes advantage of the natural characteristics of any process. All business activities can be described as specific processes with known tolerances and measurable variances. The measurement of these variances and the resulting information provide the basis for continuous process improvement. The tools presented here provide both a graphical and measured representation of process data. The systematic application of these tools empowers business people to control products and processes to become world-class competitors.

The basic tools of statistical process control are data figures, Pareto analysis, cause-and-effect analysis, trend analysis, histograms, scatter diagrams, and process control charts. These basic tools provide for the efficient collection of data, identification of patterns in the data, and measurement of variability. Figure 20–6 shows the relationships among these seven tools and their use for the identification and analysis of improvement opportunities. We will review these tools and discuss their implementation and applications.

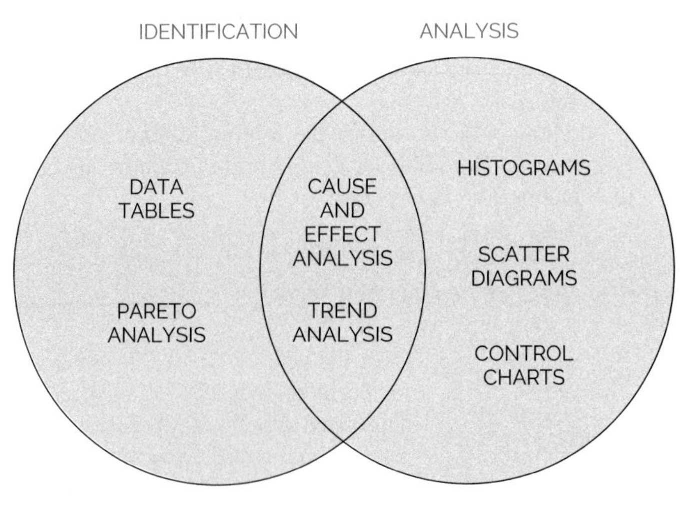

FIGURE 20-6. The seven quality control tools.

Data Tables

Data tables, or data arrays, provide a systematic method for collecting and displaying data. In most cases, data tables are forms designed for the purpose of collecting specific data. These tables are used most frequently where data are available from automated media. They provide a consistent, effective, and economical approach to gathering data, organizing them for analysis, and displaying them for preliminary review. Data tables sometimes take the form of manual check sheets where automated data are not necessary or available. Data figures and check sheets should be designed to minimize the need for complicated entries. Simple-to-understand, straightforward tables are a key to successful data gathering.

Figure 20–7 is an example of an attribute (pass/fail) data figure for the correctness of invoices. From this simple check sheet, several data points become apparent. The total number of defects is 34. The highest number of defects is from supplier A, and the most frequent defect is incorrect test documentation. We can subject these data to further analysis by using Pareto analysis, control charts, and other statistical tools.

Cause-and-Effect Analysis

After identifying a problem, it is necessary to determine its cause. The cause-and-effect relationship is at times obscure. A considerable amount of analysis often is required to determine the specific cause or causes of the problem. Cause-and-effect analysis uses diagramming techniques to identify the relationship between an effect and its causes. Cause-and-effect diagrams are also known as fishbone diagrams. Figure 20–8 demonstrates the basic fishbone diagram. Six steps are used to perform a cause-and-effect analysis.

Step 1. Identify the problem. This step often involves the use of other statistical process control tools, such as Pareto analysis, histograms, and control charts, as well as brainstorming. The result is a clear, concise problem statement.

Step 2. Select interdisciplinary brainstorming team. The team is selected based on the technical, analytical, and management knowledge required to determine the causes of the problem.

DEFECT	SUPPLIER				
	A	B	C	D	TOTAL
INCORRECT INVOICE	////	/		//	7
INCORRECT INVENTORY	/////	//	/	/	9
DAMAGED MATERIAL	///		//	///	8
INCORRECT TEST DOCUMENTATION	/	///	////	//	10
TOTAL	13	6	7	8	34

FIGURE 20-7. Check sheet for material receipt and inspection.

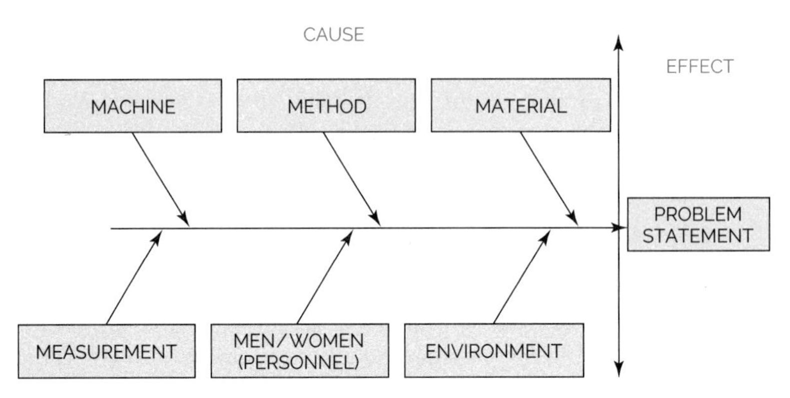

FIGURE 20-8. Cause-and-effect diagram.

Step 3. Draw problem box and prime arrow. The problem contains the problem statement being evaluated for cause and effect. The prime arrow functions as the foundation for their major categories.

Step 4. Specify major categories. Identify the major categories contributing to the problem stated in the problem box. The six basic categories for the primary causes of the problems are most frequently personnel, method, materials, machinery, measurements, and environment, as shown in Figure 20–8. Other categories may be specified, based on the needs of the analysis.

Step 5. Identify defect causes. When you have identified the major causes contributing to the problem, you can determine the causes related to each of the major categories. There are three approaches to this analysis: the random method, the systematic method, and the process analysis method.

Random method. List all six major causes contributing to the problem at the same time. Identify the possible causes related to each of the categories, as shown in Figure 20–9.

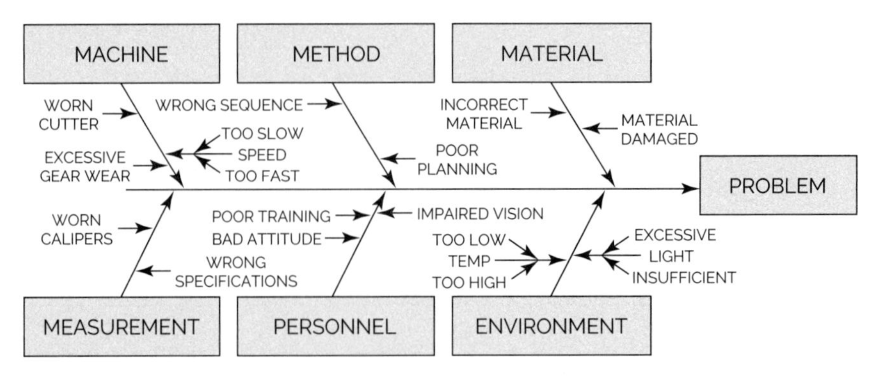

FIGURE 20-9. Random method.

Systematic method. Focus your analysis on one major category at a time, in descending order of importance. Move to the next most important category only after completing the most important one. This process is diagrammed in Figure 20–10.

Process analysis method. Identify each sequential step in the process and perform cause-and-effect analysis for each step, one at a time. Figure 20–11 represents this approach.

Step 6. Identify corrective action. Based on (1) the cause-and-effect analysis of the problem and (2) the determination of causes contributing to each major category, identify corrective action. The corrective action analysis is performed in the same manner as the cause-and-effect analysis. The cause-and-effect diagram is simply reversed so that the problem box becomes the corrective action box. Figure 20–12 displays the method for identifying corrective action.

Histogram

A histogram is a graphical representation of data as a frequency distribution. This tool is valuable in evaluating both attribute (pass/fail) and variable (measurement) data. Histograms offer a quick look at the data at a single point in time; they do not display variance or trends over time. A histogram displays how the cumulative data look *today.* It is useful in understanding the relative frequencies (percent-

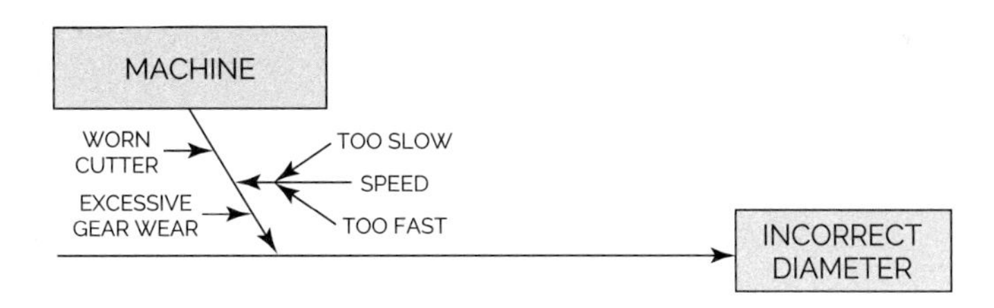

FIGURE 20-10. Systematic method.

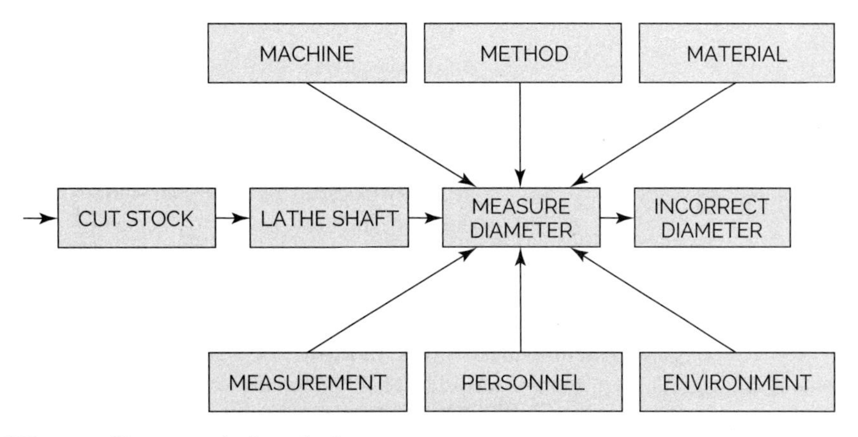

FIGURE 20-11. Process analysis method.

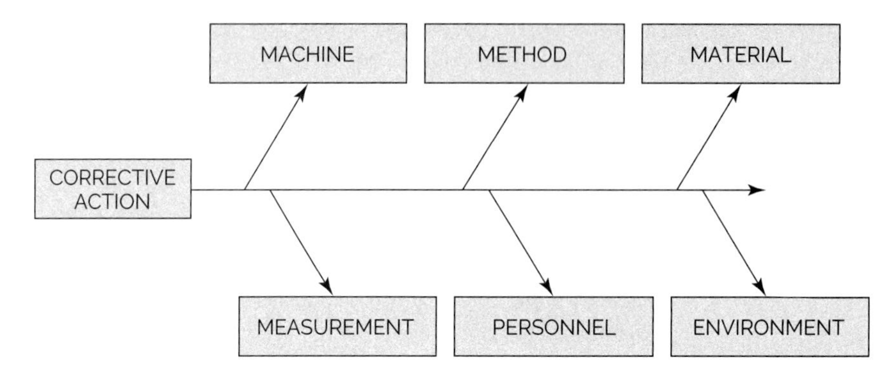

FIGURE 20-12. Identify corrective action.

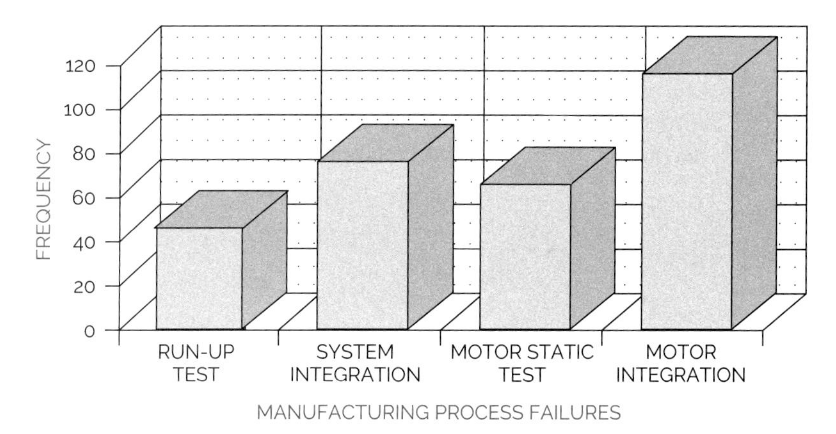

FIGURE 20-13. Histogram for variables.

ages) or frequency (numbers) of the data and how those data are distributed. Figure 20–13 illustrates a histogram of the frequency of defects in a manufacturing process.

Pareto Analysis

PMBOK® Guide, 6th Edition
8.3.2 Control Quality—Tools and Techniques

A Pareto diagram is a special type of histogram that helps us to identify and prioritize problem areas. The construction of a Pareto diagram may involve data collected from data figures, maintenance data, repair data, parts scrap rates, or other sources. By identifying types of nonconformity from any of these data sources, the Pareto diagram directs attention to the most frequently occurring element.

There are three uses and types of Pareto analysis. The basic Pareto analysis identifies the vital few contributors that account for most quality problems in any system. The comparative Pareto analysis focuses on any number of program options or actions. The weighted Pareto analysis gives a measure of significance to factors that may not appear significant at first—such additional factors as cost, time, and criticality.

The basic Pareto analysis chart provides an evaluation of the most frequent occurrences for any given data set. By applying the Pareto analysis steps to the material receipt and inspection process described in Figure 20–14, we can produce the basic Pareto analysis demonstrated in Figure 20–15. This basic Pareto analysis quantifies and graphs the frequency of occurrence for material receipt and inspection and further identifies the most significant, based on frequency.

A review of this basic Pareto analysis for frequency of occurrences indicates that supplier A is experiencing the most rejections with 38 percent of all the failures. Pareto analysis diagrams are also used to determine the effect of corrective action, or to analyze the difference between two or more processes and methods. Figure 20–16 displays the use of this Pareto method to assess the difference in defects after corrective action.

Another pictorial representation of process control data is the scatter plot or scatter diagram. A scatter diagram organizes data using two variables: an independent variable

MATERIAL RECEIPT AND INSPECTION FREQUENCY OF FAILURES			
SUPPLIER	FAILING FREQUENCY	PERCENT FAILING	CUMULATIVE PERCENT
A	13	38	38
B	6	17	55
C	7	20	75
D	9	25	100

FIGURE 20-14. Basic Pareto analysis.

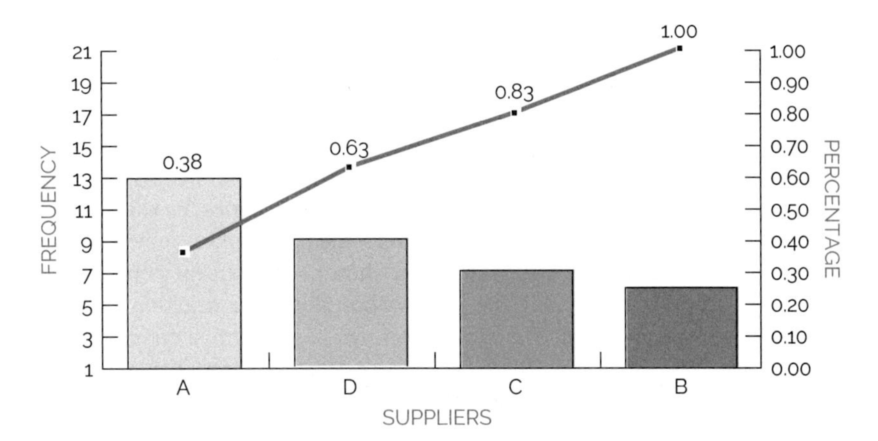

FIGURE 20-15. Basic Pareto analysis.

PMBOK® Guide, 6th Edition
8.3.2 Control Quality—Tools and Techniques

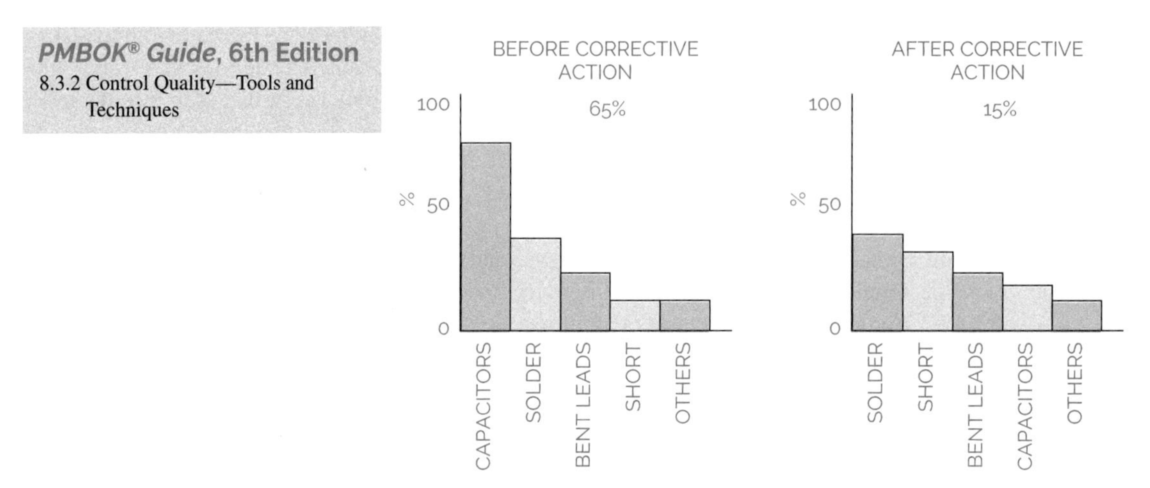

FIGURE 20-16. Comparative Pareto analysis.

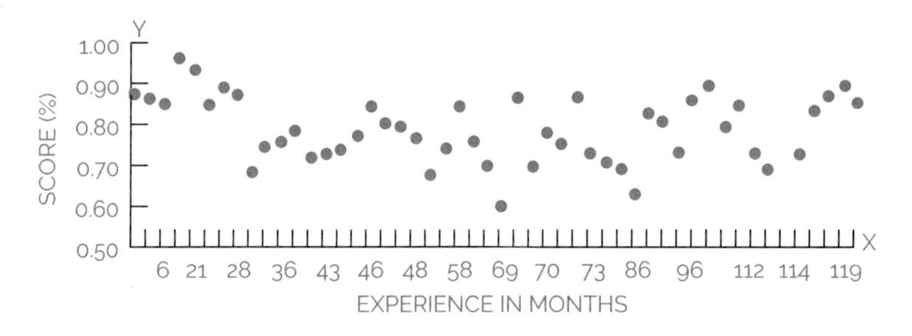

FIGURE 20-17. Solder certification test scores.

and a dependent variable. These data are then recorded on a simple graph with *X* and *Y* coordinates showing the relationship between the variables. Figure 20–17 displays the relationship between two of the data elements from solder qualification test scores. The independent variable, experience in months, is listed on the *X* axis. The dependent variable is the score, which is recorded on the *Y* axis.

These relationships fall into several categories, as shown in Figure 20–18. In the first scatter plot there is no correlation—the data points are widely scattered with no apparent pattern. The second scatter plot shows a curvilinear correlation demonstrated by the U shape of the graph. The third scatter plot has a negative correlation, as indicated by the downward slope. The final scatter plot has a positive correlation with an upward slope.

From Figure 20–17 we can see that the scatter plot for solder certification testing is somewhat curvilinear. The least and the most experienced employees scored highest, whereas those with an intermediate level of experience did relatively poorly. The next tool, trend analysis, will help clarify and quantify these relationships.

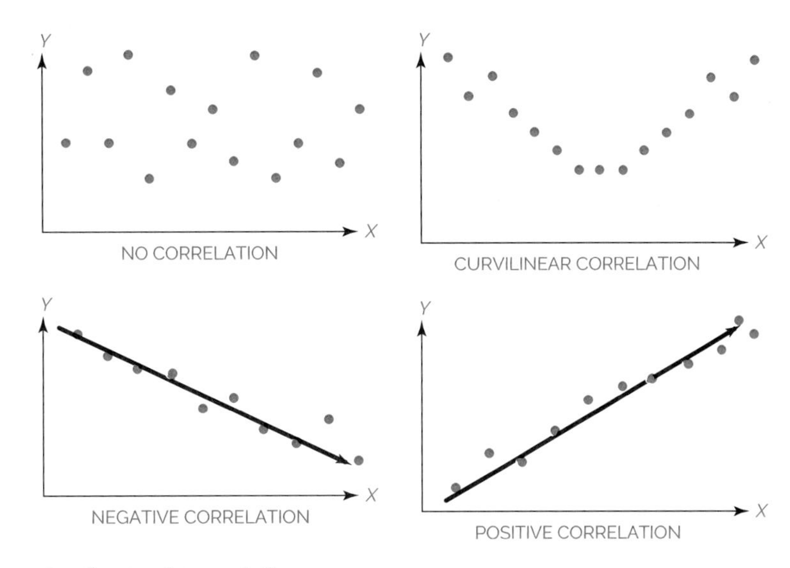

FIGURE 20-18. Scatter plot correlation.

Trend Analysis

PMBOK® Guide, 6th Edition
8.3.2 Control Quality—Tools and Techniques

Trend analysis is a statistical method for determining the equation that best fits the data in a scatter plot. Trend analysis quantifies the relationships of the data, determines the equation, and measures the fit of the equation to the data. This method is also known as curve fitting or least squares.

Trend analysis can determine optimal operating conditions by providing an equation that describes the relationship between the dependent (output) and independent (input) variables. An example is the data set concerning experience and scores on the solder certification test (see Figure 20–19).

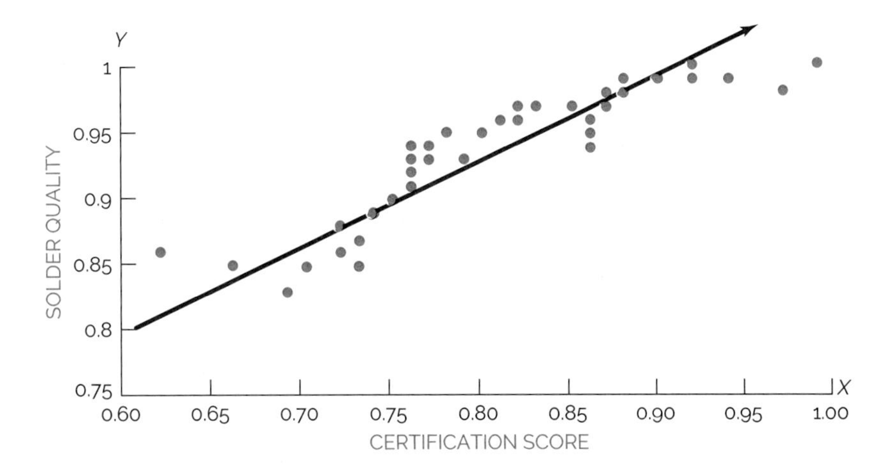

FIGURE 20-19. Scatter plot solder quality and certification score.

The equation of the regression line, or trend line, provides a clear and understandable measure of the change caused in the output variable by every incremental change of the input or independent variable. Using this principle, we can predict the effect of changes in the process.

One of the most important contributions that can be made by trend analysis is forecasting. Forecasting enables us to predict what is likely to occur in the future. Based on the regression line we can forecast what will happen as the independent variable attains values beyond the existing data.

Control Charts

PMBOK® Guide, 6th Edition
8.3.2 Control Quality—Tools and Techniques

The use of control charts focuses on the prevention of defects, rather than their detection and rejection. In business, government, and industry, economy and efficiency are always best served by prevention. It costs much more to produce an unsatisfactory product or service than it does to produce a satisfactory one. There are many costs associated with producing unsatisfactory goods and services. These costs are in labor, materials, facilities, and the loss of customers.

The cost of producing a proper product can be reduced significantly by the application of statistical process control charts.

Control Charts and the Normal Distribution

The construction, use, and interpretation of control charts is based on the normal statistical distribution as indicated in Figure 20–20. The centerline of the control chart represents the average or mean of the data ($\bar{X}$). The upper and lower control limits (UCL and LCL, respectively) represent this mean plus and minus three standard deviations of the data ($\bar{X} \pm 3s$). Either the lowercase s or the Greek letter σ (sigma) represents the standard deviation for control charts.

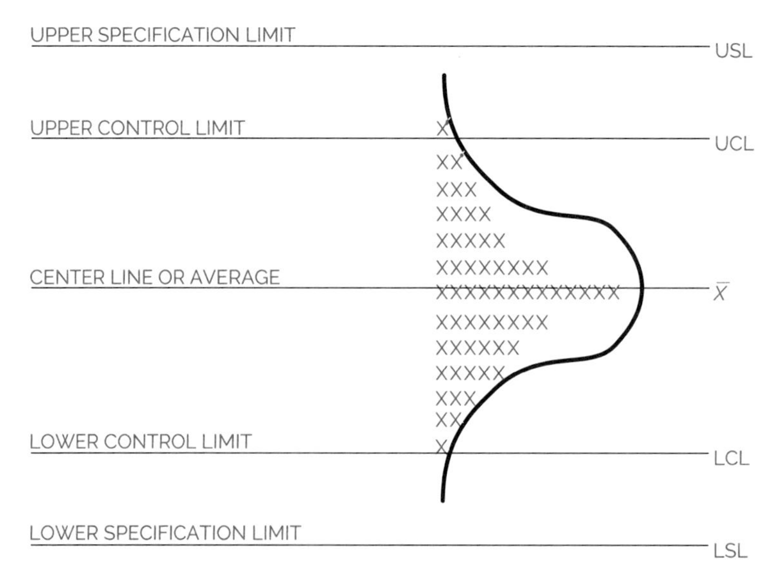

FIGURE 20-20– The control chart and the normal curve.

The normal distribution and its relationship to control charts is represented on the right of the figure. The normal distribution can be described entirely by its mean and standard deviation. The normal distribution is a bell-shaped curve (sometimes called the Gaussian distribution) that is symmetrical about the mean, slopes downward on both sides to infinity, and theoretically has an infinite range. In the normal distribution 99.73 percent of all measurements lie within $\bar{X} + 3s$ and $\bar{X} - 3s$; this is why the limits on control charts are called three-sigma limits.

Companies like Motorola have embarked upon a six-sigma limit rather than a three-sigma limit. The benefit is shown in Table 20–3. With a six-sigma limit, only two defects per billion are allowed. Maintaining a six-sigma limit can be extremely expensive unless the cost can be spread out over, say, 1 billion units produced.

Control chart analysis determines whether the inherent process variability and the process average are at stable levels, whether one or both are out of statistical control (not stable), or whether appropriate action needs to be taken. Another purpose of using control charts is to distinguish between the inherent, random variability of a process and the variability attributed to an assignable cause. The sources of random variability are often referred to as common causes. These are the sources that cannot be changed readily, without significant restructuring of the process. Special cause variability, by contrast, is subject to correction within the process under process control.

- *Common cause variability or variation:* This source of random variation is always present in any process. It is that part of the variability inherent in the process itself. The cause of this variation can be corrected only by a management decision to change the basic process.
- *Special cause variability or variation:* This variation can be controlled at the local or operational level. Special causes are indicated by a point on the control chart that is beyond the control limit or by a persistent trend approaching the control limit.

Control Chart Types

PMBOK® Guide, 6th Edition
8.3.2 Control Quality—Tools and Techniques

Just as there are two types of data, continuous and discrete, there are two types of control charts: variable charts for use with continuous data and attribute charts for use with discrete data. Each type of control chart can be used with specific types of data. Table 20–4 provides a brief overview of the types of control charts and their applications.

TABLE 20-3. ATTRIBUTES OF THE NORMAL (STANDARD) DISTRIBUTION

Specification Range (in ± Sigmas)	Percent within Range	Defective Parts per Billion
1	68.27	317,300,000
2	95.45	45,400,000
3	99.73	2,700,000
4	99.9937	63,000
5	99.999943	57
6	99.9999998	2

TABLE 20-4. TYPES OF CONTROL CHARTS AND APPLICATIONS

Variables Charts	Attributes Charts
$\overline{X}$ and R charts: To observe changes in the mean and range (variance) of a process.	p chart: For the fraction of attributes nonconforming or defective in a sample of varying size.
$\overline{X}$ and s charts: For a variable average and standard deviation.	np charts: For the number of attributes nonconforming or defective in a sample of constant size.
$\overline{X}$ and s^2 charts: for a variable average and variance.	c charts: For the number of attributes nonconforming or defects in a single item within a subgroup, lot, or sample area of constant size.
	u charts: For the number of attributes nonconforming or defects in a single item within a subgroup, lot, or sample area of varying size.

Variables Charts. Control charts for variables are powerful tools that we can use when measurements from a process are variable. Examples of variable data are the diameter of a bearing, electrical output, or the torque on a fastener.

As shown in Table 20–4, $\overline{X}$ and R charts are used to measure control processes whose characteristics are continuous variables such as weight, length, ohms, time, or volume. The p and np charts are used to measure and control processes displaying attribute characteristics in a sample. We use p charts when the number of failures is expressed as a fraction, or np charts when the failures are expressed as a number. The c and u charts are used to measure the number or portion of defects in a single item. The c control chart is applied when the sample size or area is fixed, and the u chart when the sample size or area is not fixed.

Attribute Charts. Although control charts are most often thought of in terms of variables, there are also versions for attributes. Attribute data have only two values (conforming/nonconforming, pass/fail, go/no-go, present/absent), but they can still be counted, recorded, and analyzed. Some examples are: the presence of a required label, the installation of all required fasteners, the presence of solder drips, or the continuity of an electrical circuit. We also use attribute charts for characteristics that are measurable, if the results are recorded in a simple yes/no fashion, such as the conformance of a shaft diameter when measured on a go/no-go gauge, or the acceptability of threshold margins to a visual or gauge check.

It is possible to use control charts for operations in which attributes are the basis for inspection, in a manner similar to that for variables but with certain differences. If we deal with the fraction rejected out of a sample, the type of control chart used is called a p chart. If we deal with the actual number rejected, the control chart is called an np chart. If articles can have more than one nonconformity, and all are counted for subgroups of fixed size, the control chart is called a c chart. Finally, if the number of nonconformities per unit is the quantity of interest, the control chart is called a u chart.

The power of control charts (Shewhart techniques) lies in their ability to determine if the cause of variation is a special cause that can be affected at the process level, or a common cause that requires a change at the management level. The information from the control chart can then be used to direct the efforts of engineers, technicians, and managers to achieve preventive or corrective action.

The use of statistical control charts is aimed at studying specific ongoing processes in order to keep them in satisfactory control. By contrast, downstream inspection aims to identify defects. In other words, control charts focus on prevention of defects rather than detection and rejection. It seems reasonable, and it has been confirmed in practice, that economy and efficiency are better served by prevention rather than detection.

Control Chart Interpretation

PMBOK® Guide, 6th Edition
8.3.2 Control Quality—Tools and Techniques

There are many possibilities for interpreting various kinds of patterns and shifts on control charts. If properly interpreted, a control chart can tell us much more than whether the process is in or out of control. Experience and training can help extract clues regarding process behavior. Statistical guidance is invaluable, but an intimate knowledge of the process being studied is vital in bringing about improvements. A control chart can tell us when to look for trouble, but it cannot by itself tell us where to look, or what cause will be found. Actually, in many cases, one of the greatest benefits from a control chart is that it tells when to leave a process alone. Sometimes the variability is increased unnecessarily when an operator keeps trying to make small corrections, rather than letting the natural range of variability stabilize.

20.6 ACCEPTANCE SAMPLING

PMBOK® Guide, 6th Edition
8.3.2 Control Quality—Tools and Techniques

Acceptance sampling is a statistical process of evaluating a portion of a lot for the purpose of accepting or rejecting the entire lot. It is an attempt to monitor the quality of the incoming product or material after the completion of production.

The alternatives to developing a sampling plan would be 100% inspection and 0% inspection. The costs associated with 100% are prohibitive, and the risks associated with 0% inspection are likewise large. Therefore, some sort of compromise is needed. The three most commonly used sampling plans are:

- *Single sampling:* This is the acceptance or rejection of a lot based upon one sampling run.
- *Double sampling:* A small sample size is tested. If the results are not conclusive, then a second sample is tested.
- *Multiple sampling:* This process requires the sampling of several small lots.

Regardless of what type of sampling plan is chosen, sampling errors can occur. A shipment of good-quality items can be rejected if a large portion of defective units are selected at random. Likewise, a bad-quality shipment can be accepted if the tested sample contains a disproportionately large number of quality items. The two major risks are:

- *Producer's risk:* This is called the α (alpha) risk or type I error. This is the risk to the producer that a good lot will be rejected.
- *Consumer's risk:* This is called the β (beta) risk or type II error. This is the consumer's risk of accepting a bad lot.

When a lot is tested for quality, we can look at either "attribute" or "variable" quality data. Attribute quality data are either quantitative or qualitative data for which the product or service is designed and built. Variable quality data are quantitative, continuous measurement processes to either accept or reject the lot. The exact measurement can be either destructive or nondestructive testing.

20.7 IMPLEMENTING SIX SIGMA

PMBOK® Guide, **6th Edition**
Chapter 8 Project Quality Management Introduction

Six Sigma is a business initiative first espoused by Motorola in the early 1990s. Recent Six Sigma success stories, primarily from the likes of General Electric, Sony, AlliedSignal, and Motorola, have captured the attention of Wall Street and have propagated the use of this business strategy. The Six Sigma strategy involves the use of statistical tools within a structured methodology for gaining the knowledge needed to create products and services better, faster, and less expensively than the competition. The repeated, disciplined application of the master strategy on project after project, where the projects are selected based on key business issues, is what drives dollars to the bottom line, resulting in increased profit margins and impressive return on investment from the Six Sigma training. The Six Sigma initiative has typically contributed an average of six figures per project to the bottom line. Ultimately, Six Sigma, if deployed properly, will infuse intellectual capital into a company and produce unprecedented knowledge gains that translate directly into bottom line results

Lean Six Sigma and DMAIC

Six Sigma is a quality initiative that was born at Motorola in the 1980s. The primary focus of the Six Sigma process improvement methodology, also known as DMAIC, is to reduce defects that are defined by the customer of the process. This customer can be internal or external. It is whoever is in receipt of the process output. Defects are removed by careful examination from a Six Sigma team made up of cross-functional positions having different lines of sight into the process. The team follows the rigor of the methodology of define, measure, analyze, improve, and control (DMAIC) to determine the root cause(s) of the defects. The team uses data and appropriate numerical and graphical analysis tools to raise awareness of process variables generating defects. Data collection and analysis is at the core of Six Sigma. "Extinction by instinct" is the phrase often used to describe intuitive decision making and performance analysis. It has been known to generate rework, frustration, and ineffective solutions. Six Sigma prescribes disciplined gathering and analysis of data to effectively identify solutions.

Lean manufacturing is another aspect of process improvement derived mostly from the Toyota Production System (TPS). The primary focus of lean is to remove waste and improve process efficiency. Lean is often linked with Six Sigma because both emphasize the importance of minimal process variation. Lean primarily consists of a set of tools

---

"Implementing Six Sigma" was adapted from Forrest W. Breyfogle, III, *Implementing Six Sigma* (New York: John Wiley & Sons, 1999), pp. 5–7; "Lean Six Sigma and DMAIC" was provided by Anne Foley, Director of Six Sigma, for the International Institute for Learning.

designed to assist in the identification and steady elimination of waste (muda), allowing for the improvement of quality as well as cycle time and cost reduction. To solve the problem of waste, lean manufacturing utilizes several tools. These include accelerated DMAIC projects known as kaizen events, cause-and-effect analysis using "five whys," and error proofing with a technique known as poka-yoke.

Kaizen Events. The source of the word *kaizen* is Japanese: *kai* (take apart) and *zen* (make good). This is an action-oriented approach to process improvement. Team members devote 3–5 consecutive days to quickly work through the DMAIC methodology in a workshop fashion.

Five Whys. This technique is used to move past symptoms of problems and drill down to the root causes. With every answer comes a new question until you've gotten to the bottom of the problem. Five is a rule of thumb. Sometimes you'll only need three questions, other times it might take seven. The goal is to identify the root cause of process defects and waste.

Poka-Yoke. The source of this technique is Japanese: *yokeru* (to avoid) and *poka* (inadvertent errors). There are three main principles of poka-yoke. (1) Make wrong actions more difficult. (2) Make mistakes obvious to the person so that the mistake can be corrected. (3) Detect errors so that downstream consequences can be prevented by stopping the flow or other corrective action. The philosophy behind this technique is that it's good to do things right the first time, but it is even better to make it impossible to do it wrong the first time.

When Six Sigma and lean manufacturing are integrated, the project team utilizes the project management methodology to lead them through the lean Six Sigma toolbox and make dramatic improvements to business processes. The overall goal is to reduce defects that impact the internal and external customer and eliminate waste that impact the cycle times and costs.

20.8 QUALITY LEADERSHIP

PMBOK® Guide, 6th Edition
Chapter 9 Project Resource Management

Consider for a moment the following seven items:

- Teamwork
- Strategic integration
- Continuous improvement
- Respect for people
- Customer focus
- Management-by-fact
- Structured problem solving

Some people contend that these seven items are the principles of project management when, in fact, they are the seven principles of the total quality management program at Sprint. Project management and TQM have close similarity in leadership and team-based decision making.

Section 20.8 adapted from Forrest W. Breyfogle, III, *Implementing Six Sigma* (New York: John Wiley & Sons, 1999), pp. 28–29.

Quality leadership emphasizes results by working on methods. In this type of management, every work process is studied and constantly improved so that the final product or service not only meets but exceeds customer expectations. The principles of quality leadership are customer focus, obsession with quality, effective work structure, control yet freedom (e.g., management in control of employees yet freedom given to employees), unity of purpose, process defect identification, teamwork, and education and training. These principles are more conducive to long-term thinking, correctly directed efforts, and a keen regard for the customer's interest.

To give quality leadership, the historical hierarchical management structure needs to be changed to a structure that has a more unified purpose using project teams. A single person can make a big difference in an organization. However, one person rarely has enough knowledge or experience to understand everything within a process. Major gains in both quality and productivity can often result when a team of people pool their skills, talents, and knowledge.

Teams need to have a systematic plan to improve the process that creates mistakes/defects, breakdowns/delays, inefficiencies, and variation. For a given work environment, management needs to create an atmosphere that supports team effort in all aspects of business. In some organizations, management may need to create a process that describes hierarchical relationships between teams, the flow of directives, how directives are transformed into action and improvements, and the degree of autonomy and responsibility of the teams. The change to quality leadership can be very difficult. It requires dedication and patience to transform an entire organization.

20.9 RESPONSIBILITY FOR QUALITY

Everyone in an organization plays an important role in quality management. In order for an organization to become a quality organization, all levels must actively participate, and, according to Dr. Edwards Deming, the key to successful implementation of quality starts at the top.

Top management must drive fear from the workplace and create an environment where cross-functional cooperation can flourish. The ultimate responsibility for quality in the organization lies in the hands of upper management. It is only with their enthusiastic and unwavering support that quality can thrive in an organization.

The project manager is ultimately responsible for the quality of the project. This is true for the same reason the president of the company is ultimately responsible for quality in a corporation. The project manager selects the procedures and policies for the project and therefore controls the quality. The project manager must create an environment that fosters trust and cooperation among the team members. The project manager must also support the identification and reporting of problems by team members and avoid at all costs a "shoot the messenger" mentality.

The project team members must be trained to identify problems, recommend solutions, and implement the solutions. They must also have the authority to limit further processing when a process is outside of specified limits. In other words, they must be able to halt any activity that is outside of the quality limits set for the project and work toward a resolution of the problem at any point in the project.

20.10 QUALITY CIRCLES

Quality circles are small groups of employees who meet frequently to help resolve company quality problems and provide recommendations to management. Quality circles were initially developed in Japan and have achieved some degree of success in the United States.

The employees involved in quality circles meet frequently either at someone's home or at the plant before the shift begins. The group identifies problems, analyzes data, recommends solutions, and carries out management-approved changes. The success of quality circles is heavily based upon management's willingness to listen to employee recommendations.

The key elements of quality circles include:

- They give a team effort.
- They are completely voluntary.
- Employees are trained in group dynamics, motivation, communications, and problem solving.
- Members rely upon each other for help.
- Management support is active but as needed.
- Creativity is encouraged.
- Management listens to recommendations.

The benefits of quality circles include:

- Improved quality of products and services
- Better organizational communications
- Improved worker performance
- Improved morale

20.11 TOTAL QUALITY MANAGEMENT (TQM)

***PMBOK® Guide*, 6th Edition**
8.1 Plan Quality Management

There is no explicit definition of total quality management. Some people define it as providing the customer with quality products at the right time and at the right place. Others define it as meeting or exceeding customer requirements. Internally, TQM can be defined as less variability in the quality of the product and less waste.

Section 20.11 has been adapted from C. Carl Pegels, *Total Quality Management* (Danvers, MA: Boyd & Fraser, 1995), pp. 4–27.

Figure 20–21 shows the basic objectives and focus areas of a TQM process. Almost all companies have a primary strategy to obtain TQM, and the selected strategy is usually in place over the long term. The most common primary strategies are:

- Solicit ideas for improvement from employees.
- Encourage and develop teams to identify and solve problems.

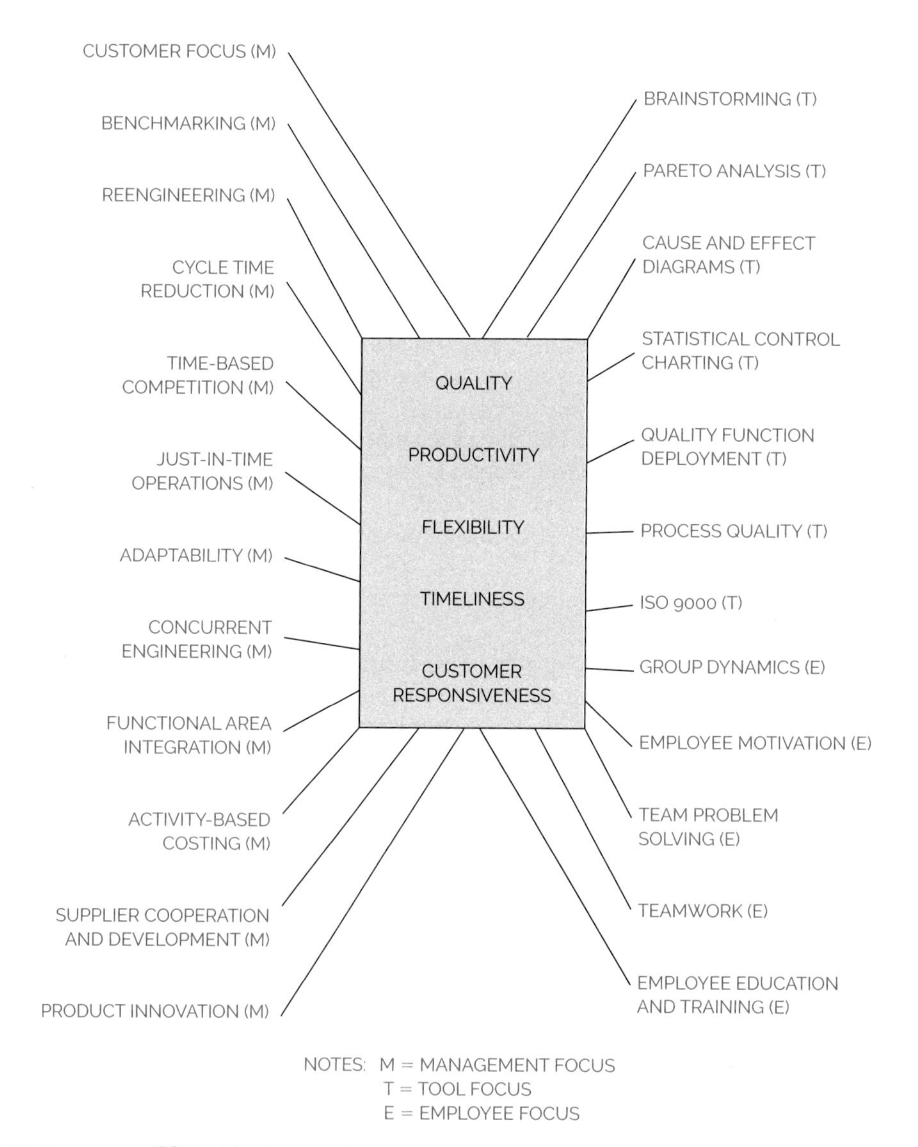

FIGURE 20-21. TQM objectives and focus areas.

Source: C. Carl Pegels, *Total Quality Management* (Danvers, MA: Boyd & Fraser, 1995), p. 6.

- Encourage team development for performing operations and service activities resulting in participative leadership.
- Benchmark every major activity in the organization to ensure that it is done in the most efficient and effective way.
- Utilize process management techniques to improve customer service and reduce cycle time.
- Develop and train customer staff to be entrepreneurial and innovative in order to find ways to improve customer service.
- Implement improvements so that the organization can qualify as an ISO 9000 supplier.

There also exist secondary strategies that, over the long run, focus on operations and profitability. Typical secondary strategies are:

- Maintain continuous contact with customers; understand and anticipate their needs.
- Develop loyal customers by not only pleasing them but by exceeding their expectations.
- Work closely with suppliers to improve their product/service quality and productivity.
- Utilize information and communication technology to improve customer service.
- Develop the organization into manageable and focused units in order to improve performance.
- Utilize concurrent or simultaneous engineering.
- Encourage, support, and develop employee training and education programs.
- Improve timeliness of all operation cycles (minimize all cycle times).
- Focus on quality, productivity, and profitability.
- Focus on quality, timeliness, and flexibility.

Related Case Studies (from Kerzner/*Project Management Case Studies*, 5th ed.)	**Related Workbook Exercises (from Kerzner/*Project Management Workbook and PMP®/CAPM® Exam Study Guide*, 12th ed.)**	***PMBOK® Guide*, 6th Edition, Reference Section for the PMP® Certification Exam**
None	• Constructing Process Charts • Constructing Cause-and-Effect • Charts and Pareto Charts • The Diagnosis of Patterns of • Process Instability, Part (A): X Charts • The Diagnosis of Patterns of Process Instability, Part (B): R Charts • Quality Circles • Quality Problems • Multiple Choice Exam • Crossword Puzzle on Quality Management	• Quality Management

PMP and CAPM are registered marks of the Project Management Institute, Inc.

20.12 STUDYING TIPS FOR THE PMI® PROJECT MANAGEMENT CERTIFICATION EXAM

This section is applicable as a review of the principles to support the knowledge areas and domain groups in the *PMBOK® Guide*. This chapter addresses:

- Project Quality Management

Understanding the following principles is beneficial if the reader is using this text to study for the PMP® Certification Exam:

- Contributions by the quality pioneers
- Concept of total quality management (TQM)
- Differences between quality planning, quality assurance, and quality control
- Importance of a quality audit
- Quality control tools
- Concept of cost of quality

The following multiple-choice questions will be helpful in reviewing the principles of this chapter:

1. Which of the following is not part of the generally accepted view of quality today?
   A. Defects should be highlighted and brought to the surface.
   B. We can inspect quality.
   C. Improved quality saves money and increases business.
   D. Quality is customer-focused.
2. In today's view of quality, who defines quality?
   A. Contractors' senior management
   B. Project management
   C. Workers
   D. Customers
3. Which of the following are tools of quality control?
   A. Sampling tables
   B. Process charts
   C. Statistical and mathematical techniques
   D. All of the above
4. Which of the following is true of modern quality management?
   A. Quality is defined by the customer.
   B. Quality has become a competitive weapon.
   C. Quality is now an integral part of strategic planning.
   D. All are true.

5. A company dedicated to quality usually provides training for:
   A. Senior management and project managers
   B. Hourly workers
   C. Salaried workers
   D. All employees
6. Which of the following quality gurus believe "zero-defects" is achievable?
   A. Deming
   B. Juran
   C. Crosby
   D. All of the above
7. What are the components of Juran's Trilogy?
   A. Quality Improvement, Quality Planning, and Quality Control
   B. Quality Improvement, Zero-Defects, and Quality Control
   C. Quality Improvement, Quality Planning, and Pert Charting
   D. Quality Improvement, Quality Inspections and Quality Control
8. Which of the following is not one of Crosby's Four Absolutes of Quality?
   A. Quality means conformance to requirements.
   B. Quality comes from prevention.
   C. Quality is measured by the cost of conformance.
   D. Quality means that the performance standard is "zero-defects."
9. According to Deming, what percentage of the costs of quality is generally attributable to management?
   A. 100%
   B. 85%
   C. 55%
   D. D 15%
10. Inspection:
   A. Is an appropriate way to ensure quality
   B. Is expensive and time-consuming
   C. Reduces rework and overall costs
   D. Is always effective in stopping defective products from reaching the customer
11. A well-written policy statement on quality will:
   A. Be a statement of how, not what or why
   B. Promote consistency throughout the organization and across projects
   C. Provide an explanation of how customers view quality in their own organizations
   D. Provide provisions for changing the policy only on a yearly basis
12. Quality assurance includes:
   A. Identifying objectives and standards
   B. Conducting quality audits

C. Planning for continuous collection of data

D. All of the above

13. What is the order of the four steps in Deming's Cycle for Continuous Improvement?

A. Plan, do, check, and act

B. Do, plan, act, and check

C. Check, do, act, and plan

D. Act, check, do, and plan

14. Quality audits:

A. Are unnecessary if you do it right the first time

B. Must be performed daily for each process

C. Are expensive and therefore not worth doing

D. Are necessary for validation that the quality policy is being followed and adhered to

15. Which of the following are typical tools of statistical process control?

A. Pareto analysis

B. Cause-and-effect analysis

C. Process control charts

D. All of the above

16. Which of the following methods is best suited to identifying the "vital few?"

A. Pareto analysis

B. Cause-and-effect analysis

C. Trend analysis

D. Process control charts

17. When a process is set up optimally, the upper and lower specification limits typically are:

A. Set equal to the upper and lower control limits

B. Set outside the upper and lower control limits

C. Set inside the upper and lower control limits

D. Set an equal distance from the mean value

18. The upper and lower control limits are typically set:

A. One standard deviation from the mean in each direction

B. 3 σ (three sigma) from the mean in each direction

C. Outside the upper and lower specification limits

D. To detect and flag when a process may be out of control

19. Which of the following is *not* indicative of today's views of the quality management process applied to a given project?

A. Defects should be highlighted and brought to the surface.

B. The ultimate responsibility for quality lies primarily with senior management or sponsor but everyone should be involved.

C. Quality saves money.

D. Problem identification leads to cooperative solutions.

20. If the values generated from a process are normally distributed around the mean value, what percentage of the data points generated by the process will *not* fall within plus or minus three standard deviations of the mean?
    A. 99.7%
    B. 95.4%
    C. 68.3%
    D. 0.3%

ANSWERS

1. B
2. D
3. D
4. D
5. D
6. C
7. A
8. C
9. B
10. B
11. B
12. D
13. D
14. D
15. D
16. A
17. B
18. B
19. B
20. D

PROBLEMS

20–1 Are all of the quality tools applicable to every project? If not, who decides which tools should be used?

20–2 Who has the ultimate responsibility for the quality of the project, and why?

20–3 You have just won a contract for an external client. Who determines the quality acceptance criteria?

20–4 On what type of projects, if any, can quality be viewed as inspection?

20–5 On what type of projects might it be necessary to perform 100% inspection rather than inspection sampling?

Modern Developments in Project Management

21.0 INTRODUCTION

As more industries accept project management as a way of life, the change in project management practices has taken place at an astounding rate. But what is even more important is the fact that these companies are sharing their accomplishments with other companies during benchmarking activities. Eight recent interest areas are included in this chapter:

- The project management maturity model (PMMM)
- Developing effective procedural documentation
- Project management methodologies
- Continuous improvement
- Capacity planning
- Competency models
- Managing multiple projects
- The business of scope changes
- End-of-phase review meetings

21.1 THE PROJECT MANAGEMENT MATURITY MODEL (PMMM)

All companies desire excellence in project management. Unfortunately, not all companies recognize that the time frame can be shortened by performing strategic planning for project management. The simple use of project management, even for an extended period of time, does *not* lead to excellence. Instead, it can result in repetitive mistakes and, what's worse, learning from your own mistakes rather than from the mistakes of others.

Strategic planning for project management is unlike other forms of strategic planning in that it is most often performed at the middle-management level, rather than by executive management. Executive management is still involved, mostly in a supporting role, and provides funding together with employee release time for the effort. Executive involvement will be necessary to make sure that whatever is recommended by middle management will not result in unwanted changes to the corporate culture.

Organizations tend to perform strategic planning for new products and services by laying out a well-thought-out plan and then executing the plan with the precision of a surgeon. Unfortunately, strategic planning for project management, if performed at all, is done on a trial-by-fire basis. However, there are models that can be used to assist corporations in performing strategic planning for project management and achieving maturity and excellence in a reasonable period of time.

The foundation for achieving excellence in project management can best be described as the project management maturity model (PMMM), which is composed of five levels, as shown in Figure 21–1. Each of the five levels represents a different degree of maturity in project management.

- *Level 1—Common Language:* In this level, the organization recognizes the importance of project management and the need for a good understanding of the basic knowledge on project management, along with the accompanying language/terminology.

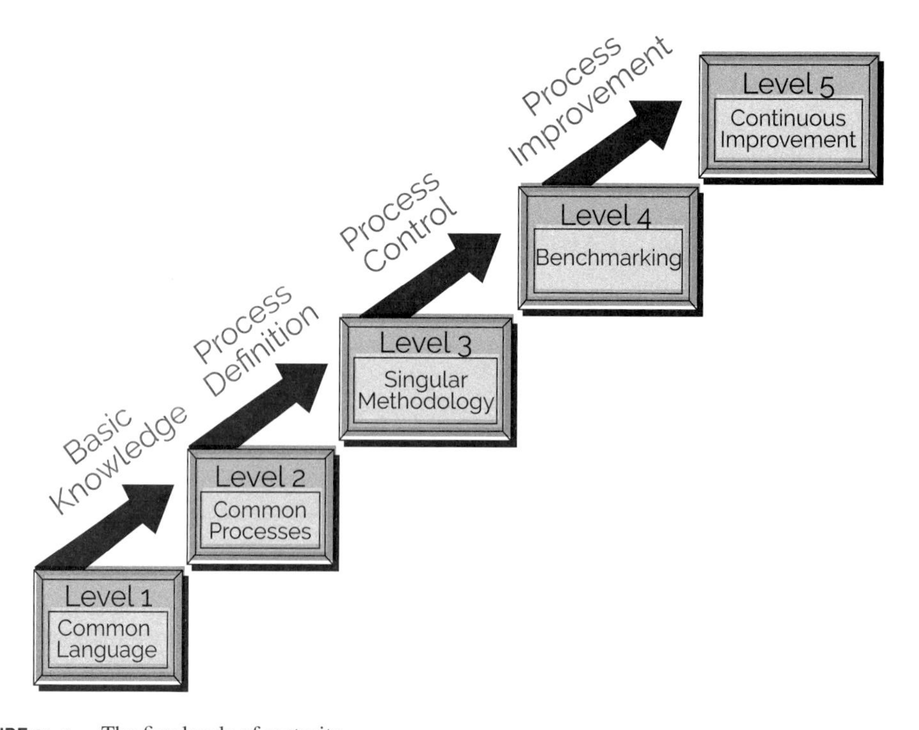

FIGURE 21–1. The five levels of maturity.

- *Level 2—Common Processes:* In this level, the organization recognizes that common processes need to be defined and developed such that successes on one project can be repeated on other projects. Also included in this level is the recognition that project management principles can be applied to and support other methodologies employed by the company.
- *Level 3—Singular Methodology:* In this level, the organization recognizes the synergistic effect of combining all corporate methodologies into a singular methodology, the center of which is project management. The synergistic effects also make process control easier with a single methodology than with multiple methodologies.
- *Level 4—Benchmarking:* This level contains the recognition that process improvement is necessary to maintain a competitive advantage. Benchmarking must be performed on a continuous basis. The company must decide whom to benchmark and what to benchmark.
- *Level 5—Continuous Improvement:* In this level, the organization evaluates the information obtained through benchmarking and must then decide whether or not this information will enhance the singular methodology.

When we talk about levels of maturity (and even life-cycle phases), there exists a common misbelief that all work must be accomplished sequentially (i.e., in series). This is not necessarily true. Certain levels can and do overlap. The magnitude of the overlap is based upon the amount of risk the organization is willing to tolerate. For example, a company can begin the development of project management checklists to support the methodology while it is still providing project management training for the workforce. A company can create a center for excellence in project management before benchmarking is undertaken.

Although overlapping does occur, the order in which the phases are completed cannot change. For example, even though Level 1 and Level 2 can overlap, Level 1 *must* still be completed before Level 2 can be completed. Overlapping of several of the levels can take place, as shown in Figure 21–2.

- *Overlap of Level 1 and Level 2:* This overlap will occur because the organization can begin the development of project management processes either while refinements are being made to the common language or during training.
- *Overlap of Level 3 and Level 4:* This overlap occurs because, while the organization is developing a singular methodology, plans are being made as to the process for improving the methodology.
- *Overlap of Level 4 and Level 5:* As the organization becomes more and more committed to benchmarking and continuous improvement, the speed by which the organization wants changes to be made can cause these two levels to have significant overlap. The feedback from Level 5 back to Level 4 and Level 3, as shown in Figure 21–3, implies that these three levels form a continuous improvement cycle, and it may even be possible for all three of these levels to overlap.

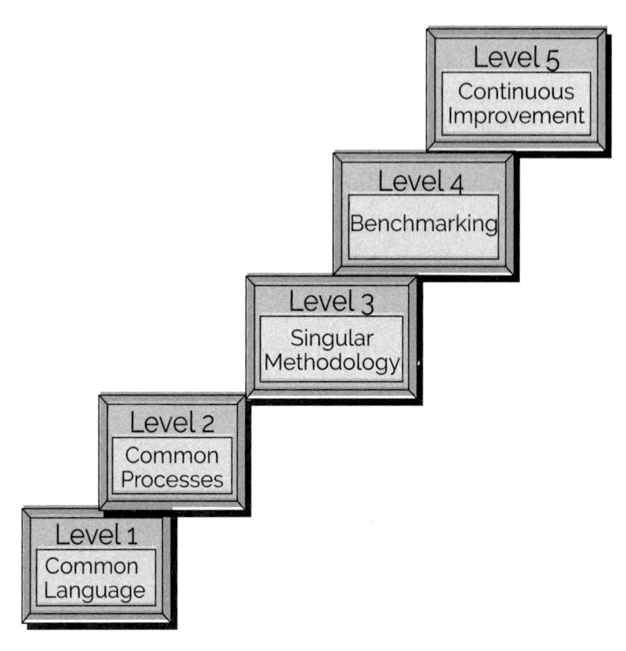

FIGURE 21–2. Overlapping levels.

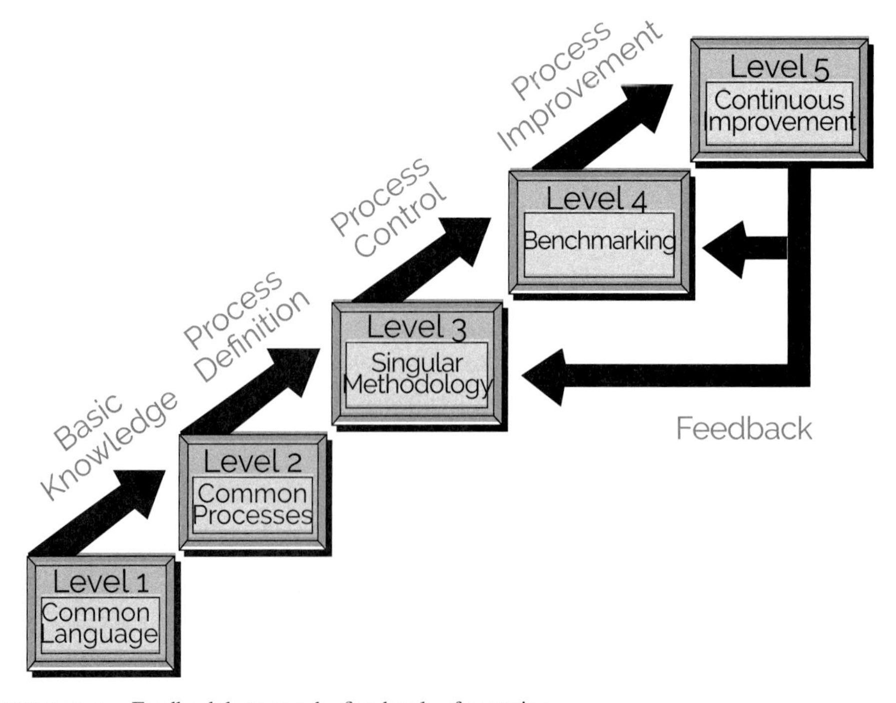

FIGURE 21–3. Feedback between the five levels of maturity.

Level	Description	Degree of Difficulty
1	Common Language	Medium
2	Common Processes	Medium
3	Singular Methodology	High
4	Benchmarking	Low
5	Continuous Improvement	Low

FIGURE 21–4. Degrees of difficulty of the five levels of maturity.

Level 2 and Level 3 generally do not overlap. It may be possible to begin some of the Level 3 work before Level 2 is completed, but this is highly unlikely. Once a company is committed to a singular methodology, work on other methodologies generally terminates. Also, companies can create a Center for Excellence in project management early in the life-cycle process, but will not receive the full benefits until later on.

Risks can be assigned to each level of the PMMM. For simplicity's sake, the risks can be labeled as low, medium, and high. The level of risk is most frequently associated with the impact on the corporate culture. The following definitions can be assigned to these three risks:

- *Low Risk:* Virtually no impact upon the corporate culture, or the corporate culture is dynamic and readily accepts change.
- *Medium Risk:* The organization recognizes that change is necessary but may be unaware of the impact of the change. Multiple-boss reporting would be an example of a medium risk.
- *High Risk:* High risks occur when the organization recognizes that the changes resulting from the implementation of project management will cause a change in the corporate culture. Examples include the creation of project management methodologies, policies, and procedures, as well as decentralization of authority and decision making.

Level 3 has the highest risk and degree of difficulty for the organization. This is shown in Figure 21–4. Once an organization is committed to Level 3, the time and effort needed to achieve the higher levels of maturity have a low degree of difficulty. Achieving Level 3, however, may require a major shift in the corporate culture.

These types of maturity models will become more common in the future, with generic models being customized for individual companies. These models will assist management in performing strategic planning for excellence in project management.

21.2 DEVELOPING EFFECTIVE PROCEDURAL DOCUMENTATION

Good procedural documentation will accelerate the project management maturity process, foster support at all levels of management, and greatly improve project communications. The type of procedural documentation selected is heavily biased on whether we wish to manage formally or informally, but it should show how to conduct project-oriented activities and how to communicate in such a multidimensional environment. The project management policies, procedures, forms, and guidelines can provide some of these tools for delineating the process, as well as a format for collecting, processing, and communicating project-related data in an orderly, standardized format. Project planning and tracking, however, involve more than just the generation of paperwork. They require the participation of the entire project team, including support departments, subcontractors, and top management, and this involvement fosters unity. Procedural documents help to:

- Provide guidelines and uniformity
- Encourage useful, but minimum, documentation
- Communicate information clearly and effectively
- Standardize data formats
- Unify project teams
- Provide a basis for analysis
- Ensure document agreements for future reference
- Refuel commitments
- Minimize paperwork
- Minimize conflict and confusion
- Delineate work packages
- Bring new team members on board
- Build an experience track and method for future projects

Done properly, the process of project planning must involve both the performing and the customer organizations. This leads to visibility of the project at various organizational levels, and stimulates interest in the project and the desire for success.

The Challenges

Even though procedural documents can provide all these benefits, management is often reluctant to implement or fully support a formal project management system. Management concerns often center around four issues: overhead burden, start-up delays, stifled creativity, and reduced self-forcing control. First, the introduction of more organizational formality via policies, procedures, and forms might cost money, and additional funding may be needed to support and maintain the system. Second, the system is seen as causing start-up delays by requiring additional project definition before implementation can start. Third and fourth, the system is often perceived as stifling creativity and shifting project control from the responsible individual to an impersonal process. The comment of one project manager may be typical: "My support personnel feel that we spend too much time planning a project up front; it creates a very rigid environment that stifles innovation. The only purpose seems to be establishing a basis for controls against outdated measures and for punishment rather than help in case of a contingency." This comment illustrates the potential misuse of formal project management systems to establish unrealistic controls and penalties for deviations from the program plan rather than to help to find solutions.

How to Make It Work

Few companies have introduced project management procedures with ease. Most have experienced problems ranging from skepticism to sabotage of the procedural system. Many use incremental approaches to develop and implement their project management methodology. Doing this, however, is a multifaceted challenge to management. The problem is seldom one of understanding the techniques involved, such as budgeting and scheduling, but rather is a problem of involving the project team in the process, getting their input, support, and commitment, and establishing a supportive environment.

The procedural guidelines and forms of an established project management methodology can be especially useful during the project planning/definition phase. Not only does project management methodology help to delineate and communicate the four major sets of variables for organizing and managing the project—(1) tasks, (2) timing, (3) resources, and (4) responsibilities—it also helps to define measurable milestones, as well as report and review requirements. This provides project personnel the ability to measure project status and performance and supplies the crucial inputs for controlling the project toward the desired results.

Developing an effective project management methodology takes more than just a set of policies and procedures. It requires the integration of these guidelines and standards into the culture and value system of the organization. Management must lead the overall efforts and foster an environment conducive to teamwork. The greater the team spirit, trust, commitment, and quality of information exchange among team members, the more likely the team will be to develop effective decision-making processes, make individual and group commitments, focus on problem-solving, and operate in a self-forcing, self-correcting control mode.

Established Practices

Although project managers may have the right to establish their own policies and procedures, many companies design project control forms that can be used uniformly on all projects. Project control forms serve two vital purposes by establishing a common framework from which:

- The project manager will communicate with executives, functional managers, functional employees, and clients.
- Executives and the project manager can make meaningful decisions concerning the allocation of resources.

Some large companies with mature project management structures maintain a separate functional unit for forms control. This is quite common in aerospace and defense, but is also becoming common practice in other industries and in some smaller companies.

Large companies with a multitude of different projects do not have the luxury of controlling projects with three or four forms. There are different forms for planning, scheduling, controlling, authorizing work, and so on. It is not uncommon for companies to have 20 to 30 different forms, each dependent upon the type of project, length of project, dollar value, type of customer reporting, and other such arguments. Project managers are often allowed to set up their own administration for the project, which can lead to long-term damage if they each design their own forms for project control.

The best method for limiting the number of forms appears to be the task force concept, where both managers and doers have the opportunity to provide input. This may appear to be a waste of time and money, but in the long run provides large benefits.

To be effective, the following ground rules can be used:

- Task forces should include managers as well as doers.
- Task force members must be willing to accept criticism from other peers, superiors, and especially subordinates who must "live" with these forms.
- Upper-level management should maintain a rather passive (or monitoring) involvement.
- A minimum of signature approvals should be required for each form.
- Forms should be designed so that they can be updated periodically.
- Functional managers and project managers must be dedicated and committed to the use of the forms.

Categorizing the Broad Spectrum of Documents

The dynamic nature of project management and its multifunctional involvement create a need for a multitude of procedural documents to guide a project through the various phases and stages of integration.

Especially for larger organizations, the challenge is not only to provide management guidelines for each project activity, but also to provide a coherent procedural framework within which project leaders from all disciplines can work and communicate with each other. Specifically, each policy or procedure must be consistent with and accommodating to the various other functions that interface with the project over its life cycle. This complexity of intricate relations is illustrated in Figure 21–5.

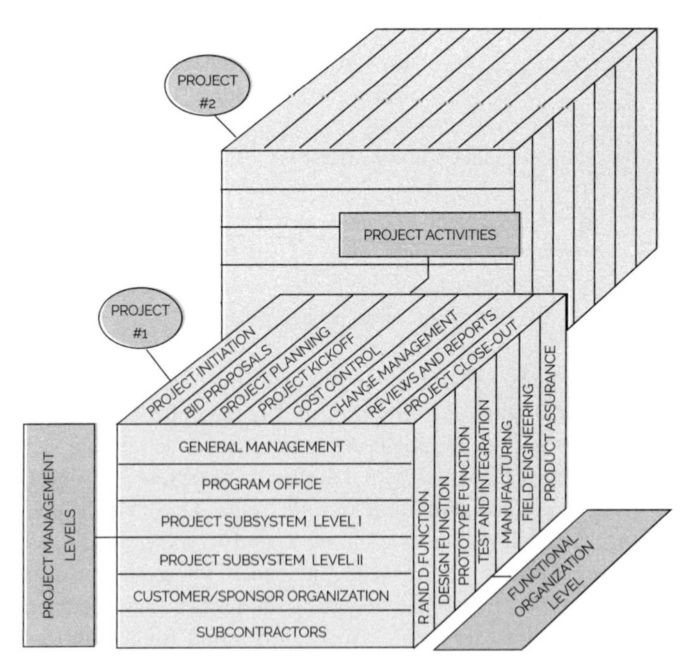

FIGURE 21–5. Interrelationship of project activities with various functional/organizational levels and project management levels.

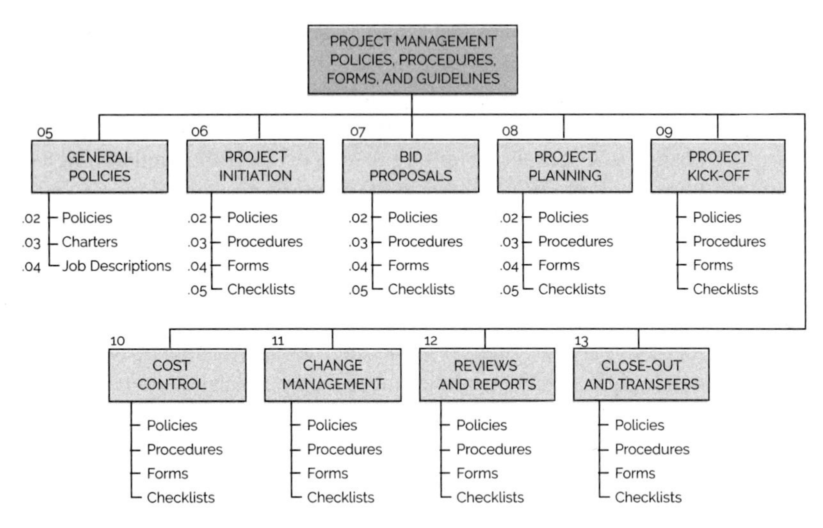

FIGURE 21–6. Categorizing procedural documents within a work breakdown structure.

One simple and effective way of categorizing the broad spectrum of procedural documents is by utilizing the work breakdown concept, as shown in Figure 21–6. Accordingly, the principal procedural categories are defined along the principal project life-cycle phases. Each category is then subdivided into (1) general management guidelines, (2) policies, (3) procedures, (4) forms, and (5) checklists. If necessary, the same concept can be carried forward one additional step to develop policies, procedures, forms, and checklists for the various project and functional sublevels of operation. Although this might be needed for very large programs, an effort should be made to minimize "layering" of policies and procedures to avoid new problems and costs. For most projects, a single document covers all levels of project operations.

As We Mature . . . As companies become more mature in executing the project management methodology, project management policies, and project management levels, procedures are disregarded and replaced with guidelines, forms, and checklists. More flexibility is provided the project manager. Unfortunately, this takes time because executives must have faith in the ability of the project management methodology to work without the rigid controls provided by policies and procedures. Yet all companies seem to go through the evolutionary stages of policies and procedures before they get to guidelines, forms, and checklists.

21.3 PROJECT MANAGEMENT METHODOLOGIES

The ultimate purpose of any project management system is to increase the likelihood that your organization will have a continuous stream of successfully managed projects. The best way to achieve this goal is with good project management methodologies that are based upon guidelines and forms rather than policies and procedures. Methodologies must have enough flexibility that they can be adapted easily to each and every project.

Methodologies should be designed to support the corporate culture, not vice versa. It is a fatal mistake to purchase a canned methodology package that mandates that you change your corporate culture to support it. If the methodology does not support the culture, it will not be accepted. What converts any methodology into a world-class methodology is its adaptability to the corporate culture. There is no reason why companies cannot develop their own methodology. Companies such as Hewlett-Packard, Johnson Controls, and Motorola are regarded as having world-class methodologies for project management and, in each case, the methodology was developed internally. Developing your own methodology internally to guarantee a fit with the corporate culture usually provides a much greater return on investment than purchasing canned packages that require massive changes.

21.4 CONTINUOUS IMPROVEMENT

All too often complacency dictates the decision-making process. This is particularly true of organizations that have reached some degree of excellence in project management, become complacent, and then realize too late that they have lost their competitive advantage. This occurs when organizations fail to recognize the importance of continuous improvement.

Figure 21–7 illustrates why there is a need for continuous improvement. As companies begin to mature in project management and reach some degree of excellence, they achieve a sustained competitive advantage. The sustained competitive advantage might very well be the single most important strategic objective of the firm. The firm will then begin the exploitation of its sustained competitive advantage.

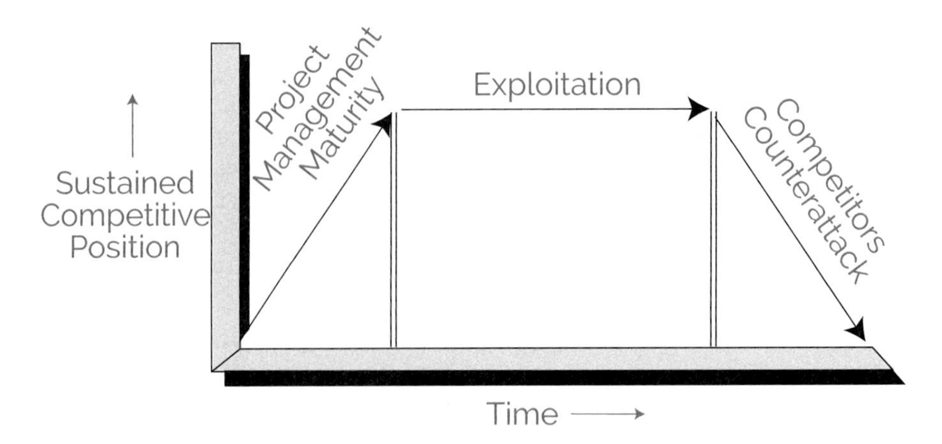

FIGURE 21–7. Why there is a need for continuous improvement.

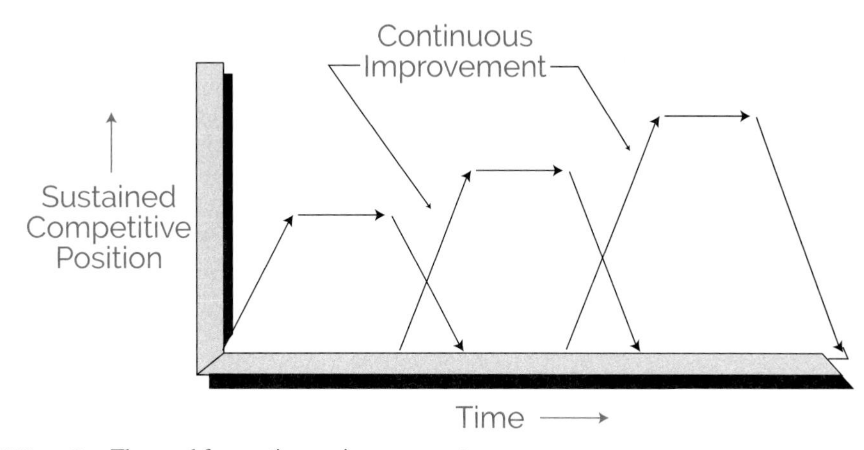

FIGURE 21–8. The need for continuous improvement.

Unfortunately, the competition is not sitting by idly watching you exploit your sustained competitive advantage. As the competition begins to counterattack, you may lose a large portion, if not all, of your sustained competitive advantage. To remain effective and competitive, the organization must recognize the need for continuous improvement, as shown in Figure 21–8. Continuous improvement allows a firm to maintain its competitive advantage even when the competitors counterattack.

21.5 CAPACITY PLANNING

As companies become excellent in project management, the benefits of performing more work in less time and with fewer resources becomes readily apparent. The question, of course, is how much more work can the organization take on? Companies are now struggling to develop capacity planning models to see how much new work can be undertaken within the existing human and nonhuman constraints.

Figure 21–9 illustrates the classical way that companies perform capacity planning. The approach outlined in this figure holds true for both project- and non–project-driven organizations. The "planning horizon" line indicates the point in time for capacity planning. The "proposals" line indicates the manpower needed for approved internal projects or a percentage (perhaps as much as 100 percent) for all work expected through competitive bidding. The combination of this line and the "manpower requirements" line, when compared against the current staffing, provides us with an indication of capacity. This technique can be effective if performed early enough such that training time is allowed for future manpower shortages.

The limitation to this process for capacity planning is that only human resources are considered. A more realistic method would be to use the method shown in Figure 21–10, which can also be applied to both project-driven and non–project-driven organizations. From Figure 21–10, projects are selected based upon such factors as strategic fit, profitability, who the customer is, and corporate benefits. The objectives for the projects selected

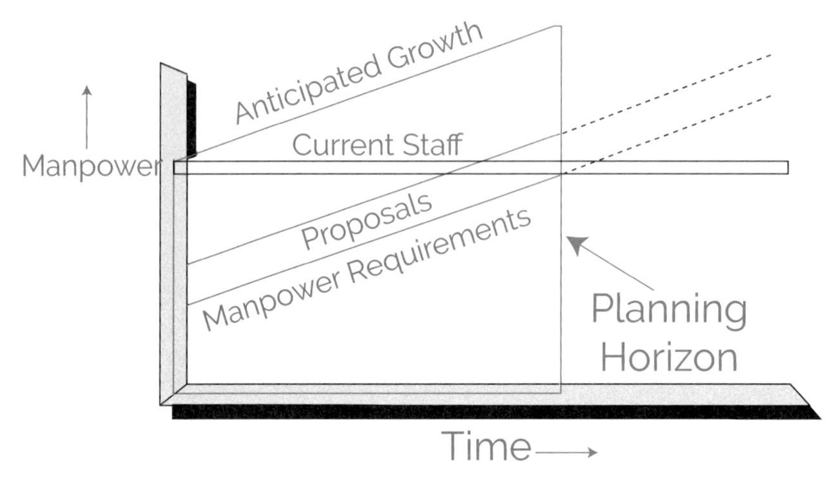

FIGURE 21–9. Classical capacity planning.

are then defined in both business and technical terms, because there can be both business and technical capacity constraints.

The next step is a critical difference between average companies and excellent companies. Capacity constraints are identified from the summation of the schedules and plans. In excellent companies, project managers meet with sponsors to determine the objective of the plan, which is different than the objective of the project. Is the objective of the plan to achieve the project's objective with the least cost, least time, or least risk? Typically, only one of these applies, whereas immature organizations believe that all three can be achieved on every project. This, of course, is unrealistic.

The final box in Figure 21–10 is now the determination of the capacity limitations. Previously, we considered only human resource capacity constraints. Now we realize that

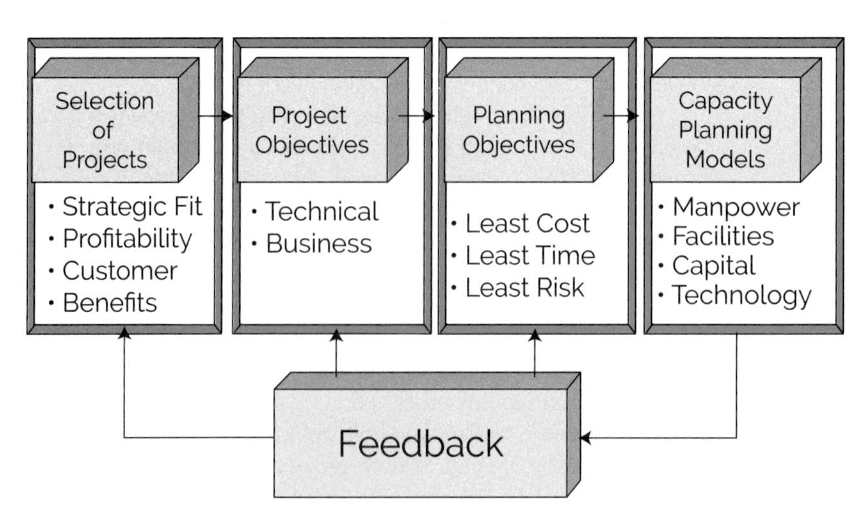

FIGURE 21–10 Improved capacity planning.

the critical path of a project can be constrained not only by time but also by available manpower, facilities, cash flow, and even existing technology. It is possible to have multiple critical paths on a project other than those identified by time. Each of these critical paths provides a different dimension to the capacity planning models, and each of these constraints can lead us to a different capacity limitation. As an example, manpower might limit us to taking on only four additional projects. Based upon available facilities, however, we might only be able to undertake two more projects, and based upon available technology, we might be able to undertake only one new project.

21.6 COMPETENCY MODELS

In the twenty-first century, companies are replacing job descriptions with competency models. Job descriptions for project management tend to emphasize the deliverables and expectations from the project manager, whereas competency models emphasize the specific skills needed to achieve the deliverables.

Figure 21–11 shows the competency model for Eli Lilly. Project managers are expected to have competencies in three broad areas[1]:

- Scientific/technical skills
- Leadership skills
- Process skills

For each of the three broad areas, there are subdivisions or grade levels. A primary advantage of a competency model is that it allows the training department to develop customized project management training programs to satisfy the skill requirements. Without competency models, most training programs are generic rather than customized.

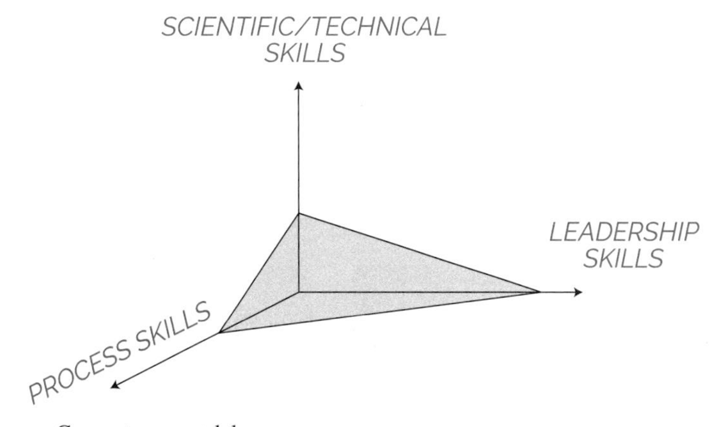

FIGURE 21–11. Competency model.

1. A detailed description of the Eli Lilly competency model and the Ericsson competency model can be found in Harold Kerzner, *Applied Project Management* (New York: John Wiley & Sons, 1999), pp. 266–283.

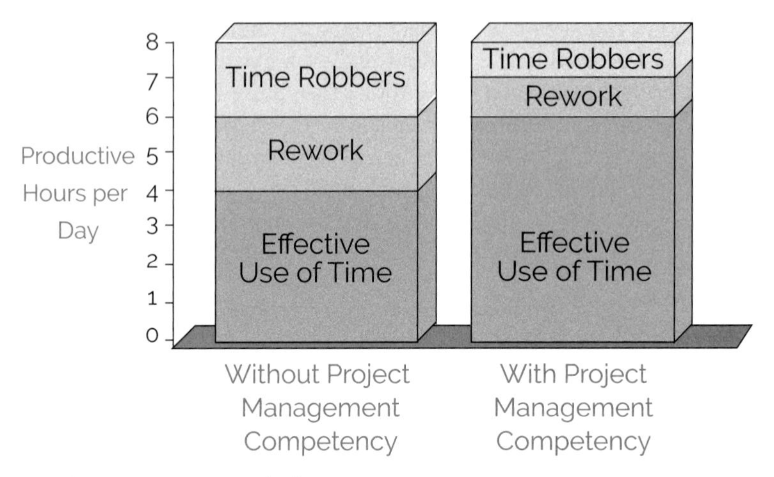

FIGURE 21–12. Core competency analysis.

Competency models focus on specialized skills in order to assist the project manager in making more efficient use of his or her time. Figure 21–12, although argumentative, shows that with specialized competency training, project managers can increase their time effectiveness by reducing time robbers and rework.

Competency models make it easier for companies to develop a complete project management curriculum, rather than a singular course. This is shown in Figure 21–13.

As companies mature in project management and develop a company-wide core competency model, an internal, custom-designed curriculum will be developed. Companies, especially large ones, will find it necessary to maintain a course architecture specialist on their staff.

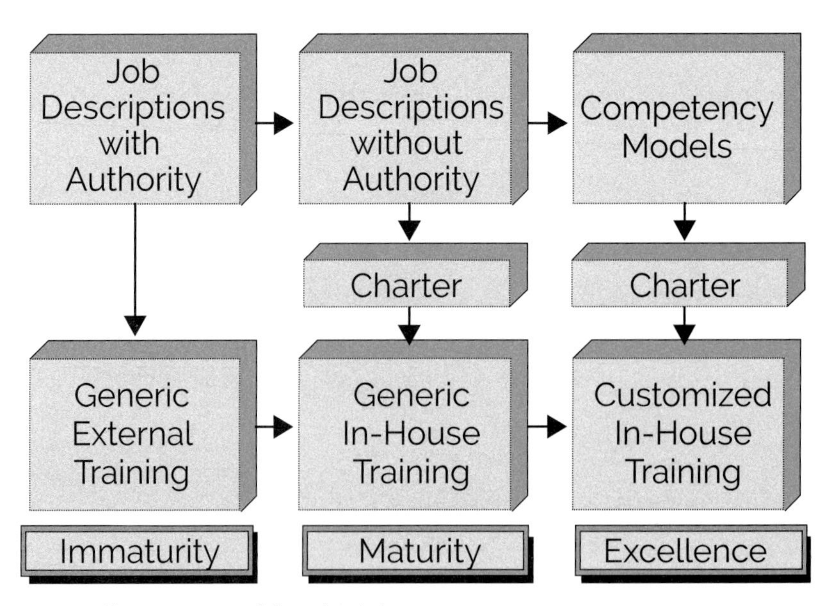

FIGURE 21–13. Competency models and training.

21.7 MANAGING MULTIPLE PROJECTS

As organizations mature in project management, there is a tendency toward having one person manage multiple projects. The initial impetus may come either from the company sponsoring the projects or from project managers themselves. There are several factors supporting the managing of multiple projects. First, the cost of maintaining a full-time project manager on all projects may be prohibitive. The magnitude and risks of each individual project dictate whether a full-time or part-time assignment is necessary. Assigning a project manager full-time on an activity that does not require it is an overmanagement cost. Overmanagement of projects was considered an acceptable practice in the early days of project management because we had little knowledge on how to handle risk management. Today, methods for risk management exist.

Second, line managers are now sharing accountability with project managers for the successful completion of the project. Project managers are now managing at the template levels of the WBS with the line managers accepting accountability for the work packages at the detailed WBS levels. Project managers now spend more of their time integrating work rather than planning and scheduling functional activities. With the line manager accepting more accountability, time may be available for the project manager to manage multiple projects.

Third, senior management has come to the realization that they must provide high-quality training for their project managers if they are to reap the benefits of managing multiple projects. Senior managers must also change the way that they function as sponsors. There are six major areas where the corporation as a whole may have to change in order for the managing of multiple projects to succeed.

- *Prioritization:* If a project prioritization system is in effect, it must be used correctly such that employee credibility in the system is realized. One risk is that the project manager, having multiple projects to manage, may favor those projects having the highest priorities. It is possible that no prioritization system may be the best solution. Not every project needs to be prioritized, and prioritization can be a time-consuming effort.
- *Scope Changes:* Managing multiple projects is almost impossible if the sponsors/customers are allowed to make continuous scope changes. When using multiple projects management, it must be understood that the majority of the scope changes may have to be performed through enhancement projects rather than through a continuous scope change effort. A major scope change on one project could limit the project manager's available time to service other projects. Also, continuous scope changes will almost always be accompanied by reprioritization of projects, a further detriment to the management of multiple projects.
- *Capacity Planning:* Organizations that support the management of multiple projects generally have a tight control on resource scheduling. As a result, the organization must have knowledge of capacity planning, theory of constraints, resource leveling, and resource limited planning.
- *Project Methodology:* Methodologies for project management range from rigid policies and procedures to more informal guidelines and checklists. When managing multiple projects, the project manager must be granted some degree of freedom.

This necessitates guidelines, checklists, and forms. Formal project management practices create excessive paperwork requirements, thus minimizing the opportunities to manage multiple projects. The project size is also critical.

- *Project Initiation:* Managing multiple projects has been going on for almost 40 years. One thing that we have learned is that it can work well as long as the projects are in relatively different life-cycle phases because the demands on the project manager's time are different for each life-cycle phase.
- *Organizational Structures:* If the project manager is to manage multiple projects, then it is highly unlikely that the project manager will be a technical expert in all areas of all projects. Assuming that the accountability is shared with the line managers, the organization will most likely adopt a weak matrix structure.

21.8 THE BUSINESS OF SCOPE CHANGES

PMBOK® Guide, 6th Edition
5.6 Control Scope

Very few projects are ever completed according to the original plan. The changes to the plan result from either increased knowledge, a need for competitiveness, or changing customer/consumer tastes. Once the changes are made, there is almost always an accompanying increase in the budget and/or elongation of the schedule.

The process for recommending and approving scope changes can vary based upon whether the client is internal or external to the organization. Scope changes for external clients have long been viewed as a source of added profitability on projects. Years ago, it was common practice on some Department of Defense contracts to underbid the original contract during competitive bidding to assure the award of the contract and then push through large quantities of lucrative scope changes.

External customers were rarely informed of gaps in their statements of work that could lead to scope charges. And even if the statement of work was clearly written, it was often intentionally misinterpreted for the benefit of seeking out scope changes whether or not the scope changes were actually needed. For some companies, scope changes were the prime source of corporate profitability. During competitive bidding, executives would ask the bidding team two critical questions before submitting a bid: (1) What is our cost of doing the work we are committing to? and (2) How much additional work can we push through in scope changes once the contract is awarded to us? Often, the answer to the second question determined the magnitude of the bid.

To make matters even worse, in the early years of project management the Department of Defense requested that the contractors' project managers have a command of technology rather than an understanding of technology. Engineers with advanced degrees were assigned as project managers, and their objective was often to exceed rather than merely meet specifications. This resulted in additional scope changes and often increased the risks in the project.

Another problem that surfaced was the downstream effect of upstream scope changes on large projects involving multiple contractors. When contractors work sequentially, as

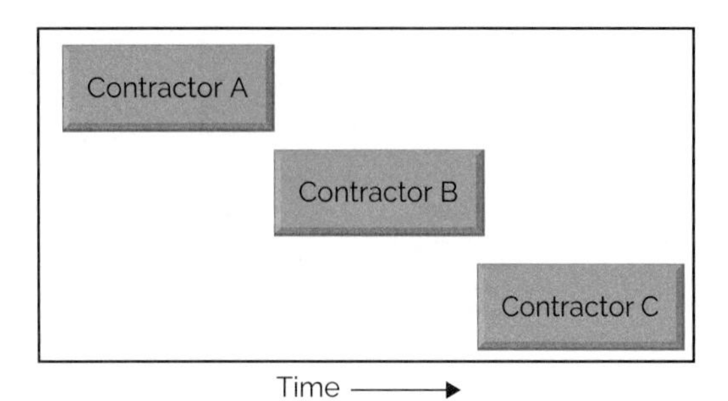

FIGURE 21–14. Sequential contractors.

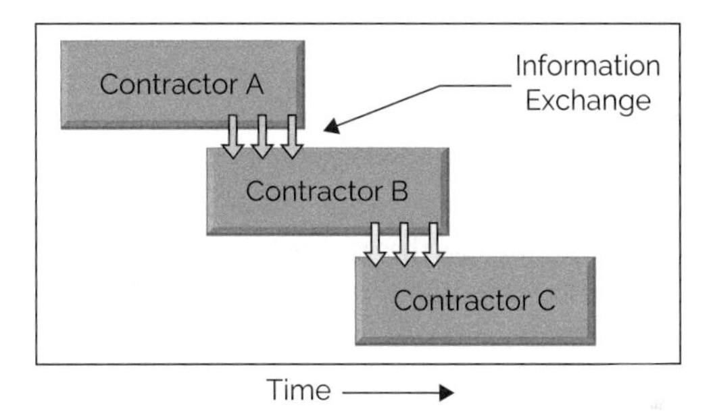

FIGURE 21–15. Overlapping contractors.

shown in Figure 21–14, a scope change in an upstream contractor might not have a serious impact on downstream contractors. Contractor B usually needs the output from contractor A to begin work. If contractor A initiates a scope change, the impact on contractor B may be minimal. If there were an impact on contractor B, the customer would incur the added costs. But if the contractors are performing work partially or completely in parallel, as shown in Figure 21–15, the downstream effect can be devastating. A relatively simple decision by an upstream contractor to change over to higher-grade or lower-grade raw materials can cause major changes in the project plans and scope baselines of downstream contractors.

Need for Business Knowledge

Using scope changes as a source of revenue has been an acceptable practice for projects funded by customers external to the organization. But for internal customers, there are numerous other reasons for scope changes, as shown in Figure 21–16. For projects that are internal, scope changes must be targeted, and this is the weakest link because it requires business knowledge as well as technical knowledge. As an example, scope changes should not be implemented at the expense of risking exposure to product liability lawsuits or safety issues. Likewise, costly scope changes exclusively for the sake of enhancing image or reputation should be avoided. Also, scope changes should not be implemented if the payback period for the product is drastically extended in order to recover the costs of the scope changes.

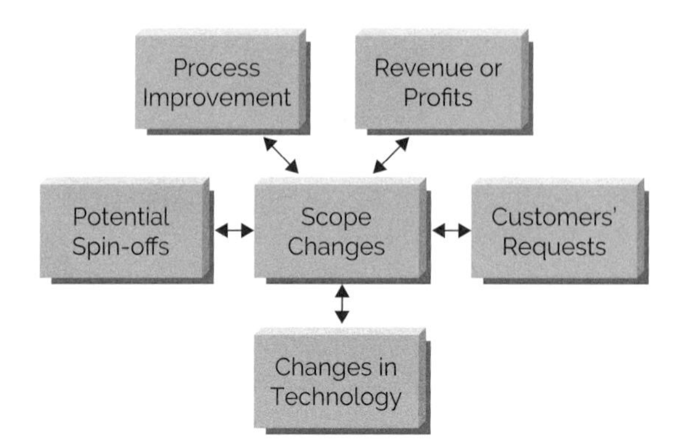

FIGURE 21–16. Factors to consider for scope changes.

Today, project managers are expected to make sound business decisions rather than merely project or technical decisions. But it wasn't that long ago when companies were nonbelievers that project management could benefit the business as a whole. Companies are slowly coming to the realization that they are managing their business by projects. Project managers are now expected to make sound business decisions rather than merely project decisions, and this includes the decision on scope changes.

Timing of Scope Changes

It is understood that the further into the project's life-cycle phases, the more costly the scope changes. As we progress through the life-cycle phases, more variables are introduced into the system such that the financial impact of a small scope change can be quite large because of the cost involved with reversing previous decisions. Scope changes in production are more costly than scope changes in technology development.

For the development of new products, it may take as many as sixty ideas to develop one successful new product. Each idea may undergo numerous scope changes before the idea is officially abandoned. Any scope changes made after capital expenditures are incurred can have a significant impact on total cost and schedule.

Another critical factor involving timing is whether the scope changes are radical scope changes. Radical scope changes require either a breakthrough in technology or the design of an entirely new platform. As an example, a competitor launches a new product that may cause the marketplace to view your product as being obsolete. In order to remain competitive, you may need to consider radical scope changes to remain competitive or to outdo the competition. Radical scope changes focus more so on creativity than execution. Radical scope changes may require a breakthrough in technology accompanied by the consumption of vast resources.

Another timing issue is whether the scope change should be done incrementally or clustered together and approved as enhancement projects. Incremental scope changes are often referred to as scope creep. These incremental changes can be done quickly and at a relatively low cost. However, if there are a significant number of incremental scope changes, such as in the case of perpetual scope creep, the project's schedule can be elongated.

Scope Creep

There are three things that most project managers know will happen with almost certainty: death, taxes, and scope creep. Scope creep is the continuous enhancement of the project's requirements as the project's deliverables are being developed. Scope creep is viewed as the growth in the project's scope. The larger and more complex the project, the greater the chances of significant scope creep.

Although scope creep can occur in any project in any industry, it is most frequently associated with information systems development projects. Scope changes can occur during any project life-cycle phase. Scope changes occur because it is the nature of humans not to be able to completely describe the project or the plan to execute the project at the start. This is particularly true on large, complex projects. As a result, we gain more knowledge as the project progresses, and this leads to creeping scope and scope changes.

Scope creep is a natural occurrence for project managers. We must accept the fact that this will happen. Some people believe that there are magical charms, potions, and rituals that can prevent scope creep. This is certainly not true. Perhaps the best we can do is to establish processes, such as configuration management systems, or change control boards to get some control over scope creep. However, these processes are designed not so much to prevent scope creep but rather to prevent unwanted scope changes from taking place.

Therefore, we can argue that scope creep is not just allowing the scope to change but an indication of how well we manage changes to the scope. If all of the parties agree that a scope change is needed, then perhaps we can argue that the scope simply changed rather than creeped. Some people view scope creep as a scope change not approved by the sponsor or the change control board.

Scope creep is often viewed as being detrimental to the success of a project because it increases the cost and elongates the schedule. While this is true, scope creep can also produce favorable results such as add-ons that give your product a competitive advantage. Scope creep can also please the customer if the scope changes are seen as additional value for the final deliverable.

Business Need for a Scope Change

There must be a valid business purpose for a scope change. This includes the following factors at a minimum:

- An assessment of the customers' needs and the added value that the scope change will provide
- An assessment of the market needs including the time required to make the scope change, the payback period, return on investment, and whether the final product selling price will be overpriced for the market
- An assessment on the impact on the length of the product life cycle
- An assessment on the competition's ability to imitate the scope change
- Is there a product liability associated with the scope change and can it impact our image?

Scope changes can be for existing products or for new products. Support for existing products is usually a defensive scope change designed to penetrate new markets with existing products. Support for new products is usually an offensive scope change designed to provide new products/services to existing customers as well as seeking out new markets.

Rationale for not Approving a Scope Change

Some scope change requests are the result of wishful thinking or the personal whims of management and not necessarily based upon sound business judgment. In such cases, the scope changes may need to be canceled. Typical rationalization for termination or not approving a scope change includes:

- The cost of the scope change is excessive and the final cost of the deliverable may make us noncompetitive.
- The return on investment may occur too late.
- The competition is too stiff and not worth the risks.
- There are insurmountable obstacles and technical complexity.
- There are legal and regulatory uncertainties.
- The scope change may violate the company's policy on nondisclosure, secrecy, and confidentiality agreements.

21.9 END-OF-PHASE REVIEW MEETINGS

For more than 20 years, end-of-phase review meetings were simply an opportunity for executives to "rubber stamp" the project to continue. As only good news was presented the meetings were used to give the executives some degree of comfort concerning project status.

Today, end-of-phase review meetings take on a different dimension. First and foremost, executives are no longer afraid to cancel projects, especially if the objectives have changed, if the objectives are unreachable, or if the resources can be used on other activities that have a greater likelihood of success. Executives now spend more time assessing the risks in the future rather than focusing on accomplishments in the past.

Since project managers are now becoming more business-oriented rather than technically oriented, the project managers are expected to present information on business risks, reassessment of the benefit-to-cost ratio, and any business decisions that could affect the ultimate objectives. Simply stated, the end-of-phase review meetings now focus more on business decisions, rather than on technical decisions.

Related Case Studies (from Kerzner/*Project Management Case Studies*, 5th ed.)	**Related Workbook Exercises (from Kerzner/*Project Management Workbook and PMP®/CAPM® Exam Study Guide*, 12th ed.)**	***PMBOK® Guide*, 6th Edition, Reference Section for the PMP® Certification Exam**
• Lakes Automotive • Ferris HealthCare, Inc. • Clark Faucet Company • Honicker Corporation* • Kemko Manufacturing*	• Project Management • Maturity Questionnaire • Multiple Choice Exam	None

*Case study also appears at end of chapter.

PMP and CAPM are registered marks of the Project Management Institute, Inc.

CASE STUDIES

HONICKER CORPORATION

Background

Honicker Corporation was well-recognized as a high-quality manufacturer of dashboards for automobiles and trucks. Although it serviced mainly U.S. automotive and truck manufacturers, the opportunity to expand to a worldwide supplier was quite apparent. Its reputation was well-known worldwide, but it was plagued for years with ultra-conservative senior management leadership that prevented growth into the international marketplace.

When the new management team came on board in 2009, the conservatism disappeared. Honicker was cash rich, had large borrowing power and lines of credit with financial institutions, and received an AA-quality rating on its small amount of corporate debt. Rather than expand by building manufacturing facilities in various countries, Honicker decided to go the fast route by acquiring four companies around the world: Alpha, Beta, Gamma, and Delta Companies.

Each of the four acquired companies serviced mainly its own geographical areas. The senior management team in each of the four companies knew the culture in their geographic areas and had a good reputation with their clients and local stakeholders. The decision was made by Honicker to leave each company's senior management teams intact provided that the necessary changes, as established by corporate, could be implemented.

Honicker wanted each company to have the manufacturing capability to supply parts to any Honicker client worldwide. But doing this was easier said than done. Honicker had an enterprise project management methodology (EPM) that worked well. Honicker understood project management and so did the majority of Honicker's clients and stakeholders in the United States. Honicker recognized that the biggest challenge would be to get all of the divisions at the same level of project management maturity and using the same corporate-wide EPM system or a modified version of it. It was expected that each of the four acquired companies might want some changes to be made.

The four acquired divisions were all at different levels of project management maturity. Alpha did have an EPM system and believed that its approach to project management was superior to the one that Honicker was using. Beta Company was just beginning to learn project management but did not have any formal EPM system, although it did have a few project management templates that were being used for status reporting to its customers. Gamma and Delta Companies were clueless about project management.

To make matters worse, laws in each of the countries where the acquired companies were located created other stakeholders that had to be serviced, and all of these stakeholders were at different levels of project management maturity. In some countries government stakeholders were actively involved because of employment procurement laws, whereas in other countries government stakeholders were passive participants unless health, safety, or environmental laws were broken.

It would certainly be a formidable task developing an EPM system that would satisfy all of the newly acquired companies, their clients, and their stakeholders.

Establishing the Team

Honicker knew that there would be significant challenges in getting a project management agreement in a short amount of time. Honicker also knew that there is never an acquisition of equals; there is always a "landlord" and "tenants," and Honicker is the landlord. But acting as a landlord and exerting influence in the process could alienate some of the acquired companies and do more harm than good. Honicker's approach was to treat this as a project, and each company, along with its clients and local stakeholders, would be treated as project stakeholders. Using stakeholder relations management practices would be essential to getting an agreement on the project management approach.

Honicker requested that each company assign three people to the project management implementation team that would be headed up by Honicker personnel. The ideal team member, as suggested by Honicker, would have some knowledge and/or experience in project management and be authorized by their senior levels of management to make decisions for their company. The representatives should also understand the stakeholder needs from their clients and local stakeholders. Honicker wanted an understanding to be reached as early as possible that each company would agree to use the methodology that was finally decided upon by the team.

Senior management in each of the four companies sent a letter of understanding to Honicker promising to assign the most qualified personnel and agreeing to use the methodology that was agreed upon. Each stated that their company understood the importance of this project.

The first part of the project would be to come to an agreement on the methodology. The second part of the project would be to invite clients and stakeholders to see the methodology and provide feedback. This was essential since the clients and stakeholders would eventually be interfacing with the methodology.

Kickoff Meeting

Honicker had hoped that the team could come to an agreement on a companywide EPM system within six months. But after the kickoff meeting was over, Honicker realized that it would probably be two years before an agreement would be reached on the EPM system. Several issues became apparent at the first meeting:

- Each company had different time requirements for the project.
- Each company saw the importance of the project differently.
- Each company had its own culture and wanted to be sure that the final design was a good fit with that culture.
- Each company saw the status and power of the project manager differently.
- Despite the letters of understanding, two of the companies, Gamma and Delta, did not understand their role and relationship with Honicker on this project.
- Alpha wanted to micromanage the project, believing that everyone should use its methodology.

Senior management at Honicker asked the Honicker representatives at the kickoff meeting to prepare a confidential memo on their opinion of the first meeting with the team. The Honicker personnel prepared a memo including the following comments:

- Not all of the representatives at the meeting openly expressed their true feelings about the project.
- It was quite apparent that some of the companies would like to see the project fail.
- Some of the companies were afraid that the implementation of the new EPM system would result in a shift in power and authority.
- Some people were afraid that the new EPM system would show that fewer resources were needed in the functional organization, thus causing a downsizing of personnel and a reduction in bonuses that were currently based upon headcount in functional groups.
- Some seemed apprehensive that the implementation of the new system would cause a change in the company's culture and working relationships with their clients.
- Some seemed afraid of learning a new system and being pressured into using it.

It was obvious that this would be no easy task. Honicker had to get to know all companies better and understand their needs and expectations. Honicker management had to show them that their opinion was of value and find ways to win their support.

QUESTIONS

1. What are Honicker's options now?
2. What would you recommend that Honicker do first?
3. What if, after all attempts, Gamma and Delta companies refuse to come on board?
4. What if Alpha Company is adamant that its approach is best and refuses to budge?
5. What if Gamma and Delta Companies argue that their clients and stakeholders have not readily accepted the project management approach and they wish to be left alone with regard to dealing with their clients?
6. Under what conditions would Honicker decide to back away and let each company do its own thing?
7. How easy or difficult is it to get several companies geographically dispersed to agree to the same culture and methodology?
8. If all four companies were willing to cooperate with one another, how long do you think it would take for an agreement on and acceptance to use the new EPM system?
9. Which stakeholders may be powerful and which are not?
10. Which stakeholder(s) may have the power to kill this project?
11. What can Honicker do to win their support?
12. If Honicker cannot win their support, then how should Honicker manage the opposition?
13. What if all four companies agree to the project management methodology and then some of the client stakeholders show a lack of support for use of the methodology?

KEMKO MANUFACTURING

Background

Kemko Manufacturing was a fifty-year-old company that had a reputation for the manufacturing of high-quality household appliances. Kemko's growth was rapid during the 1990s. The company grew by acquiring other companies. Kemko now had more than twenty-five manufacturing plants throughout the United States, Europe, and Asia. Originally, each manufacturing plant that was acquired wanted to maintain its own culture and quite often was allowed to remain autonomous from corporate at Kemko provided that work was progressing as planned. But as Kemko began acquiring more companies, growing pains made it almost impossible to allow each plant to remain autonomous. Each company had its own way of handling raw material procurement and inventory control. All purchase requests above a certain dollar value had to be approved by corporate. At corporate, there was often confusion over the information in all of the forms since each plant had its own documentation for procurement. Corporate was afraid that, unless it established a standardized procurement and inventory control system across all of the plants, cash flow problems and loss of corporate control over inventory could take its toll in the near future.

Project Is Initiated

Because of the importance of the project, senior management asked Janet Adams, director of information technology (IT), to take control of the project personally. Janet had more than thirty years of experience in IT and fully understood how scope creep can create havoc on a large project.

Janet selected her team from IT and set up an initial kickoff date for the project. In addition to the mandatory presence of all of her team members, she also demanded that each manufacturing plant assign at least one representative and that all of the plant representatives must be in attendance as well at the kickoff meeting. At the kickoff meeting, Janet spoke:

> I asked all of you here because I want you to have a clear understanding of how I intend to manage this project. Our executives have given us a timetable for this project and my greatest fear is "scope creep." Scope creep is the growth of or enhancements to the project's scope as the project is being developed. On many of our other projects, scope creep has elongated the project and driven up the cost. I know that scope creep isn't always evil, and that it can happen in any life cycle phase.
>
> The reason why I have asked all of the plant representatives to attend this meeting is because of the dangers of scope creep. Scope creep has many causes, but it is generally the failure of effective upfront planning. When scope creep exists, people generally argue that it is a natural occurrence and we must accept the fact that it will happen. That's unacceptable to me!
>
> There will be no scope changes on this project, and I really mean it when I say this. The plant representatives must meet on their own and provide us with a detailed requirements package. I will not allow the project to officially begin until we have a detailed listing of the requirements. My team will provide you with some guidance, as needed, in preparing the requirements.
>
> No scope changes will be allowed once the project begins. I know that there may be some requests for scope changes, but all requests will be bundled together and worked upon later as an enhancement project. This project will be implemented according to the original set of requirements. If I were to allow scope changes to occur, this project would run forever. I know some of you do not like this, but this is the way it will be on this project.

There was dead silence in the room. Janet could tell from the expressions on the faces of the plant representatives that they were displeased with her comments. Some of the plants were under the impression that the IT group was supposed to prepare the requirements package. Now, Janet had transferred the responsibility to them, the user group, and they were not happy. Janet made it clear that user involvement would be essential for the preparation of the requirements.

After a few minutes of silence, the plant representatives said that they were willing to do this and it would be done correctly. Many of the representatives understood user requirements documentation. They would work together and come to an agreement on the requirements. Janet again stated that her team would support the plant representatives but that the burden of responsibility would rest solely upon the plants. The plants would get what they asked for and nothing more. Therefore, they must be quite clear up front in their requirements.

While Janet was lecturing to the plant representatives, the IT portion of the team was just sitting back smiling. Their job was about to become easier, or at least they thought so. Janet then addressed the IT portion of the team:

> Now I want to address the IT personnel. The reason why we are all in attendance at this meeting is because I want the plant representatives to hear what I have to say to the IT team. In the past, the IT teams have not been without some blame for scope creep and schedule elongation. So, here are my comments for the IT personnel:

- It is the IT team's responsibility to make sure that they understand the requirement as prepared by the plant representatives. Do not come back to me later telling me that you did not understand the requirements because they were poorly defined. I am going to ask every IT team member to sign a document stating that they have read over the requirement and fully understand them.

- Perfectionism is not necessary. All I want you to do is to get the job done.
- In the past we have been plagued with "featuritis," where many of you have added in your own "bells and whistles" unnecessarily. If that happens on this project, I will personally view this as a failure by you and it will reflect in your next performance review.
- Sometimes, people believe that a project like this will advance their career, especially if they look for perfection and bells and whistles. Trust me when I tell you this can have the opposite effect.
- Back door politics will not be allowed. If any of the plant representatives come to you looking for ways to sneak in scope changes, I want to know about it. And if you make the changes without my permission, you may not be working for me much longer.
- I, and only I, have signature authority for scope changes.
- This project will be executed using detailed planning rather than rolling wave or progressive planning. We should be able to do this once we have clearly defined requirements.

Now, are there any questions from anyone?

The battle lines were now drawn. Some believed that it was Janet against the team, but most understood Janet's need to do this. However, whether it could work this way was still questionable.

QUESTIONS

1. Was Janet correct in the comments she made to the plant representatives?
2. Was Janet correct in the comments she made to the IT team members?
3. Is it always better on IT projects to make changes using enhancement projects or should we allow changes to be made as we go along?
4. What is your best guess on what happened?

APPENDIX A
Solution to Leadership Exercise

SITUATION 1

A. This technique may work if you have proven leadership credentials. Since three of these people have not worked for you before, some action is necessary.

B. The team should already be somewhat motivated and reinforcement will help. Team building must begin by showing employees how they will benefit. This is usually the best approach on long-term projects. (5 points)

C. This is the best approach if the employees already understand the project. In this case, however, you may be expecting too much out of the employees this soon. (3 points)

D. This approach is too strong at this time, since emphasis should be on team building. On long-term projects, people should be given the opportunity to know one another first. (2 points)

SITUATION 2

A. Do nothing. Don't overreact. This may improve productivity without damaging morale. See the impact on the team first. If the other members accept Tom as the informal leader, because he has worked for you previously, the results can be very favorable. (5 points)

B. This may cause the team to believe that a problem exists when, in fact, it does not.

C. This is duplication of effort and may reflect on your ability as a leader. Productivity may be impaired. (2 points)

D. This is a hasty decision and may cause Tom to overreact and become less productive. (3 points)

SITUATION 3

A. You may be burdening the team by allowing them to struggle. Motivation may be affected and frustration will result. (1 point)

B. Team members expect the project manager to be supportive and to have ideas. This will reinforce your relationship with the team. (5 points)

C. This approach is reasonable as long as your involvement is minimal. You must allow the team to evolve without expecting continuous guidance. (4 points)

D. This action is premature and can prevent future creativity. The team may allow you to do it all.

SITUATION 4

A. If, in fact, the problem does exist, action must be taken. These types of problems do not go away by themselves.

B. This will escalate the problem and may make it worse. It could demonstrate your support for good relations with your team, but could also backfire. (1 point)

C. Private meetings should allow you to reassess the situation and strengthen employee relations on a one-on-one basis. You should be able to assess the magnitude of the problem. (5 points)

D. This is a hasty decision. Changing the team's schedules may worsen the morale problem. This situation requires replanning, not a strong hand. (2 points)

SITUATION 5

A. Crisis management does not work in project management. Why delay until a crisis occurs and then waste time having to replan?

B. This situation may require your immediate attention. Sympathizing with your team may not help if they are looking toward you for leadership. (2 points)

C. This is the proper balance: participative management and contingency planning. This balance is crucial for these situations. (5 points)

D. This may seriously escalate the problem unless you have evidence that performance is substandard. (1 point)

SITUATION 6

A. Problems should be uncovered and brought to the surface for solution. It is true that this problem may go away, or that Bob simply does not recognize that his performance is substandard.

B. Immediate feedback is best. Bob must know your assessment of his performance. This shows your interest in helping him improve. (5 points)

C. This is not a team problem. Why ask the team to do your work? Direct contact is best.

D. As above, this is your problem, not that of the team. You may wish to ask for their input, but do not ask them to perform your job.

SITUATION 7

A. George must be hurting to finish the other project. George probably needs a little more time to develop a quality report. Let him do it. (5 points)

B. Threatening George may not be the best situation because he already understands the problem. Motivation by threatening normally is not good. (3 points)

C. The other team members should not be burdened with this unless it is a team effort.

D. As above, this burden should not be placed on other team members unless, of course, they volunteer.

SITUATION 8

A. Doing nothing in time of crisis is the worst decision that can be made. This may frustrate the team to a point where everything that you have built up may be destroyed.

B. The problem is the schedule slippage, not morale. In this case, it is unlikely that they are related.

C. Group decision making can work but may be difficult under tight time constraints. Productivity may not be related to the schedule slippage. (3 points)

D. This is the time when the team looks to you for strong leadership. No matter how good the team is, they may not be able to solve all of the problems. (5 points)

SITUATION 9

A. A pat on the back will not hurt. People need to know when they are doing well.

B. Positive reinforcement is a good idea, but perhaps not through monetary rewards. (3 points)

C. You have given the team positive reinforcement and have returned authority/responsibility to them for phase III. (5 points)

D. Your team has demonstrated the ability to handle authority and responsibility except for this crisis. Dominant leadership is not necessary on a continuous basis.

SITUATION 10

A. The best approach. All is well. (5 points)

B. Why disturb a good working relationship and a healthy working environment? Your efforts may be counterproductive.

C. If the team members have done their job, they have already looked for contingencies. Why make them feel that you still want to be in control? However, if they have not reviewed the phase III schedule, this step may be necessary. (3 points)

D. Why disturb the team? You may convince them that something is wrong or about to happen.

SITUATION 11

A. You cannot assume a passive role when the customer identifies a problem. You must be prepared to help. The customer's problems usually end up being your problems. (3 points)

B. The customer is not coming into your company to discuss productivity.

C. This places a tremendous burden on the team, especially since it is the first meeting. They need guidance.

D. Customer information exchange meetings are *your* responsibility and should not be delegated. You are the focal point of information. This requires strong leadership, especially during a crisis. (5 points)

SITUATION 12

A. A passive role by you may leave the team with the impression that there is no urgency.

B. Team members are motivated and have control of the project. They should be able to handle this by themselves. Positive reinforcement will help. (5 points)

C. This approach might work but could be counterproductive if employees feel that you question their abilities. (4 points)

D. Do not exert strong leadership when the team has already shown its ability to make good group decisions.

SITUATION 13

A. This is the worst approach and may cause the loss of both the existing and follow-on work.

B. This may result in overconfidence and could be disastrous if a follow-on effort does not occur.

C. This could be very demoralizing for the team, because members may view the existing program as about to be canceled. (3 points)

D. This should be entirely the responsibility of the project manager. There are situations where information may have to be withheld, at least temporarily. (5 points)

SITUATION 14

A. This is an ideal way to destroy the project-functional interface.

B. This consumes a lot of time, since each team member may have a different opinion. (3 points)

C. This is the best approach, since the team may know the functional personnel better than you do. (5 points)

D. It is highly unlikely that you can accomplish this.

SITUATION 15

A. This is the easiest solution, but the most dangerous if it burdens the rest of the team with extra work. (3 points)

B. The decision should be yours, not your team's. You are avoiding your responsibility.

C. Consulting with the team will gain support for your decision. It is highly likely that the team will want Carol to have this chance. (5 points)

D. This could cause a demoralizing environment on the project. If Carol becomes irritable, so could other team members.

SITUATION 16

A. This is the best choice. You are at the mercy of the line manager. He may ease up some if not disturbed. (5 points)

B. This is fruitless. They have obviously tried this already and were unsuccessful. Asking them to do it again could be frustrating. Remember, the brick wall has been there for two years already. (3 points)

C. This will probably be a wasted meeting. Brick walls are generally not permeable.

D. This will thicken the brick wall and may cause your team's relationship with the line manager to deteriorate. This should be used as a last resort *only* if status information cannot be found any other way. (2 points)

SITUATION 17

A. This is a poor assumption. Carol may not have talked to him or may simply have given him her side of the project.

B. The new man is still isolated from the other team members. You may be creating two project teams. (3 points)

C. This may make the new man feel uncomfortable and that the project is regimented through meetings. (2 points)

D. New members feel more comfortable one on one, rather than having a team gang up on them. Briefings should be made by the team, since project termination and phaseout will be a team effort. (5 points)

SITUATION 18

A. This demonstrates your lack of concern for the growth of your employees. This is a poor choice.

B. This is a personal decision between you and the employee. As long as his performance will not be affected, he should be allowed to attend. (5 points)

C. This is not necessarily a problem open for discussion. You may wish to informally seek the team's opinion. (2 points)

D. This approach is reasonable but may cause other team members to feel that you are showing favoritism and simply want their consensus.

SITUATION 19

A. This is the best choice. Your employees are in total control. Do nothing. You must assume that the employees have already received feedback. (5 points)

B. The employees have probably been counseled already by your team and their own functional manager. Your efforts can only alienate them. (1 point)

C. Your team already has the situation under control. Asking them for contingency plans at this point may have a detrimental effect. They may have already developed contingency plans. (2 points)

D. A strong leadership role now may alienate your team.

SITUATION 20

A. A poor choice. You, the project manager, are totally accountable for all information provided to the customer.

B. Positive reinforcement may be beneficial, but does nothing to guarantee the quality of the report. Your people may get overcreative and provide superfluous information.

C. Soliciting their input has some merit, but the responsibility here is actually yours. (3 points)

D. Some degree of leadership is needed for all reports. Project teams tend to become diffused during report writing unless guided. (5 points)

APPENDIX B

Solutions to the Project Management Conflict Exercise

PART ONE: FACING THE CONFLICT

After reading the answers that follow, record your score on line 1 of the worksheet on page 288.

A. Although many project managers and functional managers negotiate by "returning" favors, this custom is not highly recommended. The department manager might feel some degree of indebtedness at first, but will surely become defensive in follow-on projects in which you are involved, and might even get the idea that this will be the only way that he will be able to deal with you in the future. If this was your choice, allow one point on line 1.

B. Threats can only lead to disaster. This is a surefire way of ending a potentially good arrangement before it starts. Allow no points if you selected this as your solution.

C. If you say nothing, then you accept full responsibility and accountability for the schedule delay and increased costs. You have done nothing to open communications with the department manager. This could lead into additional conflicts on future projects. Enter two points on line 1 if this was your choice.

D. Requesting upper-level management to step in at this point can only complicate the situation. Executives prefer to step in only as a last resort. Upper-level management will probably ask to talk to the department manager first. Allow two points on line 1 if this was your choice.

E. Although he might become defensive upon receiving your memo, it will become difficult for him to avoid your request for help. The question, of course, is when he will give you this help. Allow eight points on line 1 if you made this choice.

F. Trying to force your solution on the department manager will severely threaten him and provide the basis for additional conflict. Good project managers will always try to predict emotional reactions to whatever decisions they might be forced to make. For this choice, allow two points on line 1 of the worksheet.

G. Making an appointment for a later point in time will give both parties a chance to cool off and think out the situation further. He will probably find it difficult to refuse your request for help and will be forced to think about it between now and the appointment. Allow ten points for this choice.

H. An immediate discussion will tend to open communications or keep communication open. This will be advantageous. However, it can also be a disadvantage if emotions are running high and sufficient time has not been given to the selection of alternatives. Allow six points on line 1 if this was your choice.

I. Forcing the solution your way will obviously alienate the department manager. The fact that you do intend to honor his request at a later time might give him some relief especially if he understands your problem and the potential impact of his decision on other departments. Allow three points on line 1 for this choice.

PART TWO: UNDERSTANDING EMOTIONS

Using the scoring table shown on page 767, determine your total score. Record your total in the appropriate box on line 2 of the worksheet on page 288. There are no “absolutely” correct answers to this problem, merely what appears to be the “most” right.

PART THREE: ESTABLISHING COMMUNICATIONS

A. Although your explanations may be acceptable and accountability for excess costs may be blamed on the department manager, you have not made any attempt to open communications with the department manager. Further conflicts appear inevitable. If this was your choice, allow a score of zero on line 3 of the worksheet.

B. You are offering the department manager no choice but to elevate the conflict. He probably has not had any time to think about changing his requirements and it is extremely doubtful that he will give in to you since you have now backed him into a corner. Allow zero points on line 3 of the worksheet.

C. Threatening him may get him to change his mind, but will certainly create deteriorating working relationships both on this project and any others that will require that you interface with his department. Allow zero points if this was your choice.

D. Sending him a memo requesting a meeting at a later date will give him and you a chance to cool down but might not improve your bargaining position. The department manager might now have plenty of time to reassure himself that he was right because you probably aren’t under such a terrible time constraint as you led him to believe if you can wait several days to see him again. Allow four points on line 3 of the worksheet if this was your choice.

E. You’re heading in the right direction trying to open communications. Unfortunately, you may further aggravate him by telling him that he lost his cool and should have apologized to you when all along you may have been the one who lost your cool. Expressing regret as part of your opening remarks would benefit the situation. Allow six points on line 3 of the worksheet.

F. Postponing the problem cannot help you. The department manager might consider the problem resolved because he hasn't heard from you. The confrontation should not be postponed. Your choice has merit in that you are attempting to open up a channel for communications. Allow four points on line 3 if this was your choice.

G. Expressing regret and seeking immediate resolution is the best approach. Hopefully, the department manager will now understand the importance of this conflict and the need for urgency. Allow ten points on line 3 of the worksheet.

	Reaction	Personal or Group Score
A. I've given you my answer. See the general manager if you're not happy.	Hostile or Withdrawing	4
B. I understand your problem. Let's do it your way.	Accepting	4
C. I understand your problem, but I'm doing what is best for my department.	Defensive or Hostile	4
D. Let's discuss the problem. Perhaps there are alternatives.	Cooperative	4
E. Let me explain to you why we need the new requirements.	Cooperative or Defensive	4
F. See my section supervisors. It was their recommendation.	Withdrawing	4
G. New managers are supposed to come up with new and better ways, aren't they?	Hostile or Defensive	4
	Total: Personal	
	Total: Group	

PART FOUR: CONFLICT RESOLUTION

Use the table shown on page 768 to determine your total points. Enter this total on line 4 of the worksheet on page 288.

PART FIVE: UNDERSTANDING YOUR CHOICES

A. Although you may have "legal" justification to force the solution your way, you should consider the emotional impact on the organization as a result of alienating the department manager. Allow two points on line 5 of the worksheet.

B. Accepting the new requirements would be an easy way out if you are willing to explain the increased costs and schedule delays to the other participants. This would certainly please the department manager, and might even give him the impression that he has a power position and can always resolve problems in this fashion. Allow four points on line 5 of your worksheet.

C. If this situation cannot be resolved at your level, you have no choice but to request upper-level management to step in. At this point you must be pretty sure that a compromise is all but impossible and are willing to accept a go-for-broke position. Enter ten points on line 5 of the worksheet if this was your choice.

	Mode	Personal or Group Score
A. The requirements are my decision and we're doing it my way.	Forcing	4
B. I've thought about it and you're right. We'll do it your way.	Withdrawal or Smoothing	4
C. Let's discuss the problem. Perhaps there are alternatives.	Compromise or Confrontation	4
D. Let me explain why we need the new requirements.	Smoothing, Confrontation, or Forcing	4
E. See my section supervisors; they're handling it now.	Withdrawal	4
F. I've looked over the problem and I might be able to ease up on some of the requirements.	Smoothing or Compromise	4
	Total: Personal	
	Total: Group	

D. Asking other managers to plead your case for you is not a good situation. Hopefully upper-level management will solicit their opinions when deciding on how to resolve the conflict. Enter six points on line 5 if this was your choice, and hope that the functional managers do not threaten him by ganging up on him.

PART SIX: INTERPERSONAL INFLUENCES

A. Threatening the employees with penalty power will probably have no effect at all because your conflict is with the department manager, who at this time probably could care less about your evaluation of his people. Allow zero points on line 6 of the worksheet if you selected this choice.

B. Offering rewards will probably induce people toward your way of thinking provided that they feel that you can keep your promises. Promotions and increased responsibilities are functional responsibilities, not those of a project manager. Performance evaluation might be effective if the department manager values your judgment. In this situation it is doubtful that he will. Allow no points for this answer and record the results on line 6 of the worksheet.

C. Expert power, once established, is an effective means of obtaining functional respect provided that it is used for a relatively short period of time. For long-term efforts, expert power can easily create conflicts between project and functional managers. In this situation, although relatively short term, the department manager probably will not consider you as an expert, and this might carry on down to his functional subordinates. Allow six points on line 6 of the worksheet if this was your choice.

D. Work challenge is the best means of obtaining support and in many situations can overcome personality clashes and disagreements. Unfortunately, the problem occurred because of complaints by the functional personnel and it is therefore unlikely that work challenge would be effective here. Allow eight points on line 6 of the worksheet if this was your choice.

E. People who work in a project environment should respect the project manager because of the authority delegated to him from the upper levels of management. But this does not mean that they will follow his directions. When in doubt, employees tend to follow the direction of the person who signs their evaluation form, namely, the department manager. However, the project manager has the formal authority to "force" the line manager to adhere to the original project plan. This should be done only as a last resort, and here, it looks as though it may be the only alternative. Allow ten points if this was your answer and record the result on line 6 of the worksheet.

F. Referent power cannot be achieved overnight. Furthermore, if the department manager feels that you are trying to compete with him for the friendship of his subordinates, additional conflicts can result. Allow two points on line 6 of the worksheet if this was your choice.

APPENDIX C

Dorale Products Case Studies

DORALE PRODUCTS (A)

***PMBOK® GUIDE* AREA**	**INTEGRATION MANAGEMENT** **SCOPE MANAGEMENT**
SUBJECT AREA	**DEFINING A PROJECT**

Background

Dorale Products was undergoing favorable growing pains. Business was good. New product development was viewed as the driving force for the company's future growth. The company was now spending significantly more money for new product development, yet the number of new products reaching the market place was significantly less than in prior years. Also, some of the products reaching the marketplace were taking longer than expected to recover their R&D costs, while others became obsolete too quickly.

Management recognized that some sort of structured decision-making process had to be put in place whereby management could either cancel a project early before massive resources were committed or redirect efforts to different objectives. David Mathews was assigned as the project manager in charge of developing a new product development (project management) methodology for Dorale Products.

David understood the benefits of a project management methodology, especially as a structured decision-making process. It would serve as a template or a repetitive process such that project success could be incurred over and over again. The methodology would contain sections for project scope definition, planning, scheduling, and monitoring and control. There would also be a section on the role of the project manager, line managers, and executive sponsors.

To make the project management methodology easy to use and adaptable to all projects, the methodology would be constructed using forms, guidelines, templates, and checklists

PMBOK is a registered mark of the Project Management Institute

rather than the more rigid policies and procedures. This would certainly lower the cost of using the methodology and make it easier to adapt to a multitude of projects. The project managers could then decide whether to implement the methodology on an informal basis or on a more formal basis.

The first draft of the new methodology was completed and ready for review by the vice president (VP) of operations, who had been assigned as the project sponsor. After a review of the methodology, a meeting was held between the sponsor and the project manager (PM).

The Meeting

VP: "I have read over the methodology. Is it your expectation that the methodology should be used on every project?"

PM: "We could probably justify using the methodology on every project. This would give us a really good structured decision-making process."

VP: "Using the methodology is costly and perhaps not all projects should require the use of the methodology. I can rationalize the use of the methodology on a $500,000 project. But what if the project is only $25,000 or $50,000? What if the project is 30 days in length rather than our usual 6- to 12-month effort?"

PM: "I guess we need to define the threshold limits on when project management should be used."

VP: "I have a concern that we should define not only when to use project management but also what a project is. If an activity remains entirely in one functional area, is it still a project according to your definition? Should we also define a threshold limit on how many functional departments must be involved before we define an activity as a project?"

PM: "I'll go back to the drawing board and get back to you in a week or so."

Questions

1. What is a reasonable definition of a project?
2. Is every activity a project or should there be a minimum number of functional boundaries that need to be crossed? If so, how many boundaries?
3. How do we determine when project management should be used and when an activity can be handled effectively by one functional group without the use of project management?
4. Do all projects need project management?
5. Since the use of a formal project management methodology requires time and money, what should be "reasonable" threshold limits for its use?

DORALE PRODUCTS (B)

DORALE PRODUCTS (B)	**INTEGRATION MANAGEMENT**
***PMBOK® GUIDE* AREA**	**SCOPE MANAGEMENT**
SUBJECT AREA	**DEFINING A PROGRAM**

Background

Dorale Products had just developed a project management methodology for the development of new products. Although the methodology was designed exclusively for new product development, the vice president of operations believed that other applications for the methodology would be possible. A meeting was held between the project manager responsible for the development of the methodology and the vice president of operations.

The Meeting

VP: "The company has invested significant time and money in the development of this methodology. It would be a shame if the methodology could not be applied elsewhere in the organization. As an example, there has to be commonality between new product development and information systems projects. Can we use this methodology, or part of it, for both new product development and information systems development?"

PM: "I'm not sure we can do that. The requirements for information systems projects are different, as are the life-cycle phases. A common project management methodology would have to be highly generic to be applicable to all types of projects."

VP: "Are you telling me that we will need to invest more time and more money to develop a family of methodologies?"

PM: "The methodology we've developed can be applied to all of our activities except information technology efforts. All of our projects are similar, or in the same domain group, except for IT. The IT people may require their own methodology, and I can understand their rationale for wanting it this way."

VP: "I assume from your comments that our existing methodology applies equally as well to programs as it does to projects. After all, isn't a program just a continuation of a project?"

PM: "I did not consider applying our methodology to programs as well as projects. Let me think about this, and I will get back to you."

Questions

1. Does it seem practical to have both a project management methodology and a systems development methodology in use concurrently?
2. What is the definition of a program? How does it differ from the definition of a project?
3. Does the project management methodology apply equally well to programs as it does to projects?

DORALE PRODUCTS (C)

***PMBOK® GUIDE* AREA**	**INTEGRATION MANAGEMENT** **SCOPE MANAGEMENT**
SUBJECT AREA	**PROJECT MANAGEMENT APPLICATIONS**

Background Dorale Products has just completed the development of a project management methodology. Although the methodology was to be used for new project development, there was hope that the methodology could be applied to other products as well.

The Meeting

VP: "Have we restricted ourselves on what type of projects we can use our methodology?"

PM: "The answer is both yes and no! Every activity in the company, regardless of the functional area, can be regarded as a project. But not all projects require the use of the methodology or even project management."

VP: "When we had these conversations months ago at the onset of this development process, you convinced me that we were managing our business by projects. Are you now changing your mind?"

PM: "Not at all. The main skill requirement for our project managers is integration management. The greater the integration requirements, the greater the need for project management."

VP: "Now I'm really confused. First you tell me that all projects need project management, and now you say that not all projects need the use of a methodology. What am I missing here?"

Questions

1. Should all projects require the use of the principles of project management?
2. What type of projects should or should not require the use of a project management methodology?
3. How does the magnitude of the integration requirements affect your answer to the previous question?
4. What conclusions can be made about the applications of project management?

DORALE PRODUCTS (D)

***PMBOK® GUIDE* AREA** — **INTEGRATION MANAGEMENT**
SCOPE MANAGEMENT

SUBJECT AREA — **PROJECT MANAGEMENT PROCESSES**

Background

Dorale Products developed a methodology for the management of projects. A vice president was assigned as the project sponsor to oversee the development of the project management methodology. It was now time for the sponsor to introduce the methodology to the executive levels of management. The vice president met with his project manager to prepare the handouts for the executive committee briefing.

The Meeting

VP: "I have looked over the methodology and am concerned that I cannot easily recognize the structure to the methodology. If I cannot identify the structure, then how can I effectively make a presentation to other executives?"

PM: "Good methodologies should be based upon guidelines, forms, and checklists, rather than policies and procedures. We must have this flexibility to adapt the methodology to a multitude of projects."

VP: "I agree with you. But there must still be some overall structure to the project management process."

PM: "Integration management involves three process areas, namely, the integration of the development of the plan, the integration of the execution of the plan, and the integration of changes to the plan. Our methodology is broken down into life-cycle phases, and these three integrative processes are included in each life-cycle phase, though not specifically addressed. I have tried to use the principles of the *PMBOK® Guide*."

VP: "Let me look at the methodology again and see if I can relate it to what you've said."

Questions

1. Is the project manager correct in his definition of the integration management process areas?
2. Can it be difficult to identify these process areas in each life-cycle phase? If so, then what can we do to make them more visible?
3. What should the vice president say in his presentation about the structure of the methodology?

DORALE PRODUCTS (E)

***PMBOK® GUIDE* AREA** **INTEGRATION MANAGEMENT**
SCOPE MANAGEMENT

SUBJECT AREA **LIFE-CYCLE PHASES**

Background The vice president made his presentation to the other senior officers concerning the methodology. Emphasis was placed upon the ten life-cycle phases. The other executives had several questions concerning the use of ten life-cycle phases. The vice president returned to the project manager for another meeting.

The Meeting

VP: "The other executives have concerns that 10 life-cycle phases are too many. You have ten end-of-phase gate reviews which require that most of our executives attend. That seems excessive."

PM: "I agree. The more I think about it, the more I believe that ten are too many. I'll be spending most of my time planning for gate review meetings, rather than managing the project."

VP: "Another concern of our executives was their role or responsibility at the gate review meetings. The methodology is unclear in this regard."

PM: "Once again, I must agree with you. We should have an established criterion for what constitutes passing the gate reviews."

Questions

1. What are the primary benefits for using life-cycle phases? Are there disadvantages as well?"
2. How many life-cycle phases are appropriate for a methodology?
3. What is the danger of having too many gate review meetings?

4. Who determines what information should be presented at each gate review meeting?
5. What questions should the information at the gate review meeting be prepared to answer?

DORALE PRODUCTS (F)

***PMBOK® GUIDE* AREA** **INTEGRATION MANAGEMENT SCOPE MANAGEMENT**

SUBJECT AREA **DEFINING SUCCESS**

Background When the executive committee made the final review of the project management methodology, they identified a lack of understanding of what would constitute project success. The recommendation was to establish some type of criteria that would identify project success.

The Meeting

VP: "We have a problem with the identification of success on a project. We need more clarification."

PM: "I assumed that meeting the deliverables specified by the customer constituted success."

VP: "What if we meet only 92 percent of the specification? Is that a success or a failure? What if we overrun our new product development process but bring in more new customers? What if the project basically fails but we develop a good customer relationship during that process?"

PM: "I understand what you are saying. Perhaps we should identify both primary and secondary contributions to success."

Questions

1. What is the standard definition of success (i.e., primary factors)? How does this relate to the triple constraint?
2. What would be examples of secondary success factors?
3. What would be a reasonable definition of project failure?
4. Should these definitions and factors be included in a project management methodology?
5. Are there any risks with inserting the primary and secondary success factors into the methodology?

DORALE PRODUCTS (G)

***PMBOK® GUIDE* AREA** **INTEGRATION MANAGEMENT**
SCOPE MANAGEMENT
HUMAN RESOURCES MANAGEMENT

SUBJECT AREA **ROLE OF THE EXECUTIVE**

Background Although senior management seemed somewhat pleased with the new methodology, there was some concern that the role of senior management was ill-defined. The vice president felt that this needed to be addressed quickly so that other executives would understand that they have a vital role in the project management process.

The Meeting

VP: "Many of our executives are not knowledgeable in project management and need some guidance on how to function as a project sponsor. Without this role clarification, some sponsors might be 'invisible' while others may tend to be too actively involved. We need a balance."

PM: "I understand your concerns and agree that some role description is needed. However, I don't see how the role description will prevent someone from becoming invisible or overbearing."

VP: "That's true, but we still need a starting point. We may need to teach them how to function as a sponsor."

PM: "If the sponsor can change based upon which life-cycle phase we are in, then we should delineate the role of the sponsor per phase."

VP: "That is a good point. Let's also make sure we define the role of the sponsor at the gate review meetings."

Questions

1. What should be the primary role for the sponsor?
2. Will the role change based upon life-cycle phases?
3. Is it advisable for the sponsor to change based upon the life-cycle phase?
4. Will role delineation in the methodology force the sponsor to perform as expected?
5. What should be the sponsor's role during gate review meetings?

DORALE PRODUCTS (H)

***PMBOK® GUIDE* AREA** — **INTEGRATION MANAGEMENT**
SCOPE MANAGEMENT
HUMAN RESOURCES MANAGEMENT

SUBJECT AREA — **ROLE OF LINE MANAGERS**

Background

The project management methodology was finally beginning to take shape. However, even though the basic structure of the methodology was in place, there were still gaps that had to be filled in. One of these gaps was a well-defined role for the line managers.

The Meeting

VP: "From what I've read about project management, it is very difficult at first to get line managers to effectively support projects. I want our line managers to become fully committed to project management as quickly as possible."

PM: "I agree with you! It's not good for a line manager to assign people to a project and then take no interest in the project at all."

VP: "I believe the line managers have the power to make or break a project. Simply stated, we need them to share in the accountability after they assign resources."

PM: "I'm not exactly sure how to do that. There is no way that I as a project manager can force a line manager to share accountability with me for the project's success or failure."

VP: "I know this will be difficult at first, but I believe it can be done. The methodology should define the expectations that the executives have on the role of the line managers in each life-cycle phase as well as the working relationships in each phase. See if you can get some of our line managers to help you in this regard."

PM: "On most of our projects, the technical direction to the employees is still provided by the line managers, even after the employee is assigned. Most of our project managers have an understanding of technology, not a command of technology. However, we do have some projects where the technical know-how resides with the project manager, who must then provide daily technical supervision. How do I cover both bases in the design of the methodology?"

VP: "It seems to me that in one situation the project manager would be negotiating with the line manager for deliverables, and in the second situation the negotiation would be for specific people. I'm sure you'll find a way to incorporate this into the methodology."

Questions

1. Should a methodology also include staffing policies? If so, what would be an example of a staffing policy?
2. When should a project manager negotiate for people, and when should the project manager negotiate for deliverables?
3. Should a staffing policy also distinguish between full-time and part-time assignments?
4. How should a company handle a situation where the line managers refuse to support project management, even though it is defined as part of the methodology?
5. Should staffing policies and the role of line management be defined in terms of policies and procedures or simply guidelines?

DORALE PRODUCTS (I)

***PMBOK® GUIDE* AREA**	**INTEGRATION MANAGEMENT** **SCOPE MANAGEMENT** **HUMAN RESOURCES MANAGEMENT**
SUBJECT AREA	**INTERPERSONAL SKILLS FOR PROJECT MANAGERS**

Background

With the role of the line manager and senior manager somewhat defined, Dorale believed that only individuals with specialized, interpersonal skills would become the best project managers. The company contemplated the preparation of a list of "universal" skills necessary to function as a project manager.

The Meeting

VP: "I would like to see a list of desired personal characteristics for project managers included in our methodology. Surely this can be done."

PM: "I think we can define knowledge areas more easily than interpersonal skills. It is easier for us to decide whether or not the project manager needs a command of technology or understanding of technology by looking at the requirements of the project. But interpersonal skills are more complicated."

VP: "I don't understand why. Please explain!"

PM: "We appoint project managers to manage deliverables, not people. Our line managers are providing significantly more daily direction to the assigned workers than do our project managers."

VP: "Are you telling me that project managers do not require any management skills or interpersonal skills while managing a project?"

PM: "That's not really what I'm saying. I just believe that the skills needed to be a project manager are probably significantly different than the skills needed to be a line manager."

VP: "I agree! See what kind of list you can develop."

Questions

1. What types of interpersonal skills are needed to be an effective project manager?
2. How do the interpersonal skills of a project manager differ from the skills needed to be an effective line manager?
3. Is your answer to the first two questions dependent upon the fact that in project management multiple-boss reporting is required?
4. Should the list that you have created be dependent upon whether or not the project manager has wage and salary responsibility (or input) for the team members?
5. Why is it often difficult for experienced line managers to become full-time project managers? (Or, in some cases, even part-time project managers?)
6. Some project managers have a command of technology while others have an understanding of technology. Can this command or understanding of technology influence the interpersonal skills needed to be a project manager?
7. Can the interpersonal skills requirements change if the project manager focuses on deliverables rather than people?
8. Can someone with very strong technical skills also have undesirable project management interpersonal skills?
9. Should a project management methodology identify the desired interpersonal skills of a project manager or should it be done on a project-by-project basis only?

DORALE PRODUCTS (J)

***PMBOK® GUIDE* AREA**	**INTEGRATION MANAGEMENT** **SCOPE MANAGEMENT** **HUMAN RESOURCES MANAGEMENT**
SUBJECT AREA	**PROJECT STAFFING POLICIES AND PROCEDURES**

Background Dorale expected conflicts to arise over the staffing of projects. There was some concern over whether or not a project management methodology should contain policies and procedures for project staffing.

The Meeting

VP: "We need some sort of direction in our methodology for the staffing of projects. If we do not have policies and procedures in this regard, then there is no guarantee that the project manager will receive adequate and timely resources."

PM: "I'm not sure I know how to do this. Right now, we are advocating that our project managers negotiate with the line managers for deliverables, rather than for people. It is then the responsibility of the line manager to provide adequate resources to get the job done."

VP: "I agree with you, but we still need direction. Project managers must make it clear what the job specifically requires so that the line manager provides the right resources. I do not want to get into a conflict situation where the project manager blames the line manager for not providing the right resources and the line manager blames the project manager for improperly defining the scope."

PM: "That seems more like accepting accountability than staffing."

VP: "Perhaps so, but it is related to staffing. I want the line managers to provide the projects with personnel with the qualification levels necessary to meet the budgetary limits. We cannot afford to have projects that are loaded with the highest-salaried workers."

PM: "That's a good idea. It might also be advisable to have some policy that mandates that the project managers release the assigned workers at their earliest convenience so that they can be picked up on other projects."

Questions

1. Is it appropriate for a project management methodology to contain policies and procedures on project staffing?
2. Should staffing policies and procedures be directed to project managers, line managers, or both?
3. Should project sponsors be involved in decisions affecting project staffing? If so, what specifically is their involvement and for the staffing of which positions?
4. How do you develop a policy that "forces" a project manager to release people to other projects, assuming they are no longer required on the existing project?
5. Is project staffing an "accountability" decision?
6. Is it the responsibility of the project manager or line manager to adequately define the skill level required to complete a task?
7. Should staffing policies be applied to full-time personnel, part-time personnel, or both?

DORALE PRODUCTS (K)

***PMBOK® GUIDE* AREA**	**INTEGRATION MANAGEMENT** **SCOPE MANAGEMENT** **HUMAN RESOURCES MANAGEMENT**
SUBJECT AREA	**THE PROJECT/PROGRAM OFFICE**

Background The methodology developed by Dorale focused on relatively small projects with time durations of less than 18 months. Could the same methodology be used on large projects?

The Meeting

VP: "Most of our projects have manpower requirements of 10–20 people with time durations of eighteen months or less. Last week, at the executive committee meeting, we approved several large projects that may run for three years or more and require more than forty people full time. How will we manage these projects?"

PM: "I assume you are talking about projects that will be managed by a project office rather than simply by a project manager."

VP: "On large projects, the project manager is more of a project office manager than a project manager. Shouldn't our methodology also discuss the role of a project office and a project office manager?"

Questions

1. What criteria should exist in deciding when to use a project office as opposed to just a project manager?
2. Are the integration management responsibilities of a project office manager different than those of a project manager?
3. What is a project office?
4. What is the role of a project office manager?
5. Can the members of a project office be part-time or must they be full-time?
6. If employees are assigned full-time to a project office, can they still report administratively to their line managers?
7. Can the assigned project office employees be full-time and yet the project manager be part-time?
8. Can project staffing policies be defined for a project office or is it more project-specific?

APPENDIX D

Solutions to the Dorale Products Case Studies

CASE STUDY (A)

1. A project is a unique activity, with a well-defined objective with constraints, that consumes resources, and is generally multifunctional. The project usually provides a unique product service or deliverable.
2. Generally, there is no minimum number of boundaries that need to be crossed.
3. Usually this is based upon the amount of integration required. The greater the amount of integration, the greater the need for project management.
4. All projects could benefit from the use of project management, but on some very small projects, project management may not be necessary.
5. Reasonable thresholds for the use of the project management methodology are based upon dollar value, risk, duration, and number of functional boundaries crossed.

CASE STUDY (B)

1. In many companies, one enterprise project methodology may be impractical. There may be one methodology for developing a unique product or service, and another one for systems development.
2. A program is usually longer in duration than a project and is comprised of several projects.
3. Project management methodologies apply equally to both programs and projects.

CASE STUDY (C)

1. All projects should use the principles of project management but may not need to use the project management methodology.
2. Projects that do not require the methodology are those that are of short duration, low dollar value and stay within one functional department.
3. Methodologies are generally required for all projects that necessitate large-scale integration. However, if the cost associated with the use of the methodology is low, or the methodology is not complex, then it could be argued that the methodology should be used on all projects.
4. This is a valid argument that the principles of project management should be applied to all projects, irrespective of constraints.

CASE STUDY (D)

1. The project manager is partially correct in his definition of the integration management process. The project manager's definition is aligned more so with the 2000 *PMBOK® Guide* rather than the 2004 version.
2. It can be difficult to identify these processes in each life-cycle phase. However, a good project management methodology will solve this problem.
3. A good project management methodology is based upon forms, guidelines, templates, and checklists, and is applicable to a multitude of projects. The more structure that is added into the methodology, the more control one has, but this may lead to the detrimental result of limiting the flexibility that project teams need to have for one methodology that can be adapted to a multitude of projects.

CASE STUDY (E)

1. The primary benefits are standardization and control of the process. The disadvantage occurs when this is done with policies and procedures rather than forms, guidelines, templates, and checklists.
2. Most good methodologies have no more than five or six life-cycle phases.
3. With too many gate review meetings, the project manager spends most of his/her time managing the gate review meetings rather than managing the project.
4. The stakeholders that are in attendance at the gate review meetings determine what information should be presented. Templates and checklists can be established for the gate review meetings as well as the stages.
5. At a minimum, the questions addressed should include: Where are we today? Where will we end up? What special problems exist?

CASE STUDY (F)

1. The standard definition of success is within time, cost, scope, or quality, and accepted by the customer.
2. Secondary success factors might include profitability and follow-on work.
3. It is more difficult to define failure as opposed to success. People believe that failure is an unsatisfied customer. Others believe that failure is a project that, when completed, provided no value or learning.
4. Absolutely, but they can be modified to fit a particular project or the needs of a particular sponsor.
5. Lack of flexibility may be the result.

CASE STUDY (G)

1. The primary role of the sponsor is to help the PM resolve problems that may be beyond the control of the PM.
2. The role of the sponsor can and will change based upon the life-cycle phase.
3. There are two schools of thought. Some believe that the same person should remain as sponsor for the duration of the project, while others believe that the sponsor can change based upon the life-cycle phase. There are advantages and disadvantages of both approaches, and it is often based upon the type of project and the importance of the customer.
4. Not necessarily, but it is a good starting point in explaining to new sponsors their role and responsibility.
5. Verify that the current phase has been completed correctly and authorize initiation of the next phase.

CASE STUDY (H)

1. Good methodologies identify staffing policies. As an example, a project manager may have the right to identify the skill level desired by the workers, but this may be open for negotiations.
2. Project managers that possess a command of technology normally negotiate for people, whereas project managers without a command of technology negotiate for deliverables.
3. This question is argumentative because it may involve an argument over effort versus duration. The line manager may carry more weight in this regard than the project manager since this may very well be based upon the availability of personnel.
4. This is why project sponsors exist; to act as a referee when there are disagreements and to make sure that line management support exists.
5. Guidelines are always better than policies and procedures, at least in the eyes of the author.

CASE STUDY (I)

1. Core skills include decision making, communications, conflict resolution, negotiations, mentorship, facilitation, and leadership without having authority.
2. Line management skills often focus on superior-subordinate relationships, whereas PM skills focus on team building where the people on the team are not necessarily under the control of the PM (and may actually be superior in rank to the PM).
3. Multiple-boss report is also a concern because the control and supervision of the worker may be spread across several individuals.
4. Wage and salary administration is an important factor. If the PM has this responsibility, the workers will adapt to the PM because he/she has an influence over their performance review and salary. Without this responsibility, the PM may be forced to adapt to the workers rather than vice versa.
5. Line managers are accustomed to managing with authority whereas project managers are not.
6. When a PM has a command of technology, he/she may align closer with the skills of a line manager rather than a PM.
7. PMs usually negotiate for deliverables when they do not have a command of technology and this can influence the interpersonal skills needed for a particular project.
8. Yes.
9. The identification should be in general terms only so that it may be applicable to a multitude of projects.

CASE STUDY (J)

1. Yes, but in general terms only.
2. Both, in order to minimize conflicts.
3. Sponsors usually take an active role in selection of the PM, but take a passive role in functional staffing so as not to usurp the authority of their line managers.
4. There is no really effective way to do this other than by closing out some of the functional charge numbers.
5. Yes, if mandated by senior management.
6. The PM can request any skill level desired, but the final decision almost always rests with the line manager.
7. Both.

CASE STUDY (K)

1. The size of the project, duration, risk, and importance of the customer.
2. They are the same and may even be more detailed.

3. A project management team.
4. The role of the PM is to coordinate and integrate the activities of the project management team.
5. They can be part-time or full-time based upon the needs of the project.
6. Yes. An example of this may be the quality specialist assigned to the project.
7. Yes.
8. Policies can be established to staffing of a project office team, but it may be company specific or client specific.

APPENDIX E

Alignment of the *PMBOK*® *Guide* to the Text

This appendix cross-lists the *PMBOK*® *Guide* Sixth Edition sections with this textbook. Not every section in the *PMBOK*® *Guide* is addressed in this textbook, only the major categories.

PMBOK® *Guide*	Page
1.0	2
1.1.3	273, 275
1.2.1	2, 71
1.2.2	47
1.2.3.2	46, 121
1.2.3.3	335
1.2.3.4	42
1.2.3.5	42
1.2.3.6	24
1.2.4.1	60, 355, 427
1.2.4.2	61, 355
1.2.4.3	65
1.2.4.5	2, 503, 582
1.2.4.7	77
1.2.5.1	77
1.2.5.2	77

PMBOK® Guide	Page
1.2.6	77
1.2.6.1	346, 488–491
1.2.6.2	346
1.2.6.4	6, 52, 299
2.0	39
2.2	348, 349
2.3	348, 349
2.3.2	77, 101
2.4	5, 89
2.4.1	22
2.4.2	18, 89, 95, 279
2.4.3	13
2.4.4	16
2.4.4.1	101
2.4.4.2	101, 103
2.4.4.3	101, 145
3.0	13, 352
3.4	4, 10, 145
3.4.2	10
3.4.4	11, 159, 160
3.4.4.1	163, 165
3.4.4.3	152, 337
4.0	13, 14, 39, 71, 76, 281, 302, 317, 386, 578
4.1	47, 391
4.1.3.2	348
4.3	379
4.4.3.1	306
4.5	381, 507
4.6	71, 396, 397
4.7	381
5.0	47, 345, 352, 360, 386, 578
5.1	358, 361
5.1.3.1	361

PMBOK® Guide	Page
5.2	346, 348, 363
5.2.3.2	395
5.3	345, 360, 363, 375
5.3.3.1	361
5.4	365, 370
5.4.2.2	365
5.4.3.1	372, 392
5.5	351, 395
5.6	326, 393, 748
6.0	168, 364, 409, 444
6.1.3.1	392
6.2.3.2	438
6.2.3.3	409
6.3	409, 411, 416, 417
6.3.2	411, 413
6.3.2.1	437, 440
6.3.2.2	417
6.3.2.3	440
6.3.2.4	442
6.4	428, 429, 453, 455, 643
6.4.2.4	428
6.4.2.5	453
6.5	417, 423, 479
6.5.2.1	416
6.5.2.2	417
6.5.2.3	423
6.5.2.5	423
6.5.2.6	432
6.6.2.1	444
7.0	501
7.1.3.1	392, 501
7.2	479
7.2.1	463, 465, 478, 484

PMBOK® Guide	**Page**
7.2.1.1	392
7.2.2	455, 474, 486
7.2.2.1	455
7.2.2.2	455
7.2.2.3	455
7.2.2.4	455
7.2.2.6	511
7.2.2.7	511
7.2.3.2	455
7.3.2.2	506
7.3.3.1	529
7.4	501, 506, 607, 511, 537
7.4.2	443, 514
7.4.2.2	517, 524, 529
7.4.2.3	524
7.4.3.5	514, 537
8.0	697–699, 703, 722
8.1	699, 725
8.1.1	697, 703
8.1.2.2	707
8.1.2.3	707
8.1.3.1	392, 706
8.2	705
8.2.2.5	706
8.3	709
8.3.2	701, 709, 714, 716–721
9.0	15, 20, 89, 115, 121, 127, 128, 133, 136, 281, 302, 303, 306, 317, 723
9.1	116
9.1.3	128
9.2	133
9.3	15, 117, 128, 131, 136
9.3.2.4	283
9.4	119, 148, 154, 266

PMBOK® Guide	Page
9.4.2.3	117, 121, 145, 148, 152
9.4.2.4	262
9.4.3.1	257
9.5	237, 257, 306
9.5.1.4	266
9.5.2.1	237, 240–242, 244
9.5.2.4	269
10.0	203, 209, 212, 214, 303, 444
10.1	205
10.1.3.1	214
10.2.2	206
10.2.2.7	210, 212
11.0	168, 479, 599, 610
11.1	601, 611
11.2	480, 604, 612, 627
11.2.3.1	611, 624
11.3	613, 615
11.4	613, 616
11.4.2	492
11.5	503, 619
11.6	603, 610, 628
11.7	621, 628, 631
12.0	593, 661, 662
12.1	662, 664
12.1.3.1	684
12.1.3.2	662
12.1.3.4	664
12.1.3.6	665
12.1.5	662
12.2	667
12.2.1.4	668
12.2.2.2	668
12.2.2.3	668

PMBOK® Guide	Page
12.2.2.4	680
12.2.2.5	668
12.2.3.1	669, 672
12.2.3.2	672, 673, 678, 680
12.3	593, 680
12.3.2.1	474
12.3.3.1	683
12.3.3.4	680
13.0	279, 317, 329

There are several ways that one can study for the PMP® Examination. The author recommends the following approach:

1. Read over a specific area of knowledge chapter in the *PMBOK® Guide*.
2. Then, read over the chapter(s) in this text that correspond to that area of knowledge
3. Then re-read the area of knowledge in the *PMBOK® Guide* for a second time. Usually things fall into place better after the second reading of the *PMBOK® Guide*.
4. Now it is time to measure what you have learned. Answer the multiple choice questions at the end of the chapter(s) where the area of knowledge information was found. Collect whatever practice questions you can find, such as with the workbook that can accompany this text. The more question you answer, the more prepared you will be to pass the exam.

Some of the sources mentioned previously for practice questions can dramatically help the learning process. For example, the software provided by the International Institute for Learning allows the user to:

- Test on all questions in an area of knowledge
- Test on all questions with a domain area, such as project initiation
- Test on all questions related to a specific *PMBOK® Guide* section such as Section 3.2.

PMP is a registered mark of the Project Management Institute.

Author Index

Aaltonen, Kirsi, 324
Abell, Derek F., 650, 651
Allport, Stephen, 33
Alt-Simmons, Rachel, 53
Archibald, Russell D., 119, 166, 503

Belack, C., 329
Bothell, T. W., 564
Breyfogle, Forrest W., III, 722, 723
Brotherton, S. A., 373
Brown, Trevor J., 33

Charvat, J., 70
Cleland, D. J., 142, 167, 205
Cohen, C., 569
Conrow, E. H., 599, 600, 602, 615
Crosby, Phillip B., 701, 703

Deming, W. Edwards, 701, 702
Duarte, D. L., 283–284

Eckerson, W. W., 559, 567, 568
Flannes, Steven W., 174
Fleming, Quentin W., 535
Foley, Anne, 722
Fried, R. T., 373
Frigon, N. L., 709

Galbraith, Jay R., 105, 153
Garrett, Gregory A., 663
Gellerman, B., 160
Gilbreath, Robert D., 54
Greiman, Virginia A., 272
Grisham, Thomas W., 31

Haimes, Y. Y., 638
Hammond, John S., 650, 651
Hirschmann, Winfred B., 651
Hodgetts, Richard M., 159
Hubbard, D. W., 551, 560
Hultman, Ken, 160

Jackson, H. K., 709
Juran, Joseph M., 701–703

Kerzner, Harold, 33, 50, 82, 111, 139, 140, 167, 174, 182, 184, 205, 228, 234, 245, 288, 312, 329, 338, 400, 445, 495, 499–500, 539, 597, 633, 657, 666, 686, 727, 745, 752
Levin, G., 174, 238, 560

Magenau, John M., 153
Mali, Paul, 366
Mantel, Samuel J., Jr., 377, 438
Mantell, Leroy H., 97
Melik, Rudolf, 79, 356
Meredith, Jack R., 377, 438
Morris, Peter W. G., 153
Mubarak, Saleh, 425
Mulcahy, Rita, 34

Norman, E. S., 373

Parmenter, David, 558
Pegels, C. Carl, 725, 726
Phillips, J. J., 564
Pinto, J. K., 153, 561

Rad, P. F., 560
Rendon, Rene G., 663
Rollins, S., 560

Schnapper, M., 560
Shewhart, W. A., 701
Sivonen, Risto, 324
Snead, G. L., 564
Snyder, N. Tennant, 283–284
Souder, William E., 374–375
Stewart, John M., 103
Stewart, R. D., 359
Thamhain, Hans J., 241, 244, 302
Venkataraman, R. R., 561
Verzuh, Eric, 352
Walker, Anthony, 120
Ward, J. LeRoy, 34
Wirick, D. W., 31
Wysocki, R. K., 217, 562

Subject Index

Acceptance sampling, 721–722
Accessibility, 303
Accommodating (in conflict resolution), 243
Accountability:
  dual, 262
  and organizational structure, 90
  shared, 16
Accounting staff, resistance to change by, 72
Active listening, 213–214
Activities, overlapping, 631–633
Activity scheduling, 383
Activity traps, 166
Actual cost for work performed (ACWP), 514–519, 523, 529, 534
Actual failure, 54
ACWP, *see* Actual cost for work performed
Added value, 32–33
Add-ons, 222
Administration, project, 17–19
Administration cycle (contracts), 680–683
Administrative closure, 65
Administrative skills (of project manager), 124
Aerospace industry, 40
Aggregate projects, 47
"Aggressor" (employee role), 136
Agile project management, 77, 287–288
Allocated baseline, 395
Alternatives:
  analysis of, 582–589
  in problem-solving, 220–221
  selection of, 589–593
  in systems approach, 80–81
  in trade-off analysis, 581–593
Ambiguity, 208
Analysis phase (systems approach), 79–82
Anxiety, 157–158
Apportioned effort technique, 525
Appraisals:
  performance, 268
  project work assignment, 259–260
Approximate estimate, 456
Arms race, 40
Aspirational standards, 276
Assertiveness (in conflict resolution), 243–244
Assumptions, 348–351
  documentation of, 350, 351
  types of, 349–350
Attribute charts, 720
Audits:
  project, 399–400
  quality, 706
Authoritarian communication style, 210
Authority, 152
  communications bottlenecks involving, 212–213
  and organizational structure, 90
  project, 148–152
  of project managers, 9
Avoiding (in conflict resolution), 244
Award fees, 675

BAs (business analysts), 562
BAC (budget at completion), 529, 534
Backup costs, 474–476
Balanced matrix structures, 101

Bar (Gantt) charts, 412, 414
Baselines, *see* Project baselines
Base pay, 265
Bathtub Period (case study), 544–545
BCWP, *see* Budgeted cost for work performed
BCWS, *see* Budgeted cost for work scheduled
Behavioralism, 90
Belief, collective, 327–328
Benchmarking, 77–78
Benefits harvesting, 76–77
Best practices, 306–311, 624
  common beliefs of, 309–310
  levels of, 308–309
  library, 310 311
  proven vs., 307–308
Best Practices audits, 400
Best-value award strategy, 669
BI (business intelligence), 569–570
Bidder conferences, 668
Bidding process (case study), 692–693
"Blocker" (employee role), 137
Boeing, 626, 627
Bonuses, 269
"Bottom-up" risk management, 623
Brainstorming sessions, 223–224
Breakthrough projects, 111, 284–285, 287
Breakthrough technologies, 222
Budgets, 511–512
Budget at completion (BAC), 529, 534
Budgeted cost for work performed (BCWP), 514–515, 518, 519, 521, 523, 529, 534, 536
Budgeted cost for work scheduled (BCWS), 507, 514–519, 523
Burnout, 170–171
Business analysts (BAs), 562
Business case, 346–348
Business intelligence (BI), 569–570
Business models, 9
Business value, 10

CACN (cost account change notice), 507, 510
Calendar project, 443
Calibrated ordinal risk scales, 615–616
Capacity planning, 743–745, 747
Capital budgeting, 488–492
  and internal rate of return, 490–491
  and net present value, 490
  and payback period, 488, 491–492
  project budget, 511–512
  and risk analysis, 492
  and time value of money, 489
Capital rationing, 494–495
Case studies:
  Bathtub Period, 544–545
  To Bid or Not to Bid, 692–693
  Conflict in Project Management, 251–256
  Corwin Corporation, 491–499
  Dorale Products, 771–783
  Estimating Problem, 499–500
  Franklin Electronics, 545–547
  Honicker Corporation, 753–755
  Invisible Sponsor, 451–452
  Irresponsible Sponsors, 341–342
  Is it Fraud?, 295–297
  Kemko Manufacturing, 755–757
  Leadership Effectiveness, 183–195
  Management Reserve, 693–695
  Mayer Manufacturing, 248–250
  McRoy Aerospace, 180–182
  Motivational Questionnaire, 195–201
  Poor Workers, 182
  Prima Donnas, 182–184
  Prioritization of Projects, 340–341
  Radiance International, 313–315
  Reluctant Workers, 184
  Risk Management Department, 641–642
  Selling Executives on Project Management, 342–344
  Teloxy Engineering, 640
  Telstar International, 250–251
  Trophy Project, 178–180
  Williams Machine Tool Company, 37–38
Cause-and-effect analysis, 711–716
CCB (change control board), 393
Ceiling price, 673
Centers for project management expertise, 101
CERs (cost estimating relationships), 453, 655
Certainty, decision-making under, 604–605
Chain of command, 26–27, 209
Champion(s):
  executive, 326
  exit, 328–329
  project, 21–22
Change:
  and corporate culture, 74–77
  management of, 71–76, 626–627
  resistance to, 43–45, 71–76, 116
Change control board (CCB), 393
Change process, 74–75
Chrysler, 697
"Clarifier" (employee role), 138
Classical management, 5

Classical organizational structure, *see* Traditional organizational structure
Closed systems, 45
Closure phase (project life cycle), 64, 65
Code of Ethics and Professional Conduct, 276
Code of Professional Conduct, 276
Collaborating (in conflict resolution), 242–243
Collective belief, 327–328
Combative communication style, 210
Commitment(s):
  of stakeholders, 330–331
  of team members, 155, 156
Committee sponsorship/governance, 19–20, 324, 330
Communication(s), 202–228
  active listening in, 213–214
  barriers to, 206–207, 214–215
  bottlenecks in, 212–213
  with customers, 205
  effective, 203, 208–209
  environment for, 207–208
  filtering of, 210
  and listening, 210
  as network of channels, 203
  between operational islands, 4
  patterns of, 166
  policy for, 211
  receiving of, 207
  styles of, 210
  and team development, 155, 157–158
  techniques for improving, 208–209
  in traditional organizational structure, 93
  traps in, 215–216
Communications management, 208
Company, responsibilities to, 278
Compaq, 742
Compensation, 262–269
  and base pay, 265
  bonuses, 269
  fixed compensation plans, 651
  and job classification, 264
  merit increases, 269
  and performance appraisals, 265–268
Competence, 276
Competency models, 745–746
Competing (in conflict resolution), 243–244
Competing constraints, 7–8
Competitive cultures, 75
Compliance audits, 399
Compression, schedule, 441–442
Compromising (in conflict resolution), 243
Conceptual phase (project life cycle), 62
Conciliatory communication style, 210
Concurrent (simultaneous) engineering, 32, 631–633
Configuration management, 397–398
Conflict(s), 237–246
  causes of, 238–239
  conflict environment, 238–239
  between line and project managers, 166
  meaningful, 239
  most common types of, 239
  and organizational structure, 91–93
  personality, 239
  within project teams, 154
  recognizing/understanding, in trade-off analysis, 578–580
  relative intensity of, 239
  schedule, 239
Conflict in Project Management (case study), 251–256
Conflict management/resolution:
  confrontation meetings for, 241–242
  and establishment of priorities, 240
  methods of, 240–241
  modes of, 242–244
  problems arising during, 166
  project manager and, 123
  role of project managers in, 239–240
  and type of conflict, 244–246
Conflicts of interest, 277
Confrontation meetings, 241–242
Confronting (in conflict resolution), 242–243
"Consensus taker" (employee role), 139
Consequence tables, 225
Constraints:
  competing, 7–8
  primary and secondary, 7
  in problem-solving, 220
  in trade-off analysis, 575–577
  triple, 7–8
Consultants, 372
Continuous improvement, 708, 742–743
Contract(s), 661
  administration cycle for, 680–683
  basic elements of, 672
  checklists for evaluation of, 684
  cost, 676
  cost-plus, 674
  cost-plus-award-fee, 594
  cost-plus-fee, 677
  cost-plus-fixed-fee, 594, 674
  cost-plus-incentive-fee, 594, 675
  cost-plus-percentage-fee, 674
  cost-sharing, 676
  definitive, 672
  fixed-price (lump-sum), 593–594, 674, 677

Contract(s) (*Continued*)
fixed-price-incentive-fee, 594, 675
fixed-price incentive successive targets, 676
fixed-price with redetermination, 676
government, 40
guaranteed maximum-share savings, 675, 677
incentive, 677, 678–679
proposal, interaction with, 684–686
and risk, 680
terminology used in, 671–674
trade-off analysis and type of, 593–594
as transition between project life-cycle phases, 663
winning new, 24–25
Contract/contractual statement of work (CSOW), 362, 394
Contract management, 662–673
activities in, 663–664
checklist for, 684
conducting procurements, 667–673
contract closure, 683
and contract types vs. risk, 680
cycle of contract administration, 680–683
definition of, 662
environment for, 662
incentive contracts, 677, 678–679
planning procurements, 664–667
procurement, 662–664
proposal-contractual interaction, 684–686
strategies for, 662
types of contracts, 673–677
Contractual closure, 65
Contract work breakdown structure (CWBS), 362–363
Control charts, 718–721
interpretation of, 721
and normal distribution, 718–719
types of, 719–721
Controlling, 146
Cooperative cultures, 75, 273
Cooper Manufacturing Company (case study), 641–642
Coordinating, 147
Corporate culture, 7, 325
and change management, 74–77
critical facets of, 76
impact on virtual project teams, 284
morality/ethics and, 273–275
Corporate governance, project governance vs., 20
Corporate procurement strategy, 662
Corwin Corporation (case study), 491–499
Cost(s):
justification of, 531–532
quality vs., 483
value-added, 644
Cost account change notice (CACN), 507, 510
Cost account codes, 506–511
Cost baseline, 529–531
Cost contracts, 676
Cost control, 501–537
and account codes, 506–511
backup costs, 474–476
budgets, 511–512
and "earned value" concept, 520–521
importance of, 501
and labor distributions, 462–463
life-cycle costing, 484–485
and logistics support, 486–487
and material costs, 534–536
and materials costs, 465–466
and MCCS, 501–503, 506, 507, 511–512, 527–528
and operating cycle, 506, 507
overhead rates, 463–465
problems with, 537–538
and project budget, 511–512
requirements for effective, 503
and status reporting, 537
and support costs, 465–466
in traditional organizational structure, 92
variance analysis for, 513–529
Cost estimating relationships (CERs), 453, 655
Cost formula, 523
Cost overruns, 525–526, 532–534
Cost performance index (CPI), 517–519, 528
Cost-plus-award-fee contracts, 594
Cost-plus contracts, 675
Cost-plus-fixed-fee contracts, 594, 674–675, 677
Cost-plus-incentive-fee contracts, 594, 675, 679
Cost-plus-percentage-fee contracts, 675
Cost reduction efforts, 222
Cost-sharing contracts, 676
Cost variance (CV), 515–516
Counseling, 147
CPI, *see* Cost performance index
CPM, *see* Critical path method
Crash times, 432–434
Creativity, in problem-solving, 221
Credibility, 303
Critical assumptions, 350
Critical path method (CPM), 409, 416, 421, 423–428, 432–437
Critical success factors (CSFs), 52–54
Crosby, Phillip B., 701
CSFs (critical success factors), 52–54

CSOW (contract/contractual statement of work), 362, 394
Culture, corporate, *see* Corporate culture
Cumulative average hours, 648–659
Cumulative total hours, 648
Customers:
  communication with, 214–215
  engagement with, 66–67
  and quality management, 697, 698
  unethical/immoral requests by, 273–274
Customer approval milestones, 357
Customer review meetings, 212
Customer Satisfaction Management phase, 66
CV (cost variance), 515–516
CWBS (contract work breakdown structure), 362–363

Dashboards, 566–569
Dashboard reporting, 51
Data gathering, 217, 219
Data tables/arrays, 711
Davis, David, 328–329
Decision making:
  under certainty, 604–605
  inappropriate influences on, 277
  meetings for, 219–220
  predicting outcome of, 224–225
  and problem solving, 215–216
  under risk, 606–607
  by teams, 157–158
  under uncertainty, 607–610
Decision trees, 608–610
Decoding, 206
De facto authority, 152
Defense industry, 40
Defensive projects, 286–287
Definitive contracts, 672
Definitive estimate, 456
De jure authority, 152
Delegation, 90
  and directing, 146
  factors affecting, 151
Deliverables, 5–6
Deming, W. Edwards, 701–702
Department of Defense (DOD), 1, 40, 282
Design freeze milestones, 356–357
Design to unit production cost (DTUPC), 477
Developmental baseline, 395
Development risks, 495
"Devil's advocate" (employee role), 137
Directing, 146–148
  difficulty of, 147
  steps of, 146–147
Discounted cash flow (DCF), 489
Discretionary dependencies, 411
Disruptive communication style, 210
Distributed budget, 512
Diversity of product lines, 105
DMAIC, 722
Documentation:
  of assumptions, 350, 351
  procedural, 737–741
  of project manager's authority, 151
DOD, *see* Department of Defense
Doing, managing vs., 166–167
"Dominator" (employee role), 137
Dorale Products case studies, 771–783
DTUPC (design to unit production cost), 477
Dual accountability, 262

EAC, *see* Estimate at completion
Earned value (EV), 520–521, 526, 534–535, 622
Earned-value measurement systems (EVMS), 512–513, 549
Economic conditions, 651
Economies of scale, 644
Education, 171, 174, 279–281
Efficiency/effectiveness, 302–303
Eli Lilly, 745
Employees:
  assignment of responsibilities to, 133
  evaluation of, 165
  functional, 17
  performance measurement with, 257–262
  problems with, 165–166
  project manager and performance of, 164
  "roles" of, 136–137
  "star," 133
Encoding, 206
"Encourager" (employee role), 138
End-of-phase review meetings, 752
Engagement project management, 66–67, 331
Engineering staff, resistance to change by, 72
Enhancements, 222
Enterprise Environmental Factors, 662
Enterprise project management methodologies, 398–399
Enterprise resource planning (ERP), 550
Entrepreneurial skills (of project manager), 124
Entry-level project managers, 9
Environment(s):
  communications, 207–208
  for conflict, 238–240
  dynamic project, 154, 156
  and organizational structure, 89–90
  problems in, 165
  review of project, 580–581
  staffing, 116–117

Equivalent units, 523
Ericsson, 742
ERP (enterprise resource planning), 550
Estimates, 453
 of activity times, 428–429
 case study, 499–500
 good information for, 455
 for high-risk projects, 479–480
 and life-cycle costing, 484–485
 and logistics support, 486–487
 and low-bidder dilemma, 474–477
 parametric, 455–456
 pitfalls with, 478
 10 percent solution with, 483
 of total project time, 429–430
 types of, 455–458
Estimate at completion (EAC), 507, 524–528
Estimated cost, 673
Estimated cost to complete (ETC), 529
Estimating manuals, 456–458
Estimating Problem (case study), 499–500
Estimative probability risk scales, 616
ETC (estimated cost to complete), 529
Ethics, 273–275
Ethical communication style, 210
EV, *see* Earned value
Evaluations, employee, 257–262
EVMS, *see* Earned-value measurement systems
Excellence, 49–50
Execution failure, 57–58
Execution risks, 495
Executives, 317
 as champions, 326
 and committee sponsorship, 324
 and decentralization of project management, 325
 defining role of, 17
 and in-house representatives, 329
 and management of scope creep, 326
 in matrix organizations, 97
 and planning, 373–379
 and program managers, 124–125
 and project management-line management relationship, 11–13
 as project managers, 125
 and project office, 132
 and project selection, 373–377
 as project sponsors, 317–326
 and risk management, 325
 selection of project manager by, 117–121
 as sponsors of multiple projects, 320
 team support from, 111, 155, 157
 in traditional organizational structure, 93
 working with, 17–19
Exit audits, 399
Exit champions, 328–329
Exit ramps, 329
Expectations:
 project, 54, 56, 303–305
 stakeholder, 331
Expected profit, 673
Expert power, 152
Explicit assumptions, 349
Extended systems, 45
External dependencies, 417
External partnerships, 279

Facilitating communication style, 210
Facilitation, 226–228
Failure, 54–60
 causes of, 57–59
 costs of, 707
 degrees of, 59–60
 of governance, 325–326
 of innovation projects, 285
 KPI, 560
 of methodologies, 70–71
 of planning, 380–381
 of public-sector projects, 31
 of stakeholder relationship management, 335
Fairness, 276
Fears, embedded, 73
50/50 rule, 522–523
Filtering, 210
Finance staff, resistance to change by, 72
Financial closure, 65
Financial risks, 495
Firsthand observations, 444
Fishbone diagrams, 711–713
Five Whys, 723
Fixed baseline, 395
Fixed compensation plans, 651
Fixed-price (lump sum) contracts, 593–594, 674, 677
Fixed-price-incentive-fee contracts, 594, 675, 677, 678
Fixed-price incentive successive targets contracts, 676
Fixed-price with redetermination contracts, 676
Follow-on orders, 655–656
Ford Motor Company, 697
Forecasting technology, 40
Formal authority-oriented leadership techniques, 160
Fragmented cultures, 75
Frameworks, 69, 285
Franklin Electronics (case study), 545–547
Fraud (case study), 295–297
Front-end analysis, 280
Functional baseline, 395
Functional employees, 17

Functional gaps, 4
Functional manager(s), 15–16. *See also* Line manager(s)
Functional organizations, 116
Future of project management, 126–127

Gantt (bar) charts, 412, 414
Gaps, organizational, 4, 5
Gates (stage-gate process), 60–61
"Gate keeper" (employee role), 60–61, 139
Gate review meetings, 65
General Electric (GE), 722
General Motors, 697
GERT (Graphical Evaluation and Review Technique), 416
Gifts, acceptance of, 278
Goal of project management, 110
"Go live" stage, 76
Governance:
  corporate, 20
  failure in, 325–326
  project, 19–20
Government contracting, 40
Graphical Evaluation and Review Technique (GERT), 416
Graphic analysis (time/cost curves), 583–588
  fixed cost, 585–586
  fixed performance levels, 583–585
  fixed time, 586–587
  learning curves, 646–657
  no constraints fixed, 587–588
Group passing technique, 291
Guaranteed maximum-share savings contracts, 675, 677

Hardware deliverables, 6
"Harmonizer" (employee role), 138
Hierarchical referral, 240–241
Hierarchy(-ies):
  considerations of, 108
  of management, 4
High-risk projects, estimating, 479–480
History of project management, 1–2, 39–45
Hodgetts, Richard M., 159
Honesty, 276
Honicker Corporation (case study), 753–755
Horizontal work flow, 4
Human behavior eduction, 171, 174
Human relations-oriented leadership techniques, 159–160
Human Resources staff, resistance to change by, 72
Hurwicz criterion, 607
Hybrid project management, 44

IFB (invitation for bid), 669
Impact implementation matrix, 226
Implementation phase (project life cycle), 63
Implicit assumptions, 349
Incentive contracts, 677, 678–679
Incentive plans, 269
Incompetency, 130
Individual projects, 47
Influence, 245–246, 277
Informal project management, 25, 40–51
Information flow, 334–335
"Information giver" (employee role), 138
"Information seeker" (employee role), 138
Information technology staff, resistance to change by, 72
In-house representatives, 329
"Initiator" (employee role), 138
Innovation, 221–222
Innovation projects, 284–287
Integrated product/project teams (IPTs), 281–283
Integrative responsibilities, 13–14
Integrity, 276
Intellectual property, 77–78
Interface management, 14
Interim deliverables, 6
Internal partnerships, 278–279
Internal rate of return (IRR), 490–491
International Institute for Learning, 398
International project management, 31–32
Interpersonal Influences, 152–154, 245–246
Interval risk scales, 615
Intimidating communication style, 210
Invitation for bid (IFB), 669
IRR (internal rate of return), 490–491
Irresponsible Sponsors (case study), 341–342
ISO 9000, 698–699
Isolated cultures, 75
Issues, 602

Japan, 702
Job classification, 264
Job descriptions, 264
Johnson Controls, 742
Judicial communication style, 210

Kaizen events, 723
Kemko Manufacturing (case study), 755–757
Key performance indicators (KPIs), 54, 334, 555–560
  characteristics of, 558–560
  components of, 557–558
  failure of, 560
  need for, 555–556
  using, 557

Kickoff meetings, 358–360
KISS rule, 147
Knowing oneself, 166
KPIs, *see* Key performance indicators

Labor distributions, 462–463
Labor efficiency, 650–653
Labor-intensive projects/organizations, 104
Lag, 440–441
Laplace criterion, 608
Large projects, 271–273, 328–329, 385
LCC, *see* Life-cycle costing
Leadership, 159–163
  definition of, 159
  elements of, 159
  organizational impact of, 163–165
  by project manager, 122, 154, 159–160, 163–165
  quality, 723–724
  in talent triangle, 10
  of team, 154, 156
  techniques for, 159–160
  transformational project management, 163
  value-based, 160–162
Leadership Effectiveness (case studies), 183–195
Lean manufacturing, 722–723
Lean Six Sigma, 722–723
Learning, 279–281
Learning curves, 456, 643–657
  as competitive weapon, 657
  cumulative average curve, 648–659
  factors affecting, 650–653
  and follow-on orders, 655–656
  graphic representation of, 646–657
  key phrases associated with, 647–648
  limitations of, 656–657
  and manufacturing breaks, 656
  and method of cost recording, 654
  selection of, 654–655
  slope measures for, 653–654
Legitimate power, 152
Lessons learned, 623–624
Letter contract (letter of intent), 672–673
Level of effort method, 524
Life cycle, project, 24, 61–66
  closure phase, 64, 65
  conceptual phase, 62
  Customer Satisfaction Management phase, 66
  implementation phase, 63
  milestones in, 356–357
  and planning, 355
  planning phase, 62
  and risk, 612
  stage-gate process vs., 61
  testing phase, 64
Life-cycle costing (LCC), 477, 484–485
  benefits of, 485
  estimates in, 485
  limitations of, 485
Linear responsibility charts (LRCs), 150–151, 385
Line manager(s):
  authority of, 148
  and communications policy, 211
  and employee evaluations, 258, 260–262
  over-the-fence management by, 39–40
  position power of, 153
  and project managers, 11–13, 166, 278
  and selection of project staff, 133
Listening, 210, 213–214
Logistics support, 486–487
Long-term projects, 5
Lot-release system, 654
Low-bidder dilemma, 474–477
LRCs, *see* Linear responsibility charts

McRoy Aerospace (case study), 180–182
Maintainability, 698
Management:
  of change, 71–76
  classical, 5
  communications, 208
  configuration, 397–398
  over-the-fence, 39–40
Management cost and control system (MCCS), 378–379, 511–512
  and cost accounting, 507
  cost data collection/reporting phase of, 507
  effectiveness of, 503
  phases of, 501, 506
  variance analysis in, 527–528
Management gaps, 4
Management pitfalls, 166–170
Management reserve, 511–512, 693–695
Managing, doing vs., 166–167
Mandatory dependencies, 417
Manpower requirements, projected, 469–470
Manufacturing breaks, 656
Manufacturing engineers, 28
Manufacturing staff, resistance to change by, 72
Marketing, 24–25
Marketing staff, resistance to change by, 72
Master production schedule (MPS), 385–386
Material costs, 465–466, 534–536
  recording, using earned value measurement, 534–535
  variances in, 535–536

Matrix organizational structure, 95–99, 104–105
development of, 95–96
functional managers in, 97
strong/weak/balanced, 101
Matrix projects, 47
Maturity, 41
definition of, 49–50
Maximax criterion, 607–608
Maximin criterion, 607
Mayer Manufacturing (case study), 248–250
MCCS, *see* Management cost and control system
Meaningful conflict, 239
Meetings:
confrontation, 241–242
effective, 210–211
end-of-phase review meetings, 752
kickoff, 358–360
for problem solving and decision making, 219–220
project review, 212
Mega projects, 271–273
Merit increases, 269
Methodologies, project management, 66–71, 628, 741–742, 747–748
creating (case study), 86–87
failure of, 70–71
and frameworks, 69
Metrics, 549–570
benefits of, 551–552
and business intelligence, 569–570
classification of, 553
dashboards and scorecards, 566–569
and failure, 552–553
identifying, 553–554
key performance indicators (KPIs), 555–560
program, 622
and role of project manager, 551
value-based, 561–566
Milestones, project life-cycle, 356–357
Milestone schedules, 364
Milestone technique, 522–523
Minimax criterion, 607–608
Modified matrix structures, 99–100
Monte Carlo process, 617–619
Morality, 273–275
Motivation, 146–148
Motivational Questionnaire (case study), 195–201
Motorola, 719, 722, 742
MPS (master production schedule), 385–386
Multiple projects, managing, 747–748
Multiproject analysis, 444
Multiproject baseline, 395

NASA, 40
Net present value (NPV), 490
Networks of channels, 203
Network scheduling techniques, 409–444
activity time estimation, 428–429
alternative models, 436–437
crash times in, 432
dependencies in, 417
lag in, 440–441
myths of schedule compression, 441–442
precedence networks, 437–440
project management software, 443–444
and replanning, 423–428
scheduling problems, 441
slack time in, 417–423
total project time estimation, 429–430
New product creation, 222
Next-generation projects, 222
Noise, 208
Nominal risk scales, 615
Noncooperative cultures, 75
Nonprofit organizations, 596
Non-project-based organizations, 22
Non-project-driven organizations, 22–23, 41, 44, 596
Normal distribution, 718–719
Normal performance budget, 512
Nortel, 742
NPV (net present value), 490

Objectives:
establishing, 360–361
quality, 704–705
reviewing, 580
unclear, 154, 157
validating, 351–352
Observations, firsthand, 444
Offensive projects, 286
Open systems, 45
Operability, 698
Operating cycle, 506, 507
Operational-driven organizations, 22
Operational islands, 4
Opportunities:
project, 25, 602–603
response options for, 621
Order-of-magnitude analysis, 456
Ordinal risk scales, 615
Organization(s):
class/prestige gaps in, 4, 5
impact of leadership on, 163–165
labor-intensive, 104
location of project manager within, 26–27
project-driven, marketing in, 24–25
project-driven vs. non-project-driven, 22–23

Organizational chart, project, 133–135
Organizational redesign, 107–109
Organizational restructuring, 90
Organizational risks, 495
Organizational skills (of project manager), 124
Organizational structure(s), 89–111
  and environment, 89–90
  matrix organizational form, 95–99
  modified matrix structures, 99–100
  pure product (projectized) organization, 93–94
  redesign of, 107–109
  selection of, 103–106
  strategic business units, 106–107
  traditional (classical), 91–93
  and work flow, 90–91
Overhead rates, 463–465
Overlapping activities, 631–633
Over-the-fence management, 39–40

Parametric estimate, 455–456
Pareto analysis, 714–716
Partnerships:
  external, 279
  internal, 278–279
Part-time project managers, 125
Payback period, 488, 491–492
Pay classes/grades, 265
Penalty power, 152
People skills, task vs., 167
Perceived failure, 54, 56
Percent complete, 523
Perception barriers to communication, 206
Performance, personnel, 116, 165
Performance appraisals, 265–268
Performance audits, 399
Performance measurement:
  with employees, 257–262
  project baseline for, 392–393
  with project managers, 266–267
  with project personnel, 268
Performance measurement baseline (PMB), 392–393
Personality conflicts, 239
Personal power, 152
Personal resistance, 71
Personal values, 165–166
Personnel, *see* Staffing
PERT, *see* Program Evaluation and Review Technique
Phaseouts, project, 381–383
P&L (profit and loss), 44
Planning:
  and configuration management, 397–398
  cycle of, 378–379
  definition of, 345
  detailed schedules/charts, use of, 383–385
  failure of, 57–58, 380–381
  and focusing on target, 354–355
  general, 352–355
  and identification of specifications, 363–364
  and life cycle phases, 355
  and management control, 396–397
  master production schedule, use of, 385–386
  and milestone schedules, 364
  and organizational level, 354
  participants in, 360
  as phase, 352–355
  for phaseouts/transfers, 381–383
  project baselines, 392–395
  project charter, use of, 391–392
  by project manager, 20–21, 123–124
  project plan use in, 386–390
  quality plan, 706
  questions to ask when, 367
  risk, 611–612
  role of executive in, 373–379
  role of project manager in, 345
  and statement of work, 361–363
  and stopped projects, 381
  subdivided work descriptions use in, 379–380
  tools/techniques for, 360–361
  validation of assumptions in, 348–351
  and work breakdown, 365–372
Planning failure, 55, 56
Planning phase (project life cycle), 62–63, 355
PMB (performance measurement baseline), 392–393
PMBOK, *see* Project Management Institute Guide to the Body of Knowledge
PMIS, *see* Project management information systems
PMMM (project management maturity model), 733–737
PMOs, *see* Project management offices
PMP (Project Management Professional), 111
POs, *see* Project offices
Poka-yoke, 723
Policies and procedures, management, 171
Policy:
  communications, 211
  conflict-resolution, 240–241
  personnel, 116
  quality, 704

Political failure, 59
Political risks, 495
Portfolio management, project, 335–337
Position power, 153
Power, 152–154
Precedence networks, 437–440
Prekickoff meeting, 359
Price-based award strategy, 669
Price ceiling, 673
Price variances (PV), 535, 536
Pricing, 453
  and backup costs, 474–476
  developing strategies for, 453–455
  and labor distributions, 462–463
  and low-bidder dilemma, 474–477
  and manpower requirements, 469–470
  and materials/support costs, 465–466
  organizational input needed for, 460–462
  and overhead rates, 463–465
  pitfalls with, 478
  process of, 458–460
  reports, pricing, 466–469
  review procedure, 471–472
  in smaller companies, 271
  special problems with, 477–478
  steps in, 466–468
  systems, 472–473
Primary constraints, 7
Primary success factors, 52–53
Priorities:
  among risks, 624–626
  conflict resolution and establishment of, 240
  project, 747
  and project success, 303
Prioritization of Projects (case study), 340–341
Problems, 215–217, 219, 602–603
Problem data, 217, 219
Problem identification, 215–216
Problem-solving:
  creativity in, 221
  data gathering for, 217, 219
  and decision making, 215–216
  evaluating alternatives in, 220–221
  by management, 164–165
  in matrix organizations, 97
  meetings for, 219–220
  project, 215–223
  systems approach to, 79–82
Procedural documentation, 737–741
Procurement, 662–664
  conducting, 667–673
  planning for, 664–667
Procurement staff, resistance to change by, 72
Produceability, 747
Product baseline, 395
Product improvements, 222
Production point, 673
Production risk, 612
Product management, project management vs., 47–48
Professionalism, 275–276
Professional resistance, 71–74
Profit and loss (P&L), 44
Profit ceiling, 673
Profit floor, 673
Programs:
  definitions of, 53–54
  projects as subdivision of, 46
  projects vs., 46
  as subsystems, 45
Program Evaluation and Review Technique (PERT), 410–416, 419–423, 434, 436
  activity time, estimation of, 428–429
  advantages of, 411
  alternatives to, 436–437
  conversion of bar charts to, 412, 414
  CPM vs., 416
  crash times in, 432
  critical path in, 416
  development of, 410–411
  disadvantages of, 411
  GERT vs., 416
  problem areas in, 436
  replanning techniques with, 423–428
  slack time in, 419–423
  standard nomenclature in, 411–412
  steps in, 430–431
  total project time, estimation of, 429–430
Program managers, PMI certification program for, 47
Program metrics, 622
Project(s):
  breakthrough, 284–285
  categories of, 47
  classification of, 25, 286–287
  defining success of, 7
  definitions of, 2, 53–54
  as "good business," 24–25
  labor-intensive, 104
  long-term, 5
  mega, 271–273
  organizational chart for, 133–135
  outcomes for, 2
  procurement strategy for, 662–663
  programs vs., 46
  scope of, 361
  short-term, 5
  technology-based, 10–11
  terminated, 381–383
Project audits, 399–400
Project-based organizations, 22–23

Project baselines, 392–395
  performance measurement, 392–393
  rebaselining, 393
  types of, 395
Project champions, 21–22
Project charter, 391–392
Project charter authority, 152
Project closure, 64, 65
Project-driven organizations, 22–23, 116
  career paths leading to executive management in, 27
  marketing in, 24–25
  resource trade-offs in, 596
Project engineers, 28
Project failure, *see* Failure
Project financing, 494–495
Project governance, 21–22
Projectized (pure product) organizations, 93–94
Project management, 48
  agile, 77, 287–288
  benefits of, 42
  controlling function of, 146
  corporate commitment to, 7
  definition of, 47
  differing views of, 27–28
  directing function of, 146–148
  driving forces leading to recognition of need for, 41
  engagement, 66–67, 331
  evolution of, 1–2, 39–45
  excellence in, 7
  future of, 126–127
  hybrid, 44
  industry classification by utilization of, 43
  informal, 25
  and integration of company efforts, 104
  international, 31–32
  matrix management vs., 95
  potential benefits from, 3
  process groups in, 2–3
  product management vs., 47–48
  and project authority, 148–152
  public-sector, 28–31
  risk management linked to, 604
  successful, 3, 17–18, 116
  transformational, 163
  ultimate goal of, 110
Project management information systems (PMIS), 549–550
Project Management Institute Guide to the Body of Knowledge (PMBOK), 514, 705–706, 789–794
Project Management Knowledge Base, 276
Project management maturity:
  criteria for, 110
  fallacies that delay, 109–111
Project management maturity model (PMMM), 733–737
Project management methodologies, *see* Methodologies, project management
Project management offices (PMOs), 101–103. *See also* Project offices
Project Management Professional (PMP), 111
Project manager(s):
  and added-value opportunities, 32–33
  administrative skills of, 124
  attitude of, 7
  authority of, 9
  and communications policy, 211
  as communicator, 208–211
  and conflict resolution, 239–240
  conflict resolution skills of, 123
  duties of, 127–128
  and employee evaluations, 257–260
  entrepreneurial skills of, 124
  entry-level, 9
  executives as, 125
  expectations of, 303–305
  integrative responsibilities of, 13–14
  leadership skills of, 122, 154, 159–160, 163–165
  and line managers, 11–13, 166, 278
  location of, within organization, 26–27
  management support-building skills of, 124–125
  in matrix organizations, 97, 104–105
  mistakes made by, 139–140
  multiple projects under single, 125
  next generation of, 126–127
  organizational skills of, 124
  part-time, 125
  performance measurement for, 266–267
  personal attributes of, 117, 119
  and planning, 345
  as planning agent, 20–21
  planning skills of, 123–124
  PMI certification program for, 47
  and problems with employees, 165–166
  professional responsibilities of, 276–278
  project champions vs., 21–22
  in project selection process, 377
  in pure product organizations, 94
  qualifications of, 127–128
  resource allocation skills of, 125
  responsibilities of, 4, 118–119, 127–128, 276–278
  and risk, 166

role of, 13–14
selection of, 117–121
skill requirements for, 121–125
team-building skills of, 121–122
technical expertise of, 117, 123
use of interpersonal influences by, 152–154, 245–246
Project milestone schedules, 269, 364
Project offices (POs), 129, 131–132, 270
communications bottleneck in, 212–213
and pricing, 471–472
scheduling by, 383
Project opportunities, 25
Project plans, 386–390
benefits of, 386
development of, 386
distribution of, 389
structure of, 389
Project portfolio management, 335–337
Project pricing model, 473
Project review meetings, 212
Project risk, 480–483, 612
Project selection process, 285, 377
Project specifications, 363–364
Project-specific baseline, 395
Project sponsors, 317–326
committees as, 19–20, 324
and decentralization of project sponsorship, 325
handling disagreements with, 327
invisible, 323, 451–452
irresponsible, 341–342
multinational, 323–324
multiple, 322
as primary stakeholders, 330
projects without, 320
responsibilities of, 321–322
role of, 317–326
termination of project by, 61
Project sponsorship, 17–20, 320
Promotional communication style, 210
Proposals, 684–686
Proven practices, best vs., 307–308
Public-sector project management, 28–31
Pure product (projectized) organizations, 93–94
PV (price variances), 535, 536

Qualitative risk analysis, 613, 615–616
Quality audits, 399
Quality circles, 725
Quality improvements, 222
Quality management and control, 697–728
acceptance sampling, 721–722
audits, quality, 706
cause-and-effect analysis, 711–716
and changing views of quality, 697–698
control charts, 718–721
costs of, 483, 707–709
as customer-driven process, 697–698
data tables/arrays, 711
and definition of quality, 698–699
leadership, quality, 723–724
objectives, quality, 704–705
Pareto analysis, 714–716
policy, quality, 704
quality assurance, 705
quality control, 705–706
quality plan, 706
responsibility for, 724–725
scatter diagrams, 716–717
Six Sigma, 722
tools for, 709–721
trend analysis, 717–718
Quality movement, 699–703
Quantitative risk analysis, 613, 616–617

Radiance International (case study), 313–315
Radical technological breakthrough projects, 287
RAM (responsibility assignment matrix), 150
Ratio risk scales, 616
R&D, *see* Research and development
Rebaselining, 393
Recession, 45
"Recognition seeker" (employee role), 137
Redesign, product, 651
Red flag, 325
Referent power, 152
Reliability, 698
Replanning:
network, 423–428
project, 593
Reports/reporting, 444
pricing, 466–469
by project managers, 26–27
software for, 443
Requests for information (RFIs), 668–669
Requests for proposals (RFPs), 461–462, 669
Requests for quotation (RFQs), 668–669
Requirements (in systems approach), 80
Requirements traceability matrix (RTM), 394
Research and development (R&D):
project management, 48, 105, 239, 484–485
resistance to change by staff of, 72

Resistance (to change), 71–76
Resource(s):
  company, 11–12
  trade-off of, *see* Trade-off analysis
Resource allocation, program managers and, 125
Resources baseline, 395
Resources Input and Review meeting, 359
Respect, 276
Responsibilities:
  to company/stakeholders, 278
  and organizational structure, 90–91
  professional, 276–278
  of sponsors, 321–322
Responsibility assignment matrix (RAM), 150
Return on investment (ROI), 564
Reviews, 471–472
Review meetings, 752
Review of Ground Rules meeting, 359
Revisable baseline, 395
Rewards, financial, 275–276
Reward power, 152
RFIs (requests for information), 668–669
RFPs, *see* Requests for proposals
RFQs (requests for quotation), 668–669
Risk(s), 599–605
  acceptance of, 620
  analysis of, 613–615
  avoidance of, 620
  causes of, 600, 601
  and concurrent engineering, 631–633
  control of, 620
  and decision-making, 604–610
  definition of, 601–603
  dependencies between, 624–628
  identification of, 610, 612
  and lessons learned, 623–624
  levels of, 614
  measuring, 430
  monitoring, 611
  prioritization of, 624–626
  procurement, 666–667
  in project financing, 495
  response options for, 620–621
  sources for identification of, 612
  tolerance for, 603–604, 628
  transfer of, 620
Risk acceptance, 620
Risk analysis, 492
Risk control, 620–621
Risk handling, 619
Risk management, 600
  and change management, 626–627
  considerations for implementation of, 622–623
  and decision-making, 604–610
  definition of, 604
  and executives, 325
  as failure component, 56
  impact of risk handling measures, 628–630
  monitoring and control of risk, 621–622
  Monte Carlo process for, 617–619
  overinvestment/underinvestment in, 629
  process of, 610
  and project management, skills, 126
  response mechanisms, risk, 619–621
  training in, 611–612
  uses of, 628
Risk Management Department (case study), 641–642
Risk Management Plan (RMP), 611, 619, 622–623
Risk monitoring, 611
Risk neutral position, 604
Risk planning, 611–612
Risk ratings, 614
Risk response strategy, 600
Risk scales (templates), 615–616
RMP, *see* Risk Management Plan
Role conflicts, with project teams, 154
Role Delineation Study (RDS), 276
RTM (requirements traceability matrix), 394

Salability, 697
Sales staff, resistance to change by, 71
Savage criterion, 607–608
SBUs (strategic business units), 106–107
Scalar chain of command, 209
Scatter diagrams, 716–717
Schedules:
  compression of, 441–442
  master production, 385–386
  preparation of, 383–385
Schedule conflicts, 239
Schedule performance index (SPI), 518, 528
Schedule performance monitoring, 622
Schedule variance (SV), 515–516
Scheduling:
  activity, 383
  network, *see* Network scheduling techniques
Scope changes, 6, 747–752
  business need for, 751
  business of, 748–749
  and need for business knowledge, 749–750
  non-approval of, 752
  timing of, 750

Scope creep, 326, 751
Scope freeze milestones, 356
Scope statement, 361
Scorecards, 566–569
Secondary constraints, 7
Secondary success factors, 52–53
Secretive communication style, 210
Self-control, 166
Selling Executives on Project Management (case study), 342–344
Sensitivity analysis, 492
Shared accountability, 16
Sharing arrangement/formula, 673
Shewhart techniques, 701
Short-term projects, 5, 521
Simultaneous engineering, *see* Concurrent engineering
Six Sigma:
  implementing, 722
  lean, 722–723
Slack time, 417–423
Slope (of learning curve), 647, 653–654
Small companies, effective project management in, 270–271
SMART rule, 351–352, 558–559
SMEs (subject matter experts), 372
Smoothing (in conflict resolution), 243
Snyder, N. T., 283–284
Social acceptability, 698
Social groups, 73
Software, project management, 443–444
  features of, 443
  reasons for using, 110
Software deliverables, 6
Solicitation package, 667–668
Solution providers, 331–332
SOOs (Statements of Objectives), 665
SOW, *see* Statement of work
Space program, 40
SPCs (statistical process controls), 701, 710
Special projects, 47
Specifications, project, 363–364
SPI, *see* Schedule performance index
Staffing, 5, 9. *See also specific job titles, e.g.:* Project manager(s)
  and directing, 146
  and employee "roles," 136–137
  environment for, 116–117
  process of, 128–131
  of teams, 154–156
Staff projects, 47
Stage-gate process, 60–61
Stakeholders:
  balancing interests of, 276
  commitments from, 330
  defined, 329
  engagement of, 334
  expectations for, 331
  failure by, 58–59
  with hidden agendas, 331
  identification of, 332
  multinational, 323–324
  of public-sector projects, 29–30
  responsibilities to, 278
  understanding issues/challenges for, 331
Stakeholder analysis, 332–333
Stakeholder mapping, 333
Stakeholder relationship management, 329–335
  and commitment, 330–331
  and engagement project management, 331
  list of expectations of stakeholders in, 331
  stakeholder engagement in, 334–335
  stakeholder interactions agreements in, 329–330
  stakeholder mapping in, 333
Standardization, product, 651
Standard Practice Manuals, 274
"Star" employees, 133
Statements of Objectives (SOOs), 665
Statement of work (SOW), 125, 361–363
  and contract statement of work, 361–362
  and contract work breakdown structure, 361–363
  misinterpretation of, 362
  preparation of, 361–363
  and requirement cycle, 664–665
  specifications in, 363–364
Statistical process controls (SPCs), 701, 710
Status, 537, 580–581
Status reporting, 537
Stonewalling, 19
Stopped projects, 381
Strategic business units (SBUs), 106–107
Strategic intelligence (SI), 569–570
Strategic project pricing model, 473
"Strawman" rating definitions, 614
Stress, 170–171
Strong matrix structures, 101
Subdivided work descriptions (SWDs), 379–380
Subject matter experts (SMEs), 372
Subsystems, 64

Success, project, 52–54, 299–305. *See also* Failure
definitions of, 52–54
degrees of, 59–60
and effectiveness of project management, 302–303
and expectations, 303–305
predicting, 299–302
Supervising, 147
Support costs, 465–466
SV (schedule variance), 515–516
SWDs (subdivided work descriptions), 379–380
"Swing" design (communication analogy), 204
Synthesis phase (systems approach), 79
System(s):
definition of, 45–47
extended, 45
open vs. closed, 45
Systems approach, 79–82
Systems engineering, 604
Systems pricing, 472–473
Systems approach, 80

Talent triangle, 4, 10
Target cost, 673
Target profit, 673
Task skills, people skills vs., 167
Teams, project, 117, 131, 133
anxiety in, 157–158
barriers to development of, 154–158
communication within, 155, 157–158
conflicts within, 154
decision making by, 157–158
expectations of/about, 345
IPTs, 281–283
leadership of, 154, 156
management of newly formed, 157–158
ongoing process of building, 158–159
performance measurement for, 266
and project manager, 121–122
support of senior management for, 155, 157
virtual, 227, 283–284
Team members, interacting with, 281
Technical expertise, 117, 123
Technical performance measurement (TPM), 622
Technical project management, 10
Technical risk dependencies, 624–628
Technology:
forecasting, 40
project managers' understanding of, 12
in pure product organizations, 94
radical breakthroughs in, 222
shifts in, 12
in traditional organizational structure, 91, 92
Technology-based projects, 10–11
Teloxy Engineering (case study), 640
Telstar International (case study), 250–251
Temporary assignments, 116
10 percent solution, 483
Terminated projects, 381–383
Testing phase (project life cycle), 64
Time management:
activity times, estimation of, 428–429
pitfalls of, 167–170
Time value of money, 489
Tip-of-the-iceberg syndrome, 23–24
To Bid or Not to Bid (case study), 692–693
Top-down estimate, 456
"Top down" risk management, 623
"Topic jumper" (employee role), 137
Total project time, estimation of, 429–430
Total quality management (TQM), 698, 725–728
Toyota Production System (TPS), 722
TPM (technical performance measurement), 622
TPS (Toyota Production System), 722
TQM, *see* Total quality management
Trade-offs, 7
Trade-off analysis, 575–597
alternatives, analyzing, 582–589
with competing constraints, 7–8
conflict, recognition/understanding of, 578–580
corrective actions, 590–591
graphic analysis, 583–588
and industry preferences, 594–596
management approval, obtaining, 593
methodology for, 578–593
objectives, review of project, 580
and project constraints, 575–577
and replanning the project, 593

selection of alternative, 589–593
status, review of project, 580–581
and type of contract, 593–594
Trade-off phase (systems approach), 79
Traditional (classical) organizational structure, 91–93
advantages of, 91–92
disadvantages of, 92
Traffic light dashboards, 566–567
Traffic light reporting system, 325
Training, 279–281
and directing, 146
for key initiatives/practices, 279–280
need for, 111
risk management, 611–612
Transfers, project, 381–383
Transformational project management leadership, 163
Translation phase (systems approach), 79
Tree diagrams, 608–610
Trend analysis, 518–524, 717–718
Trends in project management, 733–752
capacity planning, 743–745
competency models, 745–746
continuous improvement, 742–743
end-of-phase review meetings, 752
multiple projects, management of, 747–748
procedural documentation, development of, 737–741
project management maturity model, 733–737
Triple constraints, 7–8
Trophy Project (case study), 178–180
Trust, 76, 91, 108, 129

Unallocated budget, 512
Uncertainty, decision-making under, 607–610
Undistributed budget, 512
Unified Project Management Methodology (UPMM™), 398–399
United Auto Workers, 665
Unit hours, 648
Unit one, 648
UPMM™ (Unified Project Management Methodology), 398–399
Usage variances (UV), 535, 536

VAC (variance at completion), 529
Validation:
of assumptions, 348–351
of objectives, 351–352
verification and, 395–396
Value:
added, 32–33
business, 10
measurement of, 564–566
Values, personal, 165–166
Value-added costs, 644
Value-based metrics, 561–566
Value-based project leadership, 160–162
Value management methodology (VMM), 565–566
Variance analysis, 513–529
causes of variances, 526
cost variance, 515–516
and development of cost/schedule reporting system, 518
and earned value concept, 520–521
50/50 rule, 522–523
government subcontractors, 519–520
issues addressed in, 520
organization-level analysis, 527–528
price variances, 535, 536
program team analysis, 528–529
schedule variance, 515–516
thresholds, variance, 516, 517
usage variances, 535, 536
Variance at completion (VAC), 529
Variance controls, 516–517
Verification and validation (V&V), 395–396
Vertical work flow, 4
Very large projects, 271–273
Vested interest in projects, 326
Virtual project teams, 227, 283–284
Visibility, 302–303
VMM (value management methodology), 565–566
V&V (verification and validation), 395–396

Wage and salary administration, 74
Wald criterion, 607
War rooms, 385
WBS, *see* Work breakdown structure
WBS (work breakdown structure) dictionary, 372–373
Weak matrix structures, 101
What-if analysis, 444
Williams Machine Tool Company (case study), 37–38

"Withdrawer" (employee role), 137
Withdrawing (in conflict resolution), 244
Work breakdown structure (WBS), 365–373, 747
  core characteristics of, 365
  decomposition problems, 370–372
  for large projects, 367, 369
  levels of, 365–369
  preparation of, 367–370
  and pricing, 458, 460
  purpose of, 365
  setting up tasks in, 367
  WBS dictionary, 372–373
Work breakdown structure (WBS) dictionary, 372–373
Work flow, 4, 90–91
Workforce stability, 650
Work habits, 73
Work specialization, 651

Yellow flag, 325

0/100 rule, 522
Zero-based budgeting, 503

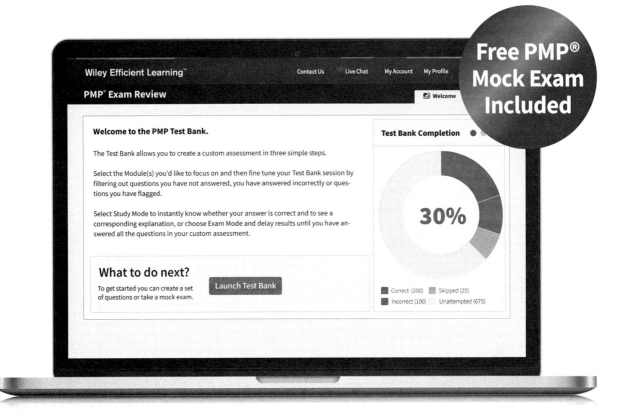